AUTONOMY AND BIOGENESIS
OF
MITOCHONDRIA AND CHLOROPLASTS

AUTONOMY AND BIOGENESIS
OF
MITOCHONDRIA AND CHLOROPLASTS

A Symposium sponsored by International Union of Biochemistry,
Australian Academy of Science and United States National Academy of Sciences

Edited by

N.K. BOARDMAN

Division of Plant Industry, CSIRO, Canberra, Australia

Anthony W. LINNANE

Department of Biochemistry, Monash University, Clayton, Victoria, Australia

and

Robert M. SMILLIE

Plant Physiology Unit, Division of Food Preservation, CSIRO,
and School of Biological Sciences, University of Sydney, N.S.W., Australia

1971

NORTH-HOLLAND PUBLISHING COMPANY – AMSTERDAM - LONDON

Library of Congress Catalog Card Number: 76—134639
ISBN North-Holland: 0 7204 4087 4

Publishers:
North-Holland Publishing Company — Amsterdam
North-Holland Publishing Company, Ltd. — London

Printed in The Netherlands

71' 3151

PREFACE

A Symposium on the "Autonomy and Biogenesis of Mitochondria and Chloroplasts" was held in Canberra on 8th-13th December, 1969. The papers presented during the Symposium are contained in this volume.

The Symposium was cosponsored by the International Union of Biochemistry, the Australian Academy of Science and the U.S. National Academy of Sciences, and arranged by a Committee comprising J.E. Falk (Chairman), N.K. Boardman, A.W. Linnane, R.M. Smillie and V.M. Trikojus. The Committee wishes to record its gratitude to a number of organizations, cited in the Chairman's opening remarks, for financial support. We also wish to acknowledge the invaluable assistance given by a Canberra Committee, comprising members of the CSIRO and the Australian National University, in the organization of the Symposium.

It is with deep regret that we record the death in October, 1970 of Dr. J.E. Falk, Chairmain of the Symposium Committee.

CHAIRMAN'S OPENING REMARKS

J.E. FALK

With great pleasure I bid you welcome.

The idea of this Symposium arose from a discussion which took place in Osaka in 1967, at a dinner given by Professor Okunuki; David Green, Tony Linnane and Keith Boardman were among those present, and all three have played their part in bringing the idea to fruition. That we are all here today is due in no small part to the vision, the enthusiasm, the stubborn determination of Tony Linnane and Keith Boardman. Many other individuals have helped, in Australia and in other countries; I could not possibly name them all now, but in expressing my gratitude to them, I am sure I speak also for all of you.

Through the good offices of Professor Trikojus, Professor Slater and Professor Handler, the Symposium gained the sponsorship of the International Union of Biochemistry. It was sponsored also by the U.S. and Australian Academies. With this backing, financial support has come from the Australian Government, through its new Ministry of Education and Science, in the form of a grant to this Academy; from the U.S. Government via the National Academy of Sciences and the National Science Foundation — and I cannot speak too highly of the sympathetic help we have had from these bodies in a period when, as we all know, funds for this kind of purpose are hard to come by. The International Union of Biochemistry also was able to give some financial help, as did two private Australian bodies: the Ian Potter Foundation and the Myer Foundation. In addition, of course, the Governments of a number of other countries have supported the travel of their countrymen, and the representatives of those Governments in Australia have all been most helpful, and have sent us their best wishes for the success of the Symposium.

A whole decade has passed since, in 1959, I sat where you are now sitting, and listened to Dr. Lemberg's opening remarks at the Symposium on Haematin Enzymes, the first scientific meeting to be held in this hall. I have a strong feeling for the family relationship between that Symposium, and the one we are commencing today, not only scientifically, but because they represent the two Symposia of the International Union of Biochemistry which have, so far, been held in Australia.

It would be tedious to try to enumerate the changes which have taken place in biological science itself over this period. During the coming week, however, we will deal in all familiarity with techniques, with methodology, with concepts, which were undreamed of ten years ago.

For ten years and longer, the importance of mitochondria and of chloroplasts has of course been appreciated. Thus it was recognised that they represent the two main energy tranducers of cells: that both contain electron-transport chains and both contain cytochromes, that both couple electron transport to ATP formation, and that in both organelles, these biochemical systems are localized on complex internal membrane structures.

In 1959, however, one could not have said most of the following, with all its more modern conceptual implications. It has become apparent that these two organelles have many more features in common: they both contain DNA, they both contain ribosomes, and they both are capable of synthesizing nucleic acids and proteins. It has thus become apparent that both organelles have a certain degree of biochemical and genetic autonomy.

Just how autonomous is each of these organelles? What is the precise role of the DNA in each of them? What is the role of nuclear DNA in their biogenesis and maintenance? These are the questions which will provide the continuous thread running through our Symposium.

I declare open this Symposium on the Autonomy and Biogenesis of Mitochondria and Chloroplasts.

CONTENTS

Fig. 1. A model of a biological membrane, showing a double layer of protein molecules, represented as spheres. The proteins in the two layers are in contact with each other, as well as with other proteins in the same layer.

recently proposed by Vanderkooi and Green [17] which satisfies all the six crucial requirements listed above and resolves the dilemma of how phospholipid can exist as a bilayer in a membrane in which it is bonded hydrophobically to protein. The model is seen in top view in fig. 1 and in cross section in fig. 2. It is assumed in this model that the protein molecules which make up the membrane continuum are of moderate molecular weight (12,000–50,000), are spheroidal and bind to neighboring protein molecules.

In this particular model each protein molecule makes 'point' contact with five other protein molecules. The model consists of a bilayer of protein molecules, with lipid bilayer filling the spaces between the proteins. The two layers of proteins are in contact with each other at the middle of the membrane. The proteins may either form a tight crystalline lattice, as shown in fig. 1, or they may form a much looser, liquid crystal array, with each protein making contact with only two or three other proteins in the same

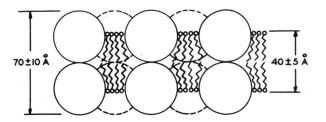

Fig. 2. Diagramatic cross section of the membrane model, showing the relationship of the lipid molecules to the proteins. The large circles are the proteins; the dashed circles are proteins behind the plane of the section. The small circles are the polar lipid heads, and the wavy lines are the nonpolar lipid tails.

layer, and none or one protein in the other layer. In such a membrane, the proteins, and the lipids between them, are in constant thermal motion in the plane of the membrane.

If there were no phospholipid, a two-dimensional protein network would be unstable. It is the insertion of oriented phospholipid molecules into the network which prevents a three-dimensional network and compels two dimensionality of the continuum [18]. Moreover, the dimensional limits of a phospholipid bilayer set a limit to the permissible thickness of the membrane continuum. In other words, membranes must be thick enough to accommodate a bilayer oriented at right angles to the direction of the continuum but not much thicker.

The polarity of the protein surface which lines the holes will determine whether the holes will be filled with phospholipid or water. If the protein surface is predominantly polar, the holes will be filled with water; if predominantly hydrophobic, the holes will be filled with phospholipid. Membranes such as the outer mitochondrial membrane which are readily penetrable by ions and water soluble molecules [19–23] may have a higher proportion of aqueous holes than membranes such as the inner mitochondrial membrane which are generally impenetrable by ions and small molecules. The corollary of this conclusion is that the penetrability of the membrane depends

upon the nature of the holes. If the holes are filled with phospholipid, then penetration of ions requires that molecular devices such as polypeptide ionophores must be available. Otherwise, ions cannot penetrate the membrane.

The model points up a feature that resolves the dilemma of the hydrophobic binding of protein to phospholipid with the presence of bilayers in the membrane. If we think in terms of a core filled with bilayer that fits into a hole, then the phospholipid molecules at the outside of the core are hydrophobically bonded to protein whereas the phospholipid molecules in the interior of the core are bonded one to another. The ultimate determinant of the binding properties of phospholipid in the membrane would be the stability of the hydrophobic binding of the phospholipid molecules to the protein. Once this binding is weakened, for example by detergents, the phospholipid is scooped out as a micellar unit. Clearly then not all the phospholipid in a membrane is hydrophobically bonded to protein – only the molecules at the periphery of the bilayer regions in the membrane continuum.

It is well known that some of the phospholipid in membranes is more easily extracted than the rest [24–26]. This is to be expected in the light of the preceding discussion. Lipids which are only in contact with other lipids should be more easily extractable than lipids which are in contact with protein. Also, the strength of interaction of the lipids with the proteins will depend upon the nature of the lipid molecule and protein molecule in question.

Reagents such as diethylstilbesterol and gramicidin [27] which dissolve in phospholipid micelles can dislodge phospholipid from membranes in micellar form. These reagents which have a high affinity for phospholipid weaken the phospholipid-protein bonds and permit the scooping out of phospholipid from the membrane.

In many biological membranes, neutral lipids such as cholesterol constitute a significant proportion of the total lipid content [28–31]. The ratio of phospholipid to cholesterol can be as high as 1:1 in membranes such as the plasma membrane of the red blood corpuscle [30]. This fact poses no problem for the model proposed in that cholesterol can interdigitate between phospholipid molecules arranged either as bilayers or monolayers. It does present a problem for models such as that of Benson [10] which depend on the interaction of single molecules of phospholipid with protein.

Although there are bilayer regions in the membrane according to the model proposed, the continuum is made up of protein, not phospholipid. Thus, it is protein-protein binding which is the primary determinant of the stability of the membrane and not protein-phospholipid binding. The presence of phospholipid in the hydrophobic holes undoubtedly contributes to the stability of the membrane but this is a second order effect.

If we examine the two surfaces of the membrane model, it is immediately obvious that both protein and the polar heads of the phospholipids are presented to the aqueous medium. The surface properties of the membrane, such as the surface charge density, are determined by contributions from both the protein and the phospholipid. A substantial part of the membrane surface area (roughly half) is provided by the protein and the rest by phospholipid. We can assume that the exposed part of the protein, i.e., the surface which fronts the solvent, is polar in nature. It has long been known that membrane proteins are not very different in amino acid composition from ordinary water soluble proteins [32], and this fact has been difficult to rationalize with the known tendency of these membrane proteins to polymerize readily to form water insoluble aggregates. The arrangement of clusters of hydrophobic groups appears to be the hallmark of membrane proteins rather than an unusually high proportion of hydrophobic groups.

The protein molecules lie at the water-lipid interface, with part of each molecule being in the water, and part in the region of the hydrocarbon tails of the lipids. This should tend to orient the proteins in a constant manner with respect to the plane of the membrane; the least polar part will be in the hydrocarbon region, and the most polar part in the water. Specific reactions between protein molecules required for their enzymic activity will be facilitated by this constant orientation.

There are a considerable number of packing arrangements for protein within a membrane continuum. The model we have presented is only one of a set of possible models. The size of the protein units is a variable, as is the exact shape of the monomer. The

protein molecules need not all be the same size, but there is a limit to the range of sizes permissible. Larger protein complexes can be built up by the specific association of several proteins. The point to be made is that while the proposed model has a certain arbitrariness, nonetheless, the general features of the membrane implicit in the model will not be significantly affected by assuming dimensions for the protein other than the dimension arbitrarily assumed.

The nature of the bonding in the membrane continuum makes it possible for the membrane continuum to be squeezed down or expanded, within limits. In view of the growing body of evidence for extensive conformational changes in the inner mitochondrial membrane which are intrinsic to energy conservation [33], this inherent plasticity of the membrane continuum is precisely the property which the conformational hypothesis would require. A rigid membrane would make it difficult to account for conformational rearrangement of the subunit proteins.

In the model shown in fig. 1, we have focussed attention primarily on the relation of phospholipid to protein and made no attempt to relate this model to the repeating unit structure of membranes [34,35]. There is basically no contradiction between the membrane model and the repeating unit concept. It would be sufficient to postulate that there are proteins within a set which have a higher affinity for one another than the affinity of one set for another. Each such set would correspond to the basepiece of a repeating unit.

The membranes and structured systems of the mitochondrion

Perhaps the simplest way to describe the mitochondrion would be in terms of two membranes, one enclosed within the other [36]. The enclosing or outer membrane (also referred to as outer boundary membrane) surrounds the enclosed or inner membrane (fig. 3). Although the inner membrane is continuous, it is made up of two readily differentiable regions — the invaginating cristae and the inner boundary membrane from which the invaginating cristae originate. The inner boundary membrane parallels the outer membrane but is separated from the

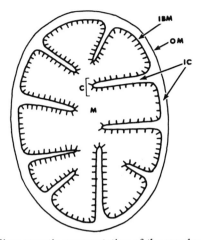

Fig. 3. A diagrammatic representation of the membranes of a mitochondrion. The enclosing outer membrane (also referred to as the outer boundary membrane) is designated OM. The inner membrane is composed of the inner boundary membrane, IBM, plus the cristae, C-bracket. The intracristal space, IC, separates the outer membrane from the inner boundary membrane as well as forms the interiors of the cristal invaginations. The matrix space, M, is enclosed by the inner membrane.

outer membrane by a space or lumen which is continuous with the lumen in the cristae (intracristal space). The space external to the cristae is known as the matrix space.

The repeating units of the cristae are the tripartite repeating units [34,35]. It has yet to be established unequivocally that the inner boundary membrane contains tripartite repeating units but if this is so, there may, nonetheless, be some qualitative difference between the tripartite repeating units in the cristae and in the inner boundary membrane as we shall discuss later [37]. As far as assembly is concerned, we have to think in terms of three membranes — the outer membrane, the inner boundary membrane, and the cristae. The outer membrane definitely does not have tripartite repeating units [38—40]. It is characterized by one dominant protein not found in the inner membrane [41]. The lipid composition of inner and outer membranes is similar but not identical [39—41]. Until outer membrane preparations are brought to an acceptable degree of purity it is not possible to press the comparison of the two membranes with respect to lipid composition too closely.

The intracristal space contains a structured system

(fig. 4) with a characteristic periodicity [42]. This system is not continuous throughout the entire space. It is found in the lumen of some cristae and also in the lumen separating the outer and inner membranes. We shall refer to this system as the intermembrane structured system (ISS). The periodicity of the ISS suggests that it is in phase with the periodicity of the repeat structure of the cristae and of the outer membrane. The center to center distance between the basepieces of the cristae is about 110 Å and the distance between the repeat structures in the ISS is also about 110 Å. The ISS shows up most clearly when the cristae are elongated and linear (fig. 5). Not all mitochondria show the ISS system while selected ones show evidence of this system in profusion. The possibility has to be entertained that the ISS system may be a structure relevant to the biogenesis of mitochondria and not necessarily relevant to the function of a completely formed mitochondrion.

Mitochondria can be subdivided into two types — a type which has relatively little matrix protein, e.g., the mitochondria of heart muscle, and a type which has a relatively high concentration of matrix protein, e.g., the mitochondria of liver and adrenal cortex [42]. The hallmarks of mitochondria with low matrix protein are two-fold: 1. the configurational changes which these undergo during the energy cycle are very precise and regular; 2. the cristal membranes can achieve very close apposition during the energy cycle. The mitochondria with high matrix protein show very irregular configurational changes, and close apposition of the cristae is almost never achieved. Moreover, there is a characteristic condensed appearance of high matrix protein mitochondria suspended in 0.25 M sucrose [43,44] (fig. 6). The matrix protein may be highly organized or amorphous depending upon the mitochondrion or on the physiological state of the cell. In a later section we shall be discussing paracrystalline structures in the matrix space.

Perhaps as a footnote to the description of the cristae, mention should be made of the variable geometries of cristae [42]. These could take the form of flattened pillowcases (heart muscle mitochondria), scalloped tubes (adrenal cortex mitochondria), triangular tubes (astrocyte mitochondria), and hexagonal tubes (mitochondria of the jumping muscle of the locust). But all cristae, regardless of geometry, are invaginations from the inner boundary membrane.

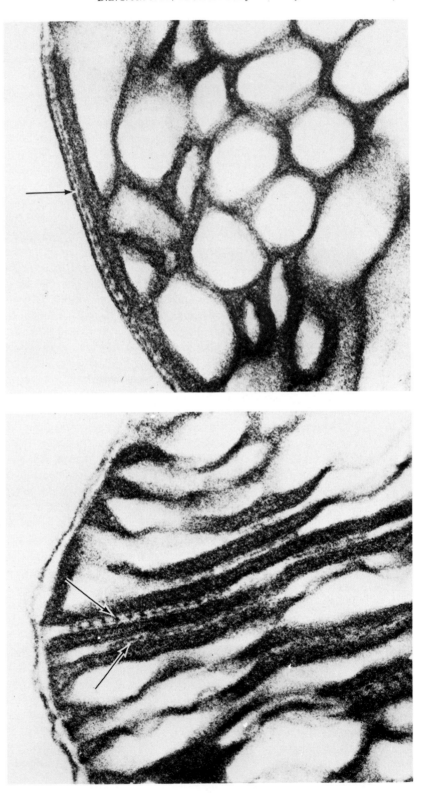

Fig. 4. Paracrystalline structures in the intermembrane space of beef heart mitochondria. ISS: Intermembrane Structural System. (A) An electron micrograph of a region of a sectioned heavy beef heart mitochondrion in the energized configuration. The intermembrane structured system a region of a sectioned heavy beef heart mitochondrion in the energized configuration. The intracristal space contains the intermembrane structured system pointed to by the arrows. (B) An electron micrograph of a region of a sectioned heavy beef heart mitochondrion in the energized configuration. The region of the intracristal space between the outer membrane and the inner boundary membrane is filled with the intermembrane structured system. See arrow. (Electron micrographs provided by Dr. T. Wakabayashi.)

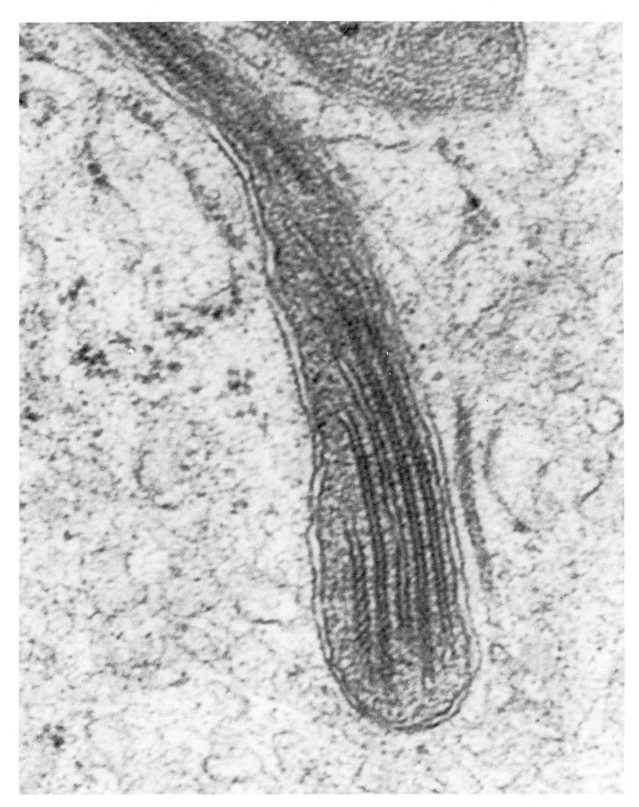

Fig. 5. Paracrystalline structures in the intracristal space of the cristae of guinea pig liver mitochondria. The intermembrane structured system is in the intracristal space of very elongated linear cristae. Electron micrograph provided by Dr. E. Valdivia.

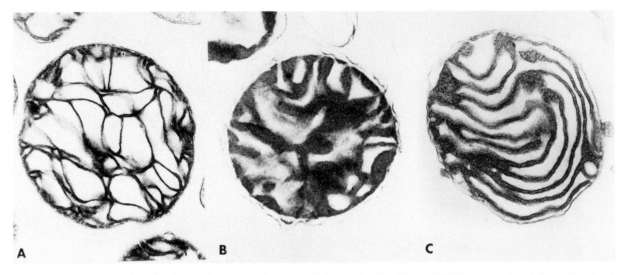

Fig. 6. The nonenergized (aggregated) configuration of various kinds of mitochondria in 0.25 M sucrose. (A) A heavy beef hart mitochondrion. The apposition of cristal membranes is very close, indicating very little matrix protein. (B) A liver mitochondron. The apposition of cristal membranes is not very close, indicating a great deal of matrix protein. (C) An adrenal cortex mitochondrion. The apposition of cristal membranes is fairly close, indicating an intermediate amount of matrix protein. (Electron micrographs provided by Dr. T. Wakabayashi.)

Localization of enzymic function in the mitochondrion

The functional proteins of the mitochondrion are either intrinsic to the membranes in the sense of being integral parts of the repeating units thereof, or are intrinsic to the structured systems which are nonmembranous and which are localized in one of the three mitochondrial spaces — the space between outer and inner membranes, the space within the cristae, and the matrix. One of the surprising developments of recent years has been the recognition that relatively few enzyme systems are intrinsic to the mitochondrial membrane. The bulk of the mitochondrial enzymes are localized in nonmembranous structured systems.

The complexes of the electron transfer chain, the ATPase complex and the coupling mechanisms are intrinsic to the tripartite repeating units of the cristal membranes [45]. The complexes of the electron transfer chain can be identified with the basepieces of the tripartite repeating units [45,46]; and the ATPase complex with the headpiece [46,47]. The stalk which connects basepiece and headpiece, i.e., the connection between the electron transfer chain and the

ATPase complex, is the determinant of the sensitivity of ATPase activity to oligomycin [48]. The stalk is intimately concerned in the conformational mechanism of energy transduction [49]. In some mitochondria, e.g., skeletal muscle mitochondria, there is an additional electron transfer complex besides Complexes I, II, III, and IV (the standard set); this complex catalyzes the oxidation of α-glycerophosphate by cytochrome c [50]. The transfer of a hydride from DPNH to TPN$^+$ is catalyzed by yet another complex which is localized in the basepieces of the cristal membrane — the complex which underlies energized transhydrogenation [51]. Thus, all the primary complexes concerned with the electron transfer chain as well as the units concerned with the coupling functions of the mitochondrion (other than substrate level phosphorylation) are localized in the tripartite repeating units of the cristal membrane.

Adrenal mitochondria are atypical in two respects. First, the cristae are tubular and scalloped (alternating spherical and flattened sections [52]); second, the cristae contain not only the systems concerned in the coupling function but also the systems concerned in steroidogenesis [53–56]. The scalloped nature of the cristae may be a reflection of this duality of function.

Perhaps the flattened sections contain the repeating units concerned with the coupling of electron transfer to synthesis of ATP whereas the spherical sections contain the repeating units concerned with steroidogenesis.

The awareness that the inner boundary membrane may not be identical in chemical composition with the cristal membranes is a recent development [40]. While there is now supporting evidence for tripartite repeating units in the inner boundary membrane [38], there is some evidence that the electron transfer chain is not in this membrane. Both in our laboratory [38] and in the laboratory of E.Racker [47], membranous particles have been isolated which have tripartite repeating units essentially devoid of electron transfer components. If such particles originated from the inner boundary, these observations would mean that the basepieces of the tripartite repeating units of the inner boundary membrane are not complexes of the electron transfer chain. Addink and Smoly [38] have found a dominant protein in these membranous preparations with tripartite repeat structures which is not present either in the outer membrane or the cristal membranes. Studies of configurational changes during the energy cycle clearly show that the inner boundary membrane participates in these configurational changes [36]. In other words, the repeating units of the inner boundary membrane must undergo conformational change during the energy cycle. This poses a dilemma if the electron transfer function is missing in these repeating units.

The outer mitochondrial membrane can be isolated free from electron transfer, ATPase and citric cycle activity [57,58]. Moreover, the isolation can be carried out under conditions in which these activities are fully recoverable in other fractions. We had previously assigned the citric cycle enzymes to the outer membrane but have since found that this association of the citric cycle enzymes with the outer membrane is a consequence of electrostatic binding [57]. When precautions are taken to eliminate such electrostatic binding, the purified outer membrane fractions show essentially no citric cycle activities [57].

Several laboratories [58—68] have found a close association of outer membrane preparations with monoamine oxidase, DPNH-cytochrome c reductase (rotenone-insensitive) and kynurenine hydroxylase (in liver mitochondria). One view to which we incline is that these activities are not intrinsic to the outer mitochondrial membrane but are referable to contamination of the outer membrane preparations by other membranes. As yet, we have not been able to eliminate either monoamine oxidase activity or DPNH-cytochrome c reductase (rotenone-insensitive) from purified outer membrane preparations and indeed these activities are concentrated as the outer membrane is purified. However, a microsomal membrane fragment can be isolated which is some 50 times more active than the outer mitochondrial membrane with respect to DPNH-cytochrome c reductase activity [69]. While this could be a happenstance, it emphasizes, nonetheless, the difficulty of deciding merely on the basis of persistent association whether the association of reductase activity with the outer membrane is intrinsic or not. The definitive proof of the association of these enzymes with the outer membrane would require the isolation of each of these activities in the form of membrane-forming complexes and the demonstration that the lipid composition of such isolated complexes matches that of the outer membrane. Until this matching of lipid composition has been demonstrated, the association of these activites with the outer membrane may be considered problematic.

The citric cycle enzymes are not intrinsic to any of the mitochondrial membranes and are all solubilizable under conditions in which the membranes are not. The, as yet unresolved, question is where in the mitochondrion these enzymes are localized — in the intracristal space or in the matrix space. At the moment there is no solid evidence to guide us in this decision. There is clear evidence of structured systems in the intracristal spaces of heart muscle mitochondria (fig. 4), but not in the matrix space [42]. Moreover, the cristae of beef heart mitochondria can expand into neighboring parallel cristae and make contact so intimate as to preclude the possibility of any protein being interposed between the two apposed membranes [42]. These two lines of evidence point to the intracristal space (this includes the intermembrane space) as the locale of the citric enzymes. Recently Smoly, Kuylenstierna, and Ernster [70] have presented a considerable body of evidence for the localization of the citric cycle enzymes in the matrix space.

The enzyme functions of the cristal membrane require at least 30–40 different proteins whereas the citric cycle and associated activites (fatty acid oxidation and elongation, phosphoryl transferases, etc.) involve at least 50 additional proteins. The assembly process, thus, has to include the synthesis and marshalling of perhaps as many as 100 different proteins within the membranes and spaces of the mitochondrion — this in addition to the incorporation of some 15 different coenzymes into one or another of these proteins and to the incorporation of a variety of ionophoric peptides as well as coenzyme Q into the phospholipid of the inner membrane.

The assembly function of the mitochondrion

In an earlier section mitochondria were subdivided into two categories — those with a protein-rich matrix space and those with relatively little protein in the matrix space. The presence of protein in the matrix space is inferred from indirect evidence of various kinds (density of staining in electron micrographs, visualization of crystalline structures, ratio of non-membraneous proteins to membraneous proteins, etc). Mitochondria such as those of heart muscle which have little matrix protein may be considered to be concerned exclusively with the energy-transducing function whereas mitochondria which have an abundance of matrix protein (liver and adrenal cortex mitochondria) may fulfill a function additional to energy transduction. In this section evidence will be presented that this additional function may be that of the assembly of structured systems in the matrix space. The assembly of a new mitochondrion within a preexisting mitochondrion is only one facet of this assembly function. Our thesis will be that mitochondria are capable of assembling a wide variety of macromolecular structures by these networks of macromolecules or membrane systems.

Evidence has accumulated of the presence in mitochondria from many different sources of a wide variety of structured systems which are readily distinguishable dimensionally from the mitochondrial membranes and which are localized in the matrix space. We shall use the term 'paracrystalline arrays' as a general catch-all description of these matrix-centered structures. Several comprehensive reviews of paracrystalline arrays in mitochondria have already been published largely from the point of view of electron microscopy and morphology [71–73].

Frog oocyte mitochondria contain hexagonal structures (fig. 7) which have been identified as paracrystalline arrays of yolk proteins [74]. In any one mitochondrion multiple arrays can be seen. The cytoplasm of the egg cell contains yolk proteins which are arrayed in precisely the same hexagonal pattern as the counterpart structures in the mitochondrion. This hexagonal pattern emerges only after the extraction of lipid from the yolk proteins in the cell. There is no direct proof that the mitochondrion assembles the yolk proteins and then delivers these assemblies to the cytoplasm but there is convincing circumstantial evidence of such a cycle. An elaborate time sequence is demonstrable electron microscopically: first the assembly of the proteins, then the addition of lipid, then the enveloping of the paracrystalline arrays within membranes, and finally the extrusion of the membranous vesicles through the mitochondrial membranes into the cytoplasm. Such evidence is strongly suggestive that the oocyte mitochondrion fulfills the function of assembling arrays of yolk proteins, interacting these arrays with lipid, and finally delivering the completed arrays to the cytoplasm [74]. It must be emphasized that there is still no information available where the yolk proteins are synthesized, and, if synthesis is external to the mitochondrion as we could expect from other data on the very limited extent of protein synthesis in the mitochondrion, how these proteins enter the mitochondrion.

Paracrystalline structures of the type found in oocyte mitochondria have been reported in mitochondria from many different sources (liver [75], kidney [71], sperm [76], etc.). These structures may not necessarily be a constant feature of mitochondria from a particular source. The episodic nature of these structured arrays may be related to pathological conditions or to cyclical physiological conditions. Thus, the paracrystalline structures seen in sperm mitochondria [76] arise only in the later stages of spermatogenesis. It is not excluded that only mitochondria of a specialized type may be implicated in assembly function. The emergence of mitochondria with atypical forms under pathological conditions is not infrequently observed and these atypical forms may be

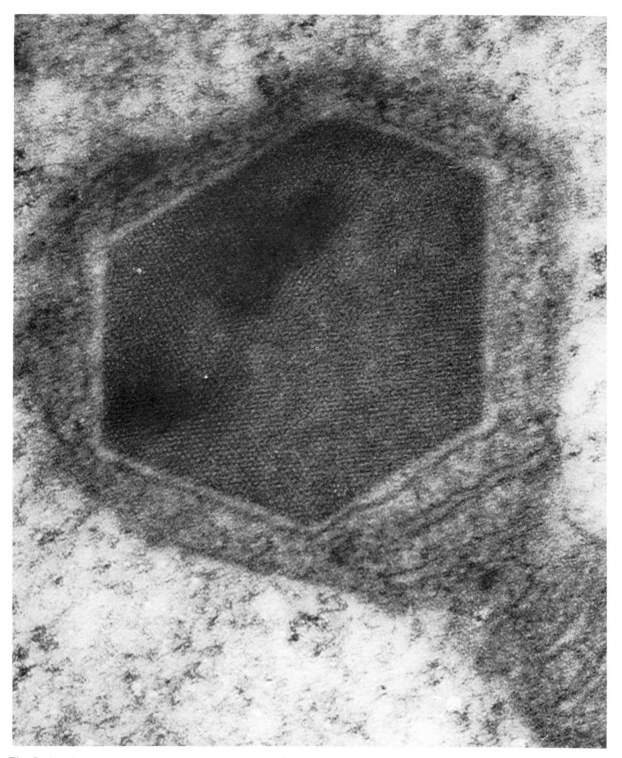

Fig. 7. An electron micrograph of a region of a sectioned mitochondrion of an oocyte of *Rana pipiens*. Note the enclosed hexagonal region with its highly ordered paracrystalline structure. Magnification: 233,000 X. Electron micrograph provided by R.T. Ward, Dept. of Anatomy, SUNY, Downstate Medical Center, Brooklyn, New York.)

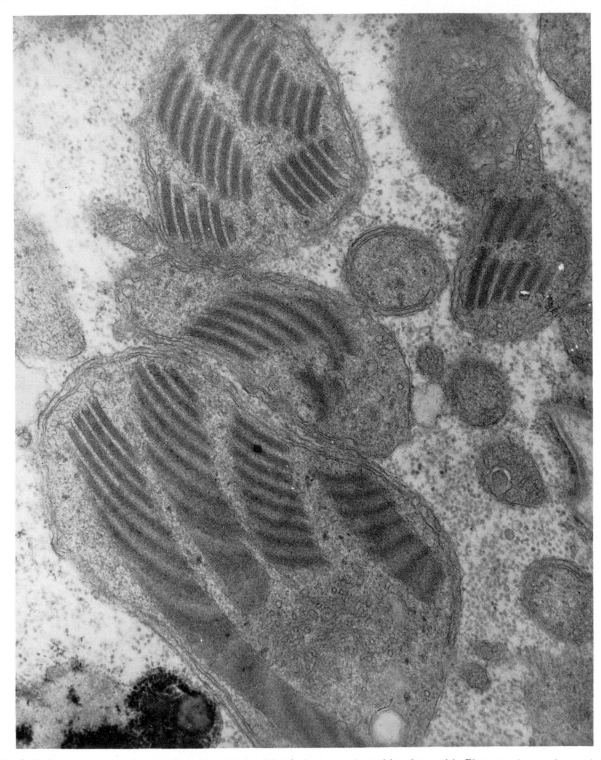

Fig. 8. Dark and light zones in skeletal muscle mitochondria of a human patient with polymyositis. Electron micrograph provided by Dr. S.M. Chou, Dept. of Pathology, W. Va. Univ.

specialized for an assembly function.

Skeletal muscle mitochondria in human patients stricken with polymyositis show a particularly interesting type of paracrystalline array [73]. We shall refer to this array as a pattern of periodic dark and light zones (fig. 8). Note that the length of each zone is fairly constant and that one dark and light array may contain 10–20 such zones. A given mitochondrion may contain two or three such arrays which are clearly independent and oriented differently. At high resolution it can be seen that the dark staining zones are made up of parallel rows of particles whereas the light zones are relatively structureless and may be considered to be spacer zones between the darkly staining structured zones. Detailed study of these dark and light arrays in skeletal muscle mitochondria has led us to the following conclusions. Each dark array eventually metamorphoses into a crista and the set of parallel dark zones within a given array emerges into a parallel set of cristae looped together to make one continuous but periodically folded sheet. Fre-

quently the dark and light zones merge into a densely staining mass with no structural pattern. This would suggest the following time sequence: first, a formless undifferentiated mass of particles; second, the formation of dark and light zones; third, the conversion of each dark zone to cristae and the looping together of the parallel cristae to form one continuous membranous sheet. We are suggesting, as did S.M.Chou [73], that the structures seen in mitochondria from the skeletal muscle of patients with polymyositis represent stages in the biogenesis of cristae. In support of this postulate is the observation that whole fields can occasionally be seen in the skeletal muscle of polymyositic patients with mitochondria crammed full with cristae either rolled up like rope or tightly looped together (fig. 9). Presumably, these mitochondria would be a later stage in the process of bioassembly of cristae.

The dark and light pattern seen in such dramatic form in skeletal muscle mitochondria is by no means unique to such mitochondria. Equally convincing

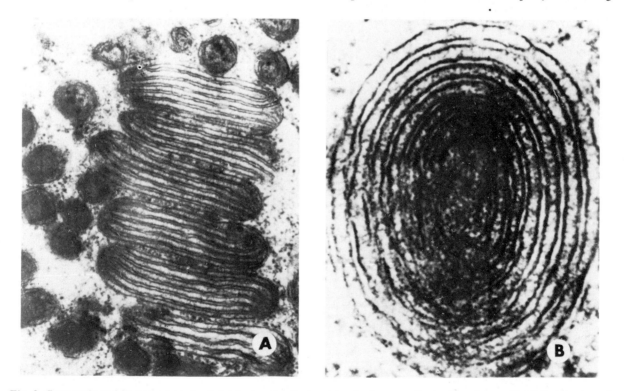

Fig. 9. Concentric and looped cristae in skeletal muscle mitochondria of a human patient with polymyositis: (A) a long, looped mitochondrion with looped cristae; (B) a cross section of a mitochondrion showing concentric cristae. Electron micrographs provided by Dr. S.M. Chou, Department of Pathology, West Virginia U.

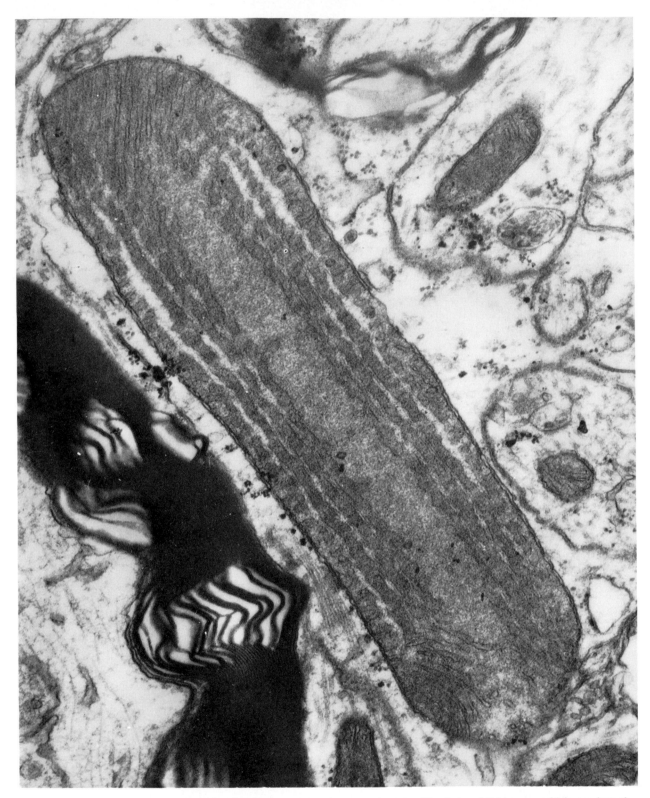

Fig. 10. Dark and light zones in the mitochondria of the Syrian hamster astrocytes. The dark and light zones, which run along the longitudinal axis of the mitochondrion undergo a transition at both ends of the mitochondrion to form normal cristae. Electron micrograph provided by Drs. R. Morales and D. Duncan, The University of Texas Medical Branch, supported by USPHS Grant NS 00690.

examples can be seen not infrequently in the mito-chondria of hamster astrocytes (fig. 10) and also in the mitochondria of the fat body cell of *Drosophila melanogaster*. Here again the transition from darkly staining arrays of parallel rows of particles to conventional cristae can clearly be traced.

It is not our intent to press this speculative interpretation much further at the present stage of our knowledge. The point that we would like to make is that mitochondrial biogenesis may be only a special case of the assembly function of the mitochondrion − a function which apparently can extend to a wide variety of macromolecular structures as well as to an organelle.

The question which obviously arises in postulating an assembly role for mitochondria is how the ingredients to be assembled (proteins, phospholipids, etc.) enter the mitochondrion and how the assembled structures leave the mitochondrion. The latter question poses less of a difficulty since extrusion of membrane-enclosed structured materials via a pino-cytotic mechanism is a well documented phenom-enon. The extrusion of viruses [77] and of golgi vesicles [78] through the plasma membrane of animal cells are two well known examples. The entry of proteins and lipid into the mitochondrion poses a more formidable problem. The clue to at least one of the mechanisms of entry may be provided by the sequence of changes which mitochondria undergo during spermatogenesis [76]. Many mitochondria coalesce to form a giant elongated mitochondrion which can wrap helically around the nuclear core. It is implicit from the sequential events observed elec-tron microscopically during spermatogenesis that sperm mitochondria are capable of fusion. During the fusion process the introduction of small and large molecules into the mitochondrion is a possibility. There is a body of evidence which points to the short term intercalation of protein molecules within the membrane structure. That is, appropriate protein molecules can pop in and out of the membrane continuum and in this way traverse the membrane without actually going through holes.

If the notion of the mitochondrion as an instru-ment of assembly has validity, the question could be raised as to the structures in the mitochondrion which control the assembly process. To what extent do the mitochondrial membranes play a role in the assembly process? Assembly involves many processes other than geometric alignment and it is difficult to believe that special devices other than the membranes of the mitochondrion will not be needed to control the assembly process. What these devices are poses a fascinating experimental challenge.

Acknowledgements

This work was supported in part by Program Project Grant No. GM−12,847 from the National Institute of General Medical Sciences (USPHS), and by NIH Grant No. 6523.

References

[1] F.S. Sjostrand, in: Regulatory functions of biological membranes, ed. J. Jarnefelt (Elsevier, Amsterdam, 1968) p. 1.

[2] D.E. Green and S. Fleischer, Biochim. Biophys. Acta. 70 (1963) 554.

[3] A.A. Benson, J. Amer. Oil Chem. Soc. 43 (1966) 265.

[4] A.A. Benson, 7th International Congress of Biochem-istry, Tokyo, Abstracts 111 (1967) 525.

[5] S. Fleischer, G. Brierley, H. Klouwen and D.B. Slautter-back, J. Biol. Chem. 237 (1962) 3264.

[6] E.D. Korn, Science 153 (1966) 1491.

[7] G.E. Palade, 7th International Congress of Biochem-istry, Tokyo, Suppl. 1 (1967) 1077.

[8] J.D.Robertson, Progr. Biophys. Biophys. Chem. 10 (1960) 344.

[9] J.F. Danielli and H.A. Davson, J. Cellular Comp. Physiol. 5 (1935) 495.

[10] A.A. Benson, this volume.

[11] D.E. Green and J.F. Perdue, Proc. Natl. Acad. Sci. U.S. 55 (1966) 1295.

[12] D. Branton, Proc. Natl. Acad. Sci. U.S. 55 (1966) 1048.

[12a] D. Branton and R.B. Park, J. Ultrastruct. Res. 19 (1967) 283.

[13] W.L. Hubbell and H.M. McConnell, Proc. Natl. Acad. Sci. U.S. 61 (1968) 12; 63 (1969) 16; 64 (1969) 20.

[14] G.S. Eisenman, M. Crane and G. Szabo, Federation Proc. 27 (1968) 1289.

[15] W. Stoeckenius and D.M. Engelman, J. Cell Biol. 42 (1969) 613.

[16] A.D. Bangham, Progr. Biophys. Mol. Biol. 18 (1968) 29.

[17] G. Vanderkooi and D.E. Green, Proc. Natl. Acad. Sci. U.S. 66 (1970) 615.

[18] D.E. Green and A. Tzagoloff, Lipid Res. 7 (1966) 587.

[19] R.L. O'Brien and G.P. Brierley, J. Biol. Chem. 240 (1965) 4527.

[20] G.R. Hunter and G.P. Brierley, Biochim. Biophys. Acta. 180 (1969) 68.

[21] E. Pfaff, in: Mitochondrial structure and compartmentation, eds. E. Quagliariello, S. Papa, E.C. Slater and J.M. Tager (Adriatica Editrice, Bari, Italy, 1965) p. 165.

[22] J.L. Purves and J.M. Lavenstein, J. Biol. Chem. 236 (1961) 2794.

[23] L.M. Birt and W. Bartley, Biochem. J. 75 (1960) 1303.

[24] A. Tzagoloff and D.H. MacLennan, Biochim. Biophys. Acta. 90 (1965) 476.

[25] A.K. Parpart and R. Ballentine, in: Modern trends in physiology and biochemistry, ed. E.S.G. Barron (Academic Press, New York, 1952) p. 135.

[26] L.L.M. Van Deenen, in: Regulatory functions of biological membranes, ed. J. Jarnefelt (Elsevier, Amsterdam, 1968) p. 72.

[27] J.M. Smoly and J. Asai, unpublished observations.

[28] D.F. Parsons and Y. Yang, Biochim. Biophys. Acta. 135 (1967) 362.

[29] S. Fleischer, H. Klouwen and G.P. Brierley, J. Biol. Chem. 236 (1961) 2936.

[30] J.S. O'Brien, J. Theoret. Biol. 15 (1967) 307.

[31] S.N. Sehgal, M. Kates and N.E. Gibbons, Can. J. Biochem. Physiol. 40 (1962) 69.

[32] H.I. Silman, J.S. Rieske, S.H. Lipton and H. Baum, J. Biol. Chem. 242 (1967) 4867.

[33] D.E. Green and R.A. Harris, FEBS Letters 5 (1969) 241.

[34] H. Fernandez-Moran, T. Oda, P.V. Blair and D.E. Green, J. Cell Biol. 22 (1964) 63.

[35] D.E. Green and D.H. MacLennan, Bioscience 19 (1969) 213.

[36] E.F. Korman, A. Addink, T. Wakabayashi and D.E. Green, J. Bioenergetics 1 (1970) 9.

[37] P.A. Berg, I.M. Roitt, D. Domack and R.W. Horne, Clin. Exptl. Pathol. 4 (1969) 511.

[38] A. Addink and J.M. Smoly, unpublished observations.

[39] D.E. Green, E. Bachmann, D.W. Allmann and J.F. Perdue, Arch. Biochem. Biophys. 115 (1966) 172.

[40] D.W. Allmann, E. Bachmann, N. Orme-Johnson, W.C. Tan and D.E. Green, Arch. Biochem. Biophys. 125 (1968) 981.

[41] J.M. Smoly, A. Senior and D.E. Green, unpublished observations.

[42] E.F. Korman, R.A. Harris, C.H. Williams, T. Wakabayashi, D.E. Green and E. Valdivia, J. Bioenergetics 2 (1970).

[43] C.H. Williams, W.J. Vail, R.A. Harris, M. Caldwell and D.E. Green, J. Bioenergetics 1 (1970) 147.

[44] C.R. Hackenbrock, J. Cell Biol. 37 (1966) 345; 40 (1968) 269.

[45] D.E. Green and A. Tzagoloff, Arch. Biochem. Biophys. 116 (1966) 293.

[46] J. Kopaczyk, J. Asai, D.W. Allmann, T. Oda and D.E. Green, Arch. Biochem. Biophys. 123 (1968) 602.

[47] E. Racker, D.D. Tyler, R.W. Estabrook, T.E. Conover, D.F. Parsons and B. Chance, in: Oxidases and related redox systems, eds. T.E. King, H.S. Mason and M. Morrison (John Wiley, New York, 1965) vol. 11, p. 1077.

[48] D.H. MacLennan and J. Asai, Biochem. Biophys. Res. Commun. 33 (1968) 441.

[49] E.F. Korman and W.J. Vail, unpublished studies.

[50] B. Saktor, L. Packer and R.W. Estabrook, Arch. Biochem. Biophys. 86 (1959) 68.

[51] D.H. MacLennan, unpublished observations.

[52] D.W. Allmann, T. Wakabayashi, E.F. Korman and D.E. Green, J. Bioenergetics 1 (1970) 73.

[53] D.G. Young and P.F. Hall, Biochem. Biophys. Res. Commun. 31 (1968) 925.

[54] K.D. Roberts, L. Bonde and S. Lieberman, Biochem. Biophys. Res. Commun. 29 (1967) 741.

[55] T. Omura, E. Sanders, R.W. Estabrook, D.Y. Cooper and O. Rosenthal, Arch. Biochem. Biophys. 117 (1966) 660.

[56] S.B. Oldham, J.J. Bell and B.W. Harding, Arch. Biochem. Biophys. 123 (1968) 496.

[57] J.M. Smoly, T. Wakabayashi, A. Addink and D.E. Green, in preparation.

[58] G.L. Sottocasa, B. Kuylenstierna, L. Ernster and A. Bergstrand, J. Cell Biol. 32 (1967) 415.

[59] D.F. Parsons, G.R. Williams, W. Thompson, D. Wilson and B. Chance, in: Round table discussion on mitochondrial structure and compartmentation, eds. E. Quagliariello, S. Papa, E.C. Slater and J.M. Tager (Adriatica Editrice, Bari, Italy, 1969) p. 29.

[60] C. Schnaitman, V.G. Erwin and J.W. Greenawalt, J. Cell Biol. 32 (1967) 719.

[61] C. Schnaitman and J.W. Greenawalt, J. Cell Biol. 38 (1968) 158.

[62] G.L. Sottocasa, L. Ernster, B. Kuylenstierna and A. Bergstrand, in: Round table discussion on mitochondrial structure and compartmentation, eds. E. Quagliariello, S. Papa, E.C. Slater and J.M. Tager (Adriatica Editrice, Bari, Italy, 1967) p. 74.

[63] D.S. Beattie, Biochem. Biophys. Res. Commun. 31 (1968) 901.

[64] M. Levy, R. Toury and J. Andre, C.R. Acad. Sci. Paris Series D. 263 (1966) 1766.

[65] M. Levy, R. Toury and J. Andre, Biochim. Biophys. Acta. 135 (1967) 599.

[66] D. Brdiczka, D. Pette, C. Brunner and F. Miller, Europ. J. Biochem. 5 (1968) 294.

[67] G. Mayer, V. Ullrich and H. Staudinger, Hoppe-Seylers Z. Physiol. Chem. 349 (1968) 459.

[68] H. Okamato, S. Yamomato, M. Nozaki and O. Hayashi, Biochem. Biophys. Res. Commun. 26 (1967) 309.

[69] N. Penn and B. Mackler, Biochim. Biophys. Acta, 27 (1958) 539.

[70] J.M. Smoly, B. Kuylenstierna and L. Ernster, Proc. Natl. Acad. Sci. (U.S.) 66 (1970) 125.

[71] T. Suzuki and F.K. Mostofi, J. Cell Biol. 33 (1967) 605.

[72] R.T. Ward, J. Cell Biol. 14 (1962) 309.
[73] S.M. Chou, Acta Neuropath. (Berlin) 12 (1969) 68.
[74] R.T. Ward, J. Cell Biol. 14 (1962) 309.
[75] E. Mugnaini, J. Ultrastruct. Res. 11 (1964) 525.

[76] J. Andre, J. Ultrastruct. Res. Suppl. 3 (1962) 1.
[77] J.J. Holland and E.D. Kiehn, Science 167 (1970) 202.
[78] A.B. Novikoff, E. Essner and N. Quintana, Federation Proc. 23 (1964) 1010.

LIPID-PROTEIN INTERACTIONS IN CHLOROPLAST LAMELLAR MEMBRANE AS BASES FOR RECONSTITUTION AND BIOSYNTHESIS

A. A. BENSON, R. W. GEE, T.-H. JI and G. W. BOWES

*Scripps Institution of Oceanography, University of California, San Diego,
La Jolla, California 92037, USA*

Cell membranes exhibit stability, activity, specificity, and adaptability which one can only describe as characteristic of Maxwell's demon [1]. Much of biology and most of physiology describes the development and remarkable function of these membranes [2]. In spite of efforts to describe them in molecular terms, there exist great differences in contemporary concepts. The mass of analytical, ultrastructural, and functional information has not dissolved present polemics. In fact they have been amplified.

The chloroplast lamellar membrane and the mitochondrial membranes are among the more specialized membranes in Nature. The retinal rod membrane system may be even more so. These paucifunctional structures should be simpler to understand and to reconstruct than the multifunctional plasma membranes of cells. Further, it is usually possible to isolate and study them in larger quantities. The classic work of David Green's laboratory in preparing mitochondrial structures on a massive scale has assured us that membrane biochemistry need not be limited to micromethodology.

The molecular structure of cell membranes has for thirty years been considered to consist of a lipid bilayer held in place by electrostatically associated protein stretched over charged surfaces. Modern NMR [3], IR, differential thermal analysis (DTA) [4], circular dichroism (CD) [5], and ultrastructural evidence [6] has been marshalled in support of this interpretation of membrane structure. The 'lipid bilayer' concept of membrane structure stemmed from Langmuir's methodology for measuring lipid interactions in monolayers. Impetus was added by Mueller and Rudin's [7] successful experimental approach to the study of a stable lipid bilayer system. Although its stability, dimensions, electrical resistance, composition, and permeability may be said to differ from those of biological membranes, the impact of their model system upon current research is profound. It is a case where experimental data speak louder than biological intuition. Evidence against the 'lipid bilayer' model for membrane structure is largely circumstantial [8]. 'Tacit understanding' [9] is a real but not widely accepted basis for persuasion in science. This paper reviews evidence for hydrophobic interaction of membrane lipid and membrane protein and presents data consistent with the lipoprotein monolayer model for membrane structure.

Lamellar and retinal membranes offer a unique opportunity for experimental study. The substrate may be added or withheld by merely turning a shutter or switch. No other membrane system demands so simple a cue to initiate its action. The several activities of the lamellar membrane are followed chemically or spectrophotometrically and many may be isolated from the total quantum conversion and electron transport processes. It is clear therefore that the lamellar membrane offers many advantages for study of membrane structure. By nondestructive methods its functional integrity may be readily assayed.

This capability is particularly sensitive in evaluating the results of membrane lipoprotein reconstitution experiments. Even when membrane function is only partially restored, there are many ways to ascertain which steps are functional and which are not.

Membrane lipoprotein structure

As a basis for designing reconstitution experiments, the concept of 'hydrophobic interaction' between amphipathic lipids, or hydrophobic pigments with membrane protein has proved useful. This concept, defined by Kauzman [10] is based upon observed entropy changes as a hydrophobic group or molecule is transferred from water to a hydrophobic medium. The entropy derives largely from the enforced organization of water by the hydrocarbon and destruction of this organization when the hydrocarbon is removed from the aqueous environment. As hydrocarbon groups approach each other, clusters of oriented water˙ must disorganize and form liquid water. This 'melting' of ice structure involves a considerable entropy increase, apparently enough to drive the equilibrium far toward micelle or folded protein structure.

$$\text{Open chain protein (hydrated)} \; \underset{\Delta S_{water} \gg 0}{\overset{\longleftarrow}{}}$$
$$\text{folded protein + water (liquid)}$$

As a consequence amphipathic lipids form micelles and (amphipathic) proteins coil to produce hydrophilic molecules with their hydrophobic groups concentrated in the interior (fig. 1).

A corollary of this relationship concerns the effect of freezing or of nonaqueous media [11] upon protein or micellar structure. When the liquid water phase is destroyed by freezing or by replacement with other solvents there no longer occurs the organized-to-disorganized water conversion in the equilibrium system and the micelle or globular protein is free to relax from its aggregated or clenched structure. Consequent loss of structural integrity, susceptibility to oxidation, or irreversible conformational changes occur. The susceptibility of many membranes to damage by freezing must be based upon the disappearance of the entropy-driven equilibrium favoring stability of these amphipathic systems.

Membrane lipoprotein stability is envisaged as a result of hydrophobic association of the hydrocarbon chains of chlorophyll or the amphipathic lipids with the hydrophobic interior of membrane structural protein. The nature of such structural protein was first revealed in the laboratories of David Green in 1960 [12] when it was noted that a 'delipidized' mitochondrial protein aggregated upon isolation but could be solubilized in the presence of appropriate detergents. It was later found that membrane structural protein possessed novel composition and characteristic molecular weights in a wide variety of membranes, low molecular weight, 23–26,000, high glutamic and aspartic acid content, low content of basic amino acids, high content of the hydrophobic amino acids, and almost no disulfide cross-linking. These proteins have been studied as detergent-solubilized 'lipoproteins'. They are exceptionally resistant to denaturation, a property resulting from their remarkable flexibility.

The phytol of chlorophyll possesses a structure superficially similar to that of polyalanine, fig. 2. It would be expected therefore to associate with the sequences of hydrophobic amino acids in the interior of the membrane lipoprotein. The resultant lipoprotein with an even more hydrophobic interior would be expected to be inaccessible to osmium or other electron-dense stains. Electron micrographs of membranes therefore reveal the un-stained membrane interior bounded by the stained surfaces of the lamellar lipoprotein monolayer. The white areas of the electron micrographs represent the lipoprotein and as such are the significant but often neglected aspects of the micrographs.

Except for chlorophyll, all the amphipathic lipids of membranes possess two hydrocarbon chains for some not yet understood reason. Lipoprotein stability may require the decreased probabilities of hydrocarbon migration over the surface of the protein.

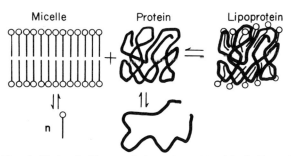

Fig. 1. Hydrophobic association of amphipathic lipids and membrane protein. Equilibria expressed by vertical arrows are associated with large changes in water entropy. Membrane lipoprotein is represented by the diagram at the right.

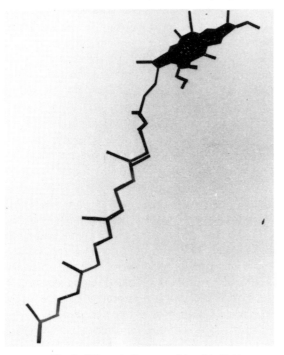

Fig. 2. Chlorophyll, an amphipathic lipid.

Lipid compositions of chloroplast and other lipoproteins have been reported [13]. As analyses of more homogeneous lipoprotein components of membrane systems become available the fatty acid compositions of their lipids become more unique. Allen and Good [14] have now reported analyses for a spinach lipoprotein fraction rich in P-700 activity which indicate by far the most specific association between membrane lipids and protein yet observed. The preparation contained no monogalactolipid but included copious amounts of digalactolipid and sulfolipid. The observed extractability of monogalactolipid with a benzene—15% acetone mixture indicates considerable difference in ease of dissociation of the glycolipid from lamellar protein.

Each of the four major amphipathic lipids of chloroplasts is associated with a characteristic limited group of fatty acids. The fatty acid content of each is distinctly different from that in the other three. Thirty-five percent of the phosphatidyl glycerol molecules are Δ^3-*tr*-hexadecenoic ester (fig. 3). This acid is found nowhere else in the plant. Of the digalactolipid molecules, almost all are dilinolenin glycosides. The monogalactolipid from which it is seemingly derived contains an appreciable amount of 16:3 acid (16:4 in *Euglena*) (fig. 4). The C_{16} acid is not found in the other lamellar lipids. The sulfolipid resembles mammalian phosphatides in that it contained equal amounts of saturated and unsaturated fatty acid esters. These specific hydrophobic 'labels' of each of the amphipathic lipids attest to the hypothesis that membrane structural protein associates best with certain hydrocarbon chains and degrees of unsaturation and that each has hydrophobic sites adapted to accommodate certain fatty chains. Recent studies in this laboratory indicate that membrane phospholipids of other organisms maintain constant fatty acid composition in spite of profound alteration in total depot fat composition. Dietary fat composition seems to have little effect upon membrane lipoprotein fatty acid composition, adding another bit of circumstantial evidence in support of specific association of hydrocarbon chains with membrane protein.

Experiments designed to evaluate the extent of association of lamellar membrane lipids with lamellar protein have revealed a stoichiometry of binding [15]. The lamellar protein associates with thirty hydrocarbon chains per molecule of protein. The same figure has been observed for chlorophyll and each of the lamellar lipids as well as for the detergent sodium dodecyl sulfate (fig. 5). This closely approximates the concentration of lipids in the native lamellar membrane. One may consider an array of lipids in a membrane lipoprotein such that the lipids are

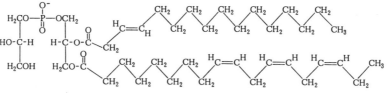

Fig. 3. Phosphatidyl glycerol, the major phospholipid of chloroplast lipoprotein.

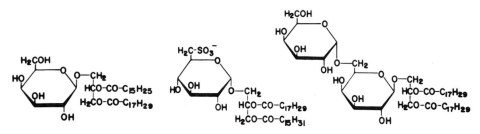

Fig. 4. The amphipathic glycolipids of chloroplasts, monogalactosyl diglyceride, the sulfolipid, sulfoquinovosyl diglyceride, and digalactosyl diglyceride.

equally distributed on the exterior and interior surfaces. This would require that seven or eight lipids and chlorophylls be associated with each surface of the 23,000 M.W. protein subunit, not an unreasonable figure. In the less likely event that membrane lipoprotein subunits are asymmetric tetrahedra [16], one face of which is exposed to the exterior of a membrane, the number of lipids on that side would be double as would the surface area.

Amphipathic molecules other than the lipids may associate with membrane protein. Xanthophylls and

long chain quinones occur naturally in chloroplasts. Their hydrophobic groups must be exposed to the aqueous environment such that their function can be performed. Cholesterol does not occur in plant membranes though the sterol glycosides certainly find sites in membranes. Certain polypeptides elicit unique activities in plant membranes. Valinomycin, antimycin A, and nigericin are compounds which, like insulin, find their way to specific sites in the membrane and act as carriers, binding sites, or inhibitors of membrane function. They must perform their

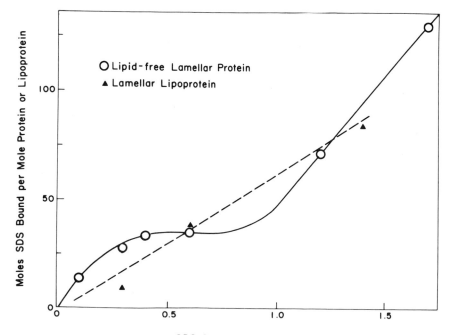

Fig. 5. Association of sodium dodecyl sulfate detergent with chloroplast lamellar protein. Stoichiometric association occurred at 0.4 mg SDS per ml.

function by assuming the structure of natural membrane components. The fact that valinomycin associates with lipid bilayer systems and facilities K⁺ transport should not preclude the possibility that it could associate hydrophobically with membrane protein to give K⁺ binding sites in mitochondrial or other membranes.

Purely hydrophobic molecules may associate with lamellar membranes. β-Carotene is such a molecule. Under certain conditions it is bound specifically by lamellar protein [17]. Phytanic acid, derived from chlorophyll's phytol associates too well with critical membrane sites in sufferers from Refsum's Disease. Certain petroleum hydrocarbon fractions act as specific toxins and can serve as weed-killers [18]. They must function by virtue of their affinity for membrane lipoprotein at especially critical sites. Even more specific are the steroid and insect hormones [19] which elicit cyclic AMP production at their sites of binding in membranes. Volatile chlorinated hydrocarbons associate with hydrophobic membrane sites with considerable non-specific damage. DDT and its derivatives bind specifically in membrane structures. Their structure and polarity lead to association at critical sites in the electron transport system and subsequent derangement or inhibition of metabolic function. Lawler and Rogers [20] studied the effects of DDT on photophosphorylation in barley and chloroplasts. We report in this paper experiments with DDT which acts rather specifically as an inhibitor of electron transport in chloroplasts. Ultimately inhibitors such as these may be useful in locating the functional sites for electron transport function. Nature, being masterful at molecular selection will probably point to novel structures like tetrodotoxin which are most selectively bound at certain sites in membranes.

Experimental methods and results

Chloroplast preparation

Spinach chloroplasts were prepared in 0.1 M Tricine buffer pH 7.8 in the presence of 5 mM $MgCl_2$ and 20 mM KCl. Class II chloroplasts were selected in order to facilitate uptake of ionic substrates. ATP production was measured as described by Avron [21] using benzene-isobutanol extraction of inorganic

phosphate as oxidized phosphomolybdate complex. ATP and esterified P^{32} in the aqueous phase was counted in the scintillation counter. Ferricyanide was measured at 420 mμ.

Chloroplasts of the green alga *Codium fragile* were prepared in the same way. Plants from deeper water yielded chloroplasts five times more active than those from the intertidal zone. Chloroplasts of the marine diatom *Ditylum brightwellii* were prepared in 0.5 or 0.1 M sodium chloride, Tricine, or 1 M sorbitol. While *Codium* chloroplasts were relatively stable with linear activity for at least ten min, those of *Ditylum* lost most of their activity after five min at 22°.

DDT and DDE (1,1-di-p-chlorophenyl-2,2-dichloroethylene) standard solutions in ethanol were added in identical volumes to each cuvette. The suspension was mixed by gentle agitation. Tungsten iodine lamps with a dilute copper sulfate solution filter gave 4,000 f.-c. at the surface of the cuvettes.

Photophosphorylation was uncoupled by ammonium sulfate. With ADP + P_i ferricyanide reduction was 320 to 355 μmoles per mg chlorophyll per hr. In the presence of 8.4×10^{-5} M ammonium sulfate the photoreduction of ferricyanide increased from 355 to 675 μmoles/mg chl/hr. The ammonium inhibited system was affected by DDT in the same manner as was ferricyanide photoreduction in the absence of ammonium ion (fig. 6, table 1).

DDE, the biological degradation product of DDT,

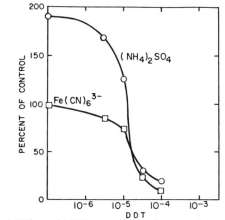

Fig. 6. Effect of DDT upon photoreduction of ferricyanide in spinach chloroplasts. $Fe(CN)_6^{-3}$ curve: Control, without added ammonium ion. $(NH_4)_2SO_4$ curve: Photoreduction of ferricyanide in the presence of 4.8×10^{-5} M ammonium sulfate as uncoupler of photophosphorylation.

Table 1
Inhibition of photoreduction of ferricyanide by DDT in spinach and *Codium* chloroplasts.

DDT Conc.	μmoles Fe(CN)$_6^{3-}$ reduced/mg chl/hr		μmoles ATP per mg chl/hr	P/2e
0	320	180*	198	1.25
10^{-6}	312	182	207	1.32
5×10^{-6}	288	155	207	1.21
10^{-5}	170	75	103	1.22
5×10^{-5}	88	–	21.4	0.5

* *Codium fragile* chloroplasts, same conditions as spinach chloroplasts.

Tricine, 250 μmoles pH 7.8; MgCl$_2$, 12.5 μmoles; KCl, 50 μmoles; Fe(CN)$_6^{3-}$, 2 μmoles; ADP, 1.0 μmole; ^{32}Pi, 10 μmoles; 55 μg chlorophyll. Total vol., 3.0 ml. 4,000 f.-c. 5 min. 22°, killed with 0.5 ml 20% TCA.

inhibited photoreduction of ferricyanide or 1,6-dichlorophenolindophenol (DPIP) reduction to the extent of 68% at 5×10^{-5} M. Under the same conditions, DDT inhibited photoreduction 72%. In comparison, we have observed concentrations of 10^{-4} M DDE in liver tissue of a local sea gull.

DDT and DDE are hydrophobic molecules which may be anticipated to associate with hydrophobic sites within membrane lipoprotein. The observed inhibition of electron transport indicates that the inhibitor has so altered the structure of some member of the electron transport chain that electron transport rates are dramatically diminished. If lamellar membrane possessed a lipid bilayer structure the DDT and DDE would be dissolved or bound within the bilayer as they would be in synthetic model bilayer. One would not, in that case, anticipate a dramatic specific effect of inhibitor upon electron transport which involves cytochromes and other lipoprotein-bound redox systems.

Freeze etch technology

A credible model for lamellar membrane structure must be consistent with the appearance of freeze etch replicas as well as with electron micrographs of their sections of stained fixed tissues. In the latter a membrane is revealed by the heavy metal adsorbed on its outer and inner surfaces. These often show individual membrane subunits as a series of white areas

spaced 39 Å apart, exactly the periodicity measured by Kreutz [22] using low angle scattering in the direction of the lamellar surface. This series of sub-units in the image of a membrane section was noted by Sjostrand [23]. The more obvious aggregation of osmium along membrane surfaces must be the result of aggregation of ion-binding groups which lead during fixation and staining to a series of aggregates of metal stain. The fact that molecules like cytochrome *c* possess most of their external carboxyl groups in certain areas and most of their cationic amino groups congregated at other sides of the molecule render plausible the proposal that certain areas of a membrane lipoprotein may bind more osmium than do other areas.

In freeze etch replicas the problem has been the assignment of inside or outside of the membrane. Interpretation of the replica micrographs has depended upon the concept of what happens when a membrane is frozen and cleaved. It is not enough to say that the cleavage follows the membrane. It has been interpreted as following one or the other side of the membrane. In a masterful experiment Deamer and Branton [24] showed conclusively that an artificial lipid bilayer cleaved in its center, through the hydrophobic plane of the lamellar micelle. They argued, further, that natural membranes do the same, a logical conclusion.

In the membrane model discussed in this paper, we see that cleavage should take place within the hydrophobic region of the lipoprotein monolayer membrane. The result is not exactly simple because the cleavage may take place on either side of the central hydrophobic region of the protein, the amphipathic lipids being 'uprooted' from the protein sites on one side of the membrane or on the other. The effect of freezing upon the lipid bilayer micelle and upon the lipoprotein monolayer membrane would be similar. Forces holding the lipids within their aggregated bilayer or within membrane protein would disappear upon freezing of the water surrounding the membrane and it would be easy to tear fatty acid or chlorophyll hydrocarbon chains from the membrane protein. The appearance of freeze etch replicas reveals features consistent with this interpretation.

In the vacuum of the etching process the two exposed surfaces of the cleaved membrane would be complementary but different. In one, the naked membrane protein would present a smooth appearance in the replica. In the other, the uprooted hydrocarbon chains would be standing, like hair, on the surface of the protein molecules previously bound electrostatically to the surface of the membrane. Electrostatic forces are not appreciably affected by freezing and hence should suffice to continue to bind the amphipathic ionic lipids to the cationic surfaces of adjacent proteins. These supralamellar proteins are the large molecules revealed so strikingly in freeze etch replicas. It is not yet possible to identify them by their appearance. Membrane-associated coupling factors, cytochromes *c*, Fraction I protein, etc., molecules which are dissociated from lamellar membrane by raising the ionic strength or removal of divalent cations, should appear as large particles associated with the membrane.

Reconstitution of lamellar membrane

Recent experiments in this laboratory and others have demonstrated the possibility that lamellar photosystems and electron transport steps can, at least in part, be reconstituted when isolated pigments and lipids are re-aggregated with their apo-lipoprotein. By application of principles outlined above, the forces determining lipoprotein stability have been controlled and directed in reconstitution of the membrane system.

A cooled suspension of cold 80% acetone-extracted lamellae was suspended and mixed with a concentrated pigment and lipid extract in 80% acetone at −10°. Careful addition of water to a concentration of 65% acetone aggregated the dispersed lipid and protein with recovery of a major fraction of the native electron transport capability. Photoreduction of methyl red (and TPN) was measured using reconstituted chlorophyll lipoprotein and an N,N′-tetramethylphenylaminediamine (TMPD)-ascorbic acid couple or DPIP-ascorbic acid couple. The reconstituted chlorophyll protein exhibited 4–6 times the rate of photoreduction of methyl red or TPN than chlorophyll alone when suspended in 2% Triton X−100. Using the DPIP-ascorbate couple a definite 2-fold stimulation in the photoreduction of methyl red or TPN could be demonstrated with addition of plastocyanin and a crude ferredoxin. These results indicate reconstitution and restoration of electron flow through Photosystem I.

Optimal activity of the reconstituted system was attained at a concentration of 2% Triton X−100. Optimal activity of a sonicated preparation of native lamellae was attained at 0.2% Triton X−100. Under such optimal conditions the rates for photoreduction of methyl red by Photosystem I were nearly equal. In both cases the activity was maximal at these Triton X-100 concentrations.

Assembly of membrane lipoprotein

Membrane structural protein may be synthesized as a water-soluble entity which associates with amphipathic lipids as they become available and which then aggregates to form membrane. The experiments of Camejo et al. [25] with serum high density lipoprotein offer an elegant example. Delipidized lipoprotein was water-soluble but adsorbed strongly at the surface. Its interaction with surface adsorbed lipids was clearly demonstrated. Although this is a possible route to lamellar lipoprotein biogenesis, the evidence of Goldberg and Ohad [26] clearly documented in greening *Euglena* the close metabolic control of biosynthesis of each of the essential membrane components, chlorophylls a and b, carotenoids, galactolipids, linolenic acid, sulfolipid, and membrane protein. It is not likely that the chloroplast accumulates membrane components to any great extent.

Examples of lipoprotein accumulation prior to membrane production have been reported. Price and Thomson [27] studied the ultrastructure of crystalline inclusions in macadamia leaf chloroplasts. These appear to be three dimensional aggregates of protein (apparently Fraction I protein, S.G. Wildman, personal communication) which, when provided with essential lipid and other components produce the two-dimensional aggregates characteristic of lamellae. Crystalline inclusions in chloroplasts of banana fruit [28] indicate similar storage of aggregated protein essential for chloroplast synthesis. Similar crystalline structures in Hydra [29] appear to be converted to mitochondrial membrane. Even the Heitz-Leyon crystalline structures of the 'prolamellar body' could be a linear aggregate of membrane lipoprotein which on addition of chlorophyll may produce lipoprotein with symmetry leading to two-dimensional lamellar aggregates. With the development of immunochemical procedures for recognizing incognito membrane proteins it will be possible to ascertain the role of these crystallites in membrane biogenesis.

A model for chloroplast lamellar structure

A three-dimensional model for the chloroplast granum is presented in fig. 7. It depicts the small 39 Å subunits of the thylakoid membrane as possessing most of the chlorophyll. As a result of contact of two of these membranes (the 'partition'), we have implicated a high concentration of chlorophyll with the possibility of dimer formation utilizing chlorophyll molecules from adjacent lipoprotein molecules. Such dimers would, of course, not be possible in the parenchyma bundle sheath type of chloroplast where the thylakoids are independent entities. The fact that chlorophyll in the two types of chloroplasts has different absorption and fluorescence properties is of considerable interest and of considerable pertinence to any structural model for the chloroplast. The packing of chlorophyll in such lamellae must affect its energy transfer and fluorescence probabilities.

The fact that the highly unsaturated linolenic acid (18:3) esters in the lamellar amphipathic lipids are stable must be interpreted. Their location in the thylakoid must preclude oxygen attack, either photosensitized or not. Isolated chloroplasts are subject to photo-induced cyclic peroxidation. This epoxide-mediated reaction leading to malondialdehyde does not occur in vivo where the polyunsaturated hydrocarbon chains must be very effectively protected from active oxidant. That these lipids can survive in an environment high in oxygen is one of the most remarkable aspects of lamellar lipoprotein structure. It is clear that the linolenic acid chains are snugly buried within a protective protein matrix. A lipid bilayer-type membrane could hardly provide this degree of protection.

The large 100 Å thick particles observed within the thylakoid stained sections and arranged upon the

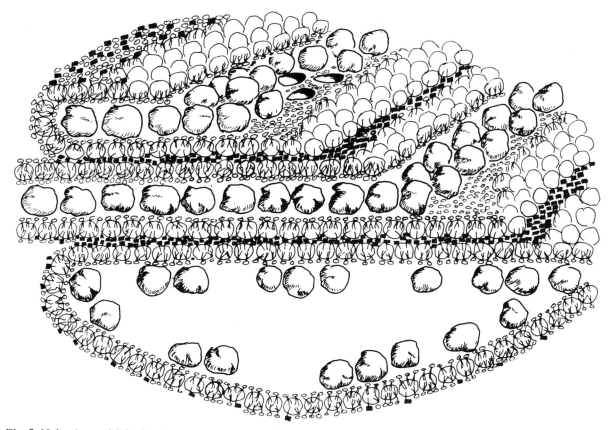

Fig. 7. Molecular model for lamellar membrane system based on the lipoprotein monolayer. Amphipathic membrane lipids and chlorophylls are associated hydrophobically with membrane proteins.

interior of the thylakoid membrane in freeze etch replicas must be associated with the membrane by virtue of electrostatic forces. In order to isolate such particles it has been necessary to rupture the thylakoids. In the work of Howell and Moudrianakis [30] it was necessary to use EDTA or other ionic bridge-breaking conditions to free particles of their ionic association with the charged membrane. The observation by Dilley [31] that poly-L-lysine is a potent uncoupler of photophosphorylation provides further convincing evidence for this point of view. It is clear that polylysine will interact with the anionic membrane and not with the larger molecules normally bound to the membrane such as phosphorylation coupling factors. A basic polyelectrolyte would displace magnesium, calcium, and any basic protein surfaces such as those of the large particles within the thylakoid. It would effectively cover the charged thylakoid surface and prevent ionic association with any large molecules. Cytochrome c appears to be such a protein (E. Margoliash, personal communication) which may be bound by its high concentration of cationic groups on certain sides of the molecule. The asymmetry of the membrane and of its associated particles is one of the important characteristics of biological systems.

Many aspects of lamellar membrane structure remain to be ascertained. The value of all types of chemical, physical, biological, and structural evidence is clear. The function of hydrophobic hormones and inhibitors such as DDT can be interpreted on the basis of their hydrophobic interactions in membrane lipoprotein. An understanding of the basic principles of lipid-protein interactions in biological systems will provide basis for experiments in reconstruction of membranes from their components as well as for investigation of the processes of biosynthesis of membranes in Nature.

Acknowledgement

This work was supported by a research grant GM-12310 from the National Institute of General Medical Science.

References

[1] J. Clerk Maxwell, in: Theory of Heat, 6th edition (Appleton and Co., New York, 1880) p. 328.
[2] R.A. Peters, Nature 177 (1956) 426.
[3] D. Chapman, Science (1968) 55.
[4] J.M. Stein, Science 162 (1968) 909.
[5] J. Lenard and S.J. Singer, Science 159 (1968) 738.
[6] W. Stoeckenius, Proc. Int. Conf. Biological Membranes, Stresa (1969), in press.
[7] P. Muller and D.O. Rudin, Nature 213 (1967) 603.
[8] A.A. Benson, Ann. Rev. Plant. Physiol. 15 (1964) 1.
[9] M. Polyani, in: The tacit dimension (Doubleday and Co, New York, 1966) p. 25.
[10] W. Kauzman, Adv. Protein Chem. 14 (1959) 1.
[11] S.J. Singer, Adv. Protein Chem. 17 (1962) 1.
[12] S.H. Richardson, H.O. Hultin and D.E. Green, Proc. Natl. Acad. Sci. (U.S.) 50 (1963) 821.
[13] C.F. Allen, O. Hirayama and P. Good, in: Lipid composition in photosynthetic systems in biochemistry of chloroplasts, Vol. I., ed. T.W. Goodwin (Academic Press, London, 1965) p. 195.
[14] C.F. Allen and P. Good, Biochem. Biophys. Res. Commun., in press.
[15] T.-H. Ji and A.A. Benson, Biochim. Biophys. Acta 150 (1968) 686.
[16] John R. Platt, personal communication.
[17] T.-H. Ji, J.L. Hess and A.A. Benson, Biochim. Biophys. Acta 150 (1968) 676.
[18] J. van Overbeek and R. Blondeau, Weeds 3 (1954) 55.
[19] J.E. Amoore, G. Palmieri, E. Wanke and M.S. Blum, Science 165 (1969) 1266.
[20] P.D. Lawler and L.J. Rogers, Biochem. J. 110 (1968) 110.
[21] M. Avron Biochem. Biophys. Acta 40 (1968) 257.
[22] W. Kreutz, Z. Naturforsch. 19B (1964) 441.
[23] F.S. Sjostrand, Nature 199 (1963) 1262.
[24] D.W. Deamer and D. Branton, Science 158 (1967) 655.
[25] G. Camejo, G. Colacicco and M.M. Rapport, J. Lipid Res. 9 (1968) 562.
[26] I. Goldberg and I. Ohad, in press.
[27] J.L. Price and W.W. Thomson, Nature 214 (1967) 1148.
[28] J.M. Bain, Austral. J. Biol. Sci. 21 (1968) 421.
[29] L.E. Davis, J. Ultrastruct. Res. 21 (1967) 125.
[30] S.H. Howell and E.N. Moudrianakis, J. Mol. Biol. 27 (1967) 323.
[31] R.A. Dilley, C.J. Arntzen and R.W. Keck, XIth Int. Botan. Congr. Abstracts (1967) p. 47; Biochemistry 7 (1968) 338.

THE RECONSTITUTION OF MICROSOMAL MEMBRANES

J. K. POLLAK, R. MALOR, Maria MORTON and K. A. WARD *

Department of Histology and Embryology, University of Sydney,
Sydney, N.S.W., Australia

In a symposium on mitochondria and chloroplasts it may seem unwarranted to include an investigation into the membranes of the endoplasmic reticulum and the derived smooth and rough microsomal fractions. Nevertheless, the mechanism of in vitro formation of microsome-like vesicles may have relevance to the assembly of membranes in other cellular organelles. The membranes of the smooth microsomal fraction have been reported to possess at least some properties in common with the outer mitochondrial membrane [1]. This superficial similarity and the possibility that the outer mitochondrial membrane has evolved from elements of the endoplasmic reticulum have been pointed out repeatedly [2,3].

Our studies of the endoplasmic reticulum and the microsomal fractions of embryonic chick liver and adult rat liver [4–8] led us to propose the presence of a non-membranous, particulate component, 'the reticulosome', within the microsomal fraction. Since reticulosomes were found to be most abundant immediately prior to membrane formation and to decrease inversely with membrane formation, it was suggested that reticulosomes were membrane precursors [7].

This concept was first formulated during studies of changes in the microsomal fraction derived from embryonic chick liver at different morphogenetic stages [4–6]. The results pointed to the presence of a non-membranous, non-ribosomal 'particulate' (i.e. sedimentable at 105,000 g (av.) for 60 min in 0.25 M sucrose) component with a high protein and low phospholipid content (protein: phospholipid ratio varied from 8:1 to 12:1).

Treatment of the rough microsomal fraction with deoxycholate (DOC) and RNase was designed originally as an analytical tool to determine the presence of 'particulate' protein other than ribosomal or membrane protein. This treatment resulted in the solubilization of membrane and ribosomal proteins [4,8]. When the results of these studies led to further investigations into the nature of the 'particulate' component, the DOC and RNase treatment became the standard method of isolation.

On investigating the amount of DOC required to solubilize membranes, consistently membrane-free reticulosomal preparations were obtained only when the DOC/protein ratio was maintained between 1–1.5 (Hallinan and Pollak, unpublished). At lower DOC concentrations, membranous vesicles contaminated the preparation, producing marked decreases in the protein/phospholipid ratios, while at higher DOC concentrations, the reticulosomes were also solubilized.

Studies with reticulosomes from embryonic chick liver (Pollak, unpublished) and subsequently with reticulosomes isolated from rat liver showed that isolated reticulosomes react with micellar microsomal phospholipid to give rise to a stable lipoprotein complex. When examined with the electron microscope, this complex was found to resemble the smooth vesicles of a microsomal fraction obtained by differential centrifugation [5].

Membrane assembly by reticulosomes and phospholipid will be described from two points of view. Firstly the properties of reticulosomes will be described in some detail, with special reference to membrane formation. Secondly the possible biological significance of reticulosomes will be discussed, bearing in mind the use of the detergent DOC during the isolation procedure of reticulosomes.

---

* Present address: Department of Zoology, University of Massachusetts, Amherst, Mass. 01002, USA.

Materials and methods

(a) Animals

Female Sprague-Dawley and Wistar albino rats of 150–180 g weight were maintained on standard laboratory rat and mouse food cubes. When fasting of various time intervals was desired, food was withdrawn from between 9–11 a.m. for the requisite time and water supplied *ad libitum.*

(b) Cell fractionation

Total microsomes and reticulosomes were prepared as previously described [8]. Ribosomes plus polysomes, rough microsomal vesicles and cell sap were isolated by the method of Bloemendal et al. [9].

(c) Preparative procedures

Microsomal micellar phospholipid was prepared by the method of Ward and Pollak [8]. Total microsomal phospholipid was prepared by sonicating the whole lipid extract and then treating it in the same way as the chromatographed phospholipids.

Structural proteins were isolated as described previously [8]. All the polyacrylamide gel electrophoresis was also carried out as described by Ward and Pollak [8].

(d) Phospholipid titration experiments

[14]C-labelled rough microsomes were prepared by injecting a female rat intraperitoneally with 10 μCi/0.31 mg of ^{14}C-labelled choline chloride in a volume of 0.2 ml, which also contained 10 mg of Difco casamino acids. After 30 min the rat was killed and a rough microsomal fraction (Fraction II, Dallner [10]) prepared.

For the preparation of ^{14}C-labelled micellar phospholipid, a rat was injected as above, a total microsomal fraction was isolated and the labelled micellar phospholipid was prepared from this fraction.

The in vitro assembled smooth membrane lipoprotein complexes were separated from the rough microsomes by the discontinuous gradient method of Dallner [10] and protein and phospholipid were determined as described previously [8].

(e) Assembly and isolation of ribosome-studded vesicles

Reticulosomes were derived from about 4 g of liver tissue and suspended in 3–5 ml 0.25 M sucrose. To this preparation, ribosomes and polysomes derived from 2 g of liver tissue and suspended in 2–3 ml 0.35 M sucrose in medium B [9] and between 1.2 mg – 6 mg microsomal micellar phospholipid suspended in 3–6.5 ml, were added simultaneously. The suspension was mixed by inverting the tube several times, left at 2°C in the dark for 20 min and then centrifuged at 105.000 g (av.) for 60 min (or for 270.000 g (av.) for 35 min). The resultant pellet was resuspended in 0.35 M sucrose in medium B and 2.5 ml portions were layered over a discontinuous gradient in a SW 39 centrifuge tube which consisted of 1.25 ml 2 M sucrose and 1.25 ml 1.5 M sucrose all made up in medium B. The tubes were then pelleted for 16 hr at 75.000 g (av.) and the layer immediately above the 2.0 M sucrose interface was removed carefully, made up to a suitable volume with 0.25 M sucrose and centrifuged either at 105.000 g (av.) for 60 min or at 270.000 g (av.) for 35 min. The pellets were then analysed for protein, phospholipid and RNA, used for amino acid incorporation studies or fixed for electron microscopy.

(f) Enzyme assays

Glucose-6-phosphatase and inorganic pyrophosphatase were assayed as described previously [6]. Pyrophosphate-glucose-phosphotransferase activity was measured by a modification of the method of Nordlie and Arion [11]. $NADH_2$-cytochrome c oxidoreductase and $NADPH_2$-cytochrome c oxidoreductase were assayed by the method described by Sottacasa et al. [12].

(g) Isotope techniques
(1) Glycerol incorporation into phospholipids

[14]C-glycerol had a specific activity of 14.3 mCi/mmol, and 140 mμmols were used per incubation tube. The precise incubation conditions are presented in the relevant table.

The incubations were carried out in flasks for 20 min and the reaction stopped by the addition of

20 volumes of chloroform-methanol (2:1 v/v). The zero time control tubes had the chloroform-methanol added before the addition of the enzyme. After extraction at room temperature for one hr the samples were filtered through glass wool and 1/5 volume of 0.9% NaCl added. The lipid phase was then washed twice with the Folch upper phase containing 0.1 M glycerol and then twice with upper phase alone.

The lipids were dried by rotary evaporation and stored under nitrogen in the cold.

The neutral lipids and phospholipids were separated by thin-layer chromatography on Kieselgel G. Two solvent systems were used. The first system was petroleum ether, ether, acetic acid (90:10:1 v/v) which removed the neutral lipids and left the phospholipids at the origin where they could be collected. The second system consisted of chloroform, methanol, water (80:35:5 v/v) which ran the entire neutral lipid fraction to the solvent front. The chromatograms were examined under ultraviolet light and the regions of the two chromatograms corresponding to total phospholipid and total neutral lipid were scraped off and counted.

(2) Amino acid incorporation into proteins

L-^{14}C-phenylalanine had a specific activity of 7 mCi/mmol, and 40 mµmols were used per incubation tube. The reactions of the incorporation experiments were terminated at 0 min and 30 min by plating 0.1 ml samples (in duplicate) onto 3 MM filter paper discs; these were exposed to a stream of hot air for about 15 sec and then immersed in ice cold 10% TCA (containing 0.05 M phenylalanine). The paper discs were then treated by the method of Mans and Novelli [13], and counted in a liquid scintillation counter.

(h) *Electron microscopy*

The fixation of liver tissue and isolated subcellular fractions and the subsequent preparative procedures as well as the negative staining techniques have been previously described [5,6]. Either a Philips 200 or an Hitachi 11b electron microscope was used for examining purposes.

Results and discussion

Electrophoretic patterns of proteins derived from the endoplasmic reticulum

When reticulosomal proteins are examined by means of polyacrylamide gel electrophoresis and compared with microsomal proteins, microsomal structural proteins and ribosomal proteins, it is apparent that distinct similarities do exist between the reticulosomal proteins and the microsomal proteins, while little overlap is shown by the bands of the reticulosomal and ribosomal proteins (fig. 1). The

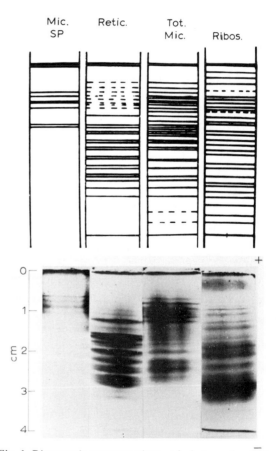

Fig. 1. Diagramatic representations and photographs of electrophoretic patterns of microsomal structural protein (Mic SP), reticulosomal protein (Retic.), total microsomal membrane proteins (Tot. Mic.) and ribosomal proteins (Ribos.).

most conspicuous difference between reticulosomal proteins and microsomal proteins shown up by gel electrophoresis is the relative paucity of structural proteins in the reticulosomes (compare figs. 1a, 1b and 1c), by corollary this would imply that reticulosomes are rich in enzymic proteins.

Titration of reticulosomes with phospholipid

Before describing some of the enzymic activities which have been measured in the reticulosomal fraction, it is convenient at this point to demonstrate the ability of rat liver reticulosomes to react with micellar microsomal phospholipid to give rise to membranous lipoprotein complexes. Micellar microsomal phospholipid was added to a suspension of reticulosomes and left for up to 20 min at either 30°C or 2°C. The mixture was centrifuged for 30 min at 105,000 g (av.), resuspended in 3 ml 0.25 M sucrose and layered over 1.8 ml 1.3 M sucrose in a SW39 Spinco tube and centrifuged 60 min at 105,000 g (av.). Under these conditions the newly formed complex was always situated at the 1.3 M − 0.25 M interface. Untitrated reticulosomes, on the other hand, always formed a pellet.

On analysis, after suitable harvesting and washing by centrifugation [5], it was demonstrated that the interface material is a lipoprotein complex with relatively constant proportions of protein and phospholipid, which are only slightly affected when protein is not present in excess (table 1). Therefore, it is stipulated that reticulosomes and micellar phospholipid combine in stoichiometric proportions of 2 parts of protein to one part of phospholipid (by weight).

When the ultrastructure of such lipoprotein complexes is examined, it is clearly seen that membranous vesicles have been formed. These assembled vesicles bear a morphological resemblance to smooth microsomal vesicles as isolated by differential centrifugation (fig. 2).

Enzymic activities of reticulosomes and assembled membranous vesicles

Reticulosomes were found to have very high specific activities of glucose-6-phosphatase. This enzyme catalyses three different reactions, being a glucose-6-phosphate phosphohydrolase, an inorganic pyrophosphate phosphohydrolase, and a pyrophosphate-glucose-6-phosphotransferase [11]. The specific activities of these three enzymically catalized reactions are greater in reticulosomes than in microsomes (table 2). Arion and Nordlie [14] have observed that in the presence of 0.2% DOC, glucose-6-phosphohydrolase and inorganic pyrophosphate-glucose-6-phosphotransferase activities are enhanced. In the present experiments the increases in the specific activities of the reticulosomal preparations are considerably greater than the increased activities obtained by incubation in 0.2% DOC as carried out by Arion and Nordlie

Table 1
Combining power of reticulosomes with micellar microsomal phospholipids

Experimental conditions	Reticulosomal protein (µg/tube)	Added phospholipid (µg/tube)	Bound phospholipid (µg/tube)	$\dfrac{\text{Protein}}{\text{phospholipid}}$ ratio
Incubation at 30°C	360	0	42	8.6
(data from	340	1500	190	2.1
Pollak et al. [5])	760	1500	360	2.3
	1160	1500	498	2.3
	1520	1500	592	2.6
Incubation at 2°C	196	0	19	10.05
	164	1600	73	2.26
	308	1600	143	2.15
	440	1600	191	2.30
	1370	1600	560	2.45

Incubation conditions and isolation and analytical procedures have been described in the methods section and in the text. Incubations were carried out for 20 min.

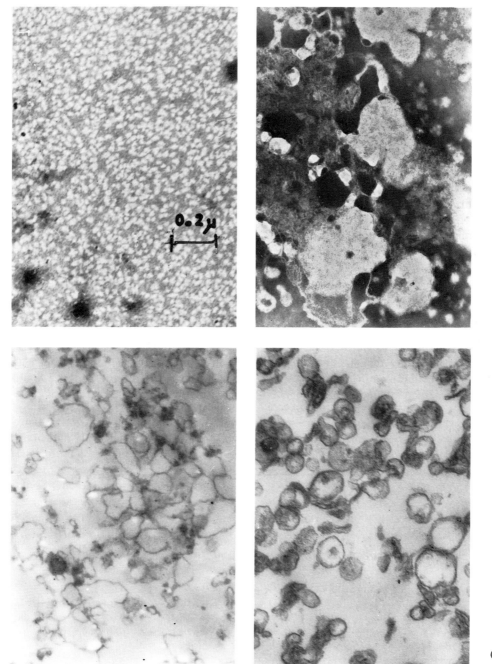

Fig. 2. (a) Reticulosomes negatively stained with potassium phosphotungstate (pH = 6.5) ($\times$ 51,000). (b) Negatively stained assembled vesicles, formed by mixing reticulosomes and micellar microsomal phospholipid ($\times$ 51,000). (c) Section through an OsO_4 – fixed pellet of assembled vesicles ($\times$ 51,000). (d) Section through an OsO_4 – fixed smooth microsomal pellet, for comparison with fig. 1c ($\times$ 51,000).

Table 4
^{14}C-Glycerol incorporation into phospholipids and neutral lipids by reticulosomes and total lipid titrated reticulosomes.

Incorporation into	Reticulosomes (cpm/mg protein)	Reticulosome-lipid complex (cpm/mg protein)
Phospholipids	18,966	90,921
Neutral lipids	776	888

The incubation mixture contained 140 mμmoles ^{14}C-glycerol (14.3 mCi/mmol), 5 μmoles MgCl$_2$, 5 μmoles palmitic acid in Tween 80/water, 0.2 μmol coenzyme A, 10 μmoles ATP, 1 μmol CDP-choline, 5 μmoles reduced glutathione, 0.2 ml cell-sap (the post microsomal cell-sap was dialysed against two changes of 0.1 M phosphate buffer pH 7.4 and 0.5 mM NaF for five hr at 2°C immediately before use) and 0.1 ml of enzyme in a total volume of 1.0 ml. Incubation was carried out at 30°C for 20 min with shaking. All other details are described in the Materials and methods section.

NADPH$_2$-cytochrome c reductase) and the proliferation of the smooth vesicles of the endoplasmic reticulum [17].

The in vitro incorporation of ^{14}C-glycerol into phospholipids and neutral lipids was also measured using reticulosomes and assembled membranes. Phospholipid synthesis as measured by the incorporation of ^{14}C-glycerol into phospholipids was significantly enhanced when membranous reticulosome-lipid complexes were used instead of free reticulosomes (table 4). Incorporation into neutral lipids was low and virtually unaffected by membrane formation (table 4). It is of interest that reticulosomes, which at least in vitro give rise to membranous vesicles when mixed with phospholipids or lipids, do contain the enzyme system which is involved in the synthesis of phospholipid, the other component, required for membrane assembly.

Formation of ribosome-studded vesicles

As an additional indication of the capacity of reticulosomes to give rise to membranous vesicles, it was attempted to assemble ribosome-studded (rough) vesicles by mixing reticulosomes, ribosomes plus polysomes and either microsomal micellar phospholipid or total lipid.

To demonstrate the formation of rough vesicles from nonmembranous precursors a method is re-quired which is able to separate smooth vesicles, ribosome-studded vesicles, reticulosomes and free ribosomes plus polysomes from one another. The discontinuous gradient method of Bloemendal et al. [9] fulfilled this requirement. After centrifugation (see Materials and Methods section) free ribosomes and polysomes formed a pellet at the bottom of the tube, isolated rough vesicles (Bloemendal fraction 3) and assembled ribosome-studded vesicles banded just above the 2.0 M sucrose interface, while smooth vesicles, either isolated or assembled from the interaction of reticulosomes and phospholipid, were situated in the upper half of the 1.5 M sucrose layer. Isolated reticulosomes when present by themselves found their equilibrium in an intermediate position between smooth vesicles and ribosome-studded vesicles. In the presence of ribosomes, most of the reticulosomes tend to aggregate with ribosomes and are sedimented along with these into the 'ribosomal' pellet. Thus during the isolation of ribosomes some reticulosomal material will be sedimented into the pellet, along with the ribosomes. Table 5 gives protein, phospholipid and RNA values of the three reactants and of the ribonucleo-lipoprotein complex of the assembled ribosome-studded vesicles at the 2.0 M sucrose interface. Reticulosomes from rat liver usually have a protein/phospholipid ratio in the vicinity of 8 (tables 1 and 5, see also [5]). The ribosomal preparations were checked for membrane contamination with the electron microscope and in spite of the presence of phospholipid (table 5) at no time were even traces of membrane detected within ribosomal preparations (fig. 4). On the other hand, the small amount of phospholipid in the ribosomal preparations may indicate the presence of reticulosomes. Since ribosomes may be regarded as consisting of approximately equal quantities of RNA and protein [18], the amount of non-ribosomal protein present in the 'ribosomal pellet' may be calculated by subtracting the RNA value from the total protein value. The high protein/phospholipid ratio of the non-ribosomal material in the pellet (last column of table 5) makes membrane contamination very unlikely, but as pointed out previously, suggests the presence of reticulosomes.

The reaction product of reticulosomes, microsomal micellar phospholipid and ribosomes, contains obviously contributions from all three components

Table 5

Chemical analyses of in vitro assembled ribosome-studded vesicles and their precursors

Fraction	Experiment no.	Protein (μg/tube)	Phospholipid (μg/tube)	RNA (μg/tube)	$\dfrac{\text{Protein minus RNA}}{\text{phospholipid}}$ ratio
Reticulosomes	1	1693	204	–	8
	2	2400	300	–	8
	3	4600	700	–	7
Microsomal, micellar phospholipid	1	–	3740	–	–
	2	–	3740	–	–
	3	–	5500	–	–
Ribosomes plus polysomes	1	1050	78	566	6
	2	1050	50	705	7
	3	1320	56	972	6
Assembled ribosome-studded vesicles	1	2070	789	435	2.07
	2	2500	917	490	2.29
	3	4500	1680	565	2.34
Isolated rough microsomal vesicles	4	4200	900	1060	3.5
	5	1400	230	570	3.6
	6	1440	394	396	2.6

Analyses were carried out as described in the Materials and methods section.

(table 5). The protein minus RNA phospholipid ratio is in all cases approximately 2, which is in accordance with the concept that the non-ribosomal components of this complex are of a membranous nature. Furthermore its position within the discontinuous sucrose gradient and the RNA content indicates the presence of ribosomes or at least RNA on or within this complex.

The (protein minus RNA) phospholipid ratio in the assembled membranes is consistently lower than that of the isolated ribosome-studded (rough) microsomal vesicles (table 5), this could well be due to the absence of structural proteins, which were not added to the assembled vesicles, but which are present in the isolated rough microsomal vesicles.

When a portion of the assembled rough vesicles was fixed with OsO_4, embedded, sectioned, stained and viewed with the electron microscope it was apparent by comparison with the electron micrographs of ribosomal and reticulosomal preparations that a complex had been formed, but the membranous and vesicular nature were not well defined (fig. 4c). Subsequently the same experiment was modified so that instead of microsomal micellar phospholipid, microsomal micellar total lipid was used. The micellar total lipid was added together with ribosomes to a suspension of reticulosomes with the

minimum amount of agitation; the resulting ribosome-studded vesicles were harvested as before and sections of such OsO_4 fixed material are shown in fig. 4d. It has been shown that the size of assembled vesicles diminishes with agitation [19], and since the demonstration of the structure of the assembled ribosome-studded vesicles is difficult if the diameter of the vesicles is of the same order as the section thickness, larger vesicles are easier to demonstrate. This is borne out by a comparison of fig. 4c and 4d.

Incorporation of ^{14}C-phenylalanine by assembled ribosome-studded vesicles

The formation of ribosome-studded vesicles per se has no biological significance, unless it can also be shown that the assembled vesicles do in fact exhibit biological activities. The results in table 6 show that assembled ribosome-studded vesicles are able to incorporate ^{14}C-phenylalanine into proteins at rates comparable to those of isolated rough microsomes or free ribosomes plus polysomes. This incorporation is energy dependent and puromycin sensitive (table 6). In the absence of an energy generating system the residual activity may be due to energy rich compounds present within the cell sap which was added to the incubation mixture. On the addition of poly-U

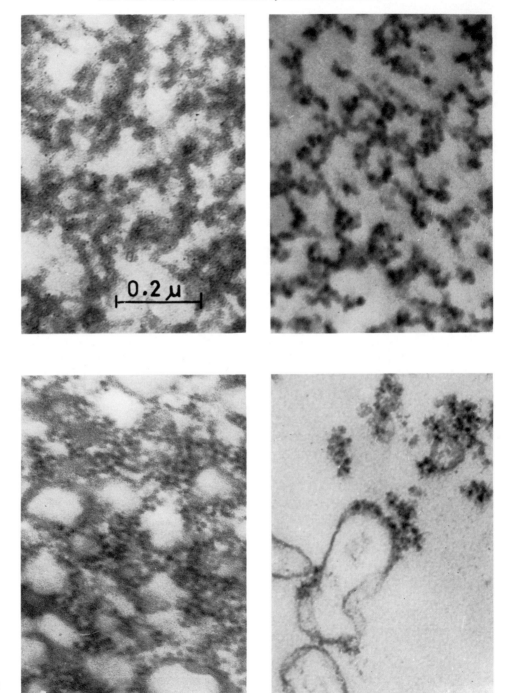

Fig. 4. (a) Section through an OsO$_4$-fixed pellet of reticulosomes ($\times$ 106,500). (Note that the reticulosomes tend to aggregate on fixation, negatively stained reticulosomes do not show this tendency.) (b) Section through an OsO$_4$-fixed pellet of ribosomes and polysomes ($\times$ 106,500). (c) Section through an OsO$_4$-fixed pellet of assembled ribosome-studded vesicles ($\times$ 106,500). For the assembly of these vesicles micellar microsomal phospholipid was vigorously stirred together with reticulosomes and ribosomes plus polysomes. (d) Section through an OsO$_4$-fixed pellet of assembled ribosome-studded vesicles ($\times$ 106,500). For the assembly of these vesicles micellar microsomal total lipid was gently mixed with reticulosomes and ribosomes plus polysomes.

Table 6

Incorporation of ^{14}C-phenylalanine into assembled ribosome-studded vesicles, isolated rough vesicles and free ribosomes plus polysomes, and the effect of addition of poly-U

Fraction	Radioactivity (counts/min/mg of RNA)				
	Complete system	+ Poly-U	+ puromycin + poly-U	+ puromycin	Minus energy and regenerating system
Assembled ribosome-studded vesicles	1620	3530 (2.3)	–	1180	1070
	1750	6875 (3.9)	–	937	–
	2000	5330 (2.7)	1332	–	–
Isolated rough vesicles	1858	18580 (10.1)	–	–	–
	5000	36700 (7.1)	–	–	–
	2722	31820 (11.7)	–	–	2085
Free ribosomes plus ribosomes	2000	43571 (29.1)	–	–	–
	550	12400 (22.5)	5550	–	–

The incubation mixture contained 40 mμmoles of L-^{14}C-phenylalanine (7 mCi/mmol), 0.5 μmol ATP, 0.25 μmol of GTP, 5 μmoles of phosphoenolpyruvate, 20 μg of pyruvate kinase, 0.06 μmoles with respect to L-isomers of the following amino acids: serine, valine, proline, lysine, tyrosine, arginine, methionine, histidine, leucine, iso-leucine, glutamic acid, aspartic acid, threonine, alanine, glycine and tryptophane, 0.1 ml of cell sap (containing 1.2 mg protein); all components were suspended in medium B [9]. Where indicated 300 μg poly-U or 0.2 mg of puromycin, or both were added. The final volume was 0.5 ml. Incubation was carried out at 37°C for 30 min with shaking. The reaction was stopped and the samples plated and counted as described in the Materials and methods section. The values in parentheses indicate the factor by which poly-U stimulated phenylalanine incorporation.

to the system ^{14}C-phenylalanine incorporation was stimulated (table 6), indicating that normal ribosomal function was retained after the in vitro assembly of ribosome-studded vesicles. This in itself was not unexpected as it had been previously shown that free ribosomes plus polysomes could be attached to microsomal vesicles which had previously been stripped of ribosomes [20].

The main difference between these experiments and the experiments in the present series is that Suss et al. [20] commenced with smooth vesicles (rough vesicles from which the ribosomes had been removed by detergent action) and free ribosomes and polysomes, while the starting material in the present study was completely non-membranous, consisting of reticulosomes, microsomal micellar phospholipid and free ribosomes plus polysomes.

As pointed out previously, in vitro assembled ribosome-studded vesicles, isolated rough microsomes and free ribosomes plus polysomes incorporate ^{14}C-phenylalanine at comparable rates (table 6). However, on the addition of poly-U, in the presence of a mixture of 17 amino acids, the in vitro assembled ribosome-studded vesicles ^{14}C-phenylalanine incorporation was stimulated only 2–4 times, as compared

to the isolated rough microsomes 7–12 times and the free ribosomes plus polysomes 23–29 times (table 6). These results lend themselves to two possible interpretations. There could be a preferential uptake of polysomes during the assembly of ribosome-studded vesicles, which are then comparatively saturated with messenger-RNA, so that any further addition of messenger-RNA in the form of poly-U causes only a relatively small stimulation to ^{14}C-phenylalanine incorporation. Alternatively, the messenger-RNA-ribosome complex may be stabilized when the ribosomes are attached to membranous vesicles, so that the half-life of messenger-RNA is longer on membrane bound ribosomes than that of messenger-RNA on free polysomes.

In the evidence presented so far it was shown that membranous vesicles may be assembled from reticulosomes in the presence of other necessary components, such as phospholipids, total lipid extracts or ribosomes plus polysomes. The vesicles formed from these components resemble microsomes as isolated by conventional methods of differential centrifugation with respect to their composition, enzymic activity and ultrastructure.

Evidence for the existence of reticulosomes in vivo

The point which still remains to be established is the existence of reticulosomes within the cell, or even within whole homogenates prior to treatment with DOC or RNase. In other words, breakdown products derived from membranous structures may produce an artefactual particulate fraction which is not present as such within the cell or the microsomal fraction.

Three different approaches have been used so far, and these provide corroborative evidence for the existence of reticulosomes in vivo.

(1) It has already been stated previously that the yield of reticulosomes is greatest when the membrane content of the endoplasmic reticulum is least. This was originally demonstrated in embryonic chick liver [4] and subsequently also in fasted rat liver (table 3). Fasted rat liver should be considered to be poised on the point of membrane formation, since in a 2–4 day fasted rat the endoplasmic reticulum, which has become either grossly disorganized or disappeared completely, is largely reconstituted within a few hours after refeeding [15]. The significance of these phenomena is twofold, firstly it is unlikely that a large yield of an artefactual membrane breakdown product is obtained from a fraction which contains less membranous components than usual, and secondly the inverse relationship between reticulosomes and membranes suggests that in vivo an equilibrium situation may exist between these two components.

(2) The main objection to the isolation procedure of reticulosomes is the treatment with DOC and RNase, thus ideally reticulosomes should be isolated by alternative means. So far it has not been possible to isolate reticulosomes by gradient centrifugation, presumably due to their aggregation with polysomes and ribosomes, a feature which has also been observed with DOC isolated reticulosomes. Hence in membrane assembly experiments lipid and ribosomes have to be added simultaneously to reticulosomes during the procedure of forming ribosome-studded vesicles.

On the other hand, if reticulosomes are present in the rough microsomal subfraction, they should react with added phospholipid to give rise to smooth membranous vesicles, which can be separated from the original rough microsomal fraction by centrifugation in a discontinuous gradient [5,10].

A rough microsomal fraction was mixed with micellar phospholipid and kept at 2°C for 20 min. The mixture was then centrifuged at 105,000 g (av.) for 60 min and the pellet rinsed several times with 0.25 M sucrose and then finally suspended in the same medium. After adding CsCl to a final concentration of 0.015 M, a Dallner [10] separation for smooth and rough microsomes was carried out; the presence of an interface fraction could be regarded as indicative of the formation of new membranes. In table 7 results are presented showing such an interaction of ^{14}C-choline labelled rough microsomes with unlabelled micellar phospholipid. When such a rough microsomal fraction was placed on a discontinuous density gradient, even in the absence of added phospholipid, and centrifuged as described above, an extremely faint interface band could be detected. On analysis it was shown to consist of protein and phospholipid, and had virtually the same specific activity as the rough microsomal preparation (table 7). After interaction with micellar phospholipid, a very definite interface band appeared after centrifugation, containing significantly greater amounts of protein and phospholipid in the form of a lipoprotein complex (table 7). Furthermore the specific activity of the interface phospholipid was reduced to about 28% of its specific radioactivity in the absence of the added phospholipid (table 7).

Further work showed that the conditions for the interaction of rough microsomes and phospholipid had to be carefully controlled, as at pH values above 7 a considerable amount of interface material was formed, even in the absence of phospholipid (table 8). On repeating the experiment described in table 7, pH was carefully controlled at 6.6 and the isotope labelling conditions were reversed, namely ^{14}C-choline labelled micellar phospholipid was added to the unlabelled rough microsomal fraction (table 9). It was shown quite clearly that the low density lipoprotein interface material derived its phospholipid entirely from the added phospholipids and only the protein of the interface lipoprotein complex came from the rough microsomal fraction. The alternative explanation, that the interface material was derived from originally rough membranes which have lost their ribosomes, is thus ruled out. On the addition of reticulosomes to this incubation mixture, there was a considerable increase in the amount of protein and

Table 7

Titration of ^{14}C-choline labelled, rough microsomes with micellar phospholipid

	Pellet (rough vesicles)			Interface material (smooth vesicles)		
	Protein (μg/g tissue)	Phospholipid		Protein (μg/g tissue)	Phospholipid	
		(μg/g tissue)	(cpm/mg)		(μg/g tissue)	(cpm/mg)
Rough microsomes	3930	1377	3839	59	9	4070
Rough microsomes + phospholipid	3930	2040	3562	220	141	1120

Rough microsomes obtained from 2 g of rat liver, which has been previously labelled with ^{14}C-choline chloride as described in the Materials and methods section were mixed with an excess of micellar phospholipid in water adjusted to pH 7.0 (5 ml of a suspension containing 60 μg phospholipid phosphorus/ml). The mixture was left in the dark at 2°C for 20 min and then centrifuged as described in the text. Pellets and interface fractions were isolated and assayed as described in the Materials and Methods section.

Table 8

Effect of pH on interface formation of a rough microsomal fraction

pH of suspension	Pellet		Interface material	
	Protein (μg)	Phospholipid (μg)	Protein (μg)	Phospholipid (μg)
6.6	2640	740	0	0
7.6	2680	755	90	50

5 ml of rough microsomal suspension (200 mg tissue equivalent/ml) were adjusted to pH values as indicated, then sedimented, resuspended and placed over a discontinuous gradient as indicated in the text. All analyses were carried out as described in the Materials and methods section.

labelled phospholipid of the interface lipoprotein complex (table 9). It was shown previously that the interface lipoprotein complex formed by the interaction of reticulosomes and phospholipid consisted of a membranous reticulum ([5] and fig. 2).

On closer examination of the data in table 9, it will be observed that the yield of the interface lipoprotein complex is significantly lower than would be expected from the amount of reticulosomes present within the rough microsomal fraction (table 3). The amount of reticulosomal protein present within the rough microsomal fraction before titration with ^{14}C-labelled phospholipid may be estimated from data in table 3, to be of the order of 0.85 mg per g of tissue or 1.36 mg per tube. The phospholipid bound in the rough microsomal fraction may be calculated

Table 9

Titration of rough microsomal fraction with ^{14}C-choline labelled micellar phospholipid

Incubation mixture	Centrifugal fraction	mg protein/ tube	mg phospholipid/ tube	cpm/ mg phospholipid
microsomal suspension + micellar phospholipid suspension	interface	0.12	0.035	5538
	pellet	7.7	3.7	8.5
microsomal suspension and micellar phospholipid suspension + reticulosomal suspension	interface	0.34	0.11	5145
	pellet	9.5	4.5	1138

8 ml of a rough microsomal suspension (200 mg tissue equivalent/ml) were mixed with 2 ml ^{14}C-choline labelled phospholipid (specific activity 5383 cpm/mg phospholipid) containing 2 mg phospholipid/ml and with 0.5 ml reticulosomes (1.2 mg protein/ml) as indicated in the table; incubated at 2°C for 20 min. All other methods as in table 7.

from the specific activity of the added phospholipid and the counts within the rough microsomal fraction to be of the order of 0.59 mg phospholipid, giving a ratio of 2.3 of reticulosomal protein to phospholipid as described previously ([5] and table 1). On the further addition of 600 mg of reticulosomal protein an additional amount of ^{14}C-labelled phospholipid was found to be protein-bound in the rough microsomal fraction (table 9). The simplest explanation for the presence of ^{14}C-labelled phospholipid within the rough microsomal fraction is the formation of ribosome-studded vesicles in the presence of ribosomes, reticulosomes and phospholipid. In fact the preferential formation of such ribosome-studded vesicles in vitro has already been demonstrated (table 5); on the other hand, exchange of phospholipids between different organelles cannot explain the appearance of ^{14}C-labelled phospholipid at the interface (at pH 6.6) in table 9, since the experiment was conducted at 2°C for 20 min and it has been shown that the exchange of phospholipid between cell organelles is temperature dependent [21].

(3) Another set of experiments was designed to establish if reticulosomes may be obtained from assembled membranous vesicles by treatment with 0.26% DOC. As shown in fig. 5, reticulosomes with characteristically high glucose-6-phosphatase activity were obtained in the usual way and pelleted through a discontinuous Dallner [10] gradient (fig. 5A). After mixing these reticulosomes with phospholipid, the membranous lipoprotein complex was isolated, and showed diminished glucose-6-phosphatase activity (fig. 5B). On treating the lipoprotein complex with 0.26% DOC, the membrane was disrupted, as evidenced by the disappearance of the interface material (fig. 5C), and on centrifugation in the discontinuous Dallner [10] gradient a pellet was obtained which showed even less glucose-6-phosphatase activity than the membrane complex (compare fig. 5B and 5C). On the addition of further phospholipid the resuspended pellet did not give rise to a lipoprotein complex since all the material again pelleted when centrifuged in the discontinuous gradient (fig. 5D). This experiment then establishes that at least assembled membranes when treated with DOC do not give rise to a granular component with high glucose-6-phosphatase activity and a facility to react with phospholipids to produce a membranous lipoprotein complex.

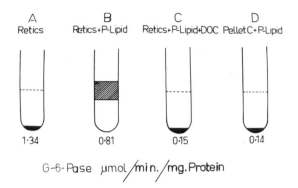

Fig. 5. Diagramatic representations of discontinuous gradient centrifugations, and glucose-6-phosphatase activities of reticulosomes (A), reticulosome-phospholipid complex (B), breakdown product of the reticulosome-phospholipid complex by DOC treatment (C) and the mixture of pellet (C) and phospholipid (D).

In conclusion it should be reiterated that two separate phenomena have been studied and described. On the one hand, evidence has been presented that a non-membranous particulate fraction was isolated by discontinuous gradient centrifugation coupled with detergent and enzymic treatment of a postmitochondrial fraction, and this particulate non-membranous fraction when mixed with micellar phospholipid gave rise to membranous vesicles. This membrane assembly was verified by chemical analysis, by electron microscopy and by measuring the concomitant changes of the enzyme kinetics in several enzyme systems.

On the other hand, some circumstantial evidence has also been presented which suggests that the reticulosomes may be present within whole homogenates or even cells, and thus provide a possible mechanism of membrane formation.

Acknowledgements

Work upon which this article is based was supported by the Australian Research Grants Committee, the National Health and Medical Research Council, the New South Wales State Cancer Council and the University of Sydney Cancer Fund. A research studentship to K.A.W. from the New South Wales State Cancer Council is gratefully acknowledged.

References

[1] G.L. Sottocasa, in: Membrane models and the formation of biological membranes, eds. Liana Bolis and B.A. Pethica (North-Holland Publishing Company, Amsterdam, 1968) p. 229.

[2] L. Sagan, J. Theor. Biol. 14 (1967) 225.

[3] D.B. Roodyn and D. Wilkie, in: The biogenesis of mitochondria (Methuen and Co. Ltd., London, 1968).

[4] J.K. Pollak and C.D. Shorey, Biochem. J. 93 (1964) 36c.

[5] J.K. Pollak, K.A. Ward and C.D. Shorey, J. Mol. Biol. 16 (1966) 564.

[6] J.K. Pollak and D.B. Ward, Biochem. J. 103 (1967) 730.

[7] J.K. Pollak and C.D. Shorey, Aust. J. Exp. Biol. Med. Sci. 45 (1967) 393.

[8] K.A. Ward and J.K. Pollak, Biochem. J. 114 (1969) 41.

[9] H. Bloemendal, W.S. Bont, M. deVries and E.L. Benedetti, Biochem. J. 103 (1967) 177.

[10] G. Dallner, Acta Pathol. Microbiol. Scand., Suppl. 166 (1963) p. 1.

[11] R.C. Nordlie and W.J. Arion, in: Methods of Enzymology, vol. IX, eds. S.P. Colowick and N.O. Kaplan (Academic Press, New York, 1966) p. 619.

[12] G.L. Sottocasa, B. Kuylenstierna and L. Ernster, J. Cell Biol. 32 (1967) 415.

[13] R.J. Mans and G.D. Novelli, Arch. Biochem. Biophys. 94 (1961) 48.

[14] W.J. Arion and R.C. Nordlie, Biochem. Biophys. Res. Commun. 20 (1965) 606.

[15] D.W. Fawcett, J. Nat. Cancer Inst. 15 (1955) suppl. 1475.

[16] P.J. Collipp, Arch. Biochem. Biophys. 118 (1967) 106.

[17] S. Orrenius, J.L.E. Ericson and L. Ernster, J. Cell Biol. 25 (1965) 627.

[18] E.L. Hess and S.E. Lagg, Biochemistry 2 (1963) 726.

[19] W. Stockenius, in: Principles of biomolecular organization, Ciba Foundation Symposium, eds. G.E.W. Wolstenholme and M. O'Connor (J. and A. Churchill Ltd., London, 1966) p. 418.

[20] R. Suss, G. Blobel and H.C. Pitot, Biochem. Biophys. Res. Commun. 23 (1966) 299.

[21] W.C. McMurray and R.M.C. Dawson, Biochem. J. 112 (1969) 91.

STRUCTURE AND FUNCTIONAL ORGANIZATION OF MICROCOCCUS LYSODEIKTICUS MEMBRANE

M.R.J. SALTON and Martin S. NACHBAR

Department of Microbiology, New York University
School of Medicine, New York, USA

The biochemical functions normally assigned to a variety of membranous organelles in higher cells are found in the 'simpler', plasma-mesosome membrane systems of the bacterial cell. Although bacteria lack discrete mitochondria, nuclear membranes, membranous endoplasmic reticulum, golgi, lysosomes and other membranous organelles, the various functions associated with these structures are performed by the bacterial membranes. Thus the division of functional labor and differentiation in the bacterial cell occurs at the membrane level. The bacterial cell is therefore deceptively 'simple', containing only two distinct membrane regions, the plasma and mesosome membranes, which must perform the functions of active transport, electron transport and oxidative phosphorylation, nuclear replication and division, wall biosynthesis, phospholipid synthesis, etc. The manner in which these various functions are organized and distributed in the 'multifunctional' bacterial membrane is a problem as challenging as the elucidation of the organization of mitochondrial membranes. We have selected the Gram-positive organism, *Micrococcus lysodeikticus,* as a suitable type of cell for membrane isolation and characterization in order to gain an insight into the structure-function relationships for a bacterial membrane system.

General features of Micrococcus lysodeikticus membranes

The anatomy of the surface and internal organization of a Gram-positive organism such as *M. lysodeikticus* is illustrated diagrammatically in fig. 1, and a thin section of the organism itself is shown in fig. 2.

The thick, outer cell-wall structure provides the cell with a rigid, protective barrier. Below the wall and closely apposed to the inner wall surface is the plasma membrane (PM in fig. 2). The plasma membrane invaginates, usually in the region of the cross-wall septum and the ingrowth of the invaginated membrane gives rise to the mesosome structures [1] which constitute the intracytoplasmic membrane region of the bacterial cell. Although the mesosomes of some bacterial cells are superficially similar to mitochondria they are not discrete organelles encased by closed outer membranes. The mesosome region can be 'outlined' in intact bacterial cells by negative staining as shown for *M. lysodeikticus* in fig. 3. The penetration

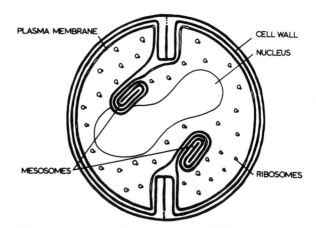

Fig. 1. A diagrammatic representation of the anatomy of the Gram-positive coccus, *Micrococcus lysodeikticus,* indicating the outer rigid cell wall and developing cross-wall (CELL WALL), the PLASMA MEMBRANE and its invagination to form the lamellated MESOSOMES, the nuclear body (NUCLEUS) lacking a limiting 'unit membrane', and RIBOSOMES.

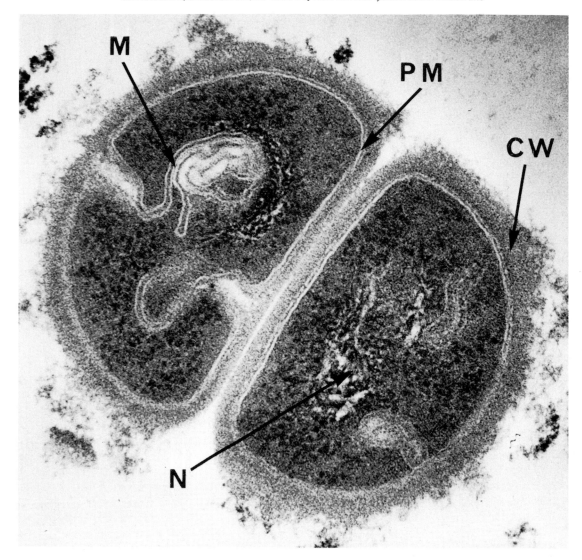

Fig. 2. Electron micrograph of a thin section of *Micrococcus lysodeikticus* illustrating the thick cell wall (CW), the plasma membrane (PM) and the invaginated region forming the mesosome (M) and the nuclear region (N) ($\times$ 137,000). (From Salton [9].)

of the negative stain into the mesosome area suggests that this region of the cell is freely accessible from the outside and is external to the plasma membrane. The extrusion of mesosome vesicles from a variety of bacteria also demonstrates the essentially 'external' location of mesosomes [2–5].

The mesosomes of *M. lysodeikticus* appear to consist of lamellated membranes or of rather open whorls of a tubulovesicular type. Salton and Chapman [6]

attempted to isolate the mesosomes of *M. lysodeikticus*, but the membranes derived therefrom co-separated with the plasma membranes. Ellar and Freer [7] have succeeded in extruding tubulovesicular mesosomes from this organism, so that a method of isolation of these structures and their separation from plasma membranes should now be feasible. At the present time, the distinction between functions carried specifically by the mesosome or the

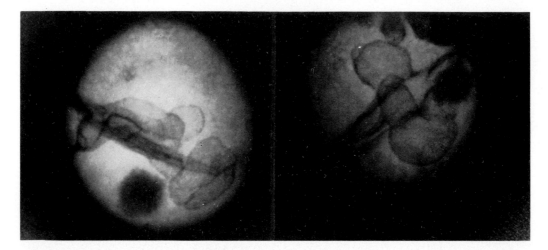

Fig. 3. *Micrococcus lysodeikticus* negatively stained with phosphotungstic acid, showing the penetration of the stain outlining the loosely-folded mesosome membrane region. The cells are about 1μ in diameter.

plasma membrane of this and other Gram-positive bacteria is still unclear [5, 8].

The plasma and mesosome membrane regions of *M. lysodeikticus* have the typical 'unit membrane' profile when viewed in thin sections of proto-plasts [9] or in isolated membranes [6] fixed by the Ryter and Kellerberger [10] procedure. The anatomy of these bacterial membranes, therefore, does not appear to differ significantly from that of other cell membranes. They show the characteristic 'double-track' profiles on fixation with osmium tetroxide and they possess an overall thickness of about 75 Å. The membranes fragment and form vesicles on isolation and their appearance in negatively stained prepara-tions is very similar to membranes derived from other types of cells. Regions of the membranes of *M. lysodeikticus* bear particles of about 100 Å in diameter (fig. 4) and the appearance of these particles along the edges of the membranes are reminiscent of the inner mitochondrial membrane structures. How-ever, these particles lack the pronounced stalked appearance seen in mitochondria [11] or in certain bacterial membrane preparations [12, 13]. The signif-icance of these particles will be discussed further below.

The determination of the chemical composition of membranes isolated from *M. lysodeikticus* indicated that they were composed predominantly of protein and lipid, the ratios being approximately 3:1 (com-pare with generally lower ratios for many other cell

membranes as summarized by Korn [14, 15]. The RNA content of the membranes can be reduced to about 1% by subjecting the membranes to six or more successive washings with buffer (0.1 M tris-HCl, pH 7.5). The functional significance of this residual RNA has not been investigated. The total membrane lipid is made up largely of phospholipid and under growth conditions used routinely in our laboratory (peptone-water-yeast extract medium, pH 7.5, vigor-ous aeration at 30° for 20—24 hr) the three phos-pholipids found in the membrane are phosphatidyl glycerol, diphosphatidyl glycerol (cardiolipin) and phosphatidyl inositol. Dimannosyl diglyceride [16], menaquinone and carotenoids [17] are also present. Vorbeck and Marinetti [18] reported that most of the cellular phospholipid of *Streptococcus faecalis* was localized in the membrane. Similarly, with *M. lysodeikticus* virtually all of the phospholipid, menaquinone and the carotenoids are localized in the membrane. The carotenoids although accounting for only approximately 1% of the total membrane lipid, have provided us with a useful, colored 'chemical marker' for the membrane of this organism. The presence of carotenoids in cytoplasmic or wash frac-tions can be used as an index of membrane break-down or fragmentation [19].

The molecular associations between the proteins and lipids in *M. lysodeikticus* membranes are un-doubtedly of the same type of weak interactions observed with other cell membranes [20]. The re-

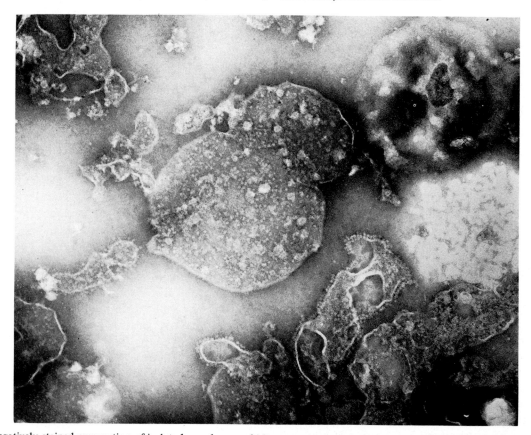

Fig. 4. Negatively-stained preparation of isolated membranes of *Micrococcus lysodeikticus*. Note the presence of small particles on the membrane surface (X 95,000) (From Nachbar and Salton [22].)

sponse of the membranes to dissociation with detergents, ultrasound, treatment with organic solvents and chelating agents [21, 22] indicates that they owe their integrity to weak cohesive forces of ionic interactions, hydrophobic associations between lipids and proteins and between the proteins themselves and possibly hydrogen bonding and London-Van der Waals forces. The asymmetric and symmetric methylene absorptions detected by infrared spectroscopy of *M. lysodeikticus* membrane fractions has also been interpreted as indicating hydrophobic interactions between proteins and lipids [23].

Thus the general properties (anatomy in the electron microscope, chemical composition and nature of membrane cohesive forces) of *M. lysodeikticus* membranes indicate a basic similarity to membrane structures derived from other types of cells or from intracellular membranous organelles.

Dissociation of functional markers

In order to gain some insight into the organization of various functions in the bacterial membrane system it is obvious that one would have to be able to recognize, localize and determine the distribution of specific functional entities of the membrane. This could be achieved by identifying specific enzymatic proteins or electron transport components by ferritin or peroxidase-labelling of antibody to individual proteins or by cytochemical staining. The investigations in our laboratory have been directed to this goal and in order to achieve this aim it has been necessary to isolate and purify individual functional components of the membrane. This has involved firstly, the recognition of specific enzymes or activities as part of the membrane and secondly, the development of suitable dissociating procedures to facilitate their purification.

As pointed out by Nachbar and Salton [22] cellular proteins, unlike the lipid components, do not always show an all or none distribution between the cytoplasmic and membrane fractions. Thus the distribution of enzyme activities varies over a wide range from those found exclusively in the cytoplasm (e.g. DPN-TPNH transhydrogenase, adenosine deaminase) to those such as phosphatidic acid-cytidyl transferase and the cytochromes found exclusively in the membrane. As shown in fig. 5 some enzymes partition between the two fractions, but it should be noted that many of these can be 'released' from the membranes by successive washings in 0.03 – 0.1 M tris-HCl buffer, pH 7.5 [22]. One of the great problems universal to all membrane studies, is the possibility that even the mildest of isolation and washing procedures may have dissociated some activities held on the membrane (in vivo) by very weak associations. However, in order to get some 'base-line' information, it has been necessary to concentrate on those functional markers firmly attached to the membranes after they have been subjected to a standardized wash procedure [24].

By the use of a variety of treatments, many of which have proven to be quite selective, we have dissociated a number of functional markers from the membrane. The mildest dissociating procedure used involves the transfer of the membranes to a low ionic strength environment (shock wash) with concommitant release of ATPase activity [26, 36]. Such a procedure would be expected to disrupt weak associations of an ionic and/or hydrophobic character. The chelating agent EDTA, resulted in the release of NADH dehydrogenase activity (Nachbar, to be published), the fraction containing both lipid and protein. Exposure to ultrasound released ATPase activity [36] without apparently disrupting the association between lipid and protein since the major 4–5S component [21] migrated as a single lipoprotein aggregate upon sucrose gradient electrophoresis [27]. More drastic procedures including the use of detergents and/or organic solvents are required to remove or displace lipid from membrane proteins and such methods frequently lead to denaturation of proteins and loss of enzymatic activities.

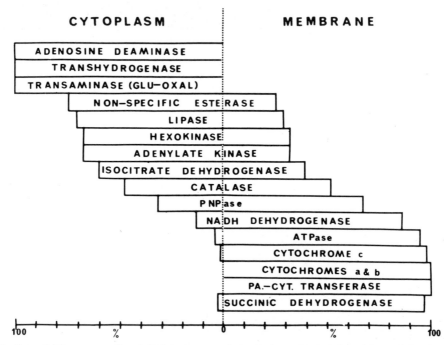

Fig. 5. The distribution of *Micrococcus lysodeikticus* enzymes between cytoplasm and membrane following separation of the fractions by centrifugation of lysates to deposit the membranes (30 min; 30,000 *g*; 0°C). Modified from Nachbar and Salton [22].)

Release of M. lysodeikticus membrane ATPase

The investigations of Munoz et al. [26, 36], in our laboratory have established the rather selective release of a Ca^{2+}-dependent ATPase from *M. lysodeikticus* membranes when they were subjected to a low ionic strength wash in 0.003 M tris-HCl buffer, pH 7.5, following four consecutive washes in 0.03 M tris-HCl buffer. The pattern of release of the ATPase activity is illustrated from the results of Munoz et al. [26] in fig. 6. The selectivity of the low ionic strength, 'shock' wash was demonstrated by polyacrylamide gel electrophoresis and the ATPase activity could be readily located and identified in the gels by the enzymatic stain method of Weinbaum and Markham [28] (Munoz et al. [26, 36]).

Examination of isolated membrane fractions before and after subjecting them to the low ionic strength wash indicated that the particles scattered over the membrane surfaces were released concomitantly with the appearance of ATPase activity in the supernatant 'shock' wash. The loss of particles

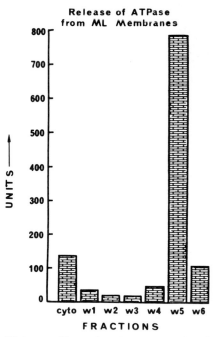

Fig. 6. Histogram illustrating the release of ATPase from *Micrococcus lysodeikticus* membranes as determined by Munoz et al. [26]. Activity was determined in the cytoplasm (cyto), membrane washes in 0.03 M tris-HCl buffer, pH 7.5 (w1-w4) and in the low ionic strength 'shock washes' with 0.003 M tris-HCl buffer, pH 7.5 (w5 and w6).

from the membranes under conditions of selective release of ATPase activity is illustrated in fig. 7. The particulate nature of the purified ATPase reported by Munoz et al. [25] is also shown in fig. 7. Thus the Ca^{2+}-dependent ATPase activity is found associated with particles of about 100 Å in diameter and each individual particle appears to consist of six peripheral subunits surrounding a central unit [25].

The ATPase can also be released from the membrane by extraction with 1% sodium deoxycholate in 0.05 M tris-HCl buffer, a method used for isolating some of the electron-transport components as a lipid-depleted, deoxycholate insoluble residue [29]. This procedure is much less selective and results in the release of many proteins, including the ATPase, into the deoxycholate (DOC)-soluble fraction. Moreover, dissociation of the membranes with this detergent results in the extraction of most of the membrane lipid into the DOC-soluble fraction. There is, however, considerable inactivation of the enzyme extracted by this procedure but the ATPase protein and residual activity can be identified by polyacrylamide gel electrophoresis of the DOC-soluble fraction.

In addition to the selective release by lowering the ionic strength of the wash buffer, its dissociation from membranes exposed to ultrasound, release into the DOC-soluble fraction, the ATPase activity can be obtained in the aqueous phase following extraction of the membranes with *n*-butanol at 0°, a method originally used by Morton [30] to release particle-bound enzymes, including ATPase. Under these conditions, most of the membrane lipid is extracted into the organic solvent layer and membrane protein and residual lipid separate as an insoluble interfacial layer, leaving the aqueous phase enriched in ATPase activity. After removal of *n*-butanol by dialysis, the ATPase can be purified further from the aqueous phase by precipitation with ammonium sulfate and/or chromatography on Sephadex G 200 (Salton and Schor, unpublished results). These results suggest that hydrophobic and/or ionic cohesive forces are involved in the association of ATPase with the membrane.

The ATPase of *M. lysodeikticus* membranes exhibits a 'latency' similar to that reported for chloroplast ATPase by Vambutas and Racker [31], low levels of hydrolysis of ATP are observed when the enzyme is in situ on the membrane but it is readily activated by trypsin in the presence of substrate [26, 36]. The

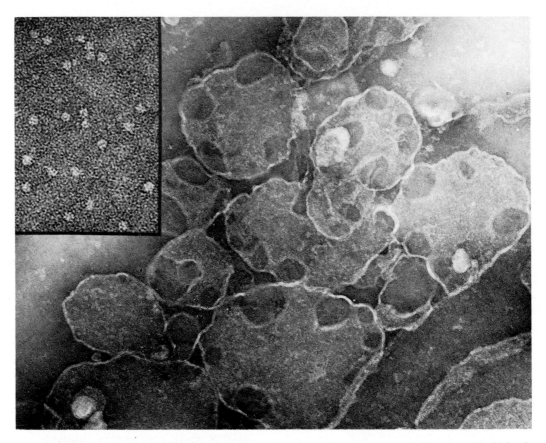

Fig. 7. A negatively stained preparation of deposit of membrane fragments after exposure to the 'shock wash' which released the ATPase activity. Note the loss of particles from the membrane (compare with fig. 4 with abundance of particles on membranes) following this treatment (X 95,000). Inset: negatively stained preparation of ATPase fraction showing the particulate nature of the enzyme (X 260,000). (From Nachbar and Salton [22]).

function of the ATPase has not been conclusively established but it is likely that it could play a role as an allotopic protein [32] in either oxidative phosphorylation [36] or in cation transport (Munoz, personal communication). It is of interest to note that Harold et al. [33] have suggested that the *Streptococcus faecalis* ATPase functions in K^+ transport. Dicyclohexylcarbodiimide (DCCD) inhibited the uptake of K^+ by intact cells and the membrane bound ATPase of *S. faecalis* but had no inhibitory effects on the solubilized enzyme. We have found that DCCD had no inhibitory effect on *M. lysodeikticus* ATPase (soluble form) at 5×10^{-5} M and only 15% inhibition was observed at 5×10^{-4} M. However, the influence of the carbodiimide on the low level of ATPase activity with whole membranes or its effects

on cation transport in *M. lysodeikticus* have not yet been investigated.

The ATPase of *M. lysodeikticus* is one of the major proteins of the membrane accounting for about 10% of the total membrane protein. It is also one of the major antigens of the membrane and its antigenic specificity and reaction with antibody is being investigated in this laboratory by Dr. Theresa Whiteside [34]. Ultimately, we hope to be able to establish the precise location of this enzyme on the bacterial membrane system.

Release of NADH dehydrogenase

The selective release of NADH dehydrogenase activity has been investigated in our laboratory by Nachbar and has been achieved by subjecting the membranes to three or four washes with 0.03 M tris-HCl buffer pH 7.5, followed by a wash in the same buffer containing 0.005 M EDTA. Treatment of the membranes with this 0.03 M tris–0.005 M EDTA wash resulted in a dramatic release of the activity as shown in fig. 8 (Nachbar, unpublished; Nachbar and Salton [22]). In addition to the large amount of activity released, the specific activity of the EDTA wash was 3 to 4 times that of the original membrane and preceding washes. Activity remaining in the membranes accounted for less than 15% of the original activity. The activity was assayed with 2,6-dichlorophenolindophenol as the electron acceptor and 2–3 bands of activity could be detected after electrophoresis of the EDTA wash fraction in polyacrylamide gel, using triphenyltetrazolium as an indicator [22].

The EDTA-wash preparation containing the NADH dehydrogenase activity had a higher lipid to protein ratio than that of the original membranes and

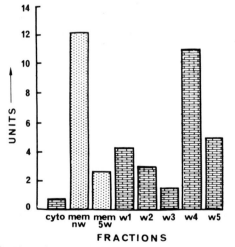

Release of NADH Dehydrogenase from ML Membranes

Fig. 8. The relase of NADH dehydrogenase activity from *Micrococcus lysodeikticus* membranes. Assay was performed with 2,6-dichlorophenolindophenol as electron acceptor and units recorded as described by Nachbar [22]. Fractions assayed were: cytoplasm (*cyto*), unwashed membranes (*memnw*), membrane after 5 washes (*mem 5w*), *w1 – w3* – consecutive washes with 0.03 M tris-HCl pH 7.5; *w4* – wash with 0.03 M tris-HCl and 0.005 M EDTA; *w5* – 0.003 M tris-HCl pH 7.5 wash.

examination of negatively stained fractions in the electron microscope showed the presence of vesicles, the smallest of which were of the order of 300 Å in diameter [22]. It is of interest to note that the ratio of cardiolipin to phosphatidyl glycerol was four times that observed for the original membrane. These results suggest that there may be wide variations in the binding specifities between particular phospholipids and certain membrane proteins.

Isolation of lipid-depleted, membranous sheets containing electron transport components

Although suitable fractionation procedures had been developed earlier in our laboratory for the selective separation of cytochromes from most of the membrane lipid, carotenoid and from about two thirds of the total membrane protein [17], the fractions were completely amorphous and showed no structural organization when viewed in the electron microscope. These procedures involved the use of detergents such as the nonionic surfactant, Nonidet P. 40, which completely 'solubilized' the membranes of *M. lysodeikticus* and recovery of the cytochromes by ammonium sulfate precipitation. Subsequent investigations led to the discovery that extraction of the membranes with 1% deoxycholate (DOC) in 0.05 M tris-HCl buffer pH 7.5, yielded a small, insoluble red-brown fraction which contained virtually all of the membrane cytochromes.

Most of the carotenoid and phospholipid was removed by 6 successive extractions of the membranes with 1% DOC in 0.05 M tris-HCl buffer. The insoluble residues deposited by centrifugation at 30,000 g for 50 min accounted for about 10–15% of the initial membrane protein and upon examination in the electron microscope they were found to be in the form of membranous sheets [29]. Analysis of the DOC-insoluble residues showed that they contained 3–5% residual lipid, including some bound DOC (compared to 23–26% total lipid in whole membranes). The appearance of these lipid-depleted membranous sheets from *M. lysodeikticus* as seen in negatively stained preparations examined in the electron microscope is shown in fig. 9.

Absorption spectra of the DOC-insoluble fraction solubilized in 0.5% sodium dodecyl sulfate (SDS) and reduced with $Na_2S_2O_4$ indicated the presence of cytochromes $a + a_3$, b and c in these lipid-depleted

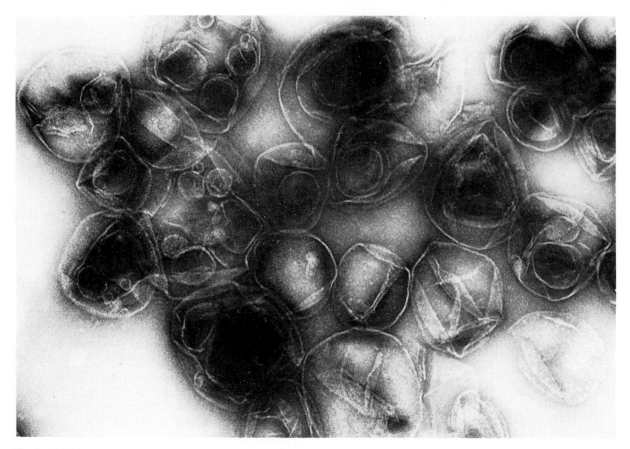

Fig. 9. Lipid-depleted, membranous sheets containing cytochromes and succinic dehydrogenase activity, isolated as an insoluble residue from *Micrococcus lysodeikticus* membranes after 6 succesive washes with 1% deoxycholate in 0.05 M tris-HCl, pH 7.5. Preparation was negatively stained with ammonium molybdate. Compare with fig. 4 for the appearance of membrane prior to extraction. (From Salton et al. [29]) ($\times$ 140,000) .

sheets [29]. That the cytochromes were almost exclusively localized in this fraction was confirmed by the detection of only trace amounts of material giving a slight peak at 560 nm when reduced spectra of DOC-soluble fractions were examined. In addition, most of the succinic dehydrogenase activity of the membrane (as determined by the method of Ells [35] was found in the DOC-insoluble residue, with less than 10% in the DOC-soluble fraction.

Although it is realized that the residual lipid and/or bound DOC may be important in binding these electron transport components in this form, it is unlikely that these membranous sheets arise as an artifact from the extraction procedure. Indeed, prefixation of the membranes with 0.5% or 1% glutaraldehyde (45 min at 0—4°) which would be expected

to cross-link adjacent proteins, has given sheets of identical appearance and size when subsequently extracted with 1% DOC. Moreover, the succinic dehydrogenase activity in the DOC-insoluble residues from membranes prefixed with 0.5% glutaraldehyde, survived storage for several months at 4°C. Fixation of the membranes with 1% glutaraldehyde before extraction with 1% DOC yielded residues which showed no succinic dehydrogenase activity. Inactivation of the latter and destruction of the cytochromes thus occurred at the higher level of glutaraldehyde treatment (1%) although the extraction of ^{32}P-labelled lipid and carotenoid by DOC was identical to that observed for unfixed membranes [29].

These results suggest that the cytochromes are more firmly bound to one another and/or the residual

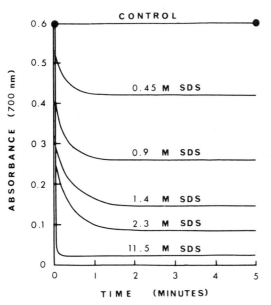

Fig. 10. Dissociation of suspensions of washed deoxycholate (DOC)-insoluble, lipid-depleted sheets isolated from *Micrococcus lysodeikticus* membranes (as seen in fig. 9) by sodium dodecyl sulfate (SDS). The dissolution was followed by measuring the absorbance at 700 mm at room temperature; the insoluble residues (0.6 mg protein/ml) were suspended in 0.05 M tris-HCl pH 7.5.

lipid than is the case for other membrane proteins. It will accordingly be of interest to determine whether the proteins of this lipid-depleted sheet are richer in hydrophobic amino acids or whether they possess longer peptide sequences of such amino acids. The DOC-insoluble residues are not dissociated appreciably by 8M urea, nor by low concentrations (0.2%) of Triton X-100. They are, however, rapidly disaggregated by SDS as indicated by the loss in turbidity of suspensions of the lipid-depleted sheets as presented in fig. 10. It would therefore appear that this region of the *M. lysodeikticus* membranes is the least sensitive to mild dissociating procedures such as the low ionic strength 'shock' wash, chelating agents and 'weaker' detergents of the deoxycholate type. Thus more drastic dissociating agents such as SDS are needed to separate the electron transport components.

Conclusions

The results of our investigations are in accord with the view that in the multifunctional bacterial membranes there is a regional distribution of specific functions rather than a random intermixing of all of the major activities found in the membrane. The organization of the membrane would thus fit a 'mosaic' or 'patch-work' arrangement of areas containing specific functional lipid-protein complexes. The distribution of both proteins and lipids in such a structure would therefore be non-random and would be in accord with the observed selective release of certain proteins and the higher ratio of cardiolipin/phosphatidyl glycerol in the EDTA-wash fraction. Our results also indicate that the strengths and probably types of cohesive forces in the membrane cover a wide range from the relatively weak association of the ATPase with the membrane to the 'firmly bound' components of the electron transport system requiring a powerful dissociating agent such as SDS to 'solubilize' this region of the membrane.

Acknowledgements

Investigations in this laboratory have been supported by a grant from the National Science Foundation (GB-17107). Studies done by M.S.N. were performed under a Public Health Service Training Fellowship (P.H.S. Grant GM00466). We wish to thank Mr. Charles Harman for the photography.

References

[1] P.C. Fitz-James, J. Biophys. Biochem. Cytol. 8 (1960) 507.
[2] A. Ryter, C. Frehel and B. Ferrandes, Comptes Rendus, Acad. Sci. Paris 265 (1967) 1259.
[3] P.C. Fitz-James, Protides Biol. Fluids 15 (1967) 289.
[4] P.C. Fitz-James, in: Microbial Protoplasts, Spheroplasts and L-forms, ed. L.B. Guze (Williams and Wilkins, Baltimore, Md., 1968) p. 124.
[5] D.A. Reaveley and H.J. Rogers, Biochem. J. 113 (1969) 67.
[6] M.R.J. Salton and J.A. Chapman, J. Ultrastructure Res. 6 (1962) 489.

[7] D.J. Ellar and J.H. Freer, J. Gen. Microbiol. 58 (1969) vii.

[8] B. Ferrandes, P. Chaix and A Ryter, Comptes Rendus, Acad. Sci. Paris 263 (1966) 1632.

[9] M.R.J. Salton, J. Gen. Physiol. 52 (1968) 227s.

[10] A. Ryter and E. Kellenberger, Z. Naturforsch. 13b (1958) 597.

[11] H. Fernandez-Moran, T. Oda, P.W. Blair and D.E. Green, J. Cell. Biol. 22 (1964) 63.

[12] D. Abram, J. Bacteriol. 89 (1965) 855.

[13] V.I. Biryuzova, M.A. Lukoyanova, N.S. Gel'man and A.I. Oparin, Dokl. Akad. Nauk SSSR 156 (1964) 198.

[14] E.D. Korn, Science 153 (1966) 1491.

[15] E.D. Korn, J. Gen. Physiol. 52 (1968) 257s.

[16] W.J. Lennarz and B. Talamo, J. Biol. Chem. 241 (1966) 2707.

[17] M.R.J. Salton, M.D. Schmitt and P.E. Trefts, Biochem. Biophys. Res. Commun. 29 (1967) 728.

[18] M.L. Vorbeck and G.V. Marinetti, Biochemistry 4 (1965) 296.

[19] M.R.J. Salton, in: Specificity of Cell Surfaces, eds. B. Davis and L. Warren (Prentice Hall, Englewood Cliffs, New Jersey, 1967) p. 71.

[20] D.F.H. Wallach, J. Gen. Physiol. 54 (1969) 3s.

[21] M.R.J. Salton and A. Netschey, Biochim. Biophys. Acta 107 (1965) 539.

[22] M.S.Nachbar and M.R.J. Salton, in: Surface Chemistry of Biological Systems, ed. M. Blank (Plenum Press, New York, 1970) p. 175.

[23] D.H. Green and M.R.J. Salton, Biochim. Biophys. Acta 211 (1970) 139.

[24] M.R.J. Salton, Trans, N.Y. Acad. Sci., Ser. II, 29 (1967) 764.

[25] E. Munoz, J.H. Freer, D.J. Ellar and M.R.J. Salton, Biochim. Biophys. Acta 150 (1968) 531.

[26] E. Munoz, M.S. Nachbar, M.T. Schor and M.R.J. Salton, Biochem. Biophys. Res. Commun. 32 (1968) 539.

[27] M.R.J. Salton, in: Microbial Protoplasts, Spheroplasts and L-forms, ed. L.B. Guze (Williams and Wilkins, Baltimore, Md. 1968) p. 144.

[28] G. Weinbaum and R. Markman, Biochim. Biophys. Acta 124 (1966) 207.

[29] M.R.J. Salton, J.H. Freer and D.J. Ellar, Biochem. Biophys. Res. Commun. 33 (1968) 909.

[30] R.K. Morton, Nature 166 (1950) 1092.

[31] V.K. Vambutas and E. Racker, J. Biol. Chem. 240 (1965) 2660.

[32] E. Racker, Federation Proc. 26 (1967) 1335.

[33] F.M. Harold, J.R. Baarda, C. Baron and A. Abrams, Biochim. Biophys. Acta 183 (1969) 129.

[34] T.L. Whiteside and M.R.J. Salton, Bacteriol. Proc. (1969) p. 43.

[35] H.A. Ells, Arch. Biochem. Biophys. 85 (1959) 561.

[36] E. Munoz, M.R.J. Salton, M.H. Ng and M.T. Schor, Europ. J. Biochem. 7 (1969) 490.

BINARY MEMBRANES IN MITOCHONDRIA AND CHLOROPLASTS

F.L. CRANE, C.J. ARNTZEN, J.D. HALL, F.J. RUZICKA and R.A. DILLEY

Department of Biological Sciences, Purdue University Lafayette, Indiana, USA
and
C.F. Kettering Research Laboratory, Yellow Springs, Ohio, USA

In this paper we will present evidence that the membranes of mitochondrial cristae and chloroplast thylakoids are formed from two layers of lipoprotein material and that different parts of the electron transport systems are contained in each layer. An understanding of the structure of these membranes is important to any consideration of the mechanisms involved in mitochondrial and chloroplast organization. Any proposed structure must allow for considerable pleomorphism which is best illustrated by conformational changes in cristae [1,2]; changes in membrane cross section thickness observed with shift of thylakoids from light to dark [3]; variation in the stoichiometry of functional components [4] and much evidence for non-polar bonding as the principal mode of association between lipid and protein in membranes [5,6].

In fig. 1 we diagram the association of subunits inherent in three different proposals for membrane structure. Unit membranes [7] would be formed from a lipid bilayer coated with protein. For 60 Å cross section usually proposed the protein would be mostly in β conformation. Evidence against this type of structure in many membranes has been thoroughly reviewed [6,8]. The second scheme advanced by Green and co-workers [9,10] for cristae envisions 115 Å × 45 Å lipoprotein complexes of the electron transport system linking together to form membrane. In this sequence the lipid functions to disperse the protein from an aggregate to a single layer. This scheme does not allow extreme variation in the stoichiometry of cytochromes. It is also inconsistent with the fact that most of the isolated subunits are smaller than 90 Å in diameter [11]. It is also not consistent with the evidence that freeze etching frac-

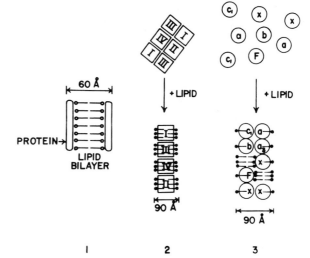

Fig. 1. Assembly of membranes from different types of subunits. 1. Lipid bilayer assembly with β configuration protein surface layers [7]. 2. Large lipoprotein subunits in monolayer with lipid bound to protein hydrophobically [9,15]. 3. Lipoprotein units in double layer with variable stoichiometry. Each layer could have mosaic lipid bilayers [8] or hydrophobic bonding of lipid to protein. See also D.E. Green et al., this volume.

tures the membrane through a central layer [12]. Primary evidence for the second scheme is the 90 Å elements or basepieces observed in the edge of negatively stained cristae. We will discuss later the evidence that basepieces are either headpieces of F_1 on the membrane surface or fibrous elements in the membrane to which headpieces are attached. The third scheme, which we present as the binary membrane, allows for a membrane made up of a diverse stoichiometry of smaller lipoprotein units. It is also consistent with a membrane which can be split through the interior during freeze fracture and with our evidence that chemically fixed membranes show three stained regions in cross section equivalent to two unit membranes [13,14].

The details of lipid protein association within the individual layers remain to be determined. As shown in the diagrams one can envision two layers with lipid cylinders between the protein as proposed by Wallach and Gordon [8] for the single layer system or two layers of lipoprotein with the lipid bound in the protein as proposed by Green and Goldberger [9] or Benson [15,16] for the single layer system. Further details are also needed about the association of core protein or non-catalytic protein with the catalytic proteins such as cytochromes and flavoproteins.

Several lines of evidence show that mitochondrial cristae and chloroplast thylakoids have two different lipoprotein layers. Part of the evidence comes from direct observation of the membrane structure using different electron microscope techniques. Part comes from fragmentation of the membrane into functional parts and determination of structure and chemical composition of the fragments. The evidence can be summarized as follows:

(1) Sections of chemically fixed membranes give the appearance of two unit membranes fused together. Fragments obtained by detergent treatment show thinner membranes equivalent to single layer structure. Recombination of fragments restores the thicker, double layer structure.

(2) Cross sections of physically fixed (freeze etch) chloroplast membranes show two globular ridges with different sized globules in each ridge. Fragments show single ridges of globules.

(3) Surfaces of physically fixed (freeze etch) chloroplast membranes show different globular structure. One surface shows 110 Å globules, the other has 170 Å globules. These surfaces must be the interior surfaces of the two membrane layers because negative staining shows that the outer surfaces are relatively smooth.

(4) Most proteins from the membranes range from 25,000 to 70,000 molecular weight which correspond to globules 30 Å to 50 Å in diameter. Each half of the membrane corresponds in thickness to a single protein layer (50 Å).

(5) Half of the electron transport system of each membrane is found in each of the separated layers. With mitochondria, recombination of the two layers restores complete electron transport.

(6) Electrophoresis shows different proteins in each half of the membranes.

(7) The amount of protein in each fragment is about half of the total membranous protein. This distribution is consistent with splitting the membrane into two halves.

(8) 90 Å globules seen at the edge of negatively stained mitochondrial membranes, which make the membrane appear to be made of 90 Å units, can be attributed either to 90 Å F_1 particles on the surface or to membranifibrils embedded in the membrane. Negative staining of the separated parts of the membrane show 40 to 50 Å globular elements and the recombined membrane with restored function still shows in surface view mostly 40 to 50 Å diameter units as the primary elements of membrane structure.

Studies on mitochondria: detergent fractionation

Mitochondrial cristae contain the complete electron transport system for oxidation of NADH and succinate by oxygen. When the membrane is treated with detergents such as bile salts, Tritons or amyl alcohol the electron transport system can be separated into two membranous parts [11]. One part contains the dehydrogenase systems for dehydrogenation of succinate and NADH by cytochrome c. The other part catalyses the oxidation of reduced cytochrome c by oxygen. The separation of parts is shown in fig. 2. Both of these fractions are in membranous form after partial removal of detergents. Cytochrome c is released as a soluble protein. This is the primary fractionation pattern in cristae upon detergent treatment. Hatefi et al. [17,18] have described further fractionation of the dehydrogenase system, which can be interpreted as further fractionation of this membrane system. The subunits so obtained have been designated I, II, III, IV and correspond respectively to NADH dehydrogenase, succinate dehydrogenase, cytochrome b and c_1 complex and cytochrome oxidase. The first three are derived by subfractionation of the original dehydrogenase or red membrane fraction.

Electron transport components account for only a part of the cristae membrane. Green et al. [19] have defined a series of white proteins, called core and structural proteins, with no known electron transport function, which make up a large part of the membrane. The relative proportion of these proteins in the membrane is shown in table 1. This data only gives a bulk distribution, based on fractionation studies, because the exact number and nature of these components is unknown. The structural protein is largely composed of protein from the headpieces of F_1 knobs which are seen projecting from the cristae membrane as 90 Å knobs [20]. The structural protein can therefore be considered not a part of the basic membrane structure. It can be removed without disruption of the membrane or loss of electron transport function. Core proteins I and II represent part of the NADH and succinate dehydrogenase. Unknown proteins represent a mostly soluble fraction released from cristae vesicles by detergent treatment.

These include many loosely attached dehydrogenases as well as small amounts of structural protein and cytochromes, including cytochrome c.

The structural elements of the membrane will be core proteins and electron transport components. From table 1 it is clear that in beef heart mitochondria each makes up about one half of the membrane after we subtract structural protein and soluble protein and add a small amount of core proteins I and II to core proteins III and IV. In other words about 50% of the ultimate membrane protein is core protein and about 50% represents electron transport carriers. Exact distribution depends on the amount of core proteins I and II present in the rather large primary dehydrogenase complexes.

When the membrane is fractionated with detergent about one half of the total membranous protein is recovered in each of the two fractions. One third of

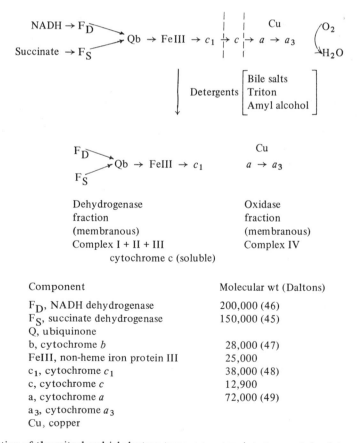

Component	Molecular wt (Daltons)
F_D, NADH dehydrogenase	200,000 (46)
F_S, succinate dehydrogenase	150,000 (45)
Q, ubiquinone	
b, cytochrome b	28,000 (47)
FeIII, non-heme iron protein III	25,000
c_1, cytochrome c_1	38,000 (48)
c, cytochrome c	12,900
a, cytochrome a	72,000 (49)
a_3, cytochrome a_3	
Cu, copper	

Fig. 2. Fractionation of the mitochondrial electron transport system into two parts by detergent treatment.

Table 1
Protein composition of mitochondrial cristae (beef heart)

	Total protein (%)
Electron transport components [a]	27
Structural protein [b]	25
Core protein III	10
Core proteins IV	10
Unknown (including soluble) [c]	28
Total protein 74%	
Total lipid 26%	

a Electron transport components include cytochromes $a + a_3$ (12%), cytochrome b (2.3%), cytochrome c_1 (0.8%) non-heme iron protein III (1.5%), succinate dehydrogenase (2.8%) and NADH dehydrogenase (7.6%) include core protein I and II. Electrophoresis of complex I and its fragments shows that about 1/3 of the protein is core protein I.

b On a yield basis. Contains a large part of the F_1 knobs or headpieces.

c Includes some protein released from vesicles by detergent treatment e.g., cytochrome c and dehydrogenases. Also includes some F_1 protein and some membranifibrils. (Data from [19,50,51].)

the total is removed as soluble protein which will not form membrane even after removal of detergent. The total weight of electron carrier protein is expected to be divided equally since the total calculated mass of cytochrome $a + a_3$, 12%, approximately equals the total mass of other carriers 15%. There is even closer to equal distribution if we consider that part of the NADH dehydrogenase protein may remain in the oxidase fraction. The core proteins III and IV have been reported to each make up 10% of the membrane protein so they are equally divided also. The actual distribution of cytochromes and total protein ob-

served in a deoxycholate fractionation is shown in table 2. Thus both theoretical and observed distribution indicates that the dehydrogenase fraction (complexes I + II + III) represents one half the membrane protein and the oxidase fraction (complex IV) represents one half the membrane protein.

Fractionation of the membrane can also be followed by gel electrophoresis. The electrophoresis shows a separation of electron carriers and core proteins as described above. The electrophoresis patterns obtained on polyacrylamide gel by the method of Takayama et al. [21] for cristae and the dehydrogenase and oxidase fractions are shown in fig. 3.

A typical separation of dehydrogenase and cytochrome oxidase activity by deoxycholate fractionation is shown in table 3. This fractionation is accomplished by 0.3 mg deoxycholate per mg protein in the presence of 1 M KCl followed by 20 hr dialysis of the fractions. The succinate dehydrogenase is concentrated in the red membrane and cytochrome oxidase is concentrated in the green membrane. NADH dehydrogenase shows a slightly different fractionation pattern in that about 50% of the activity remains in the green fraction. Part of the activity in the green fraction remains coupled to the oxidase. In other words the NADH dehydrogenase has a stronger affinity for the green fraction than does the succinate dehydrogenase. Note that the acid extractable flavin (membrane bound) is about equally divided between the two fractions. We feel that this split distribution may indicate that a part of the NADH dehydrogenase complex is actually incorporated in the green membrane layer. When we use several quinones as acceptors for the NADH dehydrogenase we can define several acceptor sites in the complex distinguished by sensitivity to several inhibitors. A diagram of the

Table 2
Distribution of protein and cytochromes by deoxycholate fractionation of cristae

	Cytochromes			Acid extracted flavin	Total protein	Percent of membrane protein
	$a + a_3$	b	$c + c_1$			
Mitochondria	9250	8560	8380		8890	
Red fraction	116	6240	5700	17%	2580	46%
Green fraction	8700	1340	650	20%	3050	54%
Supernatant	70	480	665	60%	2890	

Cytochromes as total in mμmoles.

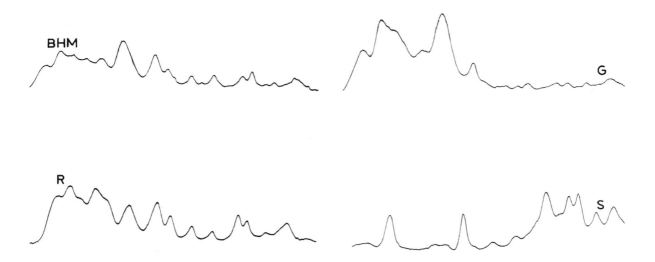

Fig. 3. Electrophoresis patterns of proteins from cristae, dehydrogenase and oxidase membrane fractions and the deoxycholate supernatant. BHM, mitochondria; R, red dehydrogenase fraction; G, green oxidase fraction; S, soluble proteins.

Table 3
Separation of dehydrogenase and oxidase activity by deoxycholate fractionation

	NADH oxidase	Succinate oxidase	Succinate cyt c reductase	NADH ferricyanide reductase	Cytochrome oxidase
Mitochondria	100	100	100	100	100
Red fraction	7	10	40	38	5
Green fraction	30	3	8	52	50
Supernatant	0	0	0	5	0

% of total activity recovered in each fraction. Succinate dehydrogenase activity is proportional to succinate cytochrome c reductase activity.

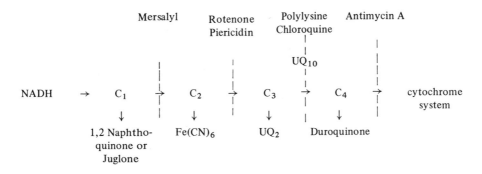

Fig. 4. Acceptor and inhibitor sites in the NADH dehydrogenase complex.
C indicates carrier site; dotted lines sites of inhibitor action.

Table 4

Distribution of NADH-quinone reductase activities in deoxycholate fractions of mitochondria

	Acceptor for NADH dehydrogenase activity [a]			
	Duroquinone	UQ$_2$	Ferricyanide	Juglone
Mitochondria	0.13	0.10	3.00	0.27
Red fraction	0.12	0.10	3.94	0.38
Green fraction	0.14	0.06	4.66	0.28
Supernatant	0.05	0.01	0.56	0.21

Data from F.J.Ruzicka, Ph.D. Thesis, Purdue University, 1969.

[a] Specific activity μmoles NADH/min/mg protein.

acceptor and inhibitor sites is shown in fig. 4. After deoxycholate fractionation a higher proportion of duroquinone reductase activity is retained in the green fraction in proportion to naphthoquinone (Juglone) reductase activity, whereas the red fraction contains less duroquinone reductase activity. Therefore the component responsible for duroquinone reduction tends to remain attached to the oxidase (table 4).

Sections of mitochondrial fractions

Cross sections of mitochondrial membranes fixed with osmium tetroxide or glutaraldehyde and post-stained with uranyl acetate or permanganate have generally been reported to show typical unit membrane structure with two dark lines enclosing an unstained region, with an overall thickness of 60 Å [7]. Sjostrand has shown further detail in the form of dark lines crossing the unstained space and has interpreted these partitions to indicate that the membrane is made up of unstained globules surrounded by a stained matrix [22]. This pattern has been interpreted as a membrane made up of globular protein units which are 50 Å in diameter.

On closer examination of many samples of fixed and sectioned cristae, especially after permanganate staining, we find many regions where there are three dark lines and two unstained globular spaces as if two unit membranes are pressed together [13,14]. Their overall thickness is 90 Å (fig. 5A). We have called this a binary membrane structure. Binary structure can be observed in many sections of mitochondria in the

literature, both in cristae and outer membrane [13,22]. Sections which show only unit membrane structure may be poorly stained or may represent conformational changes in the membrane. Sections of the green fraction show almost exclusively unit membrane structure as shown in fig. 5B. The thickness is about 50 Å. Thus it would represent one half of the original binary membrane.

The red fraction is more complicated and contains a large amount of membranifibrils as well as thin membrane sheets which we have illustrated previously [11]. There are many regions in sections of this membrane which also show unit membrane structure 40–50 Å thick, again equivalent to one half of the binary membrane (fig. 5C). In other places regions of binary membrane are found which may represent a fusion of the thinner membranes or regions of fibril structure. The details of the red fraction are under further study.

The purified NADH dehydrogenase complex (complex I) shows short fragments of membrane-like structure or clusters which in section show unit membrane structure about 70 Å thick (fig. 5D).

When the membranous complexes are recombined according to the procedure of Hatefi et al. [18] to restore succinoxidase and NADH oxidase activity the recombined membrane again shows binary structure [14]. Thus the cristae membrane can contain two layers which are separated by detergent and salt, and which can recombine into a functional structure when detergent is removed under proper conditions. Note that recombination is not achieved by simple mixing of membranes, but requires the solubilization and reprecipitation procedure. The restoration of

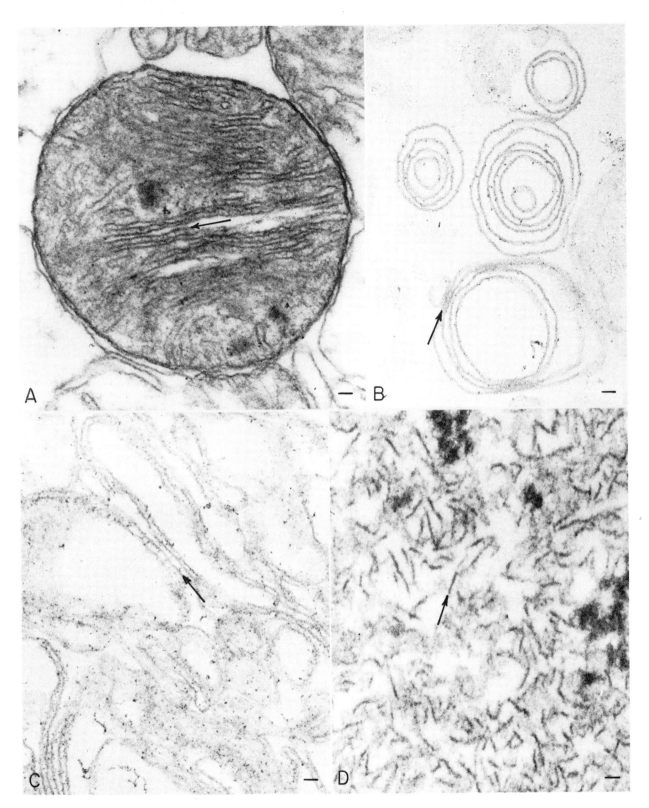

Fig. 5. Cross sections of mitochondrial cristae and fractions obtained by deoxycholate fractionation. Fixed with osmium tetroxide and post stained with uranyl acetate and lead citrate. A. Mitochondria, B. Green oxidase fraction, C. Red dehydrogenase membrane, D. Complex I prepared by further fraction of the dehydrogenase membrane [18]. Marker for all sections 500 Å. (Magnification 100,000 ×.)

Table 5
Restoration of electron transport by recombination of mitochondrial membrane fractions

	NADH oxidase	Succinate oxidase
	(μmoles/min/mg protein)	
1. Red fraction I	0.51	0.22
2. Red fraction II	0.03	0.03
3. Green fraction	0.05	0.03
4. Recombined 1 + 2 + 3	7.57	1.10

Fractions I and II are subfractions of the red dehydrogenase fraction. Fraction I has NADH cytochrome *c* reductase and II has succinate cytochrome *c* reductase. Procedure of Hatefi et al. [18] used for recombination.

activity by recombination of the red and green fractions is shown in table 6.

Freeze etching mitochondria

Freeze fracture of mitochondrial cristae membranes shows at least two types of globular structure on the exposed surfaces [23,24]. The fracture of cristae thus shows a phenomenon similar to freeze fracture layers exposed in chloroplasts which we will discuss later, and gives evidence that the cristae is actually made of two layers of globular material which are separated by the fracture. Ruska and Ruska [23] have shown freeze fracture cross sections of cristae which show two ridges. The tops of the ridges are 50 Å apart, therefore each ridge can be made up of 50 Å globules. In cross sections of cristae which we have made, we also see a detailed two ridge structure, and the overall cristae appears to be about 100 Å across. In similar sections of the dehydrogenase and oxidase fractions obtained by deoxycholate fractionation we find single ridges which are thinner than 100 Å whereas others approximate the dimensions of the cristae (fig. 6). The thinner ridges would be consistent with a splitting of the membrane into two parts but more study by the freeze etch procedure is necessary. It should be noted that the double ridge structure is not consistent with a membrane made from a single layer of 90 Å basepieces.

Negative staining: mitochondria

In some regions along the edge of cristae membranes observed with phosphotungstate negative stain one can observe 90 Å basepiece units to which the 90 Å headpieces are attached (fig. 7A). These 90 Å basepieces have been interpreted by Green et al. [10] to contain the four complexes of the electron transport chain. In other regions the basepiece region appears as a long unstained edge of the membrane. Since we have separated a fibrillar structure [25] from the membrane to which the headpieces are attached and since these fibers and headpieces preparations contain no cytochromes, very little flavin and only 10% lipid [26], we feel that the headpieces are not attached to complexes of electron transport carriers. The 90 Å basepieces can be explained as 90 Å headpieces sitting on the edge of the membrane or as areas where the fibrils run across the edge of the membrane rather than along the edge.

The green fraction shows 50 Å globules in a membrane which has full cytochrome oxidase activity (fig. 7B). The red fraction, after removal of membranifibrils [11], shows a mixture of random 70–90 Å units which may represent the larger dehydrogenases as well as a few loose headpieces (fig. 7C). Underlying these are 40–50 Å globules in the membrane. This membrane has full NADH and succinate cytochrome *c* reductase activity. Therefore the functional complexes do not have to exist as 90 Å headpiece units, but can function as units appropriate to their molecular weight. The recombined fractions with full electron transport function show a few 70–90 Å globules plus a membrane made up primarily of 40–50 Å globules and do not show a continuous row of 90 Å particles along the edges (fig. 7D). Since the recombined fractions have excellent activity it is doubtful that reformation of 90 Å complexes is necessary to restore electron transport function.

Fractionation of chloroplasts

The fractionation of the electron transport system of chloroplasts into two photosystems by digitonin or Triton treatment have been thoroughly described [27–31]. Fractionation of the system which catalyses the photo reduction of NADP by water or

Fig. 6. Cross sections and fracture surfaces of mitochondrial fractions prepared by freeze etch procedure. a. Mitochondria showing double ridges about 70 Å apart in cristae (arrow). b. Green oxidase membrane showing single membrane ridges in concentric vesicles and rough surfaces. c. Red dehydrogenase membrane showing mostly single ridges or tightly packed double ridges and smooth surfaced vesicle surfaces with scattered rodlike units. Marker for all sections 800 Å. (Magnification 75,000 ×.)

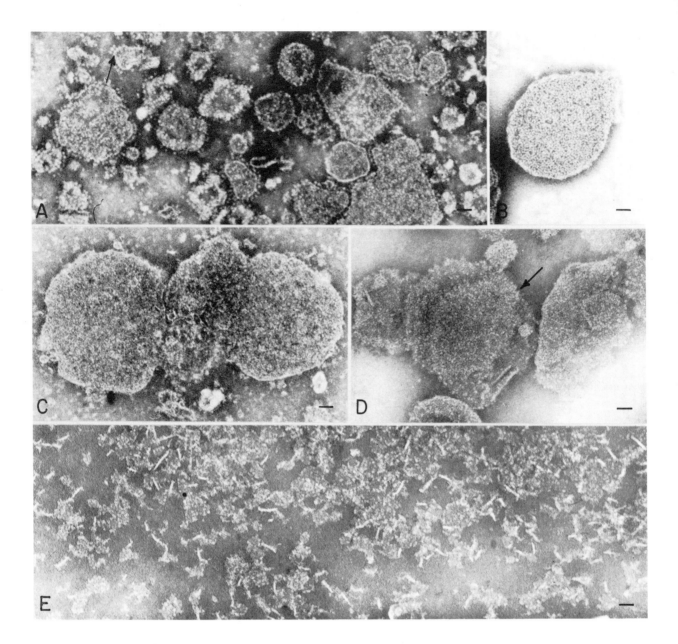

Fig. 7. Surfaces of mitochondrial fractions observed by negative staining. A. ETP shows projecting 90 Å headpieces and a few 90 Å basepiece regions (arrow). B. Cytochrome oxidase shows 50 Å globules dispersed regularly in a membrane sheet. C. Dehydrogenase membrane shows scattered large units which may be headpieces (90 Å) or NADH or succinic dehydrogenase (70 Å to 90 Å) over a membrane with smaller units (40 Å). D. Reconstituted electron transport system which still shows scattered large units and small units. Note that regular basepiece formation has not occurred (arrow) although activity is restored. E. Complex I, purified NADH dehydrogenase fraction showing 80 Å subunits. Marker for all sections 500 Å. (Magnification 100,000 ×.)

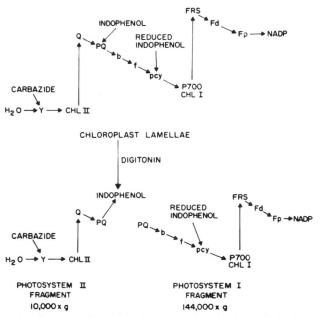

Fig. 8. Separation of photosystem I and II by digitonin fractionation. Plastocyanin (PCY) and Ferridoxin (Fd) are solublized by the treatment so must be readded to the fragments for assay. The system for interaction with water is partly lost so the substituted carbazide is used as a donor for photosystem II.

diphenyl carbazide gives two fractions which carry out partial reactions [32]. Photosystem II (PS II) catalyses reduction of indophenol by water or diphenyl carbazide whereas photosystem I (PS I) catalyses the reduction of NADP by reduced indophenol (using ascorbate as primary reducing agent). The pattern of electron transport and the separation of the parts as presently known is shown in fig. 8. The activity of each of the fragments obtained by centrifugation at 10,000 g (PS II) and 144,000 g (PS I) is shown in table VI. Both of the active fragments are membranous, although the PS I fragments are associated with chloroplast membranifibrils and F_1 knobs which must be separated [33] from the membrane. The structure of the membrane fragments indicate that they represent separate layers from a binary membrane in the original lamellae. The two fractions appear to contain about equal amounts of material when corrected for activity in the 50,000 g fraction. As with mitochondria, electrophoresis of the fragments shows a separation of the proteins of the original thylakoids (fig. 9).

Table 6
Photoreduction activities of fragments derived by digitonin fractionation of chloroplast thylakoids [43]

Fraction	Photoreduction rate (μmoles reduced/mg chlorophyll/hr)	
	Diphenyl carbazide to DCIP [a]	Ascorbate, DCIP to NADP
Orig. chloroplasts	116	150
1,000 g	27	44
10,000 g	100	36
50,000 g	8	172
144,000 g	7	272
Supernatant	5	93

[a] DCIP: 2,6-dichlorophenolindophenol

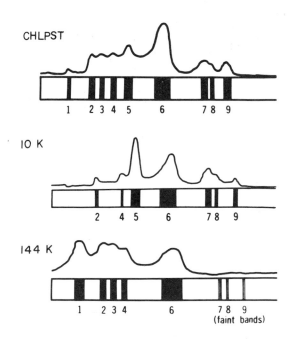

Fig. 9. Polyacrylamide gel electrophoresis patterns of chloroplast membrane fractions. Top, chloroplasts. 10K, the digitonin fraction which sediments at 10,000 g (PS II). 144K, the digitonin fraction which sediments at 144,000 g (PS I).

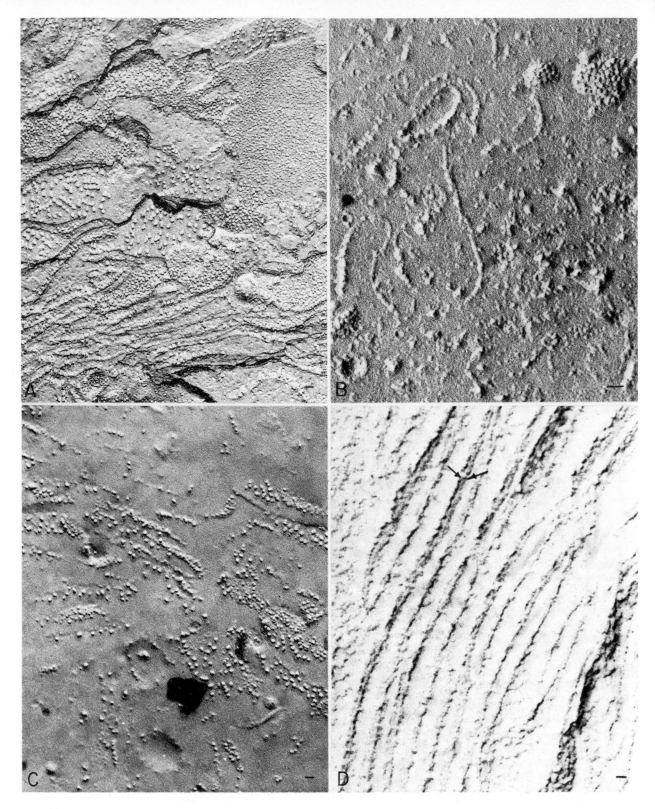

Fig. 11. Freeze fracture of spinach chloroplast fractions. A. Chloroplast membrane fracture faces showing the large widely spaced particles on one face and smaller tightly packed particles on other faces. These surfaces must be interior since globule distribution is constant in controls and in membranes from which all surface globes are removed [33]. Marker 500 Å. B. PS I fraction with small particles on surfaces and single rows of particles in cross sections. Marker 500 Å. C. PS II fraction with large particles on the surface and single rows of large particles in membrane cross section. Marker 500 Å. D. Cross section of chloroplast showing the double ridges in each membrane. Each ridge represents one layer of the binary membrane. Marker 200 Å.

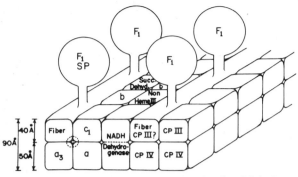

Fig. 12. Diagram of binary structure in mitochondrial cristae membranes. Alternative arrangement of cytochromes has been discussed by Kadenbach in this volume.

brane for the cytochrome oxidase layer as proposed by Wallach [8] for cell membrane. The organization of lipid in the dehydrogenase layer has not been studied although we know that cardiolipin is necessary for binding of NADP dehydrogenase [39] and loss of lecithin stops its activity without releasing it from the membrane [40].

Since mitochondrial membranes still show evidence of binary structure in sections [11] after removal of 95% of the lipid by acetone extraction [41], it would be desirable to study fractionation and freeze fracture structure in these membranes to help determine the structural significance of the lipids.

The fact that both detergent and high salt concentration are required for dispersal of cristae, which contain mostly charged lipids, whereas non-ionic detergent treatment alone is sufficient to disperse the chloroplast membrane, which contains predominantly non-ionic lipids suggest that lipid has a role in binding together the protein subunits of the membranes.

We also do not know the precise relation of core proteins to the electron carrier components. We draw them as separate entities in the mitochondria diagram in order to induce discussion, but it is equally possible that each cytochrome may have a closely associated molecule of core protein. The advantage of separation of core protein from the cytochromes is that: 1. It provides for closer interaction of the

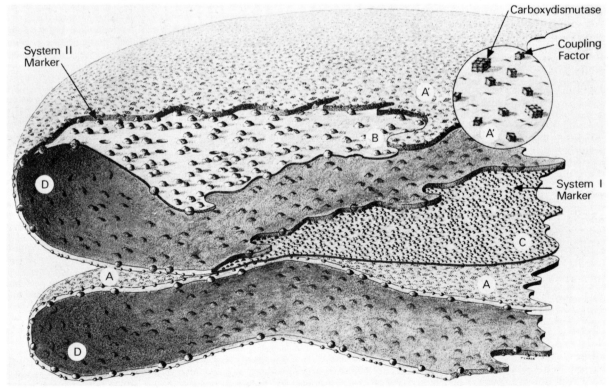

Fig. 13. Diagram of binary structure in chloroplast thylakoids.

carrier units; in a sense a redox cluster in the membrane with core proteins and membranifibrils making up the rest of the membrane bulk. 2. It allows for variation in cytochrome content in the membrane without losing membrane structure. In other words where a cytochrome is missing non-catalytic protein may replace it in the structure. Some kind of replacement of this type is necessary if we are to retain a binary membrane with equal mass on each side of the membrane.

In the chloroplast membrane we know much less about the structural or core proteins. F_1 protein is present in the knobs and membranifibrils can be separated from the membrane [25]. The fractionation studies have so far not given much help in locating the exact site of F_1 attachment and have not given clear indication of where the membranifibrils are located. The diagram shown is the relation between the large particle and small particle layers [33].

Formation of membranes

The simplest model we now have for formation of cristae is the cytochrome oxidase membrane. Clusters of lipid free cytochrome oxidase are converted to membrane sheets with tightly packed globules by addition of phospholipids and enzymatic activity is restored. Further addition of phospholipid up to a saturation level converts the sheets to vesicles with more widely spaced globules [37,38]. Heat stability of the oxidase in the model membrane is similar to its stability in mitochondria whereas the dispersed oxidase is quite unstable [38]. Phospholipids are required for membrane formation whereas detergents can restore activity without membrane formation [42].

The relatively simple formation of the single layer cytochrome oxidase membrane may be much simpler than the formation of a binary membrane where juxtaposition of functional elements and coordination of formation may become factors. The simplest approach to binary structure seems to be the reconstituted mitochondrial electron transport system.

Binary membrane formation in cristae and thylakoids may certainly be an early event in development of these structures. The earliest formed cristae

appear to be 90 Å thick although details of structure and composition are not known. In chloroplast development Arntzen [43] has found that the earliest membranes in the prolamellar body of etiolated plants seem to possess binary structure both by chemical fixation and sectioning and by the appearance of two types of surface after freeze fracture. The correlation of this structure with functional development will help to determine how the parts of the membrane are put together.

Finally we would like to point out that binary structure increases the opportunity for multiple interaction of electron carriers in three dimensions more than would be possible in a monolayer subunit system. It also is a more interesting structure in terms of conformational changes related to the juxtaposition of the two layers.

Acknowledgements

Previously unpublished work supported under grants GM10741 and AM04663 from the National Institutes of Health. Freeze etch studies at Kettering Laboratory under NSF grant B60851. F.L. Crane is supported by a Career Award (K6−21,839), F.J.R. and C.J.A. by Training Grant GM01195 and J.D.H. by Training Grant GM01392 from the National Institute for General Medical Science.

References

[1] C.R. Hackenbrock, J. Cell Biol. 37 (1968) 345.
[2] D.E. Green, J. Asai, R.A. Harris and J.T. Penniston, Arch. Biochem. Biophys. 125 (1968) 684.
[3] S. Murakami and L. Packer, Biochim. Biophys. Acta 180 (1969) 420.
[4] J.N. Williams Jr., Biochim. Biophys. Acta 162 (1968) 175.
[5] D. Chapman, ed. Biological Membranes (Academic Press, New York, 1968).
[6] L. Bolis and B.A. Pethica, eds. Membrane models and the formation of Biological Membranes (North-Holland, Amsterdam, 1968).
[7] J.D. Robertson, in: Cellular Membranes in Development, ed. M. Locke (Academic Press, New York, 1964) p. 1.
[8] D.F.H. Wallach and A.S. Gordon, in: Regulatory Functions of Biological Membranes, ed. J. Jarnefelt (Elsevier, Amsterdam, 1968) p. 87.
[9] D.E. Green and R.F. Goldberger, in: Molecular Insights

into the Living Process (Academic Press, New York, 1967) p. 247.

[10] D.E. Green, D.W. Allmann, E. Bachmann, H. Baum, K. Kopaczyk, E.F. Korman, S. Lipton, D.H. McLennan, D.G. McConnell, J.F. Perdue, J.S. Rieske and A. Tzagoloff, Arch. Biochem. Biophys. 119 (1967) 312.

[11] F.L. Crane, J.W. Stiles, K.S. Prezbindowski, F.J. Ruzicka and F.J. Sun, in: Regulatory Functions of Biological Membranes, ed. J. Jarnefelt (Elsevier, Amsterdam, 1968) p. 21.

[12] D. Branton, Proc. Natl. Acad. Sci. U.S. 55 (1966) 1048.

[13] F.L. Crane and J.D. Hall, Biochem. Biophys. Res. Commun. 36 (1969) 174.

[14] K.S. Prezbindowski, F.J. Ruzicka, F.F. Sun and F.L. Crane, Exptl. Cell Res. 57 (1969) 385.

[15] A.A. Benson, in: Membrane models and the formation of biological membranes, eds. L. Bolis and B.A. Pethica (North-Holland, Amsterdam, 1968) p. 190.

[16] T.E. Weier and A.A. Benson, in: Biochemistry of Chloroplasts, Vol. 1, ed. T.W. Goodwin (Academic Press, New York, 1966) p. 91.

[17] Y. Hatefi, A.G. Haavik and P. Jurtshuk, Biochim. Biophys. Acta 52 (1961) 106.

[18] Y. Hatefi, A.G. Haavik, L.R. Fowler and D.E. Griffiths, J. Biol. Chem. 237 (1962) 2661.

[19] D.E. Green, N.F. Haard, G. Lenaz and H.I. Silman, Proc. Natl. Acad. Sci. U.S. 60 (1968) 277.

[20] E. Racker and A. Bruni, in: Membrane Models and the Formation of Biological Membranes, eds. L. Bolis and B.A. Pethica (North-Holland, Amsterdam, 1968) p. 138.

[21] K. Takayama, D.H. MacLennan, A. Tzagoloff and C.D. Stoner, Arch. Biochem. Biophys. 114 (1966) 223.

[22] F.S. Sjostrand, in: The Membranes, eds. A.J. Dalton and F. Haguenau (Academic Press, New York, 1968) p. 151.

[23] C. Ruska and H. Ruska, Z. Zellforsch. Mikroskop. Anat. Abt. Histochem. 97 (1969) 298.

[24] G.O. Kreutziger, Proc. 27th Elect. Microsc. Soc. Amer. Meeting (1969) 294.

[25] J.W. Stiles, J.T. Wilson and F.L. Crane, Biochim. Biophys. Acta 162 (1968) 631.

[26] J.D. Hall, J.W. Stiles, Y. Awasthi and F.L. Crane, Proc. Indiana Acad. Sci. 78 (1969) 189.

[27] N.K. Boardman and J.M. Anderson, Nature 203 (1964) 166.

[28] J.M. Anderson and N.K. Boardman, Biochim. Biophys. Acta 112 (1966) 403.

[29] L.P. Vernon, B. Ke, S. Katoh, A. San Pietro and E.R. Shaw, Brookhaven Symp. in Biology 19 (1966) 102.

[30] L.P. Vernon, A. Garcia, H. Mollenhauer and B. Ke, in: Comparative Biochemistry and Biophysics of Photosynthesis, eds. K. Shibata, A. Takamiya, A.T. Jagendorf and R.C. Fuller (Univ. Tokyo Press, Tokyo, 1968) p. 3.

[31] H. Huzisige, H. Vsiyama, T. Kikuti and T. Azi, Plant Cell Physiol. 10 (1969) 441.

[32] L.P. Vernon and E. Shaw, Biochem. Biophys. Res. Commun. 36 (1969) 878.

[33] C.J. Arntzen, R.A. Dilley and F.L. Crane, J. Cell Biol. 43 (1969) 16.

[34] T.E. Weier and A.A. Benson, Am. J. Botany 54 (1967) 389.

[35] D. Branton and A.B. Park, J. Ultrastruct. Res. 19 (1967) 283.

[36] S.H. Howell and E.N. Moudrianakis, J. Mol. Biol. 27 (1967) 323.

[37] F.J. Sun, K.S. Prezbindowski, F.L. Crane and E.E. Jacobs, Biochim. Biophys. Acta 153 (1968) 804.

[38] T.F. Chuang, Y.C. Awasthi and F.L. Crane, Proc. Indiana Acad. Sci. (1969) in press.

[39] Y.C. Awasthi, R. Berezney, F.J. Ruzicka and F.L. Crane, Biochim. Biophys. Acta 189 (1969) 457.

[40] F.J. Ruzicka and F.L. Crane, Biochem. Biophys. Res. Commun. 38 (1970) 249.

[41] S. Fleischer, B. Fleischer and W. Stoeckenius, J. Cell Biol. 32 (1967) 193.

[42] T.F. Chuang, Y.C. Awasthi and F.L. Crane, Federation Proc. 29 (1970) 540 abs.

[43] C.J. Arntzen, Ph.D. Thesis Purdue Univ. 1970.

[44] K. Muhlethaler, in: Biochemistry of Chloroplasts, vol. I, ed. T.W. Goodwin (Academic Press, New York, 1966) p. 49.

[45] M.L. Baginski and Y. Hatefi, J. Biol. Chem. 244 (1969) 531.

[46] E.C. Slater ed., Flavins and Flavoproteins, (Elsevier, Amsterdam, 1966).

[47] R. Bomstein, R. Goldberger and H. Tisdale, Biochim. Biophys. Acta 50 (1961) 527.

[48] R. Goldberger, A.L. Smith, H. Tisdale and R. Bomstein, J. Biol. Chem. 236 (1961) 2788.

[49] D.G. McConnell, A. Tzagoloff, D.H. MacLennan and D.E. Green, J. Biol. Chem. 241 (1966) 2373.

[50] M. Klingenberg, in: Biological Oxidations, ed. T.P. Singer (Wiley Interscience, New York, 1968) p. 3.

[51] P.V. Blair, T. Oda, D.E. Green and H. Fernandez-Moran, Biochemistry 2 (1963) 756.

FORMATION OF PHOTOSYNTHETIC MEMBRANES
DURING CHLOROPLAST DEVELOPMENT

N.K. BOARDMAN, Jan M. ANDERSON, A. KAHN *,
S.W. THORNE and T.E. TREFFRY **

C.S.I.R.O., Division of Plant Industry, Canberra, Australia

The biogenesis of photosynthetically-active membranes in the plastids of many higher plants is controlled by light. The plastids of the dark-grown plant (termed proplastids or etioplasts) are devoid of chlorophyll, but they contain protochlorophyllide [1]. On illumination of the plants, protochlorophyllide is rapidly converted to chlorophyllide *a*, which is then esterified with phytyl alcohol. After a short lag phase, there is continued synthesis of chlorophyll *a*, and chlorophyll *b* is formed. Bean plants require several hours of illumination before the ratio of chl *a*/chl *b* reaches the value observed with fully greened plants [2]. Several hours of greening are also needed before bean plastids which exhibit significant photochemical Hill activity can be isolated [2].

The structural changes which occur during plastid development have been studied extensively [3–6]. The internal membranes of the etioplasts are organized into one or more paracrystalline, 3-dimensional tubular structures, termed prolamellar bodies. Fluorescence microscopy indicates that the protochlorophyllide is localized in centres which correspond in size to the prolamellar bodies [7]. On illumination of the dark-grown plants, the prolamellar bodies are dispersed into sheets of perforated membranes, which give rise to lamellae. Finally, there is fusion and elaboration of the lamellae in certain regions to form grana, which are a characteristic feature of mesophyll chloroplasts of higher plants.

There appears, therefore, to be continuity of membrane structures during the early stages of plastid development. This suggests that the membranes of the etioplast may be used as the building blocks for the formation of the photosynthetically-active membranes. The photosystems I and II either are an integral part of the chloroplast membranes or they are structures which are intimately accociated with the membranes.

In the present work, we have examined the spectral properties of the pigments, both of the etioplast and the developing plastid. Fluorescence spectroscopy at the temperature of liquid nitrogen was used to gain information about transfer of excitation energy between pigment molecules, which in turn gives us information on intermolecular distances. We examined the structural changes which occur during chloroplast development, in relation both to pigment synthesis and to the photochemical properties of the isolated plastids. The results suggest that the assembly of the photosynthetic membranes is a step-wise process, at least in the early stages of greening.

From studies on the formation of photosynthetic membranes during the greening of a *Chlamydomonas* mutant, it was suggested that the membranes may be assembled by a one-step process [8], but more recent work indicates that this is not so for the initial phase of greening [9].

Materials and methods

Plants

Bean plants (*Phaseolus vulgaris* L. var. Brown Beauty) were grown in vermiculite in the dark at 25° for 10–15 days before their primary leaves were

---

* Permanent address: Department of Biological Sciences, Purdue University, Lafayette, Indiana, U.S.A.
** Present address: Biochemistry Department, The University, Sheffield, S.10. U.K.

harvested. Pea plants (*Pisum sativum* L. var. Green-feast) were grown for 8 days in the dark. Plants were then illuminated either by Philips Daylight 40W fluorescent tubes (800 ft candles at level of plants or 12,000 erg cm^{-2} sec^{-1}) or by red fluorescent tubes (Philips T/L 40W) which emit primarily in the region of 600–700 nm (1900 erg cm^{-2} sec^{-1}).

Plastid isolation

Pea leaves were blended in a Servall Omnimixer for 10 sec at 85% of the line voltage in 0.05 M phosphate buffer (pH 7.2) and 0.3 M sucrose and 0.01 M KCl. The plastids were sedimented by centrifugation at 1000 g for 10 min and washed once with sucrose-phosphate buffer [10].

Hill-reaction activities were assayed with trichloro-phenol-indophenol (TCIP) as oxidant [10].

Prolamellar body membranes were prepared from bean etioplasts as described elsewhere [11].

Absorption and fluorescence spectroscopy

Absorption spectra were recorded with a Cary model 14R spectrophotometer, fitted with a scat-tered-transmission attachment. Measurements at the temperature of liquid nitrogen (77°K) were made as described previously [12]. Fluorescence emission and absorption spectra were recorded on an instrument which automatically corrected the spectra for photo-multiplier and monochromator responses and varia-tion in energy output from the light source [13]. Total chlorophyll was measured in acetone extracts [14] and chl *a*/chl *b* ratios were determined by fluor-escence spectroscopy of ethanol extracts at 77°K [15]. Chlorophyll/chlorophyllide ratios were deter-mined by a partition method [16].

Electron microscopy

Leaflets were fixed in glutaraldehyde, followed by osmium. Sections were poststained with uranyl acetate and lead citrate [17].

Spectral properties of pigments

The absorption spectra of dark-grown bean leaves both at room temperature and at 77°K are shown in fig. 1a. Two main spectroscopic forms of protochlo-rophyllide with maxima at 637 nm and 650 nm are seen in agreement with the observations of previous investigators [18]. The spectrum of a leaf which had been illuminated prior to cooling to 77°K is shown by the broken line in fig. 1a. The band at 680 nm is due to chlorophyllide *a* and the band at 628 nm is considered to be due both to a secondary band of chlorophyllide *a* and to a form of protochlorophyll which is not photoconvertible. Evidence for an in-active form of protochlorophyll absorbing at 628 nm is provided shortly by the fluorescence data. Isolated prolamellar body membranes gave similar spectra to the leaf (fig. 1b).

The form of protochlorophyllide absorbing at 637 nm (PChl–637) accounts for about 40–45% of the total protochlorophyll of the dark-grown leaf, assuming that the relative amounts of PChl–637 and PChl–650 are proportional to the peak heights at 637 nm and 650 nm at 77°K. Since about 85–90% of the protochlorophyllide of the dark-grown leaf is photoconvertible, we conclude that both PChl–637 and PChl–650 are largely photoconvertible to chloro-phyllide *a* [11].

The fluorescence emission spectra of etiolated leaves and of prolamellar body membranes at 77°K (the solid lines in fig. 2) show two main bands at 655 nm and 630 nm. These observations are in agree-ment with previous Russian work with etiolated bean leaves [19]. Fluorescence excitation spectra are shown by the broken lines in fig. 2. To record these spectra, the fluorescence monochromator was set at a slightly longer wavelength (665 nm) than the maxi-mum of the major band. The excitation spectra, both of leaves and prolamellar body membranes show peaks at 650 nm and shoulders at 638–639 nm. We conclude, therefore, that the fluorescence emitted by the major band emanates from PChl-650, but it is activated by light absorbed both by PChl–650 and PChl–637. Energy absorbed by PChl–637 is transfer-red to PChl–650 with high efficiency. An excitation spectrum for the fluorescence emitted at 630 nm showed a maximum at 628 nm [11]. Fluorescence emitted at the 630 nm band apparently originates from PChl–628. If a leaf is illuminated prior to cooling, the emission band at 655 nm is replaced by a band due to chlorophyllide *a* at 694 nm, but the 630 nm band remains [11]. PChl–628 is apparently not transformed by light. We conclude that the etiolated leaf contains three forms of protochloro-

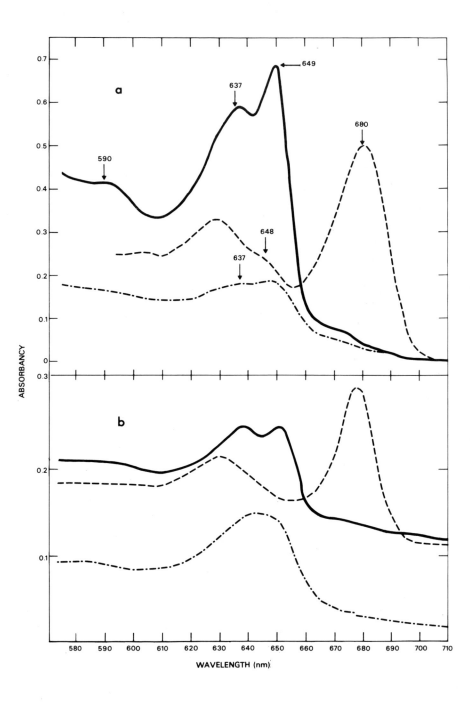

Fig. 1. Absorption spectra. (a) Leaves from 14-day old dark-grown bean seedlings. −.−.−.− 20°C; −−−− 77°K; ———— 77°K after illumination for 1 min at room temperature. Four thicknesses of leaf were used. (b) Prolamellar body membranes from dark-grown bean leaves. −.−.−.− 20°C; −−−− 77°K; ———— 77°K after illumination. The 77°K spectra are displaced by 0.12 absorbancy units.

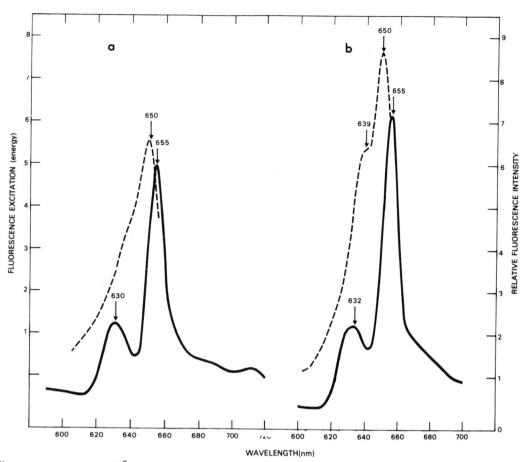

Fig. 2. Fluorescence spectra at 77°K. (a) Leaf from 15-day-old dark-grown seedlings. (b) Prolamellar body membranes. Emission spectra, obtained by exciting with 440 nm light are shown by the solid lines. The broken lines show fluorescence excitation spectra with the emission monochromator set at 665 nm.

phyll(ide), two of which (PChl–637 and PChl–650) are photoconvertible. Previous work [1] recognized two spectroscopic forms of protochlorophyll(ide), in etiolated leaves.

The fraction of fluorescence emitted at 655 nm, compared with that emitted at 630 nm varies with the age of the seedling (fig. 3). In young seedlings (4 days from planting the seed) the 630 nm band (F630) is slightly larger than F655. The ratio, F655/F630, gradually increases to a value of 2.7 at 15 days after planting and then it declines in the old etiolated plants. In the young seedlings PChl-628 may represent a precursor of the photoactive forms, whereas in the old plants it may represent a denaturation of PChl–637 and PChl–650.

We will now consider the fluorescence properties of a leaf following partial photoconversion of the photochlorophyllide. Fig. 4 shows the fluorescence emission spectrum at 77°K of a leaf in which approximately 10% of the protochlorophyllide had been converted to chlorophyllide *a* by illumination at −55°C. A high proportion of the fluorescence emission is associated with chlorophyllide *a* even though the proportion of chlorophyllide *a* is small. We conclude that some of the light energy absorbed by protochlorophyllide was transferred to newly formed chlorophyllide *a*. The efficiency of energy transfer was estimated from the fluorescence intensities at 654–655 nm of the partly converted leaf and the dark leaf [11]. We obtained a transfer efficiency of

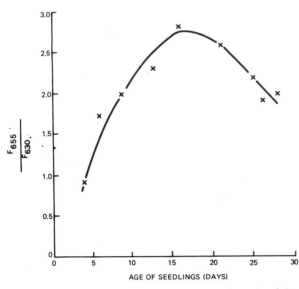

Fig. 3. The effect of age of bean seedlings on the ratio of the fluorescence emitted at 655 nm, compared with that emitted at 630 nm. Excitation wavelength, 440 nm.

40% i.e. 40% of the energy absorbed by protochlorophyllide at $77^{\circ}K$ was transferred to chlorophyllide a.

The wavelength position of the chlorophyllide a

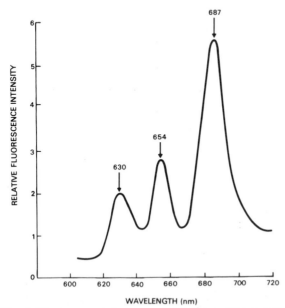

Fig. 4. Fluorescence emission spectrum of a partly converted bean leaf (ca. 10%) at $77^{\circ}K$. Excitation wavelength, 440 nm.

fluorescence varies with the amount of photoconversion. Fig. 5 compares the fluorescence emission spectra of leaves which had been illuminated for different periods at room temperature prior to cooling to $77^{\circ}K$. At low fractional conversions (ca. 5% at 3 sec), the fluorescence maximum is at 674 nm, at slightly higher fractional conversions (ca. 10% at 10 sec) it shifts to 687 nm and with maximum conversion the peak is at 694 nm.

Fig. 6 summarizes the spectroscopic intermediates which are observed on photoconversion of protochlorophyllide to chlorophyllide a. Protochlorophyllide is converted to chlorophyllide a (Chl−678, fluorescence emission at 687 nm, fluorescence excitation at 678 nm) possibly via C (fluorescence emission at 674 nm, excitation 668 nm), although C may repre-

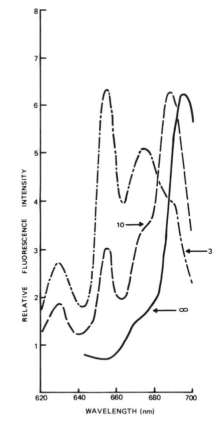

Fig. 5. Fluorescence emission spectra for 14-day etiolated bean leaves at $77^{\circ}K$, after illumination at $20^{\circ}C$ by 622 nm light, intensity 2000 erg cm^{-2} sec^{-1}. Time of illumination, −.−.−. 3 sec; − − − − 10 sec; ⎯⎯⎯ 2 min, maximal conversion.

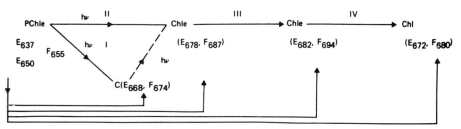

Fig. 6. Spectroscopic intermediates in the conversion of protochlorophyllide (PChle) to chl *a* at the initial photoconversion in a dark-grown bean leaf. F = fluorescence emission, E = fluorescence excitation or absorption. Transfer of excitation energy is indicated by the solid horizontal lines at the lower part of the diagram.

sent the product of a small fraction of protochlorophyllide which is converted more rapidly than the bulk of the protochlorophyllide. Chl–678 is converted fairly rapidly at room temperature to Chl–682, a shift reported previously by Grassman et al. [20] and by Bonner [21] from absorption data. Shift IV (fig. 6) in which Chl–682 is converted to Chl–672 was discovered by Shibata [18]. The Shibata shift occurs after 20–30 min and it is a dark reaction. Phytolation takes place gradually over the course of about 60 min [16], so that Chl–672 probably represents a mixture of chlorophyllide *a* and chlorophyll *a*.

The significant experimental finding we wish to stress is that if only part of the protochlorophyllide is converted, then light energy is transferred at 77°K from the remaining protochlorophyllide to the various spectroscopic forms of chlorophyllide *a*, including Chl–672 [22]. The energy transfer is in-

dicated by the horizontal lines in the lower part of fig. 6.

The observation of energy transfer at 77°K from PChl–637 to PChl–650 and from PChl–650 to chlorophyllide *a* and to chlorophyll *a* leads to some general conclusions about the organization of the pigment molecules. For efficient energy transfer, there must be good overlap between the fluorescence band of the donor and the absorption band of the acceptor, and the distance between donor and acceptor should be no greater than 30–50 Å [23]. Energy transfer, therefore, is likely to occur only if a number of protochlorophyllide molecules are organized into a unit, e.g. attached to a single apoprotein molecule. Energy transfer from protochlorophyllide to Chl–682 and to Chl–672 indicates that at least some of the newly formed chlorophyll remains in close proximity to the protochlorophyllide.

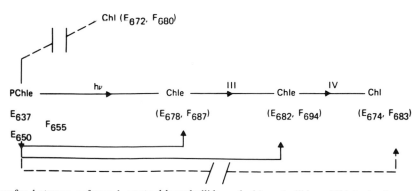

Fig. 7. Energy transfer between reformed protochlorophyllide and chlorophyllide *a* (Chle) at subsequent photoconversions. Energy transfer does not occur between the pigments connected by the broken lines. See text for explanation.

Fig. 7 summarizes the energy transfer data at subsequent photoconversions after more protochlorophyllide is allowed to accumulate. In these experiments, bean plants were illuminated for several minutes to convert their photoconvertible protochlorophyllide to chlorophyllide *a*, and the plants returned to darkness for 5 hr. During the dark period, additional protochlorophyllide was formed. Part of the new protochlorophyllide was converted by illuminating a leaf, which was then cooled to 77°K and its fluorescence properties examined. Energy transfer was observed between protochlorophyllide and Chl–678 and between protochlorophyllide and Chl–682, but not between protochlorophyllide and Chl–672 after the Shibata shift, (maximum at slightly higher wavelength, 674 nm) or to Chl–672 formed at the first transformation [24]. Following the dark period, newly-formed chlorophyllide *a* remains for a shorter period in close proximity to units containing protochlorophyllide. The photosynthetic membranes apparently are formed at an accelerated rate after the 5 hr period in the dark, compared with the rate observed during the first hour of continuous illumination.

Formation of chlorophyll during greening

When dark-grown pea seedlings are exposed to continuous white light chlorophyll accumulates, as shown in fig. 8. After turning on the light there is a lag phase of about 30 min before there is detectable synthesis of chl *a* other than the amount formed by photoconversion of pre-existing protochlorophyllide. We have developed a sensitive fluorescence method capable of detecting one molecule of chl *b* in the presence of 1000 molecules of chl *a* [15]. Chl *b* is detectable within 10 min of switching on the light. At 10 min, the chl *a*/chl *b* ratio is 300; at 20 min it is 30. The ratio remains reasonably constant between 1 and 2 hr, and then it gradually falls to 3 after about 5 hr of greening. At 5 hr, the total chlorophyll content of the leaves has increased 15-fold as compared with the protochlorophyllide content of the etiolated leaves. Phytolation of chl *a* [16] and the formation of chl *b* appear to commence soon after the photoconversion of the protochlorophyllide. Similar conclusions were reached by Rudoi et al. [25] for greening corn seedlings.

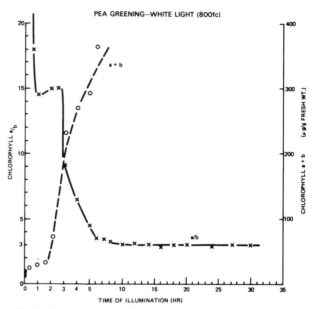

Fig. 8. Total chlorophyll ———— and chl *a*/chl *b* ratios ——————— as a function of the time of greening of 8-day dark-grown pea seedlings. White light of 12,000 erg cm^{-2} sec^{-1}.

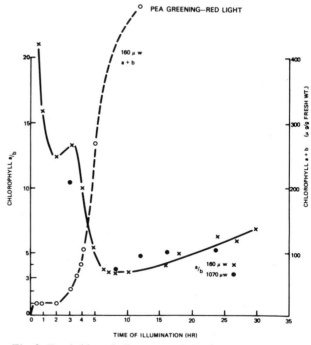

Fig. 9. Total chlorophyll ———— and chl *a*/chl *b* ratios ——————— as a function of the time of greening of 8-day dark-grown pea seedlings. Light from red fluorescence tubes (Philips TL 20W/15). —●— intensity 1600 erg cm^{-2} sec^{-1}; —X— 10,700 erg cm^{-2} sec^{-1}.

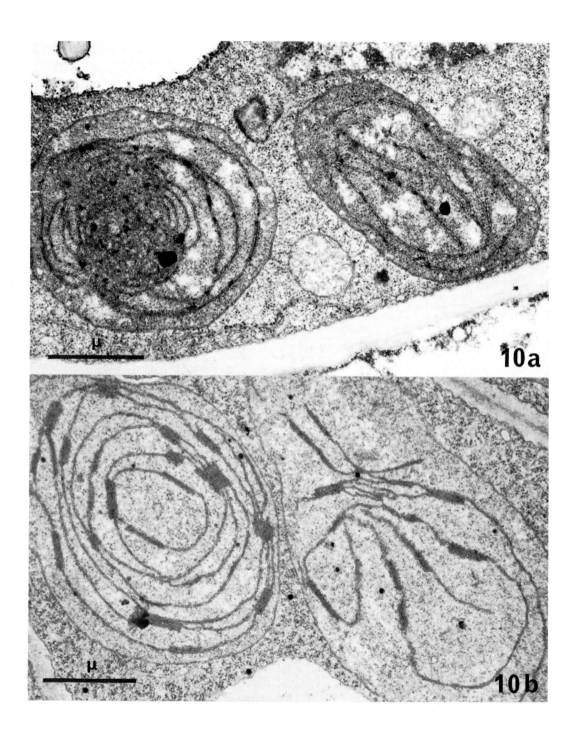

Fig. 10. Plastids in an etiolated pea seedling illuminated with white light of 12,000 erg cm^{-2} sec^{-1}, for (a) 4 hr, (b) 8 hr.

Fig. 9 shows the greening curve for etiolated pea plants after exposure to light from red fluorescence tubes, which emit most of their energy between 600 nm and 700 nm. As observed with white light, there is a 15-fold increase in chl a + chl b at 5 hr. An interesting feature of the red light treatment is the gradual increase in the chl a/b ratio from 3.5 at 7 hr to about 7 at 24–30 hr. Increasing the light intensity of the red light from 1600 erg cm^{-2} sec^{-1} to 10,700 erg cm^{-2} sec^{-1} has little effect on the chl a/chl b ratios.

Structure of developing plastids

We were interested in knowing whether there were any differences in the structure of the plastids in plants greened under red light compared with white light. Fig. 10a shows the structure of a typical plastid in plants which had been greened for 4 hr under white light. We see the fusion of vesicles to form lamellae, but there is little evidence for the fusion of lamellae to form grana. By 8 hr in white light, extensive grana formation has taken place (fig. 10b). Quantitative data on grana/plastid and partitions or lamellae/granum are shown in fig. 11. It is apparent that grana formation occurs between 4 and 8 hr of greening.

During the first 8 hr of greening in red light, the sequence of structural changes in the plastids i.e. prolamellar body dispersal, vesicle fusion and grana formation occur at approximately the same time as in white light, although grana formation may occur slightly earlier in red light. A dramatic feature at times longer than 8 hr is the appearance of prolamellar bodies. Fig. 12 shows a comparison between the

fine structure of a plastid in a plant greened in white light for 24 hr, and a plastid in a plant greened in red light for the same time. The chlorophyll contents of the plants were similar. Small prolamellar bodies are visible in the plastid from the plant greened in red light.

The chlorophyll/chlorophyllide ratio was observed to be lower in the leaves of plants greened in red light. For plants in white light for 24 hr, the chlorophyll/chlorophyllide ratio was 50, compared with 31 for plants in red light for 24 hr. Thus the formation of prolamellar bodies seems to be associated with an increase in chlorophyllide a. Protochlorophyllide could not be detected, indicating that the intensity of the red light was sufficient to convert protochlorophyllide to chlorophyllide a. Previously, it was suggested that the dispersal of prolamellar bodies in the etioplast is associated with the photoconversion of protochlorophyllide to chlorophyllide a [4], but photoconversion alone does not seem to be sufficient to cause dispersal of the prolamellar body. Fig. 13 compares the structure of a typical plastid in an etiolated pea plant which had been illuminated for 30 min at 26°C, with a plastid from a plant illuminated for 30 min at 0°C. Maximum photoconversion of protochlorophyllide to chlorophyllide occurred in both plants, but it is apparent that dispersal of the prolamellar bodies does not take place at 0° [17]. It is known that phytolation of chlorophyllide a is inhibited at 0° [17,26] and we would suggest, therefore, that the dispersal of the prolamellar body is associated with a certain minimum degree of phytolation and hence an increase in the hydrophobic character of the membrane. For some reason, phytolation is inhibited in continuous red light compared with white light, chlorophyllide a accumulates and prolamellar bodies are reformed.

Development of photochemical activity

Plastids were isolated from greening pea plants and assayed for Hill activity with trichlorophenolindophenol (TCIP) as oxidant. Reduction of TCIP is coupled to oxygen evolution and it is a measure of electron transport in photosystem II. There was no detectable Hill activity during the first 5 hr of greening and then there was a fairly rapid increase between

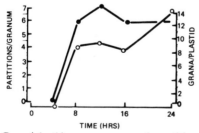

Fig. 11. Grana/plastid −o−o−o−, and partitions/granum −•−•−•− for pea plastids as a function of time of illumination of seedlings in white light.

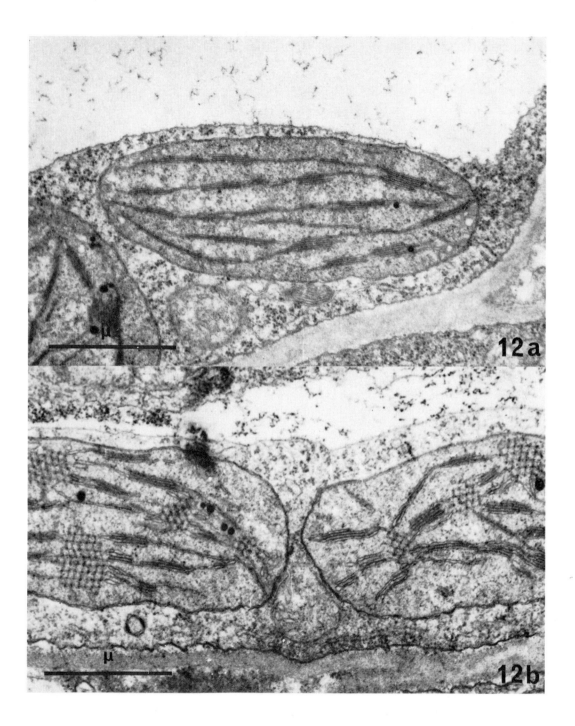

Fig. 12. Plastids in 8-day etiolated pea seedlings. The seedlings were illuminated for 24 hr in (a) white light and (b) red light.

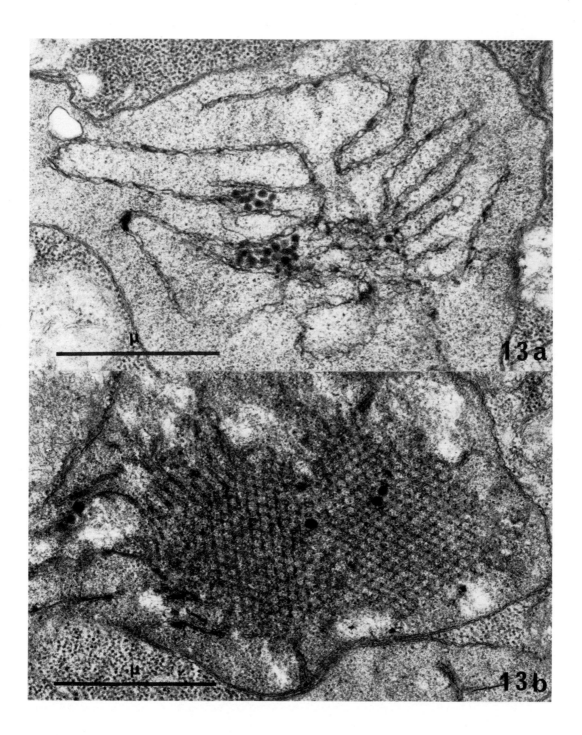

Fig. 13. Plastids in 8-day etiolated pea seedlings. The seedlings were illuminated in white light for (a) 30 min at 26° and (b) 30 min at 0°.

5 and 8 hr (fig. 14). The development of photo-chemical activity is similar for plants illuminated in white light, and in red light. Chloroplasts from mature pea plants grown under normal light conditions reduced TCIP at the rate of 200—220 μmoles/mg chl/hr.

Thus, there is a long lag phase before the onset of photosystem II activity as measured by Hill activity in isolated plastids. During this period, chlorophyll is being synthesized, and as shown in figs. 9 and 10 there is a 15-fold increase in chlorophyll at 5 hr. The onset of photosystem II activity corresponds to the formation of grana and to the decline in the chl a/chl b ratio to the value normally found in mature chloroplasts.

We have also examined the appearance of light-induced oxidation of cytochrome f during greening of pea plants. Cytochrome f is present in etiolated leaves [27,28] but we have not been able to detect any photooxidation of the cyt f on illumination of the etiolated leaves with red light in an Aminco-Chance dual wavelength spectrophotometer. This is in agreement with earlier work of Bonner and Hill [28] with etiolated bean leaves. Photooxidation with 703 nm light was first detectable in pea leaves after 30 min of greening. This is considerably sooner than the appearance of Hill activity in the isolated plastids, and suggests that photosystem I becomes photoactive before photosystem II.

Fluorescence emission spectra at 77°K

A prominent feature of the fluorescence emission spectrum of mature chloroplasts at 77°K is the band at 735 nm, which accounts for 75% of the total fluorescence emission. An examination of the fluorescence properties of the subchloroplast fragments obtained by incubation of chloroplast with digitonin indicated that the emission band at 735 nm originates mainly from photosystem I. The bands at 683 nm and 695 nm at 77°K come from photosystem II [29]. It seemed of interest to examine the appearance of the large band at 735 nm during the greening of etiolated pea plants. Plastids were isolated from plants greened in high white light, and their fluorescence emission spectra measured at 77°K.

Some representative spectra are shown in fig. 15. After 2 hr of greening, the fluorescence emitted at the 735 nm band accounts for about 25% of the total

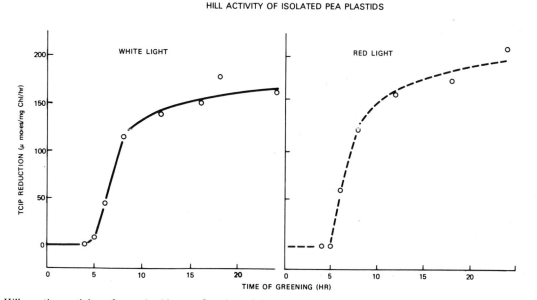

Fig. 14. Hill-reaction activity of pea plastids as a function of the time of greening of dark-grown seedlings. Seedlings illuminated with (a) white light of 12,000 erg cm^{-2} sec^{-1} and (b) red light of intensity 1600 erg cm^{-2} sec^{-1}.

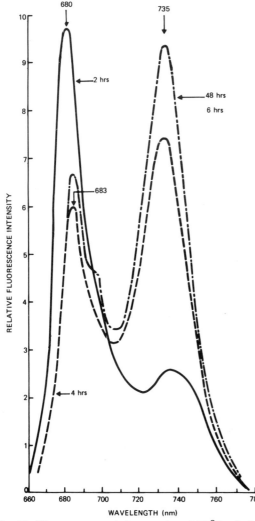

Fig. 15. Fluorescence emission spectra at 77°K of plastids from greening pea seedlings. Seedlings illuminated with white light (12,000 erg cm^{-2} sec^{-1} for ——— 2 hr, ———— 4 hr, −.−.−.− 6 or 48 hr.

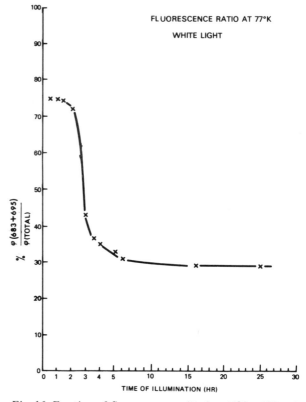

Fig. 16. Fraction of fluorescence emitted at (683 + 693 nm) by pea plastids at 77°K as a function of the time of greening of the seedlings in white light.

emission. Chlorophyll *a* in organic solvents emits a similar proportion of its fluorescence at the longer wavelength [12]. Plastids from plants greened for 6 hr show a decrease in emission at 683 nm and a marked increase at 735 nm. After 48 hr of greening the band at 695 nm is clearly visible, as a shoulder to the 683 nm band. The fraction of fluorescence emitted at the 683 nm band plus the 695 nm band is plotted as a function of the time of greening in fig. 16. The fluorescence ratio $\phi(683 + 695)/\phi$ (total)

declined sharply between 2 and 4 hr of greening and this corresponds closely to the period of rapid synthesis of chlorophyll (fig. 8). The fluorescence at 735 nm emanates from a form of chlorophyll absorbing at 705 nm. Butler [30] examined the appearance of a band at 705 nm in the excitation spectrum of long-wavelength fluorescence (> 730 nm) of greening bean leaves. An increase in excitation at 705 nm was found after 2 hr of illumination.

Discussion

Fig. 17 shows a model for the assembly of the photosystems which appears to be consistent with most of our experimental findings. The upper left-hand part of the figure shows molecules of the protochlorophyllide-protein complex (termed proto-

MODEL FOR ASSEMBLY OF PHOTOSYSTEMS

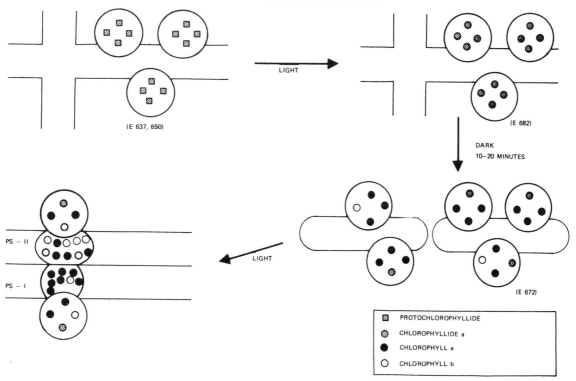

Fig. 17. Model for the assembly of photosynthetic membranes. See text for explanation.

chlorophyll holochrome) intimately associated with a prolamellar body membrane. The model shows four protochlorophyllide molecules per protein molecule, but as discussed elsewhere [11] this appears to be a minimum number.

The protochlorophyllide molecules are sufficiently close (30–50 Å) to one another to permit energy transfer between protochlorophyllide molecules, and from protochlorophyllide to chlorophyllide *a* after partial conversion. After photoconversion of protochlorophyllide, the chlorophyllide *a* is still associated with the prolamellar body membranes. After 20 min in the dark, a considerable proportion of the chlorophyllide is esterified with phytyl alcohol [16] and there is some chl *b*. The membranes of the prolamellar body have dispersed into vesicles, and the spectral shift from 683 nm to 672 nm has occurred. We believe that at least some of the molecules of chl *a* absorbing at 672 nm either are still associated with the original protochlorophyllide-containing unit, or

they are sufficiently close to permit transfer of excitation energy from protochlorophyllide. On further illumination, the molecules of chl *a* are rapidly transported from the holochrome unit to new receptor sites on the developing photosynthetic membranes. Formation of additional protochlorophyllide takes place, this is photoconverted, then phytolated and rapidly removed to the developing photosystems I and II. We do not know whether chl *b* is formed on the original holochrome units and then inserted into special receptor sites in the photosystems or whether chl *a* molecules on certain receptor sites are converted to chl *b*. Shlyk et al. [31] have suggested that chl *a* and chl *b* are formed together on a multienzyme synthetase complex. There appears to be multiple turnover of the protochlorophyllide on the holochrome before sufficient chlorophyll is incorporated into developing photosynthetic units to form fully functional units.

Our studies with the greening pea indicate that

the assembly of the photosynthetic membranes is not a single-step process. Chl *b* is first detectable after 10 min of illumination of dark-grown plants. Energy transfer from carotenoids to chl *a* is first observed after 20 min of greening, and the extent of the transfer increases during the first 2 hr [35]. Energy transfer appears to be correlated with the phytolation of chlorophyllide *a* and with the synthesis of chl *b*. Butler [32] previously reported a correlation between phytolation and energy transfer from carotenoids in greening bean leaves. The presence of chl *b* facilitates the transfer of energy from carotenoids to chl *a* [35]. Chl *b* in vivo shows an absorption maximum at 470 nm, which is in the region of carotenoid absorption.

Photo-oxidation of cytochrome *f*, a measure of activity in photosystem I is observed first after 30 min of greening. The magnitude of the absorbance changes at 554 nm increases during the first 2 hr and, like energy transfer from carotenoids, appears to be correlated with the presence of phytolated pigments. The dispersal of prolamellar bodies also seems to depend on a certain degree of phytolation. Cytochrome *b*−559, which is localized in photosystem II in chloroplasts [33] is absent from dark-grown plants [34]. It is detectable in pea plastids after 3 hr of greening. The appearance of the large fluorescence band at 735 nm at 77°K correlates with the phase of rapid chlorophyll synthesis. Hill-reaction activity with TCIP (PS-II) is further delayed and its appearance correlates with the formation of grana.

References

[1] N.K. Boardman, in: The Chlorophylls, eds. L.P. Vernon and G.R. Seely (Academic Press, New York and London, 1966) p. 437.
[2] J.M. Anderson and N.K. Boardman, Australian J. Biol. Sci. 17 (1964) 93.
[3] D. von Wettstein, Brookhaven Symposia in Biology 11 (1958) 138.
[4] H.I. Virgin, A. Kahn and D. von Wettstein, Photochem. Photobiol. 2 (1963) 83.
[5] S. Klein, G. Bryan and L. Bogorad, J. Cell Biol. 22 (1964) 433.
[6] B.E.S. Gunning and M.P. Jagoe, in: Biochemistry of Chloroplasts, Vol. II, ed. T.W. Goodwin (Academic Press, New York and London, 1967) p. 655.
[7] N.K. Boardman and S.G. Wildman, Biochim. Biophys. Acta 59 (1962) 222.
[8] I. Ohad, P. Siekevitz and G.E. Palade, J. Cell Biol. 35 (1967) 553.
[9] S. Schuldiner and I. Ohad, Biochim. Biophys. Acta 180 (1969) 165.
[10] J.M. Anderson and N.K. Boardman, Biochim. Biophys. Acta 112 (1966) 403.
[11] A. Kahn, N.K. Boardman and S.W. Thorne, J. Mol. Biol. 48 (1970) 85.
[12] N.K. Boardman and H.R. Highkin, Biochim. Biophys. Acta 126 (1966) 189.
[13] N.K. Boardman and S.W. Thorne, Biochim. Biophys. Acta 153 (1968) 448.
[14] D.I. Arnon, Plant Physiol. 24 (1969) 1.
[15] N.K. Boardman and S.W. Thorne, in preparation.
[16] N.K. Boardman, in: Harvesting the Sun, eds. A. San Pietro, F.A. Greer and T.J. Army (Academic Press, New York and London, 1967) p. 211.
[17] T.E. Treffry, Planta 91 (1970) 279.
[18] K. Shibata, J. Biochem. 44 (1957) 147.
[19] F.F. Litvin and A.A. Krasnovsky, Dokl. Akad. Nauk SSSR 117 (1957) 107.
[20] M. Gassman, S. Granick and D. Mauzerall, Biochem. Biophys. Res. Commun. 32 (1968) 295.
[21] B.A. Bonner, Plant Physiol. 44 (1969) 739.
[22] S.W. Thorne, submitted for publication.
[23] T. Forster, Discussions Faraday Soc. 27 (1959) 7.
[24] S.W. Thorne, submitted for publication.
[25] A.B. Rudoi, A.A. Shlyk and A. Yu Vezitsky, Dokl. Akad. Nauk SSSR 183 (1968) 215.
[26] J.B. Wolff and L. Price, Arch. Biochem. Biophys. 72 (1957) 293.
[27] H.E. Davenport, Nature 170 (1952) 1112.
[28] W. Bonner and R. Hill, in: Photosynthetic Mechanisms of Green Plants (Publication 1145 Natl. Acad. Sci. − Natl. Res. Council, Washington D.C., 1963) p. 82.
[29] N.K. Boardman, S.W. Thorne and J.M. Anderson Proc. Natl. Acad. Sci. U.S., 56 (1966) 586.
[30] W.L. Butler, Biochim. Biophys. Acta 102 (1965) 1.
[31] A.A. Shlyk, I.V. Prudnikova, L.I. Fradkin, G.N. Nikolayera and G.E. Savchenko, in: Progress in Photosynthesis Research, ed. H. Metzner (I.U.B.S., Tübingen, 1969) Vol. II, p. 572.
[32] W.L. Butler, Arch. Biochem. Biophys. 92 (1961) 287.
[33] N.K. Boardman and J.M. Anderson, Biochim. Biophys. Acta 143 (1967) 187.
[34] N.K. Boardman, in: Comparative Biochemistry and Biophysics of Photosynthesis, eds. K. Shibata, A. Takamiya, A.T. Jagendorf and R.C. Fuller (University of Tokyo Press, Tokyo, 1968) p. 206.
[35] S.W. Thorne and N.K. Boardman, submitted for publication.

PROPERTIES OF ETIOPLAST MEMBRANES
AND MEMBRANE DEVELOPMENT IN MAIZE

Lawrence BOGORAD, Richard H. FALK *, John M. FORGER III ** and Arnold LOCKSHIN

The Biological Laboratories. Harvard University, Cambridge, Mass. 02138, USA

Leaves of angiosperms grown in darkness lack chlorophyll and cannot carry on photosynthesis. Plastids of etiolated leaves (i.e. etioplasts) contain membranous materials organized into characteristic paracrystalline prolamellar bodies. Upon illumination these membranes are reorganized, chlorophyll is produced and is deposited in chloroplast membranes which are formed during the illumination period, and the capacity to photosynthesize develops.

The environmental controlling agent for these changes is light but virtually nothing is known about the first steps and comparatively little about intervening events in these processes. The gross differences in fine structure between the etioplast and plastid of maize as seen in the electron microscope have been described [1] and structural changes during the transformation of etioplasts into mature plastids in barley and beans have been followed by a number of different groups [2–4].

This report deals with some properties of etioplast membranes and the development of some plastid membrane functions.

Chlorophyll accumulation

The most conspicuous change during plastid maturation in the transition from etioplast to chloroplast is the accumulation of chlorophylls *a* and *b*. This process has been studied extensively and includes the conversion of protochlorophyllide (associated with the protochlorophyllide holochrome protein) to chlorophyllide *a*, the esterification of the newly formed chlorophyllide with phytol, and the deposition of the chlorophyll in its functional sites in the membranes. It also includes the synthesis of new protochlorophyllide which will be converted to chlorophyll.

Etiolated leaves accumulate many times the normal amount of protochlorophyllide if they are supplied ALA [5]; thus it seems reasonable to conclude that chlorophyll accumulation is regulated by the synthesis of ALA. It has also been shown that the synthesis of protochlorophyllide is arrested by the administration of certain inhibitors of protein and of RNA synthesis [6,7]. Furthermore, this accumulation of pigment is tightly controlled – if leaves rapidly forming chlorophyll are returned to darkness, pigment production ceases within one or two hr [7]. This has been interpreted to indicate that the production of protochlorophyllide depends upon the continued production of some relatively rapidly decaying RNA or RNAs and of some protein with the same properties [6,7]. The protein is presumably required for the production of ALA and *could be* ALA synthetase (although this enzyme has not been demonstrated in higher green plants) or some protein which regulates the activity of this enzyme.

As compared with most of the other processes to be discussed here, one of the most characteristic aspects of the chlorophyll accumulation system is its rapid shutdown when leaves which are rapidly accumulating the pigment are returned to darkness.

Membrane development

Starting with the work of von Wettstein [2] and

---

* National Institutes of Health Postdoctoral Fellow. Present address, Department of Botany, University of California, Davis, California 95616, U.S.A.
** Predoctoral Trainee of the National Institutes of Health.

continuing through the work of many others, the fate of the prolamellar body during greening has been documented. The structure is disrupted but its membraneous components seem to persist, at least as can be seen through the electron microscope. Upon illumination, the prolamellar body gives rise to dissociated vesicles which become aligned in a few rows, later these vesicles appear to fuse and become reorganized into the primary thylakoids. There is no indication that these membranes are destroyed and replaced by other membranes. Thus, if we ask whether photosynthetic membranes can be made stepwise or if they are formed only by aggregation or addition of complete functional units, we have a partial answer simply by knowing that chlorophyll is added to membranes derived from those of the prolamellar body.

Which components necessary for photosynthetic function are already present in the prolamellar body membranes and which ones are added during development? Can prolamellar body membranes carry on any parts of photosynthesis? Chlorophyll is lacking so photochemical steps cannot even begin. Consequently, we must limit ourselves to the study of partial processes not requiring light energy. Boardman [8] has shown that etiolated bean leaves contain cytochrome f and cytochrome b_6, but that cytochrome $b-559$ is absent but accumulates during the greening process.

Coupling factor for photophosphorylation in maize plastids

The present investigation began with an electron microscope study of maize etioplasts prepared by negative staining (fig. 1). Under the preparative conditions used, the prolamellar bodies dissociated and the membranes were found to be studded with stalked particles about 90 Å in diameter. Structures of the same sort appear in similar preparations of (green) chloroplasts of maize. The particles look like the photosynthetic phosphorylation coupling factor (CF) with Ca^{++}-activated ATPase activity which has been described [9–11] in spinach.

Following procedures described by students of the spinach CF [12–16], but modified as necessary for working with the maize enzyme, we have been able to show that green maize plastids contain a Ca^{++}-activated ATPase which can be solubilized partially

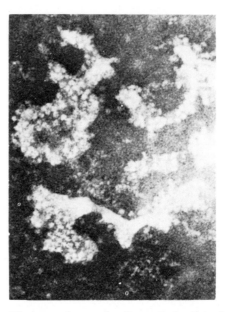

Fig. 1. Electron micrograph of negatively stained maize etioplast membranes (110,000 X).

by buffer but more effectively by NaCl-EDTA solutions. Like the Ca^{++}-activated ATPase from spinach plastids, the maize enzyme is activated by dithiothreitol (DTT) and by trypsin (table 1). In addition, extraction of maize chloroplast membranes with NaCl-EDTA greatly reduces their capacity to carry on phenazine methasulfate (PMS)-mediated photosynthetic phosphorylation and depletes them of CF particles. Addition of the NaCl-EDTA extract (which contains the Ca^{++}-activated ATPase) to the depleted membranes under appropriate conditions restores a good measure of the capacity to carry on photophosphorylation. (This is the basis for describing the material in the extract as a 'coupling' factor.)

Table 2 shows that maize etioplasts also contain a Ca^{++}-activated ATPase, and that its properties are the same as those of the enzyme from maize chloroplasts. The question of whether the etioplast ATPase is in fact a complete 'coupling factor particle' or only part of this system is answered by the data in table 3 which show that the addition of NaCl-EDTA extract of etioplast membranes (containing the Ca^{++}-activated ATPase) to green chloroplast membranes depleted of their CF results in a preparation capable of carrying on PMS photophosphorylation. The rate of photophosphorylation in the reconstituted system is pro-

Table 1

Extraction and activation of ATPase from maize chloroplasts

Fraction	Total activity [a]	Specific activity [b]	Specific activity per mg chlorophyll [c]
Minus DTT			
Chloroplasts	17.5	1.09	2.65
1st water extract	15.0	11.0	2.27 [d]
2nd water extract	15.2	4.38	2.30 [d]
1st EDTA extract	4.95 } 36.3	3.26	0.75 [d]
2nd EDTA extract	1.14		0.17 [d]
Washed chloroplast residue	6.24	0.51	0.94
1st EDTA extract and trypsin			
+ 0.05 M $CaCl_2$		9.49	16.2 [d]
+ 0.05 M $MgCl_2$		0.39	0.59 [d]
No divalent cation		0.04	0.07 [d]
Plus DTT			
Chloroplasts	21.3	1.33	3.22
EDTA extract	31.2	3.80	4.72 [d]
EDTA extract and trypsin			
+ 0.05 M $CaCl_2$		73.8	91.9 [d]
+ 0.05 M $MgCl_2$		15.7	19.5 [d]
No divalent cation		1.48	1.85 [d]

[a] μmoles ATP hydrolyzed/total fraction/hr.

[b] μmoles ATP hydrolyzed/mg protein/hr.

[c] μmoles ATP hydrolyzed/mg chlorophyll/hr.

[d] Determined on basis of amount of chlorophyll in plastid before extraction with water or EDTA solution.

Zea mays, L. (single cross WF9Tms X B37, Illinois Foundation Seeds, Inc., Champaign, Illinois) plastids were prepared by harvesting and chilling leaves which were then homogenized in a Waring Blendor with 2 ml of a solution of 0.4 M sucrose, 0.01 M NaCl, and 0.05 M Tris-HCl buffer, pH 8.0 per g of tissue. The homogenate was squeezed through muslin and then centrifuged at 3000 g for 6 min. The precipitate was resuspended in about 1/4 the volume of the homogenizing solution used initially and centrifuged at 12,000 g for 10 min. In these experiments 48 g of corn leaves were used and the pellet was suspended in 6.0 ml of 0.05 M Tris-HCl buffer, pH 8.0. 3.1 ml of the suspension was extracted and centrifuged four times consecutively: twice with water (yielding supernatants of 8.75 and 8.3 ml respectively) and twice with 0.75 mM EDTA, pH 7.0 (yielding supernatants of 10.5 and 12.0 ml respectively). After each extraction with water or EDTA, the material was centrifuged for 20 min at 35,000 g. The 4-times washed chloroplast membranes were resuspended in 4.0 ml of 0.05 M Tris-HCl.

To obtain the trypsin-activated extract, 1.2 ml of a solution containing 12 μmoles ATP, 100 μmoles Tris-HCl, pH 8.0 and 600 μg trypsin was added to 6.0 ml of the first EDTA extract. The mixture was incubated at room temperature for 6 min, and then 1.2 ml of a 2 mg/ml solution of soybean trypsin inhibitor was added to a final volume of 8.4 ml to stop the trypsin reaction.

To prepare DTT-treated chloroplast 0.5 ml of 0.25 M DTT was added to 2.0 ml of the original chloroplast suspension and the mixture was incubated at room temperature for 65 min before assay and further treatment. 1.6 ml of the DTT-chloroplast suspension was extracted once with 0.75 mM EDTA and centrifuged for 20 min at 35,000 g to yield a supernatant of 13.1 ml. The trypsin-activated DTT extract was prepared in the same manner as the non-DTT extract.

ATPase was assayed as follows: an aliquot of plastid suspension or extract was incubated at 37° and a solution containing 50 μmoles Tris-HCl, pH 8.0, 5 μmoles $CaCl_2$ was added to start the ATPase reaction. The final volume was 1.0 ml. The reaction was stopped after 20 min by adding 0.5 ml of 16% (W/V) TCA. For '0' time controls, the TCA was added before incubation. After centrifugation at 12,000 g for 10 min, the supernatant fluid was assayed for inorganic phosphate by the colorimetric method of Ames [21] modified by incubating for 5 min at 45°.

'Total Activity' is calculated on the basis of the total activity of each fraction obtained from 3.1 ml of the original chloroplast suspension. For the DTT-fractions, the corresponding calculation is on the basis of 3.83 ml of DTT-chloroplast suspension.

The specific activities related to chlorophyll concentration are based on the chlorophyll concentration of the extracts before centrifugation to remove the washed chloroplast residues.

Table 2

Extraction and activation of ATPase from maize etioplasts

Fraction	Total activity [a]	Specific activity [b]
Minus DTT		
Etioplasts	12.18	0.42
1st water extract	7.14 ⎫	3.00
2nd water extract	0.33 ⎪ 11.19	0.22
1st EDTA extract	1.41 ⎬	1.20
2nd EDTA extract	2.31 ⎭	3.78
Washed etioplast residue	5.85	0.49
1st EDTA extract and trypsin		
+ 0.05 M $CaCl_2$		8.34
+ 0.05 M $MgCl_2$		2.86
Plus DTT		
Etioplasts	16.9	0.54

[a] μmoles ATP hydrolyzed/total fraction/hr.
[b] μmoles ATP hydrolyzed/mg protein/hr.

portional to the amount of NaCl-EDTA extract of etioplast membranes which is added to the depleted green membranes.

Ca^{++}-activated ATPase activity was measured during greening of etiolated maize leaves (table 4). On the basis of activity per mg of leaf tissue, a decline occurred between 2 and 5 hr after the plants had been placed in the light, but the activity remained essentially constant over about the next 20 hr. The chlorophyll concentration increased about 13-fold during the same period. It seems reasonable to conclude tentatively that much, if not all, of the plastid Ca^{++}-activated ATPase present in a maize leaf illuminated for 25 hr was present before illumination.

Table 3

The effect of EDTA-extracts of etioplast membranes on photophosphorylation by extracted chloroplast membranes

ml of extract of etioplasts added to EDTA-extracted chloroplast membranes	cpm incorporated into ATP	
	Total	Minus boiled control
0.5 ml	22,000	6,800
1.0 ml	27,700	12,500
1.5 ml	34,500	19,300
1.0 ml (boiled extract)	15,200	(0)

Table 4

ATPase activity during greening of dark-grown maize leaves

Hours in light	μg chlorophyll per g leaf tissue	mμmoles ATP hydrolyzed in 20 min per mg leaf tissue
0	9.8 [a]	2.80
2	29.9	2.85
5	79.4	0.90
12	428.9	0.78
25.5	1065.0	0.55
Dark for 25.5 hr	8.0 [a]	2.12

[a] μg protochlorophyllide measured as chlorophyllide.

Seven-day-old dark-grown maize seedlings were harvested at times shown in the left-hand column. Illumination, where indicated, was with cool white fluorescent lamps at 200 ft-c at 28°. Crude plastids were prepared and ATPase was assayed as described in table 1. Chlorophyll concentration was determined by the method of Wintermans and DeMots [23].

To summarize: These experiments support the view that thylakoid synthesis can be stepwise and that prolamellar body membranes, though lacking chlorophyll, contain many of the things required for photosynthesis. Here, by a functional test, a coupling factor for photophosphorylation has been shown to be present in maize etioplasts. The ability of the CF from etioplast membranes to associate with CF-depleted green thylakoid membranes shows that at least parts of the surfaces of the two membranes are chemically similar although there is no basis for judging how specific the adherence to etioplast membranes may be — there is no function besides ATPase activity which can be followed in this case. Finally, if coupling factor particles are present over all or most of the prolamellar body membranes, this must be taken into account in trying to understand how the prolamellar body is organized and how it becomes disorganized when an etiolated leaf is illuminated. (Are the membranes of the prolamellar body held in place, for example, by interactions between coupling factor particles on adjacent membranes? Ionic or hydrogen-bounding interactions could be rapidly dissipated if the environment in the plastid were altered by the movement of, for example, ions into or out of the plastid.)

Acid/base phosphorylation and osmotic behavior during plastid development

Jagendorf and Uribe [17] showed that spinach thylakoid preparations produced ATP from ADP and inorganic phosphate upon transfer from a buffer solution at pH 4 to one at pH 8. This acid/base (A/B) phosphorylation can occur in darkness — the absorption of light by chlorophyll is not required. Although the mechanism of A/B phosphorylation is in dispute and the biological significance of the process is undetermined, the ability to form ATP in this way is a measure of a facet of membrane functional integrity which is independent of the capacity for photochemical activity and consequently is another interesting property which can be studied during plastid development.

We have been unable to demonstrate A/B phosphorylation in etioplast membrane preparations but plastids from leaves illuminated for 2.5 hr or more are active (table 5). On the other hand, the ability to form ATP by PMS-mediated photophosphorylation is readily detected in leaves of plants which have been illuminated for 60 to 90 min. The A/B phosphorylation activity is probably there also at that time, but the assay is much less sensitive.

The presence of osmotically responsive internal spaces and of a lamellar component(s) which can be removed by treatment with EDTA solution seem to be required for membranes to perform A/B phosphorylation.

Uribe and Jagendorf [18,19] showed that the amount of ATP made by spinach chloroplast fragments during the transition from acid to alkaline solution was correlated with the amount of succinic

Table 5
'Acid-base' phosphorylation by maize plastids

Source of plastids	cpm incorporated into ATP per g fresh weight of leaf [a]
Etiolated leaves	443
Etiolated leaves + 150 min light	1,482
Green leaves	292,015

[a] Estimated from pigment concentrations in the plastid fragments and in the leaves: cpm/μg chlorophyll added $\times$ μg chlorophyll/g tissue = cpm/g tissue.

acid taken up by the fragments into spaces inaccessible to inulin and that the tonicity of the medium in which the reaction took place affected A/B activity and organic acid uptake in the same manner. Furthermore, damaging the plastid membranes in such a way that they would become completely permeable (e.g. by freezing and thawing) abolished A/B activity. We have found that fragments of green maize plastids respond osmotically by shrinking by about half as they are transferred from 0.01 M NaCl to 0.5 M sucrose–0.01 M NaCl solution and that this volume change is reversible as they are transferred back, but that similar preparations from etiolated leaves fail to respond in this way. Thus, at least one of the reasons that preparations of etioplasts are incapable of performing A/B phosphorylation is that the membranes may be 'too permeable'.

In some of the earliest work on plastid ATPases and coupling factors for photophosphorylation Avron found [20] that photophosphorylation in Swiss chard plastid fragments was inhibited by EDTA. Jagendorf and Uribe [17] showed that A/B phosphorylation in spinach plastid fragments was also very sensitive to EDTA, but did not report any attempts to reverse this effect. Table 6 shows that some EDTA-removable membrane component(s) is required for A/B phosphorylation by maize plastid fragments. Extraction of the membranes with NaCl-EDTA solution sharply reduces their capacity to do A/B phosphorylation but activity is recovered on recombination of the two fractions in the presence of an excess of EDTA and additional Mg^{++}. These are the same conditions as are used to restore the photophosphorylating ability of CF-depleted membranes by CF solutions. The A/B activity of green membrane fragments which have been washed with NaCl-EDTA can be restored by incubating them with a similar extract from etiolated plants. We conclude that etioplasts contain at least one of the components necessary for A/B photophosphorylation — it may be identical with the coupling factor. This observation is parallel to those described for photophosphorylation reconstitution experiments described above.

Thus, etioplast membranes contain an EDTA-removable factor(s) required for A/B phosphorylation but are not responsive osmotically. It is possible that chlorophyll deposition somehow plays a role in the acquisition of A/B activity even though ATP forma-

Table 6
EDTA-removable factor(s) for A/B phosphorylation

Depleted green lamellae incubated with:	cpm incorporated/ ml reaction supernatant	% stimu-lation
EDTA in 0.01 M NaCl (control)	42,206	–
EDTA extract (etiolated)	50,909	21
EDTA extract (green)	60,375	43
Supernatants alone:		
EDTA extract (etiolated) + Mg^{++}	545	
EDTA extract (green) + Mg^{++}	(−436)	

Isolated chloroplast fragments were resuspended in 0.01 M NaCl + 0.15 mM EDTA, pH 8.0. Incubation for 15 min at room temperature partially removed the factor(s). Etioplast membranes were extracted in a similar manner. Subsequent reconstitution took place in the presence of 0.01 M EDTA, 0.1 M MgCl$_2$, and one of the following: green extract, extract of etiolated membranes, or 0.15 mM EDTA in 0.01 M NaCl; incubation: 10 min at room temperature. Chloroplast fragments containing 680 μg of chlorophyll were used per reaction mixture. Chloroplast fragments not treated with EDTA incorporated 125,000 cpm.

tion by this process does not require light − does chlorophyll or something it brings with it plug holes? (It is much more difficult to measure small changes in osmotic responsiveness than in A/B phosphorylating capacity.)

As already mentioned, chlorophyll synthesis in rapidly greening plants stops soon after leaves are returned to darkness. The object of these experiments was to determine whether the development of A/B activity stops at the same time as chlorophyll production. This would at least demonstrate whether there is an obligatory relationship between chlorophyll accumulation and the development of A/B activity. Two stages in the greening process were studied: a) the period between 5 and 12 hr after the commencement of illumination − this is a time of very rapid chlorophyll synthesis − and, b) a long period of darkness following brief illumination − when dark-grown plants are first illuminated, the protochlorophyllide present is converted to chlorophyllide (which eventually becomes chlorophyll a) and new chlorophyll is formed in an amount from roughly equal to twice as

great as the protochlorophyllide originally present. After an additional 24 hr in darkness the total amount of chlorophyll in a leaf remains at 1% or less of that in a normal light-grown leaf.

Table 7 shows that plants harvested immediately after 5 hr of illumination have about the same amount of chlorophyll per g of leaf tissue as those maintained in darkness for 13 hr after illumination was terminated. However, a large amount of additional A/B activity developed during the 13 hr in darkness following illumination. We can conclude from this experiment that the development of A/B activity does not need to parallel chlorophyll accumulation.

We have also found that illumination for 1 min followed by 24 hr in darkness results in the acquisition of some A/B activity. This, again, shows that chlorophyll accumulation and the acquisition of A/B activity can be independent. Neither of these experiments demonstrates that chlorophyll is entirely unnecessary for A/B activity but they do demonstrate that there are separate control mechanisms which regulate the production of at least two processes involved in membrane formation. We do not know the kinetics of the decay in the development of A/B activity but the present experiments show that they are clearly different from those for chlorophyll accumulation.

These observations also demonstrate that chlorophyll can be deposited in its 'correct' place in partially completed membranes. The membranes of the etioplast, since they bear coupling particles, are 'partially complete'. Furthermore, the membranes which develop after a period in darkness and become capable of A/B phosphorylation can probably later add chlorophyll and acquire the capacity for photophosphorylation. How does chlorophyll get onto or in the membrane? Are these membranes constantly breaking

Table 7
'Acid-base' phosphorylation by plastids of dark-grown maize

	μg Chl/ g fresh wt.	cpm incorporated into ATP per ml reaction mixture
5 hr light	79	2265
5 hr light + 13 hr dark	78	· 5613

Fragments containing 34.8 μg chlorophyll used per reaction mixture.

and repairing in vivo? Can the chlorophyll dissolve in the lipid phase of the lamellae and diffuse to its appropriate position, e.g. in association with some protein? Or, is there no great specificity of location for the bulk of the chlorophyll which is added?

General discussion and questions

We have seen that membranes incapable of carrying on A/B activity are also not responsive osmotically, but we have no direct evidence of a causal relationship between these two phenomena. Knowledge of the kinetics of development of osmotic responsiveness and A/B capacity might help our understanding of these relationships but changes in osmotic responsiveness cannot be measured very precisely. We would like to know what components are added to membranes to enable them to respond osmotically and to do A/B phosphorylation — are new proteins added? New lipids? What, if anything besides chlorophylls, must be added to membranes capable of A/B phosphorylation to permit them to do photophosphorylation? Furthermore, are the components needed for membrane maturation made immediately prior to their insertion into the membrane or are compounds incorporated which are present as soluble components in the etioplast?

Finally, how are the processes of photosynthetic membrane development controlled? We know that chlorophyll accumulation is blocked by inhibitors of protein and RNA synthesis. We also know that RNA metabolism of plastids changes when etiolated leaves are brought into the light. Is the development of A/B activity dependent upon light-induced RNA and protein synthesis? Are the requirements for the acquisition of osmotic responsiveness and A/B capacity by etioplast membranes precisely the same? Is there a different photoreceptor for the development of chlorophyll and for other phases of photosynthetic membrane maturation?

Acknowledgements

We are pleased to acknowledge the skilled technical services of Mrs. Mary Hegedus and Mrs. Gail Bertocchi whose assistance made the performance of this work possible.

This work was supported in part by grants from the National Institutes of Health (GM–14991 and AM–11363).

References

[1] L. Bogorad, in: Harvesting the Sun, eds. A. San Pietro, F.A. Greer, T.J. Army (Academic Press, New York, 1967) p. 191.

[2] D. von Wettstein, Brookhaven Symposia in Biology 11 (1958) 138.

[3] H.I. Virgin, A. Kahn and D. von Wettstein, Photochem. Photobiol. 2 (1963) 83.

[4] S. Klein, G. Bryan and L. Bogorad, J. Cell Biol. 22 (1964) 433.

[5] S. Granick, Plant Physiol. 34 (1959) XVIII.

[6] M. Gassman and L. Bogorad, Plant Physiol. 42 (1967) 781.

[7] M. Gassman and L. Bogorad, Plant Physiol. 42 (1967) 774.

[8] N.K. Boardman, in: Comparative Biochemistry and Biophysics of Photosynthesis, eds. K. Shibata, A. Takamiya, A.T. Jagendorf and R.C. Fuller (University of Tokyo Press, Tokyo, 1967) p. 206.

[9] R. Bronchart, C.R. Acad. Sci. Paris 260 (1965) 4564.

[10] S.H. Howell and E.N. Moudrianakis, Proc. Natl. Acad. Sci. U.S. 58 (1967) 1261.

[11] S. Murakami, in: Comparative Biochemistry and Biophysics of Photosynthesis, eds. K. Shibata, A. Takamiya, A.T. Jagendorf and R.C. Fuller (University of Tokyo Press, Tokyo, 1967) p. 82.

[12] V.K. Vambutas and E. Racker, J. Biol. Chem. 240 (1965) 2660.

[13] R.E. McCarty and E. Racker, Brookhaven Symposia in Biology 19 (1966) 202.

[14] R.E. McCarty and E. Racker, J. Biol. Chem. 242 (1967) 3435.

[15] R.E. McCarty and E. Racker, J. Biol. Chem. 243 (1968) 129.

[16] A. Bennun and E. Racker, J. Biol. Chem. 244 (1969) 1325.

[17] A.T. Jagendorf and E. Uribe, Proc. Natl. Acad. Sci. U.S. 55 (1966) 170.

[18] E.G. Uribe and A.T. Jagendorf, Plant Physiol. 42 (1967) 697.

[19] E.G. Uribe and A.T. Jagendorf, Arch. Biochem. Biophys. 128 (1968) 351.

[20] M. Avron, Biochim. Biophys. Acta 77 (1963) 699.

[21] B.N. Ames, in: Methods in Enzymology, Vol. VIII (Academic Press, New York, 1966) p. 115.

[22] M. Avron, Biochim. Biophys. Acta, 40 (1960) 257.

[23] J.F.G.M. Wintermans and A. DeMots, Biochim. Biophys. Acta, 109 (1965) 448.

RIFAMYCINS: THE INHIBITION OF PLASTID RNA SYNTHESIS IN VIVO AND IN VITRO AND VARIABLE EFFECTS ON CHLOROPHYLL FORMATION IN MAIZE LEAVES

Lawrence BOGORAD and Christopher L.F. WOODCOCK *

Harvard University, The Biological Laboratories, Cambridge, Mass, USA

Plastid ribosomes were first shown by electron microscopy in etioplasts and chloroplasts of maize [1]. Analysis of extracts of maize etioplasts by sucrose density gradient centrifugation has also shown that these organelles contain ribosomes [2]. Despite the presence of ribosomes in etioplasts, treatment of etiolated beans or maize with actinomycin D arrests light-induced chlorophyll synthesis [3–5]. This suggests that DNA-dependent RNA synthesis is required for plastid development.

In agreement with the data from experiments with actinomycin D is the observation that when dark-grown leaves of maize are illuminated, the rate of incorporation of ^{32}P-phosphate into RNA – preferentially plastid rRNA – is increased [2,6,7]. This elevated rate of incorporation may be accounted for by an increase in the activity of plastid RNA polymerase which can be detected after illumination of etiolated leaves [2]. The increase in activity of RNA polymerase is blocked by chloramphenicol; thus it seems probable that molecules of additional enzymes or some protein effector of plastid RNA polymerase form when the leaves are illuminated.

Does illumination bring about a general stimulation of protein synthesis *including* an increase in plastid RNA polymerase activity? (In which case, light-induced chlorophyll production could be independent of RNA polymerase activity.) Or, is there first an increase in the production of a single species of RNA needed for the synthesis of additional RNA polymerase followed by a general stimulation of the transcription of RNA from all unrepressed genes, in turn, followed by accelerated protein synthesis? (In the later case, chlorophyll production should be dependent upon increased RNA polymerase activity.)

Hartman et al. [8] and Wehrli et al. [9] have shown that rifamycins are potent inhibitors of *Escherichia coli* RNA polymerase; in vitro the enzyme is inhibited 50% by 2×10^{-8} M antibiotic. On the other hand, rat liver RNA polymerase was found to be relatively insensitive to rifamycin SV – 15–20% inhibition was observed at a concentration of approximately 3×10^{-4} M [9]. Unlike actinomycin D, which binds to nuclear and plastidic DNA templates, the rifamycins bind to some kinds of DNA-dependent RNA polymerase molecules [10,11].

Rifamycins seemed appropriate agents for exploring possible differences among maize RNA polymerases as well as for examining the relationship between chlorophyll formation and plastid RNA polymerase activity. Accordingly, we have examined the effects of rifamycins in vivo, on chlorophyll and plastid rRNA synthesis, and in vitro, on the activity of maize RNA polymerases.

The effects of rifamycins on chlorophyll formation

As shown in table 1 rifamycin-SV inhibits chlorophyll production by etiolated maize leaves transferred into the light. Administration of 5×10^{-5} M rifamycin-SV inhibited chlorophyll production during 5 hr in the light by 28% and at a 2-fold higher concentration 46% inhibition was observed. However, another rifamycin, rifampin (Ciba), had virtually no effect at 10^{-4} M.

* Maria Moors Cabot Foundation Fellow, Harvard University.

Table 1
Effect of rifamycin SV and rifampin on chlorophyll production by maize.

	μg Chl/g leaf	% of control
Etiolated	12	–
Light	104	100
Rifamycin–SV		
1×10^{-4} M	63	54
5×10^{-5} M	75	72
5×10^{-6} M	85	82
Rifampin		
1×10^{-4} M	99	94
5×10^{-5} M	110	105

Six-day-old plants. 0.4 ml solution administered per 12 plants. Continued in darkness for 9 hr, then illuminated for 5 hr.

These experiments were performed by growing maize plants in darkness at 28° for 6 days. Leaves from 12 plants were used for each of the four replicates. The cut bases of the leaves were recut under water and leaves from 12 plants were placed with the newly cut ends in a small vial containing 0.5 ml of water or rifamycin at the concentration noted. In the experiment shown in table 1 the leaves were placed in front of a fan until the rifamycin had been taken up and additional portions of distilled water had been administered to wash in the remaining antibiotic. (About 2 g of tissue was used in each 12-plant replicate.) Thus, if we assume that the plants were 100% water and that the antibiotic was uniformly distributed throughout the plant, the actual concentration in vivo would be approximately 1/5 of that noted (the antibiotics were probably not uniformly distributed, however). After an additional 9 hr in darkness at 28° the leaves were illuminated (about 200 ft-c), also at 28°, and samples were taken at the times indicated. Chlorophyll was measured by extracting leaves with 80% acetone and determining the pigment content of the extracts spectrophotometrically.

The differences in activity of the two rifamycins could result from differences in permeability, but this is not supported by observations on the effects of these agents on RNA synthesis in vivo.

Inhibition of ribosomal RNA synthesis by rifamycin in vivo

The effect of rifampin (the antibiotic which did not inhibit chlorophyll synthesis (table 1)) on the incorporation of ^{32}P-phosphate into ribosomal RNA of etiolated maize was examined.

Rifampin was administered to etiolated leaves in the same manner as in the chlorophyll experiments. Then, ^{32}P-phosphate was supplied in the same manner as the antibiotic. Following this, leaves were placed in the light or maintained in darkness. After the desired period of time, leaves were homogenized in buffer-phenol-bentonite and their RNAs were extracted [2]. The RNAs were separated on acrylamide gels [10] and analyzed by direct scanning in a Gilford Spectrophotometer to determine the distribution of 260 nm absorbing material. Then, after freezing, the gels were sliced into sections approximately 0.8 mm thick; the gel sections were dried and counted in a Packard Liquid Scintillation Spectrometer.

The results of one such experiment in which exposure to light was for a very long time – 18 hr – are shown in fig. 1. Illumination of etiolated maize leaves stimulates the incorporation of ^{32}P-phosphate into chloroplast rRNAs (this had been shown earlier by sucrose density gradient analyses of ribosomes and rRNAs [2]), but the promotion of incorporation by exposure to light was inhibited by rifampin.

In a series of experiments of this sort, but with shorter illumination periods, it has been found (K. Mullinix, personal communication) that there is virtually no inhibition at concentrations of rifampin up to 5×10^{-5} M but that at 10^{-4} M incorporation of ^{32}P-phosphate into chloroplast rRNAs is virtually completely abolished. Results with rifamycin-SV were about the same.

In the experiments shown in fig. 1 and in other similar experiments, incorporation into rRNAs of 80 S ribosomes seems to be stimulated in rifamycin-treated plants. The significance of this observation is not clear except that this shows that in vivo rifamycin does not inhibit whatever enzyme is responsible for synthesis of those types of RNA.

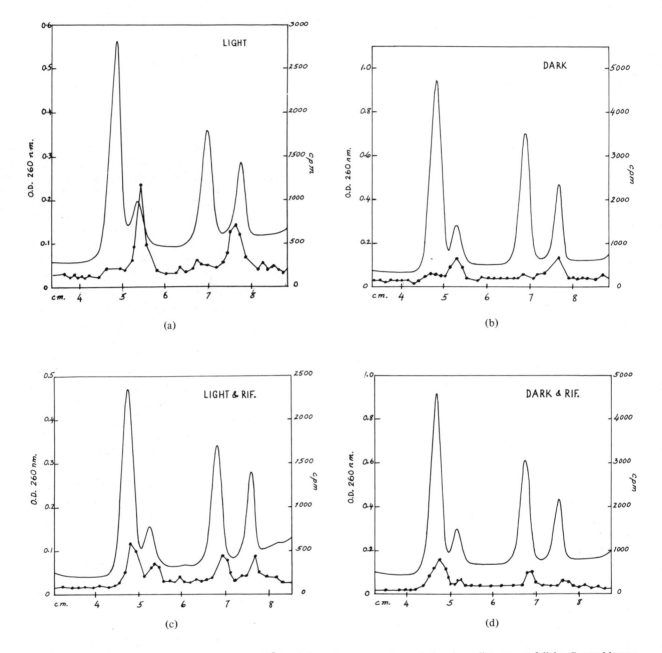

(a)

(b)

(c)

(d)

Fig. 1. Maize seedlings were grown in darkness at 28° for 6 days. Leaves were harvested under a dim green safelight. Control leaves received water, others received rifampin (0.3 ml 6.6×10^{-4} M per set of leaves from 12 plants). The rifampin was washed into the leaves by the administration to each set of leaves of 2.5 ml H_2O supplied in small aliquots. 4 hr after the beginning of treatment with rifampin , 0.018 mc ^{32}P-phosphate (carrier-free) in 0.15 ml of solution was supplied to each set of leaves and was washed in with 3 aliquots of H_2O, each 0.1 ml. Leaves were maintained in darkness or transferred into the light at 28° for 18 hr. The weights of the samples were: (a) Light control: 3.40 g; (b) Dark control: 3.46 g; (c) Light + rifampin: 3.78 g; (d) Dark + rifampin: 3.42 g. RNA was extracted [2] and 24 μg was applied per gel. Optical density at 260 nm (——) was recorded while scanning; radioactivity (●———●) of gel slices was determined as described in the text.

The inhibition of RNA polymerase in vitro by rifamycin

The experiments which we have described here show that it is possible, by the use of rifampin, to arrest incorporation of ^{32}P-phosphate into chloroplast rRNA without affecting the accumulation of chlorophyll by etiolated maize plants during illumination. Thus, under these circumstances chlorophyll production is not dependent upon all RNA synthesis in the leaf; experiments with actinomycin D showed only that some RNA synthesis probably is required.

A series of experiments was performed to try to determine whether rifamycins act on a maize chloroplast RNA polymerase or on a polymerase found in the supernatant phase of leaf extracts after chloroplasts and chloroplast fragments have been removed by centrifugation.

Washed 1,000 g plastid preparations were obtained by procedures described previously [2] and the supernatant enzyme used here was a 0—40% ammonium sulfate fraction obtained from the supernatant solution of the chloroplast preparation (by a modification of [11]). The reaction mixtures used to assay RNA polymerase were slightly modified from those previously described [2]. Incubation was for 30 min at 37°. The activity is dependent upon the presence of all four nucleotide triphosphates.

Table 2 shows the effect of two concentrations of rifamycin-SV on the activity of the supernatant enzyme. In the absence of added calf thymus DNA there was no inhibition of activity at rifampin-SV concentrations of up to 5 × 10^{-5} M. In this experiment about 10% inhibition was observed at a concentration of 5 × 10^{-4} M. Rifamycins have been shown to act on initiation of RNA synthesis but not to interfer with transcription in progress when the inhibitor is added. Consequently, it seemed possible that the failure to detect any inhibition could result from lack of initiation. To test this, the same enzyme preparation was examined in the presence of added calf thymus DNA. The data given in table 2 under the column '+ calf thymus DNA added' show the effect of the inhibitor on the fraction of the incorporation of label which was evoked by the addition of calf thymus DNA. Here, too, at a concentration of 5 × 10^{-5} M the reaction was inhibited by about 27%. By contrast, the 1,000 g plastid enzyme was inhibited

Table 2

Effect of rifamycin-SV on the activities of RNA polymerases of maize

	Supernatant enzyme		1000g enzyme
		+ Calf thymus DNA added [a]	
Control	100%	100%	100%
5 × 10^{-6} M	105%	95%	70%
5 × 10^{-5} M	104%	73%	60%

[a] In this experiment the activity of the supernatant preparation with calf thymus DNA was 191% of that without added calf thymus DNA. The calculations show the % of the added 91% of activity which was inhibited by rifamycin-SV.

appreciably (40%) at 5 × 10^{-5} M rifamycin-SV when calf thymus DNA was not supplied. The inhibition of the supernatant preparation could be due wholly or in part to contamination by a plastid enzyme.

The effect on plastid-localized RNA polymerase was examined further using plastids purified by sucrose density gradient centrifugation in a zonal centrifuge. Maize plants grown in darkness at 28° for 7 days were illuminated for 12 hr at about 200 ft-c and then harvested. Conditions for isolation of the plastids were essentially those described previously for sucrose density gradient purified organelles [2] except in this case an International BXV zonal rotor was used with a gradient of maximum concentration 56% (w/w) sucrose-buffer. The RNA polymerase of plastids in such preparations was inhibited only up to about 10% over a large rifamycin concentration range — much less inhibition than of the cruder 1,000 g plastid preparations.

Why should zonal centrifugation of plastids result in diminished sensitivity to rifamycin? Purification of the plastids, as was done here, could have resulted (a) in the preferential loss of a rifamycin-sensitive polymerase, (b) in the loss of some initiation factor, or (c) in the loss of some DNA which might have been serving as a template for initiation in the same manner as does added calf thymus DNA.

The RNA polymerase activity of zonal plastids is greatly increased by the inclusion of calf thymus DNA in the assay mixture (also see [21]). This demonstrates that the RNA polymerase in these plastids is still capable of initiating the incorporation of radioactivity from ^{14}C-8-ATP into RNA. (Requirements for the transcription of calf thymus DNA may be less stringent than for the transcription of plastid DNA by the same polymerase and thus the possible loss of a factor for initiation of transcription of plastid DNA during plastid isolation is not entirely excluded.)

The magnitude of at least one portion of RNA synthesis known to require initiation can be determined by finding the difference in activity of a preparation in the presence and absence of calf thymus DNA. Then, the effect of rifamycin on this initiation-dependent activity can be measured. In the case of zonal plastids only about 30% of the calf thymus DNA-dependent activity was inhibited by rifamycins at concentrations as high as 5×10^{-5} M. The cause (or causes) of the reduction in susceptibility to rifamycins in the absence of added calf thymus DNA in parallel with plastid purification is unclear but could arise in part from loss of DNA template and/or in part from preferential loss of some rifamycin susceptible polymerase.

Why is there so much less than 100% inhibition of $1,000 g$ or zonal plastid preparations? It is possible that there are two or more RNA polymerases in the chloroplast fraction but that not all of them are inhibited by rifamycins. The supernatant enzyme is a conspicuous rifamycin-insensitive polymerase which could be present in the plastid preparations but it seems very unlikely that purification of plastids would result in an increase in the relative abundance of such a contaminant. The number of RNA polymerases present in plastids can only be resolved unequivocally by physical separation but some of the data provided below support the possibility of multiple plastid RNA polymerases.

The effects of Mg^{++} on maize RNA polymerase

RNA synthesis by zonal plastid preparations requires either Mg^{++} or Mn^{++}. The optimal concentrations are 20 and 5 mM respectively. However, transcription from native calf thymus DNA occurs in the presence of Mg^{++} but with Mn^{++} little if any takes place. Transcription from single-stranded (heat-denatured) calf thymus DNA goes on in the presence of Mn^{++} or Mg^{++} although the rate is about 1.5 to 2

times higher with Mg^{++}. (RNA production by isolated oat plastids, like that by zonal maize plastids, is stimulated by the addition of calf thymus DNA in the presence of Mg^{++} but only slightly, if at all, with Mn^{++}.)

If there is a single RNA polymerase in the zonal plastid preparations, the Mg^{++}:Mn^{++} effects could be accounted for if Mn^{++} permitted continued transcription of endogenous DNA but little or no initiation of transcription of native (at least native calf thymus) DNA whereas continued transcription as well as initiation of native DNA occurred in the presence of Mg^{++}. On the other hand, the metal ion data could be understood easily if there are two more chloroplast polymerases.

Stout and Mans [11] have reported that Mn^{++} and Mg^{++} are equally effective in a calf thymus DNA-dependent assay of the maize supernatant RNA polymerase. Thus, the supernatant enzyme preparation contains a polymerase, which clearly differs from that (or those) in the zonal plastid preparations and the Mn^{++}/Mg^{++}/calf thymus DNA responses of RNA polymerase in plastid preparations cannot be explained entirely by assuming contamination by the supernatant enzyme.

Discussion

The in vivo experiments show that chlorophyll production in the light by dark-grown maize leaves can proceed even if light-induced plastid rRNA formation is blocked by rifampin. It might be concluded from this that although chlorophyll production in tissues of this kind appears (from actinomycin D experiments) to be dependent upon some RNA synthesis, it is independent of plastid RNA metabolism — i.e. it is dependent on nuclear RNA production. However, these results would have been obtained also if there were two or more plastid RNA polymerases — one responsible for plastid rRNA formation and sensitive to rifamycin, another responsible for the synthesis of RNA required for chlorophyll production upon illumination of etiolated maize and insensitive to rifamycins.

The failure to observe extensive inhibition of initiation of RNA synthesis by isolated maize plastids and the differences in the effects of Mn^{++} and Mn^{++} on

RNA transcription and initiation with calf thymus DNA can also be taken to indicate the presence of more than one plastid RNA polymerase. However, these data remain suggestive and not conclusive.

Summary

Rifamycin SV interferes with chlorophyll production in illuminated leaves of dark-grown maize. At the same concentration rifampin, another rifamycin, does not inhibit chlorophyll formation.

Rifamycins block the incorporation of ^{32}P-phosphate into chloroplast ribosomal RNAs in leaves of dark-grown maize plants upon illumination.

The effects of rifamycins on the activity of maize RNA polymerase in vitro were also studied. These antibiotics have a slight effect, if any, on the maize RNA polymerase which remains in solution after leaf homogenates are centrifuged at high speed; but, they markedly inhibit chloroplast RNA polymerase activity. The maximum inhibition observed is about 30—40% of newly initiated RNA synthesis, using calf thymus DNA as a template. It is suggested that the less than complete inhibition may be due, among other things, to the presence of more than one RNA polymerase in maize plastids; some data on the stimulation and control of plastid RNA polymerase activity by magnesium and manganous ions are discussed in this connection.

Acknowledgements

We are pleased to acknowledge the skilled technical services of Mrs. Mary Hegedus and Mrs. Gail Bertocchi whose assistance made the performance of this work possible. The rifampin used in these experiments was kindly provided by Dr. Justis Gelzer of Ciba, Inc. We are also indebted to Dr. J. Pene for his valuable suggestions regarding the use of rifamycins. This work was supported in part by grants from The National Institutes of Health (GM—14991 and AM—11363).

References

[1] L. Bogorad, in: Harvesting the Sun, eds. A. San Pietro, F.A. Greer and T.J. Army (Academic Press, New York, 1967) p. 191.
[2] L. Bogorad, Develop. Biology, Supplement 1 (1967) 1.
[3] L. Bogorad and A. Jacobson, Biochem. Biophys. Res. Commun. 14 (1969) 113.
[4] M. Gassman and L. Bogorad, Plant Physiol. 42 (1967) 781.
[5] M. Gassman and L. Bogorad, Plant Physiol. 42 (1967) 774.
[6] L. Bogorad, in: Molecular Organization and Biological Function, ed. J.M. Allen (Harper & Row, New York, 1967) p. 134.
[7] L. Bogorad, in: Organizational Biosynthesis, eds. H.J. Vogel, J.O. Lampen and V. Bryson (Academic Press, New York, 1967) p. 395.
[8] G. Hartmann, K.O. Honikel, F. Knüsel and J. Nüesch, Biochim. Biophys. Acta 145 (1967) 843.
[9] W. Wehrli, J. Nüesch, F. Knüsel and M. Staehelin, Biochim. Biophys. Acta 157 (1968) 215.
[10] U. Loening, Biochem. J. 102 (1967) 251.
[11] E.R. Stout and R.J. Mans, Biochim. Biophys. Acta 134 (1967) 327.

DEVELOPMENTAL INTERACTIONS AMONG CELLULAR COMPARTMENTS IN EUGLENA

Jerome A. Schiff

Department of Biology, Brandeis University, Waltham, Mass, USA

An operational distinction can be made between chloroplast development and chloroplast (or proplastid) replication in *Euglena*. * Plastid replication and development in this organism has been thoroughly reviewed elsewhere [1–4], but some background material will serve to introduce the ensuing discussion.

* *'Euglena'* will refer to both the *bacillaris* and Z strains of *E. gracilis* where there is no reason to think that the two are different; the author's work has been almost entirely with *bacillaris*.

Plastid replication

In the light, on a complete medium, the ten chloroplasts of *Euglena gracilis* var. *bacillaris* divide and are linked to cell division in such a way as to maintain the chloroplast number fairly constant from generation to generation. In the dark, on a complete medium, the same appears to be true for the proplastids. The plastids, mitochondria, and nucleus of *Euglena* contain different types of DNA distinguishable through base composition (table 1). If dividing cells are treated with streptomycin [5], temperatures above 34°C [6], or ultraviolet light (uv) [7], the

Table 1
DNA of *Euglena gracilis*

Source and method	Main band		Chloroplast satellite		Mitochondrial satellite		References
	AT	GC [a]	AT	GC	AT	GC	
Bacillaris							
Density	52	48	74	26	69	31	[8, 66, 73, 74]
Thermal denaturation	45	55	70	30	–	–	[74]
Analysis	47	53	76	24	–	–	[75]
Z strain							
Density	51	49	76	24	67	33	[62, 63]
Analysis	49	51	75	25	–	–	[62, 63]
Thermal denaturation [b]	47–52	48–53	74–79	21–26	–	–	[62, 63]
Molecular weight, as isolated (sedimentation)	$20-40 \times 10^6$		$20-40 \times 10^6$		$2.6-3.6 \times 10^6$		[75, 76]
Denaturation studies	Double-stranded		Double-stranded		Double-stranded		[8]
Density (g/cc)	1.707		1.686		1.691		See above refs.
Total DNA (%)	97.0		1.5		1.5		
pg/cell	0.485		0.0075		0.0075		[66,74]
pg/organelle	0.485		0.00075		0.000027		

[a] Includes approximately 2.3% methyl cytosine [75, 77]. Main band is assumed to represent the nuclear DNA.
[b] Calculated from data given in reference.

ability to produce plastids is rapidly lost and most of the progeny eventually lack plastids and plastid DNA [1,8]. Cell replication and viability are not affected, suggesting that these agents block *plastid* DNA replication selectively. Ultraviolet light appears to act directly to render the plastid DNA non-replicable since this agent, at doses which completely block plastid replication, does not interfere with the development of the proplastid into the chloroplast [9]. Streptomycin and temperatures above 34°C, however, also affect development [10,11].

Chloroplast development

Chloroplast development is best studied in non-dividing or resting cells of *Euglena*. Having been grown in the dark and rested, such cells contain proplastids which develop into about ten chloroplasts on exposure to light. In what follows I propose to consider our present understanding of this developmental process under three headings: (1) the structure and physiology of the proplastid and the developing chloroplast; (2) the source of building blocks for chloroplast development; and (3) the source of information for the construction of the chloroplast.

The structure and physiology of the proplastid and developing chloroplast
The proplastid

The proplastid in *Euglena* is about 1 μm in diameter (fig. 1) and is bounded by a ruffled membrane at least two layers in thickness as found by Gibbs [12] for the fully chloroplast. Ribosome-like bodies are present in the proplastid, which appear to be smaller than those of the cytoplasm, but mitochondrial ribosome-like particles seem to be rare. For the first time due to improved fixation (method of Liss et al. (unpublished) reported in Schiff et al. [13], we are able to detect prolamellar bodies which were absent when permanganate [14] or osmium without glutaraldehyde was employed [15]. In our present preparations the prolamellar body appears to be vesicular but less highly organized than that found in higher plants and is without the high order of crystallinity reported for leaves [16] (figs. 1, 2A). Several thylakoids are attached to the prolamellar body and have a rather smooth appearance compared to the outer

membrane (fig. 4A). The outer membrane itself (fig. 1) is associated with a nest of membranous material (or 'membrane depot'), but it is not clear whether this represents a true structure or an artifact of fixation. Such membranous material often seemingly connects proplastids with mitochondria or mitochondria with other mitochondria (fig. 2B).Similar whorls of membranes have been repeatedly described. They have been seen in streptomycin-treated *Euglena* [17], and in higher plants [19–24]. (I am indebted to Drs. J.V. Possingham and L. Diers for directing me to this literature.) The prolamellar bodies are frequently multiple in any particular section of a proplastid (figs. 1, 2C, 3(A,C)), and as many as three have been observed, often widely separated from each other, suggesting that they are truly distinct and not recurrent sections through a single body. This might provide an explanation for an anomaly which has troubled us for some time. In the fluorescence microscope, dark-grown *Euglena* exhibit about thirty red-fluorescing entities per cell, often connected by faint red-fluorescing strands [14]. If, as in higher plants, the protochlorophyll(ide) * and chlorophyll(ide) are confined to the prolamellar bodies and thylakoids, the red-fluorescing entities should represent prolamellar bodies and associated thylakoid materials, not proplastids as was originally thought. Thus there would be ten proplastids in the cell, each containing, on the average, three prolamellar bodies. Each proplastid would give rise to a chloroplast, ten in all as observed, obviating the need for proplastid segmentation or fusion as previously proposed [1,3], and bringing our views into agreement with those of Gibor and Granick [25], who found several red-fluorescing entities per plastid. Since the multiplicity for UV inactivation of green colony formation is 30, inferred from target analysis of inactivation curves [7,26], the number of prolamellar bodies, perhaps reflecting a fusion of three plastids in the evolution of the *gracilis* proplastid and chloroplast. As noted previously [14], there are *Euglena* species which have many more chloroplasts than the ten of *gracilis*; it would be interesting to know if these chloroplasts

---

* It is not clear whether protochlorophyllide or protochlorophyll is the photochemically transformed precursor of chlorophyll in dark grown cells. The designation here of (ide) reflects this uncertainty.

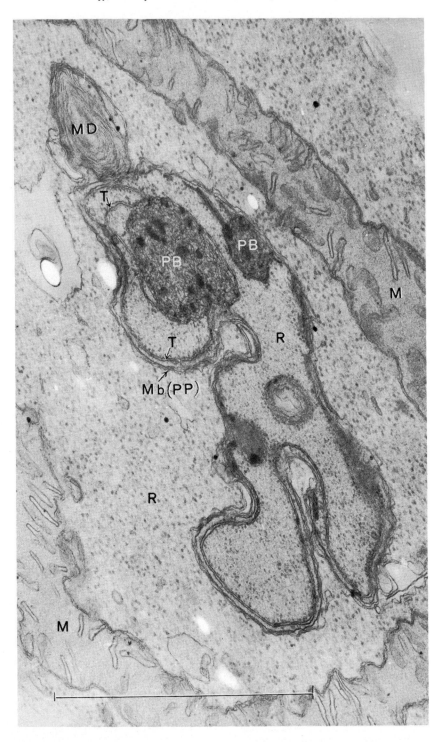

Fig. 1. Section through a dark-grown cell of *Euglena* showing a proplastid. Abbreviations here and in subsequent figures: C = chloroplast; E = endosome; ER = endoplasmic reticulum; Gu = gullet; L = lamella; M = mitochondrion; Mb = membrane(s); MD = membrane depot; N = nucleus; P = pyrenoid; PB = prolamellar body; Pe = pellicle; Pm = paramylum; PP = proplastid; R = ribosome-like particles; T = thylakoids. Scale marker indicates 1μ here and subsequently.

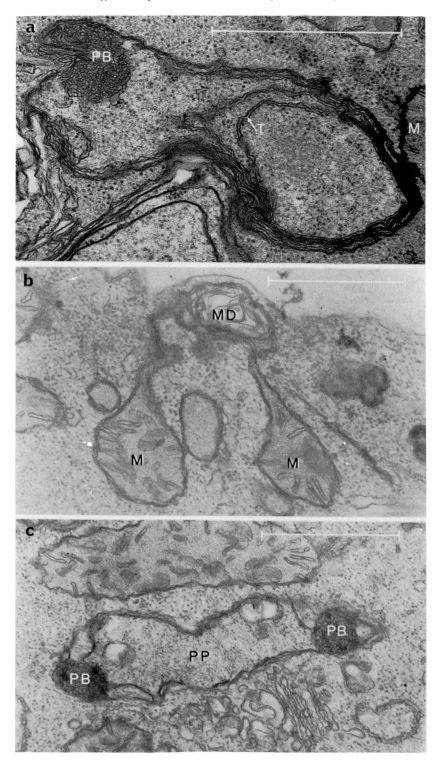

Fig. 2. Sections through dark-grown cells of *Euglena* showing: (A) proplastid, (B) mitochondria and (C) proplastid.

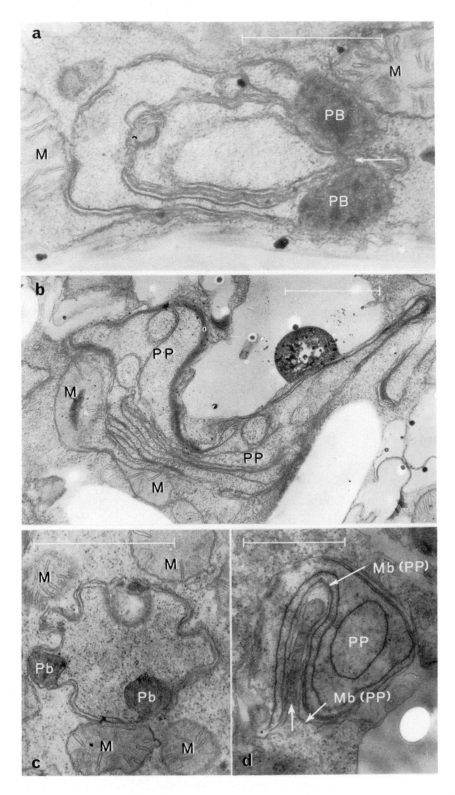

Fig. 3. Sections through dark-grown cells of *Euglena* showing: (A) proplastid, (B) proplastid with closely-apressed mitochondrion; (C) proplastid; (D) proplastid (arrow indicates cytoplasmic intrustion).

contain less DNA (perhaps one-third); perhaps their proplastids (if dark growth can be achieved) might exhibit unitary prolamellar bodies.

Two other features of the proplastid arrangement in *Euglena* should be noted, which relate to communication with the rest of the cell. Cytoplasmic intrusions into the proplastid are extensive (figs. 1, 3D) and mitochondria are frequently seen in close association (fig. 3B and other figures). The association of plastids and mitochondria has frequently been suggested, most recently by Goodenough and Levine (personal communication) for *Chlamydomonas.*

Development of the proplastid on light exposure

Improved fixation has also revealed rather extensive thylakoid formation in the proplastids of the dark-grown cells; these were not apparent with earlier fixations which yielded empty proplastids [14]. During further development membranes appear to be organized out of the ground substance of the proplastid (perhaps directly from synthesis on ribosomes) rather than from invagination of the proplastid membrane as previously supposed. By 30 min of light exposure (figs. 4B,C) membrane elaboration is clearly evident and follows the sequence already described by Ben-Shaul et al. [15] where thylakoids are formed first and increase to a maximum when formation ceases (fig. 6). Concomitantly, beginning at about 4 hr of development, thylakoids fuse into stacks called lamellae and lamellar formation is linear from about 14 hr to the completion of development somewhat beyond 72 hr. A micrograph shows the situation at 2 hr (fig. 4D). At about 14 hr the developing plastid expands tremendously in a radial direction (fig. 3A), extending the spherical or cylindrical structure into a flattened disc. The pyrenoid itself differentiates between 14 and 24 hr of development [15]. By 24–36 hr the plastid is clearly elongated (fig. 4E), but the membrane nest, seen on the proplastid, is still evident and is here surrounded with some moderately dense material. The pyrenoid is surrounded by plates of paramylum, a beta 1,3 glucan, the carbohydrate storage product of *Euglena,* deposited outside the plastid membrane. By 72 hr development is essentially complete (fig. 5) and the ribosome-like particles of the plastid remain smaller than those of the cytoplasm, although their number is probably increased over that of the proplastid.

Physiological parameters during light-induced proplastid development

Development in continuous light.

One of the first events which can be detected upon light induction of the dark-grown cells is the photoconversion of protochlorophyll(ide) to chlorophyll(ide) (fig. 7) [9]. The pattern of oxygen exchange by the cells is also altered by light exposure (fig. 8) [27]. When the dark-grown cells are initially exposed to light, there is a large increase in the rate of oxygen uptake. Most of this change is irreversible, as can be seen by turning off the light briefly as development continues, but a small part of the change is reversible, representing photorespiration. The irreversible component is interpreted as representing an increase in cellular respiration (probably mitochondrial) associated with the onset of chloroplast development. This might provide energy and intermediates needed for chloroplast development and transport would be facilitated by the close association of mitochondria and proplastids noted above. This increase in respiration might be mediated by an independent photoreceptor, but if mediated by protochlorophyll(ide) could imply a signalling system by which the proplastid informs the rest of the cell of the onset of development [28]; see also Kowallik and Gaffron [29]. As development proceeds, the light-mediated (reversible or photorespiration) component of oxygen uptake gradually becomes compensated by an increasing rate of photosynthetic oxygen evolution; by about 4 hr compensation it is almost complete and beyond this time there is an increasing net evolution of photosynthetic oxygen. The inception point for photosynthesis, then, can be taken as being somewhat before 4 hr of development, in agreement with studies of carbon dioxide fixation [30] and with the time of formation of the first lamella [15]. Chlorophyll and photosynthetic competence increase slowly between 0 and 12 hr of development but increase dramatically and maintain an approximately linear rate between 14 hr and the completion of development at 72–96 hr, which is related to the linear increase of lamellae in this period (fig. 6, 9). Chlorophyll, carotenoid, and the incoming of photosynthetic oxygen and carbon dioxide fixation all seem to be in step during this process (fig. 9). A number of proteins represented by chloroplast antigens [31] and various enzyme activities make their

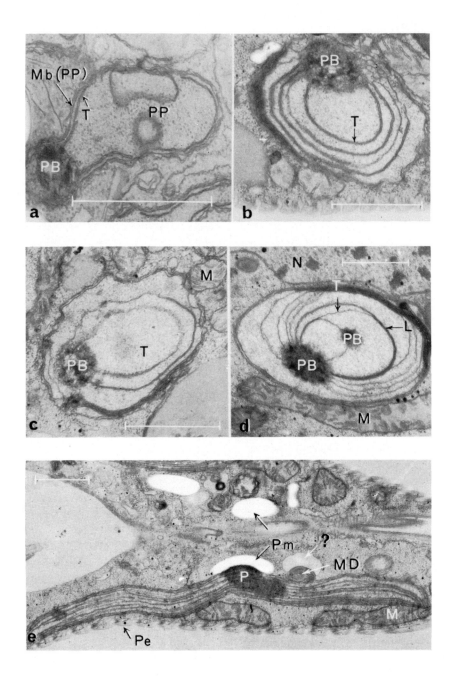

Fig. 4. (A) section through dark-grown cell showing proplastid; (B) and (C) dark-grown resting cells after exposure to 30 min of light showing developing chloroplasts; (D) section through dark-grown resting cells after exposure to 2 hr of light showing developing chloroplast; (E) section through dark-grown resting cell after exposure to 36 hr of light showing developing chloroplast.

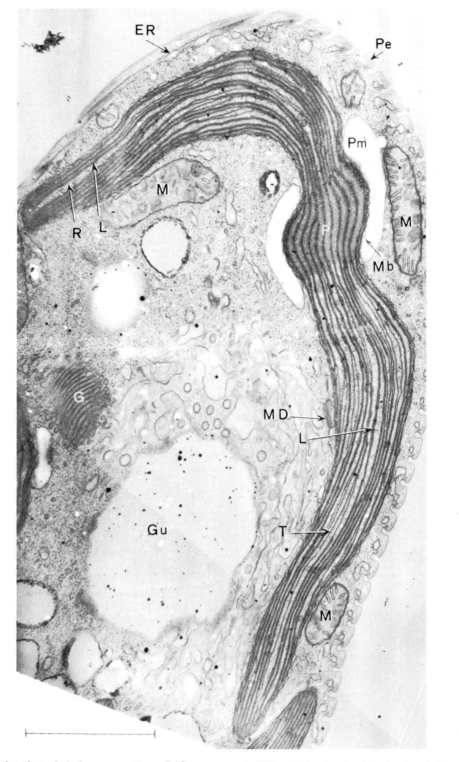

Fig. 5. Section through dark-grown resting cell after exposure to 72 hr of light showing fully-developed chloroplast.

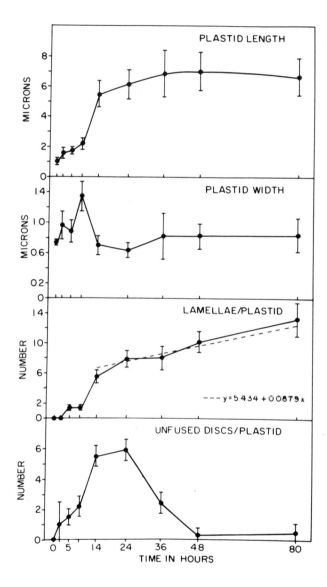

Fig. 6. Kinetics of the formation of discs or thylakoids and lamellae; and plastid lengths and widths during chloroplast development in *Euglena*. Zero time represents dark-grown cells immediately before induction of development with light. Time is measured from hr of development after dark-grown cells are exposed to light. In all cases, the points represent the means of several observations and the flags show the 95% confidence intervals of the means. The dotted line was fitted by least squares to the linear portion of lamella development [15].

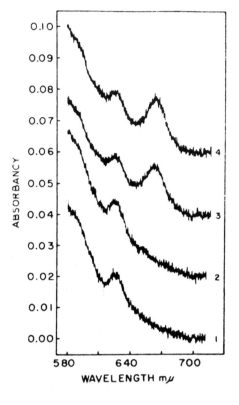

Fig. 7. Acetone extracts of equal amounts of dark-grown cells: 1) and 2) before exposure to light; 3) and 4) after exposure to 30 min of red light [9].

appearance or increase during development. TPN triosephosphate dehydrogenase [32–34], ribulose diphosphate carboxylase [32], and other enzymes of the cycle of photosynthetic carbon dioxide fixation [35] are among these, as are enzymes connected with photosynthetic electron transport such as the TPN-transhydrogenase [36,37] and cytochrome 552 [38–41]. There is every reason to believe that chloroplast development involves the synthesis of a variety of chloroplast proteins. For a list of chloroplast-localized enzymes see Smillie [42,43].

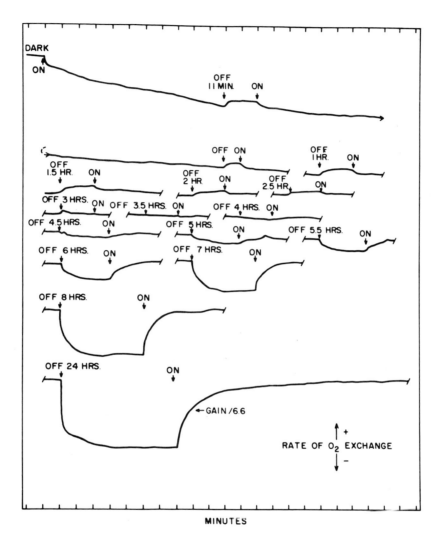

Fig. 8. Oxygen exchange of dark-grown *Euglena* cells upon light induction of chloroplast development. See text for details [27].

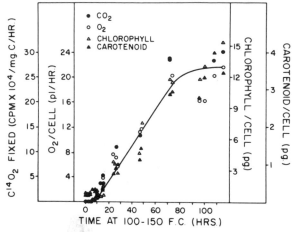

Fig. 9. Kinetics of the appearance of chlorophyll, carotenoids, photosynthetic oxygen evolution, and photosynthetic carbon dioxide fixation during chloroplast development in *Euglena*. As in fig. 6 (data on comparable cells), zero time represents measurements on dark-grown cells; time is measured from inception of light-induced chloroplast development [30].

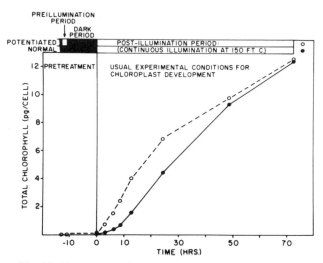

Fig. 10. Time course of chlorophyll accumulation in control (filled circles) and potentiated (open circles) cells. Zero time is taken as the beginning of the post-illumination period. Cells from a three day dark-grown resting culture were exposed to 90 min of pre-illumination (white light, 150 ft.-c) starting at −12 hr and to a dark period (potentiated cells) before exposure to continuous illumination with white light (post illumination). Control cells experienced an uninterrupted dark period until they were exposed to continuous illumination at zero hr [44,45].

Potentiation through pre-illumination

A potentially useful approach for exploring the induction and early stages of chloroplast development has been provided by the finding that a brief pre-illumination of the dark-grown cells followed by a dark period results in the elimination of the usual lag in chlorophyll formation when the cells are subsequently returned to continuous light, a phenomenon we call potentiation (fig. 10) [44,45]. Since the optimal length of the dark period for lag elimination is about 12 hr it seems reasonable to assume that a brief exposure to light triggers those events which ordinarily take place in the lag period of continuous illumination (also 12 hr in length) and permits them to occur in the dark period following pre-illumination. This should allow a separation of the inductive phase from the later consequences of induction since the latter can be studied during the dark period. The action spectrum for induction during the pre-illumination phase indicates that the photoreceptor is protochlorophyll(ide), implicating the protochlorophyll(ide) to chlorophyll(ide) conversion directly in the induction process, and not merely in chlorophyll synthesis. Unlike higher plants where a related pre-illumination phenomenon was first discovered

[46,47] and a red-far red phytochrome system was implicated, no evidence for a red-far red system mediating pre-illumination in *Euglena* has been obtained [45]. Many algae and other lower plants can make chlorophyll and chloroplasts in the dark, while what are probably single-step mutants of these organisms (e.g. *Chlamydomonas reinhardi* Y_1; Sager [48]) have lost this ability although they can still make these constituents in the light. The primitive condition, then, is probably represented by those organisms which possess two systems for carrying out the protochlorophyll(ide) to chlorophyll(ide) transformation: a dark reaction and a light-mediated one. *Euglena* seems to represent an evolutionary branch which has lost the dark system but retained the light reaction as a means of forming chlorophyll and of controlling chloroplast development. The evolutionary branch leading from the green algae to the higher plants apparently acquired the phytochrome system quite early and many morphogenetic phenomena became subjected to its control. Although the flowering plants have lost the dark system for protochloro-

phyll(ide) conversion, the light-mediated conversion of protochlorophyll(ide) seems to have become overshadowed by the phytochrome system as far as the control of early chloroplast development is concerned.

The source of building blocks for chloroplast development

Considerable chloroplast development takes place during the period before the organelle reaches full photosynthetic competence and the optimal intensity for development (150 ft-c) is far below the saturation intensity for photosynthesis (2000 ft-c) [30]. This suggests that the developing chloroplast need not obtain the energy and building blocks for development from its own photosynthetic activities.

If chloroplast development is carried out in the presence of the highly specific inhibitor DCMU, which blocks the flow of electrons from system II of photosynthesis, photosynthetic carbon dioxide fixation is completely inhibited (fig. 11). In the same

cells, chlorophyll formation is only about 20% lower than the controls. Electron micrographs of the cells developing in the presence of DCMU show chloroplasts of normal size (fig. 12) with more than the usual number of discs per lamella. Pyrenoids are rare and paramylum is greatly reduced as would be expected from the complete inhibition of carbon dioxide fixation. It has been commonly supposed that the pyrenoid serves as a centre of polysaccharide synthesis for the chloroplast where the products of photosynthesis are assembled; in this case into paramylum. The rarity of pyrenoids in chloroplasts developing with DCMU present suggests that in *Euglena* we may be dealing with a case of developmental induction of a structure by the substrates it normally acts upon. If the DCMU is washed away at the end of development at least 80% of photosynthetic ability on a chlorophyll basis can be retrieved (table 2) [13]. These data clearly indicate that the developing chloroplast can form structure, pigment and the capacity for photosynthesis in the absence of its own photosynthesis.

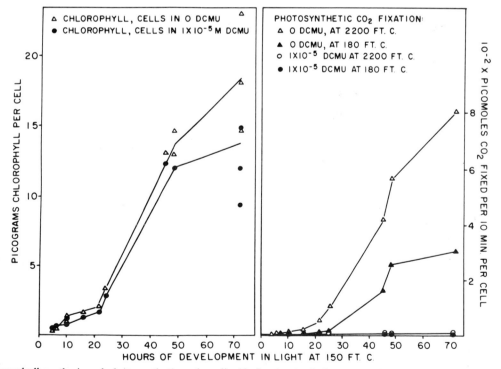

Fig. 11. Chlorophyll synthesis and photosynthetic carbon-dioxide fixation by dark-grown resting cells of *Euglena* exposed to light in the presence or absence of DCMU [13]. (a) △, 0 DCMU; •, 10^{-5} M DCMU; (b) △ 0 DCMU 2200 ft c; ▲, 0 DCMU, 180 ft c; 0, 10^{-5} M DCMU, 2200 ft c; •, 10^{-5} M DCMU, 180 ft c.

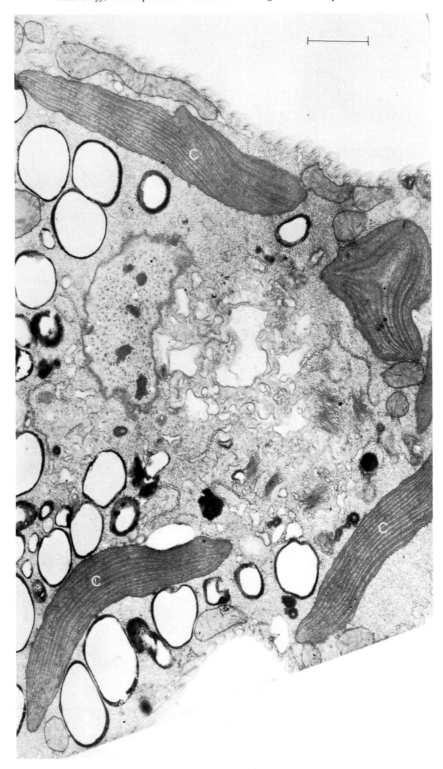

Fig. 12. Section through dark-grown resting cell which has developed chloroplasts for 72 hr in light in the presence of DCMU.

Table 2
Photosynthetic competence developed in the presence of DCMU as determined after removal of DCMU.

Time in light (hr)		$10^{-2} \times$ pmoles CO_2 fixed/pg chlorophyll/10 min			
		Before centrifugation		After centrifugation	
		Cells without DCMU	Cells with DCMU	Cells without DCMU resuspended in resting medium	Cells with DCMU resuspended in resting medium
4	A	0.11	0	0.03	0
	B	0	0	0	0
	C	0.196	0	0.196	0.065
24	A	0.31	0.004	0.27	0.08
	B	0.204	0	0.150	0.173
	C	0.339	0	0.419	0.117
48	A	0.250	0.003	0.35	0.25
	B	0.303	0.005	0.213	0.104
	C	0.390	0.010	0.332	0.255
72	A	0.340	0.001	0.350	0.310
	B	0.377	0.006	0.293	0.199
	C	0.718	0	0.824	0.400
72 [a]	A	0.47	0.007	0.460	0.400
	B	0.358	0.009	0.273	0.167

[a] Cells received DCMU after 72 hr of chloroplast development.

Cells without DCMU received ethanol and cells 'with DCMU' received DCMU in ethanol at zero time. Cells were centrifuged and were resuspended in either a 5:2 (v/v) mixture of resting medium and growth medium (entry A) or resting medium only (B and C) [13].

Since DCMU blocks photoreduction and non-cyclic photophosphorylation, much of the reducing power and ATP for development must be supplied by the rest of the cell which may also supply many of the building blocks which require energy and reducing power for their synthesis. This conclusion is supported by the fact that many organisms do not require light for chlorophyll and chloroplast development (e.g. *Chlamydomonas*, *Chlorella* and germinating pine seedlings) and mutants blocked in the photosynthetic pathway form otherwise normal chloroplast structure and pigments [48–51].

The source of information for the construction of chloroplasts

As noted above and in table 1, the nucleus, mitochondrion and chloroplast each has its own distinctive DNA. The chloroplast DNA must contain at least a crucial part of the information for the formation of chloroplast constituents since elimination of chloro-

plast DNA by streptomycin, heat or UV in *bacillaris* also results in cells incapable of forming chloroplasts [8].

RNA labelling during chloroplast development

A logical place to begin in tracing the flow of information is to study the transcription of DNA during light activation of plastid development [52]. Since assignments of RNA species to structural entities in *Euglena* is ambiguous (see below) we have contented ourselves with measuring the incorporation of radioactive phosphate into the bulk RNAs of the chloroplast and the pooled non-chloroplast fractions. Much to our surprise, light induction of chloroplast development results in incorporation of label into both the chloroplast and non-chloroplast RNAs (fig. 13) suggesting that RNA synthesis is activated in both the chloroplast and non-chloroplast compartments. This has been confirmed by Ingle [53] for radish cotyledons. These results are consistent with

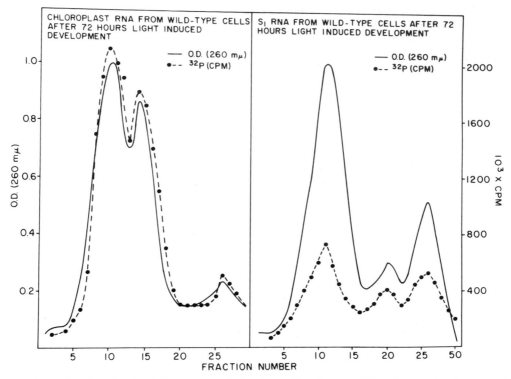

Fig. 13. Comparison of radioactive RNA from chloroplasts with radioactive non-chloroplast RNA, as separated on sucrose gradients. (a) Optical density and radioactivity of chloroplast RNA. (b) Optical density and radioactivity of non-chloroplast RNA (S_1RNA). Both were prepared from the same homogenate of wild-type cells subjected to 72 hr light-induced development on $^{32}PO_4^{3-}$. —— Optical density 260 nm; •——•, ^{32}P counts/min [52].

the finding that the chloroplast can obtain its building blocks and energy from the rest of the cell, and suggests that whatever developmental signals are necessary to mobilize the cell for chloroplast development, these must affect both the chloroplast and other sites outside the chloroplast. It is not clear whether the developmental signals arise in the chloroplast alone (on the conversion of protochlorophyll(ide) to chlorophyll(ide)) or whether other light receptors exist outside the chloroplast as well. Evidence bearing on this point which favours the latter alternative has been obtained with mutants of *Euglena* which lack plastid DNA and protochlorophyll(ide) but still show certain light-induced changes in RNA metabolism [28].

Studies of enzyme development employing streptomycin

Some extremely relevant findings bearing on the question of the origin of information for chloroplast

development have been obtained from studies of streptomycin inhibition of development. As mentioned early in this paper, streptomycin causes plastid elimination and a loss of plastid DNA in dividing cells without impairing either cell viability or cell division, suggesting that streptomycin acts on the chloroplast in a highly selective manner. In bacterial systems, this inhibitor has been shown to exert a major effect by blocking translation on ribosomes during protein synthesis and is selective for the 70 S bacterial ribosomes thought to be closely related to the chloroplast ribosomes. We surmised that streptomycin might block plastid replication in *Euglena* by selectively inhibiting translation on plastid ribosomes leading to the non-production of enzymes (such as the plastid DNA polymerase) necessary for plastid DNA replication, without affecting ribosomes external to the chloroplast. In order to determine whether streptomycin does indeed block protein synthesis in plastids, we turned to a study of the effects of this inhibitor on chloroplast development and the formation of plastid enzymes in non-dividing cells [10]. Unexpectedly, streptomycin proved to exhibit at least two patterns of action on the formation of certain chloroplast enzymes [78]; (J. Bovarnick and J. Schiff, in preparation). The syntheses of pigments and of ribulose diphosphate carboxylase (an enzyme of the photosynthetic carbon dioxide fixation cycle) and of cytochrome 552, a carrier in photosynthetic electron transport, were inhibited from 14 hr onward in development resulting in a 90% inhibition by 72 hr (fig. 14, table 3). Another enzyme of the photosynthetic carbon cycle, the TPN-linked triose phosphate dehydrogenase, behaved in an entirely different manner. Its appearance during development (fig. 14) commences without a lag and the final levels achieved demonstrate that its synthesis is not strongly inhibited by streptomycin. If streptomycin is selectively inhibiting translation on plastid ribosomes, this evidence suggests that all of the enzymes studied, except the TPN-triose phosphate dehydrogenase, are synthesized on chloroplast ribosomes. The TPN en-

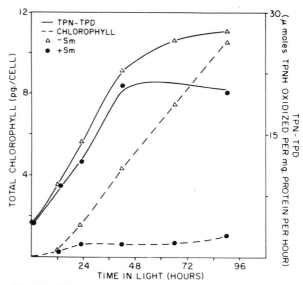

Fig. 14. Chlorophyll and TPN triose phosphate dehydrogenase appearance during light-induced chloroplast development in dark-grown resting cells of *Euglena* in the presence and absence of 0.05% streptomycin (Sm) (Bovarnick & Schiff, in preparation). ——, TPN–TPD; – – –, chlorophyll; △, –Sm; ●, + Sm.

Table 3
Enzyme formation during chloroplast development in the presence of streptomycin.

Non-dividing *Euglena*	Wild type			W$_3$BUL Light-grown
	Dark-grown	Dark-grown + light, 72–96 hr	Dark-grown + light, 72–96 hr + streptomycin	
Plastid structure present	Proplastid	Chloroplast	Chloroplast rudiment	Largely absent
Plastid DNA	+	+	+	0
Total chlorophyll (pg/cell)	0	10.50 ± 0.86	0.88 ± 0.22	0
Total carotenoid (pg/cell)	0.26 ± 0.07	2.62 ± 0.19	0.85 ± 0.12	0.13
PS CO_2 fixation (10^2 × picomoles/cell/hr)	0.08 ± 0.02	28	0.44	0
Cytochrome 552 (10^{11} × μmoles/cell)	0	2.79 ± 0.39	0.37 ± 0.6	0
Ribulose diphosphate carboxylase (μmoles CO_2 fixed/mg protein/hr)	0.14 ± 0.02	4.11 ± 0.34	0.16 ± 0.06	0
G-3-P dehydrogenase TPN (μmoles PNH DPN oxidized/mg protein/hr)	4.1 ± 0.70 43 ± 14.00	27 ± 6.83 27 ± 5.42	16 ± 6.25 32 ± 4.91	4.01 40.10

0 = below limit of detection; + = present.

zyme must be synthesized on ribosomes which are not streptomycin-sensitive and hence are external to the chloroplast. (The DPN triose phosphate dehydrogenase is not appreciably affected by streptomycin, as would be expected — it is thought to be a non-chloroplast enzyme.)

Which DNA genomes are concerned with the synthesis of these two groups of enzymes? Comparing the activities in a mutant of *Euglena* (W_3BUL) in which chloroplast DNA is undetectable (table 3) we see that those enzymes which are thought to be synthesized on chloroplast ribosomes are not detectable, suggesting that chloroplast DNA controls their synthesis. The TPN triose phosphate dehydrogenase and the DPN triose phosphate dehydrogenase are both present at the same levels as in the dark-grown cells suggesting that these enzymes are controlled by nuclear DNA. We cannot yet say that the structural genes for the first group are chloroplastic, and the structural genes for the dehydrogenases are nuclear, but further studies may well show this to be the case. Nuclear DNA, then, can influence the synthesis of a chloroplast protein; elegant studies in yeast have shown that the structural gene for the synthesis of a mitochondrial cytochrome resides in the nucleus [54].

These studies with *Euglena* suggest that the ribulose diphosphate carboxylase and the cytochrome 552 are both coded in the chloroplast DNA and are synthesized on chloroplast ribosomes. (I am indebted to Dr. K. Boardman for pointing out that if the chloroplast DNA codes for chloroplast ribosomes, the absence of these ribosomes in the mutant lacking chloroplast DNA could prevent the synthesis of various chloroplast proteins whether or not they are coded in the chloroplast DNA. Thus absence of a chloroplast enzyme from the mutant does not conclusively prove that it is coded or controlled in the chloroplast DNA.) The TPN triose phosphate dehydrogenase of the chloroplast and the DPN triose phosphate dehydrogenase of the cytoplasm are probably coded in the nucleus and synthesized on non-chloroplast ribosomes. Several models can be offered to account for the relationship between the DPN and the TPN-triose phosphate dehydrogenases. The first model assumes that both enzymes are coded in the nucleus in separate cistrons. Thus each enzyme could be regulated separately at the gene level in the nucleus. Indeed, the two enzymes in other species have proven to be so similar that it has been impossible to separate them by conventional fractionation procedures [55,56]. Since the DPN enzyme is found in photosynthetic bacteria, but not the TPN enzyme (with one exception); while the photosynthetic eucaryotic cell has both, it is possible that the TPN gene represents a duplication, during the course of evolution, of the DPN gene with minor modifications. A reciprocal relation between the levels of the activities of the two enzymes in the same cell has been suggested [34] prompted by data similar to that of table 3, where the DPN enzyme appears to fall somewhat as the TPN enzyme increases. This has been interpreted to indicate a possible interconversion between the two enzymes in certain organisms [34]. Having both enzymes coded in the nucleus would certainly facilitate such regulatory interactions. A second picture which can be made consistent with the *Euglena* data, would have only the DPN enzyme coded in the nucleus. The DPN enzyme might then be modified, in a minor way (e.g. through the addition or deletion of one or a few amino acids) or the degree of oxidation of SH groups [34] in the cytoplasm or chloroplast to yield the TPN enzyme. Evidence to support the conversion of the DPN enzyme to the TPN enzyme in the presence of high TPN concentrations has recently appeared [57]. Whatever the mechanism, it is clear that nuclear-chloroplast and nuclear-mitochondrial interactions are to be expected, and perhaps in the opposite direction as well, in eucaryotic cells. All possible permutations and combinations of informational cross-talk among the three organelles might eventually be found.

These indications of a nuclear location for the control of at least one plastid constituent might permit an explanation for the findings of Gibor and Granick [25] with the Z strain. In cells derived from growing cultures treated with streptomycin they observed red-fluorescing bodies on treatment of the cells with δ amino levulinic acid, an early precursor of haem and chlorophyll. They suggested, on the basis of this evidence, that proplastid remnants might remain in these cells. Our own data with *bacillaris* suggest that most plastid structure is missing together with the plastid DNA. Perhaps the early steps in porphyrin synthesis are catalysed by enzymes which are coded in the nuclear DNA. Elimination of the chloroplast DNA with streptomycin might leave these enzymes at low levels in the cells (as we have found for the TPN triose phosphate dehydrogenase), allowing the synthesis of precursor porphyrins, many of which show red fluorescence. The entire chlorophyll synthesis pathway should not be present in the streptomycin mutant (or, if present, the latter steps should be inactive) since this mutant cannot make protochlorophyll(ide) or chlorophyll(ide).

The level of the TPN triose phosphate dehydrogenase in the mutant is at the same level as in dark-grown wild-type cells (table 3). This indicates that the synthesis of the TPN enzyme is repressed to the same extent in both types of cells. This further suggests that, if the control is at the level of a nuclear gene, normal conditions of derepression by light involve either a signal from the proplastids to the nucleus, or a separate non-chloroplast photoreceptor as already suggested above in connection with RNA synthesis.

RNA characterization and localization in *Euglena*

The electron micrographs of the proplastid and chloroplast already presented (figs. 1, 5) indicate that the plastids contain ribosome-like particles which are somewhat smaller than those of the surrounding cytoplasm. Although chloroplast isolation methods for *Euglena* still leave something to be desired, there seems to be a current general agreement on the values for ribosomal RNAs of the chloroplast. These are thought to be 23 S and 16 S contained in a 70 S ribosome dissociating to 50 S and 30 S subunits [58—60] ; (Ledoigt, Cohen and Schiff, unpublished).

The situation with respect to non-chloroplast RNAs is in a less satisfactory condition. (We originally presented data on the *bulk*RNAs of *Euglena* cells and compared them with the findings of Brawerman [61] and Brawerman and Eisenstadt [62,63] for ribosomal RNAs [52]. Although we did not attempt to assign the RNAs we found to ribosomes or any other structure, this work has been misquoted as asserting identification of ribosomal RNAs [58,64].) One of the difficulties in assigning RNAs to cellular structures in a eucaryotic cell such as *Euglena* is the large number of compartments and ribosome-like structures present in the cell. An inspection of fig. 15 shows that ribosome-like particles are evident in the cytoplasm, and are especially rich in the nucleus and the endosome of the nucleus. As noted before (fig. 1, 12), ribosomes are also abundant in the proplastids and chloroplasts and a few can be seen in the mitochondria. It is clear from studies on other eucaryotes that the nucleus contains a spectrum of heavy RNAs, some of which serve as precursors for the cytoplasmic ribosomes [65]. To my knowledge, no satisfactory method for the preservation and isolation of nuclei from *Euglena* has been devised. When the cells are broken, the nuclei fragments and their contents undoubtedly contaminate other cell fractions. Until these problems can be resolved, the meaning of cytoplasmic ribosomes and RNAs would appear to be ambiguous. As noted above, chloroplasts can be prepared in reasonably clean form (although rather leaky) and the assignments made on the basis of such fractionation are probably a good first approximation. Although reasonably good mitochondria can also be prepared [66,67], we have not succeeded in obtaining high molecular weight RNAs from them (reflecting, perhaps, the low density of ribosomes seen in the electron micrographs (Perl and Schiff, unpublished)).

For the reasons just presented we would prefer to call cytoplasmic ribosomes 'non-chloroplast ribosomes' until an assignment can be made to the proper cellular compartment unambiguously. With this distinction in mind, there is growing agreement on the S values for non-chloroplast RNAs of the ribosomes. The major non-chloroplast ribosome seems to be 87 S and to consist of 64 S and 46 S subunits [58,59]. The RNAs show a heavy component variously measured from 24 to 27 S and another at about 20 S. A third component intermediate between these is often seen although its significance is not clear; although close to chloroplast ribosomal RNA in sedimentation it also can be obtained from mutants (such as W₃BUL) which lack conspicuous plastids and in which chloroplast DNA is undetectable [58,64,68] ; (Ledoigt, Cohen and Schiff, unpublished). These newer measurements would appear to supersede those of Brawerman and Eisenstadt [62,63], Schuit and Buetow [69] and Gnanam and Kahn [70].

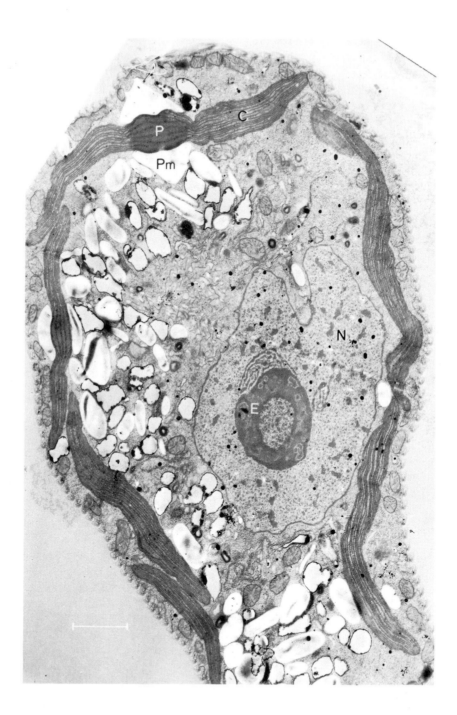

Fig. 15. Section through dark-grown resting cell after exposure to light for 72 hr showing various cellular compartments.

Conclusions

From the work presented here and other information [4], an organelle, such as the *Euglena* chloroplast, resembles a cell within a cell, to the extent that it is autonomous. No attempt has been made to review the relevant and important work done on higher plant chloroplasts [71,72]. Possessing DNA, ribosomes and the rest of the machinery for protein synthesis the plastid can synthesize many of the constituents it requires for development and maintenance. It is clear, however, that the developing chloroplast can obtain a large proportion of the energy and building blocks it requires from the rest of the cell and perhaps it lacks certain crucial synthetic capabilities. At least one enzyme, the TPN triose phosphate dehydrogenase, seems to be provided by the rest of the cell using information outside the chloroplast DNA. The various compartments of the cell undoubtedly exchange regulatory signals and evidence of cooperation on many levels seems to be emerging. It remains for the future to document further organelle interactions and to learn finally how they interact to provide a working cell and a coordinated process of cell division.

Acknowledgement

The expert technical assistance of Miss Nancy O'Donoghue in preparing the electron micrographs is gratefully acknowledged.

References

[1] J.A. Schiff and H.T. Epstein, in: Reproduction: Molecular, Subcellular and Cellular, ed. M. Locke (Academic Press, New York and London, 1965) p. 131.

[2] J.A. Schiff and H.T. Epstein, in: Biochemistry of Chloroplasts, ed. T.W. Goodwin (Academic Press, New York and London, 1966), vol. 1, p. 341.

[3] J.A. Schiff and H.T. Epstein, in: The Biology of Euglena, ed. D. Buetow (Academic Press, New York and London, 1968), vol. II, p. 285.

[4] J.A. Schiff and M.H. Zeldin, J. Cell Physiol. 72 (1968) 103.

[5] L. Provasoli, S.H. Hutner and A. Schatz, Proc. Soc. Exptl. Biol. Med. 69 (1948) 279.

[6] F. Pringsheim and O. Pringsheim, New Phytologist 51 (1952) 65.

[7] H. Lyman, H.T. Epstein and J.A. Schiff, Biochim. Biophys. Acta 50 (1961) 301.

[8] M. Edelman, J.A. Schiff and H.T. Epstein, J. Mol. Biol. 11 (1965) 769.

[9] J.A. Schiff, H. Lyman and H.T. Epstein, Biochim. Biophys. Acta 51 (1961) 340.

[10] J. Bovarnick, S.W. Chang and J.A. Schiff, Plant Physiol. 43 (1968) S-6.

[11] A. Goroll, M.H. Zeldin and J.A. Schiff, Plant Physiol 44 (1969) S-11.

[12] S.P. Gibbs, J. Ultrastruct. 4 (1960) 127.

[13] J.A. Schiff, M.H. Zeldin and J. Rubman, Plant Physiol. 42 (1967) 1716.

[14] H.T. Epstein and J.A. Schiff, J. Protozool. 8 (1961) 427.

[15] Y. Ben-Shaul, J.A. Schiff and H.T. Epstein, Plant Physiol. 39 (1964) 231.

[16] B.E.S. Gunning and M.P. Jagoe, in: Biochemistry of Chloroplasts, ed. T.W. Goodwin (Academic Press, New York and London, 1967), vol. II, p. 655.

[17] K.A. Siegesmund, W.G. Rosen and S.W. Gawlik, Amer. J. Botany 49 (1962) 137.

[18] S. Klein, J. Biophys. Biochem. Cytol. 8 (1960) 529.

[19] W. Menke, Protoplasma 51 (1960) 127.

[20] F.M. Gerola, M. Bassi and G. Belli, Caryologia 18 (1965) 567.

[21] W.A. Jensen, Amer. J. Botany 52 (1965) 781.

[22] S.Y. Zee and T.C. Chambers, Australian J. Botany 16 (1969) 37.

[23] W. Menke and B. Wolfersdorf, Planta 78 (1968) 134.

[24] L. Diers and F. Schotz, Z. Pflanzenphysiol. 60 (1969) 187.

[25] A. Gibor and S. Granick, J. Protozool. 9 (1962) 327.

[26] H.Z. Hill, J.A. Schiff and H.T. Epstein, Biophys. J. 6 (1966) 125.

[27] J.A. Schiff, Year Book, Carnegie Inst. Wash. 62 (1963) 375.

[28] M.H. Zeldin and J.A. Schiff, Planta 81 (1968) 1.

[29] W. Kowallik and H. Gaffron, Nature 215 (1967) 1028.

[30] A.I. Stern, J.A. Schiff and H.T. Epstein, Plant Physiol. 39 (1964) 220.

[31] S.C. Lewis, J.A. Schiff and H.T. Epstein, J. Protozool. 12 (1965) 281.

[32] R.C. Fuller and M. Gibbs, Plant Physiol. 34 (1959) 324.

[33] G. Brawerman and N. Konigsberg, Biochim. Biophys. Acta 43 (1960) 374.

[34] G.A. Hudock and R.C. Fuller, Plant Physiol. 40 (1965) 1205.

[35] I. Latzko and M. Gibbs, Plant Physiol. 44 (1969) 295.

[36] R.A. Lazzarini and M. Woodruff, Biochim. Biophys. Acta 79 (1964) 412.

[37] R.A. Lazzarini and A. San Pietro, Plant Cell Physiol., Tokyo (1963) spec. issue, p. 453.

[38] M. Nishimura, J. Biochem. 46 (1959) 219.

[39] J.J. Wolken and J.A. Gross, J. Protozool. 10 (1963) 189.

[40] F. Perini, M.D. Kamen and J.A. Schiff, Biochim. Biophys. Acta 88 (1964) 74.

[41] F. Perini, J.A. Schiff and M.D. Kamen, Biochim. Bio-
     phys. Acta 88 (1964) 91.
[42] R.M. Smillie, Canadian J. Botany 41 (1963) 123.
[43] R.M. Smillie, in: The Biology of Euglena, ed. D.
     Buetow (Academic Press, New York and London,
     1968) vol. II, p. 1.
[44] A. Holowinsky and J.A. Schiff, Plant Physiol. 43
     (1968) S-7.
[45] A. Holowinsky and J.A. Schiff, Plant Physiol. 45 (1969)
     339.
[46] R.B. Withrow, J.B. Wolff and L. Price, Plant Physiol. 31
     (1956) Suppl. XIII.
[47] H.I. Virgin, Physiol. Plant. 14 (1961) 439.
[48] R. Sager, Brookhaven Symp. Biol. 11 (1958) 101.
[49] L. Bogorad, Botan. Gaz. 111 (1950) 221.
[50] R.P. Levine and D. Volkmann, Biochem. Biophys. Res.
     Commun. 6 (1961) 264.
[51] N.I. Bishop, Ann. Rev. Plant Physiol. 17 (1966) 185.
[52] M.H. Zeldin and J.A. Schiff, Plant Physiol. 42 (1967)
     922.
[53] J. Ingle, Plant Physiol. 43 (1968) 1850.
[54] F. Sherman, J.W. Stewart, J.H. Parker, E. Inhaber, E.
     Shipman, J. Putterman, R. Gardinsky and E. Margoli-
     ash, J. Biol. Chem. 243 (1968) 5446.
[55] M.D. Shulman and M. Gibbs, Plant Physiol. 43 (1968)
     1805.
[56] L. Anderson, Proc. XIth Intern. Botan. Congr., Seattle,
     Washington, 1969, p. 3.
[57] B. Muller, I. Ziegler and H. Ziegler, Europ. J. Biochem.
     9 (1967) 101.
[58] J.R. Rawson and E. Stutz, J. Mol. Biol. 33 (1968) 309.
[59] J.R. Rawson and E. Stutz, Biochim. Biophys. Acta 190
     (1969) 368.
[60] R.M. Smillie, N.S. Scott, D. Graham, A.M. Grieve and
     N.F. Tobin, Plant Physiol. 43 (1968) S-6.

[61] G. Brawerman, Biochim. Biophys. Acta 72 (1963) 317.
[62] G. Brawerman and J.M. Eisenstadt, J. Mol. Biol. 10
     (1964) 403.
[63] G. Brawerman and J.M. Eisenstadt, Biochim. Biophys.
     Acta 91 (1964) 477.
[64] C. Portier and V. Nigon, Biochim. Biophys. Acta 169
     (1968) 540.
[65] M. Willems, M. Penman and S. Penman, J. Cell Biol. 41
     (1969) 177.
[66] M. Edelman, H.T. Epstein and J.A. Schiff, J. Mol. Biol.
     17 (1966) 463.
[67] D. Buetow and P.J. Buchanan, Exptl. Cell Res. 36 (1964)
     204.
[68] N.S. Scott and R.M. Smillie, Currents in Modern Biolo-
     gy 2 (1969) 339.
[69] K.E. Schuit and D.E. Buetow, Biochim. Biophys. Acta
     166 (1968) 702.
[70] A. Gnanam and J. Kahn, Biochim. Biophys. Acta 142
     (1967) 493.
[71] L. Bogorad, in: Harvesting the Sun, eds. A. San Pietro,
     F.A. Green and T.J. Army (Academic Press, New York
     and London, 1967) p. 191.
[72] S. Granick, in: Cytodifferentiation and Macromolecular
     Synthesis, ed. M. Locke (Academic Press, New York
     and London, 1963) p. 144.
[73] J. Leff, M. Mandel, H.T. Epstein and J.A. Schiff,
     Biochem. Biophys. Res. Commun. 13 (1963) 126.
[74] M. Edelman, C.A. Cowan, H.T. Epstein and J.A. Schiff,
     Proc. Natl. Acad. Sci. U.S. 52 (1964) 1214.
[75] D.S. Ray and P.C. Hanawalt, J. Mol. Biol. 9 (1964) 812.
[76] D.S. Ray and P.C. Hanawalt, J. Mol. Biol. 11 (1965)
     760.
[77] G. Brawerman, D.A. Hufnagel and E. Chargaff, Bio-
     chim. Biophys. Acta 61 (1962) 340.
[78] J. Bovarnick, Z. Freedman and J.A. Schiff, Plant Physiol.
     46 (1970) S-113.

STUDIES ON THE ASSEMBLY OF THE MITOCHONDRIAL MEMBRANES

Giorgio LENAZ, Anna Maria SECHI, Lanfranco MASOTTI,
Giovanna PARENTI-CASTELLI

Istituto di Chimica Biologica, Università di Bologna,
Bologna, Italy

and

Adriano CASTELLI, Gian Paolo LITTARRU, Enrico BERTOLI

Istituto di Chimica Biologica, Università Cattolica,
Roma, Italy

Concomitantly with progress in the understanding of membrane structure, the attention of investigators is shifting to try to elucidate the mechanisms by which the membranes of the cell are synthesized and organized. The problem of synthesis of supramolecular structures has recently received particular attention [1–3]: special types of genetic information with production of organizer and informational molecules have been invoked, together with self-assembly mechanisms which depend on the nature of the macromolecules synthesized; in the case of membranes these molecules are bimodal lipids, catalytic and noncatalytic proteins [4,5].

The problem of membrane formation can be attacked by a dual type of investigation; the first type can be directed on disassembling a membrane and studying the conditions necessary to reassemble its various pieces in order to obtain a functional membrane as close as possible to the original one; alternatively, one can follow the natural course of the formation of a membrane in vivo. Mitochondria are among the most suitable membranes for both types of investigation.

Structure of the mitochondria

The main assumption in the Danielli-Davson theory of membrane structure [6] has been the inaccessibility of a lipid bilayer (considered as the backbone of the membrane) by protein molecules. Proteins have been envisaged as apposed on the surface of the bilayer, and the linkage between lipid and protein would be assured by electrostatic interactions between the polar heads of the phospholipids and polar amino acid residues.

One main indication that at least certain membranes might not be so simple, was the finding [7] that lipid-depleted mitochondria maintain the trilaminar aspect of their inner membranes when examined at the electron microscope: lipids therefore are not essential for the microscopic structural integrity of the inner mitochondrial membrane, although it was recognized that they are essential for some of its enzymic activities [8]. A typical 'unit membrane' [9] would collapse if all of the lipids of the bilayer were removed. Green and Tzagoloff [4] have suggested that protein is contained in the core of the lipid bilayer whereas the polar residues of the phospholipids are exposed on the surface of the membrane. A similar but somewhat more detailed scheme has been proposed by Benson [10]. A subunit structure of the proteins in the membrane is an implication of the models suggested by these authors [10–12].

Since no exact knowledge on the precise localization of all the proteins of mitochondria and other membranes is available today, the problem of the relations between lipids and proteins in the structure of natural membranes should be evaluated also at the level of whole membranes.

Nature of the bonds

Fleischer and his coworkers have demonstrated a requirement for phospholipids of mitochondrial enzymic activities [cf. 8]. Several methods were effective in removing phospholipids with reversible loss of succinoxidase activity [13]. For example, extraction of beef heart mitochondria (BHM) with aqueous 90% acetone in presence of ammonia removes most of the lipids. Addition of phospholipid micelles restores the original phospholipid content. Such restoration is effective even when the incubation is accomplished in presence of neutral salts in amounts sufficient to inhibit ionic interactions [14].

Conditions affecting the binding of phospholipids to lipid-depleted mitochondria (LDM) are extensively studied in our laboratory. The association is inhibited by high concentrations of agents of the lyotropic series; in accord with the order of efficacy of this series [15], SCN^- is more effective than I^-, Br^-, and Cl^- at the same concentrations. Lyotropic agents disrupt or prevent hydrophobic associations, whose intrinsic nature is the sparing of energy deriving from the packing of nonpolar molecules in an aqueous medium [16]. Fig. 1 shows a plot of the extent of phospholipid binding against concentration of NaCl and NaSCN. There is maximum binding at low ionic strength, and a subsequent decrease, which is greater with NaSCN than with NaCl. The increased binding at low ionic strengths may be explained if there are phospholipid binding sites unavailable when the protein is folded, but available if ionic costrictions preventing unfolding are broken by salt at low concentration. Salts at high concentrations then decrease the hydrophobicity of the binding sites proportionally to their lyotropic effect.

We must assume that there are physiological binding sites (available in absence of salt) and non-physiological sites (made available by the action of salt at low concentration). The pattern of restoration of succinoxidase activity [19] by phospholipids in the presence of salts is in line with this assumption [20].

It might be argued that the interactions in vitro we have described here might be unrelated to the bonds naturally existing in the membranes. For this reason we have studied the effect of aliphatic alcohols on the extractability of lipid from intact mitochondrial membranes [20]. A direct relation holds between

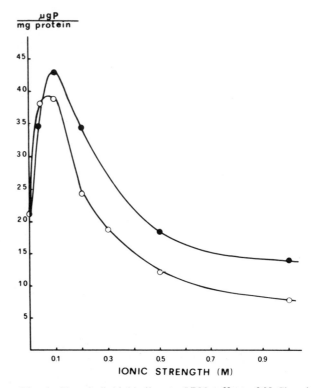

Fig. 1. Phospholipid binding to LDM; effect of NaCl and NaSCN concentration [20]. Phospholipid was Asolectin (commercial soybean phospholipids) added to LDM as an aqueous dispersion [13]. Incubation was performed as previously described [14]. Phosphorus was assayed according to Marinetti [17] and protein with a biuret method [18]. ●———● NaCl; ○———○ NaSCN.

chain length of the alcohol and efficacy of the extraction (fig. 2).

It appears from this pattern that lipids must be linked to protein by bonds which are largely hydrophobic. Table 1 summarizes some examples available for the predominance of hydrophobic bonds in mitochondria and other membrane systems.

Availability of lipids and proteins to specific reagents

Phospholipases hydrolyze membrane phospholipids [20,28]. Phospholipase C hydrolyzes phospholipid from HCl-treated mitochondria [29] and from sonic particles ETP_H [30] at a similar rate as from phospholipid micelles [20]. Protein reagents react much less effectively with intact than with delipidated membranes [20]. The interpretation of these results must be cautious, in view of the complexity

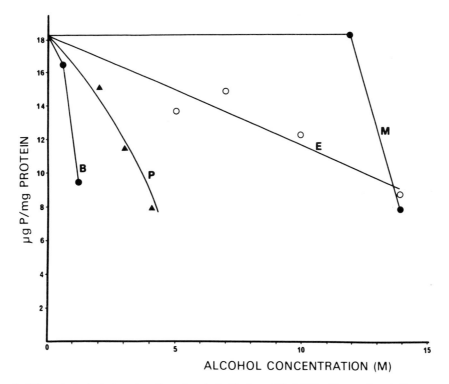

Fig. 2. Efficacy of different alcohols in extracting phospholipids from BHM [20]. M = methanol; E = ethanol; P = n-propanol; B = n-butanol. Values are μg of P per mg of particle protein after extraction with the alcohol at the indicated concentration.

and compartmentation of membranes, but the overall evidence suggests that lipids are available on the membrane surfaces, whereas many proteins are less available. Also other chemical and immunological studies are in line with this postulate [31–33].

Superficial proteins

Negatively stained mitochondrial inner membranes show typical projections towards the matrix space [34]; such 'headpieces', identified with the mitochondrial ATPase [35], complicate the pattern of lipid-protein interactions and the problem of the relative position of these constituents.

10 mM HCl extracts the structural protein fraction from mitochondria and submitochondrial particles [29]. A main component of the structural protein fraction [36] is contained in the 'headpieces' of the inner mitochondrial membrane [37]. Electron microscopy by negative staining confirms after HCl extraction the existence of membrane fragments from which headpieces are absent. About 50% of the

protein is found in the HCl extract, but all of the phospholipids remain in the membrane [20].

These membranes may be depleted of their lipids and the results on lipid-protein interactions reported in a preceding section were repeated with these lipid-depleted membranes [20]. Hydrophobic interactions are therefore the basis of the linkage of phospholipids with the 'basepieces' [11] of the membrane. The superficial position of the headpieces on the other hand could mean that they are not bound to phospholipids in vivo or that they are bound to phospholipids by bonds which are not exclusively hydrophobic.

'Headpieces' constitute a steric hindrance to the extraction by ether of a part of the phospholipids from a submitochondrial membrane (ETP$_H$) in which the headpieces are known to be exposed towards the external medium [38]. Salt is necessary in order to extract phospholipids, and this effect has been previously demonstrated for phospholipid micelles [39]; salt screens the electrostatic field of the membrane

Table 1
Lipid protein interactions in membranes

Membrane	Evidence	Type of binding	Ref.
Mainly mitochondria	Theoretical considerations	Hydrophobic	[11]
Chloroplast and others	Theoretical considerations	Hydrophobic	[10]
Mainly myelin	Theoretical considerations	Mainly intermolecular cohesive forces	[21]
Erythrocyte ghosts	Differences in extractability with solvents	Different types	[22]
Mycoplasma laidlawii	Isolation and reassembly of lipoprotein subunits	Both ionic and apolar in reassembly	
Mitochondria	Binding of structural protein to phospholipids	Hydrophobic	[24]
Halophilic bacteria	Lipoprotein persists after succinylation	Hydrophobic	[25]
Wide range of membranes	ORD studies	Hydrophobic	[26]
Wide range of membranes	NMR studies	Hydrophobic	[27]
Mitochondria	Binding of LDM to phospholipids	Hydrophobic	[20]
Mitochondria	Conditions affecting phospholipid extraction	Hydrophobic	[20]

Table 2
Phospholipid extraction from submitochondrial particles by diethyl ether [a]

Particle	− NaCl	+ 1 M NaCl
	% P extracted	
ETP_H	2	15
HCl-extracted ETP_H	20	54

[a] For experimental conditions see [20]. ETP_H, electron transport particles.

HCl from mitochondria or ETP_H (table 3). An explanation could be that cations compete with H^+ preventing protonation of acidic groups in the membrane. The extractability of protein by acid and the effect of cations suggest that these proteins are bound to the membrane when acidic groups are in the anionic forms. If endogenous divalent cations were involved in bridges, e.g. between carboxyl and phosphate anions, we would expect the observed pattern.

A tentative scheme is presented (fig. 3) suggesting that phospholipids are linked to external proteins with their polar heads which are not engaged in hydrophobic binding with the more deeply buried membrane proteins. The presence, besides weak hydrophobic and other bonds, of Mg^{++} bridges must be evaluated.

The existence of pseudomembranes made of phos-

Table 3
Effect of neutral salts on protein extraction by 10 mM HCl from BHM and ETP_H

Particle	Salt added	Protein extracted (%)
BHM	−	61
	0.01 M $MgCl_2$	45
	0.1 M $MgCl_2$	20
ETP_H	−	44
	0.1 M $MgCl_2$	17
BHM	−	58
	0.5 M KCl	15
	0.1 M $MgCl_2$	39
	0.1 M $CaCl_2$	41

BHM, beef heart mitochondria.

allowing ether to contact the lipid molecules. Little phospholipid was extracted from intact ETP_H even in presence of salt, but about 50% was extracted in presence of salt from ETP_H which had been treated with HCl [table 2]. This could also be an indication that phospholipids are linked to the acid-extractable proteins by bonds which are not affected by ether, in contrast with phospholipids linked to other mitochondrial proteins.

Neutral salts prevent the extraction of protein by

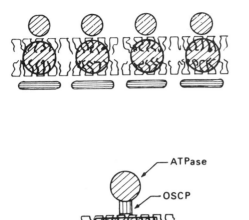

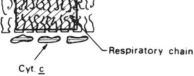

Fig. 3. A scheme of lipid-protein interactions in membranes. In the upper figure, a general membrane is drawn, composed of a lipid bilayer with globular or extended proteins on the surfaces and globular protein units in the interior. The scheme below, shows a simplified relationship in the inner mitochondrial membrane; a tripartite unit, according to Green [11] with basepiece, stalk, and headpiece, is indicated with the possible lipid to protein relationships. Headpieces are considered in an operational sense as the protein which is extracted by 10 mM HCl. Much of this protein is probably part of the ATPase complex. OSCP = oligomycin sensitivity conferring protein; Cyt. *c* = cytochrome *c*.

pholipid layers linked to the mitochondrial projections has been described previously. Kopaczyk et al. [40] have isolated an oligomycin sensitive ATPase fraction (P_2) which, combined with phospholipids, shows a peculiar membranous character: headpieces are linked by stalks to myelin figures, which have been equated with lipid bilayers. Jolly et al. [41] have obtained similar structures by lyophilization of BHM. Lyophilization must break hydrophobic bonds, because of the nature of such 'bonds' [16]: it is conceivable that after prolonged lyophilization the membrane is altered with dislocation of the hydrophobic relationships between the lipids and proteins. Since headpieces appear to be still linked to phospholipid layers, this association might not be exclusively hydrophobic.

Classes of membranes

The model presented in fig. 3, if we consider it well, has strong resemblances with a unit membrane [9]: a lipid bilayer interacts with superficial proteins (like e.g. the ATPase complex). The resemblance may be even greater if we consider that cytochrome *c* is superficially disposed on the other side of the membrane [42], most probably linked to acidic phospholipids by electrostatic interaction [8,43,44]. Different from the unit membrane model, however, there are in addition also proteins hydrophobically buried in the bilayer. As for the ATPase system, its interaction with the membrane may be complex, and involve different types of bonds.

Simple membranes like myelin may indeed be very similar to unit membranes. In more complex membranes there is however the need to accomodate an increasingly higher number of proteins having different functions. Some of these proteins can be exposed to the exterior, but more convenience and stability may be acquired by hydrophobic inclusion into the interior of the membrane within the lipid bilayer. In the inner mitochondrial membrane, the proteins of the electron transfer system obtain a more stable assembly and unique properties by disposition in a complex hydrophobic phospholipid bilayer rather than forming new unstable layers on its surface.

Classes of membranes having increasing quantities of hydrophobic proteins in their interior may be envisaged: myelin and the inner mitochondrial (and chloroplast) membranes represent the two extremes.

Composition and function of the lipids

Bimodal lipids are essential in membrane structure and function and are required for membrane assembly and reconstitution. The lipid composition and unsaturation of various membranes has been studied by several authors [45]; there is a general agreement that simple membranes like myelin contain greater amounts of lipids and these lipids are more saturated than in complex membranes like the mitochondria [46]. Unsaturation may be of far greater importance than previously recognized. Molecular models show that saturated lipids are suitable to form stable bilayers like in myelin [21] whereas unsaturated lipids are more suitable to hydrophobic interdigitation with proteins. Mitochondrial lipids contain large amounts of polyunsaturated fatty acids [45]; indeed there is large evidence for their hydrophobic binding with mitochondrial proteins. The 'unit membrane' trilaminar aspect of the mitochondrial membrane at the electron microscope could in fact even be a consequence of the disappearance of the double bonds by action of the oxidative fixatives, with enhanced stability of the bilayer after ejection of internal hydrophobically bound proteins. Findings relevant to the importance of unsaturation in membranes are not many in the literature [47–50]. Although it appears from in vitro studies that saturated phospholipids are not effective in restoring certain mitochondrial enzymic activities [47], it was shown that mitochondria from essential fatty acid (EFA) deficient rats have normal oxidative and phosphorylative capacities; however the extent of unsaturation in EFA deficiency is decreased, but the number of unsaturated fatty acids is not changed [50]. We have found that myelin lipids which are less unsaturated than mitochondrial phospholipids and have a specific fatty acid composition are less effective in restoring succinoxidase activity in lipid-depleted mitochondria (unpublished data). Conditions leading to complete saturation of phospholipids in vivo are not available in mammalian mitochondria. It is significant however that when only saturated fatty acids are present in yeast mitochondria, respiratory proteins appear to be lacking (see a later section).

Membrane assembly in vitro

Green and Perdue [11] have proposed a repeating unit structure for membranes. Repeating units are also the basis for membrane formation in vitro [51, 52]. When membranes are disrupted by detergents, the individual repeating units become separated. When detergents are functionally removed membranes reform spontaneously with restoration of the original activities. Membrane formation is effective only when lipids are present, otherwise bulky aggregates occur [4]. The conclusion has been made that lipids have the function of preventing tridimensional aggregation since their polar heads exposed at the surface of the membrane allow the repeating units to be linked together hydrophobically only in two directions with the resulting formation of bidimensional membranes [4].

A different type of membrane formation has been described by Kagawa and Racker [35]. The oligomycin sensitive ATPase, constituted of the headpiece (F_1) plus other components is functional in presence of phospholipids. The membranous character of this preparation has induced Racker to consider it as the minimal membrane, and to question whether respiratory components are necessary for membrane formation [53].

An explanation may be found for this discrepancy. In one case the complex (ATPase + phospholipids)

Table 4

Reconstitution of ETP_H from lipid-depleted ETP_H and an ATPase fraction containing phospholipids [a]

Fraction	Protein (mg)	Phospholipid		ATPase		Succinoxidase Sp. act. [c]
		$\mu gP/mg$	Total μgP	Sp. act. [b]	Total act.	
LD-ETP_H (A)	4.8	5.3	25	0	0	0.052
ATPase (B)	6.2	34.0	211	1.63	10.1	0
[A + B]-residue	10.2	19.6	200	1.23	12.5	0.123
[A + B]-supernatant	2.9	10.8	31	0.57	1.6	0

[a] ETP_H were treated with acetone and NH_3 [13] and then washed with 0.15 M KCl. The resulting fraction which has lost 40% of the total protein is referred to as lipid-depleted ETP_H (LD-ETP_H). The crude ATPase fraction has been prepared with the method of Tzagoloff et al. [55]: it has been used at the stage previous to extraction of the ATPase inhibitor, and this explains the relatively low activity. After mixing the two fractions at the concentrations indicated, they were centrifuged at 20,000 rpm, for 10 min; at this speed LD-ETP_H alone sediment, and the ATPase fraction alone remains in the supernatant.

[b] μmoles P_i/min/mg protein.

[c] μatoms O_2/min/mg protein.

should be a phospholipid bilayer linked to the knobs with a minimum of hydrophobic binding; in the other instance the electron transfer complexes plus phospholipids form exclusively hydrophobic membranes made of lipoprotein units.

Kopaczyk et al. [54] have reconstituted depolymerized electron transfer complexes with the isolated oligomycin sensitive ATPase. This reconstitution in certain conditions is effective only if Mg^{++} is present; such requirement for Mg^{++} is in agreement with the possibility outlined previously that Mg^{++} bridges may be present in the binding of the ATPase complex to the membrane.

When an ATPase preparation containing phospholipids [55] was incubated with lipid-depleted ETP_H (which are deficient in detachable proteins) the membranes reincorporated phospholipids and recovered both succinoxidase and ATPase activity, thus indicating that reconstitution had occurred (E. Boschetti and G. Lenaz, unpublished data) (table 4).

Formation of mitochondria in vivo

The origin of the mitochondria is still uncertain and controversial [3,56]. The question whether mitochondria derive from division of preexisting organelles, whether they are synthesized de novo from non-membranous constituents, or whether they originate by differentiation of simpler membranes is debated. Yeasts offer a very interesting system to study mitochondrial biogenesis. In this discussion we mainly want to emphasize certain structural aspects during mitochondrial assembly, considered by correlating the development of functional activities with the structures with which such functions are associated.

Asynchrony of development

Our experimental system [57,58] has been *Saccharomyces cerevisiae*, strain ATCC 7754, grown for 16 hr in 5% glucose medium and then transferred into a low glucose medium.

In the whole homogenate there is no respiration with succinate during the early stages of growth, in accord with the known repressive effect of glucose on respiration and cytochrome formation [59–61]; respiration however reaches normal values at the end of

the growth phase, indicating that normal mitochondria have been formed.

We have subjected to differential centrifugation cell homogenates collected at various stages of growth; the heavy particle fraction, sedimenting at 15,000 g, corresponds to the mitochondrial fraction in mature yeast cells. Electron microscopy of this fraction in the glucose-repressed cells shows vesicular structures resembling fragments of plasma membrane; at the end of the derepression stage, typical mitochondria are contained in this fraction. Vesicular structures are also observed in sections of whole cells obtained during the repression stage (A. Castelli and G. Littarru, unpublished data).

Fig. 4 shows the activities of the heavy particle fraction during growth in low glucose; the oligomycin sensitive ATPase increases from low values to a maximum around the 10th hr. Succinate oxidation begins to show up only after the 5th hr and increases to reach the highest values at the end of the growth period considered. Pyruvate dehydrogenase, and other citric cycle enzymes, have low activities at the beginning of derepression, then increase to maxima near the 10th hr.

If we however consider activities in the whole homogenate, we observe an anomalous behaviour of the oligomycin sensitive ATPase; this enzyme, which is typically mitochondrial, has high activity at all stages considered (fig. 5). This is in accord with other experimental findings [62–64].

At the 4th hr of growth, the oligomycin sensitive

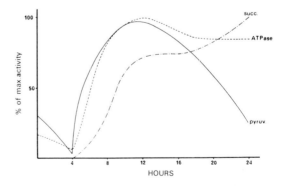

Fig. 4. Specific activities of pyruvate dehydrogenase, succinoxidase and oligomycin sensitive ATPase in the 'heavy fraction' of S. cerevisiae during growth. [58] ———— pyruvic dehydrogenase; –.–.–.–. succinoxidase; – – – – – oligomycin sensitive ATPase.

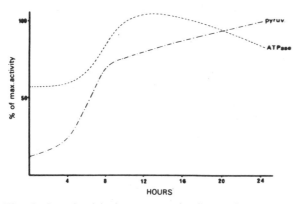

Fig. 5. Pyruvic dehydrogenase and oligomycin sensitive ATPase in whole homogenates of *S. cerevisiae* during growth. [58] –.–.–.–. pyruvic dehydrogenase; – – – – – oligomycin sensitive ATPase.

ATPase is largely recovered in the 105,000 g sediment, whereas the heavy particle fraction has lower specific activity. We cannot exclude that the selective presence of the oligomycin sensitive ATPase in the lighter fraction during repression represents a selective breakage of the fragile mitochondrial vesicles with loss of ATPase in the supernatant.

When the whole membranous fraction obtained at the 4th hr of derepressed growth was layered on a sucrose gradient, two main bands were obtained, which had maxima of ATPase activity at specific densities 1.15 and 1.21 (unpublished data); the lower

band corresponds in our condition to the position of normal mitochondria, whereas the upper band appears comparable with Schatz's promitochondria [63]. Presence of NADPH-cytochrome c reductase in the upper band seems to indicate heavy microsomal contamination.

The P:O ratios in the heavy fraction are low when respiration begins, then increase to reach a plateau around 1 with succinate as substrate after the 10th hr (table 5). Oxidative phosphorylation must be the result of the interaction of the ATPase system with the electron transfer chain. Since the appearance of the ATPase precedes electron transfer, we would expect the same efficiency of phosphorylation if the system become immediately integrated. The low efficiency of phosphorylation in the repressed stage could be the result of higher fragility and damage of mitochondria.

Development of mitochondria from repressed condition is an asynchronous process, and this must be related to asynchrony also in the morphological development. Some indication exists that also electron transfer complexes do not develop at the same rate: Antimycin-sensitive NADH-cytochrome c reductase reaches high values only at the end of the logarithmic growth phase whereas cytochrome oxidase develops more rapidly during the growth cycle [58].

In summary, soluble enzymes and ATPase develop very fast during derepression or are already present in

Table 5

Oxidative phosphorylation in the mitochondrial fraction during the growth of *Saccharomyces cerevisiae* in a low glucose medium

Time [a] (hr)	Succinate oxidation (μatoms O_2/min/mg)	Phosphate esterification (μmoles P_i/min/mg)	Oligomycin-sensitive ATPase activity (μmoles P_i/min/mg)	P : O ratio
0	0.016	–	0.501	–
4	0	–	0.179	–
5	0.028	0.013	0.611	0.48
6	0.081	0.051	1.475	0.63
8	0.160	0.099	2.554	0.62
10	0.260	0.231	2.780	0.89
12	0.285	0.316	2.930	1.11
18	0.295	0.283	2.439	0.96
24	0.440	0.374	2.504	0.85

[a] After growth on 5% glucose medium, the repressed cells were transferred at 0 hr to fresh 0.3% glucose medium.

yeast repressed by high glucose; respiratory proteins are slower to develop although the different complexes of the chain may be synthesized at different times.

Lipid composition

Several investigations have been performed on the lipid composition of yeast cells grown in different conditions and particularly in the repressed state in comparison with a normal state of the mitochondria [65—68]. These studies are relevant to the problem of mitochondrial formation. Great differences have been reported in fatty acid composition between different physiological conditions: in glucose-repressed or anaerobic yeast, saturated fatty acids are in high percentage, and unsaturated fatty acids are decreased or absent. During derepression there is a change towards a normal unsaturation of the fatty acids (which in yeast are almost completely mono-unsaturated).

The phospholipid content of whole cells and of isolated mitochondria in yeast under different culture conditions appears variable in different studies. In general, repression induces a decrease of phospholipids in yeast cells, but this is a clear indication of the decrease of total membranes and relative increase of soluble proteins. An isolated mitochondrial fraction from glucose-repressed cells appears to have a decreased phospholipid content [69]; promitochondria have normal or even increased phospholipid content [65]. It seems difficult to reconcile these differences, although the P content of the organelle, and not of the membrane, has been considered; a different proportion of soluble proteins to membranes can be on the basis of these discrepancies.

Hypotheses on the assembly of the mitochondrial membranes

Reconstitution experiments in vitro have demonstrated that membranes may be assembled in several ways; which is the way for membrane assembly followed during natural membrane biogenesis? A crucial point to understand mitochondrial assembly is the following question: which is the structure of the membrane(s) in the repressed condition?

An answer to this question does not appear easy, since there may be more than one membrane present and they may have different structures. There are at least two membrane systems in the mitochondrion (inner and outer membrane) and there are reasons to believe that the inner membrane is also distinguishable in two membrane systems (cristae and inner boundary membrane: cf. D.E. Green, this symposium). Electron microscopic and enzymatic studies point out that different relations may exist between these membrane systems in different states (there are fewer cristae in repressed yeast). Do the membranes also have a different structure? During derepression is there addition of a new different membrane, or is there a modification in complexity of a preexisting membrane? A further complication may derive from the possibility that the enzymes which appear to be lacking in repressed yeast may be present in an inactive form.

A careful study of the physicochemical structure of the membrane will be able to differentiate between various possibilities and give a clue to the understanding of membrane assembly.

Certain considerations are in line with the possibility that the mitochondrial membrane in repressed or anaerobic yeast may be a simpler system than in mature cells.

At the initial stage of mitochondrial formation, respiration is absent but the oligomycin sensitive ATPase is very active in particles which do not have significantly high amounts of the other mitochondrial activities studied; the oligomycin sensitivity suggests the membranous character of this ATPase but the lack of respiration, if associated to lack of the respiratory enzymes in the membrane, suggests incompleteness of the membrane. The described isolation in vitro from BHM of 'pseudomembranes', that is to say of membranes constituted of a headpiece-stalk sector linked to phospholipid vesicular bilayers, suggests that at this stage we might have a similar type of 'pseudomembranes' in vivo. We might postulate that a simple 'unit membrane' is formed, in which a stable bilayer of mostly saturated lipids is linked to newly synthesized or preexisting soluble proteins which dispose themselves on its surface(s). This protein in the inner mitochondrial membrane is the oligomycin sensitive ATPase.

During derepression of glucose-repressed yeast there is the synthesis of the proteins of the electron transfer chain. These proteins are accepted into the phospholipid bilayer, which is accordingly modified

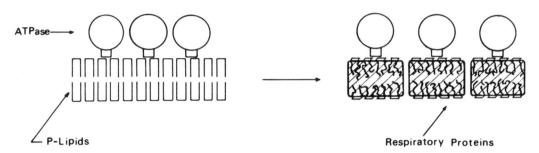

ATPase→

P-Lipids

Respiratory Proteins

Fig. 6. A scheme for the possible evolution of the inner membrane during mitochondrial formation in *S. cerevisiae*.

and also must increase in size during differentiation by acquisition of newly synthesized unsaturated phospholipids; these proteins become linked in discrete lipoprotein repeating units. This possibility is schematically outlined in fig. 6. Only when the whole mitochondrial membrane becomes fully integrated in its structure, may oxidative phosphorylation become active.

Clearly this scheme is applicable to the evolution of a simple preformed membrane. One further step will be to explore the genesis of this simple membrane. The assembly of mitochondria in vivo would be similar to certain types of reconstruction in vitro, described in the first part, where a membranous ATPase binds to depolymerized 'basepieces' or to lipid-depleted membranes.

Several considerations and experimental findings are in line with the assumptions of this model, but others are inconclusive. In order to examine these assumptions, we must obtain an exact knowledge of the lipid content and composition of authentic purified mitochondrial membranes in the repressed and normal condition, and establish a clear morphological and functional correlation of the same membranes.

The precise modalities of the assembly of the respiratory proteins and its organizational information are not known, although several hypotheses have been made on the role of informational protein molecules. Also the importance of the phospholipids must be emphasized, and one might ask if there could be a primary organizational information in the synthesis of differently saturated phospholipids.

The correspondence between changes in membrane lipids and proteins induced by repression or derepression, even if not linked by a cause to effect relationship, appears to increase the stability of the membranes in the two conditions, and could be one of the keypoints for membrane assembly in vivo.

Acknowledgement

This investigation has been supported in part by a grant of the Consiglio Nazionale delle Ricerche, Italy.

References

[1] W.B. Wood, R.S. Edgar, J. King, I. Lielausis and M. Henninger, Federation Proc. 27 (1968) 1160.

[2] D.O. Woodward, Federation Proc. 27 (1968) 1167.

[3] T.S. Work, J.L. Coote and M. Ashwell, Federation Proc. 27 (1968) 1174.

[4] D.E. Green and A. Tzagoloff, J. Lipid Res. 7 (1966) 587.

[5] D.E. Green, N.F. Haard, G. Lenaz and H.I. Silman, Proc. Natl. Acad. Sci. U.S. 60 (1968) 277.

[6] J.F. Danielli and H. Davson, J. Cell. Comp. Physiol. 5 (1935) 495.

[7] S. Fleischer, B. Fleischer and W. Stoeckenius, J. Cell Biol. 32 (1967) 193.

[8] D.E. Green and S. Fleischer, Biochim. Biophys. Acta 70 (1963) 554.

[9] J.D. Robertson, Biochem. Soc. Symp. 16 (1959) 3.

[10] A.A. Benson, J. Am. Oil Chem. Soc. 43 (1966) 265.

[11] D.E. Green and J.F. Perdue, Proc. Natl. Acad. Sci. U.S. 55 (1966) 1295.

[12] E.D. Korn, in: Theoretical and Experimental Biophysics, Vol. 2, ed. A. Cole (M. Dekker, New York, 1969) p. I.

[13] S. Fleischer and B. Fleischer, in: Methods in Enzymology, Vol. X, ed. R.W. Estabrook and M.E. Pullman (Academic Press, New York, 1967) p. 406.

[14] G. Lenaz, A.M. Sechi, L. Masotti and G. Parenti-Castelli, Biochem. Biophys. Res. Commun. 34 (1969) 392.

[15] P.H. Von Hippel and T. Schleich, in: Structure and Stability of Biological Macromolecules, ed. S.N. Timasheff and C.D. Fasman (M. Dekker, New York, 1969) p. 417.

[16] G. Nemethy, Angew. Chemie (Intern. Ed.) 6 (1967) 195.

[17] V. Marinetti, J. Lipid Res. 3 (1962) I.

[18] A.G. Gornall, C.J. Bardawill and M.M. David, J. Biol. Chem. 177 (1949) 751.

[19] R.L. Lester and S. Fleischer, Biochim. Biophys. Acta, 47 (1961) 358.

[20] G. Lenaz, Ital. J. Biochem. 19 (1970) 54.

[21] F.A. Vandenheuvel, J. Am. Oil Chem. Soc. 43 (1966) 258.

[22] L.L.M. Van Deenen, in: Regulatory Functions of Biological Membranes, ed. J. Järnefelt, BBA Library, Vol. II (Elsevier, Amsterdam, 1968) p. 72.

[23] S. Razin, H.J. Morowitz and T.M. Terry, Proc. Natl. Acad. Sci. U.S. 54 (1965) 219.

[24] S.H. Richardson, H.O. Hultin and S. Fleischer, Arch. Biochem. Biophys. 105 (1964) 254.

[25] A.D. Brown, J. Mol. Biol. 12 (1965) 491.

[26] D.F.H. Wallach and A.S. Gordon, in: Regulatory Functions of Biological Membranes, ed. J. Järnefelt, BBA Library, Vol. II (Elsevier, Amsterdam, 1968) p. 87.

[27] D. Chapman and V.B. Kamat, in: Regulatory Functions of Biological Membranes, ed. J. Järnefelt, BBA Library, Vol. II (Elsevier, Amsterdam, 1968) p. 99.

[28] G. Lenaz, A.M. Sechi, G. Parenti-Castelli and L. Masotti, 6th FEBS Meetings, 1969, Abstr. 201.

[29] G. Lenaz, N.F. Haard, A. Lauwers, D.W. Allmann and D.E. Green, Arch. Biochem. Biophys. 126 (1968) 746.

[30] M. Hansen and A.L. Smith, Biochim. Biophys. Acta 81 (1964) 214.

[31] A. Casu, G. Nanni and V. Pala, Ital. J. Biochem. 17 (1968) 301.

[32] G. Nanni, A. Casu, U.M. Marinari and I. Baldini, Ital. J. Biochem. 18 (1969) 25.

[33] A. Casu, G. Nanni, U.M. Marinari, V. Pala and R. Monacelli, Ital. J. Biochem. 18 (1969) 154.

[34] H. Fernandez-Moran, P.V. Blair, T. Oda and D.E. Green, J. Cell Biol. 22 (1964) 63.

[35] Y. Kagawa and E. Racker, J. Biol. Chem. 241 (1966) 2475.

[36] R.S. Criddle, R.M. Bock, D.E. Green and H.D. Tisdale, Biochemistry 1 (1962) 827.

[37] G. Lenaz, A. Lauwers and N.F. Haard, Federation Proc. 26 (1967) 155.

[38] C.P. Lee and L. Ernster, in: Symposium: Regulation of Metabolic Processes in Mitochondria, BBA Library, 7 (Elsevier, Amsterdam, 1966) p. 218.

[39] K.R. Bruckdorfer, P.A. Edwards and C. Green, Europ. J. Diochem. 4 (1968) 506.

[40] K. Kopaczyk, J. Asai, D.W. Allmann, T. Oda and D.E. Green, Arch. Biochem. Biophys. 123 (1968) 602.

[41] W. Jolly, R.A. Harris, J. Asai, G. Lenaz and D.E. Green, Arch. Biochem. Biophys. 130 (1969) 191.

[42] G. Lenaz and D.H. MacLennan, J. Biol. Chem. 241 (1966) 5260.

[43] K.S. Ambe and F.L. Crane, Science 129 (1959) 98.

[44] M.L. Das and F.L. Crane, Biochemistry 3 (1964) 696.

[45] G. Rouser, G.J. Nelson, S. Fleischer and G. Simon, in: Biological Membranes, ed. D. Chapman (Academic Press, New York, 1968) p. 5.

[46] J.S. O'Brien, Science 147 (1965) 1099.

[47] P. Jurtshuck, I. Sekuzu and D.E. Green, Biochem. Biophys. Res. Comm. 6 (1961) 76.

[48] G. De Pury and F.D. Collins, Chem. Phys. Lipids 1 (1966) 1.

[49] G. De Pury and F.D. Collins, Chem. Phys. Lipids 1 (1966) 20.

[50] R.C. Stancliff, M.A. Williams, K. Utsumi and L. Packer, Arch. Biochem. Biophys. 131 (1969) 629.

[51] D.E. Green and A. Tzagoloff, Arch. Biochem. Biophys. 116 (1966) 293.

[52] A. Tzagoloff, D.G. McConnell, D.H. MacLennan and D.E. Green, J. Biol. Chem. 242 (1967) 2051.

[53] E. Racker, Federation Proc. 26 (1967) 1335.

[54] K. Kopaczyk, J. Asai and D.E. Green, Arch. Biochem. Biophys. 126 (1968) 358.

[55] A. Tzagoloff, K.H. Byington and D.H. MacLennan, J. Biol. Chem. 243 (1968) 2405.

[56] M.E. Pullman and G. Schatz, Ann. Rev. Biochem. 36 (1967) 359.

[57] G. Lenaz, G.P. Littarru and A. Castelli, FEBS Letters 2 (1969) 198.

[58] A. Castelli, G. Parenti-Castelli, E. Bertoli and G. Lenaz, Ital. J. Biochem. 18 (1969) 35.

[59] P.G. Wallace, M. Huang and A.W. Linnane, J. Cell Biol. 37 (1968) 207.

[60] C. Chapman and W. Bartley, Biochem. J. 107 (1968) 455.

[61] J. Jayaraman, C. Cotman, H.R. Mahler and C.W. Sharp, Arch. Biochem. Biophys. 116 (1966) 224.

[62] A. Goffeau, H. Heslot, J. Greenawalt, W. Balcavage and J. Mattoon, 6th FEBS Meetings, Abstr. 403, 1969.

[63] R.S. Criddle and G. Schatz, Biochemistry 8 (1969) 322.

[64] K. Watson, J.M. Haslam, B. Veitch and A.W. Linnane, this symposium.

[65] F. Paltauf and G. Schatz, Biochemistry 8 (1969) 335.

[66] L. Kovac, J. Subik, G. Russ and K. Kollar, Biochim. Biophys. Acta 144 (1967) 94.

[67] D. Jollow, G.M. Kellerman and A.W. Linnane, J. Cell Biol. 37 (1968) 221.

[68] G. Serlupi-Crescenzi and S. Barcellona, Ann. Ist. Sup. Sanità 2 (1966) 431.

[69] H.B. Lukins, D. Jollow, P.G. Wallace and A.W. Linnane, Aust. J. Exptl. Biol. Med. Sci. 46 (1968) 651.

STRUCTURAL AND ENZYMIC DEVELOPMENT
OF BLOWFLY MITOCHONDRIA

L.M. BIRT

Department of Biochemistry School of General Studies,
Australian National University, Canberra, A.C.T. Australia.

Adult development in Diptera requires the formation of a new and highly-specialised group of mitochondria – the thoracic sarcosomes – which will provide ATP for flight in the mature insect. This paper presents a description of some of the processes involved in the maturation of the sarcosomes of the sheep blowfly, *Lucilia cuprina*, in which the sarcosomal protein constitutes some 20% of the total protein of the adult.

The origin of the sarcosomes

The flight muscle of the adult appears to develop on a 'template' of larval muscle, as the result of the accretion and fusion of large numbers of myoblasts derived from imaginal discs activated after the pupal/adult apolysis. These cells contain elongated mitochondria about 1μ in length and 0.2μ in diameter, with relatively few cristae. They pass into the sarcoplasm of the developing muscle where they remain among the developing fibres with little change in appearance until about 2–3 days before eclosion (fig. 1a). At that time a rapid elaboration of structural and enzymic material begins, providing the mature mitochondria about 4 days after eclosion; no information about the changes in the total number of thoracic mitochondria per insect is available for L-ucilia, though in another blowfly, Phormia, it has been reported that there is no increase in the first week of adult life [1].

Structural elaboration of the sarcosomes

While it is difficult to provide accurate estimates of increases in mitochondrial size from an inspection of electron micrographs, it is apparent (fig. 1) that there is considerable enlargement during development [2]. Thus, before emergence, fractionation of the mitochondrial population on the basis of the diameter of isolated mitochondria [3,4] reveals that few of the particles have a diameter greater than 1μ; at emergence, about 50% exceed this size, while in the mature fly up to 88% do so (fig. 2; data from [4] and Christiansen and Birt, unpublished). Comparison of micrographs of tissue before and after emergence (fig. 1, [2]) indicates that the overall increase in volume may be of the order of 10-fold. At the same time, the number of cristae per mitochondrion and the complexity of their alignment increase markedly, especially over the period of emergence, when there is a 2.5-fold increase in their frequency. At the end of the period, the characteristic pattern of aligned fenestrations [5] is clearly visible [2].

Negatively stained preparations [6] indicate that the cristal membranes detected in the sectioned material have been assembled from a relatively disorganised mass of precursor material (figs. 3–4). Before emergence, the matrix of most of the mitochondria contains diffuse bands of material without any precise structural arrangement (fig. 3); after emergence, the densely-packed cristae seen in sectioned specimens are clearly defined membranes bearing elementary particles (fig. 4).

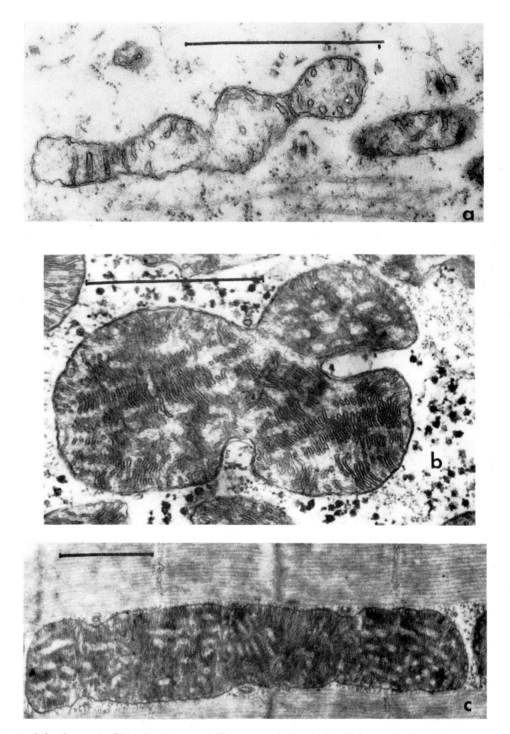

Fig. 1. Structural development of blowfly sarcosomes. Sarcosomes in developing flight muscle from (a) pharate adult three days before emergence, (b) at emergence, (c) five days post-emergence. All specimens are from longitudinal sections after osmium fixation. The length of the bar on each print represents 1 μ.

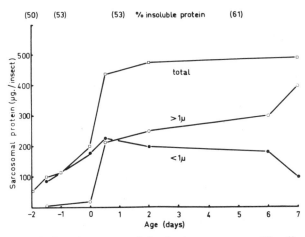

Fig. 2. Protein content of developing sarcosomes. The distribution of protein in sarcosomes isolated from thoraces of developing adults by the method of Smith [3]; basic data from [4] and Christiansen and Birt (unpublished). Day 0 is the time of adult eclosion. The % insoluble protein in sarcosomes at different stages of development was estimated with the detergent Triton-X 100 [23].

Table 1
Phospholipid content of developing mitochondria.

| | Age (days) | | | | | |
	−2	−1	0	1	4	5
	% of lipid P present in fraction					
Cardiolipin	13	12	–	17	15	–
P-ethanolamine	40	47	–	48	51	–
P-serine	10	5	–	8	7	–
P-inositol	–	11	–	15	14	–
P-choline	30	23	–	19	21	–
Total (μg/insect)	1.4	1.7	3.3	5.8	6.8	7.0
$\dfrac{\mu\text{g lipid P}}{\text{mg protein}}$	25	15	8	14	16	16

Mitochondria were prepared by a procedure involving treatment of the homogenates with Nagarse (Christiansen and Birt, unpublished) and the phospholipids extracted by the method of D'Costa and Birt [11]. Separation of classes of phospholipids was achieved by thin layer chromatography; components not identified positively (amounting to less than 10% of the total lipid P) are not tabulated.

Changes in physical and chemical properties

The rapid elaboration of mitochondrial structure at adult emergence is coincident with an abrupt increase in the mean specific gravity (from 1.184 to 1.193) of the mitochondrial population and a corresponding widening of the range of mitochondrial specific gravities [7]. These changes are explicable in terms of the preferential accumulation of protein by the particles, which in consequence become relatively poorer in phospholipids, as indicated by the decline in the ratio of phospholipid P/protein from about 20 μg lipid P per milligram protein 2 days before emergence to about 10 at emergence (table 1). The only marked variation in the composition of the phospholipid during development appears to be in the proportion of phosphatidyl inositol, which increases from a very low level before emergence to about 15% of the total at maturity. It is noteworthy that despite the considerable increases in the amount of the inner membrane [2] and in respiratory activity [4], there is little increase in the relative concentration of cardiolipin. There is no significant variation in the proportion of protein insoluble in Triton-X 100 over the period of emergence (when it remains at about

50%) though it does increase (to about 61%) in the mature fly (fig. 2). The accumulation of protein at emergence is parallelled by an increase in the amount of mitochondrial DNA (table 2, data from [8] and Christensen and Birt, unpublished). As the result of this increase, DNA becomes more concentrated at emergence in the smaller mitochondrial fraction (less than 1 μ in diameter), i.e. the DNA appears first in the fraction of the population undergoing the most rapid structural and functional development. The ratio of mitochondrial RNA/DNA is also greatest [4,8] in this fraction.

In contrast to these increases in the total amounts of DNA, protein and phospholipid is the steady level of at least one other component. The total amount of NAD present in the sarcosomal population is almost constant (at about 1.5 mμmole/insect) during development, so that its concentration declines from about 11 mμmoles per mg protein in the pharate adult to about 4 mμmoles per mg protein in the mature fly; it reaches its highest concentrations in the smaller mitochondrial fraction before emergence [10]. Thus it appears probable that the amounts of some mitochondrial components present in high concentration

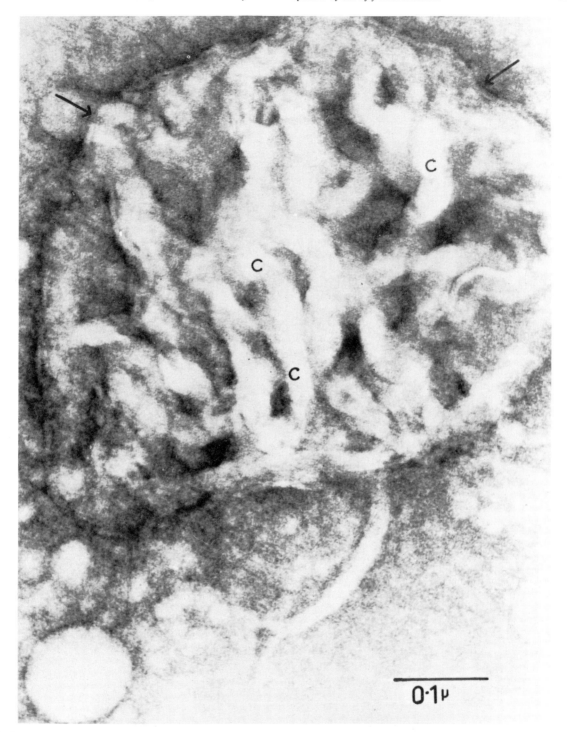

Fig. 3. Sarcosomal structure in negatively-stained preparations. An immature mitochondrion from tissue 2 days before eclosion negatively stained with potassium phosphotungstate. The outer membrane (arrows) is clearly visible in places, and there are relatively few cristae (c). The length of the bar represents 0.1 μ.

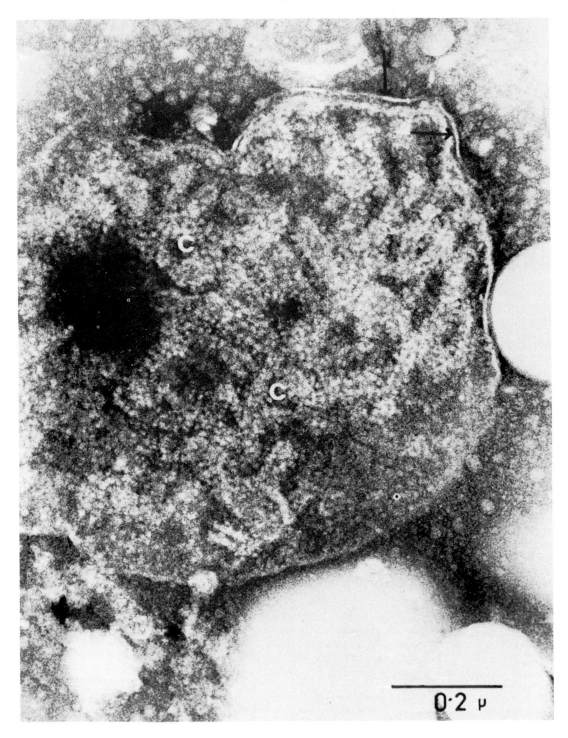

Fig. 4. Sarcosomal structure in negatively-stained preparations. A mature mitochondrion from tissue 2 days before eclosion negatively stained with potassium phosphotungstate. The outer membrane (arrows) is clearly defined and the cristae bear elementary particles. The length of the bar represents 0.2 μ.

Table 2
Nucleic acid content of developing sarcosomes

	Age								
	1 day before emergence			At emergence			1 day after emergence		
	$<1\mu$	$>1\mu$	Total	$<1\mu$	$>1\mu$	Total	$<1\mu$	$>1\mu$	Total
(a) μg DNA/mg protein	1.0	4.3	1.3	4.0	5.3	4.7	2.7	3.1	2.9
(b) μg RNA/mg protein	48	36	47	35	29	32	21	16	18
(c) μg protein/insect (calc.)	100	10	110	140	160	300	190	190	380
(d) μg DNA/insect (calc.)	0.10	0.04	0.14	0.56	0.85	1.41	0.51	0.59	1.10
(e) μg RNA/insect (calc.)	4.8	0.36	5.2	4.9	4.6	9.5	4.0	3.0	7.0
RNA/DNA	48	8.4	36.0	8.8	5.5	6.7	7.8	5.2	6.4

Sarcosomal fractions were prepared and the nucleic acids estimated as described in [8]; protein distribution is based on [8] and Christiansen and Birt (unpublished).

in the particles introduced directly from the myoblasts, are not increased during the development of the active sarcosomes.

Changes in enzyme levels

Detection of the particulate α-glycerophosphate dehydrogenase (an enzyme characteristic of the flight muscle sarcosomes) soon after pupation [4] indicates that the mitochondria of the myoblasts possess an active respiratory system of the adult type even at this early stage. However, the period of development over emergence is one of rapid increase in enzymic activity in the mitochondria [4]. At this time, a group of sarcosomal enzymes (α-glycerophosphate, succinic, malic and NAD-linked isocitric dehydrogenases and cytochrome oxidase) increase markedly (some 5- to 10-fold) in activity and more or less synchronously; it has been suggested that they constitute a 'constant proportion group' of enzymes. For all but cytochrome oxidase, the accumulation continues during the first week of adult life.

A detailed study of the changes in the specific activity of one member of this group, α-glycerophosphate dehydrogenase, indicates that incorporation of enzyme is not synchronous with the total incorporation of protein [4]. Thus, before emergence an enlargement of the mitochondria which increased the diameter of the isolated particles above 1μ produced a sharp decline in the specific activity of the dehydro-

genase; subsequently this activity rose again. As most of the mitochondrial population becomes larger than 1μ over eclosion, at which time the most striking increase in total mitochondrial protein also occurs, it is possible to represent the incorporation of enzymic and non-enzymic protein into the typical sarcosome in the diagrammatic form of fig. 5 (see also [4]). The essential feature of this representation is the extremely large and rapid incorporation of non-dehydrogenase protein at emergence, when the elaboration of cristal structures is also proceeding apace. The linkage of the dehydrogenase enzymes to a functional respira-

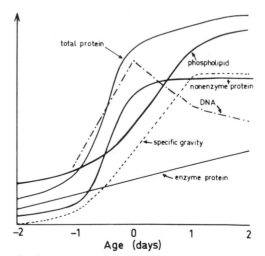

Fig. 5. Changes in sarcosomal composition at eclosion. Eclosion at day 0.

tory chain appears to be dependent on this incorporation and elaboration, as estimations of the relative specific activity of α-glycerophosphate oxidase and dehydrogenase [7] indicate that initially, dehydrogenase can be accumulated in excess of the capacity of the particles to provide a functional respiratory chain (fig. 6). In the pharate adult only, an increase in specific gravity and, as evidenced by studies of negatively stained fractions, in the structural organization of cristal membranes, produce almost equal activity of oxidase and dehydrogenase. Thereafter, the continuing accumulation of dehydrogenase protein after that of non-dehydrogenase ('structural' and 'proenzyme' — see later) protein has ceased, produces sarcosomes in which the oxidase is always of lower specific activity than the dehydrogenase. Thus the accumulation of dehydrogenase protein and of the complex of oxidase proteins can proceed independently.

In contrast, such oxidase activity as the mitochondria of the pharate adult possess appears to be tightly coupled to phosphorylation [7]. With α-glycerophosphate as substrate, the insensitivity to ADP which characterizes the sarcosomes of mature flies [7] develops over the period of emergence; whereas with pyruvate, also an important exogenous respiratory substrate, tight coupling of the two processes is maintained in the adult. The declining dependence of α-glycerophosphate oxidase on ADP is paralleled by an increasing dependence on the availability of calcium (table 3), a requirement which has an obvious physiological importance in the control of the initiation of respiratory activity in flight [7]. It seems likely therefore that the initial molecular structure of the coupled α-glycerophosphate oxidase is modified during the maturation of the mitochondria, producing a physiologically important variation in the properties of this particular oxidase system. Such a modification could result from an alteration either in the composition of the enzymatic assembly (e.g., in either the protein or phospholipid components) or in its physical conformation.

Although all the enzymes described become steadily more active as the fly matures, this does not appear to be true for at least one sarcosomal enzyme system. The energy requirements of the developing pharate adult are met almost exclusively by the oxidation of fatty acids [11]; the oxidase system for degrading these becomes detectable in thoracic mitochondria some 2–3 days before eclosion, increases rapidly in activity until emergence, and then declines so that about 4 days after emergence it is virtually undetectable [12]. This change is not the result of a dilution of the activity by increasing amounts of other mitochondrial proteins, as there is an absolute loss of activity. It has not been possible to exclude completely the possibility that the thoracic preparations contain a separate group of mitochondria (distinct from the flight muscle sarcosomes) which is responsible for these changes. However, the total mitochon-

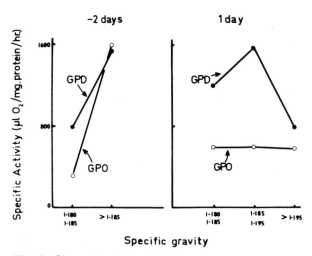

Fig. 6. Glycerophosphate dehydrogenase and oxidase of developing sarcosomes. Sarcosomal fractions were prepared in sucrose density gradients from insects before and after eclosion. The enzyme assays are described in [7].

Table 3
Dependence of glycerophosphate oxidase on calcium

Age (days)	O$_2$ uptake % maximum rate [a]
−1	33
+1	50
+8	100

[a] $\dfrac{O_2 \text{ uptake with } Ca^{++}}{O_2 \text{ uptake with ADP plus } Ca^{++}}$ as %.

Sarcosomes were isolated from insects before and after eclosion [8] and the influence of ADP (5 mM) and Ca^{++} (1 μM) tested in the presence of glycerophosphate (2 mM).

drial population cannot be fractionated on the basis of differences in specific gravity in such a way as to separate the α-glycerophosphate- and fatty acid-dehydrogenases, although a survey [2] of the ultra-structure of particles in tissue preparations indicates that a small proportion (about 3%) could be distinguished in appearance from the bulk of the mitochondria. Moreover, the activity of the fatty acid oxidase calculated on the basis of the protein content of the entire sarcosomal population (about 80 μl O_2/mg protein/hr) is directly comparable to those reported for other types of mitochondria [12] so that if it were assumed that the structural heterogeneity reflects the presence of a specialised group of particles oxidising fatty acids, the specific activity would rise very considerably (to values of about 2,500 μl O_2/mg protein/hr). These considerations suggest that in fact the sarcosomes themselves are responsible for the oxidation. The decline in activity after emergence is not a consequence of an increasing impermeability, as sonicated preparations were no more active than the original material. Thus it appears that not only does the incorporation of various mitochondrial proteins proceed independently, but also that particular enzymes can be removed when their usefulness has ended, without loss of the function of the entire mitochondrial system.

Biosynthetic capacity of the sarcosomes

The sequence of events described indicates that the developing sarcosomes might provide an extremely useful tool for the study of the biosynthetic capacities of mitochondria and accordingly, an investigation of the capacity of the sarcosomes isolated at various stages of development for protein synthesis was undertaken. Initial experiments in vitro suggested that ^{14}C-labelled amino acid could be introduced into the sarcosomal protein, but a more critical examination indicated that this apparent incorporation was the result of extremely firm binding of the labelled amino acid by the sedimented mitochondrial material. When more rigorous washing procedures were employed, no incorporation of amino acid could be demonstrated despite a great number of variations in many of the experimental conditions, e.g., in the age of the insect, the concentrations of individual

amino acids (in some experiments these were present at the same concentrations as in the intact insect [13]), the concentration and type of ^{14}C-labelled amino acid, the source of ATP in the incubation mixture, the concentrations of cofactors and coenzymes and the aeration of the incubation medium. Because of the unusually restricted permeability of the sarcosomes [14], the effect of brief sonication was also tested; it did not enhance the incorporation. (Preparations of rat liver mitochondria incubated under standard conditions with the same reagents, incorporated ^{14}C-leucine at rates almost identical to those reported by other workers.) The only possible conclusion from these experiments was that the sarcosomes have lost much of their biosynthetic capacity.

This conclusion must be viewed in the light of the available evidence about the formation of adult proteins in the entire developing organism. Other studies (Walker and Birt [15] and unpublished data) have revealed that the injection of such inhibitors as puromycin, chloramphenicol, cycloheximide and fluorophenylalanine during the period of most intense structural and enzymic development (i.e. the last three days in the puparium) has no significant effect on the emergence of the adult insect and almost none on the increase in activity of mitochondrial enzymes such as the α-glycerophosphate and pyruvic oxidases, succinic dehydrogenase and oligomycin-sensitive ATPase (e.g., see table 4). Injection of actinomycin just before emergence did inhibit the development of both α-glycerophosphate dehydrogenase and oxidase appreciably, suggesting some dependence on the synthesis of RNA and possibly protein. In contrast, injection of the inhibitors at the end of larval and the beginning of pupal life prevented normal adult development; at this time an immediate inhibition of the formation of individual enzymes which normally accumulate rapidly in the pupa (such as lipase) can also be observed. Thus, as for other blowflies (e.g., *Phormia* [16]) the entire period of rapid adult development in *Lucilia* appears to be one in which protein synthesis is quantitatively unimportant, suggesting that maturation is largely the result of a reorganisation of existing reserves. The available data suggest that the adult proteins, including those of the sarcosomes, are formed by the mature larva and the pupa and are stored in masses of

Table 4
Effect of inhibitors of protein synthesis on sarcosomal enzyme formation

Inhibitor	Enzyme	% inhibition of development after injection at		
		pupation	2 days	4 days
Puromycin	α-glycerophosphate dehydrogenase	13	2	2
	α-glycerophosphate oxidase	39	25	0
Cycloheximide plus chloramphenicol	α-glycerophosphate oxidase	82	35	2

Inhibitors were injected into batches of insects at pupation, and at 2 and 4 days after pupation; the effect of the inhibitor was assessed by comparing enzyme activities in these insects with those in sham-injected controls. The enzyme activity was estimated at eclosion by the methods described in [7].
The concentrations used were: puromycin, 1 μg/insect; cycloheximide 0.2 μg/insect; chloramphenicol, 1 μg/insect.

subcellular vesicles from which they are released at the appropriate stage in adult development.

The pattern of sarcosomal development in lucilia

The data described in this paper may be summarised as follows. Sarcosomes derived from myoblasts originating from the imaginal discs remain structurally simple and with relatively inactive enzyme systems until about three days before eclosion when rapid development is initiated, possibly by rising levels of prothoracic gland hormones [17]. These sarcosomes contain a full complement of NAD. When development is initiated, a massive incorporation of material commences, firstly of enzymic protein (dehydrogenases) then of DNA and the nonenzymic protein and phospholipid required for the assembly of the more complex oxidase and phosphorylation systems. In this phase, the accumulation of protein and phospholipid proceeds asynchronously and the characteristic ultrastructural organisation of the mature particles emerges. Further development involves the continued accumulation of adult enzymes and the loss of others whose function is primarily developmental.

The sarcosomal proteins are apparently derived almost exclusively from polypeptide units formed at an earlier stage of the insect's development and stored until the assembly of the adult commences [18]. The steady increase in the total mitochondrial enzyme protein during maturation, the absence of mitochondrial-type enzymes from the sarcoplasm, and the

dilution of activity by a massive incorporation of non-enzymic protein at emergence, suggests that the particles are acquiring both structural protein and inactive precursors of enzymes (pro-enzymes) from the reserve materials. At the earliest stages of sarcosomal development, the accumulation of this material from the sarcoplasm may be responsible for the disorganised appearance of the negatively-stained inner membranes of immature mitochondria. Its heterogeneous nature (comprising both structural and pro-enzyme protein) may be reflected in the constancy of the proportion of the detergent-soluble protein in the sarcosomes throughout the phase of rapid elaboration of cristal membranes, in which the proportion of insoluble protein might be expected to rise; if the pro-enzymes are soluble in detergent, their incorporation over emergence will contribute to the dilution of the enzyme activity and also maintain the proportion of soluble protein in the mitochondrial population.

The pattern of development also requires that the rate of activation of the pro-enzymes be slower than the rate at which they are incorporated from the sarcoplasm and also that the activation occur within the developing particle. Possible mechanisms for the activation are:

1) By the addition of a cofactor or coenzyme to the proenzyme. Such a process may be inferred from preliminary experiments, which indicate that α-glycerophosphate oxidase activity can be elicited in preparations from immature insects by the addition of a heat-stable activator present in homogenates of rapidly developing adults.

2) By the combination of polypeptide units derived directly from the larval reserves to form new quaternary protein structure [19,20]. The inhibition of enzyme formation by actinomycin suggests that protein newly synthesized in the pharate adult may also have a role in such combination, perhaps as a type of 'specifier' protein [20]; but the amount involved must be relatively small.

3) By the attachment of polypeptides to specific sites on the developed mitochondrial membranes i.e., by an 'allotopic' activation [21].

The apparent inability of the highly specialised Dipteran sarcosomes to synthesise their own proteins is consistent with the description of the evolutionary origin of mitochondria suggested by Roodyn and Wilkie [22]. These authors have discussed the evidence pointing to a bacterial origin of mitochondria, in which a host cell has established an endosymbiotic association with a bacterial cell. In the course of time, modification of the bacterial cell has produced the mitochondrion, which has lost many of the metabolic capacities of the original bacterium. In many types of mitochondria, the ability to synthesise protein has survived this evolutionary development, but, in the most advanced order of insect evolution, the Diptera, it appears that the capacity of the flight muscle sarcosomes to synthesise protein has been lost almost completely. Thus the pattern of sarcosomal elaboration in *Lucilia* may be seen as a particular reflection of the extreme specialisation of the developmental pattern of the entire insect.

References

[1] L. Levenbook and C.M. Williams, J. Gen. Physiol. 39 (1956) 497.
[2] D.W. Gregory, R.W. Lennie and L.M. Birt, J. R. Microsc. Soc. 88 (1968) 1951.
[3] D.W.E. Smith, Nature (Lond.) 192 (1961) 234.
[4] R.W. Lennie and L.M. Birt, Biochem. J. 102 (1967) 338.
[5] D.S. Smith, J. Cell. Biol. 19 (1963) 115.
[6] A.C. Walker, D.W. Gregory and L.M. Birt, J. Insect Physiol. 15 (1969) 519.
[7] A.C. Walker and L.M. Birt, J. Insect Physiol. 15 (1969) 305.
[8] R.W. Lennie, D.W. Gregory and L.M. Birt, J. Insect Physiol. 13 (1967) 1745.
[9] W.C. McMurray and R.M.C. Dawson, Biochem. J. 112 (1969) 91.
[10] L.M. Birt, Biochem. J. 101 (1966) 429.
[11] M.A. D'Costa and L.M. Birt, J. Insect Physiol. 12 (1966) 1377.
[12] M.A. D'Costa and L.M. Birt, J. Insect Physiol. 15 (1969).
[13] L.M. Birt and B. Christian, J. Insect Physiol. 15 (1969) 711.
[14] J.B. Chappell, Brit. Med. Bull. 24 (1968) 150.
[15] A.C. Walker, L.C. Barritt and L.M. Birt, Proc. Australian Biochem. Soc. (1969) 9.
[16] M.L. Dinamarca and L. Levenbook, Arch. Biochem. Biophys. 117 (1966) 110.
[17] L.C. Barritt and L.M. Birt, J. Insect Physiol. (in press).
[18] R.W. Lennie and L.M. Birt, J. Insect Physiol. 11 (1965) 1213.
[19] J.F. Donnellan and R.B. Beechey, J. Insect Physiol. 15 (1969) 367.
[20] K. Brew, T.C. Vananan and R.L. Hill, Proc. Natl. Acad. Sci. U.S. 59 (1968) 491.
[21] E. Racker, Fed. Proc. 26 (1967) 1335.
[22] D.B. Roodyn and D. Wilkie, in: The Biogenesis of Mitochondria (Methuen and Co., London, 1968).
[23] D.B. Roodyn, Biochem. J. 86 (1962) 177.

FORMATION OF MITOCHONDRIA OF
LOCUSTA MIGRATORIA FLIGHT MUSCLE

Walter KLEINOW, Walter SEBALD, Walter NEUPERT
and Theodor BÜCHER

Institute of Physiological Chemistry and Physical Biochemistry,
University of Munich, Germany

As you may remember, not long ago a well-known pioneer in chondriology called 'the tendency to erect the mitochondria to the dignity of an organelle or cell organ ... a blunder' [1]. Indeed, until the early sixties the multiplication of preexisting mitochondria was just one of several speculations concerning the cellular origin of mitochondria. Other mechanisms postulated by experienced investigators and carefully discussed in Rouiller's and Novikoff's reviews [2,3] were the synthesis de novo and the transformation of other cellular structures, such as the plasma membrane, the nuclear envelope, the endoplasmic reticulum, etc. [2,3]. With respect to multiplication, the splitting of mitochondria had been repeatedly observed [4]. The important question was whether mitochondria can grow [5], i.e., whether their mass could increase by a continuous process of insertion of specific material.

It was obvious that this problem could not be solved by morphological or biochemical investigations alone. Rather, a combination of both methods was required and that on a quantitative basis. In 1962 such studies were started, using different experimental approaches on quite different types of cells or tissues [6–8]. The current findings from our experiments with the wing depressors of the African migratory locust will be reported here. The wing depressors are a pair of strong muscle bundles, situated dorsal and longitudinal in the thorax of this orthopteran insect. They produce most of the work during flight [9].

In the days before and after the imaginal moult, the volume of chondriome in the flight muscle increases 25- to 30-fold during the formation of the muscle from a small precursor. This is mainly due to an increase in volume of the single mitochondrion (fig. 1). Simultaneously, the content of inner membrane per unit of mitochondrial volume increases by a factor of 1.4. Therefore, the mature flight muscle contains about 40 times more mitochondrial membranes than its precursor at the beginning of the preimaginal interval.

In general, the enzyme activities of the cristae and of the matrix spaces increase by this same factor. During all phases of formation, the morphologically visible mitochondrial elements are enzymatically fully equipped [8,11]. This is established by the correlation of histiometric and biochemical results, and is demonstrated here for the final phase of the development of the flight muscle (table 1), the so-called duplication phase. In the beginning of this phase, around the fourth day after the imaginal moult, all histiometric and enzymatic parameters are present in proportions identical to those of the completely formed flight muscle. However, a doubling of all these parameters occurs. This is also illustrated by the cross-sections in fig. 2.

These findings present clear evidence of the ability of pre-existing mitochondria to grow. Moreover, in the beginning as well as at the end of this phase all cellular membranes of sarcolemma, sarcoplasmic reticulum and of nuclear and mitochondrial envelopes are clearly visible and separated from each other. Therefore, any growth by transformation of extra-mitochondrial structures is excluded.

After injection of an insect with labeled amino acids (fig. 3), the label can be found in all protein fractions in about equal proportions. However, the specific activity measured in the flight muscle declines markedly during the final phase. When mito-

days before
imaginal moulting

days after imaginal moulting

8th   1st   1st   3rd   8th

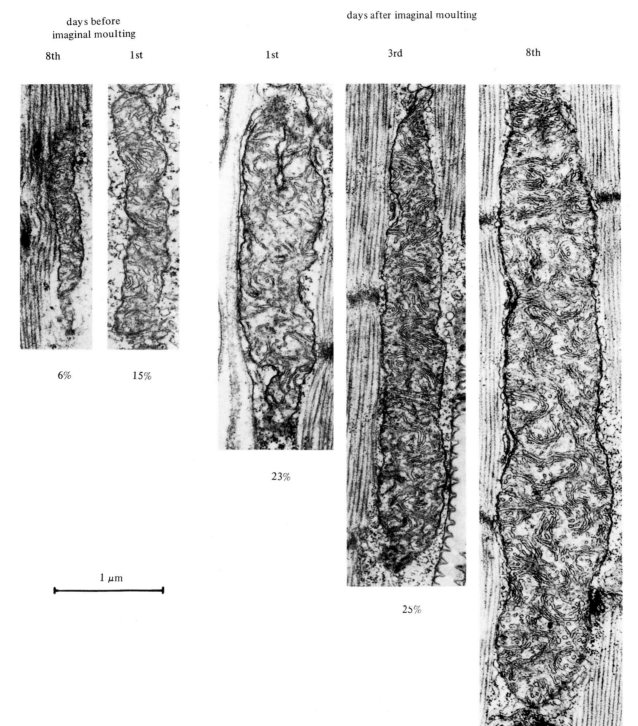

6%   15%

23%

25%

30%

1 μm

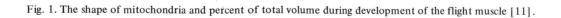

Fig. 1. The shape of mitochondria and percent of total volume during development of the flight muscle [11].

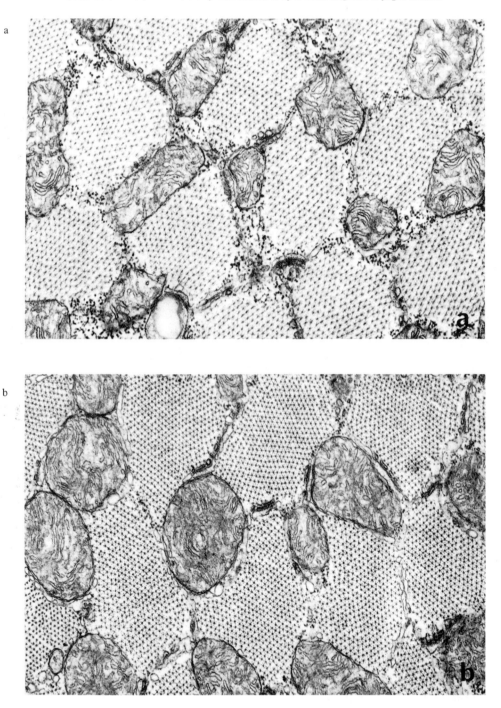

Fig. 2. Cross-sections ($\times$ 30,000) of the dorsal longitudinal muscle at the beginning and the end of the duplication phase, at 4th day (a) and at 8th day (b) after imaginal moult [8].

(a) Cross-sections of fibrils are rounded and clearly distinct. The number of filaments seen on a cross section of a fibril amounts approximately to 200.

(b) Contrary to (a), this cross-section shows a significant increase in the diameter of mitochondria, and their inner structure is more closely packed. The cross sections of fibrils are nearly unchanged, but the number of filaments therein has doubled to about 400 as compared with 200 in (a).

Table 1

Histiometric and enzymic data of the wing depressors at the beginning and the end of the final (duplication) phase of development [a]

Days after imaginal moulting	2–4	8	Increase by
Fresh weight $\times 10^3$ (g)	30	45	factor of
Mitochondria			
Volume fraction occupied by chondriome [b]	25%	30%	
Membrane profile $\times 10^{-3}$ (cm/cm^2) [c]	450	560	
Membrane area (cm^2) [d]	3,600	8,000	2.2
α-Glycerolphosphate oxidase (U) [e]	1.05	2.27	2.2
Cytochrome c (nmoles) [f]	0.75	2.5	3.3
Condensing enzyme (U) [g]	2.40	5.0	2.1
Myofibrils			
Number per muscle fiber	ca 500	800–1,000	1.6–2
Myofilaments per fibril	210±20	430±20	2.1
Glycolytic enzymes			
Glyceraldehydephosphate dehydrogenase (U) [h]	4.40	10.8	2.5
α-Glycerolphosphate dehydrogenase (U) [i]	1.68	3.83	2.3
Aldolase (U) [h]	0.68	1.44	2.1

[a] Average values per muscle pair. U denotes units of enzymic activity. Data from ref. [8].
[b] From planimetry of E-M cross-sections.
[c] Profile density in mitochondrial cross-sections.
[d] Volume % $\times 10^{-2}$ $\times$ membrane profile $\times 4/\pi \times$ fresh weight/1.2.
[e] FAD dependent activity bound on inner membrane.
[f] Protein associated to inner membrane.
[g] Extractable matrix enzyme.
[h] Extractable enzymes mainly located in the I-band regions of myofibrils (10).
[i] NAD dependent activity.

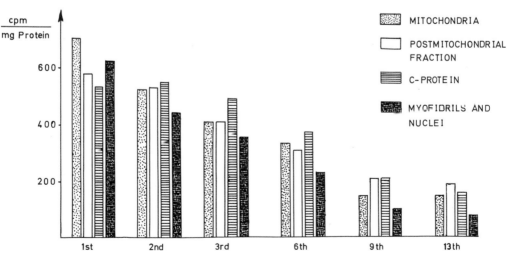

Fig. 3. Incorporation of ^{14}C-isoleucine into protein fractions of the flight muscle at different stages of imaginal development. The specific activities are shown of total mitochondrial protein (first bar), the postmitochondrial fraction (second bar), cytosolic protein (third bar) and myofibrils (fourth bar) 24 hr after the injection of 0.1 μC isoleucine per animal.

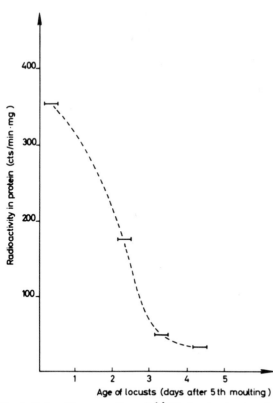

Fig. 4. Rate of incorporation of ^{14}C-isoleucine in vitro into mitochondria isolated at different stages of the developing flight muscle [12].

chondria are isolated in different phases of the muscular development and are incubated with labeled amino acids in vitro under identical conditions, the incorporation rate is observed to decrease substantially during the first days after the imaginal moult. The incorporation rate into mitochondria isolated during the duplication phase is only 10% of that found on the first day after the imaginal moult (fig. 4).

Extrinsic and intrinsic contribution to formation of mitochondria

The observations presented in the preceeding section indicate that the growth of the mitochondrial membrane is coordinated with the growth of other cellular constituents. It is harmonized with the formation of mitochondrial matrix enzymes as well as with the development of the extramitochondrial enzyme

pattern and extramitochondrial structures, such as the myofibrils. To understand the mechanism of this phenomenon we must discriminate experimentally between the extrinsic and intrinsic systems of mitochondrial membrane formation.

When isolated from the insect's flight muscle the mitochondria incorporate amino acids only into the water-insoluble fraction of the membrane proteins [12]. This is in complete agreement with the findings of Roodyn and his followers from their experiments with mammalian mitochondria incubated in vitro [13]. Furthermore, there is a reproducible separation of the insoluble membrane proteins into more than 20 bands by electrophoresis in phenol-formic acid on polyacrylamide gel. The band pattern obtained from flight muscle mitochondria is rather similar to those of mitochondrial membrane proteins from other sources (fig. 5). Only a few of these bands are labeled by incorporation of amino acids in vitro. Analogous to mitochondria of other species, the highest activity appears in band 4 [14,15].

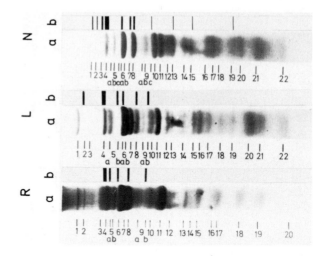

Fig. 5. Electrophoretic patterns after amido black staining (a) and in vitro labeling patterns (b) of insoluble mitochondrial proteins from *Neurospora* wild type (N), *Locusta* flight muscle (L), and rat liver (R). Mitochondria were extracted three times with phosphate buffer after sonication in order to remove soluble proteins. The resulting pellet was extracted with acetone, chloroform/methanol (2:1) and acetone and dissolved in phenol-formic acid-water (2:1:1, w/v/v). Electrophoresis was performed in 7.5% polyacrylamide plates, equilibrated with the same phenol-formic acid medium [14]. Radioactivity was determined after slicing the stained electropherograms (fig. 8).

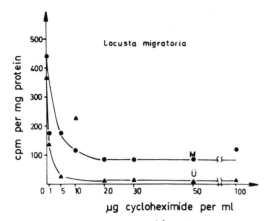

Fig. 6. Incorporation in vivo of ^{14}C-phenylalanine (0.2 μC per locust) into protein of whole mitochondria (M, ●—-—●) and 20,000 g supernatant (U, ▲—-—▲) in the presence of different amounts of cycloheximide. The locusts were killed 30 min after the injection of the ^{14}C-amino acid [15].

It may be concluded from these findings that the contribution by the intrinsic system is restricted not only to the insoluble fraction of mitochondrial proteins, but even to a quantitatively minor part of this fraction.

However, the question arises as to what extent the in vitro experiments on isolated mitochondria are representative of the situation in vivo. To answer this question, imagines of locusts were injected with cycloheximide in order to block the extrinsic (nucle-

ar-dependent) protein synthesis in vivo [15]. In the presence of cycloheximide, there is a residual amino acid incorporation into mitochondrial protein (fig. 6). The incorporation pattern shows negligible incorporation into the extractable proteins of the mitochondrial matrix (table 2), as well as into the extra-mitochondrial proteins (fig. 7b). Incorporation is observed in only a limited number of bands, the main peaks of which are similar to those found in the in vitro incorporation experiments (fig. 8). The pattern obtained in vivo in the presence of cycloheximide is distinctly sharper than that obtained in vitro. This

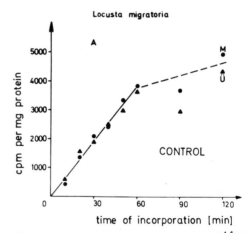

Fig. 7a. Time course of incorporation in vivo of ^{14}C-leucine plus ^{14}C-isoleucine plus ^{14}C-phenylalanine (0.2 μC each per locust) into protein of whole mitochondria (M, ●—-—●) and 20,000 g supernatant (U, ▲—-—▲). Without cycloheximide.

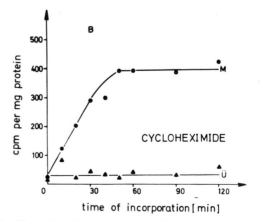

Fig. 7b. After the injection of 50 μg cycloheximide per locust. 50 μg cycloheximide was reinjected every 30 min [15].

Table 2
Specific radioactivities of submitochondrial fractions[a]

Mitochondrial fractions	Mitochondrial labelling (cpm per mg protein)		
	in vivo[b]	in vivo with cycloheximide[c]	in vitro[d]
Insoluble protein	9,200	3,640	11,000
Soluble protein	8,300	540	960

[a] See ref. [15].
[b] 1 hr with 0.4 μC leucine, 0.4 μC isoleucine, and 0.4 μC phenylalanine per locust.
[c] 1 hr in the presence of 50 μg cycloheximide per locust (50 μg cycloheximide was reinjected after 30 min) with 2 μC leucine, 2 μC isoleucine, and 2 μC phenylalanine per locust.
[d] 30 min with 1 μC leucine, 1 μC isoleucine, and 1 μC phenylalanine per ml.

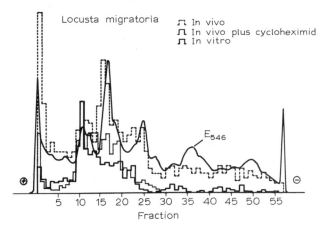

Fig. 8. Densitogram of amido black stained insoluble mitochondrial protein after gel electrophoresis (smooth line), and
distribution of incorporated radioactivity over the electropherograms (edged lines). – – –, in vivo (1 hr with 0.4 μC
leucine, 0.4 μC isoleucine, and 0.4 μC phenylalanine per
locust); ——, plus cycloheximide (1 hr in the presence of
50 μg cycloheximide per locust, with 2 μC leucine, 2 μC
isoleucine, and 2 μC phenylalanine per locust. 50 μg cycloheximide was reinjected after 30 min); ....., in vitro (30 min
with 1 μC leucine, 1 μC isoleucine, and 1 μC phenylalanine
per ml). Radioactivity in band 4 (fraction 11) was taken as
equal [15].

suggests that in the in vivo experiment the number of
incomplete peptide chains is less than in the in vitro
experiment. In principal, however, the findings on
incorporation in vitro and in vivo under action of
cycloheximide are in excellent agreement. Hence, the
conclusions drawn above are supported by two independent methods.

With respect to the coordination of the systems, it
is remarkable, that the incorporation of amino acids
by the intrinsic system continues to operate for at
least 30 min after the extrinsic system is eliminated
by cycloheximide (fig. 7b). This indicates a certain
elasticity in the coupling of the two systems [16].

RNA-pattern

A comparison of the results presented in the
preceeding section with those obtained from *Neurospora* communicated by Sebald and colleagues [17]
reveals a striking similarity in amino acid incorporation of isolated mitochondria, as well as in the mode

of cooperation between the extrinsic and the intrinsic
systems in vivo. Moreover, strictly comparable patterns and kinetics of labeling were found. These
similarities in overall performance suggest corresponding similarities of composition and function of the
intrinsic systems of both types of mitochondria.
Whereas recently the characterisation of mitochondrial ribosomes in *Neurospora* has improved greatly
[18,19], the results on animal tissues are less satisfactory. The current state of our investigations with
Locusta is presented briefly in the following.

After phenolization of the flight muscle homogenate or of the pellet and the supernatant from the
homogenate after centrifugation (fig. 9), 8 nucleic
acid bands can be differentiated by electrophoresis on
polyacrylamide gel after staining with toluidine blue.

One of the bands originating from the pellet
fraction disappears upon treatment of the homogenate with DNase. It seems to be a low molecular
weight DNA and is called 'DNA-s' (DNA small). The
other seven bands are sensitive to RNase if the
enzyme is applied to the purified nucleic acid preparation. Following centrifugation of the homogenate, a
band corresponding to tRNAs was found in both the
pellet and the supernatant. The tRNA band found in
the pellet exhibits a greater average electrophoretic
mobility than the band in the supernatant. In addition, there is a band of RNA that shows a position
characteristic of 5S ribosomal RNA, corresponding to
a molecular weight of 40,000. Under the Mg-free
conditions described below, this RNA band appears
mostly in the supernatant. It is not detectable in
RNA preparations from mitochondria from which
most of the adhering ribosomes of the extramitochondrial system have been removed.

A discrimination between cytosolic and mitochondrial contribution to the RNA patterns is difficult
because under the conditions usually employed for
the isolation of mitochondria from locust flight
muscle the RNA bands are nearly entirely associated
with the mitochondrial fraction. With the exception
of 4S RNA, no nucleic acid can be extracted from the
supernatant (fig. 10). Evidently under these conditions the extramitochondrial ribosomes are firmly
attached to the mitochondria. Separation is made
possible by homogenizing the tissue in a magnesium-
free medium in the presence of high concentrations
of EDTA [22]. As shown in fig. 10, increasing

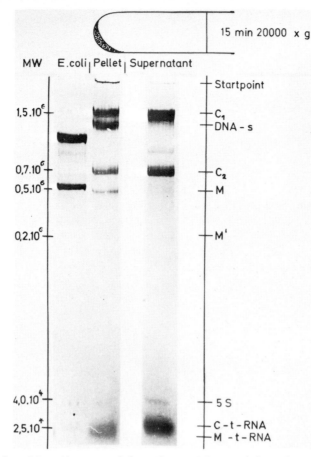

Fig. 9. Electropherograms of nucleic acids prepared from the precipitate and from the supernatant of a 15 min, 20,000 g centrifugation of a homogenate from thorax muscles of 7—day-old locusts. The homogenization was performed in 50 mM EDTA with addition of Subtilisin A.

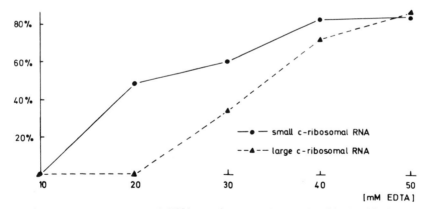

Fig. 10. Percentage of the cytosolic ribosomal RNA species appearing in the 15 min, 20,000 g supernatant at different concentrations of EDTA. Thorax muscles of 2— to 4—day-old locusts were homogenized, and the sediment and supernatant from centrifuging the homogenate 15 min at 20,000 g were examined. The homogenization medium was pH 7.2 and contained 0.3 M sucrose, 10 mM TRA and EDTA varying between 10 and 50 mM. Homogenization was accomplished by 10 strokes of a Potter homogenizer with Teflon pestle. Preparation of nucleic acids was performed by repetitive shaking with about one volume of 20% phenol:80% water and macaloid. In some cases the phenol:water was replaced by the mixture phenol:m—cresol:8—hydroxy-quinoline:water (500:70:0.5:55, w/w) following Kirby [20]. Lysis of mitochondria was accomplished by addition of 0.5 ml of 20% sodium dodecylsulphate to the first aliquot of phenol. Then the phenol was extracted by ether and the nucleic acids precipitated in the cold by two volumes of ethanol. Electrophoresis was performed according to Dingman and Peacock [21] in 2.7% acrylamide gel supported by 0.5% agarose, with Tris-borate buffer, pH 8.3, containing 2.5 mM EDTA. The relative amount of RNA corresponding to the electrophoresis band patterns was evaluated by densitographic means, from which the absolute amounts of the appropriate nucleic found in the 15 min, 20,000 g pellet and supernatant were calculated.

amounts of RNA species having molecular weights of 1.5×10^6 and of 0.7×10^6 appear in the supernatant by using increasing concentrations of EDTA. With this treatment, the smaller RNA is more easily removed from the sediment fraction than the larger one. This is similar to the action of EDTA in removing cytosolic ribosomes from liver rough endoplasmic reticulum as reported by Sabatini et al. [23]. The molecular weight of this RNA as determined from its mobility in gel electrophoresis [24,25] is in agreement with the molecular weight of cytosolic ribosomal RNA from other invertebrate animals as shown by Loening [26].

Upon treatment with 50 mM EDTA, about 10–15% of the c-ribosomal RNA remains in the mitochondrial fraction. This residue can be removed almost completely by gradient centrifugation of Subtilisin A treated homogenates (fig. 11). On the continuous gradient the flight muscle mitochondria separate into two fractions (fig. 11). The electrophoresis band patterns of the RNA isolated from these two fractions are almost identical. In addition to a prominent tRNA band with an electrophoretic mobility slightly different from that in the supernatant, there is another strong band corresponding to a molecular weight of about 500,000. Quantitatively — as deter-

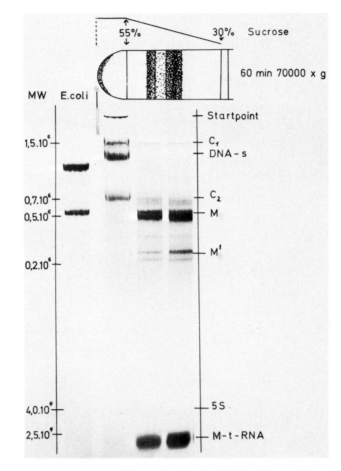

Fig. 11. Electropherograms of nucleic acids prepared from the mitochondrial fractions and from the pellet of a 60 min, 70,000 *g* gradient centrifugation. The gradient was 55–30% sucrose in 50 mM EDTA and 10 mM TRA, pH 7.2. The resuspended pellet from a 15 min, 20,000 *g* centrifugation of a thorax muscle homogenate from 7–day-old locusts, prepared in 50 mM EDTA with addition of Subtilisin A, was placed on top of the gradient. The relative amounts of RNA corresponding to the electrophoresis band patterns from the mitochondrial fractions shown, as determined by densitographic means, are approximately: C_1 3%, C_2 6%, M 35%, M' 16%, M-tRNA 37%.

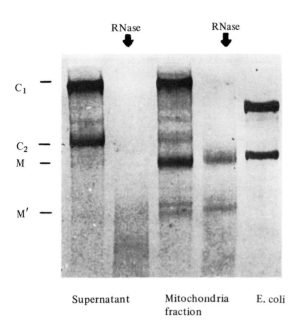

Fig. 12. Electropherograms of nucleic acids from the 15 min 20,000 g supernatant and from the mitochondrial fraction of an homogenate from thorax muscles of 2–4–day-old locusts, with and without treatment with beef pancreatic ribonuclease before phenolization. The tissue was homogenized in medium containing 50 mM EDTA, without addition of Subtilisin. The mitochondrial fraction was prepared by differential centrifugation. The resuspended pellet from 15 min, 20,000 g was centrifuged first for 10 min at 350 g and the resulting supernatant was further centrifuged 10 min at 10,000 g. The pellet from the last centrifugation was taken as the mitochondrial fraction.

mined by densitography of toluidine stain – this band amounts to up to 55% of the high molecular RNA. Besides this strong band, some minor bands, partly larger and partly smaller in molecular weight, are seen in varying proportions. Most prominent is one minor band appearing often as a double band and in a position corresponding to a molecular weight of 250,000.

Incubation with pancreas ribonuclease is another means by which the mitochondrial origin of the 500,000 and 250,000 MW bands can be demonstrated. It has been shown that ribonuclease added to suspensions of intact mitochondria, contrary to its action on cytosolic ribosomes, does not inhibit amino acid incorporation in mitochondria and leaves mitochondrial RNA unimpaired [13, 27–29]. Upon action of ribonuclease on flight muscle homogenates all

RNA bands disappear from the supernatant and also the corresponding bands in the mitochondrial fraction. The above described bands of MRNA remain with only slight diminution (fig. 12).

General remarks and conclusions

What conclusions may be drawn from the present finding? First, we must visualize, that the flight muscle is growing on an obligatorily aerobic basis, i.e. the energy required for the growth of the tissue, including that of its chondriome, is dependent upon the functional readiness of the chondriome itself. Interestingly enough, the same holds true for the growth of *Neurospora crassa* in the logarithmic phase [30]. In both tissues specific material is inserted into preexisting mitochondria in such a manner that growth in approximate constant proportions of constituents results. The preservation of constant proportions, useful in demonstrating mitochondrial growth, may be due to the special metabolic situation. In facultative aerobics and in general, the proportions of mitochondrial constituents will be subject to differentiations due to environmental circumstances and functional requirements.

Second, referring to the title of this symposium, the autonomy of mitochondria based on the results of the morphogenetic processes investigated is a rather restricted one, in that their formation is strictly coordinated with the formation of the other cellular constituents. Moreover, by far the greatest part of mitochondrial proteins is formed by the extrinsic system. Amino acid incorporation by the intrinsic system appears to be restricted to a few minor bands of the electrophoretic pattern of the insoluble mitochondrial protein fraction. The pattern of incorporation is strikingly similar to the one found in *Neurospora*. As in *Neurospora*, only a small degree of elasticity in the coupling between the extrinsic and intrinsic system could be demonstrated.

Third, the existence of RNA species in locust flight muscle mitochondria has been demonstrated. What are the functions of these RNAs? For that species, which has an electrophoretic mobility similar to cytoplasmic transfer RNA, a role as mitochondrial transfer RNA is probable. This conclusion is supported by other investigations on mammalian and *Neuro-*

spora mitochondrial tRNAs [31–34]. Considering the presence of mitochondrial tRNAs, the existence of mitochondrial ribosomes might be postulated, though recently, biosynthetic participation of tRNAs not linked to ribosomal function has been demonstrated in bacterial cell wall formation [35].

The molecular weight corresponding to the RNA-bands obtained upon gel electrophoresis are neither in that order of size observed for mammalian or insect cytoplasmic ribosmes nor in that observed for mitochondrial ribosomes from fungi, nor in that of bacterial ribosomes. In attributing the observed band pattern nevertheless to mitochondrial ribosomes, we must visualize mainly two possible explanations; either the observed bands represent the native ribosomal RNA chains or they contain specific breakdown products of these.

The first possibility would imply that the structure of locust mitochondrial ribosomes is different from that of all other types of ribosomes known at present. This difference in ribosomal structure, however, would be in contrast to the rather striking similarities in the mitochondrial amino acid incorporating systems. Therefore, attention must be drawn to the second possibility. As a matter of fact, the present standing of the problem resembles somewhat the discussion of length and continuity of bacterial ribosomal RNA occurring in the late fifties and early sixties. This matter has been reviewed by Spirin and Gavrilova [36]. Furthermore, the instability of chloroplast ribosomal RNA, as pointed out by Ingle et al. [37] seems to reflect a very similar situation.

Therefore in our case, we have to consider that at least one of the RNAs of the hypothetical large and small subunits is present as split pieces on the gel. One interpretation would be to correlate the prominent band M to the (unbroken) small subunit RNA, as well as to a major split piece of an unstable large subunit RNA.

Acknowledgement

This work was supported by the Deutsche Forschungsgemeinschaft (Schwerpunkt Biochemie der Morphogenese).

References

[1] R.R. Bensley, J. Histochem. and Cytochem. 1 (1953) 179.

[2] Ch. Rouiller, Int. Rev. Cytol. 9 (1960) 227.

[3] A.B. Novikoff, in: The Cell, eds. J. Brachét and A.E. Mirsky, (Academic Press, New York, 1961) Vol. II, p. 299.

[4] J. Frederic, Arch. Biol. 69 (1958) 167.

[5] J. André, J. Ultrastruct. Res. Suppl. 3 (1962).

[6] G.F. Bahr and E. Zeitler, J. Cell. Biol. 15 (1962) 489.

[7] D.J.L. Luck, J. Cell Biol. 16 (1963) 483.

[8] R.W. Brosemer, W. Vogell and Th. Bücher, Biochem. Z. 338 (1963) 854.

[9] W. Vogell, F.R. Bishai, Th. Bücher, M. Klingenberg, D. Pette and E. Zebe, Biochem. Z. 332 (1959) 81.

[10] H. Brandau and D. Pette, Enzymol. Biol. Clin. 6 (1966) 123.

[11] Th. Bücher, in: Aspects of Insect Biochemistry, ed. T.W. Goodwin (Academic Press, London and New York, 1965) p. 15.

[12] U. Bronsert and W. Neupert, in: Regulation of Metabolic Processes in Mitochondria, eds. J.M. Tager, S. Papa, E. Quagliariello and E.C. Slater (Elsevier, Amsterdam, 1966) p. 426.

[13] D.B. Roodyn and D. Wilkie, in: The Biogenesis of Mitochondria, (Methuen, London, 1968).

[14] W. Sebald, Th. Bücher, B. Olbrich and F. Kaudewitz, FEBS Letters 1 (1968) 235.

[15] W. Sebald, Th. Hofstötter, D. Hacker and Th. Bücher, FEBS Letters 2 (1968) 177.

[16] W. Sebald, A. Schwab and Th. Bücher, in: Inhibitors, Tools in Cell Research, eds. Th. Bücher and H. Sies (Springer Verlag, Berlin, Heidelberg, New York, 1969) p. 140.

[17] W. Sebald, G.D. Birkmayer, A.J. Schwab and H. Weiss, This Symposium.

[18] H. Küntzel and H. Noll, Nature 215 (1967) 1340.

[19] M.R. Rifkin, D.D. Wood and D.J.L. Luck, Proc. Natl. Acad. Sci. U.S. 58 (1967) 1025.

[20] K.S. Kirby, in: Methods in Enzymology, eds. S.P. Colowick and N.O. Kaplan (Academic Press, New York and London, 1968) Vol. XII, part B, p. 87.

[21] W.C. Dingman and A.C. Peacock, Biochemistry 7 (1968) 659.

[22] W. Kleinow, W. Neupert and Th. Bücher, in preparation.

[23] D.D. Sabatini, Y. Tashiro and G.E. Palade, J. Mol. Biol. 19 (1966) 503.

[24] A.C. Peacock and W.C. Dingman, Biochemistry 7 (1968) 668.

[25] D.H.L. Bishop, J.R. Claybrook and S. Spiegelman, J. Mol. Biol. 26 (1967) 373.

[26] U.E. Loening, J. Mol. Biol. 38 (1968) 355.

[27] G.D. Humm and J.H. Humm, Proc. Natl. Acad. Sci. U.S. 55 (1966) 114.

[28] D.B. Roodyn, in: Regulation of Metabolic Processes in Mitochondria, eds. J.M. Tager, S. Papa, E. Quagliariello and E.C. Slater (Elsevier, Amsterdam) 1966, p. 383.

[29] E. Wintersberger, Biochem. Z. 341 (1965) 409.

[30] D.J.L. Luck, J. Cell Biol. 24 (1965) 461.

[31] C.A. Buck and M.M.K. Nass, Proc. Natl. Acad. Sci. U.S. 60 (1968) 1045.

[32] C.A. Buck and M.M.K. Nass, J. Mol. Biol. 41 (1969) 67.

[33] M.S. Fornier and M.V. Simpson, in: Biochemical Aspects of the Biogenesis of Mitochondria, eds. E.C. Slater, J.M. Tager, S. Papa and E. Quagliariello (Adriatica Editrice, Bari, Italy, 1968) p. 227.

[34] W.E. Barnett, D.H. Brown and J.L. Epler, Proc. Natl. Acad. Sci. U.S. 57 (1967) 1775.

[35] J.L. Strominger, in: Inhibitors, Tools in Cell Research, eds. Th. Bücher and H. Sies (Springer Verlag, Berlin, Heidelberg, New York, 1969) p. 187.

[36] A.S. Spirin and L.P. Gavrilova, in: The Ribosome, (Springer Verlag, Berlin, Heidelberg, New York, 1969) p. 37.

[37] J. Ingle, R. Wells, J.V. Possingham and C.J. Leaver, this Symposium.

MONOMERS OF NEUROSPORA STRUCTURAL PROTEIN

K.D. MUNKRES, R.T. SWANK and G.I. SHEIR

Molecular Biology Laboratory, University of Wisconsin,
Madison, Wisconsin 53706, USA

An understanding of mitochondrial and chloroplast biogenesis requires some knowledge of genetics, structure, and function of membrane proteins. A priori, membrane-forming proteins must meet two requirements; (1) their amino acid sequences must lead to secondary and tertiary structure suitable for association with lipid and aggregation in two dimensional sheets, and (2) those sequences must be genetically controlled. The first requirement suggests that membrane-forming proteins possess a unique set of properties unlike other proteins. The second requirement could indicate a genetical role of organelle DNA in specifying the sequence of such proteins.

At present, it appears that all soluble proteins, readily detachable from mitochondria membranes, are encoded by nuclear genes, synthesized on cytoplasmic ribosomes, and have little role in determining primary membrane structure. Conversely, a class of proteins, called structural or non-catalytic, and ubiquitous to membranes of eucaryotes, is the only protein demonstrably synthesized by isolated mitochondria, and is the only protein for which there is evidence of functional and structural alteration as a consequence of non-chromosomal, mitochondrial mutation [18,21]. Thus, a problem central to the subject of mitochondrial biogenesis is the role of genes of those organelles in relation to membrane structural protein.

Our laboratory is currently isolating non-chromosomal mutations affecting mitochondrial structure and function and seeking to correlate those alterations with altered properties of structural protein. This paper describes experiments by electrophoretic molecular seiving methods designed to determine the minimal molecular weight of structural proteins. The results indicate a molecular weight considerably smaller than previously reported and lead to new predictions about membrane structure and genetics.

Materials and methods

Structural protein preparations

Methods for the preparation of *Neurospora* structural protein will be described in extenso elsewhere [1]. Generally, we employ a method developed in our laboratory involving initial extraction of acetone powder of mycelia [2] or mitochondria with neutral aqueous solutions containing sodium dodecyl sarcosine, ATP, and mercaptoethanol. The sarcosine detergent is preferred over sodium dodecyl sulfate employed in older methods because it is believed to be less of a protein denaturant and can be more readily removed from the protein by dialysis. Indeed, the extraction conditions are analogous to those of Stevens [3] who demonstrated the reversible dissociation of microtubule protein subunits. Following extraction and centrifugation to remove insoluble material, protein is precipitated from solution with ammonium sulfate between 36 and 48% saturation at 4° and pH 6. The water-insoluble precipitate is either resuspended in water and washed by centrifugation (undialyzed) or dialyzed for one or two days at 4° against distilled water in acetylated dialysis tubing (dialyzed). In some experiments, the proteinase inhibitor, phenylmethylsulfonylfluoride, was added to solutions throughout the preparation [4]. The overall yields of structural protein by these procedures, from either mycelia or mitochondria, have averaged about 20% of the total extracted protein.

Crystalline structural protein of *Neurospora* was prepared by the method of Kuehn et al. [5].

Structural protein of heavy beef heart mitochondria was prepared by the method of Richardson et al. as described by Criddle [6] and by the method described above employing sodium dodecyl sarcosine. Heavy beef heart mitochondria were a gift from D.E. Green.

Gel electrophoresis
Acid-urea system

Polyacrylamide gel electrophoresis was conducted at pH 3 in 35% acetic acid and 5 M urea by the method of Takayama et al. [7] with the modifications employed by Lenaz et al. [8]. The relative mobilities of proteins in gels ranging from 4 to 12% were evaluated and molecular weights were determined by the method of Hedrick and Smith [9].

Neutral SDS-urea system

Polyacrylamide gel electrophoresis at pH 7 in 0.1% SDS and 8 M urea with 10% gels was performed as outlined by Shapiro et al. [10,11] and Dunker and Rueckert [12] with slight modification allowing the determination of molecular weights as low as 2000 daltons [1].

Apparatus

Electrophoresis was performed in a Buchler Polyanalyst apparatus with twelve columns 5 mm I.D. × 65 mm long at constant temperature of 25° maintained with a refrigerated water bath and constant voltage of 4 mA per column. Densitometric tracings of stained gels were made with a Gilford recording spectrophotometer.

Chemicals

Commercial chemicals used were: horse heart cytochrome *c*, type VI; bovine carboxypeptidase, cryst.; ovalbumin, grade V, cryst.; bovine insulin, cryst.; bovine α-chymotrypsin, cryst., type II; bovine ribonuclease *A*, type XIA; bovine ribonuclease S-protein, grade XII-PR; and glucagon, cryst.; (Sigma Chemical Co.) bovine insulin, cryst.; horse heart myoglobin, cryst.; porcine adrenal corticotrophic hormone; and bovine trypsin inhibitor, cryst.; (Mann Res. Lab.)

Subtilin was a gift from Dr. George Alderton.

Electrophoresis grade acrylamide reagents were obtained from Bio-Rad Laboratories.

Enzyme grade urea and ammonium sulfate were obtained from Mann Res. Lab.

Results

Analysis of charge and size isomers in acid-urea electrophoresis

Calibration of the acid-urea electrophoresis system by the Hedrick and Smith method is illustrated in fig. 1. Apparent linearity is obtained from 3,000 to 48,000 daltons. Regression analysis reveals a standard deviation of 6% in estimating the molecular weight at the 95% confidence interval. Since the molecular

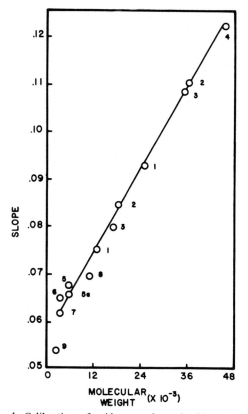

Fig. 1. Calibration of acid-urea polyacrylamide system with proteins of known molecular weight by the Hedrick-Smith method. Symbols: 1, cytochrome *c*; 2, myoglobin; 3, carboxypeptidase *A*; 4, ovalbumin; 5, insulin; 5a, insulin (carboxymethylated); 6, insulin (β-chain); 7, subtilin; 8, RNA-ase S-protein; 9, insulin (α-chain, carboxymethylated). Slope is defined as the change in the logarithm of mobilities (mm) per increment of 1% w/v gel concentration.

weights of the standard proteins are known with a precision of not less than 5%, the total error in determination is about 11%. The method is less accurate below 3,000 daltons as indicated by the marked deviation from linearity of the α-chain of insulin. In the case of ribonuclease *A*, cytochrome *c*, and myoglobin, both monomers and dimers were found. Thus, the assumption made by others (table 1) who have used this system to examine structural protein preparations, namely, that the solvent dissociates all proteins to monomers, is incorrect. Indeed, the solvent system may actually promote the aggregation of proteins that normally exist as monomers in neutral aqueous solutions.

The number of electrophoretic components observed in structural protein preparations by various laboratories in the acid-urea system are summarized in table 1. From 12 to 23 components are detected. In our laboratory, we observe 23 components by densitometry, as illustrated in fig. 2.

An analysis of the relative mobilities of nine of the 23 components as a function of gel concentration is illustrated in fig. 3. Two polymeric families of charge isomers are observed. Table 2 summarizes the ob-

served molecular weights for 18 of the 23 components. Clearly, a regular series of size isomers occurs with a monomeric weight of 3700 daltons. The weighted molecular weight average and modal molecular weight are about 22,000, a value observed by ultracentrifugal and binding analyses [13].

Various factors that may stabilize protein-protein association against acid-urea dissociation were examined. Disulfide bond formation is excluded since an identical series was obtained after carboxymethylation of sulfhydryl groups. Similarly, peroxidation of free sulfhydryl groups as a source of artifact was excluded by pre-electrophoresis to remove the excess persulfate after gel polymerization. Protein-bound detergent apparently does not contribute to stabilization of the protein aggregates since crystalline structural protein prepared without detergent yields the same molecular weight series. Protein-bound phospholipid may also be excluded as a stabilizing factor. Phospholipid analyses [14] of undialyzed preparations reveal one to two moles per 23,000 g protein. Moreover, extraction with cold 90% aqueous acetone to remove phospholipid failed to alter the electrophoretic pattern.

Table 1
Number of protein components observed in polyacrylamide electrophoresis of structural protein

Protein source	Solvent systems			
	Acid-urea		Basic	
	No. of components	Ref.	No. of components	Ref.
Rat	17	[41]	4	[31,37]
S. cerevisiae	14	[31]	5	[31]
	20	[32]		
S. carlsbergensis			5	[38]
Neurospora crassa	23	[33]	1	[5]
			1–2	[33,35,36]
Beef heart	18	[25,26]	1	[34]
	23	[33]		
	12	[34]		
	17	[40]		
Bacillus megaterium	8	[39]	1–2	[39]
Hydrogenomonas facilis	4	[5]	1	[5]
Mycoplasma laidlawii	10	[42]	1	[42]

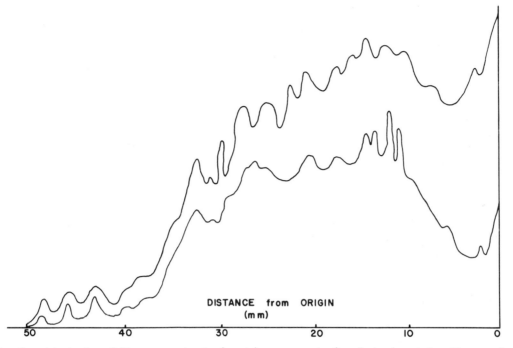

Fig. 2. Densitometric tracing of *Neurospora* structural protein components after electrophoresis in acid-urea polyacrylamide system. Conditions: total protein, 150 μg (upper curve), 75 μg (lower curve). Undialyzed type prep. Electrophoresis for 120 min at 4 mA and 25° in 7.5% gels. Stained with Coomassie blue.

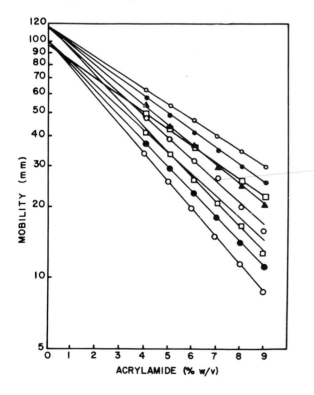

A partial demonstration that most of the electrophoretic components are members of a size isomer series can be obtained from three additional kinds of experiments. First, when the protein is allowed to stand for one week at 4° in the electrophoretic sample solvent containing phenol, urea and acetic acid, molecular weights above 26,000 disappear and 5 or 6 new small components with weights of about 3700 appear. Second, incorporation of 50% hexafluoroacetone in the sample solvent and in the gel reduces the number of components from 23 to 7. Hexafluoroacetone is superior to urea for disrupting protein hydrogen bonds and apparently does not cleave peptide bonds [15]. Finally, dialysis of the protein for one or two days at 4° against water prior to electrophoresis reduces the number of components to 7, without loss of protein. The molecular weights

Fig. 3. Electrophoretic mobilities of major components of *Neurospora* structural protein as a function of acrylamide concentration in acid-urea system. Conditions in fig. 2.

Table 2
Molecular weights of Neurospora structural protein components in acid-urea polyacrylamide electrophoresis system

| Component no. Series [a] | | Molecular weight ($\times 10^{-3}$) | | Nearest integer | Deviation (d) (obs.–calc.) |
I	II	Ob-served [b]	Calcu-lated		
23	22	3.6	3.7	1	
21	20	7.4	7.4	2	0
19	17	11.1	11.1	3	0
18	15	14.9	14.8	4	+ 100
16		17.8	18.5	5	− 700
	13	19.8	18.5	5	+1300
11		21.1	22.2	6	− 100
	10	22.5	22.2	6	+ 300
	9	25.1	25.9	7	− 800
12		29.4	29.6	8	− 200
8		33.5	33.3	9	+ 200
6		43.8	44.4	13	− 600
5		47.8	48.1	14	− 300
3		60.0	59.2	16	+ 800
	x_w [c] 21.3		21.4		Σd = 0

[a] 'Series' refers to the two charge isomer groups (cf. figs. 3 and 4).

[b] Average of seven experiments.

[c] Weighted average.

of the seven remaining components are hexamer to monomer in series I (fig. 4, table 2). At present, we think that a dialyzable component, other than phospholipid or detergent, protects the oligomers from dissociation.

The relative mobilities of all the electrophoretic components of *Neurospora* and beef heart mitochondrial structural protein in 7.5% gels as a function of molecular weight are summarized in the form of a Smithies-Connell plot [16] in fig. 4. Clearly, structural protein of both genera form similar or identical oligomeric series of two charge isomers. Previous experiments noted immunochemical and amino acid composition similarities of *Neurospora* and beef structural protein [17,18].

Analysis of molecular weight in alkaline electrophoresis systems

A few laboratories have determined the electrophoretic properties of structural protein at basic pH.

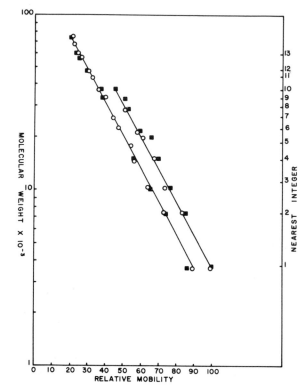

Fig. 4. Smithies-Connell plot of relative mobilities of structural protein components of *Neurospora* and beef heart mitochondria in acid-urea electrophoretic system as a function of molecular weight. Conditions: 7.5% acrylamide as in fig. 2. ○ Neurospora, □ beef.

For example (table 1) between one and five components are observed at pH 9.9 and above. The significance of the marked difference in number of components in acid and basic electrophoresis has not been analyzed. In our experience, some alkaline systems employing improper discontinuous buffer systems lead to artifactual results with all proteins migrating as a single band, regardless of charge or molecular weight. The theoretical basis for this phenomenon has been discussed elsewhere [19]. However, electrophoretic systems at basic pH can be devised which are free of such artifacts. In experiments with a continuous system at pH 10.8 and 8 M urea, we generally observe that structural protein migrates as one component at the same rate as the β-chain of insulin (mol.wt. 3300). The molecular weight, as determined by the Hedrick-Smith method, is 3500 ± 200. [1] These results indicate molecular weight homogeneity, but may fail to reveal charge heterogeneity because the protein's isoelectric point is between 6 and 7 [1].

Analysis of molecular weight in neutral SDS-urea electrophoresis

The determination of protein molecular weights by molecular seiving electrophoresis in the presence of the detergent, sodium dodecyl sulfate, has been well documented in recent years [10–12,20]. The results of analyses of over 100 proteins, whose molecular weights have been determined by classical procedures, agree with theory and practice in such systems. Fig. 5 illustrates a calibration plot of known proteins by a variation of the procedure, as devised in our laboratory, which allows determination of molecular weight as low as 2000 daltons. The standard deviation of the estimate is 5% at the 95% confidence interval with total error of about ± 10%. Fig. 6 illustrates a densitometric tracing of a typical structural protein preparation after SDS-urea electrophoresis. Virtually complete dissociation of oligomers yields monomers (3500 daltons) and dimers (7000). (A trace of material is also found with apparent molecular weight of around 1200, however this estimate requires extrapolation beyond the accuracy of the method.)

The discovery of small molecular weight components led to experiments designed to inhibit possible proteolysis during protein isolation. Identical molecular weight components are obtained from preparations made in the presence of an excess of the proteolytic inhibitor, phenylmethylsulfonylfluoride, an inhibitor of trypsin, chymotrypsin, and yeast protease [4].

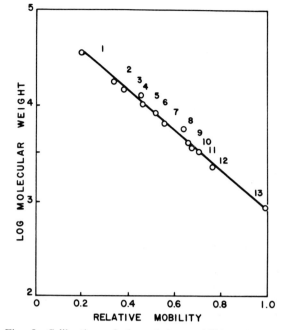

Fig. 5. Calibration of the relative mobilities of known proteins in an SDS-urea electrophoresis system. Symbols: 1, carboxypeptidase *A*; 2, myoglobin; 3, chymotrypsin (β-chain); 4, cytochrome *c*; 5, chymotrypsin (γ-chain); 6, adrenal corticotrophic hormone (dimer); 7, bovine trypsin inhibitor; 8, insulin; 9, adrenal corticotrophic hormone; 10 glucagon; 11, insulin (β-chain); 12, insulin (α-chain); 13, bromphenol blue.

Electrochromatography of Neurospora structural protein

The foregoing results indicate that analytical and preparatory methods, such as electrochromatography designed for peptides, may be more appropriate.

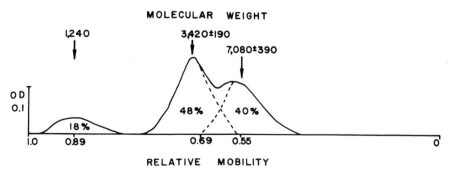

Fig. 6. Densitometric tracing of *Neurospora* structural protein components in SDS-urea electrophoretic system. Dialyzed type preparation. Stained with Coomassie blue, percentages represent fractional areas under each peak after correction for base-line absorbance.

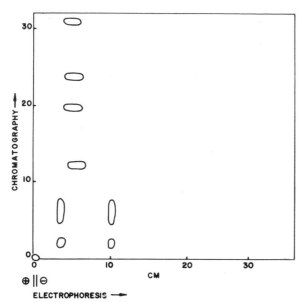

Fig. 7. Electrochromatogram of *Neurospora* structural protein monomers.

Fig. 7 illustrates a typical electrochromatogram revealing 6 to 8 major peptides differing in arginine, tryptophan, tyrosine, and cysteine content. Additional experiments are underway to determine the minimal molecular weights of those peptides.

Discussion

What are the true monomeric molecular weights of Neurospora structural protein?

The results of experiments by three independent methods all point to a minimal molecular weight of 3500 ± 400 daltons. We must still determine if the protein is degraded during isolation or analysis. The most likely source of degradation during isolation is proteolysis, a process that cannot be unequivocally excluded even after incorporating a variety of proteolytic inhibitors during the isolation. Conversely, typical proteolysis would be expected to yield a continuous array of molecular weight fragments rather than a unique monomer of 3500 daltons. However, the possibility remains that selective proteolysis occurs in vivo and/or in vitro as in the cases of: enzyme derived from proenzyme, hormones such as insulin, and certain viral structural proteins.

Artifactual degradation might occur during analysis if unusually labile peptide bonds or primary non-peptide bonds were cleaved. This possibility requires further examination but the available evidence renders it unlikely. Several groups of investigators have observed two or more amino terminal and c-terminal groups in structural protein from beef, yeast and *Neurospora* [13]. Preliminary analysis of amino terminal groups by the Edman procedure in our laboratory reveals 6 to 8 different PTH-amino acids. These end groups may correspond to the 6–8 peptides found in electrochromatography.

Several laboratories report that structural protein of *Neurospora* and beef yield about 26 moles of arginine plus lysine per 23,000 g, and that tryptic digestion yields an equivalent number of peptides [13]. These results are consistent with either a single polypeptide chain of 23,000 or several sequentially non-identical chains of lower molecular weight [18].

Previous molecular weight estimates of structural protein of *Neurospora* at pH 11 [18] and beef at pH 9 in dilute SDS [22] by sedimentation equilibrium revealed an apparent value of 23,000 but critical tests for molecular weight homogeneity were not applied nor were protein concentration effects analyzed. We observe by electrophoresis in the acid-urea system that both the weighted molecular weight average and modal molecular weight are 22,000. Moreover, in SDS-urea electrophoresis, some preparations exhibit a marked dependence of the molecular weight upon protein concentration such that at about 250 μg per column, a unique 22,000 component predominates which is progressively dissociated to the 3500 and 7000 species upon progressive dilution to 25 μg [1]. Since the minimum concentrations of protein required for sedimentation equilibrium determinations of molecular weight by interference optics are 10-fold greater than the maximum concentration employed in electrophoresis, it is unlikely that a molecular weight less than 22,000 would be obtained by ultracentrifugation. Blair et al. [23], however, reported molecular weights of 55,000 to 70,000 for beef structural protein based upon centrifugation in guanidinium hydrochloride; but Criddle [13] concluded 'that a gross heterogeneity of molecular sizes was present in the solutions analyzed in those studies'. In addition, the marked concentration dependence of molecular weight would not have been noted in those experiments. Thus, we conclude that a species of structural protein exists whose molecular weight is 22,000 as evidenced by sedimentation and electrophoretic analyses. In addition, a number of titration analyses indicate a minimal combining weight of 23,000 [13]. The latter, of course, could represent the modal or average stoichiometry of an oligomeric series as noted in acid-urea electrophoresis.

In view of the results from tryptic peptide mapping, amino terminal analysis, and electro-chromatography, oligomers of 23,000 could be either homo- or heteromultimeric with six to eight sequentially different monomers of 3500.

Relationships of structural proteins and enzymes

No enzymes or enzyme subunits are known with molecular weight as low as 3500 daltons [24]. This observation alone is sufficient to exclude the conjecture that structural proteins are derived during the isolation procedure by the denaturation of enzymic proteins and redirects attention to non-catalytic roles that such proteins may serve in membranes. However, it is recognized that detergents may denature some enzymic protein. Conversely, it is conceivable that salting out of detergent solubilized membrane protein may selectively precipitate a class of small micellar hydrophobic peptides in complex with detergent micelles. Whatever the mechanism by which structural protein is obtained, concern about the chemical relations of structural proteins to particular enzymes may not be relevant to the central issue of the nature of structural proteins per se; however, we will consider one enzyme, ATPase.

MacLennan and Tzagoloff [25] first observed that beef heart mitochondrial ATPase (F_1) subunit was found in structural protein preparations after acrylamide electrophoresis, an observation subsequently confirmed by Shatz and Saltzgaber [26]. However, the apparent minimal molecular weight of beef ATPase subunits by gel filtration in 8 M urea is about 70,000 [27]. We observe a minor component in acid-urea electrophoresis (component no. 1, fig. 2) with a molecular weight of 75,000 (fig. 4) comprising about 2% of the total structural protein as judged by relative peak areas after densitometry. A similar component of comparable magnitude has been observed in SDS-urea electrophoresis, in addition to the major components of 3500 and 7000 daltons. There-

fore, on the basis of minimal molecular weight, we conclude there may be no relationship between ATPase and structural protein. Moreover, we note that purified beef [28] and *Neurospora* [29] ATPase lack tryptophan, whereas substantial quantities occur in structural protein of various sources [13]. In addition, mutants of *Neurospora* with amino acid replacement in structural protein possess kinetically and compositionally normal ATPase [29]. Finally, Tzagoloff [30] recently demonstrated that yeast mitochondrial ATPase is synthesized by cytoplasmic ribosomes. Conversely structural proteins, as a class, are synthesized by isolated mitochondria [13] and encoded at least in part, by the mitochondrial genome [18]. Thus, by those chemical, biosynthetic, and genetic criteria there is no relationship between ATPase and structural protein.

Heterogeneity of structural protein and membrane specificity

Many enzymes and proteins exist as multiple molecular forms of isozymes. The physiological and genetical basis of such heterogeneity are in some cases obscure. Nevertheless, a large body of evidence indicates that heterogeneity is the biological rule. A priori, membrane structural protein should not be an exception. Indeed, if such proteins are to fulfill the demanding task of participating in the unique subcellular localization of perhaps 200 different enzymes and conferring unique properties on at least 10 different classes of subcellular membranes, the expectation is that a variety of multiple molecular forms of structural protein should exist.

Now, with a subclass of structural proteins consisting of sequentially different monomers of about 3500 daltons, we suggest a solution to the nature of membrane specificity which derives from principles established in the phenomena of complementation and certain isozymes. The model allows random permutations of monomers to generate a hetero- and homomultimeric series. The molecular weights of structural proteins as they occur in membranes is unknown, but isolated structural proteins in electrophoresis demonstrate the potential of forming a continuous oligomeric series from monomer to at least 16-mer. With such an oligomeric series of hetero- and homomultimers, membranes could maximally contain multiple forms of structural protein in 506 combinations and 1.7×10^8 permutations. Such organizational information may be sufficient for the task of specific localization of enzymes on various membranes and at subsites on a given membrane.

Solution of questions concerning the genetic and biosynthetic history of membrane structural protein monomers and their arrangements within membranes appear to be prerequisite to an understanding of membrane and mitochondrial biogenesis. The observations summarized here are only a hopeful beginning.

Summary

Previous minimal molecular weight estimates of mitochondrial structural protein of *Neurospora* and beef by ultracentrifugation and titration analyses were 23,000 daltons. This paper presents the results of analyses by three independent molecular seiving electrophoresis systems which confirm the weight of 23,000; but which also reveal aggregation and/or dissociation under some conditions in a regular polymeric series based upon monomers weighing 3500 ± 400 daltons. These results together with the results of various chemical analyses indicate that oligomers of 23,000 could be either homo- or hetero-multimeric with between 6 and 8 sequentially different monomers.

Acknowledgements

Paper No. 1341 from the Department of Genetics. Published with the approval of the Director of The Agricultural Experiment Station.

This research was supported by grants from The National Institutes of Health (GM–15751), The University of Wisconsin Graduate School, and The Department of Genetics, by a NIH postdoctoral fellowship (RTS) and by predoctoral fellowships to G.S. from the Wisconsin Alumni Research Foundation and NIH.

The authors thank Drs. John Garver for assistance in pilot plant operation, David E. Green for a gift of beef heart mitochondria, and Oliver Smithies for valuable discussion about electrophoresis.

Technical assistance by W. Lasher, K. Conners, and A. Rice is acknowledged.

Travel grants from the National Research Council and The University of Wisconsin Graduate School are gratefully acknowledged.

References

[1] R.T. Swank, G.I. Sheir and K.D. Munkres, in preparation.
[2] K.D. Munkres and F.M. Richards, Arch. Biochem. Biophys. 109 (1965) 466.
[3] R. Stevens, J. Mol. Biol. 33 (1968) 517.
[4] U. Kenkare, in: Methods in Enzymology (Academic Press, New York, 1966) vol. V, p. 226.
[5] G.D. Kuehn, B.A. MacFadden, R.A. Johanson, J.M. Hill and L.K. Shumway, Proc. Natl. Acad. Sci. U.S. 62 (1969) 407.
[6] R.S. Criddle, in: Methods in Enzymology (Academic Press, New York, 1967) vol. X, p. 668.
[7] K. Takayama, D.H. MacLennan, A. Tzagoloff and C.D. Stoner, Arch. Biochem. Biophys. 114 (1966) 223.
[8] G. Lenaz, N.F. Haard, H.I. Silman and D.E. Green, Arch. Biochem. Biophys. 128 (1968) 293.
[9] J.L. Hedrick and A.J. Smith, Arch. Biochem. Biophys. 126 (1968) 155.
[10] A.L. Shapiro, E. Vinuela and J.V. Maizel, Biochem. Biophys. Res. Commun. 28 (1967) 515.
[11] A.L. Shapiro and J.V. Maizel, Anal. Biochem. 29 (1969) 505.
[12] A.K. Dunker and R.R. Rueckert, J. Biol. Chem. 244 (1969) 5074.
[13] R.S. Criddle, Ann. Rev. Plant Physiol. 20 (1969) 239.
[14] K. Kopaczyk, in: Methods in Enzymology (Academic Press, New York, 1967) vol. X, p. 253.
[15] M. Goodman and I. Rosen, Biopolymers 2 (1964) 537.
[16] O. Smithies and G.E. Connell, in: CIBA Foundation on Biochemistry of Human Genetics, eds. G.E.W. Wolstenholme and C.M. O'Conner (Little, Brown and Co., Boston, 1959) p. 178.
[17] K.D. Munkres and D.O. Woodward, Biochim. Biophys. Acta, 133 (1967) 143.
[18] D.O. Woodward and K.D. Munkres, Proc. Natl. Acad. Sci. U.S. 55 (1966) 872.
[19] D.E. Williams and R.A. Reisfield, Ann. N.Y. Acad. Sci. 121 (1964) 373.
[20] K. Weber and M. Osborn, J. Biol. Chem. 244 (1969) 4406.
[21] D.O. Woodward and K.D. Munkres, in: Organizational Biosynthesis, eds. H.J. Vogel, J.O. Lamper and V. Bryson (Academic Press, New York, 1967) p. 489.
[22] R.S. Criddle, R.M. Bock, D.E. Green and H.D. Tisdale, Biochemistry 1 (1962) 827.
[23] J.E. Blair, G. Lenaz and N.F. Haard, Arch. Biochem. Biophys. 126 (1968) 753.
[24] K.M. Klotz and D.W. Darnall, Science 166 (1969) 126.
[25] D.H. MacLennan and A. Tzagoloff, Biochemistry 7 (1968) 603.

[26] G. Schatz and J. Saltzgaber, Biochim. Biophys. Acta. 180 (1969) 186.

[27] A. Senior, personal communication.

[28] H.S. Penefsky, in: Methods in Enzymology, eds. R.W. Estabrook and M.E. Pullman (Academic Press, New York, 1967) Vol. X, p. 522.

[29] S.C. Hedman and D.O. Woodward, Fed. Proc. 27 (1968) 826.

[30] A. Tzagoloff, J. Biol. Chem. 244 (1969) 5027.

[31] K. Lejsek and C.V. Lusena, Can. J. Biochem. 47 (1969) 753.

[32] H. Tuppy, P. Swetley and I. Wolff, European J. Biochem. 5 (1968) 339.

[33] This paper.

[34] G. Lenaz, N.F. Haard, A. Lauwers, D.W. Allmann and D.E. Green, Arch. Biochem. Biophys. 126 (1968) 746.

[35] L.J. Schneiderman and I.G. Junga, Biochemistry 7 (1968) 2281.

[36] C. Woodward, V. Kubik, R. Kleese and V.W. Woodward, J. Cell. Biol. (abstr.) 145a.

[37] D. Haldar, K. Freeman and T.S. Work, Nature 211 (1966) 9.

[38] S. Yang and R. Criddle, Biochem. Biophys. Res. Commun. 35 (1969) 429.

[39] R. Mirsky, Biochemistry 8 (1969) 1164.

[40] W.L. Zahler, A. Saito and S. Fleischer, Biochem. Biophys. Res. Commun. 32 (1968) 513.

[41] C.W. Cotman, H.R. Mahler and T.E. Hugli, Arch. Biochem. Biophys. 126 (1968) 821.

[42] S. Rotten and S. Razin, J. Bacteriol. 94 (1967) 349.

MITOCHONDRIAL PRECURSORS IN ANAEROBICALLY GROWN YEAST

K. WATSON, J.M. HASLAM, B. VEITCH and Anthony W. LINNANE

Biochemistry Department, Monash University, Clayton, Victoria, Australia

Under anaerobic conditions the yeast *Saccharomyces cerevisiae*, a facultative anaerobe, lacks a respiratory chain but on aeration functional mitochondria are rapidly induced [1–4]. The nature of the mitochondrial precursors in anaerobically grown yeast cells is therefore of considerable importance in a study of the biogenesis of yeast mitochondria.

The cytology of anaerobically grown yeast cells is particularly influenced by the lipid composition of the growth media [5–7] and also by catabolite repression [8–11]. When cells are grown anaerobically in the presence of lipid supplements clearly recognizable mitochondrial profiles are seen in permanganate-fixed cells [5,7]. In marked contrast, Wallace et al. [7] reported that when cells are grown anaerobically with glucose as substrate to induce catabolite repression and in the absence of lipid supplements, no obvious mitochondrial structures are visible in permanganate-fixed cells. However, Damsky et al. [12], while confirming the latter observations, have shown the presence of mitochondrial-like structures in partially disruptured lipid depleted anaerobically grown cells which had been fixed with glutaraldehyde. Recent freeze-etching studies [13] have also shown the presence of mitochondrial-like structures in glucose repressed lipid depleted cells.

The present report is concerned with the characterization of mitochondrial precursor structures from lipid depleted and lipid supplemented anaerobically grown yeast cells. Contrary to the reports of Schatz et al. [13–15], the mitochondrial precursor structures from the two types of anaerobically grown cells are morphologically distinct and differ from aerobic yeast mitochondria. Furthermore, the lipid depleted anaerobic mitochondrial structures are actually different in their biochemical properties from the structures of either aerobic or lipid supplemented anaerobically grown cells; as reported herein they apparently lack a mitochondrial protein synthesizing system. Aspects of the development of these anaerobic structures on aeration is reported. We have recently reported methods for the isolation of the morphologically intact structures from anaerobically grown cells [16].

Methods

Growth of cells

All experiments described in this paper were carried out with a prototrophic diploid strain of *S. cerevisiae* grown on 0.5% yeast extract salts medium as previously described [7]. The carbon source for strict anaerobic growth was glucose (5%) or galactose (5%); glucose (1%) was used for aerobic growth. In the case of lipid supplemented media Tween 80 (5 g/l) and ergosterol (20 mg/l) were added to the growth media.

Cultures were incubated at 28° and cells were harvested under four conditions: (a) grown to 0.8 mg dry weight/ml anaerobically on 5% glucose-yeast extract medium to give lipid depleted catabolite repressed anaerobic cells, denoted An–Glu; (b) grown to 3–4 mg dry weight/ml anaerobically in 5% glucose-yeast extract medium with lipid supplements to give lipid supplemented catabolite repressed anaerobic cells, denoted An–Glu+T+E; (c) grown to 3–4 mg dry weight/ml anaerobically in 5% galactose-yeast extract medium with lipid supplements to give lipid supplemented partially catabolite derepressed anaerobic cells, denoted An–Gal+T+E; (d) grown to 2 mg dry weight/ml aerobically in 1% glucose-yeast extract medium to give catabolite derepressed aerobic cells with a normal lipid composition, denoted Aer–Glu.

In the experiments on the aerobic induction of anaerobically grown cells, the cells were suspended at a density of 2 mg dry weight/ml in 0.5% yeast extract-salts medium containing 0.2% glucose and vigorously aerated for the times indicated in the text.

Preparation of mitochondrial fraction

Mitochondrial fractions were isolated from yeast protoplasts prepared as described by Watson et al. [16]. Precautions were taken to prevent aerobic induction of mitochondrial components during the preparation of particulate fractions from anaerobic cells and cells which had been induced aerobically for 30–120 min. Cells were chilled to 0° in ice prior to harvesting and maintained at 0–4° throughout the isolation procedure except during the incubation with snail enzyme. All solutions and cell suspensions contained cycloheximide (10 μg/ml) to prevent respiratory adaption and were deoxygenated by continual flushing with oxygen-free nitrogen.

The washed suspension of cells and protoplasts after snail enzyme digestion were broken in a French Press at a pressure of 700 psi in a medium containing 0.5 M sorbitol, 2 mM EDTA pH 7.4 (S.E. buffer). After three low speed contrifugations at 1,000 g for 5 min to remove unbroken cells and cell debris, the supernatant was centrifuged at 20,000 g for 20 min, and the particulate fraction obtained washed once in S.E. buffer. The washed pellet was layered on a discontinuous sorbitol gradient [17] consisting of 6 ml of 80% sorbitol and 3 ml each of 70, 60, 57.5, 55, 52.5 and 50% sorbitol in 20 mM tris-HCl, pH 7.4 and centrifuged at 25,000 rpm for 2.5 hr in a Spinco Model L centrifuge, SW 25.1 rotor. In all cases the main fraction formed a band at a density of 1.18–1.20 g/cm^3 which was collected.

Electron microscopy

Samples were fixed with 2% glutaraldehyde for 30 min at 0° diluted with S.E. buffer and centrifuged at 50,000 g for 15 min. The samples were washed twice in S.E. buffer then post-fixed in 1% OsO$_4$ in veronal acetate buffer, pH 7.2, for 2 hr, and collected by centrifugation. The pellets were dehydrated in acetone and embedded in araldite. Sections were cut on a LKB Ultratome with glass knives, collected on carbon-coated grids and stained with uranyl acetate-lead citrate for 15 min. Micrographs were taken in a Hitachi HU-IIA electron microscope.

Prefixation by glutaraldehyde

On occasion as detailed in the text protoplasts were prefixed with glutaraldehyde before disruption. They were incubated for 10 min at 0–4° with 2% glutaraldehyde in 0.9 M sorbitol, 2 mM EDTA, pH 7.4. The protoplasts were then washed twice in 0.9 M sorbitol, 2 mM EDTA, pH 7.4 before being broken in S.E. buffer by a French Press and the mitochondria isolated as described above.

Assays

Succinate dehydrogenase activity was measured by the method of Arrigoni and Singer [18], malate dehydrogenase by the method of Vary et al. [19] and ATP-ase activity was measured as described by Pullman and Monroy [20]. Mitochondrial DNA was purified as described by Wake [21], and its buoyant density determined as detailed by Schildkraut et al. [22] using DNA from *Micrococcus lysodeikticus*, buoyant density 1.731, as density marker. Fatty acids were determined as described by Jollow et al. [23]. Amino acid incorporation by isolated mitochondrial structures was determined as described by Lamb et al. [24].

Results

Organelle morphology
Mitochondria from aerobically grown cells

Before proceeding to the study of the nature of the mitochondrial precursor structures in anaerobic yeast cells, it was necessary to establish the morphology of aerobic yeast mitochondria under our conditions of preparation and fixation. Fig. 1 shows electron micrographs of yeast mitochondria isolated from aerobically grown cells. In fig. 1A, it is seen that the gradient purified mitochondria exist almost exclusively in the condensed conformation similar to that which has been described by Hackenbrock [25] for rat-liver mitochondria. On the other hand, non-gradient purified yeast mitochondria contain a number of different conformational forms (fig. 1B).

High magnification electron micrographs of the various conformations which can be identified in isolated yeast mitochondria are presented in fig. 2. Two extremes of conformation are apparent. One, the condensed form (fig. 2A) is characterised as

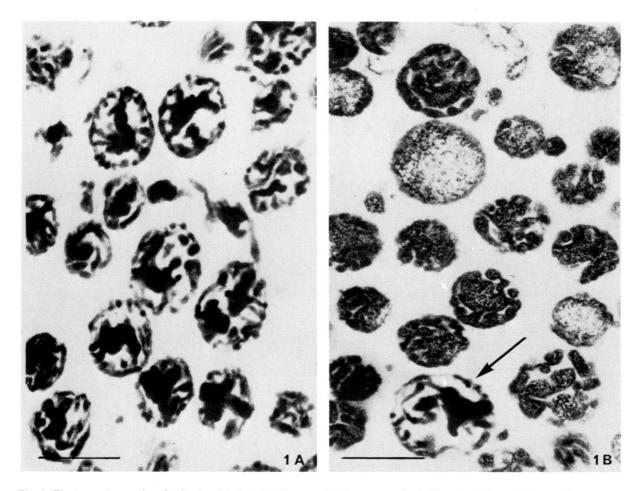

Fig. 1. Electron micrographs of mitochondria isolated from aerobically grown cells. A: Mitochondria purified on a discontinuous sorbitol gradient (50–80%). The mitochondria are exclusively in the condensed conformation. The bar corresponds to 0.5 μ. B: Non-gradient purified mitochondria. The mitochondria are mostly in the intermediate conformation. An occasional condensed mitochondria is seen (arrow). The bar represents 0.5 μ.

having an irregular inner membrane pattern, a highly condensed matrix and large intracristal spaces. The second, the so-called orthodox form (fig. 2B) is similar to the classical form of mitochondria commonly observed in situ. However, the majority of non-gradient purified mitochondria are intermediate in conformation between the two extreme types (fig. 2C). Occasionally structures are caught in a transitional stage and are observed to be partly condensed and partly orthodox in conformation (fig. 2D). Green et al. [26] have recently described the presence of such transitional forms of mitochondria in heart muscle.

Mitochondrial structures from anaerobically grown lipid supplement cells

When yeast cells are grown anaerobically in the presence of lipid supplements clearly recognizable mitochondrial profiles are seen in permanganate fixed cells [7]. Fig. 3A shows a high magnification inset and a low magnification field of the structures isolated from such anaerobically grown cells with galactose as carbon source, An–Gal+T+E cells. They have a diameter of about 0.7 μ, possess inner and outer membrane systems, and have a dense granular matrix with numerous electron transparent areas lined with

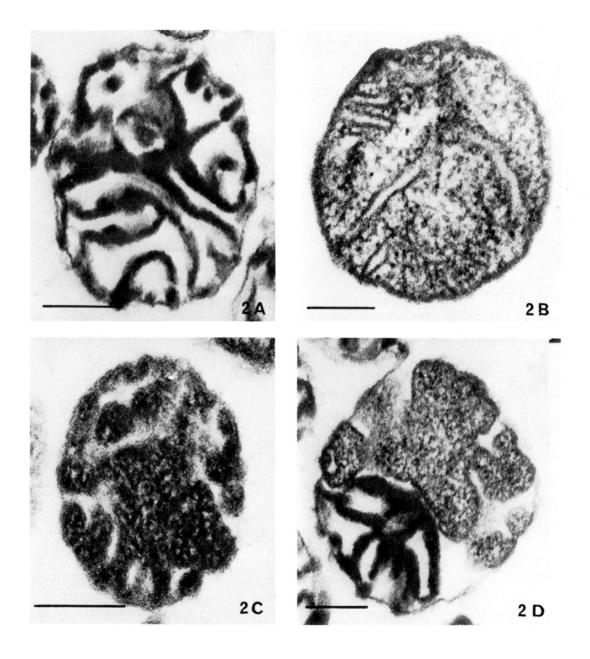

Fig. 2. Conformational forms of mitochondria isolated from aerobically grown cells. The bar corresponds to 0.25 μ.
A: Condensed form. B: Orthodox form. C: Intermediate conformation. D: Mitochondrion in a transitional state.

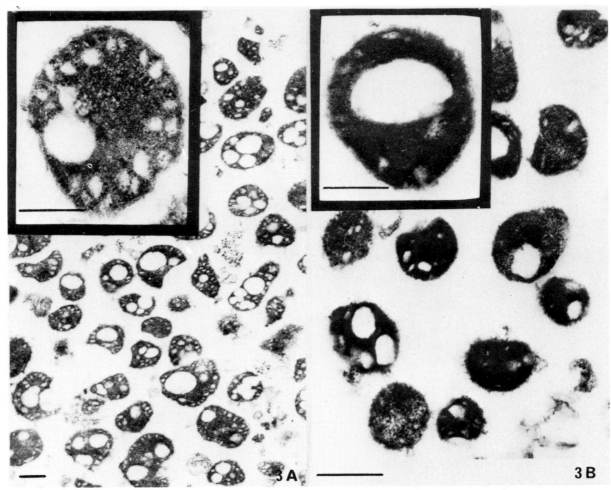

Fig. 3. Gradient purified particulate fractions prepared from yeast cells grown anaerobically in the presence of lipid supplements. A: Structures isolated from cells grown on galactose as carbon source, An–Gal+T+E cells. The structures have a characteristic honeycomb like internal membrane system. The bar represents 0.5 μ. Inset: the bar corresponds to 0.25 μ. Large field magnification $\times$ 16,000. B: Structures isolated from cells grown on glucose as carbon source, An–Glu+T+E cells. Compared with the structures in A, fewer electron transparent areas are evident within the matrix. The bar represents 0.5 μ. Inset: the bar corresponds to 0.25 μ.

an inner membrane system. These structures are distinctly different in morphology from any of the aerobic mitochondrial conformations shown in fig. 2.

The particles isolated from anaerobic cells grown on glucose under conditions of strong catabolite repression and in the presence of lipid supplements, An–Glu+T+E cells, are shown in fig. 3B. A smaller number of electron transparent areas and a lesser degree of internal membrane organization compared with that of the structures isolated from the galactose lipid supplemented cells is apparent, indicative of the effect of catabolite repression.

Mitochondrial structures from anaerobically grown lipid depleted cells

Fig. 4A shows the particulate fraction isolated from An–Glu grown cells using exactly the same procedures as for the isolation of the intact structures shown in figs. 1–3. The fraction consists mainly of membrane fragments and degraded structures, a number of outer and inner membrane bound vesicles are recognized in these fractions (fig. 4A, arrows), the majority of which appear to have been damaged and to have lost most of their contents.

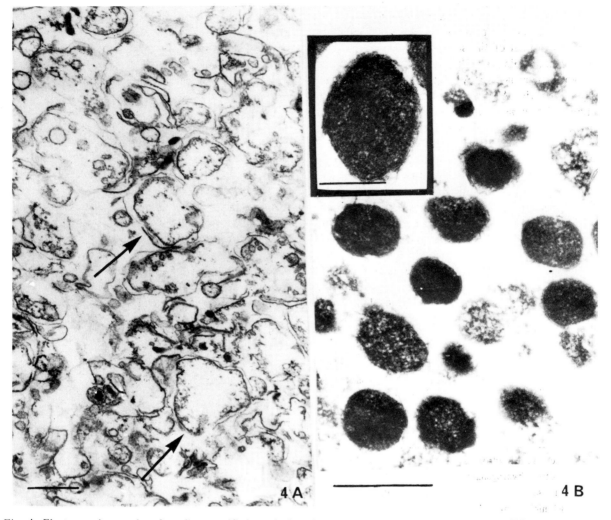

Fig. 4. Electron micrographs of gradient purified particulate fractions prepared from An–Glu cells. A: Particulate fraction isolated without prior prefixation with glutaraldehyde. The fraction consists mainly of broken membranes and degraded structures. Structures bounded by inner and outer membranes but which appear to have been damaged and to have lost most of their contents are observed in these fractions (arrows). The bar corresponds to 0.5 μ. B: Particulate fraction isolated from protoplasts prefixed with 2% glutaraldehyde (see Methods). The fraction consists of membrane bound structures enclosing a dense granular matrix. The bar represents 0.5 μ. Inset: the bar corresponds to 0.25 μ.

In order to overcome the problem of the fragility of the mitochondrial profiles present in the glucose repressed lipid depleted cells the isolation technique was modified by prefixation of protoplasts with glutaraldehyde prior to breakage and isolation (see Methods). Electron micrographs of particulate fractions isolated under these conditions show morphologically intact structures consisting of a dense granular matrix bounded by inner and outer membrane systems (fig. 4B). Under our conditions of fixation and staining there is no obvious folded inner membrane system within the dense matrix of these organelles, and we conclude that inner membrane cristae are absent or very poorly developed in these structures. The isolated structures differ considerably in morphology from any of those isolated from other cell types (cf. figs. 1–3) and are evidently the most primitive of the mitochondrial precursor structures.

Biochemical properties of isolated organelles

Biochemical investigation has confirmed the mitochondrial origin of the structures isolated from the various anaerobic cells. The particulate mitochondrial enzymes succinate dehydrogenase, and the oligomycin sensitive ATPase as well as the soluble mitochondrial enzyme malate dehydrogenase were present in the particles isolated from all anaerobic cell types (table 1). Even the extensively damaged structures isolated from An–Glu cells still contain appreciable amounts of the readily solubilized enzyme malate dehydrogenase. For comparison table 1 also includes the corresponding enzymatic activities for Aer–Glu mitochondria under our conditions of assay. The Aer–Glu mitochondria have the highest specific activities while the mitochondrial precursor structures from the An–Glu cells have the lowest specific activities.

The mitochondrial nature of the morphologically intact structures from prefixed An–Glu cells was established by measuring the activities of mitochondrial marker enzymes (table 2). The effects of prefixation on the enzyme activities of the isolated structures from An–Gal+T+E and Aer–Glu cells are included for comparison. Although prefixation led to the inactivation of 70–80% of the succinate dehydrogenase, 50–70% of the malate dehydrogenase, and 60–80% of the oligomycin sensitive ATPase, sufficient activities of these enzymes were still present in all types of structures to indicate that they were mitochondrial in nature.

DNA was extracted from the particles prepared from the cells grown under the various growth conditions and was characterized by centrifugation to equilibrium on a CsCl gradient (fig. 5). All the particulate fractions gave a similar pattern and contained DNA of buoyant density 1.683 g/cm^3 characteristic of yeast mitochondria [27]. There was some contamination by nuclear DNA (10–15%) in these preparations as shown by the presence of material banding at a density of 1.699 g/cm^3.

Table 1
Enzymic activities of purified particles from aerobic and anaerobic cells.

Enzyme	Growth conditions			
	An–Glu	An–Glu+T+E	An–Gal+T+E	Aer–Glu
Succinate dehydrogenase	0.62	1.4	1.6	6.8
Malate dehydrogenase	17	72	74	160
ATPase	0.9	2.0	1.6	4.1
% inhibition of ATPase by oligomycin	56	54	70	77

Particles were prepared and assayed for enzyme activity as described in Methods. Oligomycin sensitivity of the ATPase was assayed with 50 μg oligomycin/mg protein. Enzyme activities are expessed as μmoles substrate transformed/10 min/mg protein.

Table 2
Enzymatic activities of prefixed mitochondrial precursors and mitochondria.

Enzyme	An–Glu	An–Glu+T+E	An–Gal+T+E	Aer–Glu
Succinate dehydrogenase	0.14	0.20	0.23	2.6
Malate dehydrogenase	10	28	32	56
ATPase (oligomycin-sensitive)	0.50	0.81	0.72	2.1

Mitochondrial structures were isolated from protoplasts prefixed with glutaraldehyde. Succinate and malate dehydrogenase activities are expressed as μmoles substrate decomposed/10 min/mg protein. ATPase activity is expressed as μmoles phosphate released/10 min/mg protein.

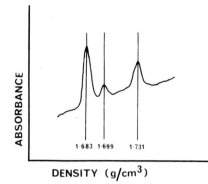

Fig. 5. Cesium chloride density gradient of a particulate fraction isolated from An–Glu cells. The peaks correspond to *Micrococcus lysodeikticus* reference DNA, buoyant density 1.731 g/cm^3, yeast nuclear DNA buoyant density 1.699 g/cm^3 and yeast mitochondrial DNA buoyant density 1.683 g/cm^3. The mitochondrial fraction contained 2.2 μg DNA/mg protein, of which about 12% was accounted for as nuclear DNA contamination.

Organelle development induced by aeration

Morphology

The recognition that An–Glu cells contain the most primitive mitochondrial structures and the ability to isolate these mitochondrial precursors provided a basis for the study of mitochondrial development induced by aeration. Fig. 6 shows high magnification electron micrographs of the morphological changes in the anaerobic structures during a 3 hr aeration experiment. The control sample at the beginning of the aeration experiment (fig. 6A) shows a dense granular matrix bounded by an inner and outer membrane but with no discernible internal membrane system. After 60 min aeration of the cells, the isolated structures show clear development of an internal membrane system as well as the presence of electron transparent areas (fig. 6B). Further development of the internal membrane system and the appearance of numerous electron transparent areas within the matrix is apparent after 120 min of oxygen induction (fig. 6C). The structures now resemble in morphology those isolated from the lipid supplemented anaerobic cell (cf. fig. 3). There is considerably more development after 180 min aeration, the structures now forming an extensive internal membrane system (fig. 6D) reminiscent of the cristae of classical aerobic mitochondria.

Biochemical properties

The changes in morphology of the mitochondrial structures during aerobic induction are paralleled by changes in the fatty acid composition (table 3) and enzymatic activities (table 4). Tables 3 and 4 also include for comparison the unsaturated fatty acid content and enzymatic activities of mitochondrial structures isolated from aerobically and lipid supplemented anaerobically grown cells.

Fatty acid composition

The An–Glu structures initially contain as much as 50% of the total fatty acids as short chain fatty acids ($C_8 - C_{14}$) and a low percentage, generally less than 7%, of long chain unsaturated fatty acids (table 3). In contrast, the An–Gal+T+E and An–Glu+T+E structures have a similar percentage amount of long chain unsaturated fatty acids to mitochondria from aerobically grown cells. They contain a low percentage (5–8%) of short chain fatty acids and a high percentage (75–80%) of long chain unsaturated fatty acids (table 3). On aeration of An–Glu cells there are extensive changes in the fatty acid composition. The most marked change is in the percentage of long chain unsaturated fatty acids which rapidly increases from 6% in the strict anaerobe to 29% after 30 min aeration. The percentage unsaturated fatty acid content of mitochondrial structures isolated after 120 min oxygen induction approaches that of the structures isolated from aerobically and lipid supplemented anaerobically grown cells (table 3). The percentage of short chain fatty acids also changes on aeration and decreases after 60 min and 120 min to 25% and 10% respectively of the total fatty acids.

Protein synthesis

The amino acid incorporation activity of the isolated mitochondrial structures from lipid depleted and lipid supplemented anaerobically grown cells and from aerobically induced An–Glu cells are presented in table 4. As reported previously [28] mitochondrial structures from An–Gal+T+E cells contain an active mitochondrial protein synthesizing system. In contrast, the incorporation of amino acids by the An–Glu structures was very small, and control experiments indicated that it could be accounted for by a slight contamination with bacteria and cytoplasmic ribosomes. However, we considered the possibility

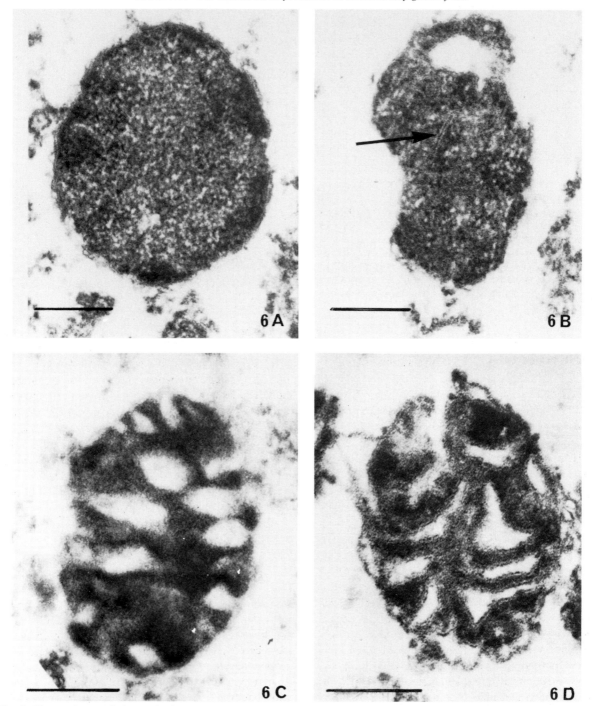

Fig. 6. Changes in morphology of mitochondrial precursor structures on aeration of An–Glu cells. The structures were isolated from protoplasts which had been prefixed with 2% glutaraldehyde. The bars correspond to 0.25 μ. A: Structure isolated from An–Glu cells. B: Structure isolated after 60 min aeration of An–Glu cells. The development of some internal membrane is apparent (arrow). C: Structure isolated after 120 min aeration of An–Glu cells. Numerous electron transparent areas are visible and the structures approach the morphology of those isolated from the lipid-supplemented anaerobic cell (cf. fig. 3). D: Structure isolated after 180 min aeration of An–Glu cells. There is now extensive development of the internal membrane characteristic of the cristae of aerobic mitochondria.

Table 3
Fatty acid composition of mitochondrial structures during aerobic induction.

Cell type and aeration time (min)	C_8	C_{10}	C_{12}	C_{14}	C_{16}	C_{18}	$C_{14:1}$	$C_{16:1}$	$C_{18:1}$	UFA [a]	SCFA [b] C_8-C_{14}
An–Glu	3	16	14	16	38	7	nil	4	2	6	49
30	1	8	16	20	18	8	4	16	9	29	45
60	<1	7	8	9	23	5	5	32	10	47	25
120	<1	2	3	4	13	4	2	45	26	73	10
An–Gal+T+E	nil	nil	<1	4	21	5	2	19	59	80	5
An–Glu+T+E	nil	<1	<1	5	20	5	2	20	52	74	7
Aer–Glu	nil	<1	1	1	11	4	1	47	36	84	3

The fatty acids are denoted by the convention; number of carbon atoms : number of unsaturated linkages.
[a] UFA – unsaturated fatty acid.
[b] SCFA – short chain fatty acid.
Fatty acid extraction and estimation was carried out essentially as described by Jollow et al. [23].

Table 4
The effect of aeration on mitochondrial precursor development.

Cell type	Aeration time (hr)	% UFA [a]	Protein synthesis	Cytochrome	NADH oxidase
An–Glu	–	5	<1 (<1)	b_1	<5
	1	60	1 (<1)	b_1	70
	2	74	3 (1)	$b_1 b c$	240
	3	78	5 (1)	$b_1 b c (c_1)$	1400
An–Gal+T+E	–	79	4 (2)	b_1	<5
Aer–Glu	–	83	18 (4)	$b_1 a b c c_1$	2000

[a] % UFA – unsaturated fatty acid expressed as a percentage of the total fatty acid of the isolated structures. Protein synthesis – pmoles leucine incorporated into protein/20 min/mg protein. Figures in parentheses are amino acid incorporation activity of structures isolated from protoplasts after prefixation with 2% glutaraldehyde. Cytochromes were identified by low temperature spectroscopy. NADH oxidase – μmoles NADH oxidized/min/mg protein.

that the absence of activity might be due to the loss of soluble cofactors from the extensively disrupted non-prefixed structures, therefore, similar experiments were also performed with the morphologically intact structures prepared from prefixed samples. Control experiments showed that mitochondria from Aer–Glu and An–Gal+T+E structures retained 20–30% of their amino acid incorporation activity after prefixation, but the intact An–Glu structures again had negligible activity, strongly suggesting the absence of a functional protein synthesizing system. Even after 60 min aeration, amino acid incorporation

by the structures was equivocal, but after 120 min aeration significant chloramphenicol-sensitive and ATP-dependent incorporation characteristic of aerobic mitochondria [24] was obtained. The specific activity of leucine incorporation by the structures after 120 min aeration was similar to that of the An–Gal+T+E structures, and increased further at 180 min.

Cytochromes and respiratory enzymes

The induction of the synthesis of cytochromes and respiratory enzymes could be approximately correlated with the appearance of amino acid incorporation activity and the changes in fatty acid composition. At the onset of aeration the anaerobic structures contained only cytochrome b_1, absorption maximum 557 mμ. However, after 120 min aeration, when the fatty acid composition of the structures approached that of the mitochondria from Aer–Glu cells, and coinciding with the appearance of clearly measurable amino acid incorporating activity, cytochromes c and b were detectable in the absorption spectrum. As a more sensitive measure of the overall activity of the respiratory chain NADH oxidase activity was determined. This activity was first measurable at 60 min and increased markedly with time of aeration, approaching the level present in mitochondria from Aer–Glu cells after 180 min aeration; traces of cytochromes $a + a_3$ and an increased amount of cytochrome c were also observed in the spectrum after this period.

Discussion

Recent studies have shown that mitochondria can exist in a number of conformational forms whose interconversion is influenced by metabolic and osmotic factors [25,29–31]. Thus in studying the nature of the mitochondrial precursors in anaerobically grown yeast cells it was important to establish whether organelle morphology reflects differences in biochemical composition or differences in metabolic state. Yeast mitochondria from aerobically grown cells were found to exist in two extreme conformational states similar to the condensed and orthodox states of rat-liver mitochondria as described by Hackenbrock [25]. However, when first isolated from non-prefixed cells, the mitochondria existed in a number of conformational forms mostly intermediate between the two extremes. The quantitative conversion of all mitochondria to the condensed conformation after density gradient contrifugation demonstrates that the various forms were different conformations of the same structure. Since a mixture of conformational forms were also observed from cells which had been prefixed with glutaraldehyde, it is clear that conformational changes also occur in vivo. The conformational transformations of yeast mitochondria both in vivo and in vitro are currently under investigation in this laboratory.

Previous studies as to the presence or absence of mitochondrial profiles in anaerobically grown yeast cells have been equivocal. Recently, however, Damsky et al. [12] have reported that mitochondrial membranes were not observed in permanganate fixed An–Glu cells, but were visible in partially disrupted aldehyde fixed cells. These profiles they described as having a dense matrix and few cristae. On the other hand, Schatz and co-workers [13] in recent freeze-etching studies have reported that the cytology of anaerobic yeast cells grown with or without lipid supplements was very similar to that of aerobic yeast cells containing mitochondrial structures with inner membranes showing typical *cristae mitochondriales*. It is nonetheless clear that lipid depleted anaerobically grown cells do contain a form of mitochondrial precursor [12,13,16]. However the present investigation on the nature of the mitochondrial precursors in anaerobic yeast cells has confirmed the observations of Damsky et al. [12] and Wallace et al. [7] that the cytology of the An–Glu cells is quite different from that of aerobic and lipid supplemented anaerobic cells. Moreover, the present report concludes that the mitochondrial structures in An–Glu cells are fundamentally different in biochemical properties from the aerobic and lipid supplemented anaerobic structures.

The mitochondrial precursors present in anaerobically grown cells differed in morphology from any of the conformations of mitochondria from aerobically grown cells. Structures isolated from An–Gal+T+E cells, while showing a clear relationship to that of mitochondria from Aer–Glu cells, had a distinctive honeycomb like internal membrane system. A somewhat similar appearance was observed in the structures isolated from An–Glu+T+E cells, but a smaller number of electron transparent areas within the matrix was noted. The presence of broken structures and membrane fragments in particulate fractions from An–Glu cells indicated that the precursor structures in these cells were very fragile in comparison to those from the lipid supplemented cells. Prefixation with glutaraldehyde prior to breakage was a prerequisite for the isolation of morphologically intact structures. The An–Glu structures were simple in morphology, possessing a dense granular matrix and no obvious folded inner membrane system. The possibility is not excluded that the method of fixation may obscure the presence in these structures of some folded inner membranes and cristae, nevertheless these structures are clearly the most primitive of the mitochondrial precursors.

Precursor structures isolated from prefixed or non-prefixed anaerobically grown cells, lipid supplemented or otherwise, contained the enzymes succinate dehydrogenase, oligomycin sensitive ATPase and malate dehydrogenase, thus confirming the conclusion from the electron microscope studies of their identification as mitochondrial precursors. The specific activities of the enzymes succinate and malate dehydrogenases were relatively low in the mitochondrial structures from An—Glu cells, significantly higher in the mitochondrial structures from the lipid supplemented cells and several-fold higher in mitochondria from Aer—Glu cells. Criddle and Schatz [14] have also recently shown the presence of similar biochemical properties in high speed particulate fractions from anaerobic yeast cells but morphological properties of the fractions were not reported.

The inability to demonstrate significant amino acid incorporation by the isolated particles from An—Glu cells raises the important question as to whether an active mitochondrial protein-synthesising system is indeed lacking in such cells in vivo. Recently, Forrester et al. [32] in our laboratory have demonstrated that whilst An—Gal+T+E and An—Glu+T+E mitochondrial precursor structures contained the two unique species of mitochondrial ribosomal RNA, these RNA species were absent from An—Glu structures. On the other hand a small amount of heterogeneous RNA was identified, corresponding in size to a mean value of about 10 S (range 6—14 S) and with a guanine + cytosine (G + C) content of 48% (32); total yeast mitochondrial RNA and yeast mitochondrial rRNA have a G + C content of about 27% [33—35]. These observations make it unlikely that the failure to detect mitochondrial rRNA was due to degradation of the RNA during isolation procedures. Taken in conjunction with the apparent lack of amino acid incorporation activity in the An—Glu structures, the complete absence of mitochondrial ribosomes is suggested. Speculations as to the possible correlation between the presence or absence of mitochondrial ribosomes and the lipid composition of the mitochondrial membrane may be made. For example, specific membrane associations,

which could be affected by lipid depletion, may be required for the transcription of mitochondrial DNA. Secondly, the inability to obtain mitochondrial ribosomes in high yield and free from mitochondrial membrane suggests an intimate association of ribosomes in the membranes. Accordingly, changes in membrane lipid composition could affect the association of ribosomes with membranes and prevent integration of the ribosomes into the organelle, and if the ribosome is assembled on the mitochondrial membrane, failure to integrate and failure to assemble could be synonymous. This in turn could lead to the inhibition of the synthesis of mitochondrial rRNA, perhaps due to a failure to remove the RNA from its template, the mitochondrial DNA [35]. A further possibility is based on the observation that the proteins of the mitochondrial ribosome are synthesized on the cytoplasmic ribosomes [28]. Consequently, the mitochondrial ribosome could be assembled in the cytoplasm, and its failure to integrate into the mitochondrion would lead to an accumulation of mitochondrial ribosomes in the cytoplasm.

The induction of cytochromes and respiratory ability of anaerobic yeast cells on aeration has long been known [2], and changes in the cytology of the cell during this period have been observed [4]. Experiments reported in this paper using isolated structures have shown that concomitant with the morphological development of the mitochondrial precursor structures into classical mitochondria there was an increase in the amino acid incorporation activity as well as the formation of mitochondrial cytochromes and NADH oxidase activity. Since the synthesis of antimycin sensitive NADH oxidase requires mitochondrial protein synthesis [36], the appearance of this enzyme activity after 60 min aeration suggests that the observed small amino acid incorporation activity by these organelles at this time is significant, and indicates that the formation of functional mitochondrial ribosomes is well advanced at this time.

References

[1] B. Ephrussi and P.P. Slonimski, Compt. Rend. 230 (1950) 685.
[2] P.P. Slonimski, La Formation des Enzymes Respiratoires chez la Levure (Masson, Paris, 1953).

[3] A. Lindenmayer and R.W. Estabrook, Arch. Biochem. Biophys. 78 (1958) 66.

[4] P.G. Wallace and A.W. Linnane, Nature 201 (1964) 1191.

[5] G. Morpurgo, G. Serlupi-Crescenzi, G. Teece, F. Valente and P. Venettacci, Nature 210 (1964) 897.

[6] E.R. Tustanoff and W. Bartley, Biochem. J. 91 (1964) 595.

[7] P.G. Wallace, M. Huang and A.W. Linnane, J. Cell Biol. 37 (1968) 307.

[8] B. Ephrussi, P.P. Slonimski, Y. Yotsuyanagi and J. Tavlitzi, Compt. Rend. Trav. Lab. Carlsberg, Sci. Physiol. 26 (1956) 87.

[9] Y. Yotsuyanagı, J. Ultrastruct. Res. 7 (1962a) 121.

[10] Y. Yotsuyanagi, J. Ultrastruct. Res. 7 (1962b) 141.

[11] A.W. Linnane, in: Oxidases and Related Redox Systems, eds. T.E. King, H.S. Mason and M. Morrison (John Wiley, New York, 1965) 2, 1102.

[12] C.H. Damsky, W.H. Nelson and A. Claude, J. Cell. Biol. 43 (1969) 174.

[13] H. Plattner and G. Schatz, Biochemistry 8 (1969) 339.

[14] R.S. Criddle and G. Schatz, Biochemistry 8 (1969) 322.

[15] H. Paultauf and G. Schatz, Biochemistry 8 (1969) 335.

[16] K. Watson, J.M. Haslam and A.W. Linnane, J. Cell Biol. 46, in press.

[17] J. Jayaraman, C. Cotman, H.R. Mahler and C. Sharp, Arch. Biochem. Biophys. 116 (1966) 224.

[18] O. Arrigoni and T.P. Singer, Nature 193 (1962) 1256.

[19] M.J. Vary, C.L. Edwards and P.R. Stewart, Arch. Biochem. Biophys. 130 (1969) 235.

[20] M.E. Pullman and G.C. Monroy, J. Biol. Chem. 238 (1963) 3762.

[21] R.G. Wake, J. Mol. Biol. 25 (1967) 217.

[22] C.L. Schildkraut, J. Marmur and P. Doty, J. Mol. Biol. 4 (1962) 430.

[23] D. Jollow, G.M. Kellerman and A.W. Linnane, J. Cell Biol. 37 (1968) 221.

[24] A.J. Lamb, G.D. Clark-Walker and A.W. Linnane, Biochim. Biophys. Acta 161 (1968) 415.

[25] C.R. Hackenbrock, J. Cell Biol. 30 (1966) 269.

[26] R.A. Harris, C.H. Williams, M. Caldwell and D.E. Green, Science 165 (1969) 700.

[27] G. Corneo, C. Moore, D.R. Sanadi, L.I. Grossman and J. Marmur, Science 151 (1966) 687.

[28] P.J. Davey, R. Yu and A.W. Linnane, Biochem. Biophys. Res. Commun. 36 (1969) 30.

[29] D.E. Green, J. Asai, R.A. Harris and J.T. Penniston, Arch. Biochem. Biophys. 125 (1968) 684.

[30] C.D. Stoner and H.D. Sirak, J. Cell Biol. 43 (1969) 521.

[31] G.R. Hunter, Y. Kamishima and G.P. Brierley, Biochim. Biophys. Acta 180 (1969) 81.

[32] I.T. Forrester and A.W. Linnane, in preparation.

[33] M. Fauman, M. Rabinowitz and G.S. Getz, Biochim. Biophys. Acta 182 (1969) 355.

[34] H.O. Halvorson, H. Morimoto, A. Scragg and J. Nikhorocheff, this symposium.

[35] E. Wintersberger and G. Viehhauser, Nature 220 (1968) 699.

[36] A.W. Linnane, D.R. Biggs, M. Huang and G.D. Clark-Walker, in: Aspects of Yeast Metabolism, ed. R.V. Mills. (Blackwell, Oxford, 1968) p. 217.

PROMITOCHONDRIA OF ANAEROBICALLY-GROWN YEAST: EVIDENCE FOR THEIR CONVERSION INTO FUNCTIONAL MITOCHONDRIA DURING RESPIRATORY ADAPTATION

H. PLATTNER, M. SALPETER, J. SALTZGABER, W. ROUSLIN and G. SCHATZ

Department of Applied Physics and Section of Biochemistry and Molecular Biology, Cornell University, Ithaca, New York, USA

It is well established that anaerobically-grown yeast cells lack a functional respiratory system but adaptively form it upon aeration [1]. Although adapting yeast is thus obviously an attractive experimental system for studying the formation of mitochondria, the fate of these organelles during anaerobic growth and respiratory adaptation has long remained uncertain. Initially, Linnane and his colleagues proposed that anaerobically-grown yeast cells were completely devoid of mitochondria and that the functional mitochondria arising during adaptation were formed either de novo [2] or by transformation of nonmitochondrial membranes [3]. However, electron micrographs of anaerobically-grown *Saccharomyces cerevisiae* by Morpurgo et al. indicated that mitochondria were still present provided the cells had been grown in a medium supplemented with Tween 80 and ergosterol [4]. Schatz subsequently isolated and characterized mitochondria-like particles from these anaerobic cells and proposed that respiratory adaptation of yeast involved the differentiation of functionally incomplete 'promitochondria' which pre-existed in the anaerobic cells [5].

In view of these results, Linnane and his collaborators modified their original hypothesis; they suggested that loss of mitochondria during anaerobic growth occurred only if the culture medium was deficient in unsaturated lipids [6–8].

Two years ago, however, Schatz et al. reported that anaerobically-grown *S. cerevisiae* contained mitochondrial particles regardless of whether the cells had been grown in the presence or the absence of added unsaturated lipids [9,10]. In either case, the nonrespiring promitochondria isolated from the anaerobic cells exhibited an ATPase activity that was inhibited by oligomycin and by a specific antiserum against purified ATPase from aerobic yeast mitochondria. The promitochondria also contained mitochondrial DNA (density = 1.685 g/cm^3) as well as several redox enzymes known to be associated with mitochondrial inner membranes [9–12]. Examination of the lipid-limited anaerobic yeast cells by the freeze-etching procedure clearly revealed numerous mitochondrial structures which closely resembled those present in aerobically-grown yeast cells [13]. The existence of promitochondria in anaerobically-grown, lipid-limited yeast was also confirmed by Damsky et al. [14]. Earlier claims about the complete loss of yeast mitochondria during anaerobic growth [2,3,6–8,15,16] must thus be discounted.

The present study attempts to define the role of yeast promitochondria during respiratory adaptation. If this process indeed involves differentiation of promitochondria [5,11], then specific components of these structures should be recovered in the functional mitochondria formed during adaptation. The label-transfer experiment described here confirms this prediction. In order to simplify the interpretation of our experiment, we divided it into several discrete steps. Moreover, we chose conditions of cell growth that permitted optimal development of the promitochondrial structures.

Methods

Yeast strains

The wild-type *S. cerevisiae* strains DT XII ($P\rho^+$, diploid) and D 273–10B ($\alpha P\rho^+$, haploid) as well as the corresponding cytoplasmic 'petite' mutants DT XIIa and D 273–10B–1 were grown aerobically in the presence of 0.8% glucose [11] or anaerobically in the presence of 0.3% glucose, Tween 80 and ergosterol [11]. Parallel labeling experiments with the two wild-type strains, or with the two 'petite' mutants gave essentially the same results; however, strain DT XII adapted to oxygen much more slowly than did strain D 273–10B.

Labeling of (pro)mitochondria in vivo

When the yeast cultures had reached the stationary phase, they were cooled to 0° and poisoned with cycloheximide (final concentration 25 μg/ml). The cells were collected by centrifugation in the cold (5 min at 1500 g) washed three times with cold 40 mM phosphate buffer pH 7.4, 25 μg/ml cycloheximide, 0.05% bovine serum albumin and suspended to 50 mg wet weight/ml in 40 mM phosphate buffer pH 7.4, 40 μg/ml cycloheximide. Three hundred ml of this suspension were mixed with 120 ml of 1 M glucose, 120 ml of 0.2 M phosphate buffer pH 7.4 and 60 ml of cycloheximide solution (375 μg/ml) and shaken for 10 min at 28° under a stream of nitrogen. 4,5-^{3}H-L-leucine (0.6 mCi; 5 Ci/mmole) was then added anaerobically and the incubation continued for 30 min. Up to this point, all steps were carried out under sterile conditions. Labeling was stopped by adding 600 ml of 0.2 M unlabeled leucine pH 7.4 and incubating for an additional 10 min.

Adaptation of the labeled cells

The labeled cells were washed six times at 0° with 40 mM phosphate buffer pH 7.4, 0.3% glucose, 20 mM unlabeled leucine and suspended to 20 mg (wet weight)/ml in ice-cold adaptation buffer (40 mM phosphate buffer pH 7.4, 20 mM unlabeled leucine, 1% ethanol and 0.3% glucose). One half of this suspension was shaken at 28° under nitrogen while the other half was shaken at 28° in air. Both incubations were carried out in the dark. At regular intervals, 2 ml samples of the aerated cell suspension were assayed for respiration in the absence and the presence of 0.5 mM KCN. Respiration was measured at 25° in a Gilson oxygen polarograph. Q_{O_2} was expressed as μl of oxygen consumed per h per mg dry weight. Only the cyanide-sensitive fraction of the observed oxygen uptake was considered in calculating the Q_{O_2}.

Fractionation of the labeled cells

The cells were collected by centrifugation and washed three times with homogenization medium (0.25 M mannitol, 20 mM Tris SO_4 pH 7.4, 20 mM unlabeled leucine, 2 mM EDTA, 25 μg/ml cycloheximide). The cells were disrupted [11] in this medium and large fragments were removed by two consecutive centrifugations for 10 min at 2000 g. The supernatant from the second centrifugation (termed 'homogenate') was divided into two aliquots. One aliquot was centrifuged for 90 min at 50,000 rpm in the Spinco no. 50 rotor to obtain the 'soluble proteins'. The second aliquot was centrifuged for 30 min at 17,500 rpm in the Spinco no. 30 rotor. The resulting pellet of 'promitochondria' was washed once by recentrifugation and further purified by 90 min centrifugation in a 20–70% sucrose gradient [11]. The aerobically-grown cells were labeled and fractionated under exactly the same conditions. In the experiments involving electron microscopy, promitochondria and mitochondria were isolated by the protoplast method of Kováč et al. [17]. All steps of the fractionation procedure were performed as close to +4° as possible.

Electron microscopy

Pellets of isolated promitochondria or of adapted mitochondria were fixed for 90 min at 0° with 1.33% (w/v) of OsO_4 in 0.8 M mannitol buffered to pH 7.2 with 70 mM s-collidine-HCl.

Staining for cytochrome oxidase was performed essentially as described by Seligman et al. [18]. Unfixed particles were suspended at a protein concentration of 2–4 mg/ml in a solution containing 0.05% 3,3'-diaminobenzidine, 2 μg/ml catalase (Boehringer), 0.1% horse heart cytochrome c (type III, Sigma), 50 mM Tris HCl pH 7.2 and 0.8 M sorbitol. (In some cases, the cytochrome oxidase reaction was blocked by adding 10 mM KCN.) The mixture (final volume 5–15 ml) was shaken in an open 400 ml beaker for 90 min at 0° and centrifuged for 30 min at 105,000 g. The sedimented particles were dispersed in 0.8 M sorbitol, 5% gelatine pH 7.2 at 38°, followed by quick cooling. This step was an attempt to achieve proper spacing of the particles for subsequent radioautography. After the preparation had solidified, it was cut into small blocks. These were washed three times with 0.8 M mannitol, 70 mM s-collidine-HCl pH 7.2, treated with OsO_4 as described above, dehydrated and embedded as described by Luft [19].

Radioautograms of the labeled and stained particles were obtained according to Salpeter and Bachmann [20]. Pale-gold ultrathin sections were mounted on collodion-coated slides, covered with a thin carbon layer, and coated with a monolayer of Ilford L-4 nuclear track emulsion (purple interference color). The specimens were exposed for 2–7 months at +4°, developed in Microdol X (Kodak), fixed in a non-hardening fixer, and photographed at random in an AE1 electron microscope whose magnification had been carefully calibrated. The autoradiograms obtained by this procedure were quantitatively evaluated according to Salpeter et al. [21].

Miscellaneous procedures

Cell growth was followed either by direct cell counts or by turbidity measurements at 550 mμ. NADH oxidase, F_1-ATPase and protein were assayed as described [11]. For the determination of protein-bound radioactivity, the samples were precipitated with an equal volume of 10% trichloroacetic acid, washed twice with 5% trichloroacetic acid, and heated in 5% trichloroacetic acid for 15 min at 90°. The residue was washed twice with ethanol:ether (3:1), dried in vacuo, dissolved in concentrated formic acid, and counted in a liquid scintillation spectrometer. Iso-1-cytochrome c was isolated from the adapted yeast cells according to Sherman et al. [22].

Results and discussion

The design of our experiment is as follows: if promitochondria are selectively labeled with ^{3}H-leucine within anaerobic yeast cells, the fate of this label during subsequent respiratory adaptation in the presence of non-radioactive leucine should identify the pathway of mitochondrial formation. Thus, functional mitochondria arising via differentiation of promitochondria should be labeled since they would receive the promitochondrial radioactivity without extensive dilution. On the other hand, respiring mitochondria formed de novo or from structures other than promitochondrial membranes should be unlabeled since the free leucine pool as well as non-promitochondrial membranes are essentially cold during adaptation.

Step 1:Selective labeling of the (pro)mitochondrial proteins within intact yeast cells. Linnane and his colleagues have shown that, in intact aerobic yeast cells, cycloheximide inhibits protein synthesis by the cytoplasmic ribosomal system without impairing mitochondrial protein synthesis [23]. Therefore, if yeast cells are labeled with ^{3}H-leucine in the presence of cycloheximide, only the mitochondrial insoluble proteins [24] should become radioactive. Experiments with aerobically-grown yeast cells substantiated this prediction (table 1). However, only three out of the five different *S. cerevisiae* strains tested by us were sufficiently sensitive to cycloheximide to yield this specific labeling pattern.

The results obtained with anaerobically-grown cells were closely similar. As seen in table 2, virtually all of the label incorporated into the cells proved to be associated with the promitochondrial fraction. In contrast, if the cells were labeled and chased in the absence of cycloheximide, the specific radioactivity of the isolated promitochondria was only about twice that of the soluble proteins.

Since the rate of labeling in the presence of

Table 1
Selective labeling of yeast mitochondria in vivo

Fraction	cpm/mg protein	% of total incorporation
Homogenate	1120	(100)
Soluble proteins	204	10
Mitochondria	7950	89

The experiment was carried out with the aerobically-grown wild-type strain D 273–10B. The concentration of mitochondria in the homogenate was determined by measuring the specific activity of NADH oxidase and F_1-ATPase in the homogenate and the purified mitochondria [11].

Table 2
Selective labeling of promitochondria in vivo

Strain	Fraction	cpm/mg protein	% of total incorporation
DT XII	Homogenate	633	(100)
	Soluble proteins	76	5.6
	Promitochondria	15,850	102
D 273-10B	Homogenate	820	(100)
	Soluble proteins	87	5.8
	Promitochondria	11,800	99

The experiment was carried out with anaerobically-grown wild-type cells. The concentration of promitochondria in the homogenates was determined by ATPase measurements [11].

cycloheximide is less than one percent of that observed in the inhibitor-free controls, the data of tables 1 and 2 may reflect labeling by contaminating bacteria or by spurious side-reactions. For

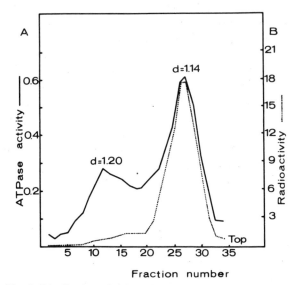

Fig. 1. Distribution of radioactivity and ATPase after centrifuging labeled promitochondria in a sucrose gradient. Promitochondria of the wild-type strain DT XII were labeled in vivo and purified as described under Methods. After gradient centrifugation, 35 fractions were collected and assayed for ATPase activity and for protein-bound radioactivity. Ordinate A: μmole ATP split per 5 min per 0.05 ml fraction. Ordinate B: cpm $\times 10^3$ per ml fraction. At least 50% of the ATPase activity equilibrating at a density of 1.20 g/ml represents nonmitochondrial ATPase insensitive to the F_1-inhibitor of Pullman and Monroy [25].

Table 3
Lack of selective promitochondrial labeling in a cytoplasmic 'petite' mutant

Fraction	Control cpm/mg protein	Plus chloramphenicol com/mg protein
Homogenate	174	174
Soluble proteins	38	39
Promitochondria	120	138

The experiment was identical to that described in table 2 except that anaerobically-grown cells of the cytoplasmic 'petite' mutant DT XIIa were used and that one aliquot of the cells was labeled in the presence of cycloheximide *and* 4 mg/ml chloramphenicol. Only the specific radioactivities are given since the impaired binding of mitochondrial ATPase to the mutant promitochondria [11] made it impossible to measure the concentration of these particles in the homogenates.

various reasons, however, this is extremely unlikely. Firstly, all steps preceding the chase with unlabeled leucine were performed under sterile conditions. Secondly, isopycnic banding of the labeled, crude promitochondria in a sucrose gradient revealed a close correlation between radioactivity and mitochondrial ATPase activity (fig. 1). Thirdly, the selective labeling of the promitochondria was approximately 90% inhibited by 4 mg/ml chloramphenicol, a specific inhibitor of mitochondrial protein synthesis [23]. Finally, analogous experiments with the two cytoplasmic 'petite' mutants (described under Methods) failed to show any comparable labeling of the promitochondrial fraction (table 3). Thus, the specific radioactivity of the promitochondria from the mutant DT XIIa was only 0.75% of that observed with promitochondria of the corresponding wild-type strain. This small residual labeling was insensitive to chloramphenicol (table 3) and probably represents a minor 'cycloheximide-leak' that is less evident in the wild-type strains.

We conclude that our experimental procedure effects a selective labeling of (pro)mitochondrial proteins.

Step 2: removal of cycloheximide from the labeled cells. Since cycloheximide inhibits respiratory adaptation [26], its removal from the selectively labeled cells was a critical step in our procedure. We found that repeated washing of the poisoned cells with buffer restored both cytoplasmic protein synthesis and adaptability to oxygen. The washing buffer was supplemented with glucose and unlabeled leucine in order to keep the cells intact and to minimize redistribution of radioactivity.

Step 3: adaptation of the labeled cells in the presence of unlabeled leucine. When the washed cells were aerated at 28°, they rapidly acquired cyanide-sensitive respiration (fig. 2). After 8 hr, the Q_{O_2} was about one-third that of aerobically-grown yeast cells. An aliquot of the cells was incubated anaerobically and, as expected, did not undergo any significant adaptation (fig. 2). These cells were used as a control for protein turnover unrelated to the adaptive process.

Step 4: fractionation of the labeled, adapted cells. Table 4 lists the distribution of radioactivity in the anaerobically-grown cells, the adapted cells, and those incubated under nitrogen. It can be seen that, even after adaptation, over 90% of the cellular radioactivity was still concentrated in the mitochondrial fraction. In contrast, iso-1-cytochrome c from the adapted cells was found to be essentially cold. Since this protein is synthesized de novo during respiratory

adaptation, the externally added unlabeled leucine must have effectively diluted the intracellular pool of ^{3}H-leucine and thus prevented re-utilization of radioactivity. Nevertheless, these results did not yet establish a physical continuity between promitochondria and mitochondria since roughly 50% of the promitochondrial label has been lost during adaptation (table 4). This loss could have occurred by either one of the following mechanisms: 1) normal turnover of (pro)mitochondrial proteins during the 8-hr incubation; 2) degradation of promitochondrial membranes and simultaneous de novo assembly of functional mitochondria; according to this alternative, the 'mitochondrial' fraction isolated from the adapted cells represented a mixture of residual (labeled) promitochondria and adapted (unlabeled) mitochondria.

The former explanation is supported by the fact that cells incubated under nitrogen lost as much label as those undergoing respiratory adaptation (table 4). Furthermore, electron microscope radioautography showed that the radioactivity was primarily associated with adapted mitochondrial membranes (cf. below).

Step 5: electron microscopy and EM radioautography. When fixed in OsO_4 immediately after isolation, the mitochondrial fraction of adapted cells is

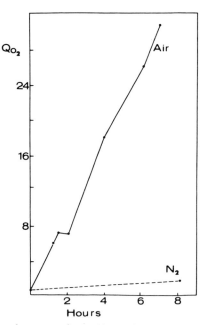

Fig. 2. Development of cyanide-sensitive respiration in the selectively labeled anaerobically-grown strain D 273-10B after removal of cycloheximide. Solid trace: aerated cells. Dotted trace: cells incubated under nitrogen.

Table 4
Fate of promitochondrial radioactivity during respiratory adaptation and during incubation under nitrogen

	Homo-genate	Soluble proteins	(Pro)mito-chondria
Anaerobic cells			
cpm/mg protein	543	83	8040
% of total radioactivity	(100)	9	103
Cells aerated for 8 hr			
cpm/mg protein	322	101	4050
% of total radioactivity	(100)	12	91
Cells incubated anaerobically for 8 hr			
cpm/mg protein	310	117	3370
% of total radioactivity	(100)	13	87

The experiment was carried out with the wild-type strain D 273–10B as described under Methods. The concentration of promitochondria in the unadapted homogenates was determined by ATPase measurements and the concentration of mitochondria in the adapted homogenates by assaying ATPase and NADH oxidase [11].

found to consist mainly of spheres (2500—3500 Å radius) with a dense matrix, but with no readily recognizable membranes (fig. 3). This picture does not change significantly after lead staining. If the unfixed pellet is first stained for cytochrome oxidase as described by Seligman et al. [18], one obtains densely stained membranes, arranged in circles (4000—5000 Å radius) with small, partly vesicular fragments inside the spheres. This reaction is eliminated by the addition of cyanide, and is not observed with promitochondria isolated from the unadapted yeast cells. It is therefore reasonable to consider the electron-dense membranes as adapted inner mitochondrial membranes. Partial fragmentation of small membrane pieces dispersed throughout the pellet could not be avoided. Aldehyde prefixation does not prevent this fragmentation and therefore was not used.

Radioautography was performed according to Salpeter and Bachmann [20] on sections from mitochondrial pellets stained for cytochrome oxidase, dispersed in gelatin, post-fixed in OsO_4 and embedded in Epon (see Methods). Due to the spread of radiation in autoradiograms, developed grains do not form immediately above a radioactive structure. The extent of this spread depends on the resolution of the preparation and has been described for a variety of specimens by Salpeter et al. [21]. One can therefore define radioactive structures in autoradiograms by comparing the experimental grain density distributions with the distributions calculated for different sources. By dispersing the pellets in gelatin, a sufficient separation of the stained particles was accomplished to make such an analysis possible. While the fragmentation of the stained membranes unfortunately made this specimen less than ideal for a clean quantitative analysis, the results were still highly significant.

On quick examination it appeared obvious that the developed grains were selectively associated with stained rings (fig 4). A quantitative analysis was performed whereby the grain density (grains/area) was plotted as a function of distance (inside and outside) from the outer edge of the stained rings. The distance was measured in units of resolution, HD, which, according to Salpeter et al. [21] is the distance from a radioactive line source within which half the devel-oped grains due to that source will lie. For the specimen used here, i.e., pale gold section, Ilford L4 emulsion, Microdol X development, the HD value is 1600 Å. The center of each developed grain was used to localize that grain and to measure its distance from the nearest stained ring. The intersection points of a grid superimposed over the autoradiogram provided a measure of relative area and were tabulated in a manner identical to that of the grain centers. (Density was thus grains/points.) To avoid areas of high grain density due to radiation spread from two adjacent stained particles, all grains and points were excluded from the tabulation if they were located less than 2 HD (3200) Å to the next nearest stained ring.

The histogram in fig. 5 presents the resultant experimental grain density distribution. Superimposed are two expected distributions, one for a circular source (of a radius equal to that of the stained rings) labeled only on the periphery, and the other for one labeled uniformly throughout. The results show that the ring structures are highly selective sites of radioactivity. The grain density rises inside the rings as would be expected if there were radioactivity distributed uniformly throughout the ring [21] and not only on the stained membrane. Stained membrane fragments are frequently seen inside the rings, and the actual amount of randomly oriented membranes is considerably underestimated even after heavy staining [27]. These may account for the occurrence of radioactivity inside the stained ring structures. The grain density outside the rings is higher than can be accounted for purely by radiation spread from the radioactive rings alone (compare histogram with expected scatter). This indicates a small amount of dispersed radioactivity. The fragmented stained particles may be the source of this radioactivity. Aggregates of unstained membranes were essentially free of radioactivity.

These results show that during respiratory adaptation promitochondrial radioactivity is transferred to respiring mitochondria. Since all the stained rings were pooled for our analysis we cannot yet assert that this is true for every individual in the entire mitochondrial population. It is quite clear, however, that the nonrespiring promitochondria of anaerobically-grown yeast cells are structural precursors of functional mitochondria.

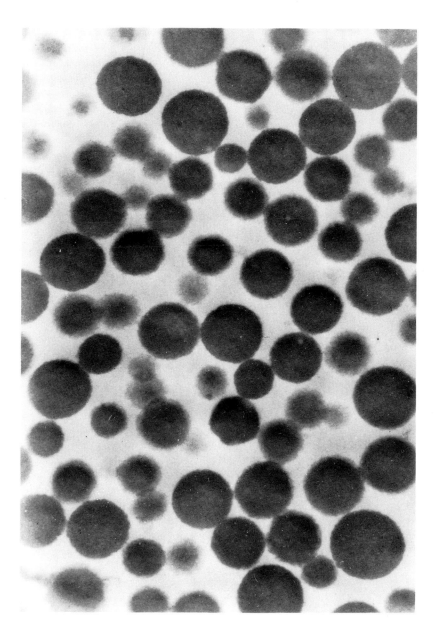

Fig. 3. Electron micrograph of mitochondria isolated from adapted yeast cells. The wild-type strain D 273-10B was grown anaerobically, labeled with leucine in the presence of cycloheximide, and adapted for 8 hr as described under Methods. Mitochondria were isolated from the cells according to Kováč et al. [17], fixed with OsO_4 and poststained with lead (27,000 X).

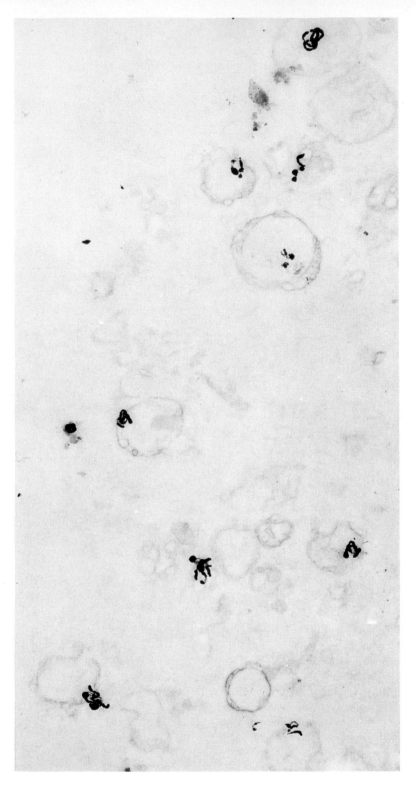

Fig. 4. Radioautogram of isolated mitochondria from oxygen-adapted yeast cells. The wild-type strain D 273-10B was grown anaerobically, labeled in the presence of cycloheximide, washed, and aerated for 8 hr. The mitochondria were isolated [17], stained for cytochrome oxidase, and prepared for electron microscope radioautography. See Methods for details. Only those membranes exhibiting cytochrome oxidase are visible in the electron micrograph. Note that most of silver grains are associated with particles limited by adapted membranes (16,000 X).

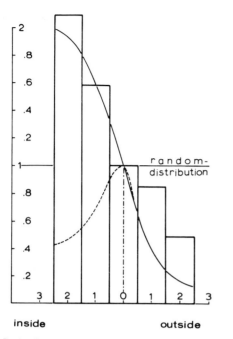

inside   outside

Fig. 5. Grain density distribution (grains/area) in radioauto-grams of mitochondria from labeled, adapted yeast D 273-10B. The experimentally observed distribution (bars) is compared with that calculated for a circular source labeled only at the periphery (broken line) and for one labeled uniformly throughout (solid line). The latter provides the best fit although there appears to be some radioactivity outside the stained structures which is not accounted for by radiation spread. Abscissa: Distance from outer edge of stained ring in units of HD (1600 Å). Ordinate: Grain density normalized to 1 at outer edge.

Summary

Promitochondria of anaerobically-grown *Saccharomyces cerevisiae* were selectively labeled in vivo by incubating the cells with radioactive leucine and cycloheximide under nitrogen. Upon aeration of the washed cells in the presence of unlabeled leucine, promitochondrial radioactivity was recovered in respiring mitochondria. These results show that respiratory adaptation of anaerobically-grown yeast involves differentiation of functionally incomplete promitochondria.

Acknowledgments

This study was supported by Grant No. GM 16320 from the U.S. Public Health Service. The experiments reported here are the outcome of a stimulating discussion with Drs. J. Johnston and Friedrich Paltauf whom we wish to thank at this point.

References

[1] P.P. Slonimski, in: La Formation des Enzymes respira-toires chez la Levure (Masson, Paris, 1953).
[2] P.G. Wallace and A.W. Linnane, Nature 201 (1964) 1191.
[3] A.W. Linnane, E. Vitols and P.G. Nowland, J. Cell Biol. 13 (1962) 345.
[4] G. Morpurgo, G. Serlupi-Crescenzi, G. Tecce, F. Valente and D. Venetacci, Nature 201 (1964) 897.
[5] G. Schatz, Biochim. Biophys. Acta 96 (1965) 342.
[6] H.B. Lukins, S.H. Tham, P.G. Wallace and A.W. Linnane, Biochem. Biophys. Res. Commun. 23 (1966) 363.
[7] P.G. Wallace, M. Huang and A.W. Linnane, J. Cell. Biol. 37 (1968) 207.
[8] G.M. Kellerman, B. Veitch and A.W. Linnane, Proc. Australian Biochem. Soc. 1969, abstract 165.
[9] G. Schatz, R.S. Criddle and F. Paltauf, in: Biochemie des Sauerstoffs, eds. B. Hess and Hj. Staudinger (Springer, Berlin, 1968) p. 318.
[10] G. Schatz and R.S. Criddle, in: Mitochondria − Structure and Function, 5th FEBS Symposium, Prague 1968, eds. L. Ernster and Z. Drahota (Academic Press, New York, 1969) p. 189.
[11] R.S. Criddle and G. Schatz, Biochemistry 8 (1969) 327.
[12] R.S. Criddle, F. Paltauf, H. Plattner and G. Schatz, J. Gen. Physiol. 54 (1967) 57.
[13] H. Plattner and G. Schatz, Biochemistry 8 (1969) 339.
[14] C.H. Damsky, W.M. Nelson and A. Claude, J. Cell. Biol. 43 (1969) 174.
[15] E.S. Polakis, W. Bartley and G.A. Meek, Biochem. J. 90 (1964) 369.
[16] C.J. Avers, J. Bact. 94 (1967) 1225.
[17] L. Kováč, H. Bednarová and M. Greksak, Biochim. Biophys. Acta 153 (1968) 32.
[18] A.M. Seligman, M.J. Karnovsky, H.L. Wasserkrug and J.S. Hanker, J. Cell Biol. 38 (1968) 1.
[19] Luft, J.H., J. Biophys. Biochem. Cytol. 9 (1961) 409.
[20] M.M. Salpeter and L. Bachmann, J. Cell Biol. 22 (1964) 469.
[21] M.M. Salpeter, L. Bachmann and E.E. Salpeter, J. Cell Biol 41 (1969) 1.
[22] F. Sherman, J.W. Stewart, E. Margoliash, J. Parker and W. Campbell, Proc. Natl. Acad. Sci. U.S. 55 (1966) 1478.
[23] R.H. Yu, H.B. Lukins and A.W. Linnane in: Biochemi-

cal Aspects of the Biogenesis of Mitochondria, eds. E.C. Slater, J.M. Tager, S. Papa and E. Quagliariello (Adriatica Editrice, Bari, Italy, 1968) p. 359.

[24] E. Wintersberger, Biochem. Z. 341 (1965) 409.

[25] M.E. Pullman and G.C. Monroy, J. Biol. Chem. 238 (1963) 3762.

[26] H. Fukuhara, Biochem. Biophys. Res. Commun. 18 (1965) 297.

[27] A.V. Loud, Proc. 25 Ann. Meet. El. Microsc. Soc. Amer. ed. C.J. Arcenaux (Claitor's Book Store, Baton Rouge, La.) 1967, p. 144.

MITOCHONDRIAL GENETICS IN YEAST:
SEGREGATION OF A CYTOPLASMIC DETERMINANT IN CROSSES AND ITS LOSS OR RETENTION IN THE PETITE

G.W. SAUNDERS, Elliot B. GINGOLD, M.K. TREMBATH,
H.B. LUKINS and Anthony W. LINNANE

Department of Biochemistry, Monash University,
Clayton, Victoria, Australia

The involvement of cytoplasmically coded genetic elements in the biogenesis of mitochondria has been recognized for many years; the cytoplasmic petite mutation, which gives rise to mitochondria deficient in a number of components necessary for respiratory activity was shown to be inherited extra-chromosomally by Ephrussi and colleagues [1]. A cytoplasmic factor, denoted the rho (ρ) factor was postulated with ρ^+ designating the factor in normal cells and ρ^- in petite, respiratory deficient cells [2]. The subsequent discovery of DNA in mitochondria suggested a possible site for the factor [3,4].

Considerable work has been done on the physical characterization of the mitochondrial DNA; however, the key question of the amount of information the mitochondrial DNA is capable of carrying has not been answered. Reports on the size of intact yeast mitochondrial DNA have varied from 5.5μ [5,6] through to very much larger pieces [7,8], intact closed circular DNA of 26μ having been recently reported [9]. It has also been suggested that there is more than one piece of DNA per mitochondrion, but it is unclear whether these are redundant or genetically distinct [10,11].

The possibility of the correspondence between ρ and the mitochondrial DNA received support from observations that the physical properties of the mitochondrial DNA often underwent large changes in ρ^- cells [12–15]. However, despite the fact that there are large numbers of pieces of mitochondrial DNA per cell, studies on the mutation to petite induced by UV-irradiation [16,17] elevated temperature [18]

and intercalating drugs [19,20] have all suggested a small number of ρ factors, although recent UV-irradiation studies of Allen and MacQuillan [21] have suggested a larger number of targets. Thus the question of genetic equivalence of all pieces of mitochondrial DNA is raised.

The use of the cytoplasmic petite mutation to elucidate the properties of the mitochondrial genetic system has provided a great deal of information, but because of its severe, irreversible, phenotypic effects it is difficult to work with. Another approach, initiated in our laboratory, has been the use of mutants resistant to antibiotics that selectively inhibit the mitochondrial system. We have previously described a mutation to erythromycin resistance which is inherited cytoplasmically in a similar manner to the petite mutation [22]. This mutation does not, however, impair the functioning of the mitochondria as does the petite and thus the use of the determinant for erythromycin resistance and sensitivity can provide information on the cytoplasmic genetic system unobtainable by study of the petite mutation alone.

This paper describes the use of the determinant for erythromycin resistance and sensitivity (denoted ER^r and ER^s) to investigate the fate of the mitochondrial genomes on cell fusion and subsequent budding from the zygote. It is demonstrated that determinants from both parents can enter each bud although the contribution made by each parent is not equal and can be influenced by the mating type. It follows that no single mitochondrial genome determines the cytoplasmic genetic character of the early progeny.

We have previously studied the retention of the erythromycin determinant in different petites and the instability of the determinant in petites [23]. Continued study of the retention of the erythromycin determinant in conjunction with the suppressiveness of petites is now reported. The data provide a new insight into suppressiveness suggesting that neutral and highly suppressive petites have the most altered mitochondrial DNA and petites of intermediate suppressiveness the least.

Methods and materials

Media

The following types of media were used in the experiments.

YEP-glucose: 1% Difco yeast extract, 2% peptone, 2% glucose, 2% agar added for solid medium.

YEP-glycerol: Same as YEP-glucose except 4% glycerol replaced glucose.

YEP-PD (petite determining): YEP-glycerol plus 0.1% glucose. Petites are only able to grow until the glucose is exhausted, whereas normal cells can continue to proliferate on the glycerol, thus a large size differentiation occurs.

Minimal-glucose: As Wickerham [24], 2% glucose as energy source.

MPD (Minimal petite determining): As minimal glucose, except 3% glycerol and 0.1% glucose replaced 2% glucose.

MPD-ERY: As MPD, with addition of 4 mg/ml erythromycin.

Mating procedures

Crosses were performed by mixing overnight cultures of the haploid parents in fresh YEP-glucose medium. After 4.5 hr the mixture was either diluted and plated, or just diluted into minimal glucose tubes and grown for a further 24 hr before plating.

For microdissection of zygotes, the initial procedure was the same, after 4.5 hr a sufficient number of zygotic figures could be observed in the mating mixture to make dissection possible. An appropriate dilution of the mixture was streaked on a YEP-glucose agar slab which was then mounted in a microdissection chamber prior to micromanipulation. Zygotes were then removed to isolated positions and

in some experiments they were grown up without further dissection. Where buds were to be dissected from the zygote, the microdissection chamber was incubated at 30°C and as mature buds formed they were detached and removed to a different position on the agar slab.

Determination of erythromycin resistance in crosses

Zygotes (which are formed after 4.5 hr in the mating mixture), their progeny (after 24 hr further growth in minimal glucose) or suspensions of colonies arising in the microdissection experiments were diluted and plated onto MPD and MPD-Ery. After 4—5 days the plates were overlayed with tetrazolium agar by the method of Ogur et al. [25]. Mixed and entirely resistant colonies are coloured on MPD and MPD-Ery plates, whereas sensitive colonies are coloured only on MPD plates. The percentage of petites present is obtained from the MPD plates and, after correction for these, the percentage of white colonies on the MPD-Ery plates gives a measure of the percentage sensitive colonies.

Determination of suppressiveness of petites

The percentage of petite zygotic colonies were determined on MPD plates as described. In calculating the percentage suppressiveness of a given petite, it is necessary to correct for the initial number of ρ^- cells in the ρ^+ parent employed in the cross [26].

Results

Vegetative segregation of cytoplasmic determinants from zygotes of an ER^S × ER^r cross

Previous work in this laboratory has shown that the diploid colonies obtained by plating the mating mixture of the cross ER^r × ER^S on minimal medium are generally comprised of a mixture of ER^r cells and ER^S cells [22]. These diploid colonies were assumed to have arisen from the vegetative growth of single zygotes. The evidence was strongly in favour of this assumption, but to rule out the possibility that the mixture arose from the clumping of cells single zygotes were isolated by micromanipulation to determine whether they formed mixed colonies similar to those obtained by plating experiments. Fig. 1 shows the degree of mixedness of 13 colonies grown from

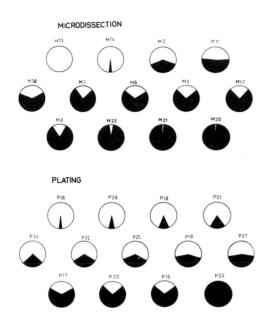

Fig. 1. Comparison of 13 clones isolated by micromanipulation of single zygotes, with 13 colonies, selected randomly, obtained by plating a mating mixture after 4.5 hr. Each clone is represented by a circle, the percentage resistant cells is represented by a black sector, the size of the black sector indicating this percentage.

single zygotes as compared with 13 colonies obtained by plating the mating mixture after 4.5 hr. Colonies obtained by both methods show a broad distribution of mixedness. Thus it can be concluded that single zygotes give rise to mixed diploid colonies, and also that the plating method is quite adequate for obtaining clones from single zygotes.

We previously reported that the resistant and sensitive cells making up the mixed colony are themselves stable, giving rise on replating to pure resistant and sensitive colonies. In addition, it has been found that if instead of plating the zygotes directly after 4.5 hr, the mating mixture is inoculated into minimal glucose medium and the progeny plated after 24 hr growth, instead of mixed colonies, pure resistant and sensitive colonies are obtained. It was concluded that an unstable state exists in the zygote, but the progeny it eventually gives rise to are stable resistant, or stable sensitive.

From a study of the genetic characteristics of the buds as they arise from the zygotes it is possible to determine whether the phenomenon of vegetative segregation involves events restricted solely to the zygote, that is whether all buds are stable and give rise only to pure clones, or whether the unstable state can be transferred to the buds which in turn give rise to mixed colonies. This question can be resolved by dissecting buds from the zygotes and analyzing their progeny. After some practice it is easy to recognize the zygotic figures although they represent only a small percentage of the total cell population in a given cross. These zygotes have a dumb-bell configuration and the first bud arises from close to the fusion bridge. The results obtained for the percentage sensitive and resistant cells in clones arising from buds separated from the zygotic figure and similarly the percentage resistant and sensitive cells in the clone arising from the remaining zygote are depicted in fig. 2. It is apparent that the zygotes and the buds all give rise to mixed clones. It follows that at least two mitochondrial genomes enter each zygotic bud. Indeed the evidence suggests a larger number due to large variation in the mixedness of the buds and the

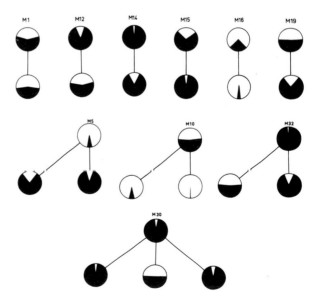

Fig. 2. Degree of mixture with respect to erythromycin resistance and sensitivity of colonies grown from zygotes and buds removed from the zygotes by microdissection. Colonies are represented in the same way as in fig. 1, the relationship between zygote and buds is shown in the form of a pedigree.

fact that no buds have been so far observed to give rise to pure colonies. Neither of these results would seem likely with only 2 or 3 determinants entering each bud.

In summary, it is clear that when two haploids, one ER^r and the other ER^s fuse to form a zygote, each zygotic bud receives determinants from both parents, and it is only in later divisions that cells containing only one type of determinant arise.

Unequal contributions from each parent to the final progeny

It was noted above that if the diploid progeny are grown for 24 hr prior to plating, then pure resistant and sensitive colonies are obtained. The ratio of resistant to sensitive colonies provides a measure of the overall average contribution each parent makes to the cytoplasmic genotype of the diploid progeny. If, for example, the contribution was equal, the ratio of stable segregants of each type should be 50:50, but this has not been observed experimentally. The crosses examined have established that the mating

type has a profound influence on the characteristics of the progeny. The following strains isogenic except for the nature of the erythromycin determinant were used in the experiments, L411 (α ER^r) derived by single step mutation from L410 (α ER^s), similarly L2201 (a ER^r) – L2200 (a ER^s), L601 (a ER^r) – L600 (a ER^s), L2001 (α ER^r) – L2000 (α ER^s). As shown in table 1 (upper part) crosses were arranged in pairs, differing only in which strain carried the resistance determinant. In all cases the cytoplasmic determinant carried by the α mating type has been found to predominate in the progeny, although the extent of the domination varies greatly, so that in some crosses 90% of the progeny have the cytoplasmic characteristic of the α mating types whereas it is only 60% in other crosses. These findings are further supported by crosses between a number of unrelated sensitive strains of different mating types with L411 (α ER^r), L2201 (a ER^r), L601 (a ER^r) and L2001 (α ER^r) (table 1, lower part). The results support a recent report by Slonimski (private communication) that the mating type is a determining factor in

Table 1
The influence of mating type on the distribution of ER^r and ER^s to vegetative diploid progeny

Mating type carrying resistance	Cross			% ER^r	% ER^s
α	L411	(α ER^r) $\times$ L2200	(a ER^s)	65	35
a	L410	(α ER^s) $\times$ L2201	(a ER^r)	28	72
α	L411	(α ER^r) $\times$ L600	(a ER^s)	90	10
a	L410	(α ER^s) $\times$ L601	(a ER^r)	29	71
α	L2001	(α ER^r) $\times$ L600	(a ER^s)	70	30
a	L2000	(α ER^s) $\times$ L601	(a ER^r)	28	72
α	L411	(α ER^r) $\times$ 101-5D	(a ER^s)	85	15
α		$\times$ L300	(a ER^s)	82	18
α		$\times$ 56-28D	(a ER^s)	61	39
α		$\times$ L2100	(a ER^s)	68	32
α	L2001	(α ER^r) $\times$ L300	(a ER^s)	65	35
α		$\times$ 45	(a ER^s)	59	41
a	656	(α ER^s) $\times$ L2201	(a ER^r)	41	59
a	C221	(α ER^s) $\times$ L2201	(a ER^r)	40	60
a	656	(α ER^s) $\times$ L601	(a ER^r)	8	92
a	C221	(α ER^s) $\times$ L601	(a ER^r)	31	69

Crosses were performed as in methods section, with the diploid progeny grown for 24 hr on minimal medium before plating on to MPD and MPD-ERY.

cytoplasmic genetics; in such a scheme α would be the female and a the male parent.

Variation of informational content of petites

The petite mutation, mainly because of its irreversibility, was originally conceived as loss of a cytoplasmic genetic particle, but recent physical studies have indicated that the mutation is characterized by a considerable modification of mitochondrial DNA rather than a complete loss. A question that arose was whether the mitochondrial DNA in the petite was able to retain some of its information, or if it was reduced to nonsense.

We have previously shown that in some petites, at least, information is retained [23], the ER^r determinant being genetically detected in the cells. Erythromycin resistance or sensitivity cannot be directly determined in petite ρ^- cells, since they are respiratory deficient. It is not, therefore, possible to test directly whether petites derived from erythromycin resistant strains retain their resistance characteristic. However, if a petite derived from a resistant cell is crossed with a respiratory competent erythromycin sensitive strain, the retention or otherwise of the erythromycin determinant by the petite can be identified in the resulting respiratory competent diploids. Thus, if a cross between $\rho^+ER^s \times \rho^-$ gives rise to some ρ^+ER^r progeny, then the ρ^- cells must contain the ER^r determinant and can be denoted ρ^-ER^r. About 33% of spontaneously arising petites, and a much smaller proportion of euflavine induced petites can be shown to be ρ^-ER^r. In these ρ^-ER^r petites, clearly the mitochondrial DNA is able to carry some information.

More recent experiments have shown that there may be three states of the erythromycin determinant. Petites derived from erythromycin resistant respiratory competent cells either retain the erythromycin resistance determinant or completely lose the determinant and may therefore be denoted ρ^-ER^r or ρ^-ER^o, respectively. Similarly petites derived from erythromycin sensitive cells can be shown to be of the genotype ρ^-ER^s or ρ^-ER^o. These conclusions are based on the experiments outlined in fig. 3. The ρ^-ER^o strains are unable to contribute resistance or sensitivity to diploid progeny on crossing with ρ^+ER^r or ρ^+ER^s cells. The ρ^-ER^o petite has lost the determinant that codes for resistance or sensitivity, that is,

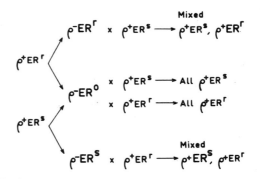

Fig. 3. The three states of the erythromycin determinant in petite cells. The figure shows the erythromycin resistance characteristics of the ρ^+ diploid derived from crossing petites with respiratory competent erythromycin resistant and sensitive strains. On the basis of these crosses, petites can be assigned the genotype ρ^-ER^r, $\rho^-ER^{(s)}$ or ρ^-ER^o, where ER^o represents the complete loss of the erythromycin determinant. Respiratory deficient ρ^- diploids are also obtained, but are not shown in the figure.

it must have undergone a loss of information from the DNA.

Suppressiveness and information content

Another characteristic that has been used to differentiate various cytoplasmic petites is that of suppressiveness, which is measured as the percentage of entirely petite zygotic colonies obtained in a cross of the petite with a ρ^+ haploid. Petites vary from the so-called neutrals, which give only ρ^+ diploids, through to highly suppressive, which may yield up to 99% ρ^- zygotes [27]. A number of authors have studied the relationship of the degree of suppressiveness to extent of the changes observed in the buoyant density of mitochondrial DNA [12,14,15]. Although the results have been equivocal the large alterations in some highly suppressive petites have led Carnevali et al. [15] to suggest that the more grossly altered the DNA, the higher would be the suppressiveness of the petite. Under such a scheme neutral petites would have the least altered mitochondrial DNA. However this model is not consistent with a number of our observations. Fig. 4 illustrates a comparison between the suppressiveness of a number of spontaneous petites and petites induced with a high concentration of euflavine (5 mg/ml); it is apparent that the two classes are quite different. The majority of the euflavine induced petites are neutral or of low suppressive-

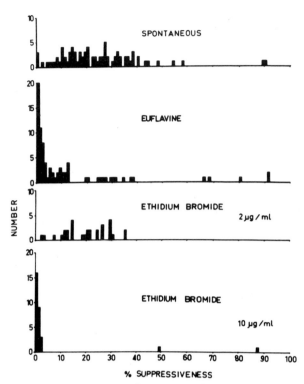

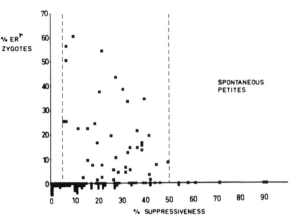

Fig. 5. The percentage suppressiveness of a large number of spontaneously arisen petites is shown plotted against the percentage ER^r zygotes, that is, the percentage of respiratory competent zygotes that show the presence of the ER^r determinant by yielding some erythromycin resistant diploids. ρ^-ER^0 petites lie along the base line; no zygotes obtained from these yield erythromycin resistant diploids.

Fig. 4. Suppressiveness of spontaneously arisen petites compared with petites obtained by euflavine induction (24 hr growth in YEP-glucose, 5 μg euflavine/ml), and induction with 2 mg/ml and 10 μg/ml ethidium bromide (24 hr growth). Histograms show number of petites found to be of each particular percentage suppressiveness.

ness, that is, less than 5% suppressive, while most of the spontaneous petites are between 10 and 40% suppressive. However, as previously reported, euflavine induced petites far more commonly lose the erythromycin determinant than do spontaneous petites [23]. For example, of the 80 spontaneous petites represented in fig. 5, 28 retain the ER^r determinant as compared with 6 out of the 80 euflavine induced petites. This would suggest that the DNA of the euflavine induced petites would be more grossly altered than that of the spontaneous petites. The suppressiveness of petites induced by two concentrations of ethidium bromide have also been compared (fig. 4). It was found that 28 of the 30 petites induced by the high concentration of ethidium bromide (10 mg/ml) were neutral or near neutral, the other 2 being highly suppressive, whereas almost all of the 30 petites induced by the low concentration were of

intermediate suppressiveness. The use of the high concentration was clearly a more severe process, none of the 30 petites induced by this concentration retaining ER^r, as compared with 12 of the 30 induced by the low concentration. Thus it can be seen that petites obtained spontaneously or by mild chemical induction as well as having the least loss of information, tend to be of intermediate suppressiveness, whereas with more severe induction processes neutral petites predominate.

The retention of the erythromycin determinant and its relation to suppressiveness has been examined in detail by the study of a large number of spontaneous petites. Fig. 5 shows the suppressiveness of a number of petites derived from respiratory competent erythromycin resistant cells plotted against the percentage of zygotes that yield some erythromycin resistant diploid progeny.

The results indicate that ρ^-ER^r clones are observed over a wide range of suppressiveness (5–50%) but to date no neutral petites or those of high suppressiveness have been observed to retain the erythromycin determinant. This conclusion also applies to the chemically induced petites, the only ethidium bromide or euflavine induced petites retaining the determinant being of intermediate suppressiveness.

Suppressiveness and instability

We have previously reported that the erythromycin determinant is unstable in ρ^- cells [23]. This phenomenon was demonstrated by sub-cloning; it was found that on plating a culture of supposedly ρ^-ER^r cells that some of the sub-clones obtained had completely lost the ability to give resistant diploids in crosses, such sub-clones had clearly arisen from ρ^-ER^o cells. This instability is the explanation for the variation between different petite clones of the percentage of ρ^+ zygotes yielding ER^r cells as seen in fig. 5. Each petite culture must be considered to contain both ρ^-ER^r and ρ^-ER^o cells and each mass cross is then a mixture of two types of cross viz. $\rho^-ER^r \times \rho^+ER^s$ and $\rho^-ER^o \times \rho^+ER^s$. The more ρ^-ER^o cells that are present in the mixture, the more zygotes will be formed that give rise only to ER^s diploids, the first type of cross yielding ER^r and ER^s diploids and the second only ER^s diploids. The ratio $\rho^-ER^r:\rho^-ER^o$ cells in the culture and thus the percentage of ρ^+ zygotes yielding ER^r cells is a function of the instability of the petite.

Additional sub-cloning experiments have now been performed to test for any correlation between the petite cells that have lost the determinant, and the suppressiveness of their sub-clones. The results of one such sub-cloning can be seen in fig. 6. Petites which retain ER^r are seen to be those with a suppressiveness close to that of the original clone, petites in which the suppressiveness has undergone a marked change have lost this determinant. The sub-clones that retain ER^r give, on crossing, a higher percentage of zygotes displaying the presence of the ER^r determinant compared with the parent clone. The increase in the zygotes containing the ER^r determinant arises from the purification of the $\rho^-ER^r - \rho^-ER^o$ mixture and the immediate use of the clone before the cells had undergone a sufficient number of divisions for many ρ^-ER^o cells to again arise. Further sub-cloning substantiates these conclusions (fig. 6) in that a small but continuing number of ρ^-ER^o cells are thrown by the purified ρ^-ER^r cells and the ρ^-ER^o sub-clones are significantly lower or higher in suppressiveness compared with the ρ^-ER^r parent.

It is important to note that except for neutral petites or those of high suppressiveness, it is not the absolute value of suppressiveness which determines the retention or loss of ER^r but the change to a new

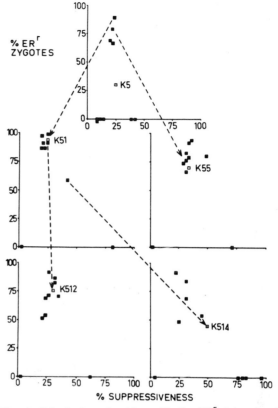

Fig. 6. Sub-cloning of petites with the ER^r determinant. Points are plotted as in fig. 5. Top graph: % ER^r zygotes versus % suppressiveness found for K5, a spontaneously arisen petite, and ten sub-clones selected as single colonies at random from a plating of K5 on YEP glucose. Middle graphs: further sub-cloning of K51 and K55, obtained as sub-clones from K5. Lower graphs: sub-cloning of K512 and K514, obtained as sub-clones from K51.

value of suppressiveness that in general produces a loss of ER^r. A study of sub-clone K5272 derived from K5 illustrates this point. Strain K5 (ρ^-ER^r, 25% suppressive) on repeated sub-cloning gave rise to many sub-clones of changed suppressiveness with genotype ρ^-ER^o, but one sub-clone, K5272, derived successively from K5, K52 and then K527 had the genotype ρ^-ER^r although only 4% suppressive. When K5272 was in turn sub-cloned, petites of lower and higher suppressiveness, all of genotype ρ^-ER^o, were obtained. One particular sub-clone K52729 was 25% suppressive, that is had the same suppressiveness as the original parent K5, yet its genotype was ρ^-ER^o.

In interpreting our results, it must be borne in mind that the erythromycin determinant is only one piece of mitochondrial information and thus a cell which has lost this determinant may still carry other information in its mitochondrial DNA, and similarly petites which retain ER^r may differ with respect to the retention of yet some other determinant. For example, K5272 has not lost ER^r during its change of suppressiveness from the 25% of K5, to 4%, but we would suggest that it is very likely to have lost other determinants. Loss or retention of ER^r can be taken as a qualitative measure of information carried, but not as a quantitative measure of it.

The results presented in this paper are consistent with the degree of suppressiveness of a petite determining the probability that the petite will carry certain information and that the degree of suppressiveness may even set an upper limit on the amount of information that can be carried by the mitochondrial DNA of each petite. However, it is important to note that there cannot be a direct relationship between the degree of suppressiveness and the total amount of information carried by a given petite. That this is so can be seen from strains K5 and K52729. Both of these strains are of the same suppressiveness yet clearly K52729 carries less information; during its isolation from K5 it has at least lost ER^r and probably other determinants as well.

A simple model may be proposed to explain the results in this section of the paper. It is proposed that mitochondrial DNA of neutral and highly suppressive petites carry little or no information in contrast to newly arising petites of intermediate suppressiveness which would have less damaged DNA. The mitochondrial DNA of petites of intermediate suppressiveness thus postulated to be able to undergo two distinct types of change, one of which causes the suppressiveness to fall, the other causing it to rise, both types of change resulting in loss of information. (These changes can be looked upon, for the purposes of illustration, as say additions to and deletions from the DNA.) In this model, the mitochondrial DNA of K5272 ($\rho^- ER^r$, 4% suppressive) would have undergone the first type of change, as compared to the original K5 ($\rho^- ER^r$, 25% suppressive) but the ER^r region would not have been involved. In the mutation of K5272 to K52729 ($\rho^- ER^o$, 25% suppressive) the mitochondrial DNA would be envisaged as having undergone an additional change of the opposite type, such that they would tend to cancel each other out in

the determination of suppresiveness. This new change, however, would result in further loss of information, this time including the erythromycin determinant.

The conclusion that newly formed petites with an intermediate suppressiveness have the least altered mitochondrial DNA is supported by an earlier observation of Ephrussi and Grandchamp [26] who reported a subcloning of a number of petites of varying suppressiveness. The results showed that while petites of intermediate suppressiveness could give rise to sub-clones of both much higher and lower suppressiveness, neutrals and highly suppressive petites were quite stable. Such a situation is fully consistent with neutrals and high suppressive petites being the most damaged.

It has been suggested by Ephrussi et al. [28] that suppressiveness acts by mutual repression between the normal factor and the suppressive factor, whereas Carnevali et al. [15] suggested that a competition of replication rates between normal and suppressive DNA was involved. In either case it would be expected where ρ^+ and ρ^- were able to compete equally, with neither predominating overall, that suppressiveness would be of intermediate value somewhat below 50% *. Our results support this conclusion in that the petites with apparently least altered DNA are of intermediate suppressiveness and thus are able to compete equally with the ρ^+ DNA in the zygotes. The change down towards neutrality or up to higher suppressiveness is a result of the DNA becoming less or better equipped to compete with the ρ^+ DNA. Furthermore, since none of the many petites examined to date has been found to contain ER, it is proposed that neutrality could be due to a complete absence of mitochondrial DNA.

---

* Although at first it might be considered that the figure would be exactly 50%, it is necessary to recall that in scoring suppressiveness only entirely ρ^- colonies are counted as petite whereas mixed colonies are scored as ρ^+ and thus the count is biased in favour of ρ^+.

References

[1] B. Ephrussi, in: Nucleo-cytoplasmic Relations in Microorganisms (Clarendon Press, Oxford, England, 1953).
[2] F. Sherman, Genetics 48 (1963) 375.

[3] M.M.K. Nass and S. Nass, J. Cell. Biol. 19 (1963) 593.

[4] G. Schatz, E. Haslbrunncr and H. Tuppy, Biochem. Biophys. Res. Commun. 15 (1964) 127.

[5] J.H. Sinclair, B.J. Stevens, P. Sanghavi and M. Rabinowitz, Science 156 (1967) 1234.

[6] L. Shapiro, L.I. Grossman, J. Marmur and A.K. Kleinschmidt, J. Mol. Biol. 33 (1968) 907.

[7] M. Guerineau, C. Grandchamp, Y. Yotsuyanagi and P.P. Slonimski, C.R. Acad. Sci. Paris 266 (1968) 1884.

[8] C.J. Avers, F.E. Billheimer, H. Hoffmann and R.M. Pauli, Proc. Natl. Acad. Sci. U.S. 61 (1968) 90.

[9] C.P. Hollenberg, P. Borst, R.W.J. Thuring and E.F.J. Van Bruggen, Biochim. Biophys. Acta 186 (1969) 417.

[10] P. Borst and A.M. Kroon, Intern. Rev. Cytol. 26 (1969) 107.

[11] M.M.K. Nass, Science 165 (1969) 25.

[12] J.C. Monoulou, H. Jakob and P.P. Slonimski, Biochem. Biophys. Res. Commun. 24 (1966) 218.

[13] G. Bernardi, F. Carnevali, A. Nicolaieff, G. Piperno and G. Tecce, J. Mol. Biol. 37 (1968) 493.

[14] B.D. Mehrota and H.R. Mahler, Arch. Biochem. Biophys. 128 (1968) 685.

[15] F. Carnevali, G. Morpurgo and G. Tecce, Science 163 (1969) 1331.

[16] D. Wilkie, J. Mol. Biol. 7 (1963) 527.

[17] N.E. Maroudas and D. Wilkie, Biochim. Biophys. Acta 166 (1968) 681.

[18] F. Sherman, J. Cellular Comp. Physiol. 54 (1959) 37.

[19] T. Sugimura, K. Okabe and A. Imamura, Nature 212 (1966) 304.

[20] P.P. Slonimski, G. Perrodin and J.H. Croft, Biochem. Biophys. Res. Commun. 30 (1968) 232.

[21] N.E. Allen and A.M. MacQuillan, J. Bacteriol. 97 (1969) 1142.

[22] A.W. Linnane, G.W. Saunders, E.B. Gingold and H.B. Lukins, Proc. Natl. Acad. Sci. U.S. 59 (1968) 903.

[23] E.B. Gingold, G.W. Saunders, H.B. Lukins and A.W. Linnane, Genetics 62 (1969) 735.

[24] L.J. Wickerham, J. Bacteriol. 52 (1946) 293.

[25] M. Ogur, R. St.John and S. Nagai, Science 125 (1957) 928.

[26] B. Ephrussi and S. Grandchamp, Heredity 20 (1965) 1.

[27] B. Ephrussi, H. Hottinguer and H. Roman, Proc. Natl. Acad. Sci. U.S. 41 (1955) 1065.

[28] B. Ephrussi, H. Jakob and S. Grandchamp, Genetics 54 (1966) 1.

MENDELIAN GENES AFFECTING DEVELOPMENT AND FUNCTION OF YEAST MITOCHONDRIA

J.C. BECK, J.H. PARKER,* W.X. BALCAVAGE and J.R. MATTOON

Department of Physiological Chemistry,
The Johns Hopkins University School of Medicine,
Baltimore, Maryland, USA

Introduction

Function as an aspect of biogenesis

The description of mitochondrial biogenesis must be in terms not only of structure, but also of function. During organelle development, as each new gene-product is integrated into the mitochondrial structure, a new component of metabolic activity emerges. In fact, we are often able to recognize development of a mitochondrial *function* long before its structural basis is known. Since our understanding of mitochondrial function is still far from complete, it seems reasonable to suppose that some of the gene-products comprising the mitochondrion are yet to be identified. Because structure and function are so closely linked, investigation of the organelle assembly process may yield unexpected insights into mitochondrial function. Conversely, elucidation of function should advance our understanding of the biogenesis process. The present report describes some new relationships between mitochondrial biogenesis and function in yeast as they are revealed through the action of various Mendelian genes.

Importance of nuclear information

Although the mitochondria themselves contain genetic information, it is now evident that the number of proteins which can be coded by mitochondrial DNA is relatively small [1,2]. Furthermore, it seems likely that a great many mitochondrial enzymes are coded by *nuclear* DNA. In fact, direct evidence is now available which indicates that structural genes

for certain mitochondrial enzymes are located on nuclear chromosomes [3–5]. Clearly, then, much of the recognized functional aspect of mitochondrial biogenesis has a nuclear basis. Table 1 is a list of Mendelian (nuclear) genes which control functions associated with the mitochondria. While recognition of a number of these genes may be made through loss of specific protein components such as cytochromes, a great many others are detected through functional changes, such as the loss of the ability to use nonfermentable substrates for growth (p genes). A striking feature of this table is the large number of Mendelian genes represented. Unquestionably the list represents only a fraction of the total. It is of interest to estimate the amount of DNA that the 63 genes listed here represent [1]. If we assume that each gene codes for a polypeptide with an average molecular weight of 40,400, 44×10^6 daltons of DNA would be required, representing an accumulated length of 23μ. It seems unlikely, of course, that these genes all represent proteins forming an actual part of mitochondrial structure. Nevertheless, the calculation emphasizes the high level of dependence of mitochondrial biogenesis upon nuclear information.

Nucleo-cytoplasmic relationships

Ephrussi [28], in his classic study of extrachromosomal inheritance of *petite yeast*, stressed the importance of the interaction of Mendelian and extrachromosomal units of heredity. Table 1 illustrates several examples of such interaction between the nuclear and mitochondrial genetic systems. Several p genes, designated as class 1 in the table, are invariably lacking in the mitochondrial determinant, ρ, while others, e.g. p_4 and p_5, exhibit a high frequency of

---

* Present address: Department of Biology, State University of New York, Buffalo, New York 14214, USA.

Table 1
Mendelian genes affecting mitochondrial structure, function, and biogenesis

Group I. Genes causing general loss of ability to assimilate nonfermentable substrates

Symbols	Class	Cytochromes [a]	Other properties	Ref.
$p_2 p_3$	1	c only	ρ^-, may have porphyrin	6, 7
p_{12}	1 (?)	(c)	ρ^-, contains pigments like b and c_1 (porphyrins?)	8
$ly_6 ly_8$	1	c	ρ^-, lysine reqd.	6, 7
glt_2	1	c	ρ^-, glutamate reqd.	9
$p_1 p_6 p_7$	2	c	ρ^+	6, 7, 10
$p_8 p_{13}$	2	(c)	ρ^+, contains porphyrins or traces of b and c_1	8
$p_{11} p_{14} - p_{17}$	3	–	no data, or genetic data only	11
gi	4	normal or c only	ρ^- when repressed	12

Group II [b]. Genes causing selective alteration in assimilation of nonfermentable substrates

Symbols	Class	Cytochromes	Other properties	Ref.
$pl_1 - pl_7$	5	$a\ b\ c_1\ c$	selective promotion of lactate assimilation	13
$glg_1 glg_2$	6	$a\ b\ c_1\ c$	selective reduction in glycerol assimilation	b

Group III. Genes altering content of one or two cytochromes

Symbols	Class	Cytochromes	Other properties	Ref.
p_4	7	$c\ c_1$ low a and b	high % ρ^-	6, 7
p_{10}	8	$a\ c$	a easily repressed	14
p_5	9	$b\ c\ c_1$	variable % ρ^-	7
$cy_1 - cy_6$	10	$a\ b\ c_1$ low c	a easily repressed in cy_3	14, 15

Group IV [b]. Genes altering nonfermentable substrate assimilation without deletion of cytochromes

Symbols	Class	Cytochromes	Other properties	Ref.
p_9 (op_1)	11	normal	abnormal oxidative phosphorylation, ρ^- lethal	16–18
$aem_1 - aem_8$	12	normal	all respire	19
$aem_9 - aem_{11}$	13	all low	low respiration	19
glt_1	13	all low	requires glutamate, lacks aconitase	3

mutation to ρ^-. Certain genes affecting biosynthesis of glutamate or lysine, glt_2, and ly_6 and ly_8, like the class 1 p genes, are always ρ^-. On the other hand, the p_9 mutation not only induces abnormalities in oxidative phosphorylation, but also creates a situation in which ρ^- is lethal [17,29].

Classification of Mendelian genes

Table 1 is a compilation of most of the known Mendelian genes which affect mitochondrial structure, function, and biogenesis. The genes are divided into 8 groups based either on the phenotypic character used to detect the gene or, whenever possible, upon the mitochondrial reaction or component which is affected. A more detailed classification is given in the legend to the table.

Group I is the largest in the table because loss of the ability to assimilate nonfermentable substrates is not a highly selective test. As more selective information is obtained, members of this group may be transferred to more specific categories. For example, since group II genes specifically affect utilization of a single nonfermentable substrate, we can already begin to focus on the type of biochemical reactions which such genes control. Likewise we see that p genes appear in groups III and IV.

Much of the work in our laboratory has been concerned with genes of group IV. Dr. Parker [19] has recently found eleven new complementation groups in this category. Our experience, particularly with cells and isolated mitochondria from the mutant containing the p_9 gene [17,30], lead us to propose the general symbol aem for this type of gene. Although mutants with *primary* lesions in oxidative

Table 1
(continued)

Group V [b]. Genes altering citric acid cycle

glt$_1$	13	all low	lacks aconitase (see group IV)	3
glt$_2$	1	c	lacks aconitase (see group I)	9
glt$_X$	14	normal	assimilates glycerol, linked to glt$_1$	b

Group VI. Genes altering porphyrin metabolism

pop$_1$ pop$_2$	15	normal	accumulates free and Zn porphyrins	20
W-1	16	none	cyt. synthesis induced by glycine or porphyrin, unstable	21, 22

Group VII. Genes altering unsaturated fatty acid formation

ol$_1$	17	a b c c$_1$	requires oleate	23
ol$_2$ –ol$_4$	18	no data	req. oleate, no glycerol assimilation, pleiotropic	23

Group VIII. Genes altering resistance to antibiotics and other chemicals affecting mitochondrial function and/or biogenesis

CAPR (1–3)	19	normal	chloramphenicol resistant, at least 3 types, 1 dominant	24
ERR	20	normal	erythromycin resistant, at least 1 type	24
or$_1$ or$_1$	21	normal	resistant to oligomycin and venturicidin	25
DNPR	22	variable	resistant to 2,4-dinitrophenol, at least 1 type	26
AAR	23	respiration high	resistant to antimycin A, at least 1 type	27

[a] Cytochrome *a* refers to the *a.a*$_3$ complex.
[b] Discussed in present report.

Classification of mutant types: 1. Multiple cytochrome deficiencies associated with a change in the ρ factor to ρ^- (petite). 2. Multiple cytochrome deficiencies even when the ρ factor is normal (ρ^+). 3. Uncharacterized biochemically. 4. Capable of assimilating nonfermentable substrates, but yields very high ρ^- when repressed, 100% ρ^- anaerobically. 5. Promote lactate assimilation selectively. 6. Promote glycerol assimilation selectively. 7. Contains all cytochromes, but *a* and *b* are low. Cyt. *a* readily repressed, and ρ^- frequent. 8. Lacks *b* and *c*$_1$. 9. Lacks Cyt. *a*. 10. Cyt. *c* reduced to varying degrees. 11. Mitochondria have low P:O except in excess ADP. 12. Abnormal energy metabolism, (won't assimilate glycerol), respire and contain all cytochromes in normal range. 13. Like class 12, but all cytochromes low. 14. Glutamate auxotroph, assimilates glycerol, cytochromes normal. 15. Accumulate Zn or free protoporphyrin, contain cytochromes. 16. Porphyrin or glycine auxotroph. 17. Respiratory competent, auxotrophic for unsaturated fatty acid (UFA). 18. No glycerol assimilation, auxotrophic for UFA. 19. Chloramphenicol resistant. 20. Erythromycin resistant. 21. Oligomycin resistant. 22. Resistant to 2,4-dinitrophenol. 23. Resistant to antimycin A.

phosphorylation would fall in this category, a variety of other alterations may yield the same general phenotype. For example, the glutamate auxotroph, *glt*$_1$, a citric acid cycle mutant, fulfills the criteria for group IV. Is seems judicious, therefore, to designate such mutants as having *abnormal energy metabolism* until more specific biochemical characterization is made.

Group V so far is represented in yeast only by the *glt* genes, both of which result in altered aconitase. A third *glt* gene, *glt*$_X$, is described in the present report. Since there are at least 10 enzymes concerned with operation of the citric acid cycle, it is obvious that group V is much larger than the table indicates. In other species, such as *Neurospora* [4] and maize [31], mutant genes for malate dehydrogenase have been reported. Isolation of other types of yeast mutants in this category should merely require application of appropriate conditions of conditional lethality.

Groups VI and VII have not been explored in depth. The probable number of possible porphyrin mutants is undoubtedly rather large, and we may anticipate major progress in our understanding of mitochondrial biogenesis after such mutants become available. Further exploration of the role of lipids in mitochondrial development and function, using group VII mutants, should likewise be highly productive.

Finally, group VIII mutants hold great promise for the future. So far, the known mutants resistant to inhibition of respiration and/or phosphorylation appear to be Mendelian. While there is great interest in the *cytoplasmic* inheritance of resistance to inhibitors of mitochondrial protein synthesis, the existence of *Mendelian* genes controlling sensitivity to these substances should not be minimized. Such genes may in fact prove to be of great consequence in analyzing some of the perplexing problems of nucleo-cytoplasmic relationships.

Interactions of genes and gene-products

The appearance of some genes in more than one group (in table 1) illustrates a highly significant aspect of the genetics of mitochondria, namely that of interaction. Interaction is precisely the theme of this communication. It is only with a broad knowledge of a great many genes and their interactions that we can hope to fully comprehend the major principles governing biogenesis. We have already noted examples of interaction between the nuclear and cytoplasmic genetic systems. As a matter of fact, we feel that pleiotropy, often considered a nuisance by geneticists, should be sought rather than shunned, if we are to elucidate these principles.

Interaction is not unique to genetics. Anyone who has studied mitochondrial function is painfully aware of the complex interlocking of activities which is the hallmark of these remarkable organelles. In a sence then, mitochondria are 'functionally pleiotropic,' since many of the proteins comprising the mitochondrion are multifunctional. Consequently, mutation in a single gene frequently induces a whole family of changes in mitochondrial activity. This is at once a blessing and a curse. On the one hand, new relationships between mitochondrial activities are revealed; on the other, exact identification of the altered gene-product is often greatly complicated. Nevertheless, it is our thesis that much genetic pleiotropy can become comprehensible through analysis of functional interactions in mitochondria isolated from mutants.

Thus, the genetic approach to the study of mito-

chondria emphasizes interaction, be it genetic or functional. We must therefore be concerned with various types of interaction: (1) interaction of biogenetic and functional systems; (2) interaction at the nucleic acid level; (3) interaction of mitochondria with metabolic systems of the cytosol, and (4) interaction of mitochondrial metabolic functions themselves.

Effects of Mendelian mutations on apparent mitochondrial numbers

In screening for mutants with altered oxidative phosphorylation, three unlinked genes have been found (aem_9, aem_{10}, aem_{11}) which cause marked reduction in the cytochrome content of whole yeast cells. In addition, a citric acid cycle mutant unable to synthesize glutamate (glt_1) has also been found to be greatly deficient in cytochromes. Table 2 illustrates this unusual effect. Strains M-33 and M-74 probably arose by mutation in the same Mendelian gene, since they do not undergo genetic complementation [19]. We may therefore designate the genes in these two mutants aem_{9-1} and aem_{9-2}. The two strains were derived from the normal parent, D311-3A. Although the growth conditions were designed so that glucose repression would be of minor importance, the mutant strains show a marked reduction in all cytochromes. Cytochrome b and $c+c_1$ are 3 to 4 times higher in the parent than in mutant M-33, while cytochrome a has been reduced 7-fold by the mutation. In the case of mutant M-74, the very low value for cytochrome $c+c_1$ is caused by an unlinked gene which lowers cytochrome c (or c_1) drastically. The reason that this second mutation occurred in this strain is unknown. As a control, a mutant with a deficiency in oxidative phosphorylation was included in the study. The mutant strain, D360-5B, contains the gene designated p_9 [16,17] or op_1 [18]. As the table shows, the cytochrome content of this mutant is of a magnitude roughly comparable to normal strain D311-3A. The relatively high ratio of cytochrome $c+c_1$ to cytochrome b appears to be more a reflection of the background genome than of the action of the p_9 gene, since it is not characteristic of other p_9 containing strains which we have examined. It is evident, therefore, that mutation affecting mitochondrial activity may or may not be accompanied by marked lowering of all cytochromes.

Table 2
Effect of Mendelian genes on cytochrome content of yeast cells

Strain No.	Symbol	Growth on YPG [a]	Glutamate requirement	Cytochrome content [b]			% ρ^- [c]
				a	b	$c+c_1$	
D311-3A	–	+	–	50.0	47.0	97.4	0
M-33	aem_{9-1}	–	–	7.1	16.7	25.8	0
M-74	aem_{9-2}	–	–	5.5	17.1	10.9	16
D360-5B	p_9 or op_1	–	–	37.8	28.2	138	–
CBS-37892	glt_x glt_1	–	+	5.0	17.5	16.6	1.2
JP293-45A	glt_1	–	+	4.1	7.0	17.9	23
JP293-45B	glt_x	+	+	76.6	62.9	173	6.2

[a] YPG refers to medium containing 1% yeast extract, 2% peptone, 3% (v/v) glycerol.
[b] nmoles/g dry wt of cells.
[c] Determined by tetrazolium overlay method [32].

It has not yet been determined whether lowering of cytochrome content is the result of a decrease in mitochondrial numbers, in mitochondrial size, or a reduction in the number of respiratory units per mitochondrion. Possibly the gene-product altered in a mutant such as M-33 is essential for mitochondrial replication or development. Although this gene-product might be a structural component made in very limited amounts in the mutant, the roughly coordinate reduction of *all* the cytochromes, suggests that a regulatory system, somewhat resembling a bacterial operon, has been altered. There is a good possibility that the altered regulatory system is the same as, or similar to, that responsible for glucose repression. In spite of the fact that both the p_9 mutant and the normal strain D311-3A were highly derepressed, it is well known that the ρ^- mutation and certain *cy* mutations greatly enhance mitochondrial sensitivity to glucose [14,33].

Another example of marked reduction of all cytochromes is also illustrated in the table, Strain CBS-37892 was described by Ogur et al. in 1964 [3]. This strain was reported to contain a single Mendelian gene, glt_1, which led to a dual phenotype: a complete nutritional requirement for glutamate, and an inability to grow on nonfermentable carbon sources. The data in the table confirm these observations. In addition, they show, that like M-33 and M-74, this mutant too has a very low content of all cytochromes. Thus, the two types of mutants share the properties of low cytochrome content and inability

to utilize glycerol, but they differ in their requirement for added glutamate. The possibility that aem_9 and glt_1 involve different portions of a single genetic locus seems unlikely, since strains bearing these genes undergo genetic complementation.

When the *glt* mutant was crossed with a normal strain, and a second generation of *glt* segregants was obtained through sporulation, the 3 properties, glutamate requirement, inability to assimilate glycerol, and reduced cytochrome, were transmitted to the glt_1 segregant, JP293-45A. However, another type of *glt* gene, illustrated by segregant JP293-45B, shows neither the low cytochrome level, nor the inability to grow on glycerol. Other properties of the latter gene are described below. Ogur et al. [3] have also reported that yeast strains bearing the glt_1 gene are almost completely lacking in aconitase. One may infer from this that strain CBS-37892 carries an altered structural gene for the enzyme. It is not immediately evident, however, what role such an alteration might play in the regulation of cytochrome synthesis and/or mitochondrial replication. The possibility that abnormal porphyrin synthesis might result from a deficiency in succinyl coenzyme A seems doubtful, since the glutamate lesion in JP293-45B causes no lowering of cytochrome content. Porphyrin deficiency resulting from a lack of reduced NADPH in mitochondria caused by lack of iso-citrate seems unlikely for the same reason.

At this point, then, one may only speculate on the basis of the reduced cytochrome. Mitochondrial DNA replication, mitochondrial protein synthesis, and membrane formation are all potential sites of regulation which may be subject to alteration. In view of the number of such possible regulation sites, our observation of several unlinked genes controlling the low cytochrome phenotype are not especially surprising. Of particular interest is the fact that these genes are *Mendelian*. It may also be possible to find similar determinants in mitochondrial DNA. If, in fact, the lowering of cytochromes by glt_1 and aem_9 has a common basis, it appears unlikely that it involves α-ketoglutarate biosynthesis, since neither of the aem_9 strains requires glutamate for growth. However, the latter strains could conceivably have alterations in the dicarboxylic acid portion of the citric acid cycle. There is a possibility, perhaps remote, that mitochondrial GTP production is deficient in both types of mutants. In the case of glt_1, it may be that exogenous glutamate is readily available for extramitochondrial protein synthesis, but it is ineffectual in driving mitochondrial GTP synthesis, via α-ketoglutarate oxidation and succinate thiokinase. The resulting lack of intra-mitochondrial GTP could hamper both nucleic acid and protein synthesis within the organelle. At this point, then, while we have an indication of function (citric acid cycle) interacting with biogenesis (apparent mitochondrial number), we have no firm basis for deciding between cause and effect. We do not know whether altered function causes altered biogenesis or vice versa.

The basis for the inability to assimilate glycerol, a property common to all 3 mutants, is also uncertain, but it may simply reflect the smaller number of respiratory assemblies per cell, indicated by the low cytochrome content. If, in fact, mitochondrial numbers are reduced in these strains, glycerol oxidation could be limited by a corresponding reduction in total mitochondrial α-glycerolphosphate dehydrogenase.

Although Reilly and Sherman [14] reported that strain CBS-37892 was not particularly sensitive to the glucose effect, Bowers et al. [34] found that another *glt* strain (37887), when grown under conditions similar to ours, was highly susceptible to repression. In electron micrographs of 37887 cells they observed only small, amorphous bodies, possibly rudimentary mitochondria, unless cells were grown under special conditions. Unfortunately, the relationship between strains CBS-37892 and 37887 was not reported; presumably both strains contain the glt_1 gene.

A final observation should be made. Ogur et al. [9] has shown that another non-allelic *glt* mutant, glt_2, like glt_1, lacks aconitase and contains abnormal mitochondria. In this case, however, the abnormality consists of a very high frequency of ρ^-, rather than a coordinate loss of *all* cytochromes. The interaction in this case seems to be more extreme, since the activity of the mitochondrial hereditary apparatus is altered irreversibly, not just inhibited.

Separation of interacting and non-interacting glt genes

In the previous section, two types of glutamate auxotrophs were distinguished; one appeared to interact strongly with the mitochondrial biogenesis system, while the other was apparently independent of this system. The present section summarizes the genetic relationships between the two types of auxotrophs. As noted above, the glutamate auxotroph, CBS−37892, lacks mitochondrial aconitase. Although there is no obvious reason to expect a block in the citric acid cycle to inhibit glycerol assimilation, the mutant is, in fact, unable to grow on this substrate. Equally puzzling were the results of our initial investigations, detailed above, showing the reduced cytochrome content of the mutant. Since it was important to ascertain whether the low cytochrome is actually determined by the glt_1 gene, the genetic study illustrated in table 3 was carried out. The results were surprising, since in 12 of 16 tetrads examined, 3 glutamate-requiring segregants were found. In only 4 tetrads was the 2:2 ratio of mutant

Table 3
Segregation of glutamate requirement from diploids derived from CBS-37892

Diploid strain	Number of tetrads		
	3 *glt* : 1 GLT	2 *glt* : 2 GLT	Total
JP293	1	0	1
JP295	4	2	6
JP296	3	1	4
JP297	4	1	5
	12	4	16

to normal, typical of single-gene inheritance, obtained. Quite obviously, at least one additional gene affecting glutamate biosynthesis was segregating. Moreover, the 3:1 ratios were found with 4 different tester strains. It should also be noted, that no tetrads containing 4 mutant spores were found.

Another important aspect of this genetic study is illustrated in table 4, which shows results obtained with a typical 3:1 tetrad. Although 3 of the 4 segregants required glutamate, only 2 of these 3 mutant strains failed to utilize glycerol. In every tetrad examined, the following relationships were obtained: (1) two segregants invariably required glutamate *and* were unable to utilize glycerol, (2) whenever 3 glutamate-requiring segregants were found, one of these was capable of assimilating glycerol.

Although the total number of tetrads examined is relatively small, the following hypothesis may be suggested: (1) the original mutant strain contained two genes for glutamate requirement, glt_1 and glt_X; (2) the glt_1 gene bestows upon the mutant not only a glutamate requirement, but also a deficiency in glycerol metabolism; (3) the glt_X gene, on the other hand, affects glutamate biosynthesis without altering glycerol assimilation. A consequence of this hypothesis is that one of the two glt_1 segregants must also contain a glt_X gene.

As a test of the above hypothesis, each of the 3 *glt* segregants was crossed with appropriate GLT tester strains and tetrad analysis performed. Table 5 summarizes the results. When the mutant parent was strain JP293-45A, segregation of the *glt* gene was always 2:2, and was always accompanied by the inability to utilize glycerol. When parent 45B was used, 2:2 segregation of the *glt* character was found, but all segregants grew on glycerol medium. Since, in the

Table 4
Glutamate requirement and glycerol assimilation in segregants from heterozygous diploid, JP293

JP293 segregant	Growth on YPG + Ad	Growth on medium lacking glutamate	Postulated genotype
45A	−	−	glt_1
45B	+	−	glt_X
45C	+	+	GLT
45D	−	−	glt_1 (glt_X)

third cross, both glt_1 and glt_X segregants appeared, it is apparent that JP293-45D carried both types of gene, as predicted by the hypothesis.

Table 6 compares the properties of the original parent CBS-37892 with a tetrad derived from a cross to a normal yeast. Both glt_1 and glt_X strains assimilate glucose much less efficiently than normal. On the other hand, while the presence of the glt_1 gene causes a drastic reduction in Q_{O_2} and cytochromes, the strain containing glt_X alone has a high cytochrome content and respires at rapid rate. Perhaps the low cell yield of glt_X indicates that a large loss of carbon source took place through citrate accumulation [3].

These studies illustrate two aspects of interaction. First, glutamate auxotrophy per se is not the cause either of reduced cytochrome or of deficiency in glycerol utilization. Thus, glt_1 might be termed an 'interacting gene' while glt_X would be 'non-interacting.' Second, there appears to be some genetic linkage between the two *glt* genes, since out of 30 tetrads derived from diploids heterozygous for both genes, only one contained 2 glt_X and 2 glt_1 segregants. The

Table 5
Determination of number and type of *glt* genes in segregants of JP293

Diploid no.	*glt* parent	No. of tetrads			
		$2glt_1:2GLT$	$2glt_X:2GLT$	$2glt_1:1glt_X:1GLT$	Total
JB-4	JP293-45A	22	0	0	22
JB-5	JP293-45B	0	17	0	17
JB-6	JP293-45D	6	0	7	13 [a]

[a] JB-6 also yielded a tetrad which was $2glt_1:2glt_X$.

Table 6
Growth, respiration, and cytochrome content of segregants from heterozygous diploid, JP293

Strain no.	Genotype	Cell [a] yield (mg/ml)	Q_{O_2} ($\mu l/hr/mg$)	Cytochrome content [b]		
				a	b	$c+c_1$
JP293-45A	glt_1	3.1	1.6	3.5	10.4	13.3
JP293-45B	glt_X	4.2	62.8	58.0	67.6	169.5
JP293-45C	GLT	11.5	28.4	32.3	31.4	66.4
JP293-45D	glt_1, glt_X	3.2	2.4	7.5	8.5	42.0
CBS-37892	glt_1, glt_X	3.9	3.3	9.6	13.8	18.2

[a] Dry weight.
[b] nmoles/g dry weight of cells.

significance of this apparent genetic linkage remains to be determined.

Perhaps the apparent discrepancy between the observations of Reilly and Sherman with *glt* strain CBS-37892 [14] and those of Bowers et al. with strain 37887 [34] is related to the action of the glt_X gene. At the present time, we do not know whether or not this gene is present in the latter strain.

Normal variation in cytochrome content and glycerol metabolism with background genome

The study of the inheritance of quantitative characters is frequently complicated because many genes are involved in determining a given phenotype. While the magnitudes of the differences in cytochrome content observed in the previous examples were rather large, the possibility exists that a given mutant gene in a different strain will give little or no effect because other genes affecting the mitochondria act in a compensatory manner. In genetic studies in which a given mutant is to be crossed with various tester strains, it is essential that the investigator view any observed quantitative variation in expression of the mutant gene within the context of normal variation.

Table 7 illustrates the effect of normal background genome on cytochrome content. A diploid was constructed by crossing two normal strains containing similar amounts of cytochrome. After sporulation, 8 tetrads of haploid recombinants were surveyed for variation in cytochrome content. Six strains, illustrating the extremes of variation encountered, are shown in the table. It may be seen that even among

the progeny of normal strains, cytochrome content may vary up to a factor of 2, depending upon the particular combination of genes in the segregant. It is therefore of critical importance in comparing quantitative characters such as cytochrome content, that the effect of background genome be assessed, since obviously there may be a high degree of interaction in some cases.

The importance of background genome effects upon the expression of quantitative characters used in studying inheritance of mitochondrial function is further illustrated by the relative efficiency with

Table 7
Normal variation in cytochrome content of segregants from normal diploid yeast strain JP292

Strain no.	Type	Cytochrome content [a]			Cell yield on YPG [b]
		a	b	$c+c_1$	
D311-3A	a parent	39.5	51.4	100	0.94
D213-1B	α parent	52.8	55.5	125	3.77
JP292-5B	Low cyt.	22.9	32.2	71.6	4.10
JP292-8C	Low cyt.	31.1	29.8	77.3	2.50
JP292-8D	Low cyt.	38.6	30.0	68.9	3.16
JP292-9B	Int. cyt.	27.0	34.0	108	1.54
JP292-8A	High cyt.	45.6	58.7	122	3.36
JP292-10D	High cyt.	62.0	56.7	151	3.61
JP273-10B	Control [c]	58.1	55.5	125	–

[a] nmoles/g dry wt.
[b] grams wet wt/100 ml.
[c] cf. ref. [35].

which various strains assimilate glycerol. As we have already seen, this capacity is commonly used to detect mutants with mitochondria altered in various ways. Fig. 1 represents the variation in growth yield of 75 segregants derived from the same cross used in the preceding cytochrome experiments. The two parent strains differ widely in the efficiency with which they assimilate glycerol. Wet weight yield of cells from strain D213-1B grown 48 hr on YPG medium is 4 times that obtained from strain D311-3A.

Although the sample is small, the data seems to fall into 4 sub-groups, I–IV. While group A (I + II) is somewhat similar than group B (III + IV), the fact that the 4 sub-groups are of roughly equal size suggests a tentative hypothesis, namely that two genes are primarily responsible for glycerol assimilation. These 'glycerol growth' genes may be viewed either as promoting assimilation (GLG) or as retarding it (glg). The relative effectiveness of GLG_1 appears to be greater than GLG_2, and the two appear to be partially additive.

As a rough test of this hypothesis diploid strains were constructed as shown in table 8. With the exception of strain JB-12, the data are consistent with the hypothesis. Although, these experiments cannot be considered as decisive in establishing the

Table 8

Growth of diploids containing different glycerol growth genes

Diploid no.	Postulated genotype	Dry wt [a] yield (mg/ml)
JB-13	glg_1 glg_2 / glg_1 glg_2	4.9
JB-16	glg_1 GLG_2 / glg_1 glg_2	16.5
JB-11	glg_1 GLG_2 / glg_1 GLG_2	19.1
JB-15	GLG_1 glg_2 / glg_1 glg_2	23.6
JB-12	GLG_1 glg_2 / glg_1 glg_2	15.5
JB-10	GLG_1 glg_2 / GLG_1 glg_2	26.6
JP292	GLG_1 GLG_2 / glg_1 glg_2	30.9
JB-14	GLG_1 GLG_2 / GLG_1 GLG_2	28.4
Controls		
D311-3A	glg_1 glg_2	3.9
D213-1B	GLG_1 GLG_2	36.7

[a] 24 hr.

number of GLG (or glg) genes, the data clearly indicate that background genome is potentially a major determinant in the expression of mutant genes which affect glycerol utilization. In considering the possible nature of the gene-products of GLG genes, extra-mitochondrial (cytosol) enzymes, e.g. glycerol kinase, should be considered. Mutations affecting such gene-products could drastically alter glycerol

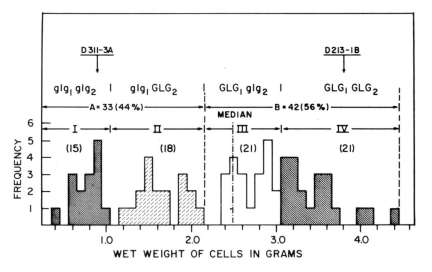

Fig. 1. Distribution of JP292 segregants according to yield on glycerol medium. Cells were grown for 48 hr at 30° on a rotary shaker. Each strain was grown in a flask containing 100 ml medium consisting of 1% Difco yeast extract, 2% Difco peptone, and 3% (v/v) glycerol; initial pH was 5.0. Cells were washed three times before weighing. A and B represent the number and percentage of segregants in the classes shown.

utilization, but have no *direct* bearing on mitochondrial structure, function, or biogenesis.

In summary, then, both cytochrome content and glycerol utilization, two frequently employed characteristics used in the study of yeast mitochondria, may be subject to wide variations resulting from interactions with background genome. Due consideration of such effects must be made in interpreting these variables in terms of mitochondrial function or development.

Correlation studies

The extent to which normal glycerol assimilation and cytochrome content are controlled by the same genes may be assessed from the data in table 9. There is no apparent correlation; in fact, JP292-5B, the strain with the lowest cytochrome content, is the most efficient in utilizing glycerol. Thus, there is no clear evidence, either in the parent strain or in the recombinants, for individual genes which control both phenotypic characteristics simultaneously. By contrast, as table 9 shows, the mutant gene in M-33 (aem_{9-1}) and the glt_1 gene show correlated alteration of cytochrome and glycerol assimilation.

From the results of the studies described above it may be concluded that we have been dealing with at least three classes of Mendelian genes which alter mitochondrial parameters (table 9): (1) genes which cause little or no change in glycerol assimilation or in cytochromes, e.g. glt_x; (2) genes which alter both cytochromes and glycerol utilization, e.g. glt_1, and aem_{9-1}, and (3) genes which affect glycerol utiliza-

tion but not cytochrome content, e.g. p_9 and *glg* genes. To determine whether the mutant gene in M-33 is of the same class as glt_1 or in a new class, a variety of reactions such as oxidative phosphorylation, ion translocation, and the citric acid cycle must be surveyed to determine what mitochondrial function, if any, has been altered.

Conclusion

The above experiments illustrate the great variety of interactions, both functional and genetic, encountered in the study of mitochondria. Although in most instances they do not represent studies in depth, they do hint at underlying principles. It is our conviction that these principles can best be discerned by developing a comprehensive viewpoint from which we may proceed to interpret the results of more specific studies.

Experimental methods

Cytochrome content was determined at 13°C in a split beam spectrophotometer constructed by one of the authors (W.X.B.). Cytochrome concentrations were calculated according to the method of Chance and Williams [36], using O_2 and sodium dithionite as oxidant and reductant respectively. Media, growth conditions, genetic analysis and Q_{O_2} determinations were carried out as described by Parker and Mattoon [19]. For cytochrome determinations cells were grown for 48 hr in medium containing 1% Difco yeast extract, 2% Difco peptone, 2% glucose and 100 mg adenine sulfate/liter. Initial pH was 4.5. Cultures were incubated at 30°C in 200 ml of medium in 2-liter Erlenmeyer flasks on a rotary shaker operated at 250 rpm. Since glucose is almost entirely consumed in 24 hr, the second 24 hr of incubation served to derepress the cells. When glutamate auxotrophs were studied, 300 mg L-glutamic acid/liter was included in the medium.

Table 9
Correlation of phenotypic characters in mutant and normal yeast

Strain no.	Gene of interest	Function affected		
		Growth on glycerol	Cytochrome content	Mitochondrial activity
JP293-45B	glt_x	−	−	+
JP293-45A	glt_1	+	+	+
D360-5B	p_9	+	−	+
M-33	aem_{9-1}	+	+	?
JP292 segregants	?	+−	±	?

Summary

(1) A scheme comprising 8 groups is proposed for classifying 63 Mendelian genes which affect mito-

chondrial function, structure, and biogenesis in yeast.

(2) Evidence is presented for classes of Mendelian genes which alter glycerol assimilation only, both glycerol assimilation and cytochrome content of cells, or neither of these.

(3) Two mutant genes, aem_9 and glt_1, are shown to lower the concentration of all cytochromes in the cell. While neither mutant assimilates glycerol, only the glt_1 mutant requires glutamate.

(4) Evidence is presented for a new Mendelian gene (glt_X) which controls glutamate biosynthesis. Although glt_X appears to be genetically linked to the glt_1 gene, its presence neither reduces cytochrome content nor limits glycerol assimilation.

(5) Certain genes are distinguished according to their influence on mitochondrial biogenesis: the glt_X gene is a *non-interacting* gene, while glt_1 and aem_9 are *interacting* genes which alter biogenesis.

(6) The influence of varying background genome on normal cell cytochrome content and glycerol assimilation is assessed. Cytochrome content may vary 2-fold, glycerol assimilation 4-fold. Preliminary evidence for 2 genes (*glg*) affecting glycerol assimilation in normal strains is reported.

Acknowledgements

This investigation was supported by U.S. Public Health Service postdoctoral research fellowships CA 30961 and CA 37087 and research grant GM 15844. The authors are indebted to Mr. R.F. Gottal, Mrs. Janice Zaron, Miss Audrey Poole, and Mr. David Barnes for their excellent technical assistance.

References

[1] J.H. Sinclair and B.J. Stevens, Proc. Natl. Acad. Sci. U.S. 56 (1966) 508.

[2] M.M.K. Nass, Science 165 (1969) 25.

[3] M. Ogur, L. Coker and S. Ogur, Biochem. Biophys. Res. Commun. 14 (1964) 193.

[4] K.D. Munkres and F.M. Richards, Arch. Biochem. Biophys. 109 (1965) 457.

[5] F. Sherman, J.W. Stewart, E. Margoliash, J.H. Parker and W. Campbell, Proc. Natl. Acad. Sci. U.S. 55 (1966) 1498.

[6] F. Sherman, Genetics 48 (1963) 375.

[7] F. Sherman and P.P. Slonimski, Biochim. Biophys. Acta 90 (1964) 1.

[8] B. Mackler, H.C. Douglas, S. Will, D.C. Hawthorne and H.R. Mahler, Biochemistry 4 (1965) 2016.

[9] M. Ogur, A. Roshanmanesh and S. Ogur, Science 147 (1965) 1590.

[10] S.Y. Chen, B. Ephrussi and H. Hottinguer, Heredity 4 (1950) 337.

[11] D.C. Hawthorne and R.K. Mortimer, Genetics 60 (1968) 735.

[12] D.B. Roodyn and D. Wilkie, in: The Biogenesis of Mitochondria (Methuen and Co., London, 1968) p. 87.

[13] P. Galzy and C. Bizeau, Arch. Mikrobiol. 52 (1965) 353.

[14] C. Reilly and F. Sherman, Biochim. Biophys. Acta 95 (1965) 640.

[15] F. Sherman, Genetics 49 (1964) 39.

[16] H.C. Douglass and D.C. Hawthorne, Genetics 49 (1964) 837.

[17] J.C. Beck, J.R. Mattoon, D.C. Hawthorne and F. Sherman, Proc. Natl. Acad. Sci. U.S. 60 (1968) 186.

[18] L. Kováč, T.M. Lachowicz and P.P. Slonimski, Science 158 (1967) 1564.

[19] J.H. Parker and J.R. Mattoon, J. Bacteriol., in press.

[20] T.P. Pretlow and F. Sherman, Biochim. Biophys. Acta 148 (1967) 629.

[21] C. Raut, Exptl. Cell Res. 4 (1953) 295.

[22] M. Yčas and T.J. Starr, J. Bacteriol. 65 (1953) 83.

[23] A.D. Keith, M.R. Resnick and A.H. Haley, J. Bacteriol. 98 (1969) 415.

[24] D.B. Roodyn and D. Wilkie, in: The Biogenesis of Mitochondria, (Methuen and Co., London, 1968) p. 100.

[25] J.H. Parker, I.R. Trimble, Jr., and J.R. Mattoon, Biochem. Biophys. Res. Commun. 33 (1968) 590.

[26] J.H. Parker and J.R. Mattoon, Genetics 60 (1968) 210.

[27] R. Butow and M. Zeydel, J. Biol. Chem. 243 (1968) 2545.

[28] B. Ephrussi, in: Nucleo-cytoplasmic Relations in Micro-orgamisms, (Clarendon Press, Oxford, 1953) p. 13.

[29] V. Kováčová, J. Irmlerová and L. Kováč, Biochim. Biophys. Acta 162 (1968) 157.

[30] J.C. Beck, Ph.D. Thesis (The Johns Hopkins University, Baltimore) 1969.

[31] G.P. Longo and J.G. Scandalios, Proc. Natl. Acad. Sci. U.S. 62 (1969) 104.

[32] M. Ogur, R.St. John and S. Nagai, Science 125 (1957) 928.

[33] J.C. Mounolou, H. Jakob and P.P. Slonimski, Biochem. Biophys. Res. Commun. 24 (1966) 218.

[34] W.D. Bowers, Jr., D.O. McClary and M. Ogur, J. Bacteriol. 94 (1967) 482.

[35] F. Sherman, H. Taber and W. Campbell, J. Mol. Biol. 13 (1965) 21.

[36] B. Chance and G.R. Williams, Advances in Enzymol. 17 (1956) 74.

THE GENIC CONTROL OF CHLOROPLAST DEVELOPMENT IN BARLEY

D. von WETTSTEIN, K.W. HENNINGSEN, J.E. BOYNTON *,
G.C. KANNANGARA and O.F. NIELSEN

Institute of Genetics, University of Copenhagen, Denmark

Genetic information for the synthesis of a chloroplast component resides in chromosomal or organelle DNA, or possibly in both. To decide between these three possibilities it is necessary to identify the structural gene for a given chloroplast component and locate it in nuclear or chloroplast DNA, respectively. When such genes have been found, we can approach the question of cooperation between the genetic information contained in the nucleus and that in the organelles.

We have tried to identify genes for chloroplast components in barley by studying both nuclear gene mutants and those with extra nuclear inheritance that have lesions in the development of the proplastid to a dark grown plastid or in the development from the dark grown plastid to a mature chloroplast. This latter part of the biogenesis of the chloroplast is dependent on morphogenic light. Even before the final aim of determining the primary action of a given gene is accomplished, the mutants can be used to elucidate pathways which have to operate for the formation of a functional chloroplast from its precursor, the proplastid. Indeed in many cases such mutants are the only possibility to provide evidence for the existence of a particular step in the formation of chloroplasts. The most interesting blocks caused by the mutations can only be identified indirectly since at present the missing metabolite such as a membrane precursor or a porphyrin cannot be supplied to the seedling from the outside as in the case of auxotrophs for low molecular weight substances. Once the specific block is characterized, the mutant becomes a highly valuable tool in the analysis of the organelle's

biogenesis as well as in studies of the interaction between biosynthetic pathways in different cell organelles.

This paper presents a survey of results which we consider relevant to the above problems and which have been obtained during the analysis of a set of induced nuclear gene mutants affecting the chloroplast

The number of chromosome loci involved

So far 198 recessive lethal chloroplast mutants have been assigned to 86 different loci by diallelic crosses (tables 1 and 2). The mutants are classified according to the system of Gustafsson [1]. It is apparent that many genes will give rise upon mutation to an *albina* mutant, whereas a restricted number of genes will mutate to produce a yellow, *xantha* phenotype. Among 33 *albina* mutants only two cases

Table 1

List of mutants in nuclear genes controlling chloroplast development in barley.

Phenotype	Gene symbols	Number of loci	Number of mutants
xantha	*xan−a* to *−q*, *xan−s* to *−u*	20	69
albina	*alb*	31	33
tigrina	*tig*	} 16	27
zonata	*zon*		
virido−albina	*vir−alb*	14	20
viridis	*vir*	5+x	49
		86	198

* Present address: Department of Botany, Duke University, Durham, North Carolina 27706, USA.

Table 2
Number of alleles at loci with more than one identified mutant allele.

Locus	xan–	a	b	c	d	e	f	g	h	i	j	m	n	o	13
Alleles		15	8	2	2	2	8	5	4	2	3	5	3	3	62

Locus	alb–	a	b
Alleles		2	2

Locus	tig–	a	b	c
Alleles		8	4 [a]	2

Locus	vir–alb–	a	b	c	d
Alleles		3	3	2	2

Locus	vir–	a	b	c	d	e
Alleles		5	2	2	2	2

[a] semidominant

of allelism have been found, which makes it likely that several hundred genes in barley can give rise to *albina* mutants. The 20 *xantha* loci found in a set of 69 mutants presumably represent a major portion of the genes that can produce seedlings with yellow color. In 13 of the *xantha* genes more than one mutant allele is available, *xan–a* comprising as many as 15 alleles. All mutants at this locus will develop a white leaf tip under green-house conditions and therefore are classified as *albo-xantha* mutants. Since this phenotype is confined to mutations in one gene, it can be used conveniently to screen for forward-mutations at the *xan–a* locus.

A group of 27 *tigrina* and *zonata* mutants, including the "infrared" mutants discussed below, are distributed among 16 different loci. About half of the mutants belong to three genes. For the *virido-albina* phenotype 14 genes have been found, whereas the analysis of a group of 49 light green *viridis* mutants has not yet been completed. So far we have determined more than one allele in five loci. It is anticipated that a large number of genes produces this phenotype upon mutation.

Survey of the lesions found in chloroplast biogenesis

The mutants are analysed with respect to their capacities during the differentiation from a proplastid to a dark grown plastid [2,3] by testing if they can form primary lamellar layers, a crystalline prolamellar body, and synthesize protochlorophyllide from δ-aminolevulinic acid [4,5]. It is further studied, which steps in the light induced and light dependent biogenesis of the mature chloroplast [2,3] a mutant can carry out. Tube transformation, dispersal of the prolamellar body, and formation of a lamellar system with grana are monitored by electron microscopy, whereas the photoconversion of protochlorophyllide to chlorophyll a_{684}, the spectral shift from chlorophyll a_{684} to chlorophyll a_{672}, and the accumulation of chlorophyll a_{678} are tested by in vivo absorption spectrophotometry [4,5]. We have tried to assess the capacity of mutant leaves to synthesize chloroplast specific lipids [6,7] as well as their capacity to accumulate soluble leaf proteins, especially ribulose-1,5-diphosphate carboxylase [8].

Fig. 1 summarizes the lesions found in such an analysis of mutants in 13 genes. In general, all members studied in a mutant allele series of a particular gene reveal the same specific pattern of defects. A closer study invariably discloses quantitative differences among the individual alleles, as expected.

The mutants in the *xantha* loci *f, g, h,* and *u* are blocked in the formation of the prolamellar body as well as in the formation of protochlorophyllide at a step between protoporphyrin IX and Mg-protoporphyrin. Locus *xan–l* seems to control a step in the synthesis of protochlorophyllide *a* subsequent to Mg-protoporphyrin. A mutant that is blocked in tube transformation or photoconversion of protochlorophyllide has not yet been found. *Albina*–17 and

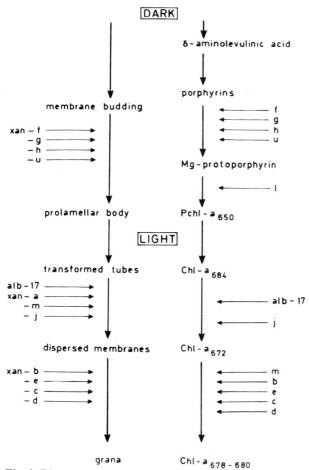

Fig. 1. Diagram of the lesions in chloroplast biogenesis (left) and chlorophyll biosynthesis (right) caused by mutations in 12 *xantha* genes and one *albina* gene.

mutants at *xan−j* are blocked in the dispersal of the prolamellar body as well as in the Shibata shift [2]. Inability to carry out the dispersal reaction is also characteristic for mutants at the loci *xan−a* and *xan−m*, but in these cases this defect is not associated with the absence of a Shibata shift, demonstrating that the parallel processes of prolamellar body dispersal and the Shibata shift can be dissected further with the aid of mutants. Mutants in many genes reveal lesions in the accumulation of chlorophyll a_{678} and the formation of grana indicating that the latter process requires the balanced synthesis of a large number of components under nuclear control.

Nuclear genes for porphyrin synthesis in plastids: *xan−f, −g, −h, −u* and *−l*

Identification of genes controlling steps from δ-aminolevulinic acid to protochlorophyllide *a* became possible with the discovery of Granick [9] that externally supplied δ-aminolevulinic acid is readily taken up by dark grown, detached barley leaves and converted into protochlorophyllide *a*. According to his experiments the dark grown seedling leaves contain all the enzymes of this pathway in non-limiting amounts and at least the steps between protoporphyrin IX and protochlorophyllide are carried out in the plastids.

A dark grown mutant seedling which as a consequence of the mutation lacks one of the enzymes in the porphyrin pathway will upon feeding with δ-aminolevulinic acid accumulate either the substrate used by the enzyme in question or one of the precursors to this substrate. At the 20 *xantha* loci 61 mutants, therefore, were tested for their ability to convert δ-aminolevulinic acid into protochlorophyllide.

An in vivo absorption spectrum of leaf pieces from 7 day old dark grown wild type seedlings reveals the absorption maximum of active protochlorophyllide *a* at 644 mμ (fig. 2). Feeding of the leaves with δ-aminolevulinic acid for 24 hr results in a large accumulation of protochlorophyllide with an absorption maximum at 636 mμ. Only a small amount of this protochlorophyllide, corresponding approximately to the amount present prior to feeding can be converted immediately to chlorophyllide a_{684} upon illumination (fig. 2). In addition to the protochlorophyllide, minor amounts of other porphyrins are accumulated in agreement with earlier observations [9]. All tested mutants at the *xan* loci *−a, −b, −c, −d, −e, −i, −j, −k, −m, −n, −o, −p, −q, −s* and *−t* respond like the wild type to δ-aminolevulinic acid feeding with the synthesis of massive amounts of protochlorophyllide *a* (fig. 3). On the other hand, most mutant alleles at the loci *xan−f, −g* and *−h* cannot accumulate detectable amounts of protochlorophyllide in the dark (fig. 2). Some mutants at these loci (e.g. *xan−f*[26] and *xan−g*[45]) as well as the only mutant at *xan−u* will form protochlorophyllide but frequently in smaller amounts than the wild type; they are thus leaky mutants. *Xantha−u*[21] accumulates

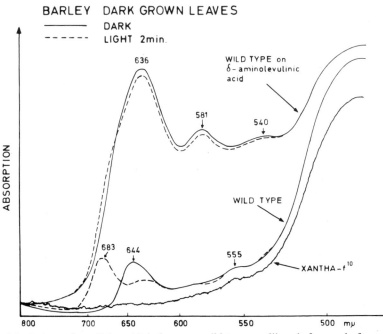

Fig. 2. In vivo spectra of leaf pieces from 7 day old dark grown wild type seedlings before and after a brief illumination with 5×10^6 erg/cm^2 of white light, giving maximum conversion of protochlorophyllide. Upper in vivo spectra are from the same number of leaf pieces after 24 hr feeding with 0.01 M δ-aminolevulinic acid in the dark. The protochlorophyllide accumulated on feeding of δ-aminolevulinic acid is not convertible. The in vivo spectrum of leaf pieces from dark grown *xantha–f*[10] seedlings reveals the absence of detectable amounts of protochlorophyllide.

an abnormal protochlorophyllide with an absorption maximum around 635 mμ which upon illumination is converted directly into a chlorophyllide a_{672}. Both the non-leaky and leaky mutants at these four loci accumulate large amounts of porphyrins upon feeding with δ-aminolevulinic acid, as can be seen from the representative spectrum reproduced in fig. 3. With the aid of thin-layer chromatography of extracts from fed mutant leaves, the major pigment accumulated was identified as protoporphyrin IX [10].

Dark grown seedling leaves of the leaky mutant *xan–l*[35] upon feeding with δ-aminolevulinic acid will accumulate large amounts of Mg-protoporphyrin IX [10] giving them an in vivo absorption spectrum characteristic for this porphyrin (fig. 3) [4].

We conclude from these results that the nuclear genes *xan–f, –g, –h,* and *–u* control steps between protoporphyrin IX and Mg-protoporphyrin IX. The nuclear gene *xan–l* controls a step between Mg-protoporphyrin IX and protochlorophyllide *a*. These genes code either for the chloroplast enzymes catalysing

these steps in protochlorophyllide synthesis or for specific structural molecules to which these enzymes and/or their substrates have to bind.

In dark grown wild type leaves which have been fed with δ-aminolevulinic acid, the accumulated inactive protochlorophyllide does not induce visible changes in the crystalline prolamellar body (fig. 4), in which the small amount of active protochlorophyllide is located. All mitochondria in such leaves, however, are highly swollen (fig. 4). Considering all mutants at the loci *xan–f, –g, –h* and *–u* the mitochondria in δ-aminolevulinic acid fed leaves are structurally indistinguishable from mitochondria in the wild type or mutant seedlings not subjected to feeding (fig. 6 vs. fig. 5; table 3). The mitochondria in mutants that accumulate protoporphyrin IX thus are protected against the structural damage caused by δ-amino-levulinic acid feeding. On the other hand, the mutant *xan–l*[35] that is blocked at a later stage in protochlorophyllide synthesis and accumulates Mg-protoporphyrin IX reacts to precursor feeding with mitochondrial swelling (table 3).

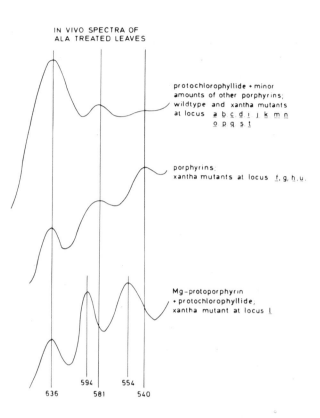

IN VIVO SPECTRA OF
ALA TREATED LEAVES

protochlorophyllide + minor amounts of other porphyrins; wildtype and xantha mutants at locus a b c d i j k m n o p q s t

porphyrins; xantha mutants at locus f, g, h, u.

Mg-protoporphyrin + protochlorophyllide, xantha mutant at locus l

636   594   581   554   540

Fig. 3. Upon feeding dark grown wild type and mutant seedling leaves with 0.01 M δ-aminolevulinic acid for 24 hr, three types of in vivo spectra are observed: The upper spectrum is found in the wild type and mutants at loci not concerned with chlorophyll synthesis and reveals accumulated protochlorophyllide. The spectrum in the center is obtained from non-leaky as well as leaky *xantha* mutants at the loci *f, g, h*, and *u* and reveals accumulated protoporphyrin IX. The lower spectrum is characteristic for the *xantha* mutant at locus *l* and reveals the accumulation of Mg-protoporphyrin.

The interaction between different cell organelles in porphyrin synthesis is further evidenced by the accumulation of abnormal sheets of cytoplasmic membranes in certain mutants upon δ-aminolevulinic acid feeding (fig. 6). These sheets appear to be derivatives of the endoplasmic reticulum and have so far been found in mutants at *xan–f, –g* and *–u*.

Heme and chlorophyll biosynthesis are controlled by feedback inhibition of δ-aminolevulinic acid synthetase [9]. Whereas heme directly represses δ-aminolevulinic acid synthetase, protochlorophyllide as such

is unable to do so: (i) The large amount of inactive protochlorophyllide accumulated in wild type seedling leaves upon δ-aminolevulinic acid feeding does not inhibit or repress the synthetase during early stages of greening [9]. (ii) The *xantha* mutants at the loci *–f, –g* and *–h* which are completely blocked in protochlorophyllide synthesis do not accumulate porphyrins unless they are fed with δ-aminolevulinic acid. Thus in these mutants the feedback inhibition of the synthetase seems operative either through protoporphyrin IX or possibly through a protoporphyrin IX holochrome.

The enzyme δ-aminolevulinic acid synthetase may be present only in mitochondria, or both in mitochondria and plastids. The subsequent four steps in porphyrin synthesis are mediated by enzymes, which may be restricted to the cytoplasm [9]. Finally, the last steps in heme and chlorophyll synthesis are conceivably compartmentalized to the mitochondria and plastids, respectively. Such a pattern requires considerable transfer of porphyrin intermediates between mitochondria and the cytoplasm as well as between plastids and the cytoplasm. If the drastic swelling of the mitochondria, which results from δ-aminolevulinic acid supply in the wild type is an expression of organelle interaction in porphyrin biosynthesis, we may speculate that the large accumulation of non-photoconvertible protochlorophyllide in the plastids tends to restrict the uptake of porphyrin intermediates made from δ-aminolevulinic acid in the cytoplasm and divert these into the mitochondria. The mitochondrial swelling would then be a direct or indirect result of their being flooded with porphyrin intermediates. In the *xantha* mutants that are blocked between protoporphyrin IX and Mg-protoporphyrin (*–f, –g, –h, .–u*), the plastids may lack the mechanism that can limit the uptake of porphyrin intermediates from the cytoplasm. The unlimited uptake of these intermediates in the plastids may protect the mitochondria from being flooded by them. Since the mutant *xan–l*[35] blocked after Mg-protoporphyrin IX shows δ-aminolevulinic acid induced mitochondrial swelling, the regulator mechanism for the uptake of porphyrin intermediates into plastids would be operating at the level of Mg-protoporphyrin IX.

Plastids of non-leaky mutants at the loci *xan–f, –g* and *–h* in the dark form widely spaced parallel discs, which are unperforated and traverse a major

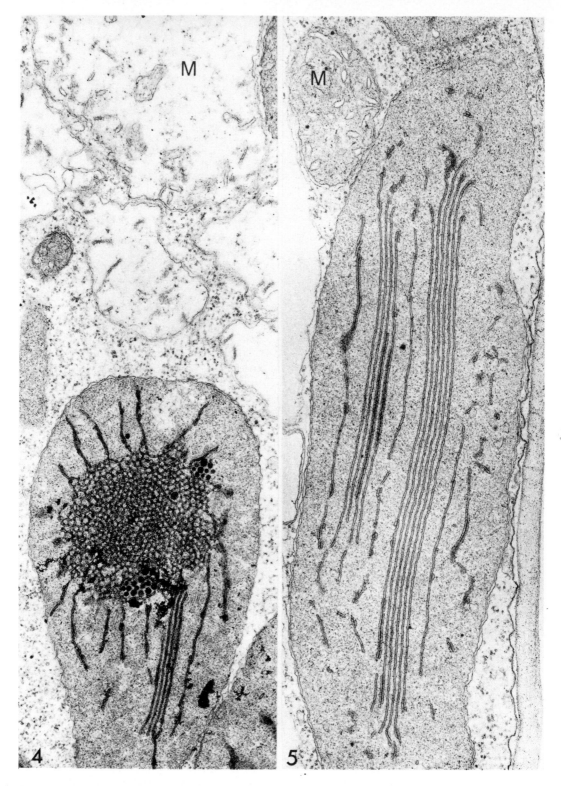

Fig. 4. Section through mesophyll cell of 7 day old dark grown wild type barley leaf after feeding with 0.01 M δ-aminolevulinic acid for 24 hr. The crystalline prolamellar body in the plastid is unchanged, but the mitochondria (M) have swollen extensively as a result of feeding with δ-aminolevulinic acid. × 30,000.

Fig. 5. Section through plastid and mitochondrion (M) in mesophyll cell of 7 day old dark grown seedling of *xan−f*[60] after 24 hr of illumination with 3200 lux of white light. The configuration of the parallel lamellar layers formed in the plastid in darkness has remained unchanged during the period of illumination. × 34,000.

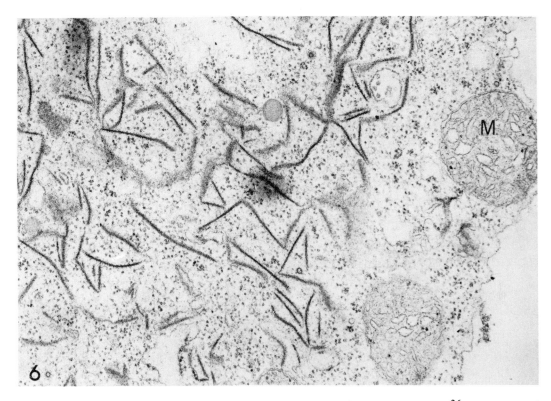

Fig. 6. Cytoplasmic area of section through mesophyll cell in 7 day old dark grown seedling of $xan-f^{26}$ after feeding with 0.01 M δ-aminolevulinic acid for 24 hr. In contrast to the wild type (fig. 4), the mitochondria (M) are not structurally affected by the feeding. Instead, abnormal sheets of cytoplasmic membranes, possibly derivatives of the endoplasmic reticulum, accumulate in the cytoplasm. × 37,000.

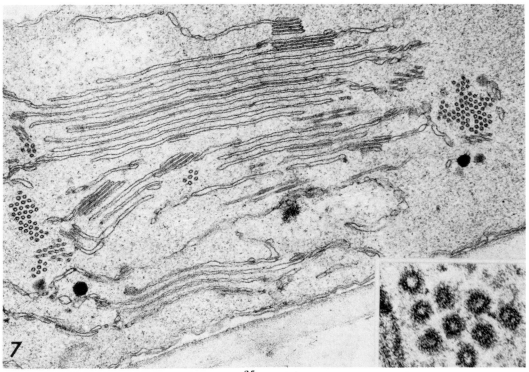

Fig. 7. Section through plastid of the leaky mutant $xan-l^{35}$. The 7 day old dark grown seedling leaf has been fed with 0.01 M δ-aminolevulinic acid for 24 hr. The feeding resulted in dispersal of prolamellar bodies into primary lamellar layers and in the formation of many groups of long tubes in hexagonal arrangement. 57,000 ×. Inset: Cross-section through group of unconnected tubes. The subunit structure of the tube walls is resolved. × 258,000.

Table 3
Pigment accumulation and structural changes resulting from feeding to 7 day old dark grown detached seedlings 0.01 M δ-aminole-vulinic acid in darkness for 24 hr (1 ml/10 seedlings).

Genotype	Pigment accumulated	Prolamellar bodies		Mito-chondria	Abnormal cytoplasmic membranes
		Before feeding	After feeding		
wildtype; $xan-b^{18}$; $-m^{48}$	Protochlorophyllide	crystalline	crystalline	swollen	none
$xan-f^{10}$; $-f^{27}$; $-f^{60}$	Protoporphyrin IX	absent	absent	normal	some
$xan-g^{37}$; $-h^{38}$; $-h^{56}$	Protoporphyrin IX	absent	absent	normal	some
$xan-f^{26}$ (leaky)	Protoporphyrin IX + some proto-chlorophyllide	crystalline, fewer than in wildtype	partially dispersed, formation of long tubes	normal	some
$xan-g^{45}$ (leaky)	Protoporphyrin IX + some proto-chlorophyllide	crystalline	partially dispersed, formation of long tubes	normal	some
$xan-l^{35}$ (leaky)	Mg-protoporphyrin IX + some proto-chlorophyllide	crystalline	partially dispersed, formation of many long tubes	swollen	none
$xan-u^{21}$ (leaky)	Protoporphyrin IX + some proto-chlorophyllide	amorphous and abnormally structured	amorphous and abnormally structured	normal	many

portion of the plastid (fig. 5) [7]. They are devoid of prolamellar bodies (table 3). Leaky mutants, such as $xan-f^{26}$, $-g^{45}$ and $-l^{35}$, make crystalline prolamellar bodies proportional to their protochlorophyllide content (table 3) in the dark [4] and spheroidel grana proportional to their chlorophyll content in the light [7]. The leaky mutant $xan-u$ with its aberrant protochlorophyllide a_{635} contains abnormally structured prolamellar bodies.

Feeding δ-aminolevulinic acid to dark grown leaves of the leaky mutants $xan-f^{26}$, $-g^{45}$, and $-l^{35}$ causes a partial dispersal of the crystalline prolamellar body (table 3). As the protoporphyrin IX or the Mg-protoporphyrin IX accumulates, many groups of long straight tubes in hexagonal arrangement are being formed (fig. 7). The diameter of the tubes and their electron scattering properties (fig. 7, inset) correspond to those composing the prolamellar body. Since the open configuration of the prolamellar body with a wide spacing of the tubes, i.e., a configuration with relatively few connections, is found in plastids with the highest amount of protochlorophyllide holochrome [3], the hexagonal arrangement of unconnected tubes may constitute a membrane conforma-

tion that is able to accommodate even more holochrome or pigment than the prolamellar body configuration with the wide spacing of the tubes. It will be interesting to see if some of the accumulated protoporphyrin IX or Mg-protoporphyrin IX in the leaky mutants is bound in these straight tubes and is substituting for protochlorophyllide in the holochrome. Alternatively, the straight long tubes may result from a slight increase in protochlorophyllide content upon δ-aminolevulinic acid feeding.

While analysing the soluble leaf proteins from non-leaky mutants at $xan-f$, $-g$ and $-h$ by gel filtration it was found that they accumulate a protein species which elutes between fraction I protein (ribulose diphosphate carboxylase) and the low molecular weight fraction II proteins (fig. 9). It is presently being studied whether this protein, which is absent in wild type leaves, is located in the plastids and there functionally connected with porphyrin synthesis or the formation of prolamellar bodies.

Besides the five nuclear genes controlling steps in the synthesis of chlorophyll a, one nuclear gene has been identified in barley that controls the synthesis of the accessory pigment chlorophyll b [11,12].

Search for nuclear genes coding for ribulose-1,5-diphosphate carboxylase

Mutation in the two genes coding for the subunits [13] of ribulose-1,5-diphosphate carboxylase will lead to mutants which either lack the enzyme or which — due to an amino acid substitution — contain an enzyme with lower specific activity than that of the wild type. The detection of such mutants requires the determination of the amount of enzyme protein and the carboxylase activity per unit enzyme protein from a mutant leaf. Ribulose-1,5-diphosphate carboxylase activity is found to be exclusively associated with fraction I protein which can be separated from

the rest of the soluble leaf proteins by Sephadex G-100 gel filtration (fig. 8) [8].

The active carboxylase accumulates slowly in the plastids of dark grown barley leaves which eventually attain an enzyme content that is comparable to leaves with fully developed chloroplasts. In leaves of 7 day old dark grown barley seedlings, the rate of accumulation of the carboxylase in the plastid stroma increases strongly upon illumination and the maximum enzyme content is obtained after about 24 hr illumination [8]. The formation of ribulose diphosphate carboxylase is independent of chlorophyll synthesis and the formation of grana.

The amount of fraction I protein and ribulose diphosphate carboxylase activity was determined in 7 day old dark grown mutant leaves illuminated for 6 and 24 hr. Such an analysis is also likely to detect

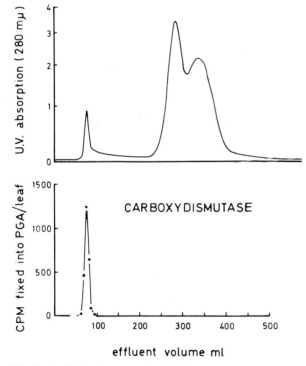

Fig. 8. Analysis of water soluble leaf proteins from 7 day old dark grown wild type seedlings illuminated for 18 hr with 3200 lux of white light. Above: Elution profile from Sephadex G-100 column consisting of three UV absorbing peaks. The first peak (left) consists of high molecular weight fraction I protein, the second peak comprises the fraction II proteins, and the third peak, lacking trichloroacetic acid precipitable material, contains various polyphenolic compounds. Below: Ribulose-1,5-diphosphate carboxylase activity was found to be exclusively associated with the fraction I protein peak.

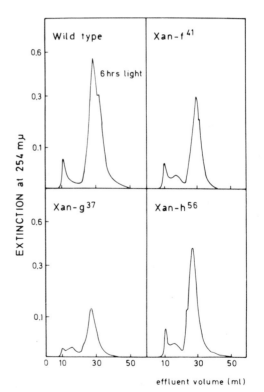

Fig. 9. Elution profiles of soluble leaf protein from 7 day old dark grown seedlings illuminated for 6 hr with 3200 lux of white light. The pattern of the wild type and one mutant each at locus xan-f, -g and -h, respectively, is shown. The mutants contain a protein species which elutes between fraction I and II proteins and which is absent in the wild type.

Table 4

Ribulose-1,5-diphosphate (RuDP) carboxylase content in wild-type and mutant leaves from 7 day old dark grown seedlings illuminated with 3200 lux for 6 and 24 hr respectively.

Genotype	6 hr light			24 hr light		
	Fraction I protein		Specific RuDP carboxylase activity [a]	Fraction I protein		Specific RuDP carboxylase activity [a]
	μg/leaf	% of wildtype		μg/leaf	% of wildtype	
wildtype	9.0	–	74.1	18.0	–	72.5
$xan-g^{45}$	8.7	97	70.0	21.7	120	70.0
$xan-d^{49}$	4.4	49	70.0	20.7	115	66.5
$xan-b^{18}$	5.9	66	70.0	7.3	41	71.4
$xan-m^3$	6.3	70	78.6	4.3	24	75.8
$alb-7$	2.1	23	70.5	2.2	12	65.4
$alb-11$	2.5	28	60.0	2.3	13	71.7
$alb-17$	4.0	44	83.2	1.8	10	66.7

[a] $\dfrac{\text{CPM incorporated into phosphoglyceric acid/20 min}}{\mu\text{g fraction I protein}}$

cases of degradation of the enzyme in mutant leaves as a consequence of secondary degeneration of the plastids in the light. Among the mutants in 3 *albina* and 21 *xantha* genes a wide variation in the fraction I protein content of the plastids was found. Selected data are given in table 4. The *albina* mutants had the lowest content in fraction I protein after 24 hr of illumination amounting to about 10 per cent of that found in the wild type. The specific activity of the fraction I protein to catalyse the reaction of one molecule ribulose-1,5-diphosphate with one molecule CO_2 to give two molecules 3-phosphoglyceric acid in vitro was found to be practically the same for all mutants tested (table 4). None of the mutants studied thus lacks the ability to synthesize the enzyme or contains serious amino acid alterations in the enzyme. This is also presumably true for the 15 *albina* mutants with low enzyme activity (1% of wild type) and the 10 *chlorina* mutants with highly variable amounts of enzyme activity studied by Kleinhofs and Shumway [14]. Under the additional assumption that the latter mutants are all the results of mutations in different genes, only 49 nuclear genes have so far been screened for the purpose of finding those coding for ribulose-1,5-diphosphate carboxylase.

The mutants surveyed in table 4 contain several representative cases: One mutant blocked in chlorophyll synthesis ($xan-g^{45}$, leaky) accumulates wild type amounts of ribulose diphosphate carboxylase at the normal rate. Another mutant ($xan-d^{49}$) reveals a delayed accumulation of the enzyme. Three mutants ($xan-b^{18}$, $xan-m^3$, $alb-17$) initially accumulate rather large amounts of ribulose diphosphate carboxylase, but then the formation ceases and the plastids may eventually be depleted of the enzyme. Finally two *albina* mutants contained low amounts of the enzyme at both times studied.

Nuclear genes controlling the synthesis of chloroplast specific lipids: *xan–a* and *–m*

Synthesis of specific lipids accompanies the formation of photosynthetic membranes and grana. Among these, monogalactosyl diglycerides with a high linolenic acid content, digalactosyl diglycerides, and sulfolipids are especially prominent (review in [6]). By incubating small leaf slices in ^{14}C labeled acetate solutions for three hours in the light, the capacities of the synthetic pathways for various lipid classes can be determined. A strongly increasing capacity to synthesize fatty acids and to insert them into phospho-, sulfo-, and galactolipids is observed as the chloroplasts develop in illuminated leaves of dark grown barley seedlings [6]. In the earliest stages of greening, 25% of the incorporated acetate appears in the phospho-, sulfo-, and galactolipids, whereas 75% is

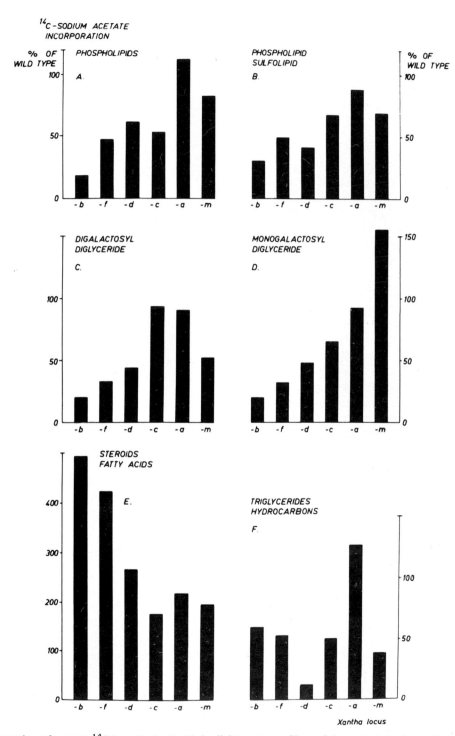

Fig. 10. Incorporation of acetate-^{14}C into the leaf lipids by light grown seedlings of the mutants in six *xantha* loci as a percentage of that incorporated by the wild type. Values given for the *xan−f* and *−m* loci are averages for three alleles.

built into steroids and other unsaponifiable lipids. At later stages of greening when large amounts of grana are being formed, the situation is reversed: then 75% of the incorporated acetate is channeled into the phospho-, sulfo-, and galactolipids, and only 25% serves as the precursor for the synthesis of steroids and other unsaponifiable lipids. Incubation of isolated barley plastids at various stages of greening in solutions containing acetate-1-^{14}C reveals that during its development the organelle itself gains an increased capacity to synthesize fatty acids and insert them into lipids [15]. Upon separating the lamellar system from the stroma the polar lipid fatty acids which have been newly synthesized by the isolated plastids are found in the photosynthetic membranes. Indeed, the amount of label in the lamellar system increases dramatically after the first grana have been formed in the plastid.

The capacity of leaf slices from light grown *xantha* seedlings to incorporate ^{14}C-acetate into various lipid classes has been compared with that of identically grown wild type seedlings (fig. 10). This capacity is assessed at the final developmental stage reached by the plastids of the mutants and compared to that of the mature wild type chloroplasts. As evident in fig. 10, a series of 6 *xantha* mutants representing the four loci *xan–b, –f, –d, –c* mimicks wild type seedlings at different greening stages with regard to the different patterns of acetate labeling of the phospho-, sulfo-, and galactolipids, on the one hand, and of the steroid-containing fraction, on the other [7]. The genetic block in chlorophyll synthesis as found in the *xan–f* mutants, for instance, leads in the light grown seedling leaves to a repression of the acetate-^{14}C incorporation into phospho-, sulfo-, and galactolipids. As in early greening stages of dark grown leaves, the acetate is channeled in these *xan–f* mutants predominately into steroids. Clearly the changes in lipid synthesis caused by mutations in *xan–b, –f, –d* and *–c* are secondary effects of the primary lesions.

A different situation is found in the mutants at *xan–a* and *–m*. In spite of the severe chlorophyll deficiency, the capacity of *xantha–a*[11] to incorporate acetate into phospho-, sulfo-, and galactolipids is about equal to wild type (fig. 10). All *xan–a* mutants assemble highly irregular and variable membrane systems in their plastids. The membrane material may

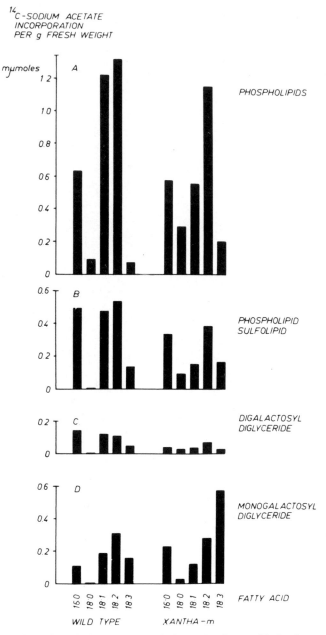

Fig. 11. Labeling patterns of the major fatty acids in the different lipid classes of the wild type compared to the average of three mutants at the *xan–m* locus.

be organized both in dark and light into a honeycomb configuration by a regular branching of the intradisc spaces (fig. 12), into individual groups of concentric

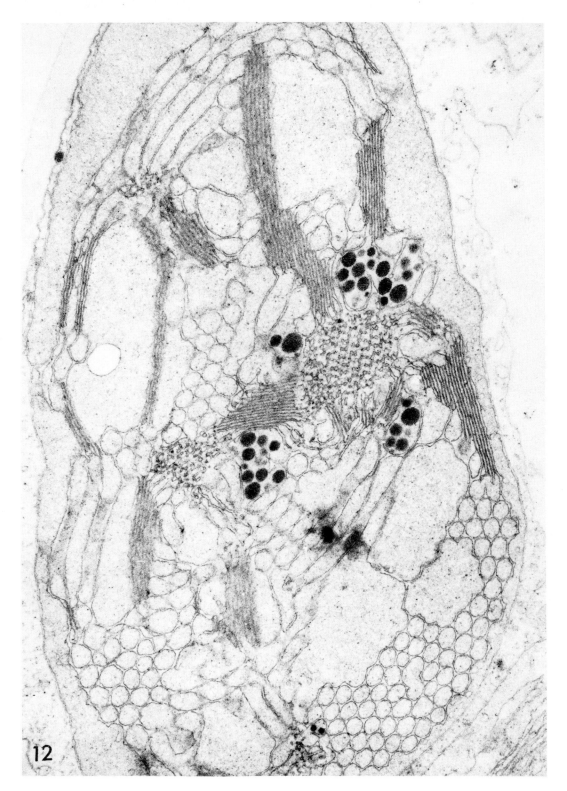

Fig. 12. Section through plastid of $xan-a^{63}$ seedling grown for 10 days at light intensity of 20 lux. The chloroplast membranes of this mutant can assume a wide variety of configurations. This plastid section shows the characteristic honeycomb arrangement of the membranes, crystalline prolamellar bodies, and grana that are formed upon accumulation of chlorophyll in dim light. × 36,000.

primary layers, into prolamellar bodies, and into large membrane bordered bodies containing lipid deposits [7]. Large aggregations of osmiophilic globuli can occur [7]. If mutants of this locus are grown in dim light for prolonged periods, they accumulate chlorophyll (up to 25% of wild type) and form grana between the complex membrane systems (fig. 12). The disc membranes as well as the intradisc spaces in the grana are continuous with the membranes and intradisc spaces of the honeycomb configuration, respectively.

Further studies on the *xan–a* mutants are necessary to find their primary lesions. However, their high capacity to synthesize chloroplast specific lipids and their prolific production of membrane material both in dark and light indicate that this gene codes for a molecule that is important in the ordered assembly of the chloroplast's lipoprotein membranes. Its absence or structural alteration allows an uncontrolled synthesis of membrane components in the dark, that after-

wards cannot be used by the plastids in the light for the construction of a normal lamellar system.

The three allelic mutants $xan–m^3$, $–m^{48}$, and $–m^{53}$ surpassed the wild type, by 50% on the average, in incorporation of acetate into monogalactosyl diglycerides (fig. 10), although incorporation into the other polar lipids was reduced. Furthermore, linolenic acid was labeled preferentially in the monogalactosyl diglycerides (fig. 11), giving a spectrum of labeled fatty acids in this lipid fraction which is more similar to that of the endogenous fatty acids than is the labeling pattern of wild type leaf slices [6,7]. The mutants at the *xan–m* locus provide the first example of higher plant leaves in which the fatty acids of the monogalactosyl diglycerides can be labeled from acetate-^{14}C in proportions corresponding roughly to the endogenous pattern of the fatty acids assembled into monogalactolipids during the formation of the lamellar system in the chloroplast.

Xan–m mutants in the dark form packages of

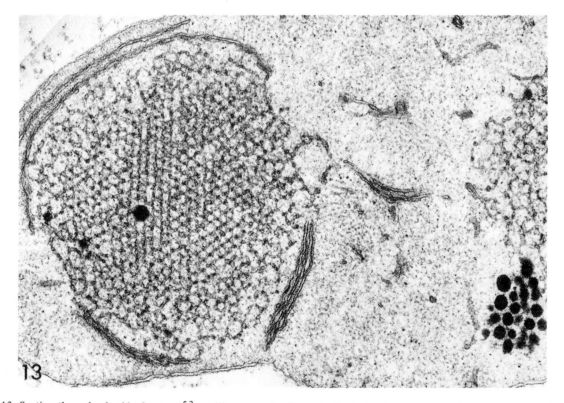

Fig. 13. Section through plastid of $xan–m^{53}$ seedling grown for 7 days in the dark. The plastid contains a crystalline prolamellar body around which packages of membrane associations and primary lamellar layers are arranged concentrically. × 62,000.

grana-like membrane associations which are frequently arranged concentrically around the prolamellar body (fig. 13). Upon exposure to light the prolamellar body is not dispersed, but considerable accumulation of chlorophyll — up to 25% of that of the wild type — can take place. Since no formation of grana or additional grana-like membrane associations takes place upon illumination, the newly synthesized pigment is possibly integrated into the grana-like membrane associations which have been formed in the dark [7]. In certain developmental stages the plastids of the mutant $xan-m^3$ accumulate large numbers of globuli in addition to the grana-like packages. The globuli contain 90% lipid by dry weight and are rich in galactolipids [16,17].

The exceptional capacity for rapid linolenic acid synthesis and its preferential insertion into monogalactosyl diglycerides exhibited by the $xan-m$ mutants indicates that these mutants synthesize monogalactolipids in a relaxed manner, both in light and dark. The accumulating monogalactolipids may coat the chloroplast discs and glue these into the observed abnormal osmiophilic membrane packages or, alternatively, become deposited as osmiophilic globuli.

Nuclear genes controlling the insertion of chlorophyll into photosynthetic membranes

A mutation that causes an alteration in the binding of the chlorophyll a molecules to the lipoprotein in the chloroplast membranes can be expected to result in a changed in vivo absorption spectrum of the leaf. More specifically one might expect a red or blue shift in the in vivo absorption maximum at 678 mμ of chlorophyll a. To see whether such mutants can be found we induced over a thousand chloroplast mutants in barley with ethyl methanesulfonate and screened these with an integrating sphere attached to a Zeiss RPQ-20A recording spectrophotometer. Several mutants were discovered which had large amounts of a pigment absorbing at 745 mμ in addition to a reduced chlorophyll a absorption maximum at 678 mμ. Tracings of the in vivo spectra from two

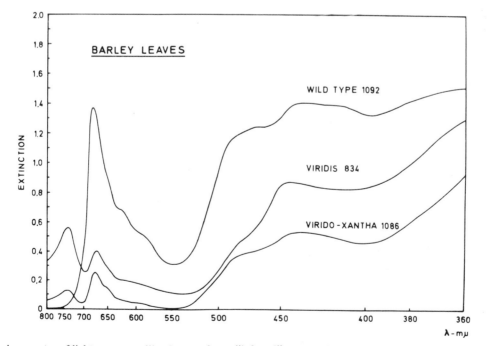

Fig. 14. In vivo spectra of light grown seedling leaves of two "infrared" mutants in comparison with the wild type. The "infrared" mutants have in addition to the chlorophyll a with an absorption maximum at 678 mμ large amounts of a chlorophyll a form with a maximum at 745 mμ.

such mutants and the corresponding wild type, as they were recorded in the screening procedure, are given in fig. 14. The mutants display a *zonata* phenotype when grown in alternating light and dark periods. We provisionally call them "infrared" mutants.

"Infrared"-5 has been studied extensively. By growing the seedlings in constant dim light but under different temperatures, the proportions of the pigment absorbing at 745 mμ and chlorophyll a_{678} can be varied over a wide range (fig. 15). With a thermoperiodicity of 16 hr at 15°C and 8 hr at 10°C, the seedling leaves produce large amounts of the 745 mμ pigment and relatively little chlorophyll a_{678}. These temperature conditions are optimal for the growth of wild type seedlings. As the temperature for seedling growth is raised, increasing amounts of chlorophyll a_{678} and correspondingly decreasing amounts of the 745 mμ pigment are produced. With a thermoperiodicity of 16 hr at 30°C and 8 hr at 25°C the infrared

absorbing pigment forms only a minor shoulder on the large chlorophyll a_{678} peak.

The pigment absorbing at 745 mμ is located in the plastids and constitutes a special form of chlorophyll *a*. If the leaves are ground in aqueous media and the cell components separated by differential centrifugation, the 745 mμ pigment is found closely associated with the chlorophylls in the chloroplast fraction. Upon extraction of the pigments with acetone, only chlorophyll *a* and chlorophyll *b* are obtained. The amount of chlorophyll *a* found corresponds to that expected from the combined in vivo absorptions at 745 and 678 mμ.

Plastids of the mutant grown under low temperatures and containing a high amount of chlorophyll a_{745} contain instead of grana with stacked discs those shown in figs. 16 to 18. In one abnormal grana configuration the membrane material is condensed into an undulated osmiophilic sheet which is bor-

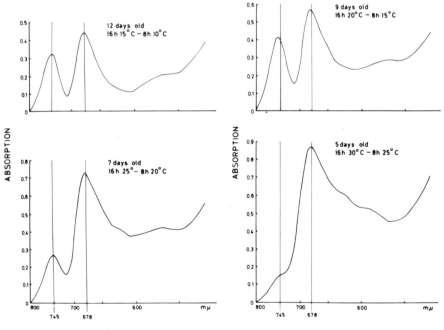

INFRARED 5: primary leaves, seedlings grown at 20 lux (white light)

Fig. 15. In vivo spectra of seedling leaves in comparable physiological age from mutant "infrared"-5. The relative amounts of the 678 mμ and 745 mμ forms of chlorophyll *a* in the plastids can be varied by using different environmental conditions during leaf development. During development in continuous dim light, the least amount of the 745 mμ form is obtained under high temperature and the largest amount under low temperature.

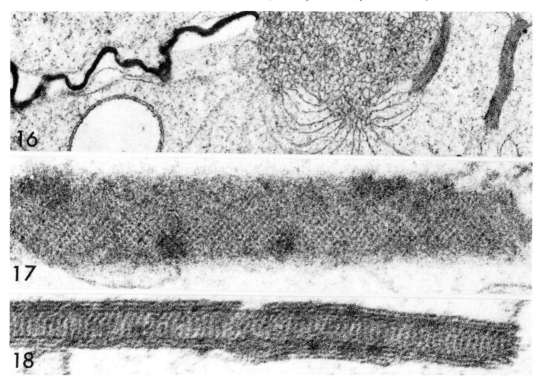

Figs. 16–18. Three types of abnormal grana found in plastids of the mutant "infrared"-5 when these contain a high amount of the 745 mμ form of chlorophyll a (20 lux, 15–10°C). × 57,000; × 156,000; × 149,000.

dered by a disc membrane. At many places this outer disc membrane is separated from the sheet by a swollen intradisc space (fig. 16). Another abnormal grana configuration consists of dense crystalloids with the two types of periodic patterns depicted in figs. 17 and 18. These abnormal grana are also bordered by disc membranes.

Membrane fractions containing these abnormal grana and both the 745 mμ and 678 mμ forms of chlorophyll a can be isolated from homogenates of the leaves by differential centrifugation. Extraction of these membrane fractions with acetone yields only chlorophyll a. On treatment with the detergent Triton X-100, the 745 mμ absorption disappears and instead a very large peak is formed at 665 mμ, characteristic of chlorophyll removed from the membrane material by the detergent micelles.

At high temperatures the mutant plastids form normal grana with stacked discs in correspondence to the preponderance of chlorophyll a_{678} present.

Not only chlorophyll a but also its precursor protochlorophyllide a becomes bound to the membrane material in an abnormal manner. The plastids in dark grown leaves accumulate considerably larger amounts of protochlorophyllide than the wild type, part of which is unconvertible to chlorophyllide a by illumination. This indicates that the membranes of the prolamellar body are also defective. An abnormal association of the protochlorophyllide with membrane components seems to lead to a partial derepression of δ-aminolevulinic acid synthetase and an accumulation of inactive protochlorophyllide.

The "infrared"-5 gene codes for a component which can determine how chlorophyll a and its precursor protochlorophyllide a are bound to the membrane. It also has a decisive role in the morphogenesis of the grana structure.

Genome versus plastome

The data presented in the preceding paragraphs demonstrate that the nuclear genes hold a very tight control over plastid biogenesis. A similar conclusion is reached by R.P. Levine, U.W. Goodenough, and co-workers through systematic studies of acetate requiring mutants in *Chlamydomonas reinhardi* [18–20]. They have identified nuclear genes in *Chlamydomonas* that control the assembly of the electron transport chain, the synthesis of chloroplast ribosomes, and the stacking of photosynthetic membranes.

At present we are still unaware of the genetic information contained in chloroplast DNA but it is likely — although not proven — that chloroplast DNA serves as the vehicle for plastome inheritance. The extensive analysis of one plastome mutant in *Nicotiana tabacum* with a maternal inheritance following the pattern of the status albomaculatus [21] in S.G. Wildman's and our laboratory has not yet given a hint on the primary lesion caused by the mutation. This plastome mutant exhibits the following characteristics: Defective white plastids can be present with normal green plastids in the same cell [21]. The structure of the mitochondria is affected in cells containing defective white plastids [21]. Wild type amounts of DNA are present in the white plastids [2,22]. The white plastids contain DNA polymerase, 70 S chloroplast ribosomes, and considerable amounts of ribulose diphosphate carboxylase [22].

This plastome mutation is transmitted only through the egg cell, not through the pollen. The plastome and thereby possibly the chloroplast DNA is thus eliminated at some stage between the formation of the pollen mother cell and the fertilization of the egg cell by the pollen tube. Using the technique of J.P. and C. Nitsch [23] to raise haploids (i.e. polyhaploids) from uninucleate immature pollen grains of the allopolyploid species *Nicotiana tabacum,* P. von Wettstein-Knowles and T. Nilsson-Tillgren have investigated the time at which this elimination of the plastome occurs [24]. Pollen grains from a white shoot of the variegated tobacco plant produce white haploid plants and those from a green shoot green haploids. Uninucleate immature pollen grains from variegated flower buds gave rise not only to all white or all green haploids but also to haploid white-green variegated plants. These plants reveal that the two distinguishable plastome factors are still present in the immature pollen grains and that the plastome in the male germ line is eliminated after meiosis at some point between the onset of the first pollen mitosis and fertilization.

Assuming that chloroplast and mitochondrial DNA are inherited maternally, we can state that organelle DNA differs in two features from the chromosomal DNA in the nucleus. (i) Organelle DNA consists of naked, histone free molecules organized into DNA plasms not bordered by an envelope (nucleoids); (ii) it is transmitted by a non-mendelian genetic system which is characterized by the presence of many copies of the nucleoids per cell and by the total or partial elimination of these nucleoids during the maturation of the male gametes. Current working hypotheses envisage that organelle DNA contains the genetic information to synthesize certain components of the organelle in which the DNA is located. Location of a given piece of genetic information in organelle DNA, however, could also be entirely independent of the place in the cell where this information is put to use. Whether a given piece of genetic information is located in the chromosomes or in organelle nucleoids is determined by the selective advantage of having it transmitted by the mendelian or the non-mendelian genetic system, respectively. The location may also be determined by the advantage of having the transcription of the genes regulated by a mechanism employing nucleohistone fibrils in the nucleus or naked DNA fibrils in a nucleoid, respectively.

Acknowledgements

We are indebted to Miss Ulla Eden for recording the large number of spectra necessary to isolate the "infrared" mutants. Financial support from the United States Public Health Service, National Institutes of Health (GM–10819), as well as the Danish Natural Science and the Swedish Agricultural Research Councils is gratefully acknowledged. The Danish Atomic Energy Commission has provided research facilities at Risö.

Summary

In our studies on barley, 86 nuclear genes have so far been identified that control chloroplast development. A group of 69 mutations in 20 of these loci give rise to *xantha* phenotypes. The 33 *albina* mutants analysed belong to 31 different loci, whereas 16 loci were found among 27 mutants in the *tigrina* and *zonata* group. Twenty *virido-albina* mutants belong to 14 loci. Five loci with more than one mutant allele have been identified among 49 *viridis* mutants.

Many of these 198 mutants have been tested for their ability to form protochlorophyllide and proplastid structures in the dark. Dark grown mutant leaves have further been analysed for their capacity to synthesize photosynthetic membranes, soluble chloroplast proteins such as ribulose diphosphate carboxylase, and chloroplast specific lipids in the light.

Five genes — *xan–f*, *–g*, *–g*, *–l* and *–u* — control steps in the biosynthesis of chlorophyll between protoporphyrin IX and protochlorophyllide. The gene *xan–m* has a controlling function in the synthesis of monogalactosyl diglycerides rich in linolenic acid. Mutants in the locus *xan–a* have a high capacity for the synthesis of all lipid classes even under conditions when these lipids cannot be incorporated into normal chloroplast membranes. Several *xantha* and *albina* genes control processes like the spectral shifts of the newly synthesized chlorophyllide *a* and the dispersal of the prolamellar body. A mutation in the gene designated as infrared 5 forms a chlorophyll membrane complex in the chloroplasts with an in vivo absorption maximum at 745 mμ. The product of this gene seems involved in the proper arrangement of the chlorophyll molecules in the grana structure.

The role of the genome and the plastome in chloroplast biogenesis is discussed briefly.

References

[1] A. Gustafsson, Kungl. Fysiografiska Sallskapets Handlingar N.F. Bd. 51 (1940) 1–40.

[2] D. von Wettstein, in: Harvesting the Sun – Photosynthesis in Plant Life, eds. A. San Pietro, F.A. Greer, T.J. Army (Academic Press, New York, London, 1967) p. 153.

[3] K.W. Henningsen and J.E. Boynton, J. Cell Sci. 5 (1969) 757.

[4] J.E. Boynton and K.W. Henningsen, Studia Biophys. 5 (1967) 85.

[5] K.W. Henningsen and J.E. Boynton, Studia Biophys. 5 (1967) 89.

[6] L.A. Appelqvist, J.E. Boynton, P.K. Stumpf and D. von Wettstein, J. Lipid Res. 9 (1968) 425.

[7] L.A. Appelqvist, J.E. Boynton, K.W. Henningsen, P.K. Stumpf and D. von Wettstein, J. Lipid Res. 9 (1968) 513.

[8] C.G. Kannangara, Plant Physiol. 44 (1969) 1533.

[9] S. Granick, in: Biochemistry of Chloroplasts, ed. T.W. Goodwin (Academic Press, London, New York, 1967) Vol. II, p. 373.

[10] S. Gough, unpublished results.

[11] N.K. Boardman and H.R. Highkin, Biochim. Biophys. Acta 26 (1966) 189.

[12] D.J. Goodchild, H.R. Highkin and N.K. Boardman, Exptl. Cell Res. 43 (1966) 684.

[13] N. Kawashima, Biochem. Biophys. Res. Comm. 38 (1970) 119.

[14] A. Kleinhofs and L.K. Shumway, Biochem. Genetics 3 (1969) 485.

[15] P.K. Stumpf, C.G. Kannangara, K.W. Henningsen and D. von Wettstein, unpublished results.

[16] A.D. Greenwood, R.M. Leech and J.P. Williams, Biochim. Biophys. Acta 78 (1963) 148.

[17] J.L. Bailey and A.G. Whyborn, Biochim. Biophys. Acta 78 (1963) 163.

[18] U.W. Goodenough and R.P. Levine, Plant Physiol. 44 (1969) 990.

[19] U.W. Goodenough, J.J. Armstrong and R.P. Levine, Plant Physiol. 44 (1969) 1001.

[20] U.W. Goodenough and R.P. Levine, J. Cell Biol. 44 (1970) 547.

[21] D. von Wettstein and G. Eriksson, in: Genetics Today, Proc. XI. Int. Congr. Genetics, The Hague 1963, ed. S.J. Geerts (Pergamon Press, Oxford, 1964) Vol. 3, p. 591.

[22] S.G. Wildman and C. Liao, personal communication.

[23] J.P. Nitsch and C. Nitsch, Science 163 (1969) 85.

[24] P. von Wettstein-Knowles and T. Nilsson-Tillgren, Nature (1970) in press.

INHIBITION OF CHLOROPLAST RIBOSOME FORMATION
BY GENE MUTATION IN *CHLAMYDOMONAS REINHARDI*

Ursula W. GOODENOUGH, R.K. TOGASAKI *, A. PASZEWSKI **
and R.P. LEVINE

The Biological Laboratories, Harvard University,
Cambridge, Massachusetts, USA

The unicellular green alga *Chlamydomonas reinhardi* contains a large cup-shaped chloroplast that occupies perhaps two-thirds of the cell's volume [1–3]. The chloroplast stroma is densely packed

* Present address: Department of Botany, University of Indiana, Bloomington, Indiana, USA.
** Present address: Institute of Biophysics and Biochemistry, Warsaw, Poland.

with ribosomes (fig. 1) that are smaller (68 S) than the cytoplasmic ribosomes (80 S) and resemble bacterial ribosomes in a number of respects [4]. Levels of chloroplast ribosomes in wild-type *C. reinhardi* remain constant whether the cells are grown phototrophically on a minimal salt medium, mixotrophically on an acetate-supplemented medium, or heterotrophically in the dark on the same acetate-supplemented medium. Chloroplast ribosome levels are also

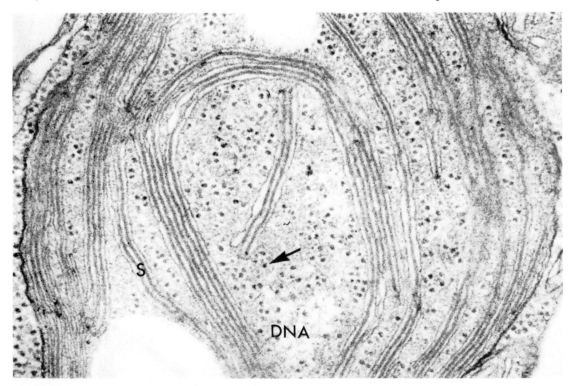

Fig. 1. Portion of the chloroplast from a phototrophically-grown wild-type cell. Chloroplast ribosomes (arrow) fill the stroma, and a region of chloroplast DNA is indicated. Chloroplast membrane is arranged in stacks (S) of from 2–6 discs. 79,000 X.

normal in 15 mutant strains of *C. reinhardi* that are either pigment-deficient or blocked in some photosynthetic reaction ([5,6] and unpublished observations). Thus chloroplast ribosome production in *C. reinhardi* does not appear to be sensitive to growth conditions nor to the photosynthetic capacity of the chloroplast.

The *ac-20* mutant strain of *C. reinhardi* is therefore exceptional, for it exhibits low levels of chloroplast ribosomes when grown phototrophically, and even more drastically reduced levels when grown mixotrophically. The *ac-20* gene has been mapped to linkage group XIII of the *C. reinhardi* genome [7]. At present we have no idea what process or processes are dictated by the wild-type gene at the *ac-20* locus, nor how a mutation of this gene can result in a 90% reduction in chloroplast ribosomes. We have, however-

er, taken advantage of the low-ribosome condition of *ac-20* to study the protein-synthesizing capacities of the *C. reinhardi* chloroplast. Full reports of this research will appear in forthcoming publications [8–10] where experimental methods are given in detail.

Demonstrations of the ribosome deficiency in *ac-20*

A comparison of electron micrographs of wild-type (fig. 1) and mixotrophically-grown *ac-20* (fig. 2) cells reveals the paucity of chloroplast ribosomes in the latter. Phototrophically-grown *ac-20* (fig. 3) possesses chloroplast ribosome levels that are intermediate between wild-type and mixotrophic *ac-20* levels. One can quantitate such impressions by cutting

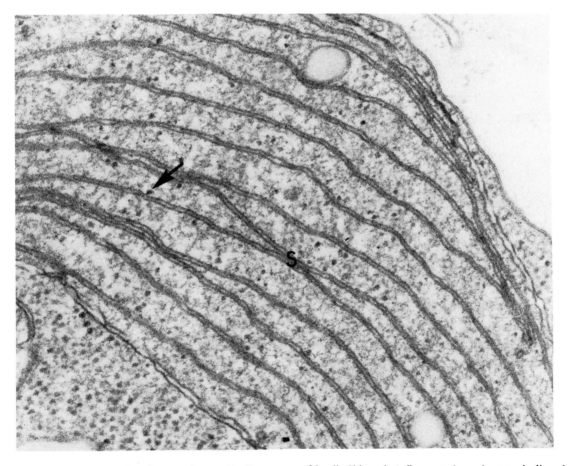

Fig. 2. Portion of the chloroplast from a mixotrophically-grown *ac-20* cell. Chloroplast ribosomes (arrow) are markedly reduced from the wild-type level (fig. 1). Chloroplast membranes in this section are predominantly in the form of single discs, with only limited regions of stacking (S). 94,000 X.

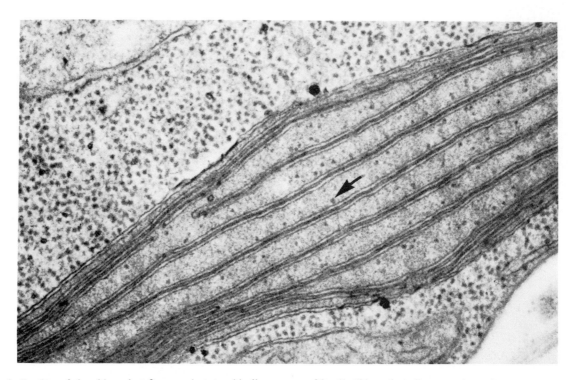

Fig. 3. Portion of the chloroplast from a phototrophically-grown *ac-20* cell. Chloroplast ribosomes (arrow) are more numerous than in mixotrophic *ac-20* (fig. 2) but less numerous than in wild-type (fig. 1). Chloroplast membranes are predominantly organized into parallel arrays of 2-disc stacks. 85,000 X.

electron micrographs with scissors so that portions of chloroplast stroma are obtained, counting the number of ribosomes in the sample, and weighing the cuttings to obtain a value of ribosomes per unit stroma weight. Such ratios are given in fig. 4. It is seen that phototrophically- and mixotrophically-grown wild-type cells contain nearly identical levels of chloroplast ribosomes by this 'cut-and-weigh' assay. The chloroplast ribosome level of mixotrophic *ac-20* is found, in 5 separate experiments, to be 10-fold less than the wild-type level, and phototrophic *ac-20*

Fig. 4. The levels of chloroplast ribosomes, as determined by the cut-and-weight assay, for the following cell types: wild-type grown on an acetate-supplemented (ac) and a minimal (min) medium; mixotrophic *ac-20* (A-cells) from 5 separate experiments; phototrophic *ac-20* (M-cells); and mixotrophic *ac-20* transferred to minimal medium and incubated in the dark for 12 hr (light-to-dark) from 3 separate experiments.

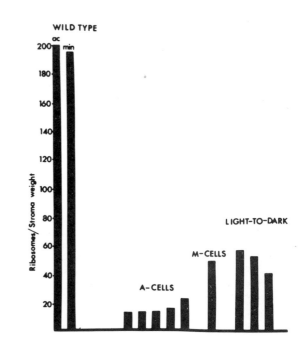

exhibits an intermediate, 4-fold reduction in chloroplast ribosome levels.

The 68 S chloroplast ribosome deficiency of mixotrophic ac-20 has also been visualized physically on sucrose gradients (fig. 5) and by polyacrylamide gel electrophoresis (fig. 6). A deficiency in the 23 and 16 S species of chloroplast ribosomal RNA has been demonstrated by polyacrylamide gel electrophoresis (fig. 7), and a deficiency in the soluble 5 S RNA associated with chloroplast ribosomes has been shown by MAK column chromatography [11]. By all of these procedures, levels of cytoplasmic ribosomes in ac-20 are found to be comparable to wild-type levels;

thus the ac-20 mutation selectively affects the production of chloroplast ribosomes. In electron micrographs, mitochondrial ribosomes also appear to be present in normal amounts in the mutant strain.

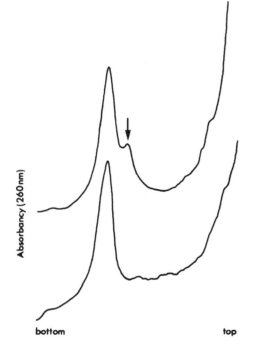

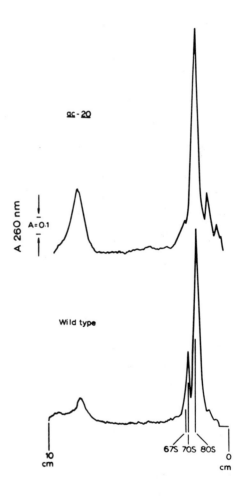

Fig. 5. Absorbancy tracings of sucrose density gradients containing ribosomes, prepared as described by Hoober and Blobel [4]. The upper profile, obtained from a mixotrophically-grown mutant strain having wild-type levels of chloroplast ribosomes [6,10], exhibits 2 distinct peaks. The smaller, slower-sedimenting peak (arrow) is identified as deriving from chloroplast ribosomes. This peak is absent from the lower profile, obtained from mixotrophic ac-20.

Fig. 6. The ribosomes of mixotrophic ac-20 (top) and wild-type (bottom) cells of C. reinhardi separated by polyacrylamide gel electrophoresis (3% acrylamide, 4 hr, 12 mA/column, buffer pH 7.5) of a broken cell suspension treated with 1% deoxycholate. (Prepared by Dr. S.J. Surzycki; methods to be published elsewhere.) The 70 S chloroplast ribosome peak of wild-type cells appears as a small shoulder, with a peak at 67 S, in mixotrophic ac-20. Note the large amount of a rapidly migrating component in the mixotrophic ac-20 gel; the material has not yet been identified.

ac—20

A=0.1

WILD
TYPE

10cm   3cm
15.5S 18S 23S 25S

Fig. 7. Polyacrylamide gel electrophoresis of ribosomal RNA from mixotrophic *ac-20* (top) and wild-type (bottom) cells of *C. reinhardi.* The 16 S and 23 S chloroplastic species are markedly deficient in the *ac-20* preparation. (Prepared by Dr. S.J. Surzycki; methods described in ref. 15.)

Chloroplast structure in *ac-20*

When figs. 1–3 are compared, differences in chloroplast membrane organization as well as ribosome levels are apparent. In the wild type (fig. 1), chloroplast membranes form long flattened sacs (discs or thylakoids), and the discs fuse together to form stacks of discs. In a typical wild-type cell, stacks of 3–4 discs predominate, and single, unfused discs are only infrequently encountered [5]. In mixotrophic *ac-20*, on the other hand, single discs predominate (fig. 2), and stacking is limited. The chloroplast membrane is also frequently found in dense vesiculate

clusters, as shown in fig. 8. Such profiles are not found in wild-type chloroplasts, nor in any other green mutant strain of *C. reinhardi* that has been examined [5,6]. They are reminiscent of the vesicles found in dark-grown, yellow mutant cells of *C. reinhardi* [2], but are more tightly aggregated. Since mixotrophic *ac-20* contains near-normal levels of chlorophyll [9] and is grown in the light, there is no reason at present to suggest that the two types of vesicles derive from a similar lesion in membrane formation.

The chloroplast membrane organization of phototrophic *ac-20* (fig. 3) is intermediate between mixotrophic *ac-20* and wild type. The vesiculate clusters are absent and stacking occurs, but the average stack size is smaller than that found in the wild type.

At the base of the *C. reinhardi* chloroplast lies a large pyrenoid from which the chloroplast disc system seems to radiate [2]. In mixotrophic *ac-20* the pyrenoid is either absent or else present in a very rudimentary form [10]. In phototrophic *ac-20* the pyrenoid is smaller than that found in wild-type cells, but it is much better developed than the pyrenoid of mixotrophic *ac-20* cells.

A pattern therefore emerges: the complex pyrenoid-disc system of the wild-type chloroplast is severely disorganized in mixotrophic *ac-20* and somewhat disorganized in phototrophic *ac-20*. This pattern, of course, is coordinate with the pattern given by chloroplast ribosome levels.

Photosynthetic properties of *ac-20*

The same pattern is found when the photosynthetic properties of *ac-20* are compared with wild type. The activity of ribulose-1,5-diphosphate (RuDP) carboxylase, an enzyme of the photosynthetic carbon reduction cycle, and the activities of cytochrome 559 and Q, two components closely associated with photochemical system II (PS-II), are dramatically reduced in mixotrophic *ac-20* and partially reduced in phototrophic *ac-20* [8,9]. As a result, mixotrophic *ac-20* is capable of very little photosynthesis, while phototrophic *ac-20* is capable of CO_2 fixation rates that are approximately 25% of the wild-type rates [8].

The photosynthetic properties of mixotrophic

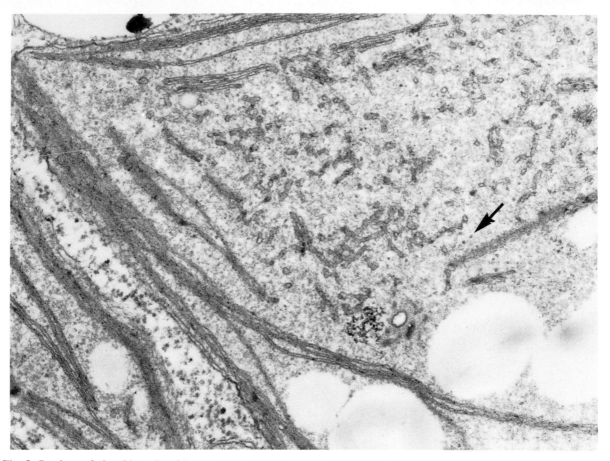

Fig. 8. Portions of the chloroplast from a mixotrophically-grown *ac-20* cell. Membrane is predominantly in the form of small, tightly-aggregated vesicles. Single, unstacked discs are also seen. Levels of chloroplast ribosomes (arrow) are low. 64,000 X.

ac-20 have been studied in detail [8,9]. These studies indicate that, with the exception of RuDP carboxylase, cytochrome 559, and Q, the enzymes and components of the photosynthetic apparatus of *C. reinhardi* are present in normal amounts in the mutant cells. These include components associated with photochemical system I (PS-I), chlorophyll, and most of the enzymes of the carbon reduction cycle. In other words, a 10-fold reduction in chloroplast ribosomes does not affect the ability of *C. reinhardi* to synthesize most of its chloroplast-specific proteins. These results are summarized in table 1. We conclude that the chloroplast components in the 'Affected' column of table 1 are dependent upon the presence of normal levels of chloroplast ribosomes for their synthesis, whereas those in the 'Not affected' column are synthesized independently of chloroplast ribosomes and thus, presumably, on cytoplasmic ribosomes.

Table 1
Chloroplast components affected and not affected in mixotropic *ac-20*

Affected	Not affected
	1. Phosphoriboisomerase
	2. Phosphoribulokinase
	3. 3-PGA kinase
1. RuDP carboxylase	4. G-3-P dehydrogenase (NAD)
	5. G-3-P dehydrogenase (NADP)
	6. Triosephosphate isomerase
	7. FDP aldolase
	8. Total quinone
	9. Plastocyanin
2. Cytochrome 559	10. Cytochrome 553
3. Q, the quencher of	11. Cytochrome 564
fluorescence of PS-II	12. P700
	13. Ferredoxin
	14. Ferredoxin-NADP reductase
	15. CHlorophyll (reduced at most by half)
	16. Carotenoid (reduced by half)
4. Membrane organization	17. Membrane formation (reduced by half)
5. Pyrenoid formation	18. Eyespot formation
	19. Starch synthesis

Experiments with antibiotics

Support for these conclusions comes from studies with two antibiotics, spectinomycin and chloramphenicol, that specifically inhibit transcription on 70 S-type ribosomes [12,4]. Experiments in this laboratory are still in progress, but thus far it has been demonstrated that if wild-type cells are grown for 3 generations in the presence of high concentrations (50–100 μg/ml) of chloramphenicol, the chloroplast membrane system is disorganized so as to be indistinguishable from that found in mixotrophic *ac-20* (fig. 9). Growth in either spectinomycin or chloramphenicol also affects photosynthetic reactions that are dependent on the operation of PS-II (Hill reactions) but has relatively little effect on reactions

that depend on PS-I alone [13,14]. Neither antibiotic affects the synthesis of chlorophyll [13,14], and spectinomycin prevents the synthesis of RuDP carboxylase [14]. In other words, it appears that the mixotrophic *ac-20* syndrome can be mimicked by growing wild-type cells under conditions where they cannot use their chloroplast ribosomes for protein synthesis.

The chloroplast ribosome content of wild-type *C. reinhardi* is not affected by growth in the presence of either spectinomycin or chloramphenicol (Fig. 9), even though the ribosomes that are produced are presumably nonfunctional. From this observation we have concluded that at least most chloroplast ribosomal proteins can be synthesized outside the chloroplast [14]. In contrast, when wild-type cells are

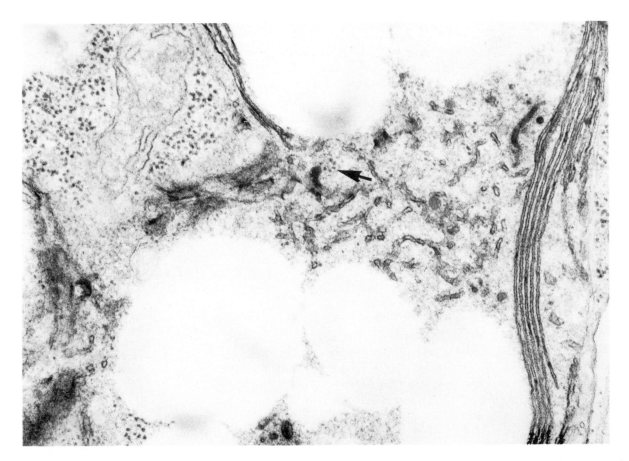

Fig. 9. Portion of a chloroplast from a wild-type cell grown mixotrophically for 3 generations (30 hr) in the presence of 100 μg/ml chloramphenicol. Membrane is predominantly in the form of small, tightly-aggregated vesicles, similar to those found in mixotrophic *ac-20* chloroplasts (fig. 8). Chloroplast ribosome levels (arrow) are normal. 64,000 X.

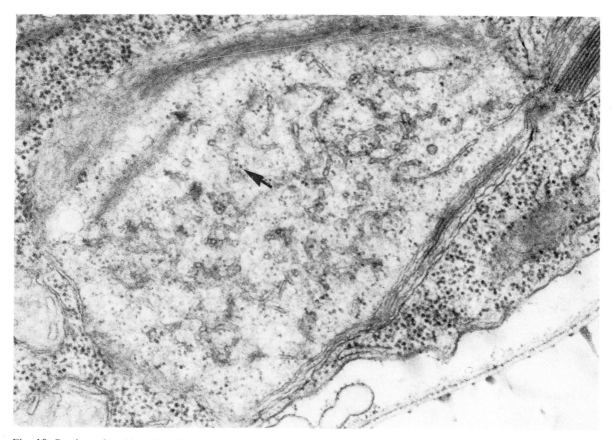

Fig. 10. Portion of a chloroplast from a wild-type cell grown heterotrophically for 4 generations (4 days) in the presence of 250 μg/ml rifampicin. Membrane is in the form of small, tightly-aggregated vesicles, similar to those found in the chloroplasts of mixotrophic ac-20 (fig. 8) and of chloramphenicol-grown wild type (fig. 9). Chloroplast ribosomes (arrow) are markedly reduced. 61,000 X.

grown in the presence of rifampicin, chloroplast ribosomes are no longer made, and after 4 generations the ribosome levels are comparable to those found in mixotrophic ac-20 cells (fig. 10). Dr. Surzycki from this laboratory has shown that rifampicin selectively inhibits the DNA-dependent RNA polymerase of the *C. reinhardi* chloroplast and thereby inhibits transcription of cistrons for chloroplast ribosomal RNA [15]. Transcription of any chloroplast messenger RNA's is presumably also blocked by the antibiotic, and thus a ribosome deficiency is probably not the only lesion suffered by rifampicin-grown cells. Perhaps for this reason, rifampicin-grown wild-type cells do not appear to resemble mixotrophic ac-20 cells in all of their photosynthetic properties [14]. Nonetheless, there is a striking structural similarity between mixotrophic ac-20 cells and wild-type cells grown in the presence of rifampicin when chloroplast membrane organization and ribosome levels are compared (figs. 8, 10); indeed it is very difficult to tell the two chloroplast types apart.

Transfer experiments

If, as is our contention, the chloroplast ribosomes of *C. reinhardi* are responsible, at least in part, for normal membrane and pyrenoid organization and for normal levels of activity of RuDP carboxylase, Q, and cytochrome 559, then it should be possible to demonstrate that the chloroplast ribosomes are necessary for the synthesis of these components. This question has been approached by transferring *ac-20* from mixotrophic to phototrophic growth conditions. Following such a transfer, ribosome levels, structural organization, and photosynthetic activities all 'recover' from their very low, mixotrophic *ac-20* level to the more intermediate, phototrophic *ac-20* level [8–10], and the time course for the recovery of these separate parameters can be monitored and compared.

In early studies of the recovery of RuDP carboxylase activity, it was discovered [8] that a transfer experiment could be performed in two ways, each giving a different result. If mixotrophic cells were transferred to minimal medium and the culture flask was returned directly to the light (a 'light-to-light' transfer), a lag of about 8 hr occurred before recovery of carboxylase activity could be detected. The rate of recovery then increased until a linear rate was established, and this rate continued until the phototrophic level of enzyme activity was attained. If, on the other hand, mixotrophic cells were transferred to minimal medium and the flask was placed in the dark (a 'light-to-dark' transfer), no recovery of carboxylase activity occurred, even after 16 hr. When a culture that had been incubated in the dark for 16 hr was returned to the light, however, recovery of activity began immediately, without any lag, and proceeded at a maximum rate until the phototrophic level was reached. These results are shown graphically in fig. 11.

When the recovery of the Hill reaction and of membrane organization is monitored in transfer experiments [9,10], similar results are obtained (fig. 11). During the 16 hr following a light-to-dark transfer experiment, levels of membrane organization and Hill activity remain low; when light is provided, recovery begins immediately. Following a light-to-light transfer, recovery does not begin until after a lag of about 4 hr, a shorter lag than that observed for the recovery of carboxylase activity. In both kinds of experiments, recovery of Hill activity and the stacking of discs appear to occur coordinately, underlining the intimate relationship between the two [6].

Studies of the recovery of chloroplast ribosomes during transfer experiments offer an explanation for the observed recovery kinetics. During a light-to-light transfer experiment, ribosome recovery begins immediately and occurs rapidly, but about 4 hr is required

before ribosome levels reach the phototrophic *ac-20* level [10]. Thus we suggest that the 4 hr lag in recovery of Hill activity and membrane organization represents the time taken for ribosomes to recover. The longer lag period taken before carboxylase recovers suggests that this recovery is a more complex process.

During a light-to-dark transfer experiment, chloroplast ribosome recovery takes more time than during a light-to-light transfer [10], but by 12 hr the level has reached that of phototrophic *ac-20* (fig. 4). Since cells growing in the dark on minimal medium are in fact starving, the slower recovery rate is perhaps to be expected. The ribosomes that are formed in the dark are not capable of participating in synthetic activity required for recovery until light is provided, but in the presence of light they are able to promote the maximum rate of synthesis of carboxylase and of components required for increased rates of Hill activity and improved membrane and pyrenoid organization. We do not, at present, understand the light requirement; light may simply provide the energy required for protein synthesis, or its role in the recovery process may be more complex and more interesting.

That net protein synthesis and not some sort of an activation process is involved in *ac-20* recovery phenomena has been demonstrated by a series of experiments using antibiotics. Recovery of Hill activity, membrane organization, and RuDP carboxylase are all blocked if chloramphenicol (50–100 µg/ml) is added at the time of transfer to minimal medium in a light-to-light experiment or just prior to turning on the lights following a light-to-dark transfer. Neither chloramphenicol nor spectinomycin inhibits the recovery of ribosomes in either type of experiment. Cycloheximide (1 µg/ml), an inhibitor of cytoplasmic ribosomes in *C. reinhardi* [4], blocks the recovery of carboxylase but not of Hill activity in a light-to-dark transfer experiment, again suggesting that the recovery of the enzyme is a more complex process, perhaps involving cytoplasmic as well as chloroplastic protein synthesis.

Summarizing, the transfer experiments indicate that the recovery of photosynthetic capacity and chloroplast membrane organization begins to occur only after chloroplast ribosome production has essentially been completed, and no recovery occurs if

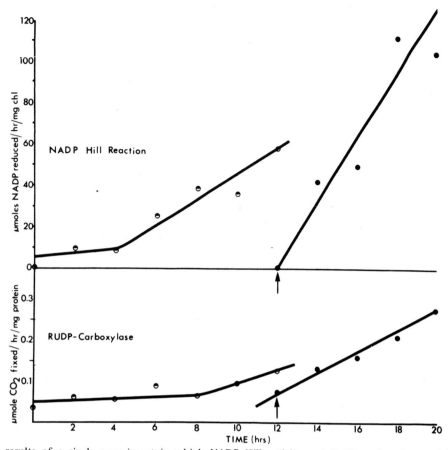

Fig. 11. The results of a single experiment in which NADP Hill activity and RuDP carboxylase activity were monitored concurrently during both a light-to-light and a light-to-dark transfer experiment. At time 0, mixotrophic *ac-20* cells were transferred to minimal medium and placed in the light (light-to-light) or the dark (light-to-dark). Results of the light-to-light experiment are plotted by the half-filled circles. A lag of 4 hr precedes the onset of Hill reaction recovery, and a longer 8 hr lag precedes the onset of recovery of carboxylase activity. Results of the light-to-dark experiment are indicated by the filled circles. After 12 hr of dark incubation, Hill activity and RuDP carboxylase activity have shown no recovery, but when the cells are returned to the light (arrow), Hill and carboxylase activity recover immediately, without any lag.

protein synthesis on these ribosomes is prevented by the absence of light or the presence of inhibitors. Taken together, these observations strongly suggest that the chloroplast components affected in the *ac-20* synchrome are synthesized on the chloroplast ribosomes.

Conclusions

Our experiments with *ac-20*, as well as the parallel and complementary experiments with antibiotics, point to a modest but important role for the protein-synthesizing system of the *C. reinhardi* chloroplast in constructing chloroplast components. The chloroplast ribosomes are apparently involved in the synthesis of component(s) involved in membrane organization, at least two components, cytochrome 559 and Q, involved in photosynthetic transport, and one enzyme, RuDP carboxylase, involved in the reduction of CO_2 to carbohydrate. Most of the other chloroplast components appear to be synthesized outside the chloroplast.

Investigators that have studied the chloroplast protein-synthetic capacity in other algae and in higher plants have, in certain cases, come to conclusions that are similar to ours [16–20] and, in other cases, to different conclusions [18, 20–22]. It is difficult to compare the extent of chloroplast autonomy from one organism to the next at the present time, but it should be stressed that there is no reason to expect that two types of chloroplasts should be identical in their genetic capacities. If chloroplasts indeed had an endosymbiotic origin, then the degree of nuclear control that has been involved in transforming the symbiont into an organelle could certainly vary greatly from one type of plant to the next.

Acknowledgements

We acknowledge the support and interest in this research given by Professor K.R. Porter and the technical assistance of Miss J.J. Armstrong. Research supported by grants GM–12336 from the NIH and GB–5005–x from the NSF to R.P.L., by a predoctoral fellowship (GM–24–306– to U.W.G. from the NIH, by a post-doctoral fellowship (92–GM–13,441) to R.K.T. from the NIH, and by training grant GM–0707 from the NIH.

References

[1] R. Sager and G.E. Palade, J. Biophys. Biochem. Cytol. 3 (1957) 463.

[2] I. Ohad, P. Siekevitz and G.E. Palade, J. Cell Biol. 35 (1967) 521.

[3] U.G. Johnson and K.R. Porter, J. Cell Biol. 38 (1968) 403.

[4] J.K. Hoober and G. Blobel, J. Mol. Biol. 41 (1969) 121.

[5] U.W. Goodenough and R.P. Levine, Plant Physiol. 44 (1969) 990.

[6] U.W. Goodenough, J.J. Armstrong and R.P. Levine, Plant Physiol. 44 (1969) 1001.

[7] P.J. Hastings, E.E. Levine, E. Cosbey, M.O. Hudock, N.W. Gillham, S.J. Surzycki, R. Loppes and R.P. Levine, Microbial Gen. Bull. 23 (1965) 17.

[8] R.K. Togasaki and R.P. Levine, J. Cell Biol. 44 (1970) in press.

[9] R.P. Levine and A. Paszewski, J. Cell Biol. 44 (1970) in press.

[10] U.W. Goodenough and R.P. Levine, J. Cell Biol 44 (1970) in press.

[11] S. Surzycki and P.J. Hastings, Nature 220 (1968) 786.

[12] P. Anderson, J. Davies and B.D. Davis, J. Mol. Biol. 29 (1967) 203.

[13] J.K. Hoober, P. Siekevitz and G.E. Palade, J. Biol. Chem. 244 (1969) 2621.

[14] S.J. Surzycki, U.W. Goodenough, R.P. Levine and J.J. Armstrong, Symp. Soc. Exptl. Biol. 24 (1970) in press.

[15] S.J. Surzycki, Proc. Natl. Acad. Sci. U.S. 63 (1969) 1327.

[16] M. Margulies and F. Parenti, Plant Physiol. 43 (1968) 504.

[17] P.G. Bartels and E.J. Pegelow, J. Cell Biol. 37 (1968) Cl.

[18] R.M. Smillie, D. Graham, M.R. Dwyer, A. Grieve and N.F. Tobin, Biochem. Biophys. Res. Commun. 28 (1967) 604.

[19] J.T.O. Kirk, Planta 78 (1968) 200.

[20] J.A. Schiff, this symposium.

[21] M. Margulies, Plant Physiol. 42 (1967) 218.

[22] M. Gassman and L. Bogorad, Plant Physiol. 42 (1967) 774.

REPLICATION, TRANSMISSION AND RECOMBINATION
OF CYTOPLASMIC DNAS IN CHLAMYDOMONAS REINHARDI

KWEN-SHENG CHIANG

Department of Biophysics, University of Chicago, Chicago, USA

DNAs of the cytoplasmic organelles, particularly of chloroplasts and mitochondria, are often distinct from the nuclear DNAs in their tertiary structures, base composition and other physio-chemical properties [1,2]. It does not follow, however, that the distinct species of DNA molecules present in the chloroplasts and mitochondria necessarily dictates the synthesis of most of the organelle components. In fact, recent studies pertaining to the structure, function and synthesis of nucleic acids and proteins in a few organisms have led to the conclusion that the interaction of the nuclear genetic system with these organelles appears to be essential to chloroplast and mitochondrial biosynthesis and function [3–7].

Whatever the mechanism that actually coordinates this apparent dual control of the biogenesis and function of the cytoplasmic organelles, the DNA molecules of chloroplasts and mitochondria must replicate, segregate and transmit into newly duplicated organelles strictly in coordination with the cell division, so that the physical conservation and functional continuity of these DNA molecules will be assured through the generations.

In this paper I intend to illustrate and discuss some aspects of cytoplasmic organelle DNA replication, transmission and recombination, using the unicellular green alga *Chlamydomonas reinhardi* as a model organism. Evidence will be presented regarding: (1) a semi-conservative mode of chloroplast DNA replication during meiosis and mitosis; (2) temporal incoordination between the replication of cytoplasmic DNA and nuclear DNA during meiosis and mitosis; (3) physical conservation of parental cytoplasmic DNA through meiosis and (4) the recombination pattern between the two parental chloroplast DNAs during the zygotic development.

The biological system and its DNA

C. reinhardi is a particularly good organism for the study of cytoplasmic organelle DNA replication, transmission and recombination because it possesses a combination of advantageous characteristics. This unicellular green alga contains a single giant chloroplast and a small number of mitochondria in the cytoplasm. It can be grown in a chemically defined medium and large numbers of cells can be obtained in a comparatively short period of time. The vegetative cells of *C. reinhardi* can be synchronized effectively by alternating dark-light cycles [8]. *C. reinhardi* is heterothallic and undergoes a zygotic, or initial, type of meiosis. This means that vegetative growth and subsequent gametogenesis of the two opposite mating type ('plus' mating type, mt^+; 'minus' mating type, mt^-) gametes can be handled separately under controlled experimental conditions so that the cytoplasmic organelle DNA of mt^+ and mt^- gametes can be labeled differentially with stable and radioactive isotopes. Thus it is possible to follow simultaneously the replication and transmission patterns of both parents throughout the zygotic developmental stages. Differential labeling of the gametes also provides an excellent chance for studying the possible recombination between the two parental organelle DNAs.

Fig. 1a shows the banding pattern of a purified DNA sample from exponentially growing vegetative cells of *C. reinhardi* in a cesium chloride density gradient. Three components were consistently observed: a major component, the nuclear or chromosomal DNA (α) with a peak density of $1.723 \, \text{g/cm}^3$, a satellite component (β) with a peak density of 1.695, and an additional small satellite component (γ) with a density of 1.715. The β satellite has been shown to be the chloroplast DNA [9,10]. The cytological origin of the γ satellite has not been clearly identified at the moment, although it is presumably from the mitochondria [11–13]. After thermal denaturation all three DNA components exhibited an increase in buoyant density of 0.015 g/cm³ (fig. 1b), indicating α, β and γ DNA all possess regular double-stranded structures. The chloroplast DNA, but not the chromosomal DNA, can be renatured readily after thermal denaturation [14]. Some of the characteristics of the DNA components in *C. reinhardi* are summarized in table 1. The methods and techniques used in handling the organism, as well as other experimental procedures, have been reported elsewhere [11–15].

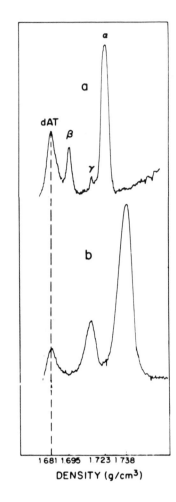

Fig. 1. Microdensitometer tracings of UV-absorption photographs of *C. reinhardi* DNA in a CsCl density gradient. Each photograph was taken after 20 hr centrifugation in 7.7 M CsCl at 44,770 rpm at 25° in a Spinco model E analytical ultracentrifuge. A Joyce-Loebl double-beam recording microdensitometer was used for tracing the photographs. Sample a, unfractionated native DNA from vegetative cells. Sample b, heat-denatured DNA of sample a. For density reference, dAT from *Cancer borealis* was used. The buoyant density of dAT, 1.681 g/cm³, was obtained by taking the density of *E. coli* DNA as 1.710 g/cm³.

Table 1
Properties of DNA components in *Chlamydomonas reinhardi*.

Growth stage	μg DNA/ cell × 10⁷	Component	Origin	Buoyant density in CsCl			% GC [d]	Relative amount of total DNA [e]
				Native	Denatured [a]	Renatured [b]		
Vegetative cells	1.24 ± 0.081 [c]	α	Nuclear	1.723	1.738	1.738	64.3	85
		β	Chloroplast	1.695	1.710	1.695	35.7	14
		γ	? (Mitochondrial)	1.715	1.730	1.715	56.1	1
Gametes	1.23 ± 0.064 [c]	α	Nuclear	1.723	1.738	1.738	64.3	89
		β	Chloroplast	1.695	1.710	1.695	35.7	7
		γ	? (Mitochondrial)	1.715	1.730	1.715	56.1	4

[a] Heating at 100° for 10 min in 0.015 M NaCl, plus 0.015 M Na citrate pH 7, then quickly cooled.
[b] Thermal denatured DNA in 0.3 M NaCl 0.03 M Na citrate pH 7 was heated to 70° for 5 hr, then slowly cooled.
[c] Standard deviation.
[d] Composition is obtained from the density, assuming no unusual bases; density (ρ) = 1.660 + 0.098 (G+C) mole percent.
[e] Estimated by integrating area under each DNA peak in the microdensitometer tracings of UV-absorption photographs.

Replication of chloroplast and γ DNA during the vegetative cell cycle

Temporal incoordination in replication

Under a light-dark cycle of alternating 12-hr periods, the vegetative cell division of *C. reinhardi* can be synchronized [8]. Under such conditions and with a selected single clone, a high degree of synchrony in cell division can thus be achieved [14]. The time course of DNA synthesis and cell division in such a culture is shown in fig. 2. The major DNA synthesis occurred shortly before cell division, which took place in the middle of the dark period. A smaller amount of DNA synthesis was detected in the ab-

sence of any cell division in the middle of the light period [14]. The biphasic curve of DNA synthesis suggests that the cytoplasmic DNA and chromosomal DNA may replicate at different times in the synchronized culture. This apparent temporal incoordination between chromosomal and cytoplasmic DNA syntheses is confirmed by the radioactive isotopic labeling experiment. If a vegetative culture that had been prelabeled with ^{32}P for a number of generations were labeled with ^{3}H-adenine only through the duration of the light period, only the chloroplast DNA and γ satellite, not the chromosomal DNA, became labeled with ^{3}H (fig. 3). On the other hand, if the synchronized vegetative culture had been labeled only through the duration of the dark period, not only was the chromosomal DNA labeled as expected, but considerable amounts of chloroplast DNA and γ satellite were also labeled (fig. 4). These results indicate that in the light period of a light-dark synchronized culture, the chloroplast DNA and γ satellite replicated coordinately with each other but independently of the chromosomal DNA replication. In the dark period during which the chromosomal DNA replicated, both chloroplast and γ DNA also underwent appreciable replication. It is of interest to note that in the cells of *C. reinhardi* grown in the dark phase and supplemented with acetate as the sole carbon source, the relative amount of chloroplast DNA (fig. 5) remained unchanged from that of the light-dark synchronized cells (fig. 1). These results suggest the presence of a light-independent replication mechanism for the chloroplast DNA; in other words, the replication of chloroplast was not necessarily dependent upon the photosynthetic activity. It is not yet certain whether there is any correlation between the light-independent chloroplast DNA replication and the chloroplast DNA replication observed in the dark period of the synchronized cycle, but it is clear that the cells can replicate chloroplast DNA in the dark. It is possible that chloroplast DNA can replicate by different mechanisms under different environmental conditions. If these implicated light-independent and light-dependent replication mechanisms for chloroplast DNA can be further substantiated, it would be extremely interesting to analyze the specific differences between these two systems and their respective control and regulatory processes.

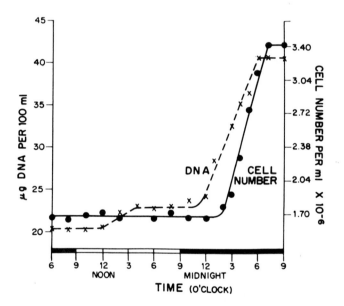

Fig. 2. DNA synthesis in a vegetative division cycle of *C. reinhardi* [14]. A clone of mt⁻ vegetative cell was selected .from a single colony that showed a high degree of synchrony in cell division in a preliminary experiment. A 2-liter medium was inoculated with this clone (10^4 cells/ml). After reaching a cell concentration of 1.67×10^6 per ml, 100-ml samples of this culture were taken at 1- or 2-hr intervals on the hour throughout the synchronous cycle. Cells obtained from these samples were subjected to DNA assay by the diphenylamine method after preliminary extraction of DNA from the cells with 0.5 N perchloric acid. Results of the DNA assay have been expressed as µg per 100 ml aliquot of the culture. The cell number was determined by quadruple countings with a haemocytometer after cells had been killed with a drop of formaldehyde.

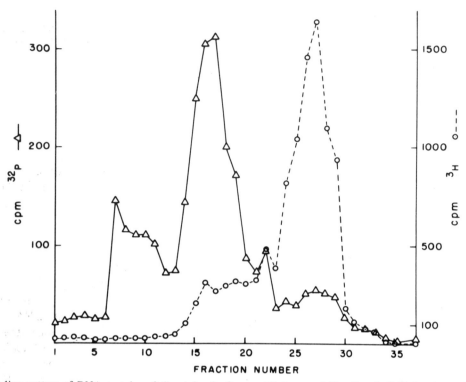

Fig. 3. The banding pattern of DNA samples of *C. reinhardi* after equilibrium centrifugation in CsCl density gradient. The DNA sample was obtained from a ^{32}P-labeled, light-dark synchronized vegetative culture, which was pulse-labeled with ^{3}H-adenine for 12 hr during the last light cycle before harvesting the cells. Four ml of CsCl solution (in 0.01 M Tris plus 0.001 M EDTA, pH 8.5) with a density of 1.703 g/cm^3 containing a purified radioactively-labeled DNA sample was centrifuged in a model B-60 International ultracentrifuge with a swinging-bucket rotor, SB-405. After 48 hr of centrifugation at a speed of 45,000 rpm (230,000 *g*) at 25°, the centrifuge tube was punctured, and between the density range of 1.735 to 1.685 g/cm^3, four drops were collected in each fraction. After dilution with SSC, each fraction was precipitated by trichloroacetic acid with the addition of carrier DNA. The precipitated DNA in each fraction was then collected onto a glass fiber filter and counted in a Packard Tri-Carb liquid scintillation counter.

The mode of cytoplasmic DNA replication

The mode of replication of chloroplast DNA and γ satellite can be analyzed by isotopic transfer experiments [14]. Details of a ^{15}N-^{14}N transfer experiment are outlined in fig. 6. The results of this experiment are illustrated in fig. 7 with microdensitometer tracings of UV absorption pictures of DNA samples. The fully ^{15}N-labeled chloroplast DNA was completely replaced by a ^{15}N-^{14}N hybrid in the middle of the light period, when a small amount of DNA synthesis was detected, as shown in fig. 2. At the same time that the hybrid β was first detected, a hybrid γ band also emerged. Two hours later, while the β band hybrid remained unchanged in density, an unlabeled light β band appeared in amounts approxi-

mately equal to the hybrid β band. The apparent synchrony in the replication of β and γ DNA became much more obvious when the fully unlabeled γ band again appeared concomitantly with the light β band (fig. 7(c)). Throughout the entire replication processes of β and γ in the light period, the chromosomal DNA α did not replicate, as indicated by its unchanged density. After the dark period, however, the chromosomal DNA replicated semoconservatively twice to give approximately equal amounts of hybrid and light α bands. At this time, estimating the relative amount of γ satellite was difficult, since the newly appeared light α band completely shaded the hybrid γ band. On the other hand, the light β band did appear now in considerably greater amounts than the hybrid

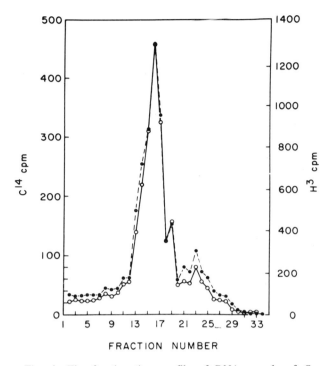

Fig. 4. The fractionation profile of DNA sample of *C. reinhardi* following equilibrium centrifugation in CsCl density gradient. The DNA sample was obtained from a ^{14}C-prelabeled synchronized vegetative culture that was pulse labeled with ^{3}H-adenine for 12 hr during the last dark cycle before harvesting the cells. See fig. 3 legend for experimental conditions.

●————●————● ^{14}C counts;   ○————○————○ ^{3}H counts.

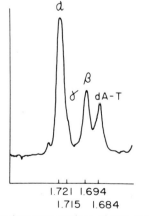

Fig. 5. Microdensitometer tracings of UV-absorption photograph of *C. reinhardi* DNA obtained from a culture grown in an acetate-supplemented minimal medium without illumination. See fig. 1 for experimental conditions.

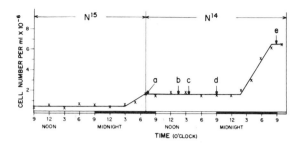

Fig. 6. The mitotic transfer experiment: growth curve and experimental scheme [14]. A clone of the mt$^+$ vegetative cell was inoculated to a ^{15}N-medium at an initial cell concentration of 1×10^4 cells/ml. This culture was subjected to a light-dark cycle to maintain synchronous division. Right after cell division resulting in a cell concentration of 1.64×10^6 per ml, cells were harvested aseptically by centrifugation, then quickly suspended in a preconditioned and adenine-supplemented ^{14}N-medium. The cell concentration was readjusted to the pretransfer concentration (1.64×10^6) by adding fresh sterile ^{14}N-medium. After this transfer, the cultures were quickly returned to the light-dark cycle. Samples of this culture were taken for DNA preparations at different times, as indicated by the a, b, c, d, and e parts of the figure. Analyses of these samples in CsCl density-gradient centrifugation are shown in fig. 7.

β band, thus indicating clearly that additional replication had occurred. This is of course consistent with the data presented in fig. 4.

From the foregoing results it becomes evident that *C. reinhardi* is particularly suitable for the study of chloroplast DNA replication. First of all, the cell division of *C. reinhardi* commands a high degree of synchrony that is a prerequisite for the isotopic labelling experiment. Secondly, there is a GC content difference of about two-fold between chromosomal and chloroplast DNA (table 1), which enables these two DNA species to be well separated in the density gradient (fig. 1). Thirdly, no complication by DNA precursor pool seems to be involved under certain experimental conditions, since no appreciable contribution of ^{15}N to chloroplast DNA was detected in the ^{15}N-^{14}N transfer experiment (fig. 7). In consequence, a discrete ^{15}N-^{14}N hybrid chloroplast DNA can be distinguished unambiguously from other DNA components (fig. 7(b)). Owing to these advantageous features, unequivocal evidence for (1) a semiconservative mode of chloroplast DNA replication and (2) difference in time of replication between chloroplast and chromosomal DNA has been observed. It was also

obvious that an apparent synchrony in replication time exists between the two satellite DNAs β and γ.

The replication, transmission and recombination patterns of cytoplasmic DNA in the sexual cycle

The mode and timing of the chloroplast DNA replication in the sexual cycle

Experimental conditions can be controlled satisfactorily to obtain a large homogeneous population of freshly conjugated zygotes of *C. reinhardi* devoid of vegetative cells and unmated gametes [11,15]. By transferring ^{15}N-labeled zygotes so obtained to an ^{14}N medium for maturation and subsequent germination, the replication mechanism of the chloroplast DNA can be elucidated [11,12]. As can be seen from fig. 8, an unexpected DNA component, the maturation band (M-band) DNA [11–13,15] replicated extensively and appeared in large quantities during maturation. A hybrid chloroplast DNA completely replaced the light chloroplast DNA in the early maturation period (fig. 8). This result indicates that the chloroplast replication in the sexual life cycle follows the familiar semiconservative mode demonstrated in the vegetative cycle. Furthermore, fig. 8 also illustrates the incoordinated replication of chloroplast and chromosomal DNA. At the completion of one round of replication of β band, the density of the fully ^{15}N-labeled α band remained unchanged, indicating that no chromosomal DNA replication had occurred in the zygotic maturation period. At the end of the germination process when chromosomal DNA appeared in hybrid density, the fully unlabeled β band DNA became the sole detectable chloroplast DNA constituent. To dilute the hybrid β band, chloroplast DNA must have replicated several times

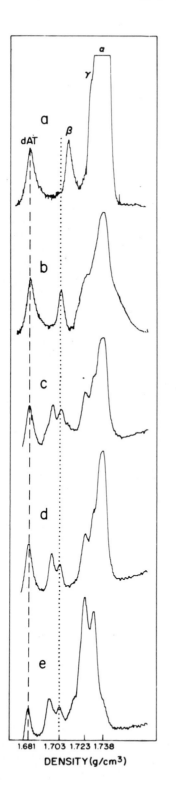

Fig. 7. The mitotic transfer experiments: analysis of CsCl density-gradient centrifugation [14]. See fig. 6 for growth curve, sampling time, cell concentration, and other experimental conditions. Density-gradient centrifugation was carried out as in fig. 1; a, immediately before the transfer; b, at 1:30 pm; c, at 3:30 pm; d, at 9:00 pm; and e at 9:00 am.

while chromosomal DNA replicated only once. This result indicates that a high degree of asynchrony existed between the replication of chloroplast and chromosomal DNA throughout the sexual life cycle of *C. reinhardi*.

Physical conservation of cytoplasmic DNA

If freshly fused gamete pairs (young zygotes) were labeled differentially with ^{3}H and ^{14}C in their DNA molecules for identification of their parental origin, the transmission pattern of each individual parental DNA to meiotic progeny can be analyzed [13]. Typical results of such meiotic DNA transmission experiments are presented in fig. 9. In the young zygotic stage, no appreciable difference between mt$^+$ and mt$^-$ cells was detected in the banding pattern (fig. 9(a)) or in the relative amounts of the three DNA components. The DNA molecules were badly sheared and gave rather diffused banding patterns of the mature zygote DNA, because of the mechanical grinding step that was necessary for the isolation of the mature zygote DNA. Nevertheless, no selective exclusion of any DNA species of either parental origin was observed in the mature zygote stage (fig. 9(b)). After meiosis, the relative amounts of ^{14}C-labeled mt$^-$, as well as of ^{3}H-labeled mt$^+$ DNA components, α, β and γ in the zoospore population, exhibited little change from the corresponding values of the gametes. The UV-absorption profile of the mature zygote sample (the one used to obtain fig. 9(c)) in the CsCl gradient is illustrated in fig. 10. Here, as could be expected from our previous work [11–15], the 'M-band' DNA appeared in large quantity, and the relative amount of chloroplast DNA, β, increased noticeably. These results indicated that nonchromosomal DNA replicated incoordinately with the chromosomal DNA during the zygote maturation period. These two DNA replication events, i.e., β and the M-band DNA, were not matched by parallel tendencies in the radioactive profile (fig. 9(b)). Therefore, no complication caused by radioactive DNA precursor pool(s) seems to be involved. These results indicate that no appreciable discrimination or selectivity was operative in the transmission or degradation of any particular parental DNA component from either parental gamete into the zoospore.

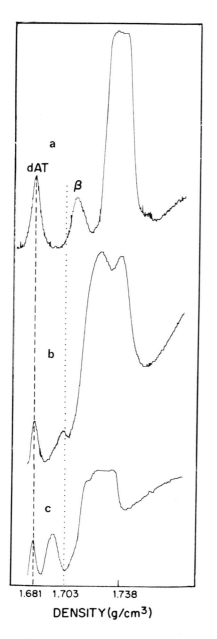

DENSITY(g/cm³)

Fig. 8. The replication of the chloroplast DNA in the sexual cycle of *C. reinhardi* [12]. Density-gradient centrifugation was carried out as in fig. 1. The analytical centrifuge cells were heavily loaded with DNA sample. (a) Freshly mated ^{15}N-labeled gamete pairs (young zygotes). (b) At termination of light period of zygote maturation in ^{14}N-medium. (c) At end of the germination of zygospores in ^{14}N-medium when eight zoospores were emerging from each germinating zygospore.

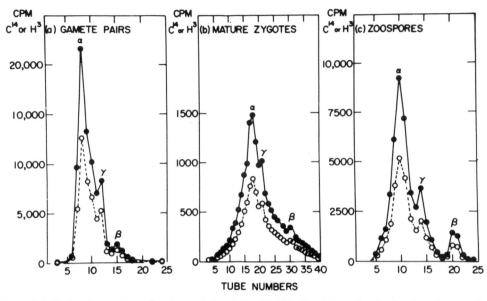

Fig. 9. Profiles of CsCl density gradient fractionated *C. reinhardi* DNA. Conditions for the fractionation are detailed in fig. 3. ●————● ^{14}C counts of ^{14}C-labeled mt$^-$ gametic DNA; ○————○ ^{3}H counts of ^{3}H labeled mt$^+$ gametic DNA.

Recombination pattern of chloroplast DNA

In addition to the differential radioisotopic labeling as described above, stable isotope ^{15}N can also be used to label either one of the two gametes within a cross. In such an isotopic labeling experiment the physical recombination of chloroplast DNA during the sexual cycle can be analyzed. These meiotic recombination experiments employed a mutant strain of the mt$^-$ carrying three uniparentally transmitted non-Mendelian genes (UP genes), streptomycin-resis-

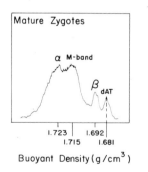

Fig. 10. Tracings of a UV-absorption photograph of *C. reinhardi* mature zygote DNA (same preparation as in fig. 9b [13]). Density gradient centrifugation was carried out as in fig. 1.

tant, neamine-resistant and spectinomycin-resistant (mt$^-$Str^{r}Nea^{r}Spe^r), obtained from N.W. Gillham.

At the gamete-pair, or young zygotic stage (fig. 11(a)), the mt$^-$ and mt$^+$ DNA in the CsCl gradient formed well-separated ^{14}C heavy (^{15}N–^{15}N) and ^{3}H light (^{14}N–^{14}N) peaks for the α and β components. The γ satellite appeared as a shoulder on the right side of the chromosomal DNA peak for both the ^{3}H-labeled mt$^+$ and ^{14}C,^{15}N-labeled mt$^-$ gametes, which carried the uniparentally inheritable non-Mendelian genes Str^r, Nea^r and Spe^r.

After meiosis and a four-fold cell increase, as shown in fig. 11(b), fraction 10, appreciable ^{3}H,^{14}N light mt$^+$ chromosomal DNA co-sedimented with ^{14}C,^{15}N heavy mt$^-$ DNA (now in its once replicated, hybrid, ^{15}N–^{14}N state), while in fraction 13, considerable ^{14}C,^{15}N heavy mt$^-$ DNA co-sedimented with ^{3}H,^{14}N light mt$^+$ DNA. This result indicates that appreciable mt$^+$ DNA became linked with mt$^-$ DNA, and vice versa. The detailed analysis of these reciprocal chromosomal DNA crossing-over events in meiosis will be reported elsewhere [16]. Here again, no selective degradation of any of the ^{3}H mt$^+$- or ^{14}C mt$^-$-labeled DNA components, chromosomal or cytoplasmic, was observed. In addition to the apparent physical conservation of both parental cyto-

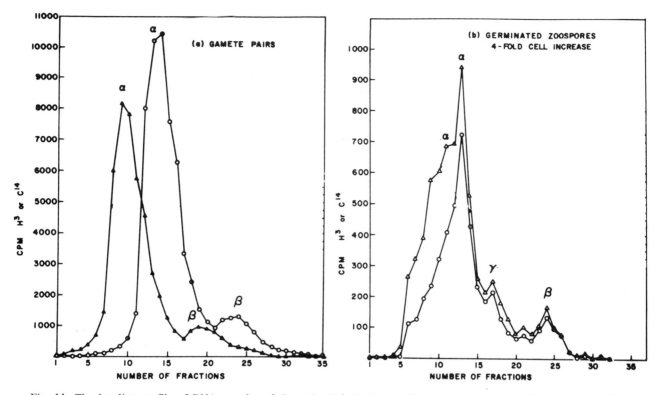

Fig. 11. The banding profile of DNA samples of *C. reinhardi* following equilibrium centrifugation in CsCl density gradient. Experimental conditions are detailed in the text and fig. 2. o————o————o ^{3}H-labeled mt$^+$ wild type DNA counts. △————⊕————△ ^{14}C-,^{15}N-labeled mt$^-$Str^{r}Nea^{r}Spe^r DNA counts.

plasmic DNAs, no resolution into hybrid and light peaks, like that of the chromosomal DNA α, was seen for either the chloroplast DNA β or the γ satellite. These patterns suggest that physical recombination in chloroplast DNA and γ DNA took place during the meiotic and/or post-meiotic mitotic divisions, and that the frequency of these cytoplasmic DNA recombinations appeared to be considerably higher than in the chromosomal DNA.

To substantiate further the above chloroplast DNA recombination result and to determine the time factor pertaining to this 'high frequency' recombination event of the chloroplast DNA during the zygotic developmental stages, meiotic recombination experiments of the following two 'reciprocal crosses' with respect to the isotopic labeling condition were carried out: ^{15}N ^{3}H mt$^-$ Str^{r}Nea^{r}Spe^r × ^{14}N ^{14}C mt$^+$ wild type; ^{15}N ^{14}C mt$^+$ wild-type × ^{14}N ^{3}H mt$^-$Str^{r}Nea^r Spe^r.

As shown in fig. 12(b), approximately 30 hr after

the fusion of mt$^+$ and mt$^-$ gametes, the heavy (^{15}N–^{15}N) ^{3}H-labeled mt$^-$ β could no longer be resolved from the light (^{14}N–^{14}N) ^{14}C-labeled mt$^+$ β peaks in the CsCl gradient, as demonstrated in the gamete pair stage (fig. 12(a)), indicating that the recombination event between the mt$^+$ and mt$^-$ chloroplast DNAs had already occurred. On the other hand, the chromosomal DNA α could still be well separated into heavy ^{3}H-labeled and light ^{14}C-labeled peaks from the mt$^-$ and mt$^+$ parents, respectively. This was of course to be expected, since it had been shown that meiotic chromosomal DNA replication does not occur until the termination of the zygotic germination period, approximately five days later, under our present experimental conditions [11,15]. No significant difference in the banding patterns of chloroplast as well as chromosomal DNA was observed in the mature zygote stage (fig. 12(c)). The ^{3}H-labeled heavy and ^{14}C-labeled light chloroplast DNA from mt$^-$ and mt$^+$ parents still appeared together, without separation, in

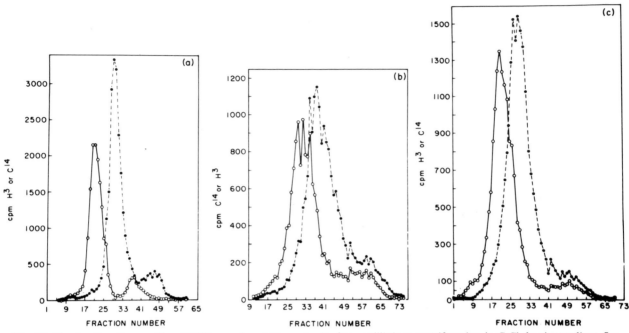

Fig. 12. The fractionation profile of DNA samples of *C. reinhardi* after equilibrium centrifugation in CsCl density gradient. See the text and fig. 2 legend for experimental conditions. (a) Gamete-pairs, (b) young zygotes 30 hr after fusion of gametes, (c) mature zygotes, 7 days after fusion of gametes. o————o————o ^{3}H counts of ^{3}H-, ^{15}N-labeled mt⁻Str^{r}Nea^{r}Spe^r DNA; ●————●————● ^{14}C counts of ^{14}C-labeled mt⁺ wild type DNA.

the CsCl gradient. This result indicates that the sedimentation property of the recombined parental chloroplast DNA, observed 30 hr after mating, was not altered significantly, even after repeated replications. Therefore, whatever the mechanism of recombination, the structure of the recombined DNA appeared to be stable and was not eliminated upon further replication of the chloroplast DNA.

In the second experiment in which the mt⁺, instead of mt⁻, was labeled with ^{15}N, the same result was again observed. Thirty hr after the fusion of gametes the heavy ^{14}C-labeled mt⁺ chloroplast DNA, but not the chromosomal DNA, could no longer be resolved from the light ^{3}H-labeled mt⁻ chloroplast DNA in the CsCl gradient after equilibrium centrifugation.

In addition to the apparent recombination pattern of the two chloroplast DNAs, no selective degradation of any of the mt⁺ or mt⁻-labeled DNA components, chromosomal or cytoplasmic, was observed in the above two isotopic labeling experiments. The generally diffused and broad appearance of the chromosomal and chloroplast DNA peaks seen here in

fig. 12b, 12c and 13b is again due to the mechanical breaking step that was used to isolate these zygotic DNAs (compare with fig. 9(b)). In addition to the rather diffused banding patterns of the badly sheared zygote DNA, as seen in figs. 12(b), 12(c) and 13(b), some fluctuation of radioactivity in the α and β peak regions is also observed. Whether this phenomenon is simply due to the heterogeneity of DNA molecules caused by the shearing or has other significance is not clear at the moment.

Unfortunately the resolution of the γ satellite in these zygote DNA samples was not very clear. The timing and general pattern of the recombination event of the γ satellite in the early zygotic developmental stages therefore cannot be determined at present.

The time at which parental gametic chloroplast DNA underwent the first semiconservative replication in zygotic maturation has not yet been precisely determined. With the same experimental design as shown in fig. 8, and taking samples at 10 to 12 hr intervals, the timing of the first chloroplast DNA

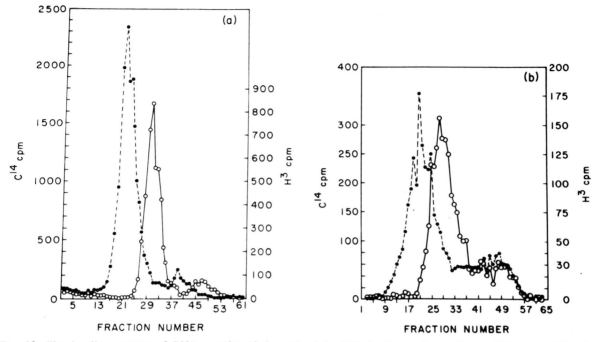

Fig. 13. The banding pattern of DNA samples of *C. reinhardi* in CsCl density gradient after equilibrium centrifugation. Experimental conditions are described in the text and fig. 2. o————o————o ^{3}H counts of ^{3}H-labeled mt⁻Str[r]Nea[r]Spe[r] DNA; ●————●————● [14]C counts of [14]C-, ^{15}N-labeled mt⁺ wild type DNA.

replication was between 36 to 48 hr after mating of the gametes [11]. Therefore, most likely prior to or around the time of the first chloroplast replication in the sexual cycle, recombination between the mt⁺ and mt⁻ chloroplast DNA occurred in such a way that after the recombination process, the heavy ([15]N–[15]N) and light ([14]N–[14]N) parental chloroplast DNA molecules could no longer be separated in the CsCl density gradient. This result was obtained, no matter which mating type parent, mt⁻ or mt⁺, was labeled with ^{15}N so that its chloroplast DNA was heavy in density.

In order to interpret the data obtained from these meiotic recombination experiments, it is important to ascertain that no radioactive DNA precursor pool(s) contributed to the cytoplasmic DNA replication that occurred after the fusion of gametes. Similar to the results presented in figs. 9(b) and 10, the 'M-band' DNA, which was present in the DNA samples shown in figs. 12b, 12c and 13b, could be detectable only by UV-absorbance but not by radioactivity (figs. 12, 13). This result indicates that under our present experi-

mental design, which employed extensive chasing steps with unlabeled adenine, no radioactive DNA precursor pool was manifested in the zygotic maturation period [13].

At present we cannot rule out the remote possibility that a strictly compartmentalized DNA precursor pool(s) existed, which could not be chased out by cold adenine and was used exclusively for chloroplast DNA synthesis. Further detailed substantiation, especially the analysis of the molecular structure of the 'recombined chloroplast DNA,' may be helpful to settle this uncertainty. With all the currently available data, it seems reasonable to assume that the chloroplast DNA recombination does occur in the sexual life cycle of *C. reinhardi*.

Two gametic chloroplasts fused in their entirety approximately 10 hr after the mating of the two parental gametes to form a single zygotic chloroplast from which all chloroplasts of meiotic progeny (zoospores) were derived [17,18]. This fusion of the two gametic chloroplasts certainly provided the necessary physical requirement for the parental chloroplast

DNA to recombine in the zygotic maturation period. The zygotic chloroplast then underwent drastic structural differentiation and dedifferentiation during the zygotic maturation and germination periods [17,18]. The first division of this single chloroplast did not occur until shortly before the first meiotic division. Moreover, during the meiotic divisions an apparent synchrony actually existed between the chloroplast division and nuclear division, i.e., in the first meiotic division the single zygotic chloroplast divided once to give rise to two chloroplasts each in one dyad. Each of these dyad chloroplasts then divided again to give rise to two chloroplasts in the tetrads [17]. In consequence, the genes on the chloroplast DNA could undergo a meiosis-like segregation similar to the Mendelian segregation pattern of the nuclear genes.

A typical set of genetic results, obtained with the same experiment whose physical result is presented in fig. 11, is summarized in tables 2 and 3. These genetic results, representative of all the above physical recombination experiments, indicate that the inheritance of the non-Mendelian UP genes followed a typical uniparental transmission pattern. Despite the physical conservation of the mt$^-$ parental chloroplast and γ DNA, as well as their physical exchange with the mt$^+$ counterparts, the three drug resistance markers originally carried by the mt$^-$t parent were not transmitted into more than 1% of the total meiotic

progeny, in accord with the usual uniparental transmission frequency of 0.1–5% for 'exceptional zygotes' [19].

Table 2
Drug resistance test for gametes [a]

Gametes	Master plates HSM	Replica plates			
		Str	Spe	Nea	SNS
mt$^+$ wild type	247	0	0	1	0
	136	0	0	0	0
	100/1729 [a]	0	0	0	0
mt$^-$ Str^r Nea^r Spe^r	145	145	145	145	144
	97	97	97	97	97
	100/1520 [b]	100	100	100	100

[a] Gametes were first plated onto non-selective plates (high salt minimal (HSM) medium with 1.5% agar) on which the gametes underwent de-differentiation to yield vegetative cells [8,11,15]. After colonies were grown to about 1–2 mm in diameter, these plates were then replica-plated onto selective plates, i.e., HSM plus streptomycin 500 μg/ml (Str); HSM plus Neamine 100 μg/ml (Nea); HSM plus Spectinomycin 100 μg/ml (Spe), and HSM plus Streptomycin 500 μg/ml, Neamine 100 μg/ml and Spectinomycin 100 μg/ml (SNS).

[b] 100 colonies randomly picked from a total of 1729 and 1520, respectively, were tested by single-clone streaking onto selective plates.

Table 3
Drug resistance test for zoospores obtained from a cross of mt$^+$ wild type × mt$^-$Str^{r}Nea^{r}Spe^r [a].

Medium	Number of cells plated	Colonies	Plating efficiency on HSM (%)	Drug resistant zoospores (%)
HSM	2,700	1606	59.4	
	1,350	627	46.4 Av. 54.9	
	1,080	555	62.3	
	540	336	51.4	
Nea	135,000	525		1.13 (0.93)
	67,500	421		0.71
Spe	135,000	236		0.32 (0.27)
	67,000	80		0.21
Str	135,000	68		0.092 (0.22)
	67,000	126		0.34
SNS	135,000	84		0.11 (0.11)
	67,000	0		–

[a] Zoospores were directly plated onto non-selective and selective plates as detailed in table 2. The average plating efficiency on non-selective, HSM plates, i.e., 54.9%, was used to calculate the percentage of drug resistant zoospores on all the different kinds of selective plates.

Concluding remarks

If chloroplast and mitochondrial DNA do contain genes whose transcription and translation are indispensable to the biogenesis and function of these organelles, then one might predict that: (1) the cytoplasmic organelle DNA may possess a replication mechanism similar to that of the chromosomal DNA, in order to maintain genetic continuity from generation to generation, (2) the replication of these organelle DNAs may be independent of, as well as temporally and physically separable from, the chromosomal DNA replication, (3) the organelle DNA may undergo regular recombination independently of the nuclear DNA recombination, as a necessary part of the evolutionary process of the organelles.

While experimental results negative to these predictions may not necessarily rule out the possibility that chloroplasts and mitochondria enjoy a certain degree of autonomy in the cell, positive experimental results obtained in the C. reinhardi system certainly lend strong support to the idea that chloroplasts and mitochondria are indeed semi-autonomous entities that possess their own independent genetic apparatus in the cell.

It is of interest to note that the degree of autonomy of the chloroplast(s) in the cell seems to vary appreciably in different organisms. In Euglena, for instance, aplastidic mutants with no detectable chloroplast DNA could be induced readily by a number of different agents such as ultraviolet light, heat and streptomycin [20]. On the other hand, aplastidic mutants that lack chloroplast DNA have not been found in Chlamydomonas. These results with aplastidic mutant isolations may be just fortuitous. It is possible, however, that the degree of autonomy of the chloroplast, or more precisely, the interdependence between the chloroplast and nuclear genetic systems, may be quite different in these two organisms.

It is not surprising that the chloroplast DNA probably possesses a recombination process, for the two gametic chloroplasts DNA probably can come together in close physical proximity, after the fusion of the two chloroplasts, to form a single zygotic chloroplast. The chloroplast DNA is no different from any other DNA in undergoing a breakage and rejoining process. From our results, the frequency of these chloroplast DNA recombinations seems higher than the nuclear recombinations (figs. 11–13). The recombination mechanism that gives rise to the high frequency of recombination is not clearly understood at present. Attempts to elucidate this mechanism are currently in progress.

Since C. reinhardi contains only one chloroplast per cell, the semiconservative replication of chloroplast DNA ensures its physical conservation and equal distribution to the progeny. Because the chloroplast does not possess a mitotic apparatus, the actual segregation of newly duplicated chloroplast DNA from the parental chloroplast to the two daughter chloroplasts may be dependent on a specific attachment site on the chloroplast membrane or other structure component, similar to bacterial DNA segregational mechanics.

From our observation that an apparent synchrony of segregation exists between the chloroplast and nucleus in meiotic divisions [17,18], it becomes clear that some chloroplast genes can segregate similarly to the nuclear genes. The implication is that at least some of the chloroplast genes can conceivably undergo a process of 'quasi-meiosis' in the zygote germination of C. reinhardi. It is therefore of importance to emphasize that one must not rule out the possibility that a particular gene is located on the chloroplast DNA simply because it possesses a segregation pattern similar to the nuclear genes. In fact it is quite possible that at least one class of mutants, i.e., the yellow-in-darkness mutants, Y_1, Y_2, Y_3 of C. reinhardi may fall into this category. These mutants have a Mendelian segregation pattern at meiosis but exhibit no linkage with any of the several dozen known Mendelian genes that have been mapped randomly on the 16 linkage groups [21]. Moreover, no linkage was detected between these Y mutants and the uniparentally transmitted non-Mendelian genes isolated by Sager and Gillham. It should be added that the particular conclusion that chloroplast genes of C. reinhardi may undergo a 'quasi-meiosis' and follow a Mendelian segregation pattern may or may not apply to other photosynthetic organisms; for those photosynthetic organisms that possess more than one chloroplast per cell, the recombination and segregation of chloroplast genes may be quite different and far more complex.

Since the transmission of chloroplast DNA is biparental in nature, i.e., both the mt^+ and mt^- gametic chloroplast DNA are physically conserved through meiosis (figs. 9, 11–13), no positive correlation seems to exist between the transmission of UP genes and chloroplast DNA from the gametes to meiotic progeny. If one assumes that the mechanism of uniparental inheritance in *C. reinhardi* is due to a selective exclusion of the mt$^-$ UP genes by physical degradation, then our present results can be interpreted to indicate that either an unidentified minute cytoplasmic DNA(s) or a rather limited portion, but not the majority, of the chloroplast DNA is responsible for the UP inheritance. Nevertheless, the mechanism of UP inheritance is totally unknown; other alternatives such as restriction of the replication or the function of the mt$^-$ UP gene carrier, regardless of its intracellular origin after the fusion of two gametes, also have to be considered. With the discovery of uniparentally transmitted non-Mendelian (UP) genes [22] together with a sum of the circumstantial evidence that these genes may be located in the chloroplast [23], little attention has been given to the investigation of other potential chloroplast mutants that do not transmit uniparentally. Critical and conclusive evidence has yet to be obtained in order to assign such UP genes to the chloroplast DNA. Even if such conclusive evidence could be secured, it is still possible that the chloroplast possesses more than one class of genes that have different transmission and segregation properties. Since the gametic chloroplast of *C. reinhardi* contains approximately 5×10^9 daltons equivalent DNA [14], it is still an open question whether all of this DNA is linked in tandem as one molecule and attached to the chloroplast only on a single site. Deliberate effort should be made to reinvestigate those suspected chloroplast mutants, such as Y_1, Y_2 Y_3 and other pigment variants, which follow biparental transmission genetics. Furthermore, a direct and unequivocal identification of some genes on the chloroplast DNA is of considerable importance. The markers used to date in the study of extra-nuclear genetics in *C. reinhardi* are not biochemically well defined. Of considerable value would be the isolation and investigation of mutants whose function at the molecular level could be rigidly defined, such as temperature-dependent photoauxotrophic mutants pertaining to the chloroplast DNA replication, repair and recombination. With these kinds of mutants the results of the biochemical and genetic analysis may be correlated directly. From these studies an unequivocal localization of some markers onto the chloroplast DNA could conceivably be made.

Acknowledgements

The author takes pleasure in acknowledging Drs. N. Sueoka, N.W. Gillham and R.P. Levine for their encouragement and criticism. The author is indebted to Mrs. Brenda Mihan, Mrs. Regina Milasius and Miss Aida Pendon for their technical assistance and to Messrs. G. Gibson, J. Hanacek, P. Kwiatkowski and G. Grofman for their mechanical, electronic and photographic work. The assistance given by Mrs. Tina Chiang in preparing the manuscript is gratefully acknowledged. The author also wishes to thank Dr. B. Strauss for the use of the Packard Tri-Carb scintillation counter and the Spinco analytical ultracentrifuge and Dr. G.B. Whifield of the Upjohn Company for the generous donation of spectinomycin sulfate (Trobicin U-18, 409E) and neamine (base) U-5214. This work was supported by U.S. Public Health Service research grant GM 15114 and by General Research Support grant FR 5367.

References

[1] M.M.K. Nass, Science 165 (1969) 25.
[2] N.H. Horowitz and R.L. Metzenberg, Ann. Rev. Biochem. 34 (1965) 527.
[3] U.W. Goodenough, R.P. Levine, S.J. Surzycki and J.J. Armstrong, this symposium.
[4] S.G. Wildman, this symposium.
[5] J.A. Schiff, this symposium.
[6] D. Von Wettstein, this symposium.
[7] J.C. Beck, J.H. Parker, W.X. Balcavage and J.R. Matton, this symposium.
[8] J.R. Kates and R.F. Jones, J. Cell. Comp. Physiol. 63 (1964) 157.
[9] R. Sager and M.R. Ishida, Proc. Natl. Acad. Sci. U.S. 50 (1963) 725.
[10] J. Leff, M. Mandel, H.T. Epstein and J.A. Schiff, Biochem. Biophys. Res. Comm. 13 (1963) 125.
[11] K.S. Chiang, Ph.D. dissertation, Princeton Univ. (1965) (Ph.D. dissertation no. 65–13,130: University Microfilms, Inc., Ann Arbor, Michigan, 1965).
[12] K.S. Chiang and N. Sueoka, J. Cell. Physiol. 70, Suppl. 1 (1967) 89.
[13] K.S. Chiang, Proc. Natl. Acad. Sci. U.S. 60 (1968) 194.

[14] K.S. Chiang and N. Sueoka, Proc. Natl. Acad. Sci. U.S. 57 (1967) 1506.

[15] N. Sueoka, K.S. Chiang and J.R. Kates, J. Mol. Biol. 25 (1967) 47.

[16] K.S. Chiang, manuscript in preparation.

[17] D. Bastia, K.S. Chiang and H. Swift, manuscript in preparation.

[18] D. Bastia, K.S. Chiang and H. Swift, J. Cell Biol. 43 (1969) 11a.

[19] N.W. Gillham, Amer. Naturalist 103 (1969) 1355.

[20] J. Mego, in: The Biology of Euglena, ed. D.E. Buetow (Academic Press, New York, 1968) p. 351.

[21] P.J. Hastings, E.E. Levine, E. Cosbey, M.O. Hudock, N.W. Gillham, S.J. Surzycki, R. Loppes and R.P. Levine, Microbiology Gen. Bull. 23 (1965) 17.

[22] R. Sager, Proc. Natl. Acad. Sci. U.S. 40 (1954) 356.

[23] R. Sager and Z. Ramanis, Proc. Natl. Acad. Sci. U.S. 61 (1968) 324.

METHODS OF GENETIC ANALYSIS OF
CHLOROPLAST DNA IN CHLAMYDOMONAS

Ruth SAGER and Zenta RAMANIS

Department of Biological Sciences, Hunter College of the
City University of New York, New York, N.Y., USA

This symposium comes at a historical moment in the development of the science of cell biology. The existence of autonomous genetic systems in chloroplasts and mitochondria, postulated for 60 years [1], has now been substantially established; and the genetic analysis of organelle systems has been put on a workable experimental basis.

The results reported in this symposium from the laboratories of Slonimski [2] and of Linnane [3] represent pioneering investigations of the organization and distribution of mitochondrial genes in yeast. In our laboratory, over the past several years, methods have been developed for the genetic analysis of non-Mendelian systems in the alga *Chlamydomonas* [4–6]. In this symposium, we summarize studies of recombination and mapping procedures which have culminated in the demonstration of an extensive linkage group or chromosome [7], probably corresponding to chloroplast DNA which we have mapped by four different quantitative methods.

The availability of genetic methods of recombination analysis and mapping opens the way for the development of an operational molecular genetics of organelles. What kinds of questions fall within the purview of molecular genetics? Broadly speaking, molecular genetics is concerned with the specificity, biosynthesis, regulation, and assembly of macromolecules into the functioning structures and reaction networks of the living cell. As a discipline, molecular genetics is the product of a tight coupling between genetics and biochemistry.

Within the past five years, great strides have been made in examining the molecular biochemistry of organelles (reviewed in this symposium): identification of organelle DNAs, fractionation of organelle RNAs and their identification by DNA-RNA hybridization, isolation of organelle-specific proteins, and studies of organelle protein synthesis. Pulse-labeling and hybridization studies have demonstrated that organelle DNAs are transcribed; ribosomal and transfer RNAs specifically associated with mitochondria and chloroplasts have been identified in a few systems. Nonetheless, not a single protein has been identified as the product of an organelle gene as yet. Thus the functions of organelle DNAs are virtually unknown, beyond their apparent contribution of some RNAs to an organelle-specific protein synthesizing apparatus.

From estimates of size and information content, organelle DNAs appear to contain substantial genomes. Mitochondrial DNA of yeast has been reported as about 5×10^7 daltons of information [8] and in *Neurospora* the 'sequence length molecular weight' has been estimated as at least 6.6×10^7 daltons [9]. Even the mitochondrial DNAs of animal cells, with only about 10^7 daltons of informational content, contain enough nucleotides to code for 25 proteins of 200 amino acids each. Chloroplast DNA of higher plants has an estimated genome size of $1.5–2.0 \times 10^8$ daltons [10]. In *Chlamydomonas* a preliminary value of 2×10^8 daltons as the information content of chloroplast DNA has been estimated from renaturation kinetics [11].

Thus the dilemma: how to uncover the functions of all these organelle genes? In general, one may predict with some confidence that the genetics of viral and bacterial systems will be of close relevance to organelle systems. It would seem fruitful, therefore, to design and evaluate organelle experiments in the light of the experiences with bacterial and viral genetics [12]. Speculation that chloroplasts and mitochondria are of bacterial origin underscores the likelihood of similarities in the genetic organization of organelle and bacterial DNAs.

Genetic analysis begins with the isolation of specific mutations, and utilizes these mutations to investigate two general classes of questions: one class having to do with the replication, organization, and distribution of the genes themselves, and the other concerned with the functions of these genes.

With respect to gene function, genetic methods can provide the following kinds of information: (1) provide critical evidence that a particular gene codes for the primary sequence of a particular protein; (2) distinguish between nuclear and organelle genes; (3) distinguish mutations of different sites within the same gene from mutations of different genes affecting the same phenotypic trait; (4) identify sets of genes linked into operons; (5) identify regulatory sequences such as operators, promotors, initiators, chain terminators; (6) distinguish between regulatory and structural genes.

More generally stated, mutations provide the raw material with which to dissect out the functions of individual genes from their integrated position in ongoing metabolism.

With respect to genetic organization, the following kinds of questions can be investigated: (1) topology of gene sequence: linear, branched, circular, terminally redundant; (2) number of linkage groups and location of particular genes on specific linkage groups; (3) number of genetic copies per cell; (4) mechanics of recombination; (5) mechanics of replication and distribution at cell division; (6) manipulation of the system, e.g., to increase the number of copies of particular genes or fragments.

As knowledge develops, increasingly precise and sophisticated questions can be examined. Perhaps the most sophisticated application of genetic analysis to morphogenesis to date has been in the elucidation of the structure and assembly of bacteriophage T_4 [13].

Organelle systems differ from bacterial cells in two important respects. In the first place of course organelle DNAs are just a part of the genome of the organism. It may be surmised that special regulatory signals exist to couple the nuclear and organelle genetic systems. These signals may involve molecular mechanisms not present in bacteria, and as such may represent a new regulatory principle developed in the evolution of eukaryotic cells.

Another consideration is the likelihood that many of the organelle genes code for 'structural' proteins, the building blocks of membranous and fibrous subcellular structures. In contrast to the bacteria, eukaryotic cells contain many classes of internal membranes and differentiated fibrous structures. It seems likely that the sequestered DNAs of organelles are concerned with the synthesis and assembly of subcellular structures, and that genetic investigation of these DNAs may provide a direct approach to the examination of subcellular assembly and specification.

Both the similarities and differences between organelle and bacterial systems encourage the systematic application of genetic principles and methods to the investigation of organelle biogenesis. The monumental discoveries in the field of molecular genetics of the past 20 years were built on the firm union of genetics and biochemistry. Substantial progress in understanding organelle formation will doubtless flow from an analogous development.

Genetic analysis of an organelle system

In what follows, we have itemized the principal steps in genetic analysis of an organelle system with detailed examples taken from our investigations of chloroplast genes in *Chlamydomonas*. A precondition of genetic analysis, of course, is the acquisition of mutant genes. To induce organelle mutations in *Chlamydomonas* we have used streptomycin as mutagen, growing cells for several doublings in the presence of a sub-lethal concentration of the drug [14]. The effectiveness of N-methyl-N′-nitro-N′-nitrosoguanidine has been reported by Gillham [15].

(1) *Identification of the genetic unit*
In Mendelian genetics, the genetic unit is identified by crossing two distinct parental types and observing the reappearance of the parental phenotypes in the F_2 generation, following their segregation out of the intervening F_1 hybrid. The molecular principle underlying this simple genetic test is the demonstration of a stable, heritable, alteration in DNA which can be transmitted unchanged, i.e., replicated precisely in a

heterozygote. One needs to show that the mutated state under consideration is stable, both in clonal multiplication and in the sexual life cycle.

An analogous test can be performed with non-Mendelian systems. With *Chlamydomonas,* for instance, new mutant strains are sub-cloned to a state of phenotypic uniformity, and then crossed to examine transmission of the trait to the sexual progeny. Characteristically uniparental transmission of non-Mendelian genes occurs, all progeny being genetically identical with the mt^+ (female) parent. Thus, mutant genes present in the mt^+ parent can be checked for stability in transmission through the hybrid zygote.

With the discovery that UV irradiation of the mt^+ parent before mating blocks maternal inheritance, we have been able to examine biparental inheritance. In this system organelle genes from both parents are transmitted to each zoospore (haploid product of meiosis), and somatic segregation of the two parental alleles occurs during vegetative multiplication of each zoospore clone.

(2) *Distinguishing organelle (non-Mendelian) from nuclear (Mendelian) inheritance*

The segregation ratios of nuclear gene pairs, typically 1:1 at meiosis, have never been seen in organelle systems. Typically, inheritance of organelles shows a sex or mating type effect, such that the organelle genome of one of the parents is totally or partially excluded from transmission to the progeny. This mating type effect or maternal inheritance (as it has been denoted in higher plants and animals) provides an excellent diagnostic criterion for initial recognition of non-Mendelian genes, distinguishing them unambiguously from the nuclear complement.

The molecular basis of this exclusion principle may not be the same in all organisms. In higher plants, the mechanism of maternal inheritance has been assumed to be simply the exclusion of male cytoplasm from the fertilized egg. In some plants, occasional (e.g., *Oenothera*), or even frequent (e.g., *Pelargonium*), transmission of non-Mendelian genes from the pollen parent does occur, and in these systems the mechanism has been postulated to be the chance transmission of male cytoplasm. No direct evidence in support of these views has been reported [16].

In *Chlamydomonas,* maternal inheritance is the rule, despite the fact that the zygote receives equal amounts of cytoplasm from both parents. The non-Mendelian genome of the mt^- is usually lost, never reappearing among the progeny or their descendants. In studies reported [17] and discussed [18] elsewhere, we have concluded that in *Chlamydomonas,* the mechanism of maternal inheritance is the destruction in the zygote of organelle DNA of the mt^- parent.

A mating type effect on the transmission of mitochondrial genes in yeast has been reported in this symposium [2,3]. The mechanism of this effect will be of great interest particularly since in yeast as in *Chlamydomonas* the gametes of opposite mating type are of equal size. Thus elimination of genes from one of the parents must occur within the zygote after fusion of gametes. In *Neurospora* maternal inheritance of mitochondrial genes has also been reported [19–21]. In that system, most of the cytoplasm is contributed by the female (protoperithicial) parent and very little by the male (conidial) parent. However, as in higher organisms, the mechanism of maternal inheritance in *Neurospora* has not been directly established.

It would appear that some evolutionary advantage underlies maternal inheritance of organelle genes, since the phenomenon occurs in isogamous species like *Chlamydomonas* and yeast as well as in the heterogamous higher forms. The most obvious consequence of maternal inheritance is the suppression of genetic recombination in the organelle genomes. Since recombination promotes variability, the suppression of recombination fosters genetic stability and non-variation, and therefore functions as a conservative force in evolution. If organelle genes play a key role in membrane formation and in the organization of complex electron transport and energy-generating systems, it may be of great importance to the cell that these genetic systems be shielded from the variability promoted by recombination. The resistance of organelle genes to many common mutagens may also reflect the presence of a mechanism, perhaps a set of very efficient repair enzymes, protecting organelle DNAs from mutation.

(3) *Allelic segregation patterns*

In Mendelian systems, the allelic segregation ratios of 1:1 seen in meiosis are the consequence of the mechanics of chromosome replication and distribution on the meiotic spindle. In classical cytogenetics, allelic pairs served as markers whose segregational patterns were correlated with the behavior of physical markers on the chromosomes, such as knobs [22] and small discontinuities [23].

In non-Mendelian systems so far studied, there has been little evidence of allelic segregation at meiosis. Typically, the pattern has been maternal; and in the exceptions to maternal inheritance, segregation has occurred during clonal or vegetative growth, i.e., somatic segregation. In the higher plant systems involving male transmission, for instance, meiotic segregation is generally rare and no evidence of an orderly pattern has been reported. Attempts to study somatic segregation in higher plants have been thwarted by the complexity of cell lineage patterns and differential cell division rates superimposed on the complexity of the organelle systems themselves [16].

In *Chlamydomonas,* by virtue of the simple sexual cycle and the clonal multiplication of zoospores, it has been possible to examine in detail the behavior of gene pairs in meiosis and in the subsequent mitotic multiplication of the progeny. Our principal findings will be briefly summarized here [4–7].

(1) Segregation rarely occurs in meiosis. Usually in biparental zygotes each of the four products of meiosis, the zoospores, is heterozygous for all of the non-Mendelian genes being studied in the cross.

(2) Each pair of alleles segregates on the average 1:1. For example, if colonies are grown from single zoospores and then subcultured to count the proportion of parental types in the colonies, those proportions are statistically close to 1:1 for each of the marker pairs in the cross.

(3) However, if individual zygotes are plated at the 16- to 32-cell stage and colonies grown and classified, we find that individual gene pairs do not segregate 1:1 in these first few doublings. However, the deviation from 1:1 is only seen in the first few cell divisions of zoospore clones.

(4) The explanation of 1:1 segregation and of the deviation from 1:1 was discovered by pedigree analyses of zoospore clones. We found three patterns of segregation at cell division. Type 1, the commonest type, produces no segregant offspring; the hybrid cell gives rise to two daughter cells both hybrid. Type 2 segregation produces one daughter cell which is hybrid and the other which is a pure type like either one or the other of the parents. Type 3 gives rise to two segregants, one resembling each parent. It is the type 2 events that may distort the 1:1 ratio temporarily in individual zygotes. However, the type 2 events produce equal numbers of the two parental alleles on the average.

(5) In multi-factor crosses involving several genes, the frequency of type 2 events is approximately the same for each gene. However, the frequency of type 3 segregation shows a definite polarity. That is, different genes show different and characteristic frequencies of type 3 segregation. This polarity provides an order on the basis of which genes can be mapped.

A schematic model of type 3 segregation based upon the genetic data is shown in fig. 1. The polarity of type 3 segregation is taken as evidence of a centromere-like 'attachment point' (*ap*) governing the distribution of molecules to daughter cells at cell division. In the figure, a reciprocal exchange occurring between genes *ac* and *ery* results in type 3 segregation for the genes *ery* and *sm2* which are beyond the exchange point with respect to *ap*. Mapping by the frequency of type 3 segregation is based on the assumption that exchanges occur at random along the molecule and that the further a gene lies from *ap*, the more frequently will it undergo type 3 segregation.

The type 3 polarity map agrees well in order and relative spacing of genes with maps based upon recombination and upon cosegregation frequencies (discussed below). It seems likely therefore that type 3 segregation is the result of a recombination or exchange event occurring between the gene and *ap*. The occurrence of type 3 polarity among linked genes also provides evidence that type 3 segregation is not the result of a sorting out process but rather is the result of a true recombinational event.

(6) We have inferred that type 2 segregations also are not the result of sorting out. Progeny of multifactor crosses often show a pattern of type 2 and type 3 segregational events occurring on the same strand and presumably at the same doubling. Intervening linked genes may still be heterozygous. Thus it seems likely that the cells remain permanently diploid for this linkage group whether heterozygous or homozygous for individual marker genes. Formally type 2 segregation resembles 'gene conversion' [23a], a kind

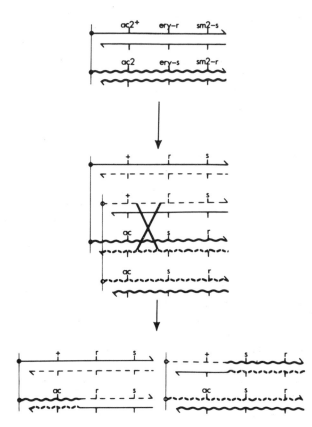

Fig. 1. Proposed scheme of reciprocal recombination. Hetero-zygous chloroplast DNA shown before replication, after replication, and after cell division. An exchange is shown, occurring at the 4-strand stage (after replication) between the genes *ac2* and *ery*. Daughter cells are heterozygous for *ac2* and homozygous for genes *ery* and *sm2* located beyond the exchange point. No molecular mechanism is implied in the figure. Solid lines: one parent; wavy lines: other parent; dotted lines: new replicated strands; ● old attachment points (*ap*); ○ new attachment points.

of miscopying process. We recover 3 copies of one allele and one copy of the other, instead of two and two; the molecular mechanism is not known.

(4) *Recombination between marker genes*

In early studies zygotes were allowed to germinate and form colonies of 10^6-10^7 cells, a sample of wich was then scored for parental and recombinant types. Subsequently we discovered that the frequency of exchange events was so high that data obtained from cells which had undergone many doublings were virtually uninterpretable in terms of recombination analysis. As an example of the difficulty, consider the genes *ac* and *sm2*. Initially we reported that these genes were unlinked and even in studies with cells which had undergone only 4—6 doublings, no evidence of linkage was found. However, by restricting our analysis to the first two doublings and with the aid of intervening markers, it became apparent that the two loci are indeed members of the same linkage group [7].

Current mapping procedures make use of multi-factor crosses analyzed after two doublings of each zoospore. Thus, we examine 16 progeny from each zygote. In each cross three pairs of unlinked nuclear genes are used to distinguish the four products of meiosis. Thus at the 16-cell stage, we can tell which four cells came from each of the four zoospores. In order to distinguish between type 2 and type 3 segregation patterns, it is necessary to identify the pairs of sister cells arising at each doubling. We do this in the following way. Zygotes are picked at the 8-cell stage and spread individually onto whole agar plates. After one doubling, each of the eight cells is respread by a method of limited respreading so that the two daughter cells can be identified. The 16 cells are then allowed to grow to form colonies which are subsequently classified for all markers. Since the segregation frequencies are substantial (10—30% per doubling), an extensive amount of data becomes available by this procedure.

The data may be analyzed in a number of ways. There are three possibilities for each marker: homozygous for one parental allele or the other or still heterozygous. The heterozygotes are identified by subculturing: they give rise to both parental types in later rounds of segregation. In addition to considering the progeny as individuals, we derive information from the pedigree relations. We can distinguish between events occurring at the first and second doubling. We can distinguish recombinations based upon type 2 from those based upon type 3 events. We can check for any segregation or recombination occurring in meiosis. We can score either recombination events or recombinant progeny.

We have developed four mapping procedures which are compared in table 1 and fig. 2. Mapping by polarity of type 3 segregation has been described above. Mapping by recombination frequency is a

Table 1 *

Recombination analysis of progeny from cross 2
(mt+) ac2+ ery-r sm2-s X (mt-) ac2 ery-s sm2-r

	P	R	%R$_A$	%R$_B$	%R$_C$	%R$_D$
ac2-ery	100	36	25.9	6.4	4.08	4.6 (ac2)
ery-sm2	140	27	13.4	4.9	3.3	7.9 (ery)
ac2-sm2	62	35	37.3	6.7	5.56	10.7 (sm2)

* Definitions:

P = parental

R = recombinant

R_A = P/(P + R)

R_B = R/(total progeny (551))

R_C = (total progeny (551))/(P + R)

R_D = (no. type III segregations/total progeny (551))

conventional genetic procedure; one simply counts up recombinants considering all pairwise combinations of genes segregating in the cross, computes the fraction of recombinant progeny, and uses those fractions as a basis for mapping. In our system, this method is

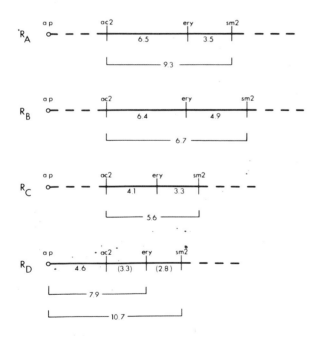

Observed values ÷ 4

Fig. 2. Mapping of three genes, *ac2*, *ery* and *sm2* by four methods based on data of table 1. Values for R_A divided by 4 for purposes of comparison. The attachment point, *ap*, mapped only by method R_D, frequency of reciprocal recombination.

complicated by alternatives in the choice of a suitable denominator. Those progeny that are still heterozygous for the markers in question cannot be classified as either recombinants or parental progeny. Thus the denominator can be restricted to include only progeny which have already segregated (R_A) or all progeny can be included (R_B). The two procedures are compared in table 1 and fig. 2.

We have also developed a mapping procedure based upon the pairwise frequency of cosegregation. In this procedure we utilize the frequency with which two genes become homozygous at the same doubling as a basis of measuring their linkage. Map distance is taken as the reciprocal of the percent of linkage. The result of mapping by cosegregation frequency is shown in fig. 2 (R_C).

Cross 2 is one of a set of five crosses on the basis of which the first chloroplast linkage map was proposed [7]. The principal difference to be seen between methods R_A and R_B is the better additivity which we find using segregated progeny (R_A) rather than total progeny (R_B) as the denominator. In R_C when cosegregants (parental and recombinant) are evaluated as a fraction of the total population, the additivity is poor. Presumably the reason for the low values seen over long intervals in R_B and R_C is the loss resulting from multiple exchanges. In R_A those losses affect both parental and recombinant alike and consequently the additivity is good. In R_D the map itself·is a measure of polarity; with this method we cannot examine addivity of intervals. (The values of 2.8 and 3.3 in R_D are calculated by subtraction from the observed values.) The four methods can be compared by examining the ratio of intervals *ac2-ery* and *ery-sm2* in table 1. They show remarkably good agreement.

Mapping of organelle genes is in a preliminary state. It seems advisable to continue evaluating the data by a number of different methods, rather than attempting to choose one preferred method. At this stage the similarity in gene order and relative map distances seen in all four methods is particularly reassuring. Not only in cross 2 shown here as an example but also in other crosses, both published [7] and unpublished, we have seen a reasonably good agreement in the maps generated by these different procedures.

In conclusion, studies of segregation and recombination occurring in the first few mitotic divisions after meiosis in *Chlamydomonas* have provided evidence in support of the following points: the non-Mendelian linkage group or chromosome under observation undergoes a regular pattern of replication, recombination, and distribution to daughter cells at cell division. This linkage group is genetically diploid; extensive recombination both reciprocal and nonreciprocal occurs during vegetative multiplication; homozygosis results from somatic recombination; reciprocal recombination occurs at a '4-strand' stage, after replication but before chloroplast division.

Comparison of chloroplast and mitochondrial genetic systems

Although the mitochondrial genetic system of yeast is at an earlier stage of investigation than the chloroplast system of *Chlamydomonas,* some comparisons can already be made. Of first importance is the identification of the genetic units themselves. In *Chlamydomonas* we have used the phenotypic stability of mutant traits in the sexual cycle, and particularly in segregation from heterozygotes, as an essential criterion of their identity as unit genes. Maternal inheritance in crosses has distinguished them as non-Mendelian.

In yeast, an important distinction has been demonstrated between the mutations to drug-resistance on the one hand, and the heterogeneous class of mutations known as 'cytoplasmic petites', or *rho⁻* mutations. The non-Mendelian mutations to drug-resistance so far described, are phenotypically stable, and the parental types, i.e., drug-resistance and sensitivity, segregate out of heterozygous diploids [2,3]. Thus far, meiosis has not been studied with these heterozygotes, since they become homozygous before sporulation occurs. Thus, their identification as non-Mendelian units rests on their somatic segregation in diploid clones, which on the basis of present knowledge, is a reliable criterion.

The situation with respect to *rho⁻* is quite different. Ephrussi et al. [24] described neutral petites in which the petite phenotype was never transmitted to diploids in crosses or to progeny in meiosis. Subsequently suppressive petites were discovered [25] in which the petite phenotype was transmitted to some fraction of the diploid clones arising from crosses. Different suppressive petites were shown to exhibit different degrees of suppressiveness, as measured by the percent of petite colonies produced by zygotes from crosses of petite × wild type. Suppressiveness was also shown to be transmitted in meiosis to a variable fraction of the progeny. This evidence established the non-Mendelian nature of these petite mutations but left the mechanism of suppressiveness unknown.

Subsequently Slonimski et al. discovered that some (if not all) *rho⁻* mutations involve extensive changes in mitochondrial DNA, seen as alterations in buoyant density and base composition [26]. As described in this symposium [2,3], some mutations to *rho⁻* wipe out the mitochondrial drug-resistance mutations initially present (i.e., *rho⁻* ER^r becomes *rho⁻* ER^o) while in other strains, the *rho⁻* still carries ER^r, as shown by suitable crosses. Thus it seems likely that few if any mutations to *rho⁻* are simple point mutations, and that different *rho⁻* mutations involve more or less extensive losses of genetic information. Furthermore, some *rho⁻* strains, while remaining phenotypically petite, undergo increasingly drastic shifts in base composition and buoyant density of mitochondrial DNA as reflected in changes in suppressiveness [27] and loss of ER^r [3].

In view of these findings, further studies of *rho⁻* petite strains, while of the greatest interest, will require additional precautions to establish homogeneity and stability for biochemical investigations. Perhaps the presence of a set of marker genes of known location along the DNA will be necessary to define the state of the mitochondrial DNA. In order to develop such strains, as well as to investigate segregation and recombination of mitochondrial DNAs, it seems essential to use stable mutations in *rho⁺* parental strains. Some crosses of this type have been reported in this symposium, and the results are of great interest for comparison with the *Chlamydomonas* data.

The analysis of somatic segregation in *Saccharomyces cereviseae* is complicated by the process of budding. A single zygote may give rise to as many as 20–25 successive buds, during which process the early buds proceed to bud again, leading to the overlapping of generations in zygote clones whether grown in liquid or on agar. Thus the cell lineages in a growing culture of budding yeast are much more difficult to assess than in organisms like *Chlamydomonas* which regularly divide in two at each cell division.

Of prime concern in genetic analysis is the evidence that the gene pairs from the two parental strains do not segregate 1:1, but that the allele from the parent of mating type α predominates among the progeny [2,3]. Does this deviation from 1:1 reflect a difference in the number of copies present in each parent or a difference in the survival and replication of genes coming from the two parents? Since the mating type effect is not all-or-none, different genes along the DNA might have different chances of survival, resulting in a polarity which could be utilized for mapping. In any event, the deviations of allelic segregation from 1:1 need to be understood in order to provide a basis for recombination analysis.

Another fundamental question concerns the mechanism of segregation. In *Chlamydomonas*, segregation appears to be entirely the result of exchange events, whether reciprocal (type 3) or non-reciprocal (type 2). We have found no evidence of a sorting-out process, leading to reduction in the number of copies. However, we found it essential to limit our observations to the first two doublings in order to keep the system comprehensible. In the yeast system, analysis is complicated by the presence of many molecules of mitochondrial DNA, and of many mitochondria per cell, as well as by the budding process. Nonetheless, it will be necessary to distinguish between molecular exchange events and sorting-out of DNA in order to analyze the observed segregational and recombinational events.

Despite the complexities of the system, Slonimski et al. [2] have demonstrated the occurrence of recombination between erythromycin and chloramphenicol resistance markers giving stable recombinant molecules. They have proposed a mapping procedure based on the frequency of recombinants

found in a set of 2-factor crosses between a number of different erythromycin- and chloramphenicol-resistant mutants.

Concluding remarks

This article has as its theme the role of organelle genetics in the overall investigation of organelle structure, function, and biogenesis. In the opening section we considered some classes of information which can be investigated uniquely by applying the methods of molecular genetics to the organelle system. To use genetics in the study of organelles, of course, depends upon the availability of a well-developed genetic analysis of organelle DNAs.

In the second (and major) section of this article, we discussed the methods of genetic analysis and mapping procedures developed for chloroplast DNA in *Chlamydomonas*. We have demonstrated that the chloroplast system is diploid in cells which are haploid for their nuclear genome as judged by genetic criteria; and that recombinational events occur continuously between the DNAs from the two parents during vegetative clonal multiplication of haploid cells. Normally these cells are homozygous for their organelle genes because of maternal inheritance and consequently the recombination events are not expressed phenotypically. By suppressing maternal inheritance, we have been able to introduce multiple genetic differences into the progeny of test crosses, and to follow the segregation and recombination of these marker genes during clonal growth.

The rules of segregation and recombination, as we now understand them, have been summarized, together with a brief description and comparison of four mapping procedures which we have developed. All of the stable non-Mendelian mutants so far investigated fall into one linkage group which we consider to be located in chloroplast DNA [7,18]. In addition

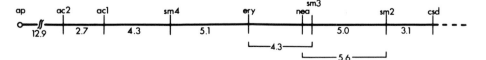

Fig. 3. Composite map based on published data [7]. Symbols are: *ap*, attachment point; *ac2* and *acl*, acetate requirement; *sm4*, streptomycin dependence; *ery*, erythromycin resistance; *nea*, neamine resistance; *sm3*, low level streptomycin resistance; *sm2*, high level streptomycin resistance; *csd*, conditional streptomycin dependence.

to those markers already reported [7], and shown in fig. 3, we have evidence of linkage for two additional *ac* mutants as well as spectinomycin-resistance, kanamycin-resistance, and a temperature sensitive mutant [28].

Comparison of the *Chlamydomonas* system with the mitochondrial genetic system in yeast indicates certain fundamental similarities which must be of considerable significance to the organism. Both systems have the capacity for genetic recombination. Molecular exchanges between DNAs from the two parents have been demonstrated in both systems by genetic methods. In *Chlamydomonas,* the high frequency of exchanges seen at every cell doubling at which heterozygous genes are still present, shows that pairing of chloroplast DNAs and molecular exchanges are regular occurrences in vegetative growth. In the yeast system, it is not yet clear whether exchange events occur only in the zygote or whether they continue in the buds.

In nuclei, genetic recombination is largely limited to meiosis, with rare exceptions occurring in mitosis. In organelles, on the contrary, it seems that the frequency of recombination is not regulated in this way. Perhaps the mating type effect, i.e., the preferential transmission to progeny of organelle genes from one of the parents, acts to limit genetic recombination. In *Chlamydomonas,* maternal inheritance is usually all-or-none with 0.01−1.0% exceptions, while in yeast the mating type effect is less extreme. Nonetheless in both systems the consequence is to decrease the opportunity for genetic recombination.

From the viewpoint of developing a workable system for genetic analysis and mapping, it is the recombination feature which is essential. Thus molecular exchanges in DNAs provide the backbone upon which the organelle genetic methodologies are being constructed. With the developments reported in this symposium, it would appear that organelle genetics is rapidly becoming available to play its part in a concerted investigation of structure, function, and biogenesis.

Added in proof

Mapping of additional mutants has provided the first evidence of genetic circularity [28]. On the basis of distance from the attachment point *ap*, the gene *tr1* (temperature sensitive) is close to the acetate region and *spc* (spectinomycin resistance) is close to *sm2*. On the basis of recombination frequencies however, *tr1* is not linked to *ac*, and *spc* is loosely linked to *sm2*; *tr1* is closely linked to *spc*. These findings suggest a 2-armed chromosome with *tr1* and *spc* on one arm across the attachment point from all the other markers. The evidence for circularity comes from the close linkage of a set of intervening drug-resistance markers between *spc* and *sm2*.

References

[1] C. Correns, Z. Induktive Abstammungs-Vererbungslehre 1 (1908) 291; Erwin Baur, Z. Induktive Abstammungs-Vererbungslehre 1 (1908) 330.

[2] Slonimski et al., this symposium.

[3] G.W. Saunders, E.B. Gingold, M.K. Trembath, H.B. Lukins and A.W. Linnane, this symposium.

[4] R. Sager and Z. Ramanis, Proc. Natl. Acad. Sci. U.S. 53 (1965) 1053.

[5] R. Sager and Z. Ramanis, Proc. Natl. Acad. Sci. U.S. 58 (1967) 931.

[6] R. Sager and Z. Ramanis, Proc. Natl. Acad. Sci. U.S. 61 (1968) 324.

[7] R. Sager and Z. Ramanis, Proc. Natl. Acad. Sci. U.S. (1970) in press.

[8] P. Borst, this symposium.

[9] D.D. Wood and D.J.L. Luck, J. Mol. Biol. 41 (1969) 211.

[10] R. Wells and M. Birnstiel, Biochem. J. 112 (1969) 777; K.K. Tewari, Symp. Soc. Exptl. Biol. 24, in press.

[11] R. Wells and R. Sager, in preparation.

[12] W. Hayes, The Genetics of Bacteria and their Viruses, 2nd edition (John Wiley & Sons, New York, 1968).

[13] R.S. Edgar, Harvey Lectures 63 (1969) 263.

[14] R. Sager, Proc. Natl. Acad. Sci. U.S. 48 (1962) 2018.

[15] N.W. Gillham, Genetics 52 (1965) 529.

[16] J.T.O. Kirk and R.A.E. Tilney-Bassett, in: The Plastids (W.H. Freeman & Co., London and San Francisco, 1967).

[17] R. Sager and D. Lane, Federation Proc. 38 (1969) 347.

[18] R. Sager, Symp. Soc. Exptl. Biol. 24, London, in press.

[19] M.B. Mitchell and H.K. Mitchell, Proc. Natl. Acad. Sci. U.S. 38 (1952) 442.

[20] L. Garnjobst, J.F. Wilson and E.L. Tatum, J. Cell Biol. 26 (1965) 413.

[21] A.M. Srb, Symp. Soc. Exptl. Biol. 17 (1963) 175.

[22] H.S. Creighton and B. McClintock, Proc. Natl. Acad. Sci. U.S. 17 (1931) 492.

[23] C. Stern, Biol. Zbl. 51 (1931) 547.

[23a] R. Holliday, Genet. Res. 5 (1964) 282.

[24] B. Ephrussi, H. Hottinguer and J. Tavlitski, Ann. Inst. Pasteur, 76 (1949) 419.

[25] B. Ephrussi, H. Margerie-Hottinguer and H. Roman, Proc. Natl. Acad. Sci. U.S. 41 (1955) 1065.

[26] J.C. Mounolou, H. Jakob and P.P. Slonimski, Biochem. Biophys. Res. Comm. 24 (1966) 218.

[27] B. Ephrussi, H. Jakob and S. Grandchamp, Genetics 54 (1966) 1.

[28] R. Sager and Z. Ramanis, in preparation.

SIZE, STRUCTURE AND INFORMATION CONTENT
OF MITOCHONDRIAL DNA

P. BORST *

Department of Medical Enzymology, Laboratory of Biochemistry,
University of Amsterdam, Amsterdam, The Netherlands

In a sense the organelle DNA is the most important subject in a Symposium like this. The most characteristic feature of the biogenesis of mitochondria and chloroplasts is the fact that they contain a genetic system of their own with the entire enzymic apparatus to transcribe this DNA and translate RNA into protein. This fact distinguishes the assembly of mitochondria and chloroplasts from the assembly of less aristocratic, albeit complex, organelles, like lysosomes and peroxisomes. This fact is also the basis for the three main questions that dominate the discussion on the biogenesis of mitochondria and chloroplasts: (1) What is the organelle DNA coding for? (2) How are the contributions of nucleus and organelle DNA to the biogenesis of the organelle coordinated? (3) What is the evolutionary advantage for a cell to have 2 separate, and in the case of a plant cell even 3 separate genetic systems? Essential for our understanding of the 'what and why' of organelle DNA is a thorough knowledge of its genetic information content. In this paper I shall discuss the evidence on this point for mitochondrial DNA. Other aspects of mitochondrial biogenesis, studied in my laboratory, have been discussed in detail elsewhere [1,2].

Size and structure of mitochondrial DNA

As a first approximation to the genetic information content of mitochondrial DNA, one can measure its size. Table 1 presents a summary of the data reported up till now. All cases where some uncertainty remained whether the entire genome had been measured or only fragments derived from it, have been omitted. It is clear from this table that the

Table 1
Size and structure of mitochondrial DNAs *

Species	Structure	Size (μ)
ANIMALS		
Chordata		
Mammalians	Circular	4.7–5.6
Birds	Circular	5.1–5.4
Amphibia	Circular	4.9–5.9
Fish	Circular	5.4
Echinodermata		
Echinoidea	Circular	4.6–4.9
Arthropoda		
Insects	Circular	5.2
PROTOZOA		
Tetrahymena pyriformis	Linear	17.6
FUNGI		
Saccharomyces	Circular	25

* Taken from [2].

simple generalization that mitochondrial DNA is always circular and always about 5 μ long is not correct, because *Tetrahymena* mitochondrial DNA is not circular, whereas the mitochondrial DNAs from *Tetrahymena* and *Saccharomyces* are not 5 μ long, in fact the DNA of *Saccharomyces* mitochondria is five times this size.

A five-fold difference in size does not necessarily imply a five-fold difference in genetic information content. It is conceivable that the mitochondrial

* Postal address: Jan Swammerdam Institute, 1e Constantijn Huygensstraat 20, Amsterdam, The Netherlands.

DNA of animal tissues is heterogeneous in base sequence, though not in base composition or size, resulting in a genetic information content equivalent to a multiple of 15,000 base pairs, the number present in a DNA molecule 5 μ long [3]. Since animal mitochondria contain on the average more than one DNA molecule [1] and since fusion and fission in a mitochondrial population may well occur (see discussion in refs. [1 and 4], effective complementation could be envisaged for this heterogeneous population [5]. On the other hand it is also conceivable that the size of yeast mitochondrial DNA represents an overestimate of its genetic information content because the DNA is highly redundant or contains sections of pure dA:dT or dAT:dAT. In principle, these possibilities can be tested by quantitative renaturation studies of mitochondrial DNA and these are discussed in the next section.

Genetic information content of mitochondrial DNA

The renaturation of denatured DNA is a second-order reaction and one can therefore expect that the renaturation rate of any DNA under standard conditions will be inversely proportional to its genetic complexity, defined as the number of base pairs in this DNA, disregarding all repeated sequences. This expectation was experimentally verified by Britten and Kohne [6] and Wetmur and Davidson [7] and found to be correct.

Thuring, Hollenberg and I have used this approach to study the genetic complexity of mitochondrial DNAs from several sources. The DNAs were fragmented by sonication to an average single-stranded fragment size of 2–3 X 10^5, as determined by alkaline band sedimentation. They were then denatured at 100° in low salt and after raising the salt concentration to 0.15 M NaCl, 0.015 M sodium citrate (pH 7.0) the renaturation rate was determined at the optimal temperature by following the decrease in absorbancy at 260 nm as a function of time. The renaturation rates obtained were compared with those of a number of reference DNAs of known complexity.

Fig. 1 shows a comparison of the renaturation of rat mitochondrial DNA and DNA from the *Bacillus subtilis* bacteriophage ϕ 29. ϕ 29 DNA is a linear

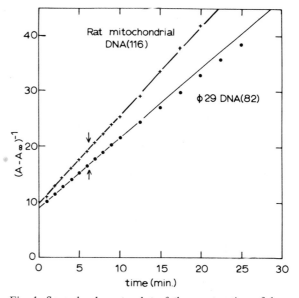

Fig. 1. Second-order rate plot of the renaturation of denatured DNA of bacteriophage ϕ 29 and rat-liver mitochondria. The data are presented according to Wetmur and Davidson [7]. In this plot the slope of the line is only dependent on the renaturation constant and not on the DNA concentration used. $(A-A_\infty)^{-1}$ is the reciprocal absolute value of the absorbance at 260 nm of the DNA at time = t, minus the absorbance of the DNA at time = infinity. Renaturation was measured in 0.15 M NaCl, 0.015 M sodium citrate (pH 7.0) at a temperature 25° below the melting temperature of the DNA in this solvent. For further experimental details, see text. The values between parentheses are the second-order renaturation constants calculated from the slope of the line by a procedure slightly modified from Wetmur and Davidson [7]. The arrow indicates the point at which 50% renaturation was reached.

DNA with a length of 5.8 μ [8], that is only 20% more than that of rat-liver mitochondrial DNA [1]. It has about the same density in CsCl and melting temperature as rat mitochondrial DNA, so presumably the base composition of both DNAs is the same *. If rat mitochondrial DNA is not heterogeneous one would expect it to renature slightly faster than ϕ 29 DNA. If it also lacks redundancy one

---

* In his excellent review Szybalski [9] gives a buoyant density for ϕ 29 DNA of 1.694 g/cm^3 (*Escherichia coli* DNA = 1.710 g/cm^3) and a base composition of 34 mole % GC. I presume that this is a printing error since we find a buoyant density of 1.701 g/cm^3 and a midpoint of the (very sharp) thermal transition profile in 0.15 M NaCl, 0.015 M sodium citrate (pH 7.0) of 86°. These values suggest that the base composition is 42% GC.

would expect, in addition, that the renaturation reaction will follow second-order kinetics without components renaturing faster than the bulk of the DNA. This is exactly the result obtained in the experiment presented in fig. 1. The absence of rapidly renaturing components can be studied more effectively when the renaturation rate is lowered by decreasing the salt concentration as was done in the experiment presented in fig. 2. Again there is no indication for heterogeneity in any of the DNAs studied. (It should be noted that the renaturation constants calculated for mitochondrial DNA and ϕ 29 DNA in this experiment are less reliable than those of fig. 1 because the renaturation rate measured was too low for accurate measurements.)

Experiments on other mitochondrial DNAs and additional reference DNAs are summarized in table 2. In all cases renaturation followed second-order kinetics without evidence of heterogeneity. For yeast mitochondrial DNA the results varied more than for the other DNAs studied and therefore the range of our values is given rather than the average value. The interpretation of these data is complicated by the fact that the renaturation constants of our reference DNAs are not exactly proportional to genome size. This is apparently due to a small effect of base composition on renaturation rate as shown by the positive correlation between GC content and renaturation rate (table 2). A similar, though less pro-

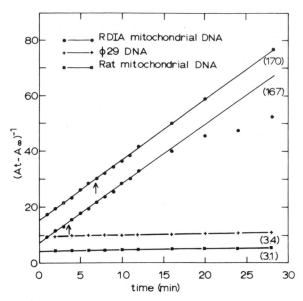

Fig. 2. Second-order rate plot of the renaturation of the denatured DNA of the 'petite' mutant RDIA of *Saccharomyces cerevisiae* and two reference DNAs. Renaturation was measured in 0.045 M NaCl, 0.0045 M sodium citrate (pH 7.0) as specified in the legend of fig. 1.

nounced, effect has been observed by others [7,13]. If the effect of base composition is taken into account, it is clear from table 2 that the kinetic complexity of the bulk of rat and guinea-pig mitochondrial DNA is equivalent to its size. Since no

Table 2

Renaturation rates of mitochondrial DNAs and reference DNAs related to genome size and base composition

Source of DNA	Renaturation constant [a] ($l.mole^{-1}.sec^{-1}$)	Size of genome [b] (μ)	(a) $\times$ (b) ($\times 10^{-2}$)	Mole % G+C [c]
Escherichia coli	0.8	1100	8.7	50
Phage lambda	50	17.2	8.6	49
Guinea-pig mitochondria	100	5.6	5.6	43
Rat mitochondria	120	4.9	5.9	42
Phage ϕ 29	80	5.8	4.6	42
Phage T_4	6	56	3.4	34
Yeast mitochondria	6–10	25	1.5–2.5	17
Yeast 'petite' mitochondria	>4000	?	–	<3 ?

[a] From [2] and [10] and unpublished experiments.
[b] From [11] (*E. coli*), [12] (lambda and T_4, assuming that the lengths of T_2 and T_4 are equal), [8] (ϕ 29), [1] (rat and guinea-pig mitochondria), [10] (yeast mitochondria).
[c] From [12] (lambda and T_4), [1] (rat and guinea-pig mitochondria), this paper (ϕ 29 and yeast 'petite' mitochondria) and a personal communication from G.Bernardi (yeast mitochondria).

reference DNAs are available with such a low GC content as yeast mitochondrial DNA we cannot estimate its genetic complexity with certainty. However, also in this case the kinetic complexity observed is fully compatible with the size of the mitochondrial genome.

Renaturation studies that complement ours have been done by Flavell and Jones [14] with mitochondrial DNA from *Tetrahymena*. Also in this case they find a renaturation rate equivalent to the molecular size of 17.6 μ and without major gene repetitions.

The last line of table 2 presents preliminary experiments with the low-density DNA of a petite mutant induced by ethidium and studied about 150 generations after induction by Hollenberg in our laboratory. This DNA replaces the wild-type mitochondrial DNA and we assume, like others [15—17] have done previously, that it represents the defective mitochondrial DNA of the mutant. The density of this DNA in CsCl is 1.671 g/ml (*E. coli* = 1.710) which is significantly lower than the density of alternating dAT:dAT of 1.678—1.679 g/cm^3 but much higher than the density of dA:dT of 1.647 g/ml [9]. The melting point of the mutant DNA is 67—68° in 0.2 M Na$^+$. This is intermediate between the melting points of dAT:dAT and dA:dT [18] and it is about the melting point expected for a random double-stranded polymer containing only A and T. Its base composition has not been determined but it is likely that it contains less than 3 mole percent GC, like the low-density DNA of the mutants studied by Tecce and co-workers [15] and Mehrotra and Mahler [16]. The renaturation of the low-density DNA of the mutant was so fast that it could only be studied in 0.06 M Na$^+$. The results of quantitative renaturation in that solvent, shown in fig. 2, have two interesting aspects: (1) the renaturation rate shows the kinetics and the concentration dependence of a bimolecular reaction; this suggests that notwithstanding its apparent low GC content, the separated strands are unable to renature by the formation of hairpin-type structures and that long stretches of pure dA or dT or alternating dAT are absent in this DNA. (2) By comparing the renaturation rate with that of ϕ 29 and rat mitochondrial DNA in the same solvent, we can calculate an approximate genetic complexity equivalent to a molecular weight of about 100,000 daltons. This is even less than the fragment size used in these

experiments and therefore an upper limit. Hence, we conclude that the bulk of the low-density DNA of this mutant contains little, if any, genetic information. This conclusion provides experimental support for our previous suggestion [1] that the mitochondrial enzymes still present in this type of mutant must be specified by nuclear genes.

Discussion

The renaturation studies reviewed in this paper show that the kinetic complexity of all normal mitochondrial DNAs studied is equivalent to the size of these DNAs. Before discussing the implications of this finding it may be useful to point out some of the limitations of this experimental approach. The limited analytical precision of the determination of renaturation constants and the uncertainty introduced by the effect of base composition on renaturation rate are illustrated in table 2. Hence, the discrepancies between size and kinetic complexity by less than a factor two, should not be considered significant unless the DNA can be prepared completely free of contaminants and only when a suitable reference DNA of about the same size and base composition is available. Lack of analytical precision also prevents the detection of a limited degree of repetition, affecting less than 10% of the DNA. The most unfortunate limitation, however, is that renaturation experiments cannot detect a limited degree of heterogeneity. This is obvious for point mutations, small insertions and deletions, but it also holds for the two models for mitochondrial DNA, shown in fig. 3. The limited degree of heterogeneity depicted in fig. 3 will show up in renaturation experiments as a deviation from second-order kinetics in the last part of the renaturation reaction. Unfortunately, deviations from second-order kinetics are a constant feature of the later stages of the renaturation reaction with all DNAs [19], including DNAs of undisputed homogeneity, and it is not possible to distinguish between this 'normal' deviation and the deviation due to heterogeneity, if this heterogeneity is limited to a small part of the DNA.

Although we cannot exclude the models shown in fig. 3 on the basis of renaturation experiments, it should be stressed that they represent a type of

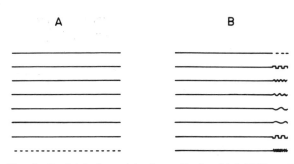

Fig. 3. Far-fetched models for mitochondrial DNA not excluded by renaturation experiments. The separate lines represent separate DNA molecules present in one cell. The continuous lines indicate DNA molecules with the same base sequence; the broken line in model A represents a DNA molecule with a different base sequence from the other molecules. The broken, resp. wavy lines in model B represent a variable gene in otherwise identical molecules.

complicated genome organization without precedent, not supported by any evidence and requiring elaborate control systems in the cell to ensure the maintenance of the heterogeneity. Hence, I think it reasonable to conclude that the information content of the normal mitochondrial DNAs studied, is equivalent to their length. This conclusion has two interesting implications. (1) One cannot extrapolate results obtained for the biogenesis of mitochondria in yeast and *Tetrahymena* to higher organisms because most of the genes present on yeast and *Tetrahymena* mitochondrial DNA must be absent in the DNA of animal mitochondria. (2) Our results show that the total genetic information available in the DNA of animal mitochondria indeed does not exceed the information available in 15,000 base pairs. We can, therefore, now return to the question: what are these 15,000 base pairs coding for to justify the maintenance in animal cells of a complete second system for the duplication and transcription of DNA and the biosynthesis of protein?

Knowledge about the genetic function of mitochondrial DNA is unfortunately still limited, but the evidence as it stands is rather consistent. Mitochondrial RNA components with the characteristics of ribosomal RNA specifically hybridize with mitochondrial DNA in *Tetrahymena* [20], yeast [21], *Neurospora* [22] and *Xenopus laevis* [23]. In addition, 4 S RNA hybridizes with mitochondrial DNA in *Xenopus* [23] and rat liver [24] and in the latter case

it has been shown that at least some of this 4 S RNA represents mitochondrial tRNA species.

In my opinion no protein coded for by mitochondrial DNA has been identified as yet with certainty. The cytoplasmic inheritance of resistance to erythromycin, chloramphenicol and other antibiotics in yeast [25–29] suggests that the structural genes for some mitochondrial ribosomal proteins are present on mitochondrial DNA. It has not been excluded, however, that a change in ribosomal RNA could lead to resistance to these antibiotics, e.g. because it secondarily affects the tertiary structure of the ribosomal proteins. Since the ribosomal RNA is coded for by mitochondrial DNA, this could explain the cytoplasmic inheritance of some cases of drug resistance even if the ribosomal proteins were coded for by nuclear DNA.

Evidence for the origin of the proteins of the mitochondrial inner membranes is even less satisfactory [30]. Although it is clear that biogenesis of functional mitochondria requires an intact mitochondrial system for protein synthesis, which in turn requires functional mitochondrial DNA, I do not know of any experiment that rigorously excludes the possibility that all inner membrane proteins synthesized on mitochondrial ribosomes are specified by nuclear messenger RNAs imported into the mitochondrion.

In view of the fragmentary state of the evidence, I shall limit my comments on the question 'why is there mitochondrial DNA at all?' to animal cells, where the need for a sensible answer is most pressing in view of the low genetic information present in mitochondrial DNA. Recent work by Dawid [23] strongly suggests that at least in *Xenopus laevis* egg mitochondria, the specification of stable mitochondrial RNA components takes up a large fraction of the available genetic information. Dawid's estimate of the combined molecular weights of the two ribosomal RNAs of *Xenopus* mitochondria is 1.1–1.3 million daltons. Since the molecular weight of *Xenopus* mitochondrial DNA is about 10 million daltons, one expects to find a hybridization plateau of about 12% for the ribosomal RNA, if this is coded for by mitochondrial DNA and if one copy of the ribosomal cistrons is present per DNA molecule. This is exactly what Dawid found. In addition, Dawid observed that 4 S RNA hybridized with 3% of the mitochondrial

DNA. Therefore, a total of $2 \times 15\% = 30\%$ of the mitochondrial DNA is already used up for the synthesis of stable RNA components and this is a minimal estimate because in other systems ribosomal RNAs are not synthesized as such but as part of a larger precursor [31–33]. Therefore it is probably more realistic to put the estimate for the part of the DNA coding for stable RNA components at 40%. The remaining 60% is equivalent to about 3,000 amino acids or 20 very modest proteins. This is not even a complete set of ribosomal proteins.

On the basis of these results Dawid [23] has suggested that the function of mitochondrial DNA in animal cells is restricted to the elaboration of a specialized protein synthesizing machinery, all the information specifying the sequences of product proteins being provided by the nucleus. A genetic system providing only some components of a protein synthesizing system does not seem very useful to the cell and equipment that is not useful does not survive a few hundred million years of evolutionary pressure. Nevertheless, mitochondrial DNA is still with us. I see only 3 ways to rationalize this paradox:

(1) Mitochondrial DNA represents an effective form of gene amplification, because it provides multiple copies of genes for the synthesis of ribosomes at the site where they are required, in the mitochondria. This idea is not very attractive because the amplified ribosomal genes in amphibian eggs [34,35] demonstrate that nature can come up with less gene-consuming remedies than the elaboration of a complete second protein-synthesizing system, if the synthesis of ribosomal RNA becomes limiting.

(2) The assembly of the inner membrane of double-membraned organelles like chloroplasts or mitochondria requires a number of proteins that can only be put in place from the inside and that cannot pass the membrane. Hence, they must be synthesized inside and this in turn requires a protein-synthesizing system inside. According to this idea mitochondrial DNA would be primarily required to provide those components of the system that cannot be imported. The defects of this idea are obvious, but it serves to emphasize the importance of finding out, whether nuclear messengers are present in the mitochondrial matrix space. This is investigated in my laboratory.

(3) This brings me to a third explanation that I have put forward on previous occasions [1,2] and

that has proved to be extremely repulsive to geneticists. This explanation says that eukaryotes in general are not happy with their mitochondrial DNA and that they are trying to get rid of it without simultaneously losing their mitochondria. According to this idea the mitochondrion started out as a nearly completely autonomous endosymbiotic bacterium. In the course of evolution more and more functions originally encoded in the symbiont DNA have been taken over by the nucleus. What we see in an animal mitochondrion are the last remnants of this bacterial genome. The use of this hypothesis is that it emphasizes that the evolutionary advantage of having an organelle DNA, which is usually taken for granted, may not exist at all. Hence, we may still find an organism (superman?) that manages to synthesize its mitochondria without mitochondrial DNA.

Acknowledgements

I am greatly indebted to Professor E.C. Slater for his interest and advice. The renaturation experiments were carried out in collaboration with Mr. C.P. Hollenberg, Mr. R.W.J. Thuring and Dr. G.J.C.M. Ruttenberg with expert technical assistance by Mrs. F. Fase-Fowler, Miss H.M. Moerdijk and Mr. H.D. Batink. Stocks of bateriophage ϕ 29 and its host were kindly provided by Dr. E. Viñuela and Dr. G. Venema. I am grateful to Dr. E. Viñuela for sending me his unpublished procedure for the purification of bateriophage ϕ 29 DNA. The experimental work in my laboratory is supported (in part) by the Netherlands Foundation for Chemical Research (S.O.N.) with financial aid from the Netherlands Organization for the Advancement of Pure Research (Z.W.O.).

References

[1] P. Borst, Int. Rev. Cytol. 26 (1969) 107.
[2] P. Borst, in: The Development and Interrelationship of Cell Organelles, ed. P.L. Miller (Cambridge University Press, Cambridge, 1970) p. 201.
[3] S.B. Leighton and J. Rubenstein, J. Mol. Biol. 46 (1969) 313.
[4] S. Nass, Int. Rev. Cytol. 25 (1969) 55.
[5] Y. Suyama and K. Miura, Proc. Natl. Acad. Sci. U.S. 60 (1968) 235.

[6] R.J. Britten and D.E. Kohne, Science 161 (1968) 529.
[7] J.G. Wetmur and N.J. Davidson, J. Mol. Biol. 31 (1968) 349.
[8] D.L. Anderson and E.T. Mosharrafa, J. Virol. 2 (1968) 1185.
[9] W. Szybalski, in: Methods in Enzymology, eds. L. Grossmann and K. Moldave (Academic Press, New York, 1968) vol. 12B, p. 330.
[10] C.P. Hollenberg, P. Borst and E.F.J. van Bruggen, Biochim. Biophys. Acta 199 (1970) 373.
[11] J. Cairns, Cold Spring Harbor Symp. Quant. Biol. 28 (1963) 43.
[12] C.A. Thomas, Jr. and L.A. McHattie, Ann. Rev. Biochem. 36, part 2 (1967) 485.
[13] K.J. Thrower and A.R. Peacocke, Biochem. J. 109 (1968) 543.
[14] R.A. Flavell and I.G. Jones, private communication.
[15] G. Bernardi, F. Carnevali, A. Nicolaieff, G. Piperno and G. Tecce, J. Mol. Biol. 37 (1968) 493.
[16] B.D. Mehrotra and H.R. Mahler, Arch. Biochem. Biophys. 128 (1968) 685.
[17] J.C. Mounolou, H. Jakob and P.P. Slonimski, Biochem. Biophys. Res. Commun. 24 (1968) 218.
[18] Handbook of Biochemistry. Selected Data for Molecular Biology, ed. H.A. Sober (The Chem. Rubber Co., 1968) p. H-16.
[19] J.A. Subirana, Biopolymers 4 (1966) 189.
[20] Y. Suyama, Biochemistry 6 (1967) 2829.

[21] E. Wintersberger and G.L. Viehhauser, Nature 220 (1968) 699.
[22] D.D. Wood and D.J.L. Luck, J. Mol. Biol. 41 (1969) 211.
[23] J.B. Dawid, in: The Development and Interrelationship of Cell Organelles, ed. P.L. Miller (Cambridge University Press, Cambridge, 1970).
[24] M.M.K. Nass and C.A. Buck, Proc. Natl. Acad. Sci. U.S. 62 (1969) 506.
[25] D.Y. Thomas and D. Wilkie, Genet. Res. 11 (1968) 444.
[26] A.W. Linnane, G.W. Saunders, E.B. Gingold and H.B. Lukins, Proc. Natl. Acad. Sci. U.S. 59 (1968) 903.
[27] A.W. Linnane, A.J. Lamb, C. Christodoulou and H.B. Lukins, Proc. Natl. Acad. Sci. U.S. 59 (1968) 1288.
[28] G. Saunders, E. Gingold, K. Trembath, H.B. Lukins and A.W. Linnane, this volume.
[29] P.P. Slonimski, this volume.
[30] H.R. Mahler, P. Perlman and B.D. Mehrotra, this volume.
[31] U.E. Loening, K.W. Jones and M.L. Birnstiel, J. Mol. Biol. 45 (1969) 353.
[32] N.B. Hecht and C.R. Woese, J. Bacteriol. 95 (1968) 986.
[33] M. Adesnik and C. Levinthal, J. Mol. Biol. 46 (1969) 281.
[34] D.D. Brown and I.B. Dawid, Science 160 (1968) 272.
[35] M. Birnstiel, J. Speirs, I. Purdom, K. Jones and U.E. Loening, Nature 219 (1968) 454.

WILL THE REAL CHLOROPLAST DNA PLEASE STAND UP

John T.O. KIRK

C.S.I.R.O. Division of Plant Industry, Canberra, A.C.T., Australia

It is difficult to state precisely when the idea that chloroplasts contain their own DNA gained general acceptance: 1963 was probably the turning point, and by the end of 1964 people were already forgetting that the very supposition that DNA existed outside the nucleus had formerly been heretical. In the wave of enthusiasm for extranuclear DNA that swept through the biological sciences in the next few years it was not generally noticed that there was something of a discrepancy between the two original reports [1–3] on the isolation and analysis of higher plant chloroplast DNA, as to its actual base composition. In both cases it had been reported that there was a difference between the composition of the nuclear DNA and that of the corresponding chloroplast DNA, but the findings were in disagreement as to both the size and the direction of this supposed difference.

In the years since these first reports, many papers on chloroplast DNA have appeared, some supporting one side, some the other, with the result that there is now a great deal of confusion as to which of the various DNA species that have been described, really originate in the chloroplast. In the present article I shall attempt to clarify the situation by tracing out the way opinion in this field has developed and changed from 1963 onwards, and I shall also attempt to reach a conclusion on the basis of what I consider to be the best evidence, as to which DNA is really the chloroplast DNA.

The first reports

In my own work, with broad bean (*Vicia faba*) leaves [1,2], I found the chloroplast DNA to have a base composition close to, but not identical with,

that of the nuclear DNA (table 1), the actual values for GC content * being 39.4% for the nuclear DNA, and 37.4% for the chloroplast DNA. The difference, although small, was undoubtedly real, being statistically significant at the 0.1% level: this particular finding was important at the time because it was the main piece of evidence indicating that the chloroplast preparations had a DNA of their own, which was not merely contaminating nuclear DNA. However, for our present purposes, now that the existence of chloroplast DNA is no longer in doubt, it is sufficient to note, first, that the two base compositions were very close together, and second, that such difference as did exist consisted of a lower GC content in the chloroplast DNA (table 1). In view of the differences

* By GC content is meant the proportion of guanine + cytosine (+ 5-methylcytosine, where present) in DNA, expressed as a percentage of the total number of moles of purine and pyrimidine bases present.

Table 1
Base compositions of broad bean chloroplast and nuclear DNAs [2]

	Adenine/guanine molar ratio	
	Chloroplast DNA	Nuclear DNA
	1.66	1.57
	1.72	1.57
	1.63	1.54
	1.65	1.50
	1.67	1.54
Mean	1.67	1.54
Calculated GC content	37.4%	39.4%

Pooled estimate of standard deviations of A/G ratios, $S = 0.03$. Difference between means = 0.13. Probability that the means are not different, $p < 0.001$.

between these results and those of other workers, some note should be taken of the methods used. Leaves of young, greenhouse-grown broad bean seedlings were used. The chloroplasts were isolated first by differential centrifugation and then further purified by high speed density gradient centrifugation. The base compositions were determined by chemical analysis. For this purpose a new method had been devised involving liberation of the purine bases by gentle acid hydrolysis, followed by separation of the adenine and guanine by ion-exchange chromatography. Thus the actual parameter measured was the adenine/guanine molar ratio, from which, assuming Watson-Crick base pairing, the GC content can readily be calculated.

The other report of isolation of DNA from higher plant chloroplasts to appear in 1963 was that of Chun et al. [3] who claimed that spinach (*Spinacia oleracea*) and beet (*Beta vulgaris*) chloroplasts both contained two kinds of DNA, β and γ, of 46% and 60% GC content, respectively. Both these DNAs, in contrast to the broad bean results, had a very different base composition from that of the nucleus, the difference consisting of much higher GC contents in the chloroplast DNAs than in the nuclear DNA (36% GC in both species). The chloroplasts were isolated by simple differential centrifugation from commercially obtained spinach and beet leaves. The conclusions as to the base compositions of the various DNAs were based on analysis by caesium chloride equilibrium density gradient centrifugation. The nuclear DNA contained a single component of buoyant density *, 1.695 g cm^{-3}. The chloroplast DNA preparations still consisted predominantly of a 1.695 g cm^{-3} component but also contained 5–15% of a DNA with density 1.705 g cm^{-3} (β DNA) and 5–40% of a DNA with density 1.719 g cm^{-3} (γ DNA). The base compositions of the various DNAs were calculated from the buoyant densities using the relationship [4] ρ = 1.660 + 0.098 (G+C). The presence of the 1.695 g cm^{-3} component in the chloroplast DNA preparations was attributed to nuclear contamination and it was considered that the two dense DNA bands constituted the true chloroplast DNA.

* The values for buoyant densities of DNAs in CsCl given in this review relate to a value of 1.710 g cm^{-3} for *Escherichia coli* DNA [4].

The satellites continue to appear

The difference between the broad bean results and the spinach and beet results was puzzling but it seemed at the time that it could plausibly be attributed simply to a species difference: since the nuclear DNAs of different plant species can have different base compositions, it seemed entirely feasible that the GC content of chloroplast DNA could similarly vary from one species to another. However, the results reported from other laboratories on other plant species, in the next few years, seemed to be in better agreement with the findings of Chun et al. than with the broad bean results. Kislev et al. [5] reported that swiss chard (*Beta vulgaris* var. *cicla*) chloroplasts contained a DNA of a density (1.700 g cm^{-3}) which was 0.011 g cm^{-3} higher (indicating 11% higher GC content) than that of the nuclear DNA (1.689 g cm^{-3}). Shipp et al. [6] claimed that tobacco (*Nicotiana tabacum*) chloroplasts contained a DNA of density 1.703 g cm^{-3}, 0.013 g cm^{-3} more dense (indicating 13% higher GC content) than the nuclear DNA (1.690 g cm^{-3}) *. Also working with tobacco, Green and Gordon [7] claimed that the chloroplasts contained a DNA of density (1.706 g cm^{-3}) which was 0.010 g cm^{-3} higher (indicating 10% higher GC content) than that of the nuclear DNA (1.696 g cm^{-3}). Similarly, Nilsson-Tillgren et al. [8] reported that tobacco chloroplasts contained a DNA of density 1.707 g cm^{-3}, 0.009 g cm^{-3} more dense (indicating 9% higher GC content) than the nuclear DNA (1.698 g cm^{-3}). Green and Gordon [9] extended their observations to several other plants – marigold (*Tagetes patula*), snapdragon (*Antirrhinum majus*), carnation (*Dianthus caryophyllus*), buttercup (*Ranunculus repens*) and horsetail (*Equisetum* sp.) – and concluded that in every case the chloroplast DNA was substantially more dense than the corresponding nuclear DNA the chloroplast values ranging from 1.702 to 1.713 g cm^{-3}, and the nuclear values from 1.692 to 1.697 g cm^{-3}. It should be noted that all these different groups of workers found in the chloroplast DNA preparations a DNA component – indeed

* Comparison of the data for nuclear DNAs with those obtained by other workers on the same species suggests that all the buoyant density values obtained by Kislev et al., and by Shipp et al., may be 0.005–0.006 g cm^{-3} too low.

the major band in every case except possibly the results of Shipp et al. – of buoyant density close to that of the nuclear DNA. But the presence of this component was in every case attributed to nuclear contamination.

By this stage the broad bean data was out of step, not only with the results obtained with spinach and beet, but also with the findings reported for Swiss chard, tobacco, marigold, snapdragon, carnation, buttercup and horsetail; the supposition that the discrepancy was merely a species difference was beginning to look increasingly untenable. However, two papers had appeared in this period which, while not precisely in agreement with the broad bean data, were not so clearly in disagreement as were the other papers discussed above. Suyama and Bonner [10] in a study primarily concerned with mitochondrial DNA, noted that the major peak in a DNA preparation from turnip (*Brassica rapa*) chloroplasts, had a density of 1.695 g cm^{-3}. Although there was some indication that a small amount of a more dense satellite DNA was also present in the chloroplast DNA preparation, and despite the fact that the major chloroplast DNA band had a density close to that of the nuclear DNA (1.692 g cm^{-3}), Suyama and Bonner did not, unlike other workers, jump to the conclusion that the major band in the chloroplast DNA preparations was merely contaminating nuclear DNA. The other paper was that of Tewari and Wildman [11], working with tobacco who found, in disagreement with other reports, only one band in the DNA from purified chloroplasts. This had a density of 1.702 g cm^{-3} (indicating 43% GC content). The Tewari and Wildman value for density of chloroplast DNA was still higher than that for the nuclear DNA (1.697 g cm^{-3}) but by only about half as much as the other workers had reported *. Nevertheless despite these straws in the wind, the balance of the evidence was against the broad bean data, and by the autumn of 1967 anyone wishing to generalize about higher plant chloroplast DNA would probably have come to the conclusion that it was a DNA with a density in the range $1.702-1.713$ g cm^{-3}, but usually around 1.706 g cm^{-3}, with a base composition in the range

43–54% GC, usually around 47% GC, and which was in all plant species substantially more dense (and therefore with a higher GC content) than the corresponding nuclear DNA.

The satellites disappear

However, also in 1967, there began to appear a series of papers, which changed the whole picture. The first of these, by Beridze et al. [12], reported studies by the CsCl technique on DNA from nuclei and chloroplasts of *Phaseolus vulgaris,* the French bean. Both kinds of DNA gave a somewhat irregular band in CsCl, with a suggestion of a shoulder on the high density side. The Russian workers attempted to interpret the shape of these bands as representing in each case a mixture of two kinds of DNA, predominantly DNA of density 1.694 g cm^{-3}, with a minor component of density 1.703 g cm^{-3}. Putting aside the possible dual nature of each of these DNAs (the question of interpretations of 'satellite' DNA is discussed later) the average densities work out to be about 1.697 g cm^{-3} for the nuclear DNA and 1.696 g cm^{-3} for the chloroplast DNA. That the DNA in the chloroplast fraction was not merely contaminating nuclear DNA was shown by the fact that the chloroplast DNA, unlike that from the nucleus, underwent extensive renaturation after heat denaturation (this criterion for chloroplast DNA was first established by Tewari and Wildman [11]). Thus the chloroplast DNA in French bean seemed to be much less dense than the values previously reported for other plants, and in fact had a density close to that of the nuclear DNA. Late in 1967, Ruppel [13] reported that *Antirrhinum majus* chloroplast DNA consisted of a single band in CsCl with a density of 1.698 g cm^{-3}: the density of nuclear DNA was 1.689 g cm^{-3}. Although the density of this chloroplast DNA was much lower than the currently accepted values, Ruppel apparently saw his results as being essentially in agreement with those of Chun et al. and others, in so far as the *Antirrhinum* data apparently supported the generalization that in higher plants the chloroplast DNA is more dense than the corresponding nuclear DNA.

The results of the Russian workers and of Ruppel were on species which had not previously been

---

* The more recent results of Tewari and Wildman have narrowed the density gap between chloroplast and nuclear DNAs still further – see later.

examined and so did not directly contradict the findings of Chun et al. and others, although the new data did make generalization about the properties of higher plant chloroplast DNA more difficult. The results which really caused a rethinking of the nature of chloroplast DNA were those of Wells and Birnstiel [14,15] in Edinburgh and Whitfeld and Spencer [16] in Canberra. These workers studied (amongst others) two of the species – spinach and tobacco – which had been examined previously, and their results failed to confirm the earlier reports. Wells and Birnstiel looked at four species – spinach, lettuce (*Lactuca sativa*), sweet pea (*Lathyrus odoratus*) and broad bean. In every case they found that the DNA from the chloroplast fraction yielded a single component in CsCl, of density $1.697 \, g \, cm^{-3}$: there was no sign of the satellite bands, densities $1.705 \, g \, cm^{-3}$ and $1.719 \, g \, cm^{-3}$, reported by Chun et al., and others. The nuclear DNAs had densities close to, but not identical with, that of chloroplast DNA, being $1.694 \, g \, cm^{-3}$ for spinach and lettuce, and $1.695 \, g \, cm^{-3}$ for broad bean and sweet pea. In the case of lettuce the DNA obtained from chloroplasts was shown to lack 5-methylcytosine (a property of chloroplast DNA which distinguishes it from nuclear DNA; first established for *Euglena gracilis* [17,18]) and was renaturable, confirming that it was not nuclear DNA.

Essentially the same results were obtained by Whitfeld and Spencer [16] with spinach and tobacco. The chloroplast DNA from both species gave a single band in CsCl, with no satellites. The density was $1.696 \, g \, cm^{-3}$ in the case of spinach, and $1.697 \, g \, cm^{-3}$ in the case of tobacco; the corresponding nuclear DNAs had densities of $1.694 \, g \, cm^{-3}$ and $1.697 \, g \, cm^{-3}$, respectively. Both chloroplast DNAs lacked 5-methylcytosine and were renaturable, showing that they were not contaminating nuclear DNAs.

Confirmation of some of these new findings on the density of chloroplast DNAs has now appeared. Wolstenholme and Gross [19] have examined nuclear and chloroplast DNAs of *Phaseolus vulgaris*. Like Beridze et al. [12] earlier, they find both DNAs to give somewhat irregular bands in CsCl. The nuclear DNA seemed to have its main peak at a density of $1.693 \, g \, cm^{-3}$ with a shoulder at $1.702 \, g \, cm^{-3}$. The chloroplast DNA had its main peak at $1.695 \, g \, cm^{-3}$ with a suggestion of a shoulder in the $1.702 \, g \, cm^{-3}$

region. An approximate graphical analysis of the UV tracings in the paper of Wolstenholme and Gross [19] indicates that both the chloroplast and nuclear DNAs have an average buoyant density in the region of $1.696 - 1.697 \, g \, cm^{-3}$ in good agreement with the data of Beridze et al. [12]. The presence of a satellite band in the nuclear DNA probably represents a genuine heterogeneity of the DNA in the nucleus. The presence of a much less definite shoulder in the chloroplast DNA band, I suggest is likely to be due to the presence of a certain amount of contaminating nuclear DNA. This explanation may also account for the suggestion of heterogeneity in the DNA isolated from *Phaseolus* chloroplasts by Beridze et al. [12] and in the DNA isolated by Suyama and Bonner [10] from turnip chloroplasts (in this plant too the nuclear DNA is distinctly heterogeneous – more so than the chloroplast DNA). In the case of broad bean, Kung and Williams [20] have shown that DNA from chloroplasts put through a very exhaustive purification procedure, which removes the last traces of nuclear DNA, gives a single band in CsCl of density $1.696 \, g \, cm^{-3}$, the same as the density of the nuclear DNA; the chloroplast DNA, however, unlike the nuclear DNA, was renaturable. The data of Kung and Williams thus essentially confirm those of Wells and Birnstiel [15] for this plant.

In all this recent work, however (not to mention the earlier work) there was a shortage of accurate chemical analyses, as opposed to density measurements, of these DNAs. It therefore seemed desirable to, once again, measure the base composition of the chloroplast and nuclear DNAs of a suitable plant. The species we chose was *Phaseolus vulgaris*, partly because there was buoyant density data available for comparison purposes. The method used for analysis was a much improved version [21] of the procedure used in the original broad bean DNA analyses [2]. With the new method, which as before involves measuring the adenine/guanine ratio, it is possible to determine the base composition of DNA substantially more accurately than by any of the other methods in common use. The analyses [22] indicated 37.7% GC content for the chloroplast DNA (table 2) which is in very good agreement with my original value of 37.4% GC obtained with broad bean chloroplast DNA [1,2]. In French bean, unlike the broad bean, the nuclear DNA has a base composition (37.3% GC) not signifi-

Table 2
Base compositions of french bean chloroplast and nuclear DNAs [22]

Chloroplast DNA		Nuclear DNA	
Adenine/ guanine molar ratio	%GC calc. from A/G ratio	Adenine/ guanine molar ratio	%GC calc. from A/G ratio
1.668	37.47	1.703	37.00
1.622	38.14	1.686	37.23
1.637	37.92	1.665	35.52
1.673	37.40	1.652	37.71
–	–	1.685	37.24
Mean value			
1.650	37.73	1.678	37.34

cantly different from that of the chloroplast DNA. Proof that the chloroplast DNA analyzed was not merely contaminating nuclear DNA was provided by the fact that the chloroplast DNA – unlike the nuclear DNA – did not contain 5-methylcytosine. The average buoyant density of *P. vulgaris* chloroplast DNA is, as we have seen, 1.696–1.697 g cm^{-3}. This would indicate a GC content of 37–38%, and so it turns out that the chemical data and the density data are in very good agreement.

Let us now look at all the data together. Table 3 contains all the results which I consider to be essentially correct. I have listed, in addition to accurate chemically-determined base compositions for two of the species, the observed buoyant densities for nuclear and chloroplast DNAs, together with the GC

Table 3
Buoyant densities and base compositions of higher plant chloroplast and nuclear DNAs

Plant species	Nuclear DNA		Chloroplast DNA		Reference
	Density in CsCl (g cm^{-3})	%GC [a,b] content	Density in CsCl (g cm^{-3})	%GC [a] content	
Broad bean	–	39.4	–	37.4	[1,2]
Broad bean	1.695	38.8	1.697	37.8	[14,15]
Broad bean	1.696	39.8	1.696	36.8	[20]
Turnip	1.694 [c]	37.8	1.695	35.7	[10]
Spinach	1.694	37.8	1.696	36.8	[16,25]
Spinach	1.694	37.8	1.697	37.8	[15]
Lettuce	1.694	37.8	1.697	37.8	[15]
Sweet pea	1.695	38.8	1.697	37.8	[15]
Tobacco	1.697	40.8	1.697	37.8	[16]
Tobacco	1.697	40.8	1.700	40.8	[24]
Tobacco	1.695	38.8	1.697 (2)	38.0	[25]
Snapdragon	1.689	32.7	1.698	38.8	[13]
French bean	1.697	40.8	1.696	36.7	[12]
French bean	1.696–1.697	40	1.696–1.697	37–38	[19]
French bean	–	37.3	–	37.7	[22]
Swiss chard	1.694	37.8	1.696 (5)	37.3	[25]
Onion	1.691	34.7	1.696 (4)	37.2	[25]
Wheat	1.702	47.0	1.697 (6)	38.4	[25]

a) Except for the two chemical determinations (broad bean – line 1, French bean – line 15) all GC contents have been calculated from the buoyant densities in CsCl.

b) In order to correct for the lowering of the buoyant densities of the nuclear DNAs caused by methylation of some of the cytosine residues [23], a value of 0.003 g cm^{-3} (0.004 g cm^{-3} in the case of wheat) was added to all the observed buoyant densities of the nuclear DNAs before calculating the GC content.

c) Since the turnip nuclear DNA consisted of a main band, density 1.692 g cm^{-3} with a marked shoulder at 1.700 g cm^{-3}, it was necessary to obtain this approximate value for the average density by a rough calculation based on examination of the published UV tracing.

contents calculated from these values. It may seem puzzling at first that in some cases the table gives the same densities for chloroplast and nuclear DNAs, but different base compositions. This is because higher plant nuclear DNAs invariably have a proportion – of the order of one quarter – of their cytosine residues methylated, i.e., as we have seen, the nuclear DNAs contain 5-methylcytosine. Since it seemed likely that the presence of these methyl groups might affect the buoyant density of DNA, I carried out a calculation [23] of the effect of a given degree of methylation on the volume and on the weight of a hydrated molecule of DNA in CsCl solution. The calculations led to the conclusion that methylation lowers buoyant density by an amount which, for most higher plant DNAs will be about $0.003-0.004$ g cm^{-3}, the precise effect depending on the base composition (for a given proportion of methylated cytosines, the effect increases with GC content). This is quite significant since it introduces errors of $3-4\%$ GC in determinations of base composition. To allow for this effect, I added 0.003 g cm^{-3} to each of the nuclear DNA buoyant densities before calculating the GC content. (In the case of wheat, 0.004 g cm^{-3} was added.)

Before we attempt to see the pattern running through all these data, it is instructive to recall the earlier stages ideas on chloroplast DNA have passed through. In the 1963–67 period when various workers thought they were finding high-density, high-GC, chloroplast DNAs, the results seemed to be leading people to the generalization (despite the contrary evidence from the broad bean work) that chloroplast DNA in any higher plant had a GC content about 10% higher than that of the corresponding nuclear DNA. However, when, in the 1967–69 period it began to look as if these high density satellites were not the chloroplast DNA at all, and that it was really the DNA of similar density to the nuclear DNA which had really been the chloroplast DNA all along, the idea seemed to be emerging that perhaps it was generally true that in higher plants chloroplast DNA had a density and GC content very close to, although not always quite identical with, those of the corresponding nuclear DNA. However, even though our own *Phaseolus* data were in particularly good agreement with this generalization, it seemed to us that the results were possibly misleading

because it just happened that nearly all the plants which had been looked at had nuclear DNA GC contents in the region of 37–40%. It seemed to us that the chloroplast DNA base composition was more constant from one species to another than was the nuclear DNA base composition. As can be seen by examining lines 1–15 of table 3, all but one of the (accepted) values for chloroplast DNA GC content up to that time lay between 35.7% and 38.8%. The odd one out is the result of Tewari and Wildman: their most recent value [24] for tobacco chloroplast DNA density – 1.700 g cm^{-3} (indicating 40.8% GC) – is lower than their original [11] estimate, of 1.702 g cm^{-3}, but is still higher than the value of 1.697 g cm^{-3} obtained by others [16,25]. On the basis of these data we therefore put forward the hypothesis [22] that chloroplast DNA in all higher plants has a rather constant base composition in the region of 37–38% GC. We further suggested that this hypothesis should be tested by studying plants with a widely varying nuclear DNA GC content. This has now been done. Wells and Ingle [25] have now shown (lines 17 and 18, table 3) that in onion, *Allium cepa,* (which has a low nuclear DNA GC content) the chloroplast DNA is of substantially higher density than the nuclear DNA, whereas in wheat, *Triticum vulgare* (which has a high nuclear DNA GC content) the chloroplast DNA is significantly less dense than the nuclear DNA. The chloroplast DNAs have much the same density in both plants. In this very recent work Wells and Ingle have also studied Swiss chard and find, contrary to an earlier report [5] that the chloroplast DNA in this species too, consists of a single band in CsCl, of density 1.695 g cm^{-3}.

Thus, the data on wheat and onion are in agreement with our hypothesis as to the relative constancy of chloroplast DNA density and base composition throughout the higher plants, although further evidence from many more species is desirable to test the hypothesis thoroughly. In summary, then, the best evidence available at the moment leads us to the generalization that higher plant chloroplast DNA has a base composition of about $37.5 \pm 1\%$ GC, and a buoyant density in CsCl of about 1.697 ± 0.001 g cm^{-3}. The chloroplast DNA is more constant in composition and density than nuclear DNA which can vary markedly from one species to another.

What went wrong?

Assuming that our new generalization is correct, what were the other, dense, DNA satellites that people were finding in their chloroplast preparations in the 1963–67 period? The very dense satellite, density 1.719 g cm^{-3}, observed by Chun et al. [3], but by no other workers, can perhaps most plausibly be attributed to bacterial contamination. Unlike most researchers in this field, Chun et al. used commercially obtained spinach and beet leaves, instead of growing their own. But App and Jagendorf [26] reported in the same year that commercially obtained spinach leaves, which may well have been stored for some time, are likely to be extensively contaminated with bacteria. However, the other satellite found by Chun et al. density 1.705 g cm^{-3} or thereabouts, was also observed by several other groups [5–9], using less suspect plant material, and so is less likely to be due to bacterial contamination. I believe the most plausible explanation to be that this was mitochondrial DNA. Suyama and Bonner [10] reported that the mitochondrial DNA from four different plant species – mung bean (*Phaseolus aureus*), turnip, sweet potato (*Ipomoea batatas*) and onion – had a density in CsCl of 1.706 g cm^{-3}. Other workers have found mitochondrial DNA to have a density in CsCl of 1.707 g cm^{-3} in French bean [19], 1.705 g cm^{-3} in broad bean [15], 1.706 g cm^{-3} in lettuce [15], 1.706 g cm^{-3} in onion [25] and 1.705 g cm^{-3} in wheat [25].

It may seem unlikely that there could be such extensive contamination of chloroplast fractions by mitochondria. However, this assumption, although very often true, is by no means always true. To give an example, we have recently found that in homogenates of mustard (*Sinapis alba*) cotyledons, most of the mitochondria (as judged by cytochrome oxidase activity) sediment together with the chloroplasts at 800 g. Another factor which might possibly give rise to DNA with a significant proportion of mitochondrial DNA, is the probably greater sensitivity of DNA in chloroplasts (which so readily rupture their outer membrane) to endogenous nucleases, than DNA in mitochondria. In this way, although there might be only a small amount of mitochondrial contamination to begin with, there could, during the isolation procedure, be an increase in the proportion of mitochondrial DNA resulting from a selective loss of chloroplast DNA.

Other possibilities that should be borne in mind when seeking for an explanation for the presence of dense satellites in DNA preparations, are that there might be bands of partially or completely denatured DNA, or DNA-RNA hybrids, or bands of polysaccharide [29].

It is perhaps also relevant to ask ourselves just what we mean when we say that a DNA preparation contains a satellite band. The implicit meaning often attached to such a statement is that the DNA preparation in question consists of two distinct populations of DNA molecules, each with the classical Gaussian distribution of densities (and GC contents) around its mean density. The two distributions will be considered as overlapping to a greater or lesser extent, accordingly as the 'satellite' is seen as a shoulder on the main band or as a distinct peak. However, there is no obvious reason why a particular kind of DNA (especially DNA from the nucleus of a eukaryotic organism, with all its different chromosomes, and perhaps chloroplast DNA too) should have a simple Gaussian distribution of densities. Thus if a DNA preparation is seen to have an irregularity such as a shoulder, in CsCl, this should not automatically be attributed to the presence of a 'satellite' DNA (with the likelihood, often implied, of a different origin within the cell). It may just be that that particular population of DNA molecules has a non-Gaussian distribution of densities.

Is the matter finally closed?

I hope that by now I have made it clear that I support the view that the typical higher plant chloroplast DNA has a GC content of about 37–38%, and a density in the region of 1.697 g cm^{-3}. Nevertheless, tempting though the prospect is, it is perhaps too early to allow ourselves to stop worrying about this problem. Although an explanation of the 1.705 g cm^{-3} satellite in terms of mitochondrial DNA is entirely feasible, it is still a little puzzling that some workers in the 1963–67 period should have succeeded in achieving quite so much mitochondrial contamination of their chloroplast preparations. Also a very recent paper by Bard and Gordon [27] claims that there are two distinct chloroplast DNAs. These workers now accept, contrary to the earlier opinion from the same laboratory [9], and in agreement with more recent views, that the major chloroplast DNA has a density of 1.696 g cm^{-3} and a GC content in the 36–39% region. But their chloroplast DNA preparations show, in CsCl, another band comprising 35% of the total, of density 1.706 g cm^{-3}, and about 46% GC content, and this too they regard as an authentic chloroplast DNA. However, an examination of the methods used suggests alternative explanations. Contrary to normal practice, no attempt was made to separate the chloroplasts from mitochondria by a preliminary low speed centrifugation before the final purification. A whole spinach leaf homogenate (which would contain mitochondria as well as chloroplasts) was placed on a sucrose gradient and centrifuged at high speed in a zonal rotor. The band containing the chloroplasts comprised one sixth of the actual gradient and was assumed not to be contaminated with mitochondria. In view of the fact that mitochondria as well as chloroplasts were placed on the gradient, some evidence for the absence of mitochondria from the chloroplast fraction would be desirable. A further criticism that can be levelled at this work is that commercially obtained spinach was used, with a consequent danger of bacterial contamination. In short, I believe that this 1.706 g cm^{-3} satellite too can be explained in terms of contamination, most probably by mitochondria. Bard and Gordon could best clarify the matter by repeating the experiment with young leaves, freshly harvested (i.e., non-commercial), carrying out a preliminary partial purification of the chloroplasts by low speed centrifugation prior to the density gradient run, and by measuring (using cytochrome oxidase activity) the proportion of the mitochondria that end up in the chloroplast fraction.

Another possible exception to the apparently uniform picture that has recently been building up is the report by Mache and Waygood [28] that wheat (*Triticum aestivum*) chloroplasts contain a single DNA species, density 1.714 g cm^{-3}, compared to 1.702 g cm^{-3} for the nuclear DNA. This claim is in direct contradiction to the report by Wells and Ingle [25] that wheat chloroplast DNA has a density of 1.697(6) g cm^{-3}. The fact that the 1.697(6) g cm^{-3} value is in good agreement with the values obtained for so many other plants, one of which (onion) is also a monocotyledon, rather tends to support the Wells and Ingle data against the results of Mache and Waygood. One possibility is that the 1.714 g cm^{-3} satellite is a band of polysaccharide; the danger of confusing such substances with DNA in CsCl gradients has been well documented by Edelman et al. [29]. However, the findings of Mache and Waygood can not be dismissed, and further work on wheat, and preferably other members of the *Graminae* too, would be highly desirable.

In summary, although I believe the simple generalization about higher plant chloroplast DNA given at the beginning of this section to be probably true, there are still some unexplained flies in the ointment, and while we may be surprised, we should not be astonished, if it turns out either to be only partially true for all plants (e.g. if there really is another chloroplast DNA) or to be entirely untrue for some plants (e.g. if the chloroplast DNA of some species is really quite different, as has been suggested for wheat).

Chloroplast DNA in the algae

The algal field was to begin with relatively free of disagreement about which DNA is really the chloroplast DNA. However, some confusion has now arisen as to the identity of chloroplast DNA in *Chlorella* and also in *Acetabularia*. So far as *Chlorella* is concerned this problem may be said to have started when Iwamura [30] observed that the DNA in a homogenate of *C. ellipsoidea* was mainly found in the 20,000 g pellet and the 105,000 g supernatant, and that there appeared to be a difference in base composition between the DNAs of the two fractions, the supernatant DNA having a higher GC content. Iwamura concluded that there are two kinds of DNA in *Chlorella,* and that the pellet DNA might possibly be a chloroplast DNA and the supernatant DNA might be the nuclear DNA. With hindsight we can now say that the 20,000 g pellet probably contained chloroplast and mitochondrial DNA, in addition to nuclear fragments. It is rather surprising that the 105,000 g supernatant contained so much DNA; this may well have consisted largely of very small fragments of nuclear DNA, but the possibility of the presence of some chloroplast and/or mitochondrial DNA in this fraction can not be a priori eliminated with certainty.

The next stage was the report by Chun et al. [3] that the whole cell DNAs of *Chlamydomonas reinhardi* and *Chlorella ellipsoidea* each contained, in CsCl, in addition to the main — presumed nuclear — DNA bands of densities 1.723 g cm^{-3} and 1.716 g cm^{-3}, respectively, a small satellite band of density 1.695 g cm^{-3}. They had no evidence as to the cellular location of this band, but shortly afterwards Sager and Ishida [31] were able to show that isolated chloroplasts of *Chlamydomonas reinhardi* were greatly enriched in this satellite DNA, indicating that it originated in the chloroplasts. Chemical analysis indicated about 39% GC content for the chloroplast DNA, compared to 62% GC for the major (presumed nuclear) DNA.

Iwamura and Kuwashima [32] have now attempted to more rigorously determine the cellular location of the different DNAs in *Chlorella ellipsoidea*. They were able to show that the nuclei contained only the 1.716 g cm^{-3} component. In the chloroplast fraction, however, purified by density gradient centrifugation, two DNA bands, a large one of density 1.717 g cm^{-3} and a small one of density 1.692 g cm^{-3}, were observed. Iwamura and Kuwashima concluded that not only was the 1.692 g cm^{-3} band a chloroplast DNA (presumed to be identical with the 1.695 g cm^{-3} satellite found by Chun et al.) but so was the 1.717 g cm^{-3} band. They rejected the superficially more obvious possibility that the 1.717 g cm^3 band was merely contaminating nuclear DNA. One of their main pieces of evidence for this was that when they deliberately contaminated the chloroplast preparations with a radioactive 100,000 g supernatant of *Chlorella* homogenate (which would presumably contain nuclear DNA), subsequent purification of the chloroplasts removed 97—98% of the contaminating supernatant DNA. However, it must be pointed out that while this establishes that the chloroplasts are not likely to be contaminated with the very small nuclear fragments such as occur in the 100,000 g supernatant, it does not show that there has been removal of the large nuclear fragments that would have been present in the initial 20,000 g pellet from which the chloroplasts were normally purified. Therefore, while I believe that Iwamura and Kuwashima have now presented good evidence that the 1.692—1.695 g cm^{-3} component is really a chloroplast DNA, I do not consider that their evidence for

the 1.717 g cm^{-3} DNA being also a chloroplast DNA is satisfactory. Valuable controls here would be to isolate the 1.717 g cm^{-3} component from the chloroplast fraction, and show that it differed from the nuclear DNA in being renaturable and having no 5-methylcytosine. Until such evidence is available, the supposition that the 1.717 g cm^{-3} DNA in the chloroplast fraction is contaminating nuclear DNA, will continue to seem marginally more plausible.

In the case of *Acetabularia* too there is a lack of unanimity concerning the identification of chloroplast DNA. Gibor [33] reported that chloroplasts from bacteria-free, enucleated cells of *Acetabularia mediterranea* contained DNA which gave a single band of density 1.695 g cm^{-3} in CsCl. Green et al. [34] however, claim that DNA from chloroplasts of *A. mediterranea* has a density in CsCl of 1.704 g cm^{-3}. The published data do not permit a conclusion as to which of these two claims is nearer the truth: here again is a problem on which more work is needed.

It is something of a relief to note that in the case of *Euglena gracilis,* the only other alga whose chloroplast DNA has been characterized, there is a gratifying conformity of opinion amongst the different workers as to which DNA is the chloroplast DNA. It was shown independently in three laboratories in 1964 that the chloroplasts contain a DNA with a density in CsCl of about 1.685 g cm^{-3} and 25% GC content, compared to 1.707 g cm^{-3} and 52% GC content for the main cellular (presumed nuclear) DNA [17,18,35], in agreement with an earlier prediction [36] based on the disappearance of a DNA band, density 1.685 g cm^{-3}, from the whole cell DNA of a mutant which had lost the ability to form chloroplasts.

It is interesting to note that the chloroplast DNA in *Chlorella* and *Chlamydomonas* is rather close in density and hence base composition to the chloroplast DNA of higher plants (1.692—1.695 g cm^{-3} compared to 1.696—1.698 g cm^{-3}). This may of course be pure coincidence, but it could be relevant that both these algae are members of the *Chlorophyta,* and according to one school of taxonomic thought, the higher plants could be included, together with the green algae, in the *Chlorophyta.* The *Euglena* chloroplast DNA, on the other hand has a distinctly lower density (1.685 g cm^{-3}) and a correspondingly lower GC content (25% as opposed to 37%). This need not surprise us. It fits in perfectly well with the supposition, that the *Euglenophyta* are taxonomically, and evolutionarily, farther removed from the green algae and higher plants than these two latter groups are from each other. Thus we have the possibility that the chloroplast DNAs are very much the same (although not identical) not only throughout the higher plants but in the green algae as well. In the other major groups of algae, however, − the *Euglenophyta, Rhodophyta, Cryptophyta, Phaeophyta, Pyrrophyta* and *Chrysophyta* − the chloroplast DNA may have quite a different base composition and density, and perhaps information content. Clearly, then, there is a great need for research on the chloroplast DNAs of many different kinds of algae, from all the different major taxonomic groups. However, in view of all the troubles that workers in the higher plant chloroplast DNA field have encountered, it is to be hoped that anyone wishing to carry out similar studies on these algae will make quite sure that it is really chloroplast DNA that has been isolated before describing it in the literature.

References

[1] J.T.O. Kirk, Biochem. J. 88 (1963) 45P.

[2] J.T.O. Kirk, Biochim. Biophys. Acta 76 (1963) 417.

[3] E.H.L. Chun, M.H. Vaughan and A. Rich, J. Mol. Biol. 7 (1963) 130.

[4] C.L. Schildkraut, J. Marmur and P. Doty, J. Mol. Biol. 4 (1962) 430.

[5] N. Kislev, H. Swift and L. Bogorad, J. Cell. Biol. 25 (1965) 327.

[6] W.S. Shipp, F.J. Kieras and R. Haselkorn, Proc. Natl. Acad. Sci. U.S. 54 (1965) 207.

[7] B.R. Green and M.P. Gordon, Science 152 (1966) 1071.

[8] T. Nilsson-Tillgren, D. Luck and D. von Wettstein, quoted by D. von Wettstein, in: Harvesting the Sun, eds. A. San Pietro, F.A. Greer and T.J. Army (Academic Press, New York, 1967) p. 153.

[9] B.R. Green and M.P. Gordon, Biochim. Biophys. Acta 145 (1967) 378.

[10] Y. Suyama and W.D. Bonner, Plant Physiol. 41 (1966) 383.

[11] K.K. Tewari and S.G. Wildman, Science 153 (1966) 1269.

[12] T.G. Beridze, M.S. Odintsova and N.M . Sissakian, Mol. Biologiya 1 (1967) 142.

[13] H.G. Ruppel, Z. Naturforschg. 22b (1967) 1068.

[14] R. Wells and M. Birnstiel, Biochem. J. 105 (1967) 53P.

[15] R. Wells and M. Birnstiel, Biochem. J. 112 (1969) 777.

[16] P.R. Whitfeld and D. Spencer, Biochem. Biophys. Acta 157 (1968) 333.

[17] D.S. Ray and P.C. Hanawalt, J. Mol. Biol. 9 (1964) 812.

[18] G. Brawerman and J. Eisenstadt, Biochim. Biophys. Acta 91 (1964) 477.

[19] D.R. Wolstenholme and N.J. Gross, Proc. Natl. Acad. Sci. U.S. 61 (1968) 245.

[20] S.D. Kung and J.P. Williams, Biochim. Biophys. Acta 169 (1968) 265.

[21] J.T.O. Kirk, Biochem. J. 105 (1967) 673.

[22] R. Baxter and J.T.O. Kirk, Nature 222 (1969) 272.

[23] J.T.O. Kirk, J. Mol. Biol. 28 (1967) 171.

[24] K.K. Tewari and S.G. Wildman, in: The Development and Interrelationship of Cell Organelles, symp. Soc. Exptl. Biol. 24 (1970) in press.

[25] R. Wells and J. Ingle (1970), in preparation; see also J. Ingle, in: Symp. Soc. Exptl. Biol. 24 (1970), in press.

[26] A.A. App and A.T. Jagendorf, Plant Physiol. 39 (1964) 772.

[27] S.A. Bard and M.P. Gordon, Plant Physiol. 44 (1969) 377.

[28] R. Mache and E.R. Waygood, FEBS Letters 3 (1969) 89.

[29] M. Edelman, D. Swinton, J.A. Schiff, H.T. Epstein and B. Zeldin, Bact. Revs. 31 (1967) 315.

[30] T. Iwamura, Biochim. Biophys. Acta 42 (1960) 161.

[31] R. Sager and M.R. Ishida, Proc. Natl. Acad. Sci. U.S. 50 (1963) 725.

[32] T. Iwamura and S. Kuwashima, Biochim. Biophys. Acta 174 (1969) 330.

[33] A. Gibor, in: Biochemistry of Chloroplasts, ed. T.W. Goodwin (Academic Press, London, 1967) vol. II, p. 321.

[34] B. Green, V. Heilporn, S. Limbosch, M. Boloukhere and J. Brachet, Proc. Natl. Acad. Sci. U.S. 58 (1967) 1351.

[35] M. Edelman, C.A. Cowan, H.T. Epstein and J.A. Schiff, Proc. Natl. Acad. Sci. U.S. 52 (1964) 1214.

[36] J. Leff, M. Mandel, H.T. Epstein and J.A. Schiff, Biochem. Biophys. Res. Commun. 13 (1963) 126.

CHARACTERIZATION OF EUGLENA GRACILIS
CHLOROPLAST SINGLE STRAND DNA

Erhard STUTZ

Department of Biological Sciences, Northwestern University,
Evanston, Illinois, USA

Chloroplasts and mitochondria from lower and higher organisms have their own sets of double-stranded DNA. According to our present understanding, such organellar DNA encodes some, but not all the essential information necessary for growth and development of the organelle. Nuclear genes definitely control the synthesis of important organellar constituents. The obvious problem is to understand the complex interaction of the various gene pools within the eucaryotic cell.

Chloroplasts of lower and higher plants contain between 10^{-16} g [1] and 10^{-14} g [2,3] DNA. Unless such DNA is reiterated in base sequences, the information content of the average chloroplast is similar to that of a bacterial cell and considerably higher than that of a mitochondrion. However, it was recently shown that chloroplast DNA from lettuce is about 20 times less complex than *E. coli* DNA and extensive base sequence reiteration was postulated [4].

Chloroplast DNA isolated from higher plants is rather similar to the corresponding nuclear DNA in base ratio and therefore neutral buoyant density in CsCl (for reference see paper from J.T.O. Kirk, this volume). This brings about analytical and preparative problems, e.g. cross-contamination is not easily detected. In contrast, *Euglena gracilis* chloroplast DNA (density, 1.685 g/cc, neutral CsCl) differs sufficiently from the nuclear DNA (density, 1.708 g/cc, neutral CsCl) to be easily separated and purified in a preparative CsCl density gradient [2,3,5].

One important function of chloroplast DNA seems to be the coding for chloroplast specific ribosomal RNA (rRNA). *Euglena* chloroplast rRNA differs from the cytoplasmic rRNA in molecular weight [6,7] and base composition. Table 1 summarizes recent data. A

Table 1

Chemical composition of *Euglena* chloroplast and cytoplasmic rRNA [a]

Nucleo-tide	Cytoplasmic rRNA mole %		Chloroplast rRNA mole %	
	X	$\sigma(n=6)$	X	$\sigma(n=5)$
C	24.1	0.7	20.4	1.6
A	21.6	0.6	25.5	1.3
G	32.0	1.3	31.1	0.8
U	22.3	0.8	23.0	0.7

[a] See ref. [6].

necessary, though not conclusive experiment to demonstrate the existence of rRNA-cistrons on chloroplast DNA is to measure the hybridization specificity between rRNA and the DNA. During such experiments we noted that alkaline-denatured *Euglena* chloroplast DNA yields two bands in an alkaline CsCl density gradient. The observation seemed worthwhile to pursue especially in view of recent reports about successful separation of complementary single strands from mitochondrial DNA in alkaline CsCl density gradients [8,9].

Some characteristics of *Euglena* chloroplast single strand DNA

A sufficient bias in the content of the titratable bases G + T between the complementary strands of a DNA results in a proportional difference in alkaline buoyant density [10]. In fig. 1 we show a few

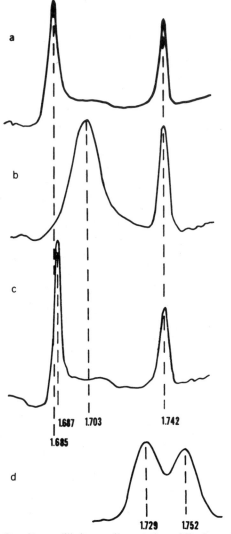

Fig. 1. Density equilibrium sedimentation of *Euglena* chloroplast DNA in CsCl. (a) 2 μg native DNA; (b) 5 μg DNA denatured in 0.13 ml 0.1 × SSC (0.15 M NaCl, 0.015 M Na-citrate) adjusted to pH 12, 10 min, room temperature, chilled in ice and rapidly neutralized; (c) 3 μg DNA renatured in 0.13 ml, 2 × SSC, pH 7, 1 hr, 63°C; (d) 5 μg DNA denatured in 0.13 ml 0.1 × SSC adjusted to pH 12, 10 min, room temperature. Centrifugation: Spinco Model E, An-D rotor, 44,770 rpm, 18–20 hr, 20°C; tracings a,b,c are aligned with respect to the reference DNA peak, ^{15}N-DNA of *Ps. aeruginosa* (density 1.742 g/cc). The densities in the neutral CsCl gradients (a,b,c) are calculated relative to the reference according to Sueoka [20]. The refractive indices of the neutral CsCl solution were kept in the range of 1.4000 ± 0.002. The densities of the samples in the alkaline gradient (d) were calculated according to Vinograd et al. [21].

banding characteristics of *Euglena* chloroplast DNA. A single, sharp and symmetric band is obtained when native DNA is equilibrated under neutral conditions. Alkaline denatured DNA yields a broadened band with an average neutral buoyant density of 1.703 g/cc (fig. 1b). Heat denatured chloroplast DNA gives a double band (not shown) with neutral buoyant densities of 1.704 g/cc and 1.696 g/cc respectively. The two bands are not equally strong. This observation shall not be discussed any further in this paper. Alkaline-denatured DNA, equilibrated in an alkaline CsCl gradient, reproducibly gives two separate bands (fig. 1d). The denaturing process is reversible to approximately 65–70% as measured by the change in hyperchromicity during renaturing. This considerable renaturing capacity is also reflected in the shifting of the neutral buoyant density back to 1.687 g/cc which is close to the density of the native DNA (fig. 1c).

Exploiting this analytical observation, we prepared a light and a heavy single-strand DNA fraction following basically the procedure of Flamm et al. [11]. A detailed description of these experiments is given elsewhere [12]. The two single-strand DNA preparations were characterized as shown in the graphs of fig. 2. The data may be summarized as follows: the heavy single-strand DNA fraction which was shown to have an average alkaline buoyant density of 1.752 g/cc equilibrates in a neutral CsCl gradient at a mean density of 1.705 g/cc (fig. 2a). This DNA, when incubated under renaturing conditions, slightly changes its overall density; the respective tracing (fig. 2b) becomes asymmetric with a shoulder on the lighter side. The light fraction (alkaline buoyant density, 1.729 g/cc) bands at a neutral buoyant density of 1.698 g/cc (fig. 2c). The neutral buoyant density values of the two single-strand DNA fractions are bracketing the value of total denatured DNA (1.703 g/cc). Renatured light single-strand DNA has an average buoyant density of 1.693 g/cc (fig. 2d). Neither of the single-strand DNA preparations displays a significant renaturing capacity, taking the density shift as the criterium (other parameters such as shift in hyperchromicity have not been measured). However, when equimolar amounts of light and heavy single strand DNA are co-renatured, a new type of DNA results with a density of 1.687 g/cc, which is close to the density of the native DNA and identical with the density of directly renatured total chloro-

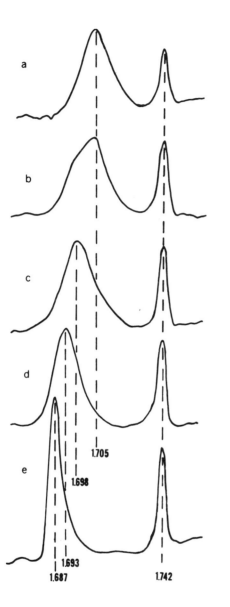

plast DNA. The tracing of this co-renatured DNA preparation (fig. 2e) is slightly skewed towards the heavier side. From these results we conclude that the heavy and light fraction contain single-strand DNA (fragmented) essentially complementary to each other.

The relationship of chloroplast single-strand DNA to chloroplast rRNA

Using the filter technique according to Gillespie and Spiegelman [13] we studied the affinity of chloroplast rRNA to the various DNA preparations. *Euglena* nuclear DNA and *E. coli* DNA were included and table 2 shows hybridization percentages expressed in μg rRNA bound per 100 μg denatured DNA. On the average, 1.2% of the chloroplast DNA hybridizes with rRNA. This value is close to earlier reports from Scott and Smillie [14]. Annealing experiments between chloroplast rRNA and various fractions of chloroplast DNA are given in table 3. Under these experimental conditions, the heavy single-strand DNA binds approximately nine times more rRNA than the light single-strand DNA fraction. The counts from the heavy fraction and light fraction together match reasonably well the counts retained by total denatured chloroplast DNA, indicating that we did not have significant or selective losses during the separation steps. We must add that saturation curves for chloroplast total denatured DNA had previously been made and the experiments presented here were run under saturation conditions. From these data we may tentatively conclude that the heavy single-strand DNA fraction contains the majority of base sequences coding for 23S/16S rRNA.

Fig. 2. Density equilibrium sedimentation of fractionated *Euglena* chloroplast DNA. (a) 6 μg heavy fraction DNA; incubated in 0.11 ml 0.1 × SSC adjusted to pH 12, 10 min, room temperature, chilled in ice, rapidly neutralized; (b) 6 μg heavy fraction DNA incubated in 0.13 ml 2 × SSC, pH 7, 6 hr, 63°C; (c) 6 μg light fraction DNA, treated as mentioned sub a); (d) 6 μg light fraction DNA, treated as mentioned sub b); (e) 3 μg heavy fraction plus 3 μg light fraction DNA in 0.13 ml 2 × SSC, pH 7, jointly incubated for 6 hr, 63°C. Centrifugation, DNA reference and density calculation as mentioned in legend to fig. 1.

Table 2

Hybridization between *Euglena* chloroplast rRNA and various DNAs

Source of DNA	% hybridization
Euglena chloroplast	1.2
Euglena nuclei	0.2
E. coli	0.07

Incubation: 16 hr, 64°C, 2 × SSC (SSC = 0.15 M NaCl, 0.015 M Na-citrate).

Table 3
Hybridization between *Euglena* chloroplast rRNA and different fractions of *Euglena* chloroplast DNA

DNA fraction	µg DNA filter	^{32}P-rRNA [a] µg	cpm/filter charged	blank	Per cent hybridization
Chloroplast total	22	8	402	25	1.2
			364	21	1.0
		16	407	19	1.2
			407	21	1.2
Chloroplast heavy	10–11	8	347	14	2.3
			302	17	1.9
		16	335	17	2.2
			312	19	2.0
Chloroplast light	9–10	8	39	14	0.1
			45	16	0.2
		16	45	18	0.2
			38	20	0.1

[a] 145 cpm/0.1 µg chloroplast rRNA.
[b] Incubation conditions as in table 2.

Discussion

Separation of individual intact complementary DNA strands has been reported in several instances, e.g. for phage DNA [15], mitochondrial DNA [9], or nuclear satellite DNA [16]. Without exception so far, it is the heavier of the two complementary strands (higher ratio in pyrimidines) or the heavier strand segment at any given region which is transcribed. It was postulated that pyrimidine rich clusters are involved in the transcription regulation [17] and special proteins (e.g. σ-factor) seem to be involved in the initiation step [18]. The strand separation achieved so far with *Euglena* chloroplast DNA is still rather unspecific but better standardized and more sophisticated separation techniques might lead in the future to very characteristic single-strand DNA products.

In evaluating our data, we have to keep the following points in mind: 1) *Euglena* chloroplasts contain approximately 6×10^9 daltons of double-stranded DNA [2,3]. The average molecular weight of our purified native DNA preparations (table 4) is in the range of 10×10^6 daltons and the DNA must be considerably degraded, even when we assume that chloroplasts contain several 'chromosomes'. 2) The

rRNA-cistrons per se must have a G + C content of 52% (see base ratios in table 1) which is quite different from the base ratio of the total DNA (G + C = 26%). Such extensive fragmentation and remarkable base sequence heterogeneity bring about specific preparative problems. This is demonstrated by data given in table 4. We hybridized 23S/16S rRNA with nuclear DNAs (density 1.708 g/cc, neutral CsCl) which we had isolated either from a nuclei pellet

Table 4
Hybridization between *Euglena* chloroplast rRNA and different types of *Euglena* DNA

DNA source	µg DNA filter	^{32}P-rRNA [a] µg	cpm/filter [b] charged	blank	$S_{20,w}$
Chloroplast	20–21	16	516	43	24.6
			563	37	
'Nuclear'	20–21	16	199	15	28.0
			189	15	
Nuclear	20–21	16	101	18	28.1
			112	33	

[a] 243 cpm/0.1 µg rRNA.
[b] Incubation conditions as in table 2.

heavily contaminated by chloroplasts ('nuclear') or from pure nuclei (nuclear). Otherwise both DNA preparations were processed identically. The two preparations are identical with respect to neutral buoyant density and $S_{20,w}$-value, but they bind different amounts of rRNA. We believe that the higher hybridization percentage of 'nuclear' DNA is due to contamination by G + C-rich fragments, rRNA-cistrons being part of it. If this assumption is correct, our hybridization results with chloroplast rRNA and chloroplast DNA represent minimal values and scanning of the whole density gradient will be necessary to get the maximum hybridization percentage.

One faces a somewhat similar problem when isolating and purifying the single-strand DNA from the alkaline density gradient. First, hidden single-strand nicks become apparent and the molecular weight in the average drops by a factor of ten. Second, the separating principle is the ratio of G+T/C+A. The G+T content of the sense-strand complementary to the 23S/16S rRNA is 45% and the respective anti-strand sequence therefore 55%. Unless T-rich clusters are contiguous to these cistrons the coding strand fragment bands under alkaline conditions with a lower buoyant density than the anti-strand fragment. Unfavorable cuts in the two strands may be the reason for the 10% rRNA bound to the lighter fraction. It is noteworthy that in neutral CsCl, the coding base sequence for rRNA has the higher buoyant density than the corresponding anti-strand. Adjacent T-rich clusters would of course enhance that difference.

Our hybridization and cross-hybridization experiments with *Euglena* chloroplast rRNA and chloroplast and nuclear DNA do not allow answering the important question whether the nuclear DNA has separate cistrons coding for 23S/16S RNA, as was suggested for tobacco nuclear DNA [19]. We have done some preliminary DNA/RNA hybrid melting studies and competition experiments and according to these results we tentatively conclude that no separate 23S/16S RNA cistrons exist on nuclear DNA. The affinity found between chloroplast rRNA and nuclear DNA and vice versa is of lower specificity than the affinity to the homologous rRNA and cytoplasmic and chloroplast rRNA compete for the same sites, which is to be expected if remnant similarities in base sequences still persist.

Acknowledgement

The *P. aeruginosa* ^{15}N-DNA was a gift from Dr. N. Welker. Dr. J.R. Rawson helped in the preparation of RNA and DNA. I thank Mrs. K. Patel for skillful technical help. This investigation was supported by Biomedical Sciences Support Grant FR–07028 from the National Institutes of Health. I am grateful to Dr. H. Noll for additional support.

References

[1] A. Gibor and M. Izawa, Proc. Natl. Acad. Sci. U.S. 50 (1963) 1164.
[2] G. Brawerman and J.M. Eisenstadt, Biochim. Biophys. Acta 91 (1964) 477.
[3] M. Edelman, C.A. Cowan, H.T. Epstein and J.A. Schiff, Proc. Natl. Acad. Sci. U.S. 52 (1964) 1214.
[4] R. Wells and M. Birnstiel, Biochem. J. 112 (1969) 777.
[5] D.S. Ray and P.C. Hanawalt, J. Mol. Biol. 2 (1969) 339.
[6] J.R. Rawson and E. Stutz, Biochim. Biophys. Acta 190 (1969) 368.
[7] N.S. Scott and R.M. Smillie, Currents in Mod. Biol. 2 (1969) 339.
[8] G. Corneo, L. Zardi and E. Polli, J. Mol. Biol. 36 (1968) 419.
[9] P. Borst and G.J.C.M. Ruttenberg, Biochim. Biophys. Acta 190 (1969) 391.
[10] J. Vinograd, J. Morris, N. Davidson and N. Dove, Proc. Natl. Acad. Sci. U.S. 49 (1963) 12.
[11] W.G. Flamm, H.E. Bond and H.E. Burr, Biochim. Biophys. Acta 129 (1966) 310.
[12] E. Stutz and J.R. Rawson, Biochim. Biophys. Acta 209 (1970) 16.
[13] D. Gillespie and S. Spiegelman, J. Mol. Biol. 12 (1965) 829.
[14] N.S. Scott and R.M. Smillie, Biochem. Biophys. Res. Commun. 28 (1967) 598.
[15] W. Doerfler and D.S. Hogness, J. Mol. Biol. 33 (1968) 635.
[16] W.G. Flamm, M. McCallum and P.M.B. Walker, Proc. Natl. Acad. Sci. U.S. 57 (1967) 1792.
[17] W. Szybalsky, H. Kubinsky and P. Sheldrik, Cold Spring Harbor Symp. Quant. Biol. 31 (1966) 123.
[18] R.R. Burgess, A.A. Travers, J.J. Dunn and E.K.F. Bautz, Nature 221 (1969) 43.
[19] K.K. Tewari and S.G. Wildman, Proc. Natl. Acad. Sci. U.S. 58 (1967) 689.
[20] N. Sueoka, J. Mol. Biol. 3 (1961) 31.
[21] J. Vinograd and J.E. Hearst, Prog. Chem. Org. Nat. Prod. 20 (1962) 395.

COMPARISON OF THE PROTEIN SYNTHESIZING SYSTEMS FROM MITOCHONDRIA AND CYTOPLASM OF YEAST

H. MORIMOTO, A.H. SCRAGG, J. NEKHOROCHEFF,
V. VILLA and H.O. HALVORSON

*Laboratory of Molecular Biology, University of Wisconsin,
Madison, Wisconsin, USA*

Numerous observations in recent years have suggested that yeast mitochondria possess an autonomous protein synthesizing system. Inhibitors of bacterial protein-synthesizing systems (chloramphenicol, lincomycin, erythromycin, carbomycin, spiramycin and oleandomycin) inhibit amino acid incorporation into mitochondria of *Saccharomyces cerevisiae* in vivo and in vitro but are without effect on the cytoplasmic protein-synthesizing system [1–3]. Mutants resistant to these antibiotics show non-Mendelian inheritance (this symposium). Cycloheximide, on the other hand, inhibits only the latter. f-Met-tRNA, the chain initiating tRNA in bacteria, has been detected in mitochondria of yeast, rat liver [4] and Hela cells [5]. These observations suggest that yeast, like *Neurospora* [6], contain a bacterial-like protein-synthesizing system in the mitochondrion, but a mammalian-like system in the cytoplasm [7]. This diversity is further supported by the isolation of several unique ribosomal-like RNA species from yeast mitochondria [8,9].

Isolation of ribosomes from mitochondria

If the protein-synthesizing system in mitochondria were strictly analogous to that in bacteria, one might expect mitochondria to contain 70 S ribosomes. Although ribosomes have been isolated from organelles of a number of species, there is some controversy over their sedimentation values. 70 S type ribosomes have been isolated from chloroplasts of *Chlamydomonas* [10] and tobacco [11] and from mitochondria of *Neurospora* [12]. On the other hand 80 S

type ribosomes were reported in mitochondria of yeast [13,14] and *Neurospora* [15]. O'Brien and Kalf [16] reported the isolation of 55 S particles from rat liver mitochondria.

To further characterize the properties of ribosomes from yeast mitochondria, the following procedure was developed for their isolation (fig. 1). Freshly harvested cells of a homothallic diploid of *Saccharomyces cerevisiae* were transformed into spheroplasts by digestion of cell walls by glusulase as described by Lamb et al. [3] or disrupted by passing the cells suspended in 20% glycerol and 1 M sorbitol through an Eaton press at 5,000 psi. Following two 15 min centrifugations at 2,500 g at 4° with a GSA rotor (Sorvall) and one 5 min centrifugation at 5,000 g, the majority of the debris (whole cells, unbroken protoplasts and cell walls) was removed. The supernatant was centrifuged at 15,000 g for 30 min to pellet the mitochondria. This mitochondrial pellet was repeatedly (3–4 times) washed with 0.6 M sorbitol, 1 mM EDTA, 0.02 M Tris-HCl buffer pH 7.2. After gentle resuspension by a hand homogenizer, the mitochondrial suspension was layered on a 12 to 26% linear renografin gradient containing 0.6 M sorbitol and centrifuged as described by Schatz et al. [17]. Alternatively a discontinuous sorbitol gradient may be employed [18]. The mitochondrial fractions between the densities of 1.2–1.15 were pooled, pelleted by centrifugation, washed twice to remove the high UV absorbing renografin, quick-frozen in dry ice and alcohol and stored at −20° in order to facilitate the isolation of mitochondrial ribosomes. The purified mitochondria were thawed and carefully resuspended by homogenization in

ISOLATION OF MITOCHONDRIA AND THEIR COMPONENTS

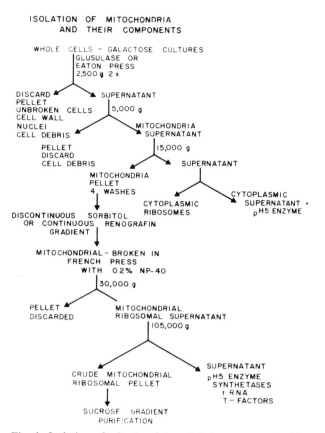

Fig. 1. Isolation of mitochondria and their components. All procedures were carried out at 4°. Mitochondria were isolated in TKM buffer (0.05 M Tris, pH 7.4, 25 mM KCl, 10 mM MgCl$_2$) containing 0.9 M sorbitol. Ribosomal fractions were maintained in a TKM containing 10 mM Cleland reagent.

TMK buffer (0.05 M Tris-HCl pH 7.4, 0.025 M KCl and 0.01 M MgCl$_2$) containing 0.01 M Cleland reagent, and 0.2% Nonidet. To release the ribosomes, purified mitochondria were disrupted by slowly passing the suspension through a French press at 8,000 psi. The homogenate was centrifuged in a Ti 50 rotor (Spinco L2–65B) at 30,000 g for 30 min; the clear bright yellow supernatant was recentrifuged for 1.5 hr at 105,000 g. The supernatant was retained for isolation of various components in the protein-synthesizing system.

Sedimentation properties of ribosomes

The 105,000 g pellet (fig. 1) was resuspended by homogenization in TMK buffer and subjected to a 10–40% linear sucrose gradient centrifugation in a SW 25.1 (Spinco) rotor at 23,000 rpm for 9 hr at 4° (fig. 2). A single prominent peak, with a sedimentation coefficient of 80 S was observed in agreement with the value reported by Schmitt [13]. The absorbancy at the top of the tube is due to Cleland reagent. The primary ribosome peaks were pooled, ribosomes collected by centrifugation and recentrifuged on a similar sucrose gradient. The pooled ribosomal material was then washed twice in TMK buffer, and the final purified mitochondrial ribosomes were quick-frozen and stored at −80°.

During the early stages of purification the mitochondrial ribosome fraction is darkish red-brown in color, and after one sucrose gradient centrifugation, slightly yellowish. The ratio of absorbance ratio (260 mμ/280 mμ) of pure mitochondrial ribosomes is 1.94, a value very similar to that observed for purified cytoplasmic ribosomes (1.92). The absorption spectrum of purified cytoplasmic and mitochondrial ribosomes is identical and typical of ribosomal particles. The absorption maximum and minimum for both are 258 and 242 mμ.

The thermotransitions of the cytoplasmic and

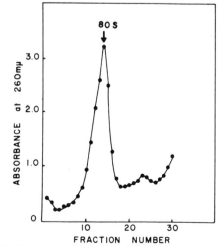

Fig. 2. Sucrose gradient centrifugation of mitochondrial ribosomes. Linear 10–40% sucrose gradient. Centrifugation: 9 hr at 23,000 rpm, 4°, in a SW 25.1 rotor.

mitochondrial ribosomes, however, differ [19]. In 5 mM MgCl$_2$ in TMK buffer, mitochondrial ribosomes have a maximal hyperchromatic change at 49.5° whereas the maximal change of the cytoplasmic ribosomes occurs at 60.5°.

Cytoplasmic ribosomes and their subunits were used to calibrate the mitochondrial ribosomes. Fig. 3 shows the Schlieren pattern for purified cytoplasmic and mitochondrial ribosomes. Since the sedimenta-

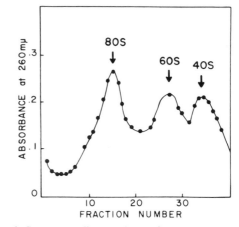

Fig. 4. Sucrose gradient analyses of mitochondrial ribosomal subunits. Subunits were formed suspending mitochondrial ribosomes in TKM buffer containing 5×10^{-6} M MgCl$_2$. Centrifugation on a 10–40% exponential sucrose gradient for 4 hr at 40,000 rpm. 4° in a SW 40 rotor.

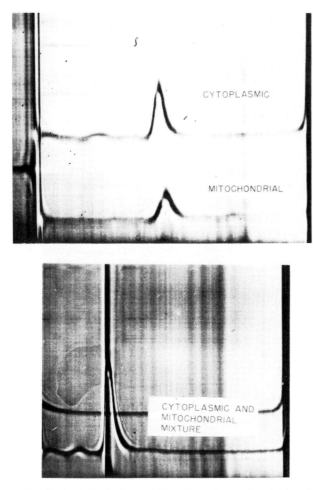

Fig. 3. Analytical ultracentrifugation of mitochondrial and cytoplasmic ribosomes. Centrifugation using Schlieren optics with 30 mm cells at 31,410 rpm, 20°. Photographs taken at 4 min intervals after speed was attained. Ribosomes (approximately 1 mg/ml) were suspended in 0.02 M Tris pH 7.2, 0.3 M NH$_4$Cl, and 0.005 M Mg acetate (TMA). A. Cytoplasmic ribosomes. Bar angle 65°, 8 min after speed attained. B. Mitochondrial ribosomes. Bar angle 65°, 8 min after speed attained. C. Mixed solution of mitochondrial and cytoplasmic ribosomes. Bar angle 80°, 8 min after speed attained.

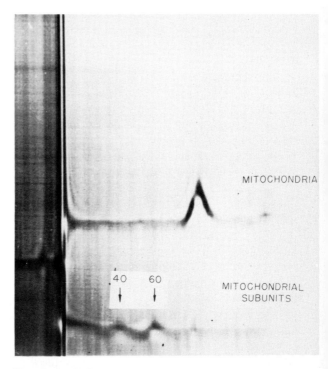

Fig. 5. Analytic ultracentrifugation of mitochondrial subunits. Subunits were formed by suspending mitochondrial ribosomes in 0.5 M KCl-TMA buffer. Centrifugation in a 30 mm cell using Schlieren optics, 31,410 rpm, 20°. Photographs taken at 12 min intervals after speed attained. Concentration 0.8 mg/ml, bar angle 65°.

tion velocity of the two was identical, we assume the mitochondrial ribosomes are 80 S. An identical S value was obtained when *E. coli* 70 S ribosomes were used as a standard.

Mitochondrial ribosomes can be dissociated into their subunits by either of two procedures. a) When these particles were suspended in buffer containing 5×10^{-5} M $MgCl_2$ and centrifuged in an exponential sucrose gradient (fig. 4), a dissociation of 80 S monosomes to two particles was observed. Based on the sedimentation of the main monosome peak, the subunits were calibrated as 60 S and 40 S respectively. Kuntzel [12] observed that in *Neurospora* mitochondrial ribosomes require a higher Mg^{++} concentrations to maintain their integrity than do cytoplasmic ribosomes. At 10^{-6} M Mg^{++} mitochondrial ribosomes dissociate to 60 S and 37 S subunits. b) A similar dissociation occurs when mitochondrial ribosomes are suspended in 0.5 M KCl (fig. 5), in agreement to that reported by Schmitt [13].

Characterization of mitochondrial rRNA

Table 1 summarizes the base composition of RNA extracted from pure cytoplasmic and mitochondrial ribosomes. The GC composition of total cytoplasmic rRNA is 45.2% in agreement with previous observations that both 28 S and 18 S rRNA from yeast contain approximately equal proportions of each nucleotide [20]. In sharp contrast, mitochondrial RNA contains approximately equal proportions of purine and pyrimidine bases but has considerably lower GC ratio (30.2%). These findings provide for the first time proof that the bulk of the mitochondrial RNA previously isolated (8,21) is ribosomal.

From differences in base composition one would expect dramatic differences in homology between the two types of ribosomal RNAs. To test this, ^{32}P-labeled RNA was isolated from mitochondrial and cytoplasmic ribosomes and hybridized with nuclear and mitochondrial DNA as previously described [20]. rRNA-DNA hybridization is highly specific. Mitochondrial rRNA shows very weak hybrid formation with nuclear DNA and cytoplasmic rRNA fails to hybridize with mitochondrial DNA [9,20]. Cytoplasmic rRNA hybridized with about 2.3–2.8% of the nuclear DNA [8,20,22]. In our experience, the

Table 1

Base composition of mitochondrial RNA and cytoplasmic tRNA

	CPM (%)	AMP (%)	GMP (%)	UMP (%)	GC (%)
Cytoplasmic [a]					
4 S	24.2	18.8	30.5	26.5	54.7
18 S and 26 S	17.0	26.4	28.2	28.4	45.2
Mitochondrial					
4 S	15.6	26.5	23.0	35.1	38.6
15 S	12.9	33.3	17.4	36.4	30.3
25 S	16.1	32.6	17.2	34.0	33.3

Base composition of cytoplasmic and mitochondrial ribosomal RNA. 10 μg of each RNA was completely hydrolyzed by a combined mixture of T_1 and T_2 RNase. The relative amounts of each nucleotide were analyzed by chromotography on a Picker nucleotide analyzer. *a* see ref. [20].

saturation plateaus for hybridization of the two mitochondrial rRNA with mitochondrial DNA is about 2.3% (fig. 6). Wintersberger and Viehhauser [9] found a higher plateau of about 4.3%. To determine homologies between mitochondrial and cytoplasmic rRNA, the competition experiment shown in fig. 7 was carried out. Filters containing mitochondrial or nuclear DNA were first prehybridized with varying concentrations of unlabeled cytoplasmic rRNA and then with a low concentration mitochondrial ^{32}P-rRNA. As can be seen, cytoplasmic rRNA fails to compete with mitochondrial rRNA for sites on mitochondrial DNA. The reverse competition is complete and suggests that the apparent presence of nuclear DNA sequences homologous to mitochondrial rRNA observed by us and others [9,20,22] could be explained by a small contamination with cytoplasmic rRNA.

In addition to difference in base composition, rRNA from pure mitochondrial and cytoplasmic ribosomes can be separated on sucrose gradient (fig. 8) or on polyacrylamide gels according to the method of Peacock and Dingman [24]. As shown in gel scans of the electropherograms (fig. 9), mitochondrial and cytoplasmic ribosomes each contain in addition to 4 S RNA, two major RNA species. When the two RNA preparations were mixed, only the major RNA species migrated separately. Size estimates based on sucrose gradient and model E ultracentrifuge analysis

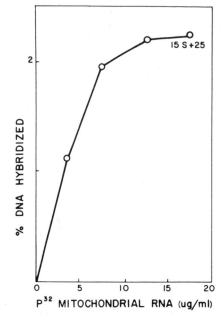

Fig. 6. Hybridization of mitochondrial rRNA to mitochondrial DNA. Mitochondrial DNA bound on nitrocellulose membrane filters was first prehybridized with cold cytoplasmic ribosomal RNA. Mitochondrial ^{32}P-rRNA at various concentrations was incubated with mitochondrial DNA membrane filters and the percent hybrid formed determined according to Schweizer and Halvorson [21].

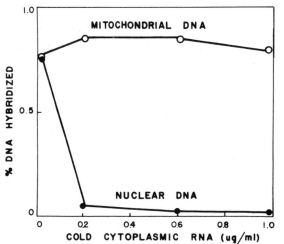

Fig. 7. Competition for binding by cold cytoplasmic RNA to nuclear and mitochondrial DNA. Both nuclear and mitochondrial DNA bound membrane filters were preincubated at various concentrations of cold cytoplasmic RNA. The prehybridized papers were incubated with 3.55 μg/ml of mitochondrial ^{32}P-rRNA and the percent hybrid formation determined.

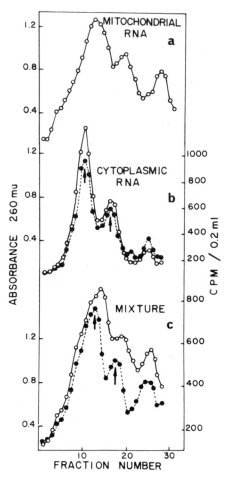

Fig. 8. Sucrose gradient centrifugation of mitochondrial RNA. Linear 5–20% sucrose gradients. Centrifugation 15 hr at 23,000 rpm, 4° in a SW 25.1 rotor. a. Mitochondrial RNA. b. Cytoplasmic ^{3}H-uracil labeled RNA. c. Mixture of a and b.

have given various results. Rogers et al. [8] reported S values of mitochondrial RNA of 12.7, 17.8 and 22.4 and of cytoplasmic ribosomes of 16.2 and 24.6. Leon and Mahler [21] calculated the S values of mitochondrial RNA to be 11.3, 15.1 and 21.9. Wintersberger [23] reported two RNA species in yeast mitochondria having S values of 23 S and 16 S. Thus the larger RNA species from mitochondria seems to be smaller than the corresponding cytoplasmic RNA. In the polyacrylamide gels, on the other hand, the two mitochondrial rRNA species consistently migrate slower than the cytoplasmic rRNAs at pH 8.3, and presumably have larger molecu-

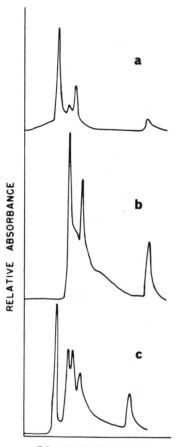

Fig. 9. Acrylamide gel electrophoresis of cytoplasmic and mitochondrial RNA. RNA was electrophoresed according to the method of Peacock and Dingman [24] in 2.5% gels polymerized in quartz tubes. The migration of the RNA species was determined by UV scanning of the gels in a Cary Spectrophotometer. a. Mitochondrial RNA. b. Cytoplasmic RNA. c. Mixture of a and b.

lar weights. However since these differences in migration are not seen at pH 7.0, more detailed studies are required to determine the actual molecular weight differences.

Comparison of ribosomal proteins

It is now recognized that each ribosomal subunit contains numerous distinct protein species [25]. To determine whether ribosomal proteins, as well as

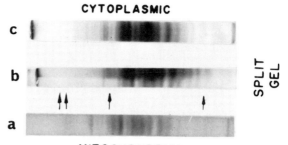

Fig. 10. Polyacrylamide gel electrophoresis of mitochondrial and cytoplasmic ribosomal proteins. Dialyzed LiCl split proteins of ribosomes were electrophoresed according to the method of Gestland and Staehelin [26]. a. Mitochondrial ribosomal proteins alone. b. Split gels one side containing mitochondrial and the other side cytoplasmic ribosomal proteins. c. Cytoplasmic ribosomal proteins alone.

rRNA, differ between cytoplasm and mitochondria, the predominant protein species present in monosomes were compared by polyacrylamide gel electrophoresis. Pure mitochondrial and cytoplasmic ribosomes were washed twice with 0.5 M NH_4Cl and 0.1 M $MgCl_2$ in TMK. Ribosomal proteins were prepared from both according to the method of Gestland and Staehelin [26]. After dialysis of ribosomal proteins against 4 M urea and TMK buffer containing 0.01 M mercaptoethanol, the ribosomal proteins were subjected to the acrylamide gel electrophoresis according to Gesteland and Staehelin [26], as shown in fig. 10. Many of the bands are common to both ribosomal preparations. Of significance, however, is the finding of a number of bands unique to either mitochondrial or cytoplasmic ribosomes. Kuntzel [6], based on chromatography on carboxy cellulose, reported numerous differences between ribosomal proteins from cytoplasm and mitochondria of *Neurospora*.

Comparison of aminoacyl-RNA synthetases and tRNA

tRNA and aminoacyl-RNA synthetases have been reported in mitochondria from a number of organisms [28,29]. Comparisons of mitochondrial and cytoplasmic tRNAs and synthetases have revealed marked differences [29,33] supporting the view of an unique protein-synthesizing system in mitochondria. To further define the mitochondrial system in yeast heterologous comparisons of the activity of mitochondrial and cytoplasmic synthetases on mitochondrial and cytoplasmic tRNA were carried out (table 2). As also reported in *Neurospora*, mitochondria contain a high leucyl RNA synthetase activity which shows a preference for mitochondrial tRNA at 10 mM magnesium [29]. In contrast the cytoplasmic phenylalanyl synthetase under the same conditions is more active in the homologous test system than in the heterologous one.

One basis for the apparent difference in the rate of acylation could be due to species differences and the optimum reaction conditions, in particular the Mg^{++} concentration. That these in fact can widely vary is shown for 5 synthetases in table 3. With the exception of phenylalanine, the cytoplasmic synthetases require higher Mg^{++} concentrations than do the mitochondrial enzymes.

Table 2
Transfer RNA specificities of mitochondrial and cytoplasmic synthetases

| Amino acid | $\mu\mu$moles aminoacyl tRNA formed/ mg tRNA/10 min | | | |
| | Mitochondrial synthetases | | Cytoplasmic synthetases | |
	Mito. tRNA	Cyto. tRNA	Mito. tRNA	Cyto. tRNA
Phenylalanine	101	77	154	205
Leucine	6600	2550	2200	1070
Serine	57.5	76	24.5	9.4

The reaction mixture contained: 5 mM Tris-Cl pH 7.6 containing 10 mM KCl, 10 mM dithiothreitol and 10 mM magnesium acetate; 10–100 μg of tRNA, 50–150 μg of enzyme, 4 μmoles ATP and 0.25 μc of ^{14}C-amino acid. The total volume was 0.25 ml. The mixture was incubated at 30°C, 50 μl samples were taken at 0, 5, 10, 15 min and assayed by a filter paper disc method of Bollum [54]. The values at 10 min were taken for the table. The cytoplasmic synthetases and tRNA were extracted from the post-mitochondrial supernatant.

Table 3
Magnesium optima for aminoacylation

| Amino acid | Magnesium acetate concentration (mM) | |
	Cytoplasmic system	Mitochondrial system
Methionine	40	8
Phenylalanine	12	22
Leucine	40	10
Serine	40	15
Isoleucine	40	10

The reaction mixture and assay was as for table 2 except the volume was 0.125 ml, the magnesium concentration varied and two 50 μl samples were taken after 15 min incubation at 30°.

Polymerization factors from cytoplasm and mitochondria

A variety of enzymes and factors are required for protein biosynthesis of which T and G [34] catalyze the reactions of peptide chain elongation. Three factors have been isolated from bacterial sources [35,36] and two or three from yeast [37–39] and other eucaryotic systems [40]. The bacterial-type ribosome will not function in poly U directed protein synthesis with eucaryotic supernatant enzymes [41–43], the specificity being due to the T and G factors [43]. Thus a comparison between these factors from mitochondria and cytoplasm would further define any differences in their protein-synthesizing systems.

GTP-binding material was found in the mitochondrial supernatant and was extracted by the procedures outlined for *E. coli* T and G factors [44,45]. Chromatographically on DEAE-Sephadex, the mitochondrial T factors were similar to those of *E. coli*. However, in agreement with Richter et al. [37], the elution pattern of the cytoplasmic factors differed markedly from that of the *E. coli* factors. Upon heat inactivation, the GTP-binding material from both cytoplasm and mitochondria were similar (fig. 11).

E. coli Ts can be separated from Tu or a Tu-Ts complex by DEAE-Sephadex chromatography [46]. We have observed a similar separation for mitochondria but not for the cytoplasm factors by chro-

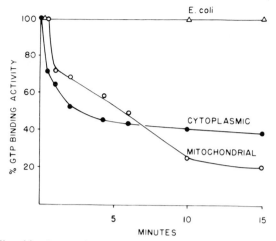

Fig. 11. Ammonium sulphate extracts of mitochondrial, cytoplasmic (80%) and E. coli (40–60%) supernatants were heated at 50°C. 20 μl samples were removed, cooled and assayed according to Allende and Weissbach [52] for GTP-binding. – o – o – mitochondrial, – ● – ● – cytoplasmic, – △ – △ – E. coli.

matography and heat inactivation at 50° for 15 min. In addition, E. coli Ts will stimulate the binding of GTP by Tu [44]. A similar result was found for the mitochondrial system (table 4) although the low stimulation does not rule out another component required for stimulation or that Tu is in fact still a Tu-Ts complex. In E. coli the Tu-Ts complex has recently been crystallized [47].

In addition to binding GTP the T factors are required for the binding of tRNA to ribosomes [48–50]. Using ammonium chloride-washed E. coli ribosomes [48] the mitochondrial factors have been assayed for the binding of E. coli tRNA to ribosomes (table 5). As can be seen, both mitochondrial Tu and Ts can partially replace the E. coli factors.

To test further the ability of mitochondrial factors to replace those of E. coli, their activity in the formation of polypeptides from phenylalanine tRNA in the E. coli system was determined. As can be seen in table 6, mitochondrial Tu and Ts show little or no incorporation although Ts appears to be able to replace E. coli Ts. This is in marked contrast to the results of Ciferri et al. [51] in which it was found that E. coli Ts could not be replaced by yeast cytoplasmic factors. Thus in the present experiments, the mitochondrial factors do not appear to be significantly contaminated with cytoplasmic factors. The

Table 4
^{14}C-GTP binding by mitochondrial fractions

	μμmoles ^{14}C-GTP bound
Tu	1.1
Ts	0.04
Tu + Ts	3.35

The GTP binding was assayed as described by Allende and Weissbach [52], Tu and Ts were prepared as described by Ertel et al. [44]. Where specified, 11.5 μg of Tu and 0.27 μg of Ts were added.

Table 5
Binding of ^{14}C-phenylalanine tRNA to E. coli ribosomes

Addition	μμmoles ^{14}C-phenylalanine tRNA bound/mg ribosomal protein
E. coli Tu	0.54
E. coli Ts	0.195
E. coli Tu + Ts	1.56
E. coli Tu + Mito Ts	0.98
Mito Tu + E. coli Ts	0.74

The assay used the method of Lucas-Lenard and Haenni [48]. The total volume was 0.5 ml containing 460 μg of ribosomes. Where specified, 59 μg of E. coli Tu, 4 μg of E. coli Ts, 11.5 μg of mitochondrial Tu and 2.7 μg of mitochondrial Ts were added.

Table 6
Incorporation of ^{14}C-phenylalanine from ^{14}C-phenylalanine-tRNA by mitochondrial and E. coli T factors

Addition	μμmoles ^{14}C-phenylalanine incorporation 30 min at 30°/mg ribosomal protein
E. coli Ts	0.69
E. coli G	0.63
E. coli Tu + Ts	14.9
E. coli Tu + Ts + G	22.5
E. coli Tu + G + Mito. Ts	18.6
E. coli Ts + G + Mito. Tu	1.10
Mito Tu + Mito. Ts	0.70

The ribosomes were prepared as described by Lucas-Lenard and Haenni [48] and the cell-free system as described by Ciferri et al. [51]. The total volume was 0.25 ml containing 920 μg of ribosomes. Where specified 59 μg of E. coli Tu, 4 μg of E. coli Ts, 11.4 μg of E. coli G, 11.5 μg of mitochondrial Tu and 0.27 μg of mitochondrial Ts were added.

discrepancy between these results in table 6 and the tRNA-binding (table 5) can perhaps be due to a) binding of the tRNA to the wrong sites on the ribosome, b) an incompatibility between the bound T-tRNA-GTP complex and the *E. coli* translocase G, which does not occur in the cytoplasmic system [51], or c) differences in the optimum Mg^{++} requirement for tRNA binding.

Characteristics of the cell-free incorporating system from mitochondria

Employing supernatant and ribosomes isolated from yeast mitochondria, some of the characteristics of basal and poly U dependent phenylalanine incorporation have been determined (table 7). Amino acid incorporation is dependent upon ribisomes and an ATP-generating system. It is stimulated by poly U and is sensitive to RNase. Rapid freezing, which has proven useful for storing cytoplasmic ribosomes of yeast, leads to a 85% loss of incorporating activity

Table 7
Mitochondrial cell-free system

Additions	$\mu\mu$moles ^{14}C-phenylalanine incorporated at 30°C for 30 min/mg ribosomal protein
Complete system	4.0
– poly U	2.7
+ 100 μg poly U	14.4
– ribosomes	0
– ATP, UTP, CTP, GTP	1.02
– nucleotide and PEP, PEP kinase	0
+ 5 μg ribonuclease	1.18

The reaction mixture (0.20 ml) contained: 28.5 μmoles Tris-HCl pH 7.6, 1.0 μmoles magnesium acetate, 2.0 μmoles KCl, 2.0 μmoles dithiothreitol, 0.2 μmoles spermidine, 0.5 μmoles ATP, 0.15 μmoles each of UTP, GTP and CTP, 0.6 μmoles PEP, 5 μg PEP kinase, 25 μg of poly U, 20 μg of mitochondrial tRNA, 0.5 μc of ^{14}C-L-phenylalanine, 50–100 μg of supernatant enzyme and 100–300 μg of ribosomal protein. After incubation at 25° for 45 min, 0.2 ml of ice-cold 10% TCA was added and the mixture was retained in the cold for at least 1 hr. The mixture was then heated at 90° for 15 min and treated as described by Bretthauer et al. [7]. The stored ribosomes had been frozen rapidly and stored at –20° for 24 hr.

Table 8
The effect of various inhibitors upon poly U directed cell-free synthesis from the cytoplasm and mitochondria of yeast

Inhibitor		% inhibition	
		Cytoplasmic	Mitochondrial
Cycloheximide	0.1 mM	85	0
Chloramphenicol	5.0 mM	17	47
Streptomycin	0.2 mM	25	47
Erythromycin	0.1 mM	25	52

The reaction mixture for the mitochondrial system is described in the legend to table 7 except that the magnesium concentration was 7 mM. The cytoplasmic system was run as described by Bretthauer et al. [7]. In both cases, two 75 μl samples were taken after 30 min incubation at 30° and assayed using a filter paper disc method of Bollum [54].

after 24 hr storage at –20°. The incorporating activity of the mitochondrial incorporation system (1–4 $\mu\mu$moles phenylalanine/30 min/mg ribosomal protein) is about 1/5 or less of the activity of the cytoplasmic system. These differences could be due to insufficient knowledge of the optimum conditions for the mitochondrial system or to possible dependency of activity of ribosomes on the physiological age of the culture [7]. We have found, for example, that the optimum Mg^{++} concentration for the mitochondrial system (about 15 mM) is higher than that for the cytoplasmic system (5 mM) in the poly U directed assay.

A more critical test of the uniqueness of the cell-free mitochondrial protein synthesizing system is its sensitivity to antibiotics. Table 8 compares the effect of cycloheximide, chloramphenicol, streptomycin and erythromycin on the cytoplasmic and mitochondrial yeast systems in the presence of poly U. In the absence of inhibitors, the activities of the mitochondrial and cytoplasmic systems were 4.5 and 50 $\mu\mu$moles phenylalanine/30 min/mg ribosomal protein respectively. In agreement with the findings of Kuntzel [6] in *Neurospora,* only the cytoplasmic ribosomes are sensitive to cycloheximide. On the other hand, the mitochondrial ribosomes are more sensitive to chloramphenicol, streptomycin and erythromycin. Cross contamination of the two robosome preparations could contribute to the partial sensitivities observed.

Conclusions

Do mitochondria, as has been suggested by several workers [1–3,10–12,15], contain a bacterial-type protein-synthesizing system? The preparation of various components from yeast mitochondria described here, permits, for the first time, an evaluation of this possibility in yeast.

The mitochondrial protein-synthesizing system from yeast resembles in several respects that from bacteria. These include a) the pattern of sensitivity to antibiotics (table 8), b) chromatographic behavior as well as the specificity of T factors (tables 4,5 and 6), c) magnesium optimum and d) the presence of f-met tRNA [4]. A similar relationship probably also exists in *Neurospora* since Kuntzel [53] found that the cell-free system from mitochondria could be interchanged with the *E. coli* system but not with the cytoplasmic system. One significant difference is that the ribosomes of yeast mitochondria [13] (fig. 2,3) are 80 S particles rather than the 70 S particles found in bacteria.

The uniqueness of the mitochondrial system is further evident by comparison of this to the cytoplasmic system. Although the size of the monosomes and subunits are similar in each, they differ in their stability [13], thermotransitions [19], ribosomal protein patterns (fig. 10), and the composition [8,20,21] (table 1), homology to DNA [9,20] (fig. 6) and size of RNA [8,21,23; fig. 8,9]. This is further supported by differences in the specificity and optimum conditions for amino acyl synthetases (tables 2 and 3) and T factors [51] and sensitivity to antibiotics (table 8). The latter confirm previously reported differences between whole mitochondria and the cell-free cytoplasmic system [1–3]. These differences provide a focus for identifying at the molecular level cytoplasmic mutations as well as defining possible interdependence of the mitochondrial and cytoplasmic protein-synthesizing systems.

Acknowledgements

This investigation was supported by grants from the National Science Foundation (GB–6993X), a Public Health Service research grant AI–01459 from the National Institute of Allergy and Infectious Diseases, CA–18203 from the National Institute of Cancer and GM–01874 from the National Institutes of General Medical Sciences. A NIH Research Career Professorship was awarded to one of us (H.O.H.). The authors wish to express their appreciation to Mr. Timothy Johnson for assistance in the determination of base compositions.

References

[1] M. Huang, D.R. Briggs, D.G. Clark-Walker and A.W. Linnane, Biochim. Biophys. Acta 114 (1966) 434.

[2] G.D. Clark-Walker and A.W. Linnane, Biochem. Biophys. Res. Comm. 25 (1966) 8.

[3] A.J. Lamb, G.D. Clark-Walker and A.W. Linnane, Biochim. Biophys. Acta 161 (1968) 415.

[4] A.E. Smith and K.A. Marker, J. Mol. Biol. 38 (1968) 241.

[5] J.B. Galper and J.E. Darnell, Biochem. Biophys. Res. Commun. 34 (1969) 205.

[6] H. Kuntzel, Nature 222 (1969) 142.

[7] R.K. Bretthauer, L. Marcus, J. Chaloupka and H.O. Halvorson, Biochemistry 2 (1963) 1079;
A.G. So and E.W. Davie, Biochemistry 2 (1963) 132;
H.O. Halvorson and C. Heredia, in: Aspects of Yeast Metabolism, ed. A.K. Mills (Blackwell Sci. Pubs, Oxford, 1968) p. 107.

[8] P.J. Rogers, B.N. Preston, E.B. Titchener and A.W. Linnane, Biochem. Biophys. Res. Commun. 27 (1967) 405.

[9] E. Winterberger and G. Viehhauser, Nature 220 (1968) 699.

[10] J.K. Hoober and G. Blobel, J. Mol. Biol. 41 (1969) 121.

[11] N.K. Boardman, R.I.B. Francki and S.G. Wildman, J. Mol. Biol. 17 (1966) 470.

[12] H. Kuntzel, J. Mol. Biol. 40 (1969) 315.

[13] H. Schmitt, FEBS Letters 4 (1969) 234.

[14] P.V. Vignais, J. Huet and J. Andre, FEBS Letters 3 (1969) 177.

[15] M.R. Rifkin, D.D. Wood and D.J.L. Luck, Proc. Natl. Acad. Sic U.S. 58 (1967) 1025.

[16] T.W. O'Brien and G.F. Kalf, J. Biol. Chem. 242 (1967) 2180.

[17] G. Schatz, H. Haselbrunner and H. Tuppy, Biochem. Biophys. Res. Commun. 15 (1964) 127.

[18] D.J. South and H.R. Mahler, Nature 218 (1968) 1226.

[19] H. Morimoto and H.O. Halvorson, in preparation.

[20] E. Schweizer, C. MacKechnie and H.O. Halvorson, J. Mol. Biol. 40 (1969) 261.

[21] S.A. Leon and H.R. Mahler, Arch. Biochem. and Biophys. 126 (1968) 305.

[22] J. Retal and R.S. Plauta, Europ. J. Biochem. 3 (1967) 248.

[23] E. Wintersberger, Z. Physiol. Chem. 348 (1967) 1701.

[24] A.C. Peacock and C.W. Dingman, Biochemistry 6 (1967) 1818.

[25] S.J.S. Hardy, C.G. Kurland, P. Voynow and G. Mora, Biochemistry 8 (1969) 2897.

[26] R.F. Gestland and T. Staehelin, J. Mol. Biol. 24 (1967) 129.

[27] E. Wintersberger, Biochem. Z. 341 (1965) 409.

[28] W.D. Barnett, D.H. Brown and J.L. Epler, Proc. Natl. Acad. Sci. U.S. 57 (1967) 1775.

[29] W.E. Barnett and D.H. Brown, Proc. Natl. Acad. Sci. U.S. 57 (1967) 452.

[30] Y. Suyama and J. Eyer, Biochem. Biophys. Res. Comm. 28 (1967) 746.

[31] D.H. Brown and B.G.D. Novelli, Biochem. Biophys. Res. Commun. 31 (1968) 262.

[32] J.L. Elper, Biochemistry 8 (1969) 2285.

[33] C.A. Buck and M.M.K. Nass, J. Mol. Biol. 41 (1969) 67.

[34] Y. Nishizuka and F. Lipmann, Proc. Natl. Acad. Sci. U.S. 55 (1966) 212.

[35] J. Lucas-Lenard and F. Lipmann, Proc. Natl. Acad. Sci. U.S. 55 (1966) 1562.

[36] J.E. Allende, N. Seeds, T. Conway and H. Weissbach, Proc. Natl. Acad. Sci. U.S. 58 (1967) 1566.

[37] D. Richter, H. Hameister, H.G. Peterson and F. Klink, Biochemistry 7 (1968) 3753.

[38] M.S. Ayuso and C.F. Heredia, Europ. J. Biochem. 7 (1968) 111.

[39] U. Albrecht, K. Prenzel and D. Richter, Biochemistry, in press.

[40] R. Arlinghaus, G. Favelukes and R. Schweet, Biochem. Biophys. Res. Commun. 11 (1963) 92; P. Rao and K. Moldave, Biochem. Biophys. Res. Commun. 38 (1967) 909; R. Ibuki and K. Moldave, J. Biol. Chem. 243 (1968) 44.

[41] D. Nathans and F. Lipmann, Proc. Natl. Acad. Sci. U.S. 47 (1961) 497.

[42] C.F. Heredia and H.O. Halvorson, Biochemistry 5 (1966) 946.

[43] B. Parisi, G. Milanesi, J.L. Van Etten, A. Perani and O. Differri, J. Mol. Biol. 28 (1967) 295.

[44] R. Ertel, N. Brot, B. Redfield, J. Allende and H. Weissbach, Proc. Natl. Acad. Sci. U.S. 59 (1968) 861.

[45] D.L. Miller and H. Weissbach, Arch. Biochem. Biophys. 132 (1969) 146.

[46] R. Ertel, B. Redfield, N. Brot and H. Weissbach, Arch. Biochem. Biophys. 128 (1968) 331.

[47] A. Parmeggiani, Biochem. Biophys. Res. Commun. 30 (1968) 613.

[48] J. Lucas-Lenard and A.L. Haenni, Proc. Natl. Acad. Sci. U.S. 59 (1968) 554.

[49] J.M. Ravel, Proc. Natl. Acad. Sci. U.S. 57 (1967) 1811.

[50] Y. Ono, A. Skoultchi, J. Waterson and P. Lengyel, Nature 222 (1969) 645.

[51] O. Ciferri, B. Parisi, A. Parani and M. Grandi, J. Mol. Biol. 37 (1968) 529.

[52] J.E. Allende and H. Weissbach, Biochem. Biophys. Res. Commun. 28 (1967) 82.

[53] H. Kuntzel, FEBS Letters 4 (1969) 3.

[54] F.J. Bollum, J. Biol. Chem. 234 (1959) 2733.

PROPERTIES OF MITOCHONDRIAL RNA IN HELA CELLS

G. ATTARDI, Y. ALONI, B. ATTARDI, M. LEDERMAN,
D. OJALA, L. PICA-MATTOCCIA and B. STORRIE

Division of Biology, California Institute of Technology,
Pasadena, California, USA

Twenty years after the discovery of respiratory deficient mutants in yeast [1,2] and *Neurospora* [3], the biological significance of mitochondria-associated inheritance is today still elusive. Paradoxically, the discovery and characterization of mitochondrial DNA (see [4] for a review), while providing the physical basis for the postulated cytoplasmic determinants, have placed strong constraints on their genetic role. Thus, although mitochondrial DNA controls in a still unknown way the assembly and division of mitochondria, it is now clear that its informational content is sufficient to specify only a minor fraction of the constituents of these organelles. One would like to know why mitochondria have DNA at all, why it is advantageous for the eukaryotic cell to maintain this family of DNA molecules confined in the cytoplasm.

To the molecular biologist interested in the organization and expression of the eukaryotic genome, the mitochondrial system is uniquely attractive. Apart from the specific problem of the role of mitochondrial DNA (mit-DNA) in controlling the growth and reproduction of mitochondria, the availability of a set of genes which can be fairly easily isolated and studied in their structural properties, replication and pattern of expression, offers the possibility of analyzing in a relatively simple system some of the basic problems concerning replication and transcription of DNA, processing, transport and translation of RNA and the regulation of these phenomena.

Animal cell systems, especially cells growing in culture, are at present only limitedly accessible to genetic analysis; however, they are particularly suitable for a biochemical approach to the problems mentioned above.

We shall summarize in this paper the work that we have carried out in the last two years to characterize the molecular and biological properties of mitochondrial RNA in a human cell line grown in vitro, HeLa cells.

Rapidly labeled heterogeneous mitochondrial RNA

When exponentially growing HeLa cells are exposed to a labeled RNA precursor, radioactivity is incorporated linearly without any detectable lag (after correction for pool equilibration) into RNA associated with a crude mitochondrial fraction isolated by differential centrifugation [5,6]. This fraction is contaminated by smooth membrane components and by elements of rough endoplasmic reticulum (ER) [7]. Buoyant density or sedimentation velocity centrifugation in sucrose gradient (fig. 1) of the crude mitochondrial fraction reveals that after a short pulse of ^{3}H-uridine (up to 20 to 30 min) all, or the great majority, of the labeled RNA is in structures with a size and density distribution which follows fairly closely that of mitochondria (recognized from the cytochrome oxidase activity), while being significantly different from that of the rough ER elements (revealed by the 2 hr labeling profile which is about 50% due to extramitochondrial ribosomal RNA (rRNA) [6]. This fact is even more clearly illustrated in experiments of buoyant density fractionation in dextran-sucrose gradients (fig. 1), in which 50 to 60% of the rough ER bands at a different density from that of mitochondria and the rest of rough ER elements (Storrie and Attardi, in preparation). These experiments suggest strongly that the bulk of the pulse-labeled RNA of the crude

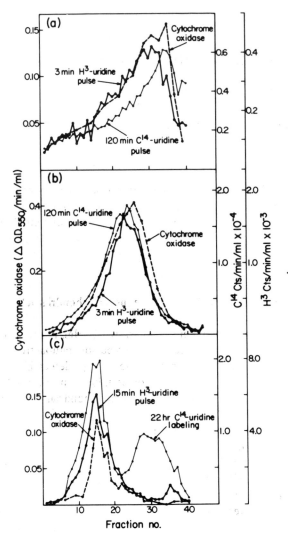

Fig. 1. Fractionation of a crude mitochondrial fraction from HeLa cells by sedimentation velocity (a) or buoyant density (b) centrifugation in sucrose gradient, or by buoyant density centrifugation in a dextran-sucrose gradient (c). The crude mitochondrial fraction was prepared as described previously [6] from a mixture of cells subjected to a short ^{3}H-5-uridine pulse and cells labeled for 2 hr or 24 hr with ^{14}C-uridine. For details of the conditions of fractionation see [6,7].

mitochondrial fraction is associated with mitochondria. That the labeled RNA is indeed intramitochondrial is indicated by the fact that 70 to 80% of it is resistant to digestion by RNase of the isolated subcellular fraction: this figure is obviously a

minimum one since a certain damage of mitochondria during the isolation procedure probably cannot be avoided.

Fig. 2 shows the sedimentation pattern of RNA extracted by phenol-SDS from the isopycnically separated crude mitochondrial fraction of HeLa cells exposed to short ^{3}H-uridine pulses. The O.D.$_{260}$ profiles reveal the two rRNA components pertaining to the ribosomes of the rough ER which contaminates the mitochondrial fraction (in (a) carrier 28 S RNA was added). The 3 min pulse-labeled RNA (fig. 2(a)) is represented by heterogeneous components with sedimentation constants ranging from 4 S

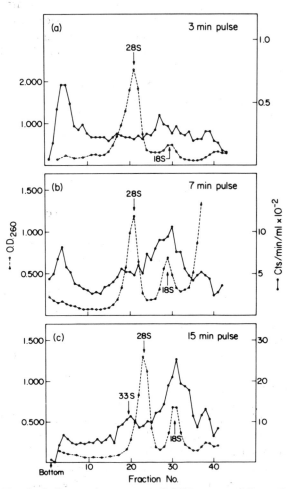

Fig. 2. Sedimentation pattern of RNA extracted from the crude mitochondrial fraction, isopycnically separated in sucrose gradient, from HeLa cells exposed for various times to ^{3}H-5-uridine [8].

to 50 S and more, with the heaviest components accumulating against a dense sucrose cushion at the bottom of the tube. After long pulses (fig. 2(b) and 2(c)) the proportion of the heavier components decreases and there is a progressive accumulation of radioactivity in components sedimenting slower than 28 S. Already after a 7 min pulse and even more after a 15 min pulse one recognizes discrete components emerging over the background of heterogeneous RNA: a fairly broad peak at about 33 S, a pronounced peak at about 17 S and a shoulder at about 12 S, and a peak at 4 S. The discrete components sedimenting slower than 28 S will be better seen in the long sedimentation runs (see below). The nature of the heavy (> 30 S) components of heterogeneous RNA is now being investigated in our laboratory. The kinetics of labeling of the various size components is consistent with the idea that the heavy material contains the nascent form of mitochondrial RNA, which is progressively converted to the slower sedimenting components.

Fig. 3 shows that ethidium bromide, an intercalating phenanthridine dye, at a concentration of 1 μg/ml, inhibits uniformly by about 85% the labeling

of different-size components of heterogeneous mitochondria-associated RNA; this inhibition is about 90% at a concentration of the drug of 2 μg/ml. At 1 μg/ml, on the other hand, ethidium bromide has no appreciable effects on nuclear RNA synthesis, in agreement with published evidence [9]. The selective inhibition by ethidium bromide, in short term experiments, of the labeling of mitochondria-associated heterogeneous RNA may reflect a direct effect of this drug on the transcription of closed circular mit-DNA: in fact, this intercalating dye is known to distort the structure of supercoiled circular DNA [10–12]. If this interpretation is correct, the above discussed results provide strong support for the view that mit-DNA is the template of the heterogeneous RNA, previously suggested by the linear kinetics of labeling and the intramitochondrial location of this RNA fraction. Consistent with this view is the observation that concentrations of actinomycin D which inhibit completely rRNA synthesis and almost completely the arrival in the cytoplasm of mRNA of free polysomes do not affect appreciably the pulse-labeling of the heterogeneous mitochondria-associated RNA (Attardi and Ojala, in preparation).

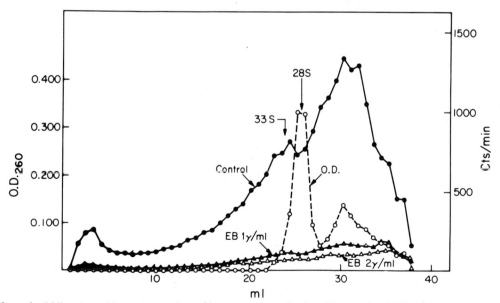

Fig. 3. Effect of ethidium bromide on the labeling of heterogeneous mitochondria-associated RNA. The RNA was extracted from the crude mitochondrial fraction, which had been treated with 3×10^{-2} M EDTA [7], and isopycnically separated in sucrose gradient, from HeLa cells exposed to ^{3}H-5-uridine for 15 min in the absence or in the presence of 1 μg/ml, or 2 μg/ml ethidium bromide.

Decisive evidence in favor of the idea that mit-DNA is the template involved comes from RNA-DNA hybridization experiments utilizing closed circular mit-DNA purified by two cycles of cesium chloride density gradient centrifugation in the presence of ethidium bromide [10,13]. As shown in table 1, after a 7 or 15 min ^{3}H-uridine pulse, both the RNA components sedimenting slower than 28 S and the heavier components hybridize with great efficiency with mit-DNA and only to a very limited extent with total cell or nuclear DNA. More surprising is the observation that a similar efficiency of hybridization with mit-DNA is exhibited by the heterogeneous RNA of the crude mitochondrial fraction isolated from cells labeled for 120 min with ^{14}C-uridine; in fact, after this time of labeling, a substantial portion (estimated to be about 50%) of the radioactive heterogeneous RNA of this fraction is extramitochondrial and presumably represented by messenger RNA (mRNA) of contaminating ER-bound polysomes [6]. This observation suggests that this mRNA

fraction may have sequence homology to mit-DNA. We shall return to this problem later.

Table 2 shows the ^{32}P-nucleotide composition of pulse-labeled mitochondria-associated RNA from cells exposed for 30 min to ^{32}P-orthophosphate [5]. After this time of labeling the great majority of the radioactive RNA in the crude mitochondrial fraction still consists of heterogeneous components. One can see that both the heavier (> 25 S) and lighter (9 to 25 S) components have a base composition strikingly different (especially for the high A and low G content) from that of free polysome mRNA. The base composition appears to be complementary, as concerns the A and U content, to that of the heavy strand of mit-DNA: this suggests that mitochondrial RNA is transcribed mostly, if not exclusively, from the heavy strand [8]. RNA-DNA hybridization experiments carried out between separated mit-DNA strands and mitochondrial RNA (see below) have led to the same conclusion.

The rate of synthesis of mitochondrial RNA, as

Table 1
Homology with mitochondrial DNA of RNA from a crude mitochondrial fraction of HeLa cells

Expt. no.	Pulse (min)	Fraction	DNA	cpm in hybrid	% input cts/ min in hybrid
1	7	10–26 S	Mitochondrial	194	13.3
	7	10–26 S	Total cell	5	0.3
	7	33–50 S	Mitochondrial	284	25.6
	7	33–50 S	Nuclear	10	0.9
2	15	8–35 S	Mitochondrial	30	11.4
	15	8–35 S	Total cell	< 1	< 0.4
	15	35–50 S	Mitochondrial	51	14.1
	15	35–50 S	Total cell	< 1	< 0.3
3	120	10–26 S	Mitochondrial	138	10.3
	120	10–26 S	Total cell	8	0.6
	120	33–50 S	Mitochondrial	172	16.6

The crude mitochondrial fraction was isolated, as in fig. 1, from HeLa cells exposed for 7 or 15 min to ^{3}H-5-uridine or for 120 min to ^{14}C-uridine, and the RNA was phenol-SDS extracted and run through sucrose gradients as in fig. 2. RNA components from different portions of each gradient were precipitated with ethanol and dissolved in SSC/10. The incubation mixtures, containing 2 to 3.6 μg heat denatured mit-DNA and 10 to 22 μg RNA in a total volume of 2 ml 2 X SSC, were incubated at 70°C for 5 hr. After cooling, the mixtures were treated with 10 μg/ml heated pancreatic RNase for 60 min at 22°C and the RNA/DNA hybrids isolated by Sephadex chromatography [14]. (Washing of the nitrocellulose membranes with 0.5 M KCl, 0.01 M Tris buffer was at 56°C.) The data are corrected for nonspecific background estimated with E. coli DNA. For the 120 min labeled RNA the percentage of input cts/min in hybrid refers to heterogeneous RNA only, estimated as described by Girard et al. [15]. Control experiments showed that labeled rRNA from ER-bound ribosomes does not contribute to any detectable extent to hybrid formation.

Table 2

Nucleotide composition of pulse labeled mitochondrial RNA from HeLa cells

Fraction	Moles per cent					
	A	C	U(T)	G	GC%	A/U(T)
Mitochondrial RNA						
9–25 S	33.9	24.5	22.6	18.9	43.4	1.50
26–48 S	31.4	23.9	25.3	19.4	43.3	1.24
Free polysome mRNA						
10–38 S	24.8	21.4	27.9	25.8	47.2	0.89
Total DNA [a]	29.8	20.0	30.1	20.1	40.1	0.99
Mitochondrial DNA [b]						
Heavy strand	22.5		31.5		46	0.71
Light strand	31.5		22.5		46	1.46

[a] Huberman and Attardi [16]
[b] Hallberg and Vinograd (personal communication).

The crude mitochondrial fraction was isolated as described in fig. 1, and free polysomes as described in Attardi et al. [7] from HeLa cells labeled for 30 min with ^{32}P-orthophosphate; the RNA was phenol-SDS extracted from the two fractions and run through sucrose gradients as in fig. 2.

estimated from the initial rate of labeling of the RNA of the mitochondrial fraction is very high: it is about two-thirds of the steady-state rate of arrival in the cytoplasm of newly synthesized free polysome mRNA and about 30% of the rate of arrival of free polysome 28 S RNA [8]. As shown in table 3, the rate of transcription of mit-DNA, when referred to unit length, is 10 to 20 times as high as the average rate of transcription of nuclear DNA, although not quite so high as that of the rRNA genes. On the other hand, it is one-third to one-sixth the average rate of transcription of *E. coli* DNA. In spite of its high rate of synthesis, mitochondrial RNA does not accumulate in the cytoplasm of HeLa cells, where it represents only 1% or less of the RNA: this is due to the fact that it is metabolically unstable. Addition of actinomycin D at a concentration of 7.5 μg/ml immediately stops the labeling of mitochondrial RNA, and the heterogeneous RNA synthesized prior to addition of the drug decays with an estimated half-life of about 30 min [5].

Table 3

Comparative rate of transcription of various DNAs

DNA type	Total mol wt/cell	Rate of transcription	
		Nucleotides/ min/cell	Nucleotides/min/ 10^9 daltons
HeLa mitochondrial DNA	1.2–2.4×10^{10} [a]	4.1×10^6 [e]	1.7–3.4×10^5
HeLa total DNA	1.2×10^{13} [b]	2.2×10^8 [f]	1.8×10^4
HeLa rDNA	1.3×10^{10} [c]	4.2×10^7 [g]	3.2×10^6
E. coli DNA	5.6×10^9 [d]	5.6×10^6 [h]	1.0×10^6

[a] The minimum estimate was derived from the reported maximum yield of closed circular mit-DNA from HeLa cells, the maximum estimate from the DNA content of a HeLa cell [17], by assuming that 0.2% of it is mit-DNA as in L cells [18].
[b] McConkey and Hopkins [17].
[c] Estimated as equal to twice (for the asymmetry of transcription) the fraction of HeLa DNA complementary to 45 S RNA [14].
[d] Corresponding to two genome equivalents/cell [19], each with a mol.wt. of about 2.8×10^9 daltons [20].
[e] This estimate was obtained from the initial rate of labeling of mitochondria-associated RNA relative to the rate of arrival in the cytoplasm of mRNA or 28 S RNA of free polysomes.
[f] Estimated as equal to twice the rate of synthesis of mature rRNA (to account for the half of the 45 S RNA which is lost during maturation [14]).
[g] Refers to the rate of synthesis of rRNA precursors and heterogeneous nuclear RNA (the rate of synthesis of the latter was estimated to be about four times as high as that of 45 S RNA [21].
[h] Mueller and Bremer [19].
(For details concerning the data of this table see [8].)

Discrete RNA species present in crude mitochondrial preparations from HeLa cells

Recently, several discrete RNA species have been identified by polyacrylamide gel electrophoresis and sedimentation analysis of bulk RNA extracted from a crude mitochondrial fraction of hamster, human (HeLa) and *Xenopus* cells [8,13,22–24]. A complication in these studies has been the presence of a large excess of rRNA of contaminating rough ER elements [7], which can mask the presence of minor mitochondrial RNA components. In some of the above quoted investigations low doses of actinomycin D were used to suppress selectively rRNA synthesis without affecting mitochondrial RNA synthesis [22,23]. In our investigations we took advantage of the fact that the bulk of the ER-bound ribosomes can be removed from the crude mitochondrial fraction by EDTA treatment. In fact, as previously shown for rat liver microsomes [25], this EDTA-stripping procedure releases from the rough ER elements almost all the small ribosomal subunits and about 70% of the large subunits, the residual membrane-stuck 50 S subunits being presumably those which carry the more complete polypeptide chains [7]. Figs. 4 and 5 show the labeling of RNA components sedimenting slower than 28 S from an EDTA-treated crude mitochondrial fraction of cells exposed to ^{3}H-uridine pulses of various duration. One can recognize in the O.D.$_{260}$ profile a prominent 28 S RNA peak at the bottom of the gradient (partially pelleted), an 18 S RNA peak with a shoulder at about 16 S, two somewhat smaller peaks at about 21 S and 12 S and a pronounced peak at 4 S. A minor component not always well resolved and possibly in variable amount can be seen at about 23 S. As previously mentioned, after a 15 min ^{3}H-uridine pulse (fig. 4), discrete peaks of radioactivity are already clearly recognizable over a high background of heterogeneous RNA. One can see a labeled peak at 17 S with a shoulder at about 16 S in correspondence with the shoulder in the O.D. profile. The 12 S and 4 S components appear also to be labeled at this time; by contrast, no radioactivity appears to be associated with 21 S RNA. After a 30 min pulse (fig. 4), the three peaks of radioactivity observed earlier have become more pronounced. The radioactive peak sedimenting behind 18 S has its center now at about 16 S, suggesting a conversion of

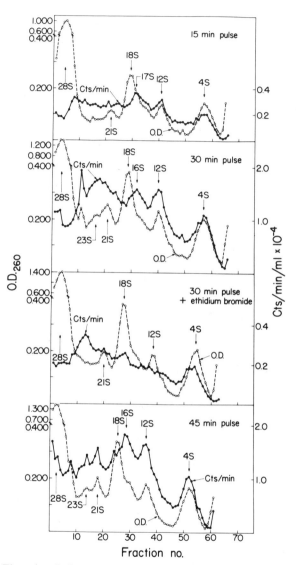

Fig. 4. Sedimentation pattern of RNA components sedimenting slower than 28 S RNA from the EDTA-treated and isopycnically separated crude mitochondrial fraction of HeLa cells exposed to ^{3}H-5-uridine pulses of various duration in the absence or in the presence of 1 μg/ml ethidium bromide.

the early labeled 17 S RNA to somewhat slower sedimenting molecules; a shoulder at 18 S suggests the initial labeling of this component. If the 30 min ^{3}H-uridine pulse is carried out in the presence of 1 μg/ml ethidium bromide, the labeling of the 16 S component is completely inhibited, unmasking the initial labeling of 18 S RNA (fig. 4). It should be

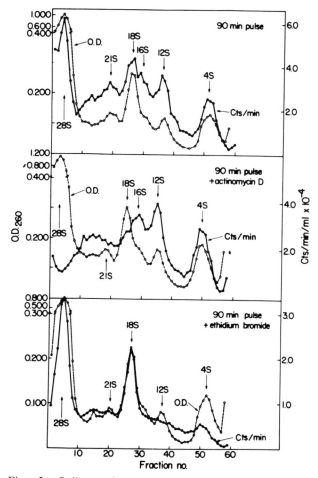

Fig. 5. Sedimentation pattern of RNA components sedimenting slower than 28 S RNA from the EDTA-treated and isopycnically separated crude mitochondrial fraction of HeLa cells exposed to ^{3}H-5-uridine for 90 min in the absence or in the presence of 0.04 μg/ml actinomycin D or 1 μg/ml ethidium bromide.

the 16 S component: this is recognizable as a shoulder on the light side of 18 S RNA. If the 90 min ^{3}H-uridine pulse is carried out in the presence of 0.04 μg/ml actinomycin D to block selectively rRNA synthesis [26], the 18 S RNA labeling is abolished and the 16 S labeled peak stands out clearly (fig. 5). Under the same conditions, the labeling of the 12 S and 4 S peaks is also not affected to any appreciable extent; by contrast, the labeling of the 21 S component appears to be greatly reduced. A different pattern is observed if the 90 min ^{3}H-uridine pulse is carried out in the presence of 1 μg/ml ethidium bromide. Fig. 5 shows that this drug strongly suppresses the labeling of the heterogeneous RNA, and abolishes that of the 16 S, 12 S and of most of the 4 S components. The labeling of the 21 S and 23 S RNA seems to be on the contrary only partially inhibited.

The results of the kinetics and inhibition experiments described above strongly suggest that the 16 S, 12 S and most of the 4 S RNA components are synthesized on a mit-DNA template. Furthermore, they provide evidence in favor of the idea that the residual 18 S RNA of the EDTA-treated crude mitochondrial fraction (estimated to correspond to 10 to 15% of that originally present in this fraction) pertains to a small amount of 30 S subunits of nuclear origin not removed by the chelating agent. Finally, they suggest a non-mitochondrial origin of the 21 S and 23 S RNA components.

The conclusions stated above have been fully corroborated by RNA-DNA hybridization tests. In these experiments, the sedimentation distribution of long-term labeled RNA from a crude mitochondrial fraction of HeLa cells was scanned for capacity to hybridize with mit-DNA [13]. As shown in fig. 6, the sedimentation profile of the components homologous to mit-DNA (as determined by RNA-DNA hybridization experiments carried out at a low RNA to DNA ratio (1:6) after normalization for the differences in amount of RNA in various fractions of the gradient) shows a broad band between 9 and 15 to 16 S, with a peak at 12 S, and a pronounced 4 S peak. The shoulder of RNA homologous to mit-DNA at 15 to 16 S presumably corresponds to the ethidium bromide-sensitive 16 S component. A small amount of faster sedimenting components hybridizable with mit-DNA can also be seen in the region 16 to 30 S.

noticed that the 18 S RNA of free ribosomes starts becoming labeled at about the same time. Ethidium bromide also inhibits the labeling of the 12 S component and to a great extent (by about 80%) that of 4 S RNA; the labeling of the heterogeneous RNA components is likewise strongly suppressed by the drug, as previously discussed.

After a 45 min pulse, the 21 S and 23 S components are clearly labeled (fig. 4). After a 90 min ^{3}H-uridine pulse (fig. 5), the now considerable labeling of the 18 S RNA species masks in part

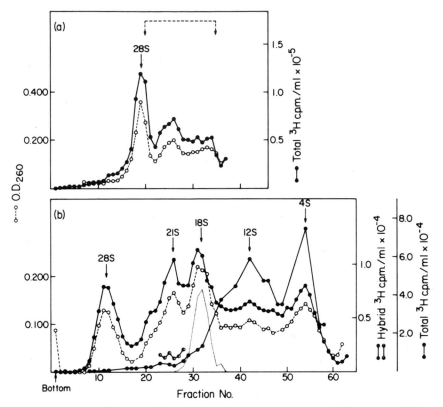

Fig. 6. Sedimentation profiles of total RNA and RNA hybridizable with mitochondrial DNA (at a low RNA to DNA input ratio) from the EDTA-treated and isopycnically separated crude mitochondrial fraction of HeLa cells labeled for 48 hr with ^{3}H-uridine. The components corresponding to 4 to 28 S in the sedimentation pattern of fig. 6(a) (indicated by arrows) were pooled, precipitated with ethanol, redissolved in SDS buffer and, after addition of a small amount of ^{14}C-labeled 18 S RNA from free polysomes, run through a 15 to 30% sucrose gradient in the Spinco SW 25–3 rotor at 25,000 rpm for 25 hr at 20°C. Components in individual or pooled fractions were collected by ethanol precipitation and centrifugation, and subjected to DNase digestion and Sephadex chromatography. An aliquot of each sample was tested for capacity to hybridize with ^{14}C-labeled closed circular mit-DNA (at a ratio of 1 to 6) by the Gillespie-Spiegelman method [27]. The hybridization values are corrected for non-specific background determined with *E. coli* DNA (<10%) and normalized for the differences in amount of RNA in various fractions of the gradient (○–––○). One set of hybridization experiments (●———●) was carried out at an RNA to DNA ration of 1:20 [13].

No, or very little, hybridization seems to occur between the 21 S or 23 S RNA component and mit-DNA, as judged from the absence of a peak of hybridizable RNA at 21 S or 23 S emerging over the low background of hybrids in this region of the gradient. When the RNA-DNA hybridization experiments were carried out at a relatively high RNA to DNA input ratio (5:1), a different distribution in the gradient of the RNA hybridized with mit-DNA was observed (fig. 7): in comparison with the sedimentation profile of the RNA hybridizable at low RNA to DNA input ratio there is in fact, under these conditions, a higher amount of hybrids formed by

components heavier than 16 S and a uniformly high level of hybridization between 4 and 16 S. These differences are not surprising, since at the relatively high RNA to DNA ratio used in this experiment, one would expect that the contribution of various RNA species to the hybrids would, within certain limits, reflect more the fraction of DNA complementary to each species than the concentration of the latter. The relatively high levels of hybridization occurring with components in the region of the gradient corresponding to S values between 16 and 28 S is presumably due to the small amounts of heterogeneous RNA sedimenting in this region, detected also in the

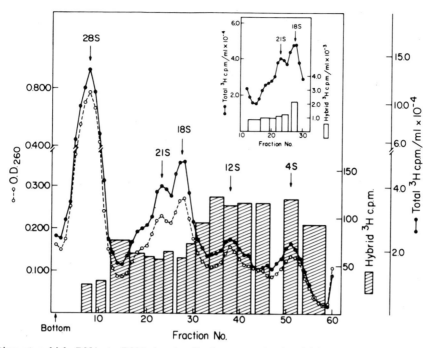

Fig. 7. Hybridization at a high RNA to DNA input ratio between mitochondrial DNA and different components of the sedimentation pattern of RNA from the EDTA-treated and isopycnically separated crude mitochondrial fraction of HeLa cells labeled for 24 hr with ^{3}H-5-uridine. The RNA-DNA hybridization experiments were carried out at a ratio of RNA to DNA of 5:1. The data are corrected for non-specific background as in fig. 6. In the insert, the results of hybridization assays carried out at a low RNA to DNA input ratio (1:17) are plotted (after correction for non-specific background and normalization for differences in amount of RNA in various fractions of the gradient) (from Attardi and Attardi [8]).

hybridization at a low RNA to DNA ratio. Hybridization with heterogeneous components is also predominant in the region of 4 to 16 S at high RNA to DNA ratio, and tends to cover here the contribution to the hybrids by 12 S and 4 S RNA.

The 16 S and 12 S mit-DNA-coded RNA species detected in the present work probably correspond to the unresolved '17 S' RNA found by Dubin [22] in the crude mitochondrial fraction from hamster cells treated for a long time with a low dose (0.1 μg/ml) of actinomycin D, and which has been recently shown to consist of two components [28].

An analysis by polyacrylamide gel electrophoresis of individual components isolated by sucrose gradient centrifugation has shown (Attardi and Attardi, in preparation) the RNA species sedimenting at 12 S corresponds to the '12 S' component recognizable in the gel electrophoresis pattern of RNA extracted from a crude mitochondrial fraction of HeLa cells [23] and Xenopus cells [24]. On the other hand, the

RNA species sedimenting at 16 S has been found to move through the gel slower than expected, i.e. slightly behind the 18 S rRNA marker. It is very likely that the 16 S RNA corresponds to the '21 S' electrophoretic component previously described [23,24]. The reason for the higher electrophoretic mobility of this species in the present experiments is not known. A discrepancy between sedimentation and electrophoretic properties of discrete mitochondrial RNA components has also been observed by others [23,29,48], and is presumably due to the less compact configuration of these species as compared to the rRNA markers.

Fraction of mitochondrial DNA which is transcribed in HeLa cells

In order to obtain direct evidence as to the fraction of mit-DNA which is expressed in HeLa cells.

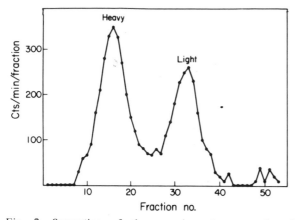

Fig. 8. Separation of the complementary strands of ^{14}C-labeled HeLa mitochondrial DNA in an alkaline CsCl gradient. A solution containing 13 μg closed circular mit-DNA in 4.0 ml 0.055 M K_3PO_4, 0.01% SDS, was brought to a refractive index of about 1.405 with solid CsCl and to pH 12.4 with KOH. The mixture was centrifuged in a polyallomer tube in the Spinco 65 angle rotor at 42,000 rpm for 42 hr. Eleven drop fractions were collected from the bottom of the tube and assayed for radioactivity.

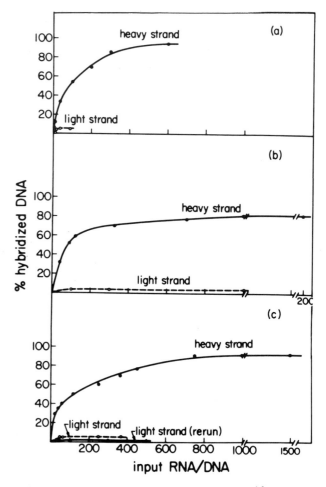

RNA-DNA hybridization experiments were carried out between RNA from an EDTA-treated crude mitochondrial fraction of cells labeled for 48 hr with ^{3}H-uridine and ^{14}C-labeled mit-DNA. Since the great tendency of mit-DNA to renature would have made the quantitation in these experiments difficult, advantage was taken of the fact that the two complementary strands of human mit-DNA can be separated in alkaline CsCl gradients [30] due to their different GT/CA ratio [31]. Fig. 8 shows an example of such a separation. As appears from fig. 9, HeLa mitochondrial RNA hybridizes exclusively with the heavy strand, in agreement with the above discussed complementarity of their base compositions (table 2). The low level of hybridization observed with the light strand (about 5% of that obtained with the heavy strand) is presumably due in part to a small amount of contaminating heavy strand: in fact, this level was considerably reduced by using light strand run twice through alkaline CsCl gradient (fig. 9(a)). The exclusive transcription of the heavy strand of mit-DNA previously reported for rat liver [32] thus holds true also for exponentially growing cells. As appears in fig. 9, the RNA components sedimenting slower than 21 S and faster than 32 S saturate the heavy strand of mit-DNA at a level corresponding to more than 90% of the DNA; the RNA components

Fig. 9. Hybridization of separated strands of ^{14}C-labeled HeLa mitochondrial DNA with increasing amounts of RNA from the EDTA-treated and isopycnically separated crude mitochondrial fraction of HeLa cells labeled for 48 hr with ^{3}H-5-uridine. The RNA components from the EDTA-treated crude mitochondrial fraction corresponding to S values > 32, between 32 and 21, and < 21, separated by sucrose gradient centrifugation, were collected by ethanol precipitation and centrifugation, subjected to DNase digestion and Sephadex chromatography, and used for hybridization tests by the Gillespie and Spiegelman procedure [27]. The specific activity of the RNA was 28,000 cts/min/μg for the heavy cut (a), 31,000 cts/min/μg for the intermediate cut (b) and 31,000 cts/min/μg for the light cut (c). Appropriate controls showed that the specific activity of the unstable RNA components was approximately equal to that of stable RNA (in particular rRNA). Aliquots of 0.0125 μg (for the heavy RNA cut) or 0.025 μg (for the intermediate and light RNA cuts) of ^{14}C-thymidine labeled heavy or light strand of mit-DNA were immobilized on nitrocellulose membranes and incubated with the RNA samples in 2 ml 4 X SSC containing 0.01 M Tris buffer pH 7.8 for 24 hr at 68°C. The data are corrected for non-specific background determined without DNA or with SV 40 DNA (less than 10% of the hybrid level).

sedimenting between 21 and 32 S give a slightly lower saturation value (about 80%). These results indicate that the whole or almost whole length of the heavy strand of mit-DNA is transcribed in HeLa cells (Aloni and Attardi, in preparation). It is possible that this transcription takes place in the form of a continuous RNA chain: this would be in agreement with the occurrence in the sedimentation profile of pulse labeled mitochondrial RNA of heavy RNA molecules ethidium bromide sensitive and homologous to mit-DNA. We are at present investigating this possibility.

Informational role of mitochondrial RNA for protein synthesis

An important question to which we have given our attention is whether any of the transcription products of mit-DNA is utilized as mRNA for protein synthesis. A priori one can conceive that translation of mitochondrial messages can occur either inside or outside the mitochondria. Furthermore, ignoring the improbable event that an unusual mechanism of protein synthesis operates here, we have to assume that any utilization of mitochondrial mRNA will involve its association with ribosomes. In yeast and *Neurospora* mitochondria there is convincing evidence for the presence of specific ribosomes containing distinctive RNA species [8], which appear to be coded by mit-DNA [29,33,34]. *Neurospora* mitochondria have been shown to contain a whole set of tRNA species and amino acyl-tRNA synthetases, some of which have different properties from those of the cytoplasmic counterparts [35,36]; furthermore, the mitochondrial protein synthesizing system in yeast exhibits both in vivo and in vitro a pattern of sensitivity towards various inhibitors which is different from that of the extramitochondrial machinery and strikingly resembles that of bacterial systems [37]. It is likely that the mitochondrial protein synthesizing apparatus in yeast and *Neurospora* utilizes at least in part mit-DNA-coded messages, although no conclusive evidence has been presented as yet in this respect. In animal cells, although there is considerable evidence in the literature of an intrinsic mitochondrial protein synthesizing system [38–40], the search for

mitochondrial ribosomes has so far been elusive. In HeLa cells, it has been shown that the great majority of the ribosomes found in crude mitochondrial preparations pertains to contaminating rough ER elements [7]. The 28 S RNA which cannot be removed from these preparations by EDTA treatment (about 30% of the original amount) belongs mostly if not exclusively to ER-stuck 50 S subunits, as in rat liver microsomes [25]. The high resolution sedimentation analysis used in the experiments described earlier has revealed that a small amount of 18 S RNA (estimated to correspond to 10 to 15% of the original) is also consistently not removed by EDTA: this residual RNA presumably pertains to small ribosomal subunits, which may derive in part from dissociation of intramitochondrial ribosomes. The EDTA-resistant 18 S RNA has been shown to be identical in sedimentation properties and kinetics of labeling to 18 S RNA of cytoplasmic ribosomes and, furthermore, has no base sequence homology to mit-DNA; therefore, it is presumably of nuclear origin.

As to the other discrete RNA components which are found in a crude mitochondrial fraction of HeLa cells, no evidence exists as yet that they are constituents of mitochondria-specific ribosomes. Clearly, more work is needed to answer the question of the nature and origin of mitochondrial ribosomes in HeLa and other animal cells. On the other hand, we have evidence which suggests that mit-DNA-coded mRNA does indeed occur in polysomal structures present in the crude mitochondrial fraction. This is illustrated in the experiment shown in figs. 10 and 11. In this experiment, polysomes of the crude mitochondrial fraction from cells labeled for 30 min with ^{3}H-uridine were released by deoxycholate from their association with membrane components and purified by two cycles of sucrose gradient centrifugation (fig. 10). The RNA was extracted from the polysome peak and separated on a sucrose gradient. As shown in fig. 11, the labeled mRNA components hybridize with extraordinary efficiency with mit-DNA: more than 20% of the labeled RNA was specifically bound to mit-DNA in this experiment, and the components homologous to this DNA did not appear to be exhausted yet. This experiment does not give any definite information as to the location of the polysomes containing the mit-DNA-coded mRNA. Since, however, the great majority if not all of the

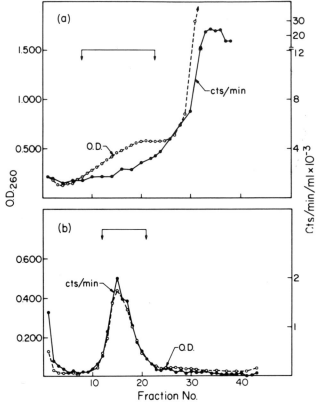

Fig. 10. Isolation of polysomes from the crude mitochondrial fraction of HeLa cells labeled for 30 min with ^{3}H-uridine. (a) The crude mitochondrial fraction isolated by differential centrifugation was lysed with 1% NaDOC, run on a 15 to 30% sucrose gradient in TKM (0.01 M Tris buffer pH 7.1, 0.01 M KCl, 0.00015 M MgCl$_2$) for 120 min at 24,000 rpm in the SW 25.1 Spinco rotor. (b) The fractions indicated by arrows in (a) were pooled, dialyzed against TKM, and centrifuged through a sucrose gradient consisting of 5 ml 23% (w/w) sucrose, 24 ml 23 to 55% sucrose gradient and 10 ml 55% sucrose, all in TKM (SW 25.2 Spinco rotor, 20,000 rpm, 11.5 hr). From Attardi and Attardi [8].

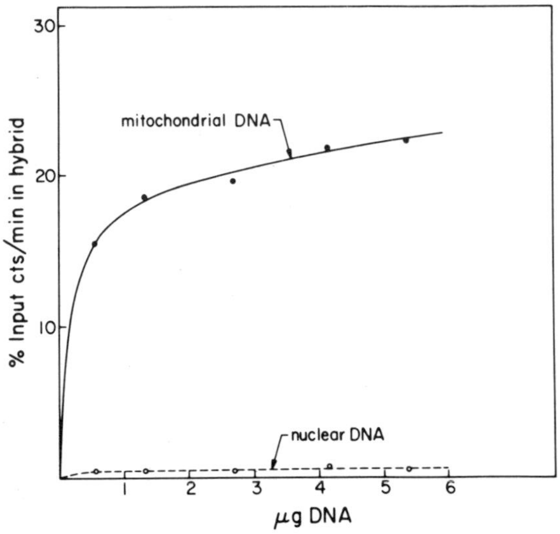

Fig. 11. Hybridization of mRNA from membrane-associated polysomes with mit-DNA and nuclear DNA. RNA was released by SDS from the polysomal band shown in fig. 10 and run in a sucrose gradient in SDS buffer. After 30 min labeling, essentially all radioactivity in these polysomes is associated with heterogeneous RNA, which is presumably mRNA. RNA components between 5 and 33 S were collected by ethanol precipitation and centrifugation, and tested for capacity to hybridize with increasing amounts of denatured ^{14}C-labeled mit-DNA or nuclear DNA bound to nitrocellulose membranes.

polysomes present in a crude mitochondrial fraction from HeLa cells pertain to contaminating rough ER [7], these results suggest the possibility that mit-DNA-coded RNA is exported to the rough ER, where it becomes associated with ribosomes of nuclear origin. This possibility is in agreement with the previously mentioned results of hybridization with mit-DNA of the RNA extracted from the crude mitochondrial fraction of cells labeled for 2 hr with ^{14}C-uridine. Experiments carried out with RNA from purified rough ER components should be able to test this possibility.

In order to obtain direct evidence as to the existence and properties of a mitochondrial protein synthesizing system in HeLa cells and, eventually, on the nature of the proteins synthesized in mitochondria and in the ER-bound polysomes under the direction of mit-DNA, a cell-free protein synthesizing system utilizing a mitochondria-ER fraction has been investigated in this laboratory (Lederman and Attardi, in preparation). Table 4 illustrates the results obtained with a system utilizing the whole unfractionated 8,100 g membrane fraction and 229,000 g supernatant. One can see that the amino acid incorporation is totally dependent on an added ATP generating system and is not inhibited by dinitrophenol or oligomycin: this indicates that, under the conditions employed in these experiments, exogenous ATP satisfies the total energy requirements of the system without appreciable participation of oxidative

Table 4
Amino acid incorporating system utilizing a membrane fraction, or free polysomes, and the soluble fraction from HeLa cells

Cell fraction	Incubation mixture	Incorporation	
		cpm/mg protein	% of control
Experiment 1			
8100 *g*	Complete, no addition	15,090	100
membrane fraction [a]	Same minus creatine phosphate and creatine phosphokinase	252	2
	Same plus RNase, 40 μg/ml	968	6
	Same plus dinitrophenol, 2 $\times$ 10^{-4} M	16,080	107
	Same plus oligomycin, 25 μg/ml	14,050	93
	Same plus puromycin, 150 μg/ml	455	3
	Same plus chloramphenicol, 50 μg/ml	13,390	89
	Same plus chloramphenicol, 100 μg/ml	11,840	78
	Same plus cycloheximide, 150 μg/ml	3,120	21
	Same plus cycloheximide, 300 μg/ml	2,480	16
Experiment 2			
5000 *g*	Complete, no addition	6,570	100
membrane fraction [b]	Same plus RNase, 40 μg/ml	2,410	37
	Same plus dinitrophenol, 2 $\times$ 10^{-4} M	7,740	118
	Same plus chloramphenicol, 100 μg/ml	3,360	51
	Same plus cycloheximide, 300 μg/ml	4,230	64
Experiment 3			
Free polysomes	Complete, no addition	7,150	100
	Same plus RNase, 40 μg/ml	344	5
	Same plus puromycin, 150 μg/ml	232	3
	Same plus chloramphenicol, 50 μg/ml	6,650	93
	Same plus chloramphenicol, 100 μg/ml	6,940	97
	Same plus cycloheximide, 300 μg/ml	526	7

In experiments 1 and 2, the complete incubation mixture contained, in 0.5 ml, 350 mM sucrose, 44 mM Tris buffer pH 7.3, 35 mM KCl, 8 mM MgCl$_2$, 1 mM CaCl$_2$, 1 mM dithiothreitol, 2 mM ATP, 0.4 mM GTP, 15 mM creatine phosphate, 50 μg/ml creatine phosphokinase, 0.1 mM of each amino acid (minus leucine), 5 μg/ml ^{3}H-leucine (40 mC/μM), 8,100 *g* membrane fraction (700 μg protein) or 5,000 *g* membrane fraction (280 μg protein) and 0.1 ml of 229,000 *g* supernatant (300 μg protein).
In experiment 3, the complete incubation mixture contained, in 0.5 ml, 0.2 ml of the 15,800 *g* supernatant of a cytoplasmic extract, containing the bulk of free polysomes (900 μg protein), and all the components specified above at the same final concentrations, with the exception of the 229,000 *g* supernatant.
All incubations were for 60 min at 37°C. For experiments 1 and 2 the incorporation data are referred to milligram of protein of the membrane fraction only. 8,000 cts/min correspond to at least 10^{-4} μmoles of leucine incorporated.
[a] Prepared according to Attardi et al. [7].
[b] Prepared according to Kroon [38].

phosphorylation processes. Puromycin inhibits leucine incorporation almost completely. RNase also inhibits it to a great extent, the RNase resistant fraction varying between 6 and 16% in different experiments: this RNase-resistant amino acid incorporation is presumably real polypeptide synthesis, because it is almost totally abolished by puromycin or chloramphenicol (not shown), though not by cycloheximide. The system is only slightly sensitive to chloramphenicol, the observed degree of inhibition for a concentration of the drug of 50 to 100 μg/ml being always less than 25%, with a certain variability from experiment to experiment. Cycloheximide inhibits amino acid incorporation by 80 to 90%.

Protein synthesis by a cell-free system utilizing free polysomes, apart from a more complete inhibition by cycloheximide and an apparent lack of sensitivity to chloramphenicol, has properties indistinguishable from those of the protein synthesizing activity of the 8,100 g membrane fraction (table 4). These results suggest that the majority of the amino acid incorporation by the latter system is supported by ER-bound polysomes.

Banding the 8,100 g membrane fraction in a 1.0 to 1.7 M sucrose gradient so as to eliminate the smooth membrane components does not change the properties of the system appreciably. On the contrary, fractionation in a dextran-sucrose gradient separates a rough ER fraction, which supports amino acid incorporation with a decreased sensitivity to chloramphenicol, and a mitochondria-enriched fraction, which has a protein synthesizing activity more resistant to RNase (about 35%) and cycloheximide (45%) and more sensitive to chloramphenicol (about 30% for a concentration of the drug of 50 μg/ml) than the unfractionated 8,100 g membrane components. These observations suggest that a considerable enrichment in a protein synthesizing system with properties different from those of free polysomes and similar to those reported for mitochondria [38,39] can be achieved by this fractionation. Likewise, preparation of a crude mitochondrial fraction by a procedure involving homogenization in 0.25 M sucrose instead of hypotonic medium and a lower speed centrifugation (5,000 g) to pellet the membrane components [32] yields a fraction which appears to be enriched in mitochondria, with respect to the 8,100 g membrane fraction, as judged from the increased RNase resistance and sensitivity to chloramphenicol and the decreased sensitivity to cycloheximide of the amino acid incorporation which it supports (table 4). In all the experiments described above, the 229,000 g supernatant had been added to the in vitro amino acid incorporating system. As shown in table 5, omission of the soluble fraction causes a considerable change in the properties of the system utilizing the 8,100 g membrane fraction. The total level of incorporation apparently does not change; however, the incorporating activity observed under these conditions is to a great extent (up to about 80%) resistant to RNase and completely insensitive to cycloheximide; the system is, on the other hand, only partially (about 50%) sensitive to chloramphenicol. The incorporating activity is completely dependent on external ATP and is not affected by dinitrophenol, suggesting that under these conditions oxidative phosphorylation does not play any significant role as energy yielding mechanism for protein synthesis. If, instead of an ATP generating system, a respiratory substrate (succinate), ADP and phosphate are added, the incorporation becomes in part dependent on respiratory-chain phosphorylation, as suggested by the partial sensitivity to dinitrophenol (table 5). The ATP needs are presumably satisfied in most part, under these conditions, by dinitrophenol-insensitive substrate level phosphorylation [41]. Under the same conditions, the amino acid incorporation is almost completely sensitive to chloramphenicol and resistant to cycloheximide. Similar results were obtained, in the absence of an external ATP generating system and in the presence of added substrate, with the 5,000 g membrane fraction. A reasonable interpretation of these results is that elimination of the supernatant factors suppresses drastically the amino acid incorporation by elements of rough ER: the apparent lack of effect on the measured level of labeled leucine incorporation is presumably due to the fact that the relatively small fraction of amino acid incorporation which is not dependent on supernatant utilizes leucine of a much greater specific activity, as a result of the elimination of the soluble leucine pool. This supernatant-independent incorporation appears to take place to a great extent in an RNase-resistant compartment, presumably mitochondria. The complete resistance to cycloheximide and almost complete sensitivity to chloramphenicol of the amino acid incorporation also argues in favor of a predominant role of an intramitochondrial protein synthesizing system under these conditions. The basis for the different sensitivity to antibiotics of this system as compared to the free and ER-bound polysomes needs further investigation. Whether mitochondrial protein synthesis in HeLa cells utilizes mit-DNA-coded mRNA and whether any of the protein synthesis by ER-bound polysomes is directed by mit-DNA are questions at present being investigated in our laboratory.

Table 5
Amino acid incorporation by membrane fractions from HeLa cells in the absence of the soluble fraction

Expt. No.	Energy source	Incubation mixture	8100 g membrane fraction		5000 g membrane fraction	
			cpm/mg protein	% of control	cpm/mg protein	% of control
1	Exogenous ATP generating system	Complete, no addition	11,740	100		
		Same minus ATP, creatine phosphate and creatine phosphokinase	112	1		
		Same plus RNase, 40 μg/ml	9,250	79		
		Same plus dinitrophenol, 2 × 10⁻⁴ M	5,940	51		
		Same plus chloramphenicol, 100 μg/ml'	14,230	121		
		Same plus cycloheximide, 300 μg/ml	12,710	108		
2	Oxidative phosphorylation	Complete, no addition	14,440	100	43,520	100
		Same plus RNase, 40 μg/ml	9,410	65	36,640	84
		Same plus dinitrophenol, 2 × 10⁻⁴ M	10,170	70	33,170	76
		Same plus puromycin, 150 μg/ml	479	3	–	–
		Same plus chloramphenicol, 100 μg/ml	1,735	12	3,850	9
		Same plus cycloheximide, 300 μg/ml	13,080	91	50,190	116

The 8,100 g membrane fraction (390 μg protein) was incubated, in experiment 1, under the same conditions described in the legend of table 4, except for the omission of the 229,000 g supernatant; in experiment 2, ATP, creatine phosphate and creatine phosphokinase were omitted, and succinate (30 mM), ADP (2 mM) and potassium phosphate (20 mM) were added.
The 5,000 g membrane fraction (280 μg protein) was incubated in 0.5 ml medium containing 350 mM sucrose, 50 mM Tris buffer pH 7.4, 60 mM KCl, 7 mM MgCl$_2$, 1 mM EDTA, 30 mM succinate, 2 mM ADP, 20 mM potassium phosphate, 0.1 mM of each amino acid (minus leucine), and 5 μC/ml ³H-leucine (40 mC/μM).
An assay for bacterial contamination by surface viable cell count revealed that no more than 10³ bacteria could be present per ml of incubation mixture for either membrane fraction: this maximum level of contamination would not contribute to any detectable extent to the measured incorporation [47].

Relationship of mitochondrial DNA replication to the cell cycle

Information useful for an understanding of the informational role of mit-DNA can be derived from an analysis of the transcription of this DNA in relationship, on one hand, to the cell cycle, on the other, to the assembly and division of mitochondria. As a preliminary for this type of analysis the replication of mit-DNA in different stages of the cell cycle was investigated (Pica-Mattoccia and Attardi, in preparation). Physiologically synchronized populations of HeLa cells were obtained by a technique based on the selective detachment of mitotic cells from their solid substrate [42,43]. Cell populations at various stages of the cell cycle were exposed to a ³H-thymidine pulse and the labeling of the closed circular mit-DNA (isolated from the mitochondrial fraction by CsCl-ethidium bromide density gradient centrifugation) was analyzed. In these experiments a

constant amount of unsynchronized HeLa cells labeled with ¹⁴C-thymidine was added to the ³H-thymidine pulse-labeled cells in order to provide an internal standard, so that the results could be normalized for variations in the recovery of mit-DNA. Fig. 12 shows the results of a typical experiment. It is evident that the labeling of mit-DNA in early G1 phase is about 30% of that in S phase and about two-thirds of that in G2 phase. Differences in the specific activity of the precursor pool could conceivably account for these results. Therefore, an analysis of the total cell pool of thymidine derivatives was carried out. The specific activity of this pool, after a ³H-thymidine pulse, showed indeed considerable variations in different phases of the cell cycle. Since the relative rate of uptake of exogenous ³H-thymidine into the intramitochondrial pool in various phases of the cell cycle was found to parallel the relative rate of uptake of the precursor into the total cell pool, it was considered likely that the

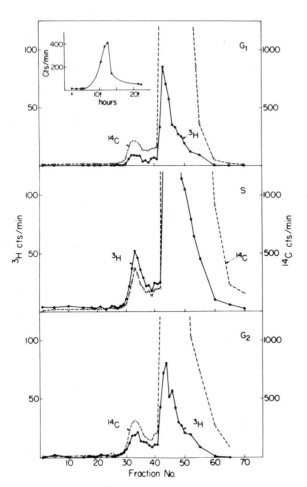

specific activities of the thymidine triphosphate in the intra- and extramitochondrial compartments should vary proportionally during the cell cycle. Therefore, the data of labeling of mit-DNA with ^{3}H-thymidine were corrected on the basis of the determined specific activity of the total cell precursor pool. Table 6 summarizes the results of several experiments after correction for pool differences. It appears that in the first few hours of the G1 phase there is relatively little replication of mit-DNA, with the residual level possibly being accounted for by the 20 to 35% of unsynchronized cells present in the cell

Fig. 12. Labeling of closed circular mitochondrial DNA from HeLa cells subjected to a ^{3}H-thymidine pulse at different stages of the cell cycle. Physiologically synchronized HeLa cells were obtained by the Terasima and Tolmach method [41]. Samples at different stages of the cell cycle were exposed for 1 hr to ^{3}H-thymidine, and mixed with a constant amount of unsynchronized cells labeled for 24 hr with ^{14}C-thymidine: a crude mitochondrial fraction was prepared from the mixed population, lysed with SDS and run in a CsCl-ethidium bromide density gradient [10,13]. In the insert, the rate of labeling with ^{14}C-thymidine of total cell DNA in the synchronized HeLa cell population at various stages of the cell cycle is shown. Arrows point to times at which samples were removed from the synchronized population for measurement of rate of mit-DNA synthesis.

Table 6

Relative rates of mitochondrial and nuclear DNA synthesis in different phases of the HeLa cell cycle

	G_1		S		G_2	
	1−5 hr	5−10 hr	10−12 hr	14−16 hr	18−21 hr	21−24 hr
Nuclear DNA	1.0 [a]	1.3	10.1	16.5	7.5	7.1
Mitochondrial DNA	1.0 [a]	2.0	2.5	4.1	4.6	5.0

[a] Arbitrary unit.

population. The overall rate of mit-DNA replication increases in the late G1-early S phase and reaches a maximum in the late S-G2 phase. The data are compatible with the idea that mit-DNA replication starts after initiation of the nuclear DNA replication in each cell and proceeds asynchronously over the S and G2 phases.

Discussion

From the results which have been discussed above it is clear that a crude mitochondrial fraction from HeLa cells contains both heterogeneous RNA components ranging in sedimentation constant from 4 S more than 50 S, which are rapidly labeled during exposure of the cells to a radioactive RNA precursor, and discrete RNA species which become labeled more slowly. While the rapidly labeled heterogeneous RNA appears to be totally or in its majority of mitochondrial origin, only some of the discrete species are transcribed from mit-DNA. In particular, the 16 S, 12 S and most of the 4 S RNA components appear to be specified by mit-DNA, as judged from their kinetics of labeling, ethidium bromide sensitivity and homology to mit-DNA.

As to the nature of the 16 S and 12 S components, further work is needed to clarify their biological significance. The 4 S RNA homologous to mit-DNA and ethidium bromide sensitive which has been detected in the present work presumably contains at least some of the mitochondria-specific tRNA species which have been described in animal cells, including HeLa cells [44–46]; sequence homology of mitochondrial leucyl-tRNA from rat liver to mit-DNA has been recently reported [18]. The relatively small fraction of 4 S RNA (15 to 20%) which is resistant to ethidium bromide presumably represents cytoplasmic 4 S RNA, possibly bound to the EDTA-resistant membrane-stuck 50 S subunits.

As to the 21 S and 23 S RNA components detected in the present work, their kinetics of labeling, sensitivity to actinomycin and lack of homology to mit-DNA indicate a non-mitochondrial site of synthesis. If these components are intramitochondrial, they must be imported from some other cellular site.

The elucidation of the biological significance of the discrete RNA species present in the crude mitochondrial fraction of animal cells will be of utmost importance for understanding the informational content of mit-DNA and the role of mitochondria in protein synthesis.

Another important question which needs to be considered is that of the nature of the rapidly labeled heterogeneous RNA and of the possible relationship between this and the mit-DNA coded discrete RNA species. The early kinetics of labeling mitochondrial RNA of different sedimentation constant (fig. 2) is consistent, as was mentioned earlier, with a precursor to product relationship between the fast sedimenting mitochondrial RNA components and the discrete RNA species. If the heavy strand of mit-DNA is transcribed in the form of a continuous long RNA chain, this precursor to product relationship would be unescapable. The heterogeneous components spread in the region from 4 to 50 S could represent intermediates or waste products in the processing of the large precursor and/or incomplete nascent mitochondrial RNA chains. The heterogeneous RNA might also include mit-DNA-coded mRNA species destined to be utilized inside the mitochondria or exported. The transcription of DNA in the form of gigantic RNA molecules destined to be processed to smaller size functional molecules, which is a well established phenomenon in the eukaryotic nucleus, may prove to be true also for mit-DNA, and thus represent the fundamental way of expression of the eukaryotic genome whether in the nucleus or in the cytoplasm.

Added in proof

Recent experiments have shown that the 16 S and 12 S RNA components are contained in ribonucleoprotein particles which for their structural and functional properties appear to be mitochondria-specific ribosomes (Attardi and Ojala, Nature, in press).

Acknowledgements

These investigations were supported by a grant from the National Institutes of Health (GM–11726) and by fellowships from the World Health Organization (Y.A.), the National Institutes of Health and the National Science Foundation. The capable technical assistance of Mrs. La Verne Wenzel and Mrs. Benneta Keeley is gratefully acknowledged.

References

[1] B. Ephrussi, H. Hottinguer and J. Tavlitzki, Ann. Inst. Pasteur 76 (1949) 419.

[2] P. Slonimski and B. Ephrussi, Ann. Inst. Pasteur 77 (1949) 47.

[3] M.B. Mitchell and H.K. Mitchell, Proc. Natl. Acad. Sci. U.S. 38 (1952) 442.

[4] P. Borst and A.M. Kroon, Intern. Review of Cytology 26 (1969) 107.

[5] B. Attardi and G. Attardi, Proc. Natl. Acad. Sci. U.S. 58 (1967) 1051.

[6] G. Attardi and B. Attardi, Proc. Natl. Acad. Sci. U.S. 61 (1968) 261.

[7] B. Attardi, B. Cravioto and G. Attardi, J. Mol. Biol. 44 (1969) 47.

[8] G. Attardi and B. Attardi, in: Internat. Symp. on Problems in Biology, 1. RNA in Development, ed. W.E. Hanly, Park City (Academic Press, 1969) in press.

[9] E. Zylber, C. Vesco and S. Penman, J. Mol. Biol. 44 (1969) 195.

[10] R. Radloff, W. Bauer and J. Vinograd, Proc. Natl. Acad. Sci. U.S. 57 (1967) 1514.

[11] L.V. Crawford and M.J. Waring, J. Mol. Biol. 25 (1967) 23.

[12] W. Bauer and J. Vinograd, J. Mol. Biol. 33 (1968) 141.

[13] B. Attardi and G. Attardi, Nature 224 (1969) 1079.

[14] Ph. Jeanteur and G. Attardi, J. Mol. Biol. 45 (1969) 305.

[15] M. Girard, H. Latham, S. Penman and J.E. Darnell, J. Mol. Biol. 11 (1965) 187.

[16] J.A. Huberman and G. Attardi, J. Cell Biol. 31 (1966) 95.

[17] E.M. McConkey and J.W. Hopkins, Proc. Natl. Acad. Sci. U.S. 51 (1969) 1197.

[18] M.M.K. Nass and C.A. Buck, Proc. Natl. Acad. Sci. U.S. 62 (1969) 506.

[19] K. Mueller and H. Bremer, J. Mol. Biol. 38 (1968) 329.

[20] J. Cairns, Cold Spr. Harb. Symp. Quant. Biol. 28 (1963) 43.

[21] R. Soeiro, M.H. Vaughan, J.R. Warner and J.E. Darnell, J. Cell Biol. 39 (1968) 112.

[22] D.T. Dubin, Biochem. Biophys. Res. Commun. 29 (1967) 655.

[23] C. Vesco and S. Penman, Proc. Natl. Acad. Sci. U.S. 62 (1969) 218.

[24] I. Dawid, Fed. Proc. 28 (1969) 349.

[25] D.D. Sabatini, Y. Tashiro and G.E. Palade, J. Mol. Biol 19 (1966) 503.

[26] S. Penman, C. Vesco and M. Penman, J. Mol. Biol. 34 (1968) 49.

[27] D. Gillespie and S. Spiegelman, J. Mol. Biol. 12 (1965) 829.

[28] D.T. Dubin and B.S. Montenecourt, J. Mol. Biol. 48 (1970) 279.

[29] H.O. Halvorson, A. Morimoto, A. Scragg and J. Nikhorocheff, this symposium.

[30] G. Corneo, L. Zardi and E. Polli, J. Mol. Biol. 36 (1968) 419.

[31] J. Vinograd, J. Morris, N. Davidson and W.F. Dove, Proc. Natl. Acad. Sci. U.S. 49 (1963) 13.

[32] P. Borst and C. Aaij, Biochem. Biophys. Res. Commun. 34 (1969) 358.

[33] E. Wintersberger and G. Viehauser, Nature 220 (1968) 699.

[34] D.D. Wood and D.J.L. Luck, J. Mol. Biol. 41 (1969) 211.

[35] W.E. Barnett and D.H. Brown, Proc. Natl. Acad. Sci. U.S. 57 (1967) 452.

[36] W.E. Barnett, D.H. Brown and J.L. Epler, Proc. Natl. Acad. Sci. U.S. 57 (1967) 1775.

[37] A.W. Linnane, in: Biochemical Aspects of the Biogenesis of Mitochondria, eds. E.C. Slater, J.M. Tager, S. Papa and E. Quagliariello (Adriatica Editrice, Bari, Italy, 1968) p. 333.

[38] A.M. Kroon, Biochim. Biophys. Acta 91 (1964) 145.

[39] L.W. Wheeldon and A.L. Lehninger, Biochemistry 5 (1966) 3533.

[40] T.S. Work, J.L. Coote and M. Ashwell, Federation Proc. 27 (1968) 1174.

[41] H.W. Heldt, H. Jacobs and M. Klingenberg, Biochem. Biophys. Res. Commun. 17 (1964) 130.

[42] T. Terasima and L.J. Tolmach, Exptl. Cell Res. 30 (1963) 344.

[43] E. Robbins and P.I. Marcus, Science 144 (1964) 1152.

[44] C.A. Buck and M.M.K. Nass, J. Mol. Biol. 41 (1969) 67.

[45] J.B. Galper and J.E. Darnell, Biochem. Biophys. Res. Commun. 34 (1969) 205.

[46] A.E. Smith and K.A. Marcker, J. Mol. Biol. 38 (1968) 241.

[47] A.M. Kroon, M.J. Botman and C. Saccone, in: Biochemical Aspects of the Biogenesis of Mitochondria, eds. E.C. Slater, J.M. Tager, S. Papa and E. Quagliariello (Adriatica Editrice, Bari, Italy, 1968) p. 439.

[48] M. Edelman, I.M. Verma and U.Z. Littauer, J. Mol. Biol. 49 (1970) 67.

EFFECTS OF THYROXINE ON ^{14}C-LEUCINE INCORPORATION AND MITOCHONDRIAL PROTEINS, RESPIRATORY CONTROL AND tRNA

D. RAO SANADI, Suresh S. KAPLAY and Stringner S. YANG

*Department of Bioenergetics Research, Retina Foundation,
Boston, Massachusetts, USA*

Our experiments in 1961 on the turnover rates of certain mitochondrial components raised the question whether the mitochondrion turns over as a unit [1]. Similar results have been obtained by some workers [2–5]. In a more recent study the turnover rate of mitochondrial DNA was determined in several organs of the rat [5]. The turnover rate was the highest in the heart ($t_{1/2}$ = 6.7 ± 1 days) and lowest in the brain (31 ± 6 days). The biochemical studies have been supported by electron microscopic observations that mitochondria are engulfed as a whole by lysosomes and digested [6,7]. On the other hand differential replacement of certain components has been revealed by other studies [8–13]. Thus, Neupert [11] and Beattie [12] have noted that the inner and outer mitochondrial membranes have different half lives. Swick et al. [8] obtained half life values of 0.78–1.03 days and 0.73 to 0.92 days respectively for the two mitochondrial enzymes, ornithine aminotransferase and alanine aminotransferase. In these experiments the enzymes were induced with high protein diets or prednisolone injection. In another experiment [9] a half life of as low as 72 min was observed for rat liver δ-aminoevulinatesynthetase induced under the influence of a porphyrogenic drug. These latter results obtained by enzyme activity determinations give rise to the possibility that although the mitochondrion as a whole may turnover as a unit, a small complement of inducible protein components in it could be replaced more rapidly. The metabolic state of the animal may be the important determinant in these replacements and may affect the estimates of gross turnover rates if measured without adequate control of physiological parameters. In order to probe deeper into this question, the effect of thyroxine on mitochondrial function and biogenesis have been examined. These results also have important bearing on the function of thyroxine.

In our recent studies [14], rats were injected intraperitoneally with ^{14}C-L-leucine (5 μc/100 g body weight) and its incorporation in different liver mitochondrial fractions was measured at the intervals of 15, 30, 60 and 180 min. The mitochondrial protein suspension in 0.15 M KCl was sonicated for 2 min and then centrifuged at 100,000 g for 1 hr. The soluble protein was passed through Sephadex G-25 and then fractionated on a DEAE-cellulose (DE-52) column. Five fractions were eluted with 5, 20, 40, 50 and 250 mM potassium phosphate pH 7.5. It was observed that the incorporation reached a maximal value in all fractions in 30 min and then remained constant. Some of the soluble protein fractions (20, 40 and 50 mM − group 1 proteins) were labeled 2–3 times more rapidly than the total mitochondria or the other (5 and 250 mM) fractions, consistant with earlier findings that the appearance of certain enzyme activities could be rapidly induced in mitochondria. These group 1 proteins represented about 6% of the total mitochondrial proteins. In similar incorporation studies with thyroidectomized animals, the labeling in all the fractions was found to be depressed to about 50% of control, but the increased incorporation in group 1 proteins was still apparent. When the tyroidectomized rats were treated with a single dose of L-thyroxine (0.52 μg/g body weight) the labeling in only the group 1 proteins was increased significantly within 60 min after thyroxine treatment. The radioactivity in the total mitochondria and in the rest of the fractions, however, was unaffected by the treatment. The incorporation in the insoluble protein fraction (100,000 g pellet) was not significantly different from the normal, either in thyroidectomy or after thyroxine treatment. Fractionation of the membrane proteins could of course reveal differences in incorporation rates but this has not been carried out. Six mitochondrial enzymes (malate, lactate and glutamate dehydrogenases, glutamate-oxalocetate and glutamate-pyruvate transaminases and the soluble glycerol-1-phosphate dehydrogenase) have been assayed in the fractions. They were localized in the 5 or 250 mM fractions but the group 1 proteins had no detectable activity. Simultaneous studies of the respiratory activities of mitochondria showed that state 3 (active) and state 4 (controlled, ADP limiting) respiration were both lowered to almost the same extent by thyroidectomy, but state 4 respiration alone was increased after thyroxine treatment leading to a decrease in respiratory control, as reported by Hoch [15]. The effect of these group 1 proteins on the mitochondrial activity and their site of synthesis are described here.

Results and discussion

Direct effect of thyroxine and effect via protein synthesis on respiration

Effect of group 1 proteins on mitochondrial activities
The protein fractions eluted from the DE-52 column by 20, 40 and 50 mM potassium phosphate were obtained from thyroidectomized rats that had been treated with a single injection of thyroxine three hours before sacrifice. The fractions were precipitated with ammonium sulfate and dialyzed against 20 mM K-phosphate pH 7.5. It was found that addition of this 20 mM fraction to mitochondria from thyroidectomized rats resulted in a striking stimulation of state 4 respiration (table 1). If the effect of in vivo injection of thyroxine on state 4 in these experiments was taken as 100%, then the increase in state 4 after the in vitro addition of different levels of 20 mM fraction was 50 to 70%. The 40 and 50 mM eluate fractions at the concentrations used in these experiments, did not have any effect on the mitochondrial activity. According to Hoch [15] the effect of injected thyroxine on the controlled (state 4) respiration of mitochondria is reversed by in vitro addition of bovine serum albumin. However, the effect of the

Table 1
Effect of the rapidly labeled '20 mM protein' fraction on state 4 respiration.

Mitochondrial source	20 mM protein µg	µatoms O_2/hr/mg		
		State 3	State 4	RCR
T_X		1.5	0.52	2.9
$T_X + T_4$ (3 hr)		1.7	1.0	1.7
T_X	112($T_X + T_4$)	1.6	0.75	2.1
T_X	280($T_X + T_4$)	1.8	0.90	2.0
T_X		2.0	0.65	3.1
T_X	200 (normal)	1.9	0.65	2.9
T_X	400 (normal)	2.0	0.70	2.9

The 20, 40 and 50 mM fractions were obtained from the thyroidectomized rats treated with L-thyroxine (0.52 µg/g body weight) for 3 hr. The proteins were added to the polarographic reaction chamber directly. The experimental details were as described previously [14].
No effect was observed with the 40 and 50 mM protein fractions. The 40 mM fraction was used in the range of 50 to 200 µg and the 50 mM fraction 4 to 40 µg protein. The P/O was the same in these experiments.

20 mM eluate fraction on state 4 respiration of thyroidectomized rat liver mitochondria was not reversed by bovine serum albumin. The 20 mM fraction obtained from normal rat liver mitochondria did not have any effect under the same conditions on the thyroidectomized mitochondrial activity (table 1). Nor did state 4 respiration of mitochondria from a normal rat show any change with the 20 mM fraction which was active with deficient mitochondria. The effect is seen only when mitochondria from thyroidectomized rats are combined with the 20 mM fraction from treated rat liver mitochondria. No other combination produced a change in state 4 respiration.

The above results together with the stimulation of labeling of this protein fraction by a single dose of thyroxine indicated a close relationship between the effect of thyroxine and synthesis of the 20 mM protein fraction which amounts to 1 to 2% of the mitochondrial protein. It was of interest therefore to identify the site of synthesis of these proteins.

Effect of chloramphenicol and cycloheximide on ^{14}C-leucine incorporation into different mitochondrial fractions
It is apparent from the work of several laboratories

that mitochondrial proteins are synthesized at two different sites in the cell. The small part, probably a membrane component, synthesized inside mitochondria in in vitro experiments is sensitive to chloramphenicol [16–22]. The bulk of the protein is synthesized extramitochondrially in a cycloheximide sensitive system [22–26] and then integrated into the mitochondrial structure [16]. Administration of cycloheximide (1 mg/100 g body weight one hour before sacrifice) inhibited all but a small part of the incorporation of ^{14}C-leucine in the insoluble fraction (table 2). Thus, the group 1 proteins are synthesized extramitochondrially.

Administration of chloramphenicol in a single dose (53 mg/100 g body weight) four hours before sacrifice did not show any effect on ^{14}C-leucine incorporation into mitochondria. But four such doses administered as shown in table 3 produced two effects. One was an approximately 30% decrease in mitochondrial protein per g tissue. The decrease was entirely due to a decrease in the insoluble protein content of mitochondria, which is consistent with the observations of Firkin and Linnane [32] on the inhibition of cytochrome synthesis in rat liver by high doses of chloramphenicol. Secondly, the incorporation of ^{14}C-leucine was inhibited in all the soluble fractions but to a variable degree. This inhibition is in agreement with the findings of the earlier workers [27–29]. * It would appear from these results that the presence of newly synthesized membrane protein is essential for binding and retention of extramitochondrially synthesized soluble protein. The difference in inhibition of the 5, 20, and 40 mM fractions compared to the 50 and 250 mM fractions is of interest. The former are apparently more dependent on the presence of the newly synthesized membrane protein than the 50 and 250 mM fractions. An alternative explanation has been proposed by Beattie [16] for the inhibition by

---

* Similar inhibition of incorporation into the soluble fraction by high levels of chloramphenicol was observed earlier by Kadenbach [35] in liver slice experiments. However, no inhibition was observed in his in vivo experiments with a single dose of chloramphenicol administered 2 hr before sacrifice, although there was 50 to 60% inhibition of incorporation into cytochrome $a + a_3$. The differences in the dose of chloramphenicol and duration of exposure to the antibiotic could account for the differences in the results.

Table 2

Effect of cycloheximide on ^{14}C-leucine incorporation into mitochondrial proteins

Fraction	cpmg/mg protein	
	Control	Treated
Mitochondria	150	10
Soluble protein	150	0
Insoluble protein	100	20
5 mM	95	0
20 mM	300	0
40 mM	215	0
50 mM	320	0
250 mM	115	0
Microsomes	1800	15

Cycloheximide (1 mg/100 g body weight) was injected i.p. 1 hr before sacrifice and 5 μc ^{14}C-leucine/100 g 30 min before sacrifice.

excess chloramphenicol, namely, that the antibiotic inhibits mitochondrial respiration, which in turn could result in the decreased amino acid incorporation in the soluble protein fraction seen in table 3. If this were solely responsible for the inhibition, all fractions may be expected to be equally inhibited. However, the above difference between the 5, 20 and 40 mM fractions compared to the 50 and 250 mM fractions would favor at least a combination of the above two interpretations.

Effect of cycloheximide on the thyroxine induced respiratory change

The inhibition of leucine incorporation into the 20 mM and other fractions by cycloheximide offered a way of distinguishing between a direct effect of injected thyroxine on respiration and an indirect effect exerted through the newly synthesized 20 mM protein fraction. In these experiments cycloheximide was injected one hour before administration of thyroxine and ^{14}C-leucine was given 30 min before sacrifice. It was observed that the effect of thyroxine on controlled respiration was still present in the cycloheximide treated animals whereas the ^{14}C-leucine incorporation in the 20 mM as well as other soluble proteins was completely inhibited (table 4).

Bronk [30] had reported that the effect of triiodothyronine (T_3) on mitochondrial state 3 activity was still present in the presence of actinomycin D

Table 3
Effect of chloramphenicol administration on mitochondrial protein

	Control		Experimental	
	mg protein/g liver	cpm/mg	mg protein/g liver	cpm/mg
Mitochondrial	9.1	188	6.1	166
Soluble protein	3.4	235	3.3	163
Insoluble protein	6.3	640	3.8	500
5 mM fraction		125		54
20 mM fraction		305		120
40 mM fraction		305		125
50 mM fraction		412		331
250 mM fraction		140		115

The values represent the averages of two experiments. 53 mg chloramphenicol-succinate/100 g body weight was injected intraperitoneally in 4 doses at 48, 40, 24 and 4 hr before sacrifice. The control group of animals received similar amounts of Na-succinate. ^{14}C-leucine incorporation was measured as described previously [14].

Table 4
Effect of L-thyroxine in the presence of cycloheximide

	^{14}C-leucine incorporation cpm/mg protein	Controlled respiration (μatoms O_2/mg/hr)	
		Succinate	Glutamate-maleate
	Mitochondria		
Normal	150 ± 10	3.0 ± 0.5	2.0 ± 5
T_X	70 ± 5	2.0 ± 0.5	0.8 ± 0.2
$T_X + T_4$			
1 hr	68 ± 2	2.5 ± 0.3	1.2 ± 0.1
3 hr	69 ± 5	–	1.5 ± 2
$T_X + T_4$ + cycloheximide			
1 hr T_4	10	2.6 ± 0.3	1.2 ± 0.1
3 hr $_4$	10	–	1.4 ± 0.2

Cycloheximide (1 mg/100 g body weight) was injected 1 hr before thyroxine (0.52 μg/g body weight). L-^{14}C-leucine (5 μc/100 g body weight) was injected 30 min before sacrifice. State 3 (active) respiration was not affected by T_4 treatment with or without cycloheximide.

even though the incorporation of ^{14}C-leucine was inhibited. But in view of the observed effect (on state 4) of a very small fraction (20 mM) — about 1 to 2% — of the total mitochondrial protein on the state 4 respiration, it was of importance to decide whether its synthesis could be inhibited independently of the effect on respiration. The experiments presented here show that thyroxine may have two different rapid effects on mitochondrial respiration, one independent of protein synthesis [14,15] and the other dependent on the synthesis of a small amount

of soluble protein. It should be recognized that this protein synthesis is different from that concerned with the formation of entire respiratory assemblies as described by Tata and coworkers [31], since the latter showed a lag of over 36 hr.

Effects of thyroxine on the distribution of isoaccepting forms of tRNA

The rapidity of the stimulation of incorporation of

the labeled amino acid into mitochondrial protein after thyroxine administration suggested involvement of translational control of mitochondrial protein synthesis in the early effects of the hormone. We have chosen to examine first the changes in tRNA induced by thyroxine [33].

The nucleic acids were isolated from the mitochondrial, microsomal and soluble protein fractions of normal, thyroidectomized and thyroxine treated rats by phenol extraction, and tRNA was separated by chromatography on a methylated albumin kieselguhr (MAK) column [33,34]. The samples were deaminoacylated and recharged with the particular labeled amino acid together with 19 other unlabeled amino acids. The charging was carried out with a mixed 1:1 synthetase preparation from normal and thyroidectomized rats or thyroidectomized and thyroxine treated rats depending upon what we were comparing in this experiment. ^{14}C or ^{3}H labeled amino acids were used to distinguish between the tRNA from the two sets of animals, and the samples were chromatographed on the same column.

So far, only the samples from the 100,000 g supernatant have been examined. When the tRNA from thyroidectomized and normal rats was charged with glycine, valine, leucine, proline, serine, methionine, tyrosine, glutamate and tryphophane no significant difference in the kinetics of labeling or extent of charging was observed. Lysine, as observed previously by other workers, showed two isoaccepting peaks of tRNA, and phenylalanine showed three. Fig. 1 shows that $tRNA_I^{Lys}$ (peak I) was larger than $tRNA_{II}^{Lys}$ (peak II) in normal rats, the ratio of the amounts being 70:30. In thyroidectomized rats the ratio was altered to 35:65, form II being much greater than form I. With phenylalanine three peaks were obtained, the first two being difficult to separate from each other (fig. 2). In normal rats the combined amount of $tRNA_I^{Phe}$ and $tRNA_{II}^{Phe}$ was less than that of $tRNA_{III}^{Phe}$, the ratio being 16:84. In thyroidectomized rat livers, a decrease in peak III was observed, with the ratio of 63:37. Intraperitoneal administration of 0.5 μg thyroxine/g body weight 30 min before sacrifice produced a reversal in the distribution ratio to 32:68.

Pending analysis of the tRNA derived from the particulate fractions, a preliminary in vivo charging experiment was carried out. Liver slices from thyroid-

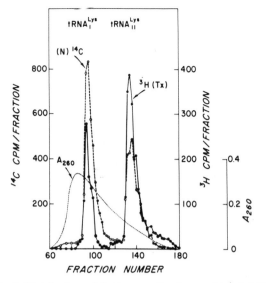

Fig. 1. Distribution of the isoaccepting forms of lysyl tRNA in normal (N) and thyroidectomized (T_X) rats. Methylated albumin kieselguhr chromatography was used to separate the forms and the experiments were carried out as described previously [33].

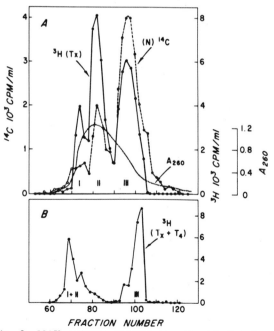

Fig. 2. MAK column cochromatography of phenylanyl tRNA. A. Comparison of tRNA from normal (N) and thyroidectomized (T_X) rats. B. tRNA from thyroidectomized rats which had been treated with 0.52 μg thyroxine/g body weight 30 min before sacrifice. The experimental details have been reported [33].

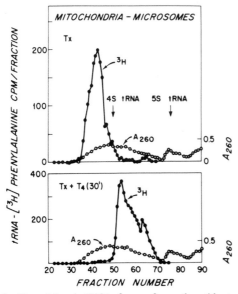

Fig. 3. Phenylalanyl tRNA forms from thyroidectomized (T_X) and thyroxine treated ($T_X + T_4$) rat liver slices. Approximately 15 g of livers from two groups of thyroidectomized rats – one injected with saline, other with thyroxine (0.5 μg/g body weight – were sliced into 1–2 mm thickness at 0–2°. The liver slices were then incubated in a medium containing 25 mM phosphate, pH 7.4, 0.25 M sucrose, 0.01 M $MgCl_2$, 0.2 M glucose, 300 μc ^{3}H-phenylalanine, and 19 other amino acids at 37° with vigorous shaking for 15 min. The activation reaction was stopped by dropping the pH to 6.0 and by quick chilling. The liver slices were rinsed and homogenized in a similar buffer except that Mg^{++} was 1 mM, it contained 0.1 mM EDTA and the pH was 6.0. After removal of nuclei by centrifugation, the extranuclear supernatant was spun at 105,000 g for 60 min to separate mitochondria and microsomes from soluble cytoplasm. Extraction of tRNA from the combined particulate fraction was carried out at pH 6.0 as in the in vitro experiment. Analysis of ^{3}H-phenylalanyl tRNA chromatography was carried out as before [33].

ectomized and treated rats were incubated with ^{3}H-phenylalanine as described under fig. 3 and the aminoacyl tRNA was isolated. Chromatography revealed that the radioactivity peak preceded the absorbance peak in the experiment with the mitochondria plus microsomal fraction from thyroidectomized rat livers. In the thyroidectomized rats which had been treated with thyroxine before sacrifice, the radioactivity peak followed the absorbance peak. The results indicate a change in the distribution pattern of organelle-bound tRNA also, although con-

firmation with the double labeling technique would be needed to establish the finding on a firm basis.

It is somewhat premature to speculate on the significance of the changes in the distribution of the isoaccepting forms of tRNA as a function of the thyroid state of the animal. The possibility of the multiple peaks being artifacts has not been completely excluded. If the results turn out to be meaningful, they might give a clue to at least one aspect of translational control of protein synthesis in mammalian systems.

Acknowledgement

This work was supported by a research grant from the American Cancer Society (P–473).

References

[1] D.S. Fletcher and D.R. Sanadi, Biochim. Biophys. Acta 51 (1961) 356.
[2] D.S. Beattie, R.E. Basford and S.B. Koritz, J. Biol. Chem. 242 (1967) 4584.
[3] J.E. Wilson and J.L. Dove, J. Elisha Mitchell Scientific Society 81, supplement 1 (1965) 21.
[4] E. Bailey, C.B. Taylor and W. Bartley, Biochem. J. 104 (1967) 1026.
[5] N.J. Gross, G.S. Getz and M. Rabinowitz, J. Biol. Chem. 244 (1969) 1552.
[6] A.B. Novikoff and E. Essner, J. Cell Biol. 15 (1962) 140.
[7] D. Brandes, D.E. Buetow, F. Bertini and D.B. Malkoff, Exptl. Mol. Path. 3 (1964) 583.
[8] R.W. Swick, A.K. Rexroth and J.L. Stange, J. Biol. Chem. 243 (1968) 3581.
[9] H.S. Marver, A.Collins, D.P. Tschudy and M. Recheigl, J. Biol. Chem. 241 (1966) 4323.
[10] M. Pascaud, Biochim. Biophys. Acta 84 (1966) 528.
[11] W. Neupert, D. Bradiczlla and W. Sebald, in: Biochemical Aspects of Biogenesis of Mitochondria, eds. E.C. Slater, J.M. Tager, S. Papa and E. Quagliariello (Adriatica Editrice, Bari, Italy, 1968) p. 395.
[12] D.S. Beattie, Biochem. Biophys. Res. Commun. 35 (1969) 721.
[13] B. DeBernard, G.S. Getz and M. Robinowitz, Biochim. Biophys. Acta 193 (1969) 58.
[14] P. Volfin, S. Kaplay, D.R. Sanadi, J. Biol. Chem. 244 (1969) 5631.
[15] F.L. Hoch and Michelle V. Motta, Proc. Natl. Acad. Sci. U.S. 59 (1968) 118.
[16] D.S. Beattie, J. Biol. Chem. 243 (1968) 4027.

[17] P. Borst, A.M. Kroon, G.J.C.M. Ruttenberg, in: Genetic Elements: Properties and Functions, ed. D. Shuger (Academic Press, New York, 1967).

[18] M. Huang, D.R. Biggs, G.D. Clark-Walker and A.W. Linnane, Biochim. Biophys. Acta 114 (1966) 434.

[19] G.D. Clark-Walker and A.W. Linnane, J. Cell Biol. 34 (1967) 1.

[20] H.R. Mahler, P. Perlman, C. Henson and C. Webber, Biochem. Biophys. Res. Commun. 31 (1968) 474.

[21] L.W. Wheeldon and A.L. Lehninger, Biochemistry 5 (1966) 3533.

[22] A.J. Lamb, G.D. Clark-Walker and A.W. Linnane, Biochim. Biophys. Acta 161 (1968) 414.

[23] A.C. Trakatellis, M. Montjar and A.E. Axelrod, Biochemistry 4 (1965) 2065.

[24] D.S. Beattie, R.E. Basford and S.B. Koritz, Biochemistry 6 (1967) 3099.

[25] W. Sebald, T. Bucher, B. Olbrich and F. Kaudewitz, FEBS Letters 1 (1968) 235.

[26] W. Neupert, W. Sebald, A.J. Schwab, P. Massinger and T. Bucher, Europ. J. Biochem. 10 (1969) 589.

[27] D.B. Roodyn, J.W. Suffie and T.S. Work, Biochem. J. 83 (1962) 29.

[28] D.E.S. Truman, Biochem. J. 91 (1964) 59.

[29] B. Kadenbach, Biochim. Biophys. Acta 134 (1966) 430.

[30] J.R. Bronk, Science 153 (1966) 638.

[31] J.R. Tata, in: Mechanisms of Hormone Action, ed. P. Karlson, (Thieme Stuttgart, 1965) 173.

[32] F.C. Firkin and A.W. Linnane, Exptl. Cell Res. 55 (1969) 68.

[33] S.S. Yang and D.R. Sanadi, J. Biol. Chem. 244 (1969) 5081.

[34] S.S. Yang and D.G. Comb, J. Mol. Biol. 31 (1968) 139.

[35] B. Kadenbach, in: Biochemical Aspects of Biogenesis of Mitochondria, eds. E.C. Slater, J.M. Tager, S. Papa and E. Quagliariello (Adriatica Editrice, Bari, Italy, 1968) p. 355.

MITOCHONDRIOGENESIS IN ANIMAL CELLS:
STUDIES WITH DIFFERENT INHIBITORS

A.M. KROON and H. DE VRIES

Laboratory of Physiological Chemistry, State University, Groningen, The Netherlands

With respect to the biosynthetic properties of mitochondria the following general conclusions may be drawn from the data presented in the literature so far (for reviews see refs. [1–4]): (1) isolated mitochondria are completely equipped for the synthesis of DNA, RNA and proteins; (2) the amount of genetic information present in mitochondrial DNA from animal cells is limited to about 15,000 basepairs; (3) much of the genetic information for mitochondrial proteins resides within the nucleus; (4) evidence accumulates that the components of the nucleic acid and protein synthesizing machineries of mitochondria are of the bacterial type.

We are aware, of course, that the generalisations made may introduce an oversimplification. For instance the experiments by Birt with blowfly mitochondria do not fit the above statements for as yet unknown reasons [5]. Furthermore Attardi and co-workers studying the properties of mitochondrial RNA in HeLa cells were unable to gather any experimental support for the contention that the mitochondria of these cells contain specific ribosomes. This led them to consider seriously the possibility that all mitochondrial messenger-RNAs are translated outside the mitochondria [6,7]. The sharp delineation of the contribution of the synthetic activities of mitochondria to the structure and function of the eukaryotic cell is further hampered by the fact that the nature of the organelle transcription- and translation-products is established very poorly, especially in higher animals.

Inhibitors of different types are often used as a tool to distinguish, both qualitatively and quantitatively, between the contribution of the different synthetic machineries present in the eukaryotic cell to the formation of fully equipped and active mito-

chondria. In table 1 we have summarized some of our experiments with a number of antibiotics and intercalating dyes. Mitochondrial protein synthesis in vitro was studied by following the incorporation of amino acids into acid-precipitable proteins, that were not further identified. We have, moreover, investigated the effects of most of the inhibitors listed on the formation of cytochrome c oxidase in rat heart cells in tissue culture and in regenerating rat liver. As indicated in table 1, chloramphenicol and oxytetracycline are potent inhibitors in all three systems. We have discussed these experiments recently on different occasions [8–10]. Moreover, similar results for chloramphenicol have been obtained independently by others [11]. For these reasons we want to concentrate in this paper on two other groups of inhibitors: (1) lincomycin and the macrolides, (2) the intercalating dyes.

Lincomycin and the macrolides

Experiments with isolated mitochondria

The antibiotics lincomycin and the macrolides such as erythromycin, oleandomycin and carbomycin interfere with bacterial protein synthesis at the level of the ribosomes [12]. They bind to the large subunit of these ribosomes and prevent either translocation as has been suggested for erythromycin [13,14] or peptidebond formation as has been shown for lincomycin [14]. Also protein synthesis in isolated mitochondria from yeast has been shown to be inhibited by these antibiotics [15,16] and these inhibitions have been interpreted to mean that the ribosomes from bacteria and yeast mitochondria are of the same type. With respect to the ribosomes from animal

Table 1
Effects of various inhibitors on the biosynthetic activities of mitochondria of rat tissues in vivo and in vitro.

Inhibitor	(^{14}C) leucine incorporation in vitro (rat-liver)		Cytochrome c oxidase formation in	
	Intact mitochondria	Swollen mitochondria	Cultured rat-heart cells	Regenerating rat-liver
Oxytetracycline	++	++	++	++
Chloramphenicol	++	++	++	++
Lincomycin	–	+	–	0
Erythromycin	–	++	–	0
Oleandomycin	–	++	0	0
Carbomycin	++	++	++	0
Ethidium bromide	++	++	++	+
3,6-diamino-10-methyl acridine (euflavin)	++	++	++	++
3,6-diamino acridine (proflavin)	+	+	cytotoxic	0
9-amino acridine	–	–	cytotoxic	0

++ = 50–95% inhibition; + = 20–50% inhibition; – = no inhibition; 0 = not tested.
The details concerning the different experimental systems are given in the legends to fig. 2 and tables 2, 3 and 4.

mitochondria the situation is less clear. There are several reports that the mitochondrial ribosomes from animal cells are characterised by a sedimentation value of 55 S [17–19], suggesting the presence of a new and unique class of ribosomes in animal mitochondria. In our opinion, however, the characterisation of these particles and its constituents has not yet made enough progress as to justify the final decision that these 55 S particles are the intact and undamaged ribosomal monomers of animal mitochondria [cf. ref. 4]. Moreover, in animal cells mitochondrial protein synthesis is affected in the same way as bacterial protein synthesis by a reasonable number of antibiotics (chloramphenicol, oxytetracycline, etc.). This points to similarities rather than to extreme differences between the ribosomes of both systems.

In an attempt to extend the number of reagents interfering with both bacterial and mitochondrial protein synthesis we have investigated the effects of lincomycin and erythromycin on protein synthesis by intact isolated mitochondria. No inhibitions were found, even at very high concentrations. Resistance of mitochondrial protein synthesis to erythromycin has also been observed for certain mutant strains of yeast [20]. Two types of resistance were distinguished: one due to impermeability for erythromycin

of the mitochondrial membranes and genetically determined by the nucleus, the other caused by an insensitivity to the drug of the mitochondrial ribosomes and specified by cytoplasmic genes. In order to decide whether the lack of inhibition of rat-liver mitochondrial protein synthesis was due to a permeability barrier at the level of the mitochondrial membranes or to insensitivity of the mitochondrial ribosomes themselves, experiments with 'swollen' mitochondria were performed. In addition two other macrolides, oleandomycin and carbomycin, were tested. The results of a number of experiments are shown in table 2. It can be seen that protein synthesis is indeed inhibited by lincomycin, erythromycin and oleandomycin if the mitochondria are 'swollen'. This swelling or at least this altered permeability was obtained by incubating the mitochondria for 15 min at 4° in a hypotonic sucrose solution (0.025 M) in the experiment with erythromycin. In the experiments 1 and 3 of table 2, osmotic shock was omitted but instead 0.05% (w/v) digitonin was added to the medium for amino acid incorporation. In a number of experiments hypotonic or digitonin treatment led to a serious decrease in the incorporation activity already in the absence of the inhibitor. In these experiments the inhibition was either much smaller

Table 2
Effects of lincomycin and some macrolides on mitochondrial protein synthesis.

Expt. number	Additions	Mitochondria	(^{14}C) leucine incorporation (pmoles/mg protein/h)
1	None	intact	27
	100 μg lincomycin/ml	intact	26
	None	swollen	26
	100 μg lincomycin/ml	swollen	15
2	None	intact	36
	250 μg erythromycin/ml	intact	35
	None	swollen	33
	50 μg erythromycin/ml	swollen	12
3	None	intact	23
	100 μg oleandomycin/ml	intact	25
	None	swollen	22
	100 μg oleandomycin/ml	swollen	11
4	None	intact	32
	2 μg carbomycin/ml	intact	11
	10 μg carbomycin/ml	intact	5

Mitochondria were prepared from the livers of male rats weighing about 125 g after fasting overnight. Incubations were carried out at 30° for 60 min. The incubation medium contained 50 mM sucrose, 50 mM Tricine-buffer, 20 mM KCl, 30 mM NH$_4$Cl, 5 mM MgCl$_2$, 1 mM EDTA, 20 mM potassium phosphate buffer, 2 mM ADP, 30 mM sodium succinate, 20 μg ribonuclease (EC 2.7.7.16), 50 μg of a synthetic amino acid mixture [21], 0.04 mM (^{14}C)leucine (10 μC/μmole), 2–3 mg mitochondrial protein plus the additions specified in the table. The final volume was 1 ml, the final pH 7.4. The reactions were stopped by the addition of 5 ml 5% trichloroacetic acid. The washed and defatted precipitates were solubilized in 1 ml soluene 100 (Packard Instrument Co.) and counted in a liquid scintillation counter using a toluene PPO/POPOP scintillation mixture. The counting efficiency was 60–65% in the different experiments. The incorporation data given are corrected for zero time activity which was 1 pmole/mg protein/hr or less. All experimental details have been described previously [22,23]. The procedures used for the preparation of swollen mitochondria are described in the text.

on a percentage basis or absent. The inhibitions obtained were never complete and specially in the case of oleandomycin were somewhat variable. Swelling the mitochondria in phosphate buffer [24] led to a complete loss of incorporation activity. From these experiments we have concluded that the insensitivity of rat-liver mitochondrial protein synthesis to lincomycin and some macrolides is due to the fact that the mitochondrial membranes are impermeable to these drugs, but that the mitochondrial ribosomes are themselves still susceptible to the action of the inhibitors. This point of view is strengthened by the observation (expt 4, table 2) that carbomycin is a very strong inhibitor of mitochondrial protein synthesis. Carbomycin belongs also to the macrolide antibiotics. The different macrolides are supposed to have a common mechanism of action [12]: cross-

sensitivity and cross-resistance may, therefore, be expected.

In a similar approach using sonic vibrations to break up the mitochondrial membranes instead of swelling by hypotonic or digitonin treatment, Firkin and Linnane [25] have drawn an opposite conclusion. These authors found no inhibition by lincomycin and erythromycin under their experimental conditions. Unfortunately these conditions are quite different from ours. It is, therefore, difficult to account for this apparent discrepancy. However, it should be noticed that the incorporation activity expressed as pmoles/mg protein is very low in the experiments of the latter authors [25] and only at the level of 10% of the activity in the experiments given in table 2. It is of course important that indeed the ribosomal processes are rate-limiting if one is

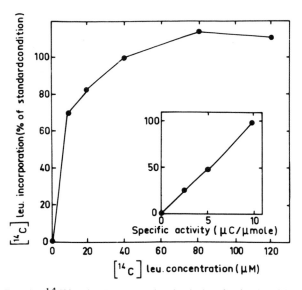

Fig. 1. (^{14}C)leucine incorporation by isolated mitochondria from rat-liver as a function of the concentration of added (^{14}C)leucine (10 μC/μmole) and of (insert) the specific activity of the added leucine under our standard incubation conditions. With the exception of (^{14}C)leucine, the incubation medium contained all the components specified in the legend to table 2. Incubations were carried out at 30° for 60 min.

looking for inhibitions at the ribosomal level, especially if these inhibitions may be only partial. Although it can not be assumed with certainty that this rate-limitation exists under our experimental conditions, we can exclude from the data given in fig. 1 that the supply of radioactive amino acid will be the rate-limiting step. It can be easily calculated from this figure that a tenfold increase of the specific activity of the added (^{14}C) leucine at a constant input of radioactivity will drop the incorporation measured to about 10% of the value under our standard conditions.

Experiments with beating rat-heart cells in tissue culture

It has already been indicated in table 1 that lincomycin and erythromycin do not inhibit the formation of cytochrome c oxidase in rat-heart cells in tissue culture. We have used this primary cell-line in earlier studies [4,8,10,26] and shown that chloramphenicol added to the tissue culture medium evokes a decrease in the specific activity of cytochrome c oxidase in these heart cells. The lack of inhibitory effect by

lincomycin and erythromycin on the cytochrome c oxidase formation in cultured rat-heart cells is very easily to reconcile with the impermeability of intact mitochondrial membranes to these antibiotics as we have discussed above. In this connection it was of course interesting to look for the effect of carbomycin on the heart cell. A typical experiment is given in fig. 2. It can be seen that the formation of cytochrome c oxidase in the heart-cell cultures is specifically inhibited. This seems to substantiate the conclusion drawn in the preceding paragraph that mitochondrial protein synthesis is the primary point of attack for the inhibition by carbomycin. However, based on the lower protein content of the carbomycin-treated cultures, we have to assume that carbomycin gives some inhibition of cell growth too. Since a similar effect was not observed in our experiments with choramphenicol [26], the reason for this inhibition needs further clarification. So far we have no indication that carbomycin interferes with mitochondrial energy-conserving processes. It is, therefore, unlikely that the inhibition of protein synthesis in isolated mitochondria is due to an insufficient energy supply as in the presence of 2,4-dinitrophenol or cyanide [22,23].

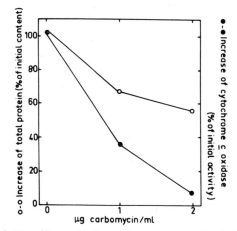

Fig. 2. The effects of carbomycin on growth and cytochrome c oxidase activity of beating rat-heart cells in tissue culture. Age of cultures: 4 days. Carbomycin was present during the whole experiment. Each value is the mean of 3 different cultures. Cell cultures were prepared and maintained as described previously [26]. Cytochrome c oxidase activity was measured using the method of Cooperstein and Lazarow as modified by Borst et al. [27]. Protein was measured by the Lowry method.

Intercalating dyes

It has been known for a long time that euflavin is a potent mutagen in yeast. It causes preferentially cytoplasmic mutations [28,29]. More recently ethidium bromide has been shown to exert a similar mutagenic activity [30]. It is well established that both euflavin and ethidium bromide react with double-stranded DNA by intercalation of their flat polycyclic rings between two adjacent basepairs. Other acridine derivatives may act in the same way [31,32].

We have reported [33] that euflavin at very low concentrations is a strong inhibitor of amino acid incorporation in isolated rat-liver mitochondria. Similar inhibitions by *high* concentrations had been described before [24,34]. On the basis of the linearity of the amino acid incorporation and an inhibition by actinomycin D we have proposed earlier [35] that mitochondrial protein synthesis in vitro is continuously depending on transcription. This working hypothesis was backed by the finding that euflavin at the 1 μM level inhibits the synthesis of proteins in isolated mitochondria. However, it will be necessary to exclude the possibility that the inhibition by euflavin affects this process at the level of translation. We have, therefore, compared the effects of euflavin and some other intercalating dyes in mitochondrial, microsomal and bacterial systems for protein synthesis.

Effects of intercalating dyes on protein synthesis

There are two well-known and distinct systems for protein synthesis: the 80 S-ribosomal system from the cytoplasm of eukaryotic cells and the bacterial system operating with 70 S-ribosomes. We have studied the effects of increasing concentrations of ethidium bromide, euflavin, proflavin and 9-amino acridine on the incorporation of amino acids by rat-liver microsomes plus pH 5 enzymes as an example of the former system and by a supernatant system from *E. coli* as an example of the latter. The results of these experiments as compared with the effects on mitochondrial protein synthesis are given in fig. 3. It can be seen that both euflavin and ethidium bromide are very strong inhibitors of the mitochondrial process and give no inhibition at all in the bacterial system. In the case of ethidium bromide only there is partial

inhibition of microsomal incorporation but at a concentration much higher than that necessary for complete inhibition in mitochondria. Proflavin is much less active in inhibiting the mitochondria, and partially affects the bacterial and microsomal systems. 9-Amino acridine is slightly inhibitory in the mammalian systems and gives some (in different experiments quite variable) stimulation in the S 30 supernatant fraction of *E. coli*. After treatment of the S 30 fraction in a way that the endogenous messenger-RNA is removed [36] the effects of all four inhibitors were tested in a polyuridylic acid directed system for phenylalanine incorporation. No inhibitions were obtained, even with proflavin. In a separate set of experiments the microsomes and pH 5 enzymes and the 'S 30' fraction were preincubated for 30 min at 4° with a high concentration of either of the dyes and then tested for amino acid incorporation activity. The patterns of inhibition (not shown) remained grossly the same: no inhibition whatsoever by ethidium bromide, euflavin and 9-aminoacridine and partial inhibition by proflavin. To assure ourselves that the inhibition of the mitochondrial incorporation had not to be ascribed primarily to inhibition of either oxidation or phosphorylation, all four inhibitors were tested for possible interference with these processes by measuring respiratory rates, respiratory control indices, and P/O ratios with different substrates and at different concentrations of the drugs. Only proflavin gave a consistent but far from complete inhibition of coupled respiration at the concentrations used to inhibit mitochondrial protein synthesis. The other dyes were inactive.

In view of the experiments described above we tend to ascribe the inhibitions of mitochondrial protein synthesis by ethidium bromide and euflavin to the interaction of these dyes with mitochondrial DNA. This implies that we expect protein synthesis in mitochondria to be firmly coupled to a DNA-dependent process. It has been reported in the literature that DNA-directed RNA-synthesis in mitochondria is indeed inhibited by ethidium bromide and euflavin [7,37,38]. Furthermore, as is shown in fig. 4, the incorporation of labelled deoxyribonucleotides into mitochondrial DNA is inhibited by the same low concentrations of ethidium bromide as the incorporation of amino acids into protein. For reasons that will not be discussed in detail here the DNA

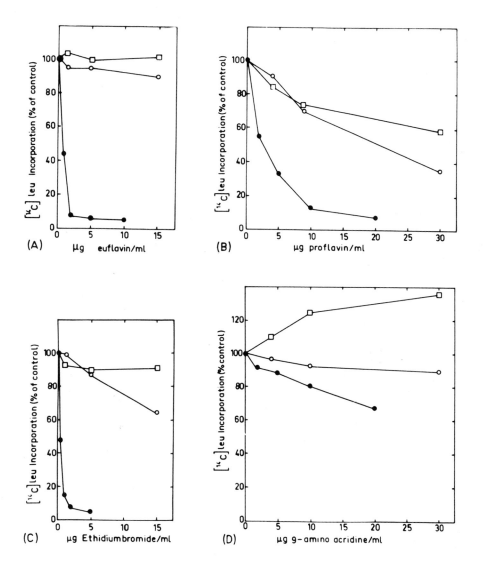

Fig. 3. Comparison of the effects of intercalating dyes on the (^{14}C)leucine incorporation activity of isolated rat-liver mitochondria, rat liver microsomes and an *E. coli* system for protein synthesis. (A): euflavin; (B): proflavin; (C): ethidium bromide; (D): 9-amino-acridine. For experimental conditions for the mitochondrial system and for counting procedures see table 2. Microsomes and pH 5 enzymes from rat liver were isolated as described previously [22]. The medium for incorporation contained 50 mM sucrose, 50 mM Tricine buffer, 50 mM KCl, 5 mM MgCl$_2$, 1 mM EDTA, 3 mM ATP, 0.3 mM GTP, 4 mM sodium phosphoenolpyruvate, 3 mM glutathione (reduced), 50 μg of a synthetic amino acid mixture [21], 0.04 mM (^{14}C)leucine (10 μC/μmole), pyruvate kinase, about 1 mg protein of pH 5 enzyme fraction and about 2 mg microsomal protein. The final volume was 1 ml, the final pH 7.8. Incubations were carried out at 37° for 15 min. The *E. coli* system was prepared according to Nirenberg [36]. The incubation medium contained 50 mM Tricine, 80 mM KCl, 1 mM EDTA, 20 mM MgCl$_2$, 2 mM ATP, 0.3 mM GTP, 5 mM phosphoenolpyruvate, 0.1 mg pyruvate kinase, 5 mM mercaptoethanol, 0.04 mM (^{14}C)leucine (10 μC/μmole), 50 μg of a synthetic amino acid mixture [21] and about 5 mg of bacterial protein (deoxyribonuclease-treated supernatant (S 30), centrifuged at 30,000 g for 30 min, and containing ribosomes, tRNA, endogenous mRNA and enzymes for transcription and translation). The final volume was 1 ml, the final pH 7.8. Incubations were for 20 min at 37°C. Symbols used: ●———●: mitochondria; ○———○: microsomes; □———□: coli system.

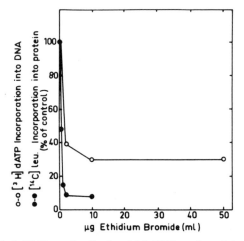

Fig. 4. Inhibition of mitochondrial DNA and protein synthesis by ethidium bromide at different concentrations. DNA synthesis was measured by following the incorporation of tritiated dATP into DNA by chick-liver mitochondria. The mitochondria were prepared at a large scale as described elsewhere [27]. Incubations were carried out at 37° for 10 min in shaking Warburg vessels. The incubation medium contained 50 mM Tris HCl, 4 mM KCl, 7 mM $MgCl_2$, 20 mM sodium phosphate buffer, 20 mM sodium succinate, 0.015 mM dCTP, 0.015 mM dGTP, 0.015 mM dTTP, 0.004 mM (^{3}H)dATP (3070 dpm/pmole) and about 10 mg of mitochondrial protein. The final volume was 1 ml, the final pH 7.4. Further experimental details are described elsewhere [39]. The experiment was performed by Mr. J. ter Schegget, Laboratory of Medical Enzymology, University of Amsterdam. The curve representing (^{14}C)leucine incorporation into mitochondrial protein is the same as given in fig. 3B.

synthesis measured in this type of experiment is thought to be mainly repair synthesis [39].

Effects of intercalating dyes on cells and tissues

We have tested the four intercalating dyes discussed in the preceding paragraph also in the heart-cell system. As already indicated in table 1, proflavin and 9-amino-acridine had a cytotoxic effect; the heart cells did not stick to the culture dishes, did not form a monolayer and refrained from beating if one of these inhibitors were present in a concentration of 1 μg per ml of culture medium. At lower concentrations some growth and beating activity were obtained, but in these cases there was no *specific* inhibition of the formation of cytochrome c oxidase. On the contrary very low concentrations of ethidium

bromide (0.25 μg/ml) or euflavin (0.5 μg/ml) completely blocked the formation of this enzyme, whereas the increase in total protein in the dye-treated cultures was not significantly different from the controls. The effects of ethidium bromide and euflavin are, therefore, phenomenologically similar to those obtained previously with chloramphenicol and oxytetracycline [10,26]. There is, however, one major difference. For the inhibition by chloramphenicol or oxytetracycline the presence of the antibiotics during the whole period of culture is absolutely required. As soon as the antibiotic is withdrawn, the formation of cytochrome c oxidase resumes [26]. In the case of ethidium bromide and euflavin this requirement does *not* exist. If the dye is present in the culture medium only during the first two days, the formation remains blocked in the subsequent period of culture. This means that either the damaging effects of ethidium bromide and euflavin are irreversible or that they are bound irreversibly to their target of inhibition.

In view of the cytotoxicity of proflavin and 9-amino-acridine only euflavin and ethidium bromide were tested on liver regeneration [cf. refs. 10,41]. Both drugs were injected only once, at the time of the operation.

Tables 3 and 4 show that as a result of the treatment with small quantities of ethidium bromide or euflavin the increase of the total cytochrome c oxidase activity is much smaller than the increase of total protein. Also in this system, therefore, a clear and specific inhibition of cytochrome c oxidase formation is induced by the intercalating dyes. This inhibition is also visible in spectra of mitochondria; the molar ratio of cytochrome aa_3 to cytochromes $c + c_1$ is significantly lower in mitochondria from drug-treated animals than from controls. Just as is the case for the cultured heart cells, ethidium bromide and euflavin evoke the same effect in regenerating liver as chloramphenicol and oxytetracycline [10,41]. Since we do not have a sensitive method for measuring the rate of disappearance of the dyes from blood and tissues, it is as yet not possible to decide whether these effects depend on the continuous presence of the inhibitor or whether they are caused by some irreversible change in the mitochondrial biogenetic machineries shortly after administration.

Table 3
Effect of ethidium bromide on cytochrome c oxidase formation in regenerating rat liver.

Drug treatment	Liver homogenates				Liver mitochondria
	Number of animals	Total cytochrome c oxidase activity (percent)	Total (percent)	Specific cytochrome c oxidase activity (percent)	Molar ratio of cytochrome aa_3 to cytochromes $c + c_1$
None	7	64.9	65.6	99.0	0.37
Ethidium bromide, 10 μg per g body weight, intraperitoneally	3	33.9	57.0	59.6	0.30
Ethidium bromide, 10 μg per g body weight, intravenously	4	46.2	65.0	70.5	0.29

The table gives total cytochrome c oxidase activity, total protein content and specific cytochrome c activity of liver homogenates, 46 hr after partial hepatectomy in percent of the amounts present in the original liver before the operation. The amount of protein and the total cytochrome c oxidase activity in the liver part remaining after the removal of 70% of the liver weight are taken as 30%. Specific activities at the time of operation are 100%. Two groups of partially hepatectomized rats were studied: control animals and rats, injected at the time of the operation with one injection of ethidium bromide, as indicated. Molar ratios of cytochrome aa_3 to cytochromes $c + c_1$ in mitochondria prepared from these homogenates are given in the last column. For the calculation of the ratios we have used a value of 24.0 as the extinction coefficient of cytochrome aa_3 (605–635 nm) [40] and 17.8 as the extinction coefficient of cytochromes $c + c_1$ (551–540 nm) (J. Berden, personal communication). The experimental conditions are as described previously [41].

Table 4
Effect of euflavin on cytochrome c oxidase formation in regenerating rat liver.

Drug treatment	Liver homogenates				Liver mitochondria
	Number of animals	Total cytochrome c oxidase activity (percent)	Total protein (percent)	Specific cytochrome c oxidase activity (percent)	Molar ratio of cytochrome aa_3 to cytochrome.$c + c_1$
None	7	64.9	65.6	99.0	0.37
Euflavin, 5 μg per g body weight, intraperitoneally	2	49.8	60.0	83.7	0.26
Euflavin, 5 μg per g body weight, intravenously	1	39.7	60.6	65.5	0.29
Euflavin, 10 μg per g body weight, intraperitoneally	2	39.4	59.2	67.0	0.25
Euflavin, 10 μg per g body weight, intravenously	1	40.3	68.1	59.3	0.22

The table gives total cytochrome c oxidase activity, total protein content and specific cytochrome c activity of liver homogenates, 46 hr after partial hepatectomy. Molar ratios of cytochrome aa_3 to cytochromes $c + c_1$ in mitochondria prepared from these homogenates are given in the last column. To the drug-treated rats one injection of euflavin was given at the time of the operation, as indicated. For further details, see table 3.

General remarks

Inhibitions by ethidium bromide and euflavin have been observed for mitochondrial DNA synthesis [39], mitochondrial RNA synthesis [7,37,38] and mitochondrial protein synthesis [8,10,16]. The concentration of the dyes necessary to obtain half maximal inhibition is strikingly similar in all cases. Because of this similarity it is inviting to speculate that the basis for the inhibitory effects on the different synthetic reactions is a common point of attack. The phenomenon of intercalation into DNA may well represent this common causal relation, provided that mitochondrial protein synthesis is continuously dependent on a DNA-directed process. Besides the correlation in the sensitivity of the three synthetic reactions just mentioned, there is also a quite interesting correlation with the sensitivity to ethidium bromide and euflavin of other mitochondrial DNA-dependent processes like the adaptation to oxygen of anaerobically grown yeast and the rate of formation of cytoplasmic mutants in yeast. Some of these data are compared in table 5.

However, in spite of the attractiveness of the above speculation, alternative explanations have to be considered. In the first place it has been shown that mitochondrial DNA-polymerase from rat-liver is more sensitive to ethidium bromide than the corresponding nuclear DNA-polymerase [42]. This difference in sensitivity is independent from the type and source of the DNA used as the template and points, therefore, to specific interference with the enzyme rather than with the DNA. In the second place, in order to let the hypothesis hold, ethidium bromide and euflavin may

not interfere directly with mitochondrial translation. The experiments described in this paper show that the well-known bacterial and eukaryotic ribosomal systems for translation are not affected by these drugs. Since the similarities between the components of the mitochondrial and bacterial protein synthesizing systems are numerous, it seems justified to assume that the mitochondrial translation machinery is bacterial in its properties and evolutionary origin [43]. However, this assumption does not exclude *minor* differences. Another possibility that has to be seriously considered is that the incorporation of amino acids within mitochondria is limited by other factors than both of the reconstituted systems used in these studies to trace the possible inhibitory effects of the intercalating dyes at the level of translation. Unlike proflavin [44,45], neither ethidium bromide nor euflavin are well-known inhibitors of protein synthesis. The recent observations that ethidium bromide is able to inhibit amino acid activation in rat liver [46] and to form complexes with transfer-RNA from *E. coli* [47] were made at concentrations of the drug exceeding those necessary for the effects described in this paper by at least one order of magnitude. For this reason we doubt that our observations can be explained by an interaction of the dyes with transfer-RNA or a transfer-RNA dependent process but further experiments to substantiate this will be required. With respect to the in vivo effects it should be stressed that the inhibitions appear to be irreversible. This irreversibility is hard to reconcile with an inhibition of one of the translation reactions but points rather to a damaging effect at a basic level. The most

Table 5

Quantitative comparison of the minimal molarities, necessary for different inhibitory effects of the intercalating dyes ethidium bromide, euflavin and proflavin.

Parameter	Ehidium bromide	Euflavin	Proflavin
50% inhibition of mitochondrial protein synthesis [a]	1 μM	6 μM	‘14 μM
50% specific inhibition of cytochrome *c* oxidase formation in cultured rat-heart cells [a]	0.2 μM	1 μM	cytotoxic
50% inhibition of adaptation to O_2 in yeast [b]	–	0.7 μM	3 μM
50% ρ^- mutant formation in yeast [b]	1 μM	1 ‧ μM	not reached

[a] Data presented in this paper.
[b] Literature data [28–30].

basic level we can think about is the mitochondrial DNA. In that case it would appear that mammalian cells do not behave differently from lower eukaryotic organisms such as yeast in this respect. Cell systems like the beating heart-cells discussed in this paper may then offer a nice tool to investigate the primary events that occur if cells are exposed to mutagens such as ethidium bromide or euflavin.

Acknowledgements

We wish to thank Prof. F.J. Loomeijer for his valuable criticism and interest and Miss A.J. Arendzen and Miss H.J.L. Schuuring for expert technical assistance.

References

[1] Biochemical Aspects of the Biogenesis of Mitochondria, eds. E.C. Slater, J.M. Tager, S. Papa and E. Quagliariello (Adriatica Editrice, Bari, Italy, 1968).

[2] T.S. Work, J.L. Coote and M.A. Ashwell, Federation Proc. 27 (1968) 1174.

[3] P. Borst and A.M. Kroon, Int. Rev. Cytol. 26 (1969) 107.

[4] A.M. Kroon, in: Handbook of Molecular Cytology, ed. A. Lima-de-Faria (North Holland Publishing Company, Amsterdam, 1969) p. 943.

[5] L.M. Birt, this volume.

[6] B. Attardi and G. Attardi, Nature 224 (1969) 1079.

[7] G. Attardi, B. Attardi, B. Storrie, L. Mattocia, M. Lederman and Y. Aloni, this volume.

[8] A.M. Kroon, in: Inhibitors, Tools in Cell Research, eds. Th. Bucher and H. Sies (Springer Verlag, Heidelberg, 1969) p. 159.

[9] A.M. Kroon and H. de Vries, FEBS Letters 3 (1969) 208.

[10] A.M. Kroon and H. de Vries, in: The Development and Interrelationships of Cell Organelles, ed. P.L. Miller (Cambridge University Press, Cambridge, 1970) p. 181.

[11] F.C. Firkin and A.W. Linnane, Exptl. Cell Res. 55 (1969) 76.

[12] J.M. Wilhelm, N.L. Oleinick and J.W. Corcoran, in: Antimicrobial Agents and Chemotherapy, (American Soc. for Microbiol. 1968) p. 236.

[13] E. Cundliffe and K. McQuillen, J. Mol. Biol. 30 (1967) 137.

[14] K. Igarashi, H. Ishitsuka and A. Kaji, Biochem. Biophys. Res. Commun. 37 (1969) 499.

[15] A.W. Linnane, A.J. Lamb, C. Christodoulou and H.B. Lukins, Proc. Natl. Acad. Sci. U.S. 59 (1968) 1288.

[16] G.M. Kellerman, D.R. Biggs and A.W. Linnane, J. Cell Biol. 42 (1969) 378.

[17] T.W. O'Brien and G.F. Kalf, J. Biol. Chem. 242 (1967) 2172, 2180.

[18] I.B. Dawid, in: The Development and Interrelationships of Cell Organelles, ed. P.L. Miller (Cambridge University Press, Cambridge, 1970), in press.

[19] M.A. Ashwell and T.S. Work, Biochem. Biophys. Res. Comm. 39 (1970) 204.

[20] D.Y. Thomas and D. Wilkie, Genet. Res. 11 (1968) 33.

[21] D.B. Roodyn, P.J. Reis and T.S. Work, Biochem. J. 80 (1961) 9.

[22] A.M. Kroon, Biochim. Biophys. Acta 72 (1963) 391.

[23] A.M. Kroon, Biochim. Biophys. Acta 91 (1964) 145.

[24] D. Neubert, H. Helge and H.J. Merker, Biochem. Z. 343 (1965) 44.

[25] F.C. Firkin and A.W. Linnane, FEBS Letters 2 (1969) 330.

[26] A.M. Kroon and R.J. Jansen, Biochim. Biophys. Acta 155 (1968) 629.

[27] P. Borst, G.J.C.M. Ruttenberg and A.M. Kroon, Biochim. Biophys. Acta 149 (1967) 140.

[28] P. Slonimski, in: Adaption in Micro-organisms, eds. E.F. Gale and R. Davis (Cambridge University Press, Cambridge, 1953) p. 76.

[29] P. Slonimski and H. de Robichon-Szulmajster, in: CIBA Foundation Symp. on Drug Resistance in Micro-organisms, (Churchill, London, 1957) p. 210.

[30] P.P. Slonimski, G. Perrodin and J.H. Croft, Biochem. Biophys. Res. Commun. 30 (1968) 232.

[31] M.J. Waring, Biochim. Biophys. Acta 114 (1966) 234.

[32] L.S. Lerman, J. Cellular Comp. Physiol. 64 (1964) suppl. 1, p. 1.

[33] P. Borst, A.M. Kroon and G.J.C.M. Ruttenberg, in: Genetic Elements. Properties and Function, ed. D. Shugar (Academic Press and P.W.N., London and Warsaw, 1967) p. 81.

[34] E. Wintersberger, Biochem. Z. 341 (1965) 409.

[35] A.M. Kroon, Biochim. Biophys. Acta 76 (1963) 165.

[36] M.W. Nirenberg, in: Methods in Enzymology, vol. 6, eds. S.P. Colowick and N.O. Kaplan (Academic Press, New York, 1963) p. 17.

[37] A.M. Kroon, C. Saccone and M.J. Botman, Biochim. Biophys. Acta 142 (1967) 552.

[38] E. Zylber, C. Vesco and S. Penman, J. Mol. Biol. 44 (1969) 195.

[39] J. ter Schegget and P. Borst, in press.

[40] R.F. van Gelder, Biochim. Biophys. Acta 118 (1966) 36.

[41] H. de Vries and A.M. Kroon, Biochim. Biophys. Acta 204 (1970) 531.

[42] R.R. Meyer and M.V. Simpson, Biochem. Biophys. Res. Commun. 34 (1969) 238.

[43] S. Nass, Intern. Rev. Cytol. 25 (1969) 55.

[44] I.B. Weinstein and I.H. Finkelstein, J. Biol. Chem. 242 (1967) 3757.

[45] Th. Finkelstein and I.B. Weinstein, J. Biol. Chem. 242 (1967) 3763.

[46] J.H. Landez, R. Roskoski and G.L. Coppoc, Biochim. Biophys. Acta 195 (1969) 276.

[47] R. Bittman, J. Mol. Biol. 46 (1969) 251.

AMINO ACID INCORPORATION INTO MITOCHONDRIAL RIBOSOMES OF NEUROSPORA CRASSA WILD TYPE AND MI-1 MUTANT

W. NEUPERT, P. MASSINGER and A. PFALLER

Institut für Physiologische Chemie und Physikalische Biochemie
der Universität München, Germany

Up to the present time the best-characterized mitochondrial ribosomes are those from *Neurospora crassa*. As first shown by Küntzel and Noll [1] and by Rifkin et al. [2] they can be isolated in a rather pure state. They differ from cytoplasmic ribosomes in many respects; namely, in sedimentation constants, dissociation behaviour, electrophoretic and chromatographic patterns of their proteins, properties and composition of their RNAs and sensitivity to different inhibitors of protein synthesis [1−5,7,9].

In order to give an impression about the significance of these differences and to demonstrate the purity of the mitochondrial ribosomes used in the experiments to be described here, the two types of ribosomes will be compared with respect to their density gradient profiles and the properties of the ribosomal RNAs on gel-electrophoresis.

In fig. 1 sucrose density gradient profiles of mitochondrial and cytoplasmic ribosomes isolated from *Neurospora* wild-type cells (WT) in the late logarithmic growth phase are shown. With the cytoplasmic ribosomes a monosome peak and three distinct polysome peaks can be seen. No free subunits are present. The mitochondrial ribosomes also show the monosome peak, which moves slightly slower on the gradient, but they exhibit no clear polysome formation, and free subunits are present. The degree of dissociation into subunits of the mitochondrial ribosomes under our isolation procedure is dependent on the age of the *Neurospora* cells. The dissociation into subunits is greater the further they come into the stationary phase (cf. fig. 8). The S_{20} values of the cytoplasmic and mitochondrial ribosomes as determined by Küntzel and Noll [1] and Küntzel [3] are 77 S and 73 S for the monomers, respectively, and

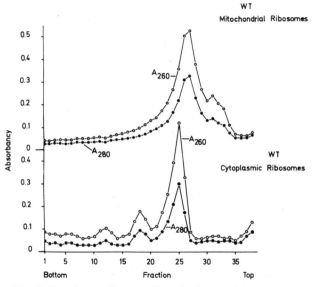

Fig. 1. Density gradient profiles of mitochondrial and cytoplasmic ribosomes from *Neurospora* wild type hyphae. Mitochondrial and cytoplasmic ribosomes were prepared from hyphae grown for 18 hr. The ribosomes were centrifuged for 2 hr at 41,000 rpm in the Spinco rotor SW 41 through a convex sucrose gradient according to Noll [24]. Fractions were collected and monitored at 260 and 280 mμ. For further details see [8].

57 S/37 S and 50 S/37 S for the subunits, respectively.

Küntzel and Noll [1] and Rifkin et al. [2] have also characterized the mitochondrial ribosomal RNA. The sedimentation coefficients vary, depending upon the salt concentration in the density gradient, from values similar to those of bacterial ribosomal RNA to those of cytoplasmic ribosomal RNA.

In fig. 2 a polyacrylamide gel electrophoretic run

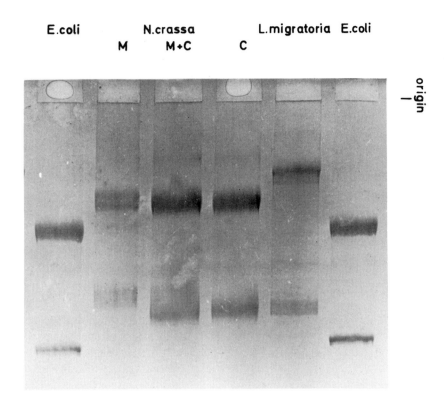

Fig. 2. Polyacrylamide gel electrophoresis of RNA samples. RNA was prepared from *E. coli* ribosomes, mitochondria (M) and 27,000 *g* supernatant (C) from *Neurospora* wild type hyphae, and from the cytoplasm of the flight muscle of *Locusta migratoria.* Electrophoresis on polyacrylamide gel (2.7% with 0.5% agarose) was carried out essentially as described by Dingman and Peacock [11,12]. The gels were stained with toluidine blue. For further details see [8].

of both types of RNA is presented. The large subunit RNAs of cytoplasmic and mitochondrial ribosomes have the same electrophoretic mobility, whereas the small subunit RNA of mitochondria has a lesser mobility than the cytoplasmic counterpart. To the same gel, RNA preparations from *E. coli* ribosomes and cytoplasmic ribosomes from locust flight muscle were applied as standards. According to Dingman and Peacock [13] and Loening [14], the electrophoretic mobility on polyacrylamide gel is a measure of the molecular weight of the ribosomal RNA species. The molecular weights estimated by this method are 1.28 and 0.67 million for cytoplasmic ribosomal RNA, and 1.28 and 0.72 million for mitochondrial ribosomal RNA. Mitochondrial and bacterial ribosomal RNA appear to be very dissimilar in size. A similar

electrophoretic mobility for cytoplasmic and mitochondrial transfer RNA is obvious in pictures of the whole gel slab (cf. fig. 10). It should be noted that under our preparative conditions no 5 S RNA can be observed in mitochondrial ribosomes, in contrast to cytoplasmic and bacterial ribosomes.

Synthesis of peptide chains on mitochondrial ribosomes of Neurospora wild type

Are these mitochondrial ribosomes active in synthesizing peptide chains? In order to answer this question the following experiment was performed. Mitochondria were isolated from *Neurospora* hyphae and incubated with radioactive amino acids under conditions which were determined to be optimal [6]. After 20 min incubation, the suspension was divided into two equal portions. One was immediately cooled to 0° and served as a control. To the other portion puromycin was added to a final concentration of 0.4 mM and incubation was continued for 7 min. This portion also was then cooled to 0°, and the ribosomes were isolated from both preparations in exactly the same way.

In table 1 the specific radioactivities of the different fractions of the isolation procedure of the ribosomes are shown. The control ribosomes have a very high specific radioactivity compared to the whole mitochondria, but only 7–10% of the total radioactivity incorporated is found in the ribosomal fraction. This percentage is dependent on the efficiency of incorporation and becomes higher when the rate of incorporation becomes lower. This observation indicates that, at least under certain conditions, the radioactive amino acids are incorporated into complete peptide chains which are released from the ribosomes and integrated into the insoluble membrane protein. There, as shown by Sebald et al. [6], they can be detected as definite bands by means of polyacrylamide gel electrophoresis.

In the preparation treated with puromycin, the specific radioactivity of the whole mitochondria is lower, probably because of loss of trichloroacetic acid

soluble peptides released from mitochondria. Similar results were obtained for rat liver mitochondria by Wheeldon and Lehninger [10]. Most interesting is the finding that the specific radioactivity of the ribosomes is very low compared to the control, indicating a substantial loss of radioactive amino acids from the ribosomes.

Fig. 3 represents sucrose density gradients of the ribosomes of these preparations. In the control ribosomes, radioactivity is found at the polysome region and at the monosome peak. The polysomes have the highest specific radioactivity. The radioactivity near the top of the gradient probably represents light mitochondrial particles not destroyed by the lysing agent, Triton X–100. In the preparation treated with puromycin, all of the radioactivity associated with poly- and monosomes has disappeared. This demonstrates that peptide chains are synthesized on the ribosomes, and that they can be released by puromycin, as is known for the bacterial system and for the cytoplasmic system of eucariotic cells.

To provide further proof that the label in the gradient below the monosome peak actually corresponds to peptide chains on polysomes, i.e., to demonstrate the existence of polysomes in mitochondria, another experiment was carried out. Isolated mitochondria were labelled with radioactive amino acids in the same way as described for the puromycin

Table 1

Specific radioactivities of mitochondrial fractions after incorporation of labelled amino acids into isolated *Neurospora* wild type mitochondria.

Fraction	Control		Puromycin incubated	
	Specific radioactivity cpm/mg protein	Total radioactivity cpm	Specific radioactivity cpm/mg protein	Total radioactivity cpm
Mitochondria after incubation	7,000	525,000	4,900	343,000
Mitochondrial lysate	7,210	392,000	4,960	238,000
Sediment of mitochondrial lysate 30 min, 35,000 g	7,910	13,050	3,650	6,930
Crude ribosomes	17,660	33,500	4,160	5,830
Supernatant after sedimentation of ribosomes	6,185	180,700	7,110	242,000

Isolated mitochondria were labelled for 20 min with 1-^{14}C-leucine, 1-^{14}C-isoleucine and 1-^{14}C-phenylalanine (0.1 μC/ml each) in a medium described by Sebald et al. [6]. For further details see text and ref. [8].

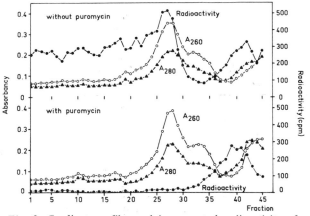

Fig. 3. Gradient profiles and incorporated radioactivity of ribosomes isolated from mitochondria labelled in vitro with and without successive incubation with puromycin. The crude ribosomal preparations of table 1 were used. For further experimental details see [8].

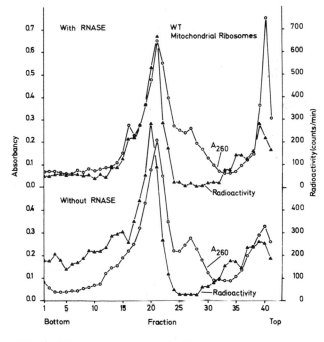

Fig. 4. Ribonuclease treatment of ribosomes isolated from *Neurospora* wild type mitochondria labelled in vitro with radioactive amino acids. Labelling of mitochondria, isolation of ribosomes and density gradient centrifugation were performed as described for table 1 and fig. 3. After isolation, one half of the ribosomes were incubated for 1 hr at 4°C with 1 μg/ml of pancreatic ribonuclease, essentially according to Rich et al. [18] prior to density gradient centrifugation.

experiment. After 20 min, the incubation was stopped by cooling and the ribosomes were isolated. These ribosomes were divided into two equal portions. One portion served as a control and was put on the gradient without further treatment. The second portion was incubated for 60 min at 4° with 1 μg/ml pancreatic ribonuclease essentially according to Rich et al. [18]. Both preparations were submitted to density gradient centrifugation, the result of which is demonstrated in fig. 4. When compared to the control, most of the radioactivity at the polysomes has disappeared in the ribonuclease experiment. The radioactivity removed from the polysome region cannot be found quantitatively at the monosome peak. The reason for this may be that under the conditions of the experiment part of the aminoacyl-transfer-RNA at the monosomes is degraded and removed.

It cannot be determined from our experiments whether peptide chains are synthesized only by polysomes, nor can it be determined how many of the polysomes present in the mitochondria, in vivo, might be converted to monosomes during the isolation.

One more interesting experimental observation was the inability of isolated mitochondria to incorporate amino acids into the structural ribosomal proteins. However, this provides no evidence that mitochondria are unable to synthesize their ribosomal proteins, since the protein-forming system as well as the system for the synthesis of the ribosomal ribonucleic acids may be damaged.

Biosynthesis of the mitochondrial ribosomal proteins

A more definite answer to the question of where the mitochondrial ribosomal proteins are synthesized can be expected from experiments in vivo with specific inhibitors of cytoplasmic and mitochondrial amino acid incorporation.

Table 2 represents experiments in which *Neurospora* cells were labelled with and without preincubation with cycloheximide and chloramphenicol, respectively, followed by a chase of unlabelled amino acids to make certain that the amino acids in the peptidyl-transfer-RNA on the ribosomes are not radioactively labelled. In the control experiment (first column) all fractions have a very similar specific radioactivity. In the experiment with cycloheximide

Table 2
Influence of inhibitors on the incorporation of labelled amino acids in vivo into fractions of *Neurospora* hyphae.

Fraction	Specific radioactivity (cpm/mg protein)		
	Control	Cycloheximide preincubated	Chloramphenicol preincubated
Mitochondrial lysate	25,010	2,380	13,280
Sediment of mitochondrial lysate 30 min, 30,000 *g*	27,700	1,460	10,830
Crude mitochondrial ribosomes	21,100	750	10,400
Cytoplasmic ribosomes	8,700	234	5,700
Supernatant of cytoplasmic ribosomes	17,700	95	11,810

Neurospora wild type hyphae grown for 18 hr were labelled for 20 min with 1-^{14}C-leucine, 1-^{14}C-isoleucine and 1-^{14}C-phenyl-alanine (6.7 mμC/ml). Then a 20-min chase period with unlabelled amino acids (2 mM each) followed. The inhibitors (cyclo-heximide 100 μg/ml, chloramphenicol 4 mg/ml) were added 10 min prior to the addition of the radioactive amino acids. Separa-tion of the crude mitochondrial ribosomes is shown in figs. 5–7 and ref. [9].

preincubation (second column) incorporation into cytoplasmic soluble proteins and into the cytoplasmic ribosomes is inhibited by 99%, while whole mito-chondria still show 10% of the incorporation of the control. However, incorporation into mitochondrial ribosomes is inhibited by 99%. In the experiment with chloramphenicol preincubation (third column), incorporation into all fractions is lower, if compared to the control. No distinct specific inhibition of any fraction can be observed. Similar experiments and results were described by Küntzel [4]. In the follow-ing figures sucrose density gradient profiles of the

mitochondrial ribosomes isolated in these experi-ments are presented.

In the control experiment (fig. 5) a heavy labelling of the monosome peak takes place; polysomes do not have a higher specific radioactivity. There is also no label at the subunits. Fig. 6 represents the mito-chondrial ribosomes from the experiment in which the *Neurospora* cells were preincubated with cyclo-heximide before adding the labelled amino acids. Virtually all radioactivity above the experimental limit of error has disappeared from the ribosomes. In fig. 7, representing the chloramphenicol experiment, the monosome peak is labelled to approximately the

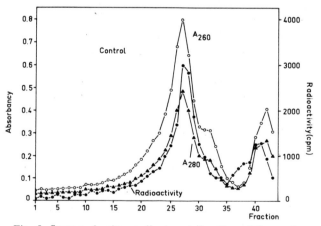

Fig. 5. Sucrose density gradient centrifugation of the crude ribosomal preparation from the control experiment in table 2.

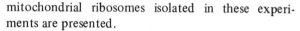
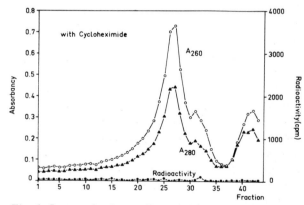

Fig. 6. Sucrose density gradient centrifugation of the crude ribosomal preparation from the cycloheximide preincubation experiment in table 2.

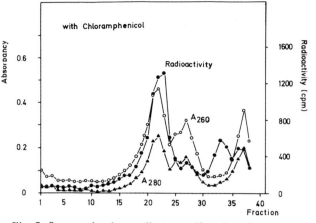

Fig. 7. Sucrose density gradient centrifugation of the crude ribosomal preparation from the chloramphenicol preincubation experiment in table 2.

same specific radioactivity as in the control experiment.

To summarize, amino acid incorporation into the mitochondrial ribosomal proteins in vivo can be blocked by cycloheximide, a specific inhibitor of cytoplasmic protein synthesis [6,21,23], but not by chloramphenicol which specifically inhibits mitochondrial amino acid incorporation [15,20–22]. These results demonstrate that the synthesis of mitochondrial ribosomal proteins is dependent on the functioning of the cytoplasmic protein synthesis.

On the basis of the very strong degree of inhibition by cycloheximide the most probable explanation is that the vast majority, if not all, of the mitochondrial ribosomal proteins are synthesized by the cytoplasmic system. Of course, these experiments do not exclude the possibility that a minor part of these proteins, a few out of 40 or 50, are synthesized by the mitochondrial system. Our conclusions implicate a cooperation of mitochondrial and extramitochondrial systems in the biogenesis of mitochondrial ribosomes, since experimental evidence suggests that mitochondrial ribosomal RNA is coded for and synthesized within the mitochondria [25–28].

Other less probable interpretations of our results will be discussed in the following. (1) It could be that just one or a few protein components essential for the formation of the mitochondrial ribosome are synthesized by the cycloheximide-sensitive system. This explanation implies that the average time be-

tween the synthesis of these components and their integration into the mitochondrial ribosomes is much shorter than 10 min. In other words, only a small amount of these components would exist in the cell in a free state, not yet integrated into the mitochondrial ribosomes, so that this precursor protein pool is exhausted after the 10 min incubation with cycloheximide. (2) The mitochondrial ribosomal proteins could be all of mitochondrial origin, but a 'trigger' protein could be synthesized on the cytoplasmic ribosomes and then transferred into the mitochondria where it would regulate specifically the synthesis of the mitochondrial ribosomes or one of their components. For the same reason as stated above, this protein would have an extremely short half-life, i.e., it would be degraded or inactivated within less than 10 min. Such a high turnover for a protein seems improbable.

Mitochondrial ribosomes of Neurospora mi-1 (poky) mutant

We should like to turn now to experiments with the cytoplasmic mutant mi-1 (poky) of *Neurospora crassa*.

It is well known from the early work of Mitchell and Mitchell [16] that the mutant character of this slowly growing strain does not obey Mendelian genetics but is maternally inherited. It is also known from these authors that the hyphae possess mitochondria with an altered respiratory chain, with greatly decreased amounts of cytochromes a and b and an elevated amount of cytochrome c. The composition of the mitochondrial membrane, as shown by gel electrophoresis, is different from the wild type. Also the gel electrophoretic pattern of the mitochondrially labelled proteins is drastically changed as compared to the wild type [6]. The rate of amino acid incorporation into isolated mitochondria under optimal conditions amounts to about 20% of that of the wild type [6]. Reich and Luck [19] did not find a difference in the buoyant densities of poky and wild type mitochondrial DNA. Finally, the instability of the poky character should be mentioned, i.e., as the poky cells approach the stationary growth phase the more they aquire more of the cytochromes [17].

Fig. 8 shows a comparison of sucrose density

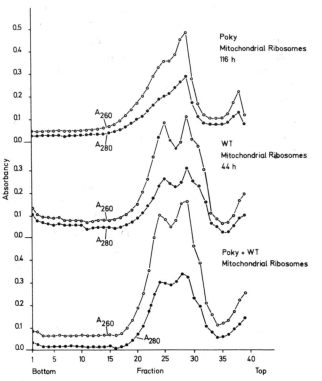

Fig. 8. Density gradient profiles of mitochondrial ribosomes isolated from *Neurospora* mi-1 mutant (poky) and wild type (WT). Poky hyphae were grown for 116 hr, wild type hyphae for 44 hr. For experimental details of poky growth see [6]. In the mixture poky + WT, equal amounts of each type of ribosomes were applied.

gradient profiles of mitochondrial ribosomes from wild type and poky *Neurospora* hyphae. The wild type mitochondrial ribosomes were isolated from hyphae grown for 44 hr, i.e. from hyphae in the stationary growth phase. As mentioned above, a large part of the wild type mitochondrial ribosomes in the stationary growth phase are present as free subunits. One can distinguish the monosome peak, the large subunit peak and the small subunit present as a clear shoulder. Since the difference of the S values of the mitochondrial ribosomal subunits is relatively small as compared to the cytoplasmic ribosomes it is difficult to resolve the small subunit as a peak. In this context we refer to the excellent characterization of the mitochondrial ribosomal subunits by Küntzel [3].

Poky mitochondrial ribosomes isolated from the late logarithmic growth phase show a peak corresponding to the monosomes, and a stronger peak corresponding to the large subunit. However, no shoulder corresponding to the free small subunit can be detected. Absence of the small subunit shoulder was observed in some 30 density gradients without an exception. When wild type and poky mitochondrial ribosomes are mixed in equal proportions and subjected to gradient centrifugation the small subunit shoulder can be seen clearly. These results strongly suggest that in the case of the poky mitochondrial ribosomes, the small subunits are not equivalent in number to the large subunits, and that only in the relatively small proportion of monosomes are small subunits present. In contrast, ribosomes from the cytoplasm of wild type and poky *Neurospora* do not differ. This is shown in fig. 9. Gradient profiles of both these types of cytoplasmic ribosomes and of a mixture of the two types are very similar.

This non-equivalence of large and small subunits in mitochondrial poky ribosome preparations should be reflected in the relation of large and small subunit ribosomal RNA. In fig. 10 polyacrylamide gel electrophoresis is shown in which RNA preparations from isolated whole wild type and poky mitochondria were run in parallel. Only a very small amount of small subunit RNA is present in poky mitochondria. As in the wild type, no 5 S RNA can be detected. This, and the observation that ribonuclease does not preferentially destroy the large subunit RNA when added to whole mitochondria, prove the mitochondrial localisation of the large subunit isolated from mitochondria. The small amount of small subunit RNA is presumably derived from the monosomes.

Are these poky ribosomes active in synthesizing peptide chains? The experiments reported by Sebald et al. [6] which demonstrate the ability of isolated poky mitochondria to incorporate radioactive amino acids, suggest that they are. However, as already mentioned, the electrophoretic pattern of the radioactively labelled bands is appreciably changed compared to that of the wild type. We have carried out the following experiment to check the function of mitochondrial ribosomes in whole poky cells. Two equal portions of poky hyphal suspensions were first incubated for 10 min with cycloheximide, then radioactive amino acids were added to each and the incubation continued for a further 10 min. One portion was then immediately harvested and cooled; the other was chased with approximately a 10,000-fold

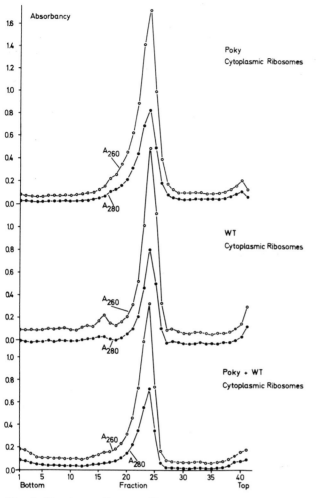

Fig. 9. Density gradient profiles of cytoplasmic ribosomes from *Neurospora* mi-1 mutant (poky) and wild type (WT). Poky hyphae were grown for 137 hr wild type hyphae for 47.5 hr. Other experimental conditions as described before.

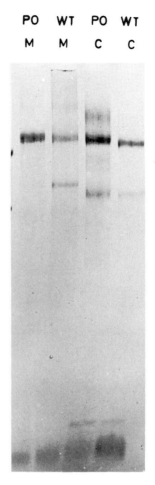

Fig. 10. Polyacrylamide gel electrophoresis of RNA samples from *Neurospora* mi-1 mutant and wild type. Abbreviations: PO, poky; WT, wild type; M, mitochondria; C, cytoplasm (27,000 g supernatant). The ages of hyphae from which RNA was isolated were: C (poky) 137 hr, C (WT) 44 hr, M (poky) 71.5 hr, M (WT) 47.5 hr. Preparation of RNA samples and conditions of electrophoresis as described for fig. 2.

excess of unlabelled amino acids for 20 min, then harvested. In figs. 11 and 12 density gradient centrifugations of the mitochondrial ribosomes from this experiment are shown. In the experiment without chase, a radioactivity peak is found at the monosomes and also some radioactivity at the polysomes. Also, an appreciable amount of radioactivity is located in the upper part of the gradient probably representing, as already pointed out for the wild type experiments, light mitochondrial particles. However, compared to the wild type gradients, the proportion of this frac-

tion is much higher. The obvious explanation is the much lower labelling of the poky ribosomes. In the experiment with chase, the radioactivity in the polysome and monosome region has disappeared, whereas the label in the upper part of the gradient is still present.

These experiments suggest that poky ribosomes are also able to synthesize polypeptide chains, but they do not indicate whether or not these monosomes are 'normal'. Experimental evidence indicates that they are much more labile than those from the

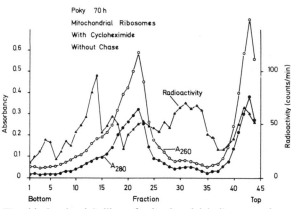

Fig. 11. In vivo labelling of mitochondrial ribosomes from *Neurospora* poky hyphae with radioactive amino acids after preincubation with cycloheximide. Poky cells grown for 70 hr were incubated first with cycloheximide (100 μg/ml) for 10 min, then 1-^{14}C-leucine, 1-^{14}C-isoleucine and 1-^{14}C-phenylalanine (12.5 mμC/ml each) were added and after 10 min the hyphae were harvested. Then mitochondrial ribosomes were isolated and subjected to density gradient centrifugation as described.

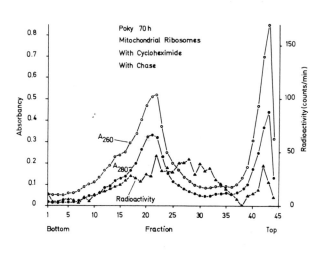

Fig. 12. In vivo labelling of mitochondrial ribosomes from *Neurospora* poky hyphae with radioactive amino acids after preincubation with cycloheximide followed by a chase with unlabelled amino acids. Poky cells were grown and labelled identically to those described in fig. 11. In this experiment after the incubation period with the ^{14}C-amino acids, un-labelled amino acids were added to a final concentration of 2 mM and incubation was continued further for 20 min. Then the hyphae were harvested and fractions were isolated in parallel to those of fig. 11.

wild type. Similar conclusions can be drawn from experiments in which puromycin was used to release newly synthesized peptide chains from poky ribosomes after incorporation of labelled amino acids into isolated poky mitochondria.

It has already been mentioned that several properties of poky cells and mitochondria change during the growth of a poky culture. Therefore, it seemed

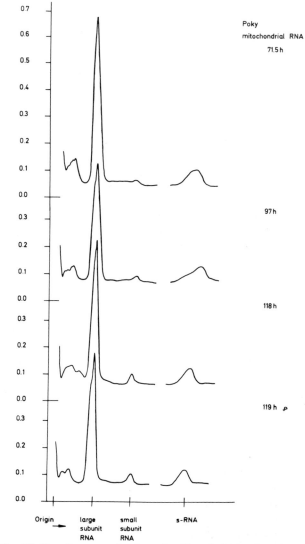

Fig. 13. Densitograms of polyacrylamide gel electrophoresis of RNA samples isolated from mitochondria of poky hyphae of different age (71.5 hr, 97 hr, 118 hr and 119 hr). Isolation of RNA and gel electrophoresis were performed as described for fig. 2, and the toluidine blue stained gel was submitted to densitometry at 546 mμ.

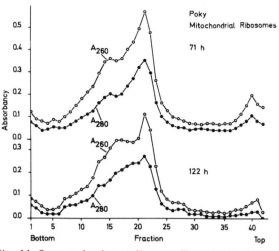

Fig. 14. Sucrose density gradient profiles of mitochondrial ribosomes isolated from *Neurospora* poky hyphae of different age (71 hr and 122 hr). Isolation and density gradient centrifugation were performed as described before.

worthwhile to determine whether the properties of the mitochondrial ribosomes might change also. In fig. 13 densitograms are shown of polyacrylamide gel electrophoreses of RNA preparations extracted from whole mitochondria isolated from poky cells of different ages. Hyphae were taken from 71.5, 97, 118 and 119 hr cultures. The first ones correspond to the early logarithmic phase, the last ones to the very late logarithmic phase. The proportion of small subunit RNA is relatively low throughout the whole period investigated. However, it can be clearly seen that it increases with the age of the hyphae. If this is true there should also be an increase of the proportion of the small subunit in density gradient profiles of ribosomal preparations, either as appearance of free small subunits or as an increase of the proportion of monosomes. This is illustrated in fig. 14. Ribosomes of the early and late logarithmic phase are shown. In both cases no small subunit shoulder can be seen. However, the shoulder corresponding to the monosomes is increased in the ribosome preparation from the late logarithmic growth phase. In order to explain the described changes of ribosomal composition in poky mitochondria, the following possibilities are considered: (1) the change in ribosomal composition is caused by a genetic change affecting the structure or the synthesis of one or several components of the mitochondrial ribosomes. (2) The genetic alteration

does not concern the information for structure or synthesis of a ribosomal component, and the change of the ribosome is a secondary effect. For instance, there could be an increased amount of endomitochondrial nuclease which specifically attacks and destroys the small subunit. Or conditions inside the poky mitochondria, such as alteration of membrane structure or the absence of normal messenger RNA, could favour the dissociation into subunits and the small subunit could be lost because of a peculiar instability. It is difficult to exclude these possibilities, but the reproducibility of the results and the dependence of ribosomal composition on the growth do not favour such an explanation. In this context it is interesting to note that the cytoplasmic mutant *mi-3*, which in many respects resembles *mi-1*, has large and small subunits in equal amounts [29]. Furthermore, in vivo pulse labelling experiments with radioactive amino acids of poky hyphae in the logarithmic phase show that the label is present almost entirely in the large subunit. We have to assume that either the newly synthesized small subunit is broken down immediately upon formation or the amount of small subunit synthesized is very small compared to that of the large subunit. Therefore the first explanation is preferred, namely that the small subunit is not formed because of a genetic defect in the mitochondrial DNA. Further speculations, such as whether the small subunit RNA is affected, should be based on further experimental data.

Acknowledgements

This work was supported by the Deutsche Forschungsgemeinschaft (Schwerpunktsprogramm 'Biochemie der Morphogenese'). The authors wish to thank Prof. Th. Bücher for continuous advice and support, and Dr. A.J. Schwab for stimulating discussions. They also wish to thank Dr. and Mrs. G.D. Ludwig for help in preparing the manuscript.

References

[1] H. Küntzel and H. Noll, Nature 215 (1967) 1340.
[2] M.R. Rifkin, D.D. Wood and D.J.L. Luck, Proc. Natl. Acad. Sci. U.S. 58 (1967) 1025.

[3] H. Küntzel, J. Mol. Biol. 40 (1969) 315.

[4] H. Küntzel, Nature 222 (1969) 142.

[5] H. Küntzel, FEBS Letters 4 (1969) 140.

[6] W. Sebald, Th. Bücher, B. Olbrich and F. Kaudewitz, FEBS Letters 1 (1968) 235.

[7] W. Sebald, A.J. Schwab and Th. Bücher, FEBS Letters 4 (1969) 243.

[8] W. Neupert, W. Sebald, A.J. Schwab, A. Pfaller and Th. Bücher, Europ. J. Biochem. 10 (1969) 585.

[9] W. Neupert, W. Sebald, A.J. Schwab, P. Massinger and Th. Bücher, Europ. J. Biochem. 10 (1969) 589.

[10] L.W. Wheeldon and A.L. Lehninger, Biochemistry 5 (1966) 3533.

[11] A.C. Peacock and C.W. Dingman, Biochemistry 6 (1967) 1818.

[12] C.W. Dingman and A.C. Peacock, Biochemistry 7 (1968) 659.

[13] C.W. Dingman and A.C. Peacock, Biochemistry 7 (1968) 668.

[14] U.E. Loenig, J. Mol. Biol. 38 (1968) 355.

[15] M. Huang, D.R. Biggs, G.D. Clark-Walker and A.W. Linnane, Biochim. Biophys. Acta 114 (1966) 434.

[16] M.B. Mitchell and H.K. Mitchell, Proc. Natl. Acad. Sci. U.S. 38 (1952) 442.

[17] F.A. Haskins, A. Tissieres, H.K. Mitchell and M.B. Mitchell, Biochem. J. 200 (1953) 819.

[18] A. Rich, J.R. Warner and H.M. Goodman, Cold Spring Harb. Symp. on Quant. Biol. XXVIII (1963) p. 269.

[19] E. Reich and D.J.L. Luck, Proc. Natl. Acad. Sci. U.S. 55 (1966) 1600.

[20] E. Wintersberger, Biochem. Z. 341 (1964) 409.

[21] D.G. Clark-Walker and A.W. Linnane, Biochem. Biophys. Res. Commun. 25 (1966) 8.

[22] A.G. So and E.W. Davie, Biochemistry 2 (1963) 132.

[23] D.S. Beattie, J. Biol. Chem. 243 (1968) 4027.

[24] H. Noll, Nature 215 (1967) 360.

[25] E. Wintersberger, Z. Physiol. Chem. 348 (1967) 1701.

[26] H. Fukuhara, Proc. Natl. Acad. Sci. U.S. 58 (1967) 1065.

[27] Y. Suyaina, Biochemistry 6 (1967) 2829.

[28] D.D. Wood and D.J.L. Luck, J. Mol. Biol. 41 (1969) 211.

[29] W. Neupert, unpublished results.

INCORPORATION OF AMINO ACIDS INTO ELECTROPHORETIC AND CHROMATOGRAPHIC FRACTIONS OF MITOCHONDRIAL MEMBRANE PROTEINS

W. SEBALD, G.D. BIRKMAYER, A.J. SCHWAB and H. WEISS

*Institut für Physiologische Chemie und Physikalische Biochemie
der Universität München, Germany*

Intrinsic and extrinsic contribution to the biogenesis of mitochondria

Two systems contribute to the biogenesis of the mitochondrial proteins. One is localized outside the mitochondria and may be called the extrinsic system. The other one is localized inside the mitochondria and may be called the intrinsic system.

There are various experimental approaches to differentiate between the extrinsic and the intrinsic origin of the mitochondrial proteins [1]. One possibility is to label specifically the products of the intrinsic system by incorporation of radioactive amino acids. This can be achieved by incorporation of radioactive amino acids into isolated mitochondria in vitro, or by incorporation in vivo in the presence of cycloheximide.

Neurospora grind mill

The experiments to be described here have been carried out with mitochondria from *Neurospora crassa*. This organism has the advantage that intact mitochondria can be prepared rapidly and in large amounts [2]. We have constructed a simple device, which we call a '*Neurospora* grind mill', to disrupt the cell walls of the filamentous hyphae between two grinding wheels (fig. 1). The filtered hyphae are suspended homogeneously in the sucrose isolation medium and poured into the upper reservoir cylinder. They pass through the space between the grinding surfaces of the two rotating wheels. At the outlet tube the broken cells flow out ready for differential centrifugation. Approximately 25% of the mitochondria present in the cells are recovered.

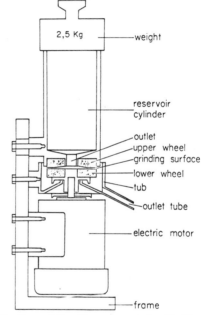

Fig. 1. Section of the *N. crassa* grind mill. The grinding wheels consist of corundum (8350 R, Kleinschleifscheiben, Edelkorund, Fa. Kratsch, München, Germany). The lower wheel is rotated by an electric motor at 1350 rpm and 135 watt.

In vitro labelling pattern of mitochondrial membrane proteins from wild type and mi-1 mutant of N. crassa

Isolated mitochondria from *N. crassa* wild type incorporate under conditions determined to be optimal 300 $\mu\mu$moles of leucine per hr per mg protein [3]. Nearly all of the radioactivity incorporated is associated with the mitochondrial membrane proteins, as has been found by other workers with mitochondria from many different sources.

These membrane proteins were dissolved in phenol-formic acid medium and gel electrophoresis was performed in the same medium [4]. In fig. 2 the protein pattern of the mitochondrial membrane proteins after separation is shown. Twenty-five protein bands can be distinguished. Only a few of them are labelled by in vitro incorporation of ^{14}C-amino acids. This is demonstrated by the autoradiograph taken from the dried gel (fig. 2). The highest specific radioactivity is obtained in the 'band 4' protein. High incorporation is also found in the proteins of bands 3, 6a, 7 and 8. In addition minor components were labelled, as indicated by the small amounts of radioactivity present in other protein bands.

Identical experiments have been done with mitochondria from the cytoplasmic *mi-1* mutant (poky). The *mi-1* mitochondria show optimal incorporation under the same conditions as the wild type mitochondria. The rate of incorporation, however, is found to be only 20% of that of the wild type mitochondria. The protein pattern of the membrane proteins isolated from *mi-1* mitochondria shows after gel electrophoretic separation, bands analogous to those of the wild type mitochondria. Less protein is found in bands 4, 8 and 10. It is especially noteworthy that the labelling pattern is markedly altered. The most significant change is that there is no labelling of band 4. Thus, by direct evidence, it is shown that the cytoplasmically inherited alterations in the *mi-1* mutant can be correlated with alterations

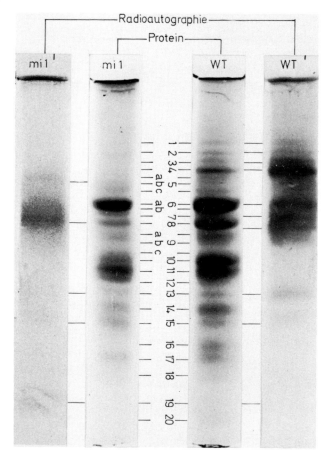

Fig. 2. Electrophoretic pattern of insoluble mitochondrial protein from *N. crassa* wild type (WT) and cytoplasmic *mi-1* mutant (*mi-1*). Inner gels, amidoblack staining; outer gels, autoradiographs corresponding to the pherograms after amino acid incorporation in vitro. For experimental conditions see ref. [3].

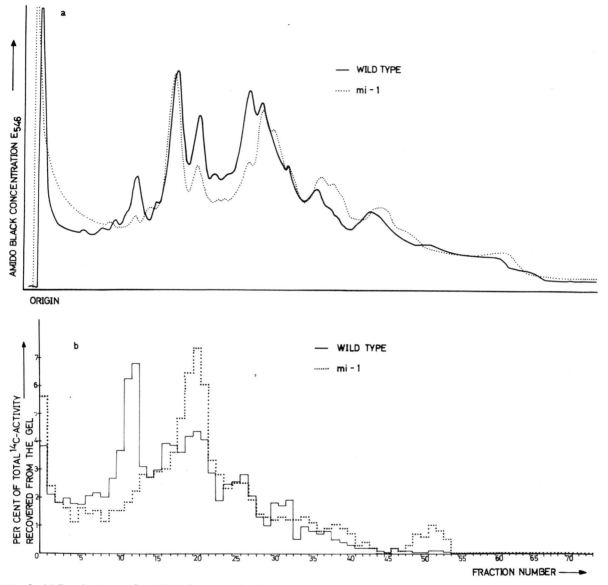

Fig. 3. (a) Densitograms of amidoblack stained insoluble mitochondrial proteins from *N.c.* wild type and *mi-1* mutant after gel-electrophoresis. (b) Distribution of radioactivity in the pherograms shown in (a) (total as 100%).

in the biosynthetic activities of the isolated mitochondria.

Fig. 3 shows a quantitative evaluation of the protein patterns presented in fig. 2. The continuous curve represents a densitometric tracing of the stained wild type proteins and the dotted curve that of the *mi-1* proteins. The distribution of the radioactivities (discontinuous curves) was determined by cutting the

gel and counting the single slices. An attempt was made to calculate the specific radioactivity of the 'band 4 protein'. However, as shown in the figure, there is still an appreciable amount of protein present in band 4 of the *mi-1* mitochondria, compared to the small amount of radioactivity. This indicates that band 4 in the wild type may contain both labelled and unlabelled proteins. Therefore, the specific radioactivity of the protein which is missing in the band 4 of the *mi-1* mitochondria was calculated from the differences in the protein and labelling patterns. It is found that this protein constitutes only 1% of the total membrane proteins, although it contains 10% of the total radioactivity incorporated. This suggests that the intrinsic mitochondrial protein synthesis contributes 10% to the proteins of the mitochondrial membranes, assuming that all proteins labelled in vitro achieve the same specific radioactivity. Since in *N. crassa* half of the mitochondrial proteins are membrane proteins, 5% of the total mitochondrial protein would be synthesized in the mitochondria.

In vivo labelling of mitochondrial proteins under the influence of cycloheximide

It might be asked to what extent these in vitro results are artifacts due to the isolation of the mitochondria from their natural environment. Hence, it is desirable to devise an experimental approach to specifically label in vivo the protein synthesized by the intrinsic mitochondrial system. Cycloheximide, which inhibits protein synthesis in eukaryotic cells but not in prokaryotic cells, such as bacteria, or in mitochondria [5], was used. Fortunately, in vivo, extrinsic and intrinsic protein synthesis is not strongly coupled in *N. crassa*. As shown in fig. 4, cycloheximide, in concentrations in excess of 10 μg per ml,

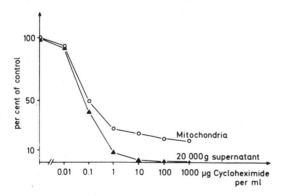

Fig. 4. Effect of different amounts of cycloheximide on the incorporation in vivo of ^{14}C-leucine (0.5 μC per 50 ml culture) into the proteins of whole mitochondria (○—○) and of the 20,000 *g* supernatant (▲—▲) of *N. crassa*. Hyphae were preincubated for 5 min with cycloheximide before the 10 min labelling period.

inhibits incorporation of ^{14}C-amino acids into extra-mitochondrial proteins (marked S) by more than 99%. In contrast, approximately 20% of the incorporation into the whole mitochondrial proteins (marked M) is found to be resistant to cycloheximide [6].

Rate of protein synthesis in the presence of cycloheximide

The effect of cycloheximide on the labelling of the total mitochondrial proteins was investigated in greater detail using different times of preincubation with cycloheximide and a constant labelling period of 10 min. The incorporation of ^{14}C-leucine decreases with increasing times of cycloheximide preincubation (fig. 5). Simultaneously, the cellular leucine pool increases as measured in cultures grown in parallel. This appears to be reasonable since the bulk of cellular protein synthesis is inhibited, while leucine production still proceeds. The actual rates of protein synthesis within the mitochondria can be calculated from these experimental data for the various periods of cycloheximide preincubation. Extrapolation of this curve to zero time yields a rate of protein synthesis of 8% as compared to that without cycloheximide. This indicates that about 8% of total mitochondrial protein is synthesized by the intrinsic mitochondrial system. This is a maximal value and it is in good agreement with the value of 5% which was obtained in the electrophoretic experiments following incorporation in vitro.

From the experiment shown in fig. 5 the additional information is obtained that the mitochondrial protein synthesis is still active for at least 60 min after addition of cycloheximide. It can be calculated that the products of the mitochondrial protein synthesis increase about 10% within this time. In the following experiments the hyphae were preincubated with cycloheximide for 5 min. The labelling period thereafter was 10 to 30 min. The ^{14}C-amino acids were then diluted by the corresponding unlabelled amino acids, and the hyphae were incubated for a further 30 min. This chase was effective since no labelled growing peptide chains could be detected attached to the mitochondrial ribosomes [7]. Hence it may be assumed that the labelling pattern represents predominantly complete peptide chains.

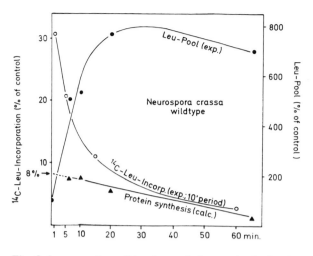

Fig. 5. Incorporation of leucine and changes in the leucine pool in *N. crassa* in the presence of cycloheximide. ○─○, ¹⁴C-leucine incorporation in vivo in a 10-min labelling period (0.5 μC per 50 ml culture). The hyphae were preincubated with 100 μg cycloheximide per ml for the times plotted in the abscissa. The radioactivity incorporated without cycloheximide was taken as 100%. ●─●, leucine pool per 50 ml culture after different times of cycloheximide (100 μg per ml) incubation. The leucine pool without cycloheximide was taken as 100%. For the incorporation and the pool experiment parallel grown cultures were used (one 50 ml culture per experimental point). The hyphae were grown from an inoculum of 2×10^6 conidia per ml in 50 ml of Vogels minimal medium plus 2% sucrose at 25°C for exactly 18 hr before addition of cycloheximide. ▲─▲, actual mitochondrial protein synthesis after different times of cycloheximide preincubation as calculated from the pool size and the leucine incorporation. The following assumptions were made: that the mitochondrial proteins contain 7% (w/w) leucine and that without cycloheximide all proteins of the exponentially growing hyphae enhance proportionally with a doubling time of 4 hr. This rate of synthesis was set as 100%. The protein synthesis of the control cannot be calculated from leucine incorporation because the added label is diluted very rapidly. (The turnover of the cellular leucine pool is less than 2 min.)

In vivo labelling pattern of the mitochondrial membrane proteins

Under these in vivo conditions, incorporation which is resistant to cycloheximide is found to occur exclusively into the mitochondrial membrane proteins (table 1). The labelling of the soluble mitochondrial proteins is inhibited to the same large

Table 1

Effect of cycloheximide on the in vivo incorporation of amino acids into mitochondrial subfractions of *N. crassa*.

Protein fraction [a]	Control cpm per mg protein	Cycloheximide cpm per mg protein
Cytoplasmic (20,000 *g* supernatant)	130,000 (93%)	3,000 (5%)
Mitochondria		
total	120,000 (86%)	28,000 (47%)
extractable	105,000 (75%)	3,100 (5%)
residual [b]	140,000 (100%)	60,000 (100%)

Percentages refer to the insoluble mitochondrial proteins (residual) in each experiment. Hyphae grown for 18 hr in five 50 ml cultures were incubated for 30 min with 0.1 μC per ml of each ¹⁴C-leucine, ¹⁴C-isoleucine and ¹⁴C-phenylalanine, followed by a 30 min chase (2 μmoles of each ¹²C-leucine, ¹²C-isoleucine, ¹²C-phenylalanine). In the cycloheximide experiment the hyphae were preincubated for 5 min with 100 μg cycloheximide per ml prior to the addition of the labelled amino acids.

[a] For fractionation conditions see ref. [3].

[b] The further separation of these fractions is shown in fig. 6.

extent as the labelling of the extramitochondrial proteins. The electrophoretic separation (fig. 6) of the mitochondrial membrane proteins indicates that the radioactivity incorporated in vivo under the influence of cycloheximide is greatest in fraction 11 and 18, while minor radioactivity is concentrated in fractions 8, 10, 16 and 23. These fractions correspond to the protein bands 2, 3, 4, 6, 8 and 10. These are essentially the same as were labelled following in vitro incorporation. In the control experiment without cycloheximide all bands are labelled nearly proportional to the amido black staining. Hence, by two independent methods it is shown that only a few fractions of the mitochondrial membrane proteins are formed by the intrinsic mitochondrial protein synthesis, and that these proteins amount to 5–8% of the whole mitochondrial proteins. Consequently, for isolation of the products of the mitochondrial protein synthesis, utilizing the label as a marker, the highest enrichment of radioactivity which can be expected for the pure proteins is about 20-fold, starting with whole mitochondria, and about 10-fold starting with the mitochondrial membrane proteins.

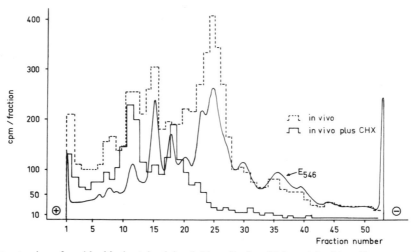

Fig. 6. Densitometer tracing of amido black stained insoluble mitochondrial protein from *N. crassa* after gel electrophoresis (smooth curve) and distribution of radioactivity (edged lines) after incorporation in vivo (– – –) and in vivo plus cycloheximide (———). (For labelling conditions see table 1.)

Molecular sieve chromatography in phenol-acetic acid medium

A fractionation of the mitochondrial membrane proteins was attempted on a preparative scale using molecular sieve chromatography on Biogel P–200. The same phenol medium in which the electrophoresis was performed served as solvent, with the exception that the formic acid was replaced by acetic acid. In the experiment shown in fig. 7, hyphae were incubated for 1 hr with tritiated leucine. Thereafter the usual labelling in the presence of cycloheximide was performed with ^{14}C-leucine. Hence, all mitochondrial proteins are labelled with tritiated leucine, but only the products of the intrinsic protein synthesis contain ^{14}C-leucine.

The proteins containing ^{14}C-leucine leave the column with the first fractions. This suggests that the products of the intrinsic system belong to the largest of the mitochondrial membrane proteins. However, at present the possibility that these proteins are still aggregated even in the strong dissociating phenol medium cannot be excluded.

The separation of the proteins by this method is not very effective. Nevertheless, the proteins which migrate slowly during electrophoresis, emerge first

from the column and are separated from those proteins which migrate rapidly in electrophoresis. This may be useful in future experiments for the separation of protein fractions which are obtained by another method which is described below.

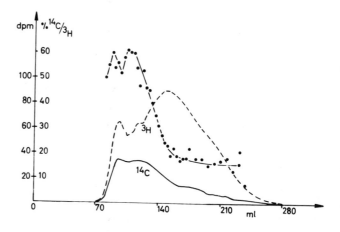

Fig. 7. Elution profile of mitochondrial membrane proteins after separation on Biogel P–200 in phenol/acetic acid/water (2/1/1, w/v/v). The hyphae were labelled for 1 hr with ^{3}H-leucine before the incorporation of ^{14}C-leucine under the influence of cycloheximide (5 min cycloheximide, 20 min ^{14}C-leucine, 30 min chase).

Solubilization of labelled proteins in Triton X–100 medium

The hyphae used in these experiments had been labelled with ^{3}H- and ^{14}C-leucine as described in the previous paragraph. Submitochondrial particles were prepared and dissolved in 1% Triton X–100 plus 100 mM phosphate buffer pH 7.2 at a protein concentration of 1 mg per ml 95% of ^{14}C-leucine incorporated in vivo under the influence of cycloheximide could be solubilized and remained in solution even after centrifugation 1 hr at 100,000 g. The mitochondrial ribosomes sedimenting under these conditions were not significantly labelled as already shown by Neupert et al. [7].

The solubilized proteins were fractionated with ammonium sulphate at 35%, 45% and 55% saturation.

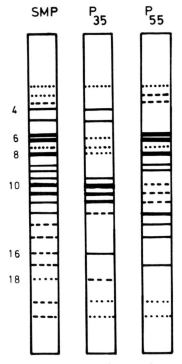

Fig. 8. Electrophoretic pattern of ammonium sulphate fractions of Triton X–100 solubilized submitochondrial particles. For abbreviations see table 2.

Table 2
Specific radioactivity of ammonium sulphate fractions of Triton X-100 solubilized submitochondrial particles.

Mitochondrial fractions [a]	Specific radioactivity (cpm per mg protein) [b]	
	^{3}H	^{14}C
M	20,000	1,350
SMP	21,500	2,100
P_0	19,500	2,100
P_{35}	20,400	4,300
P_{45}	21,200	320
P_{55}	18,600	100
S_{55}	19,600	60

[a] Abbreviations: M = mitochondria; SMP = submitochondrial particles; P_0 = Triton X-100 insoluble fraction; P_{35}, P_{45} and P_{55} = ammonium sulphate fractions precipitating at 35%, 45% and 55% saturation, respectively; S_{55} = supernatant of the P_{55} fraction.
[b] For labelling conditions see fig. 7.

Distinct protein fractions were obtained as shown by the electrophoretic patterns in fig. 8. In one fraction (P_{35}) the specific ^{14}C-activity is more than twice that of the submitochondrial particles. In all other fractions specific ^{14}C-activity is strongly reduced (table 2).

References

[1] W. Sebald, A. Schwab and Th. Bücher, in: Inhibitors, Tools in Cell Research, eds. Th. Bücher and H. Sies (Springer Verlag, Berlin, Heidelberg, New York, 1969) p. 140.
[2] H. Weiss, G. von Jagow, M. Klingenberg and Th. Bücher, Europ. J. Biochem. 14 (1970) 75.
[3] W. Sebald, Th. Bücher, B. Olbrich and F. Kaudewitz, FEBS Letters 1 (1968) 235.
[4] G. Braunitzer and G. Bauer, Naturwissenschaften 54 (1967) 70.
[5] G.D. Clark-Walker and A.W. Linnane, Biochem. Biophys. Res. Commun. 25 (1966) 8.
[6] W. Sebald, A.J. Schwab and Th. Bücher, FEBS Letters 4 (1969) 243.
[7] W. Neupert, W. Sebald, A.J. Schwab, P. Massinger and Th. Bücher, Europ. J. Biochem. 10 (1969) 589.

THE PROTEIN SYNTHETIC CAPACITY OF YEAST MITOCHONDRIA AND THE ROLE OF THE MITOCHONDRIAL GENOME IN THE ECONOMY OF THE CELL

G.M. KELLERMAN, D.E. GRIFFITHS *, J.E. HANSBY,
A.J. LAMB and A.W. LINNANE

Biochemistry Department, Monash University, Clayton, Victoria, Australia

We have previously reported that a number of antibacterial antibiotics, including chloramphenicol, tetracycline, lincomycin and a number of the macrolide series such as erythromycin reversibly and specifically inhibit mitochondrial protein synthesis both in vivo and in vitro. Growth in the presence of these antibiotics was reported to produce cells which resembled the cytoplasmic mutant petite organism in that they lacked cytochromes a, a_3, b and c_1, and the amount of inner mitochondrial membrane was greatly reduced [1−4]. At that time the results were interpreted to indicate that cytochromes a, a_3, b and c_1, together with some uncharacterized structural protein(s) of the inner mitochondrial membrane, were coded for by mitochondrial DNA and formed by the mitochondrial protein synthesizing system. It followed that the major portion of the mitochondrion was synthesized under nuclear genetic control by the cytoplasmic ribosomes. This paper is a continuation of the earlier studies and is concerned with the quantitation and the attempted identification of the proteins formed by the organelle. It is reported that the contribution of the mitochondrial protein synthesizing system to the total organelle protein is quite small, ranging from 4−13% depending upon the degree of catabolite repression of the cells. The study of chloramphenicol inhibited wild type cells and petite cells in the absence of catabolite repression reconciles the large morphological changes previously reported with the present quantitative estimate of the protein formed by the mitochondria. Sub-fractiona-

tion of mitochondria selectively radioactively labelled in vivo and in vitro yield different distributions of label and the results are not inconsistent with part of the cytochrome complexes being synthesized by the mitochondria.

Preliminary reports of the work on quantitation of the contribution of the mitochondrial protein synthesizing system [5] and on the fractionation studies [6] have been made at meetings of the Australian Biochemical Society.

Methods

A locally isolated diploid strain of *Saccharomyces cerevisiae* [7] denoted strain M, a haploid strain, denoted L410, and a cytoplasmic erythromycin resistant mutant derived from L410, denoted L411 [8] were used for the experiments. Cytoplasmic petite mutants were induced by euflavine, and some spontaneously arising petites from strain L411 were also examined. Cultures were maintained on slopes and grown in liquid 0.5% Difco yeast extract−salts−glucose media [7]. The medium for chemostat culture was supplemented with 1% Difco yeast extract and 0.1% Difco peptone. Liquid media for strains L410 or L411 were routinely supplemented with 25 mg uracil and 50 mg histidine per litre. For each experiment the appropriate initial glucose concentration in the growth medium was chosen, and this is specified in the text.

Growth conditions

Cells were grown aerobically in batch culture as

---

* Present address: Department of Molecular Sciences, University of Warwick, Coventry, U.K.

previously described [7], or in chemostat culture with glucose as the limiting substrate [9]. In the chemostat the dilution rate was in the range of 0.15–0.2 volume per hr, which results in a mean generation time 3.5–4 hr. The glucose concentration of the inflowing medium was chosen, according to the type of cell being grown and the presence or absence of antibiotic, so that the cell density was in all cases maintained at about 2–3 mg/ml. Thus respiratory competent cells capable of utilizing the products of glucose fermentation required 0.5% glucose, whereas the respiratory incompetent chloramphenicol grown or petite cells, which are completely dependent on fermentation reactions, required 2% glucose. In order to ensure maximal derepression, especially with petite cells, cultures were routinely maintained under equilibrium conditions for at least 44 hr before being employed for an experiment. The equilibrium concentration of glucose in the medium was measured by a glucose oxidase method [10] and found to be from 1.5 to 3×10^{-4} M. Isolation of mitochondria, measurement of in vitro protein synthesis and assay of adenosine triphosphatase activity of isolated mitochondria were carried out as previously described [3].

Effects of antibiotics on protein synthesis in vivo
    Cells were grown aerobically in batch culture on a

1% glucose medium to the desired cell density (0.5 mg/ml for glucose repressed cells, about 1.7 mg/ml for derepressed cells). Aliquots of 150–200 ml were then transferred to warmed flasks containing the required amount of chloramphenicol or erythromycin glucoheptonate. The flasks were then shaken for 15 min at 28° to allow solution of the antibiotic and recovery from the transfer procedure. 2 μCi of carrier-free (^{14}C) leucine was then added to each flask and incubation continued for the appropriate time. Cycloheximide in the form of a 1 mg/ml solution was added exactly 60 sec before the (^{14}C) leucine, a period found to be adequate with the strain used for maximal inhibition of protein synthesis to have occurred. In view of the much reduced rate of protein synthesis in the presence of cycloheximide, 5 μCi (^{14}C) leucine was added to each flask containing this antibiotic.

    The methods used for isolation of mitochondria, and the determination of the specific radioactivities of the mitochondrial and cytoplasmic proteins were as previously described [3]. Cytoplasmic protein was defined as that protein remaining in solution after centrifugation of the post-mitochondrial supernatant at 100,000 g for 1 hr.

Labelling and fractionation of mitochondria
    Cells were derepressed by growth to a density of

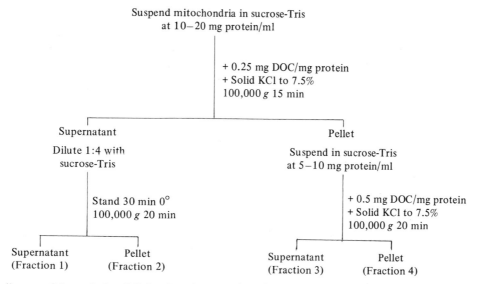

Fig. 1. Flow diagram of deoxycholate-KCl fractionation procedure. Sucrose-Tris = 0.25 M sucrose, 0.05 M Tris-HCl, pH 7.8. DOC = sodium deoxycholate, pH 7.8

1.7 mg/ml in 1% glucose medium. For in vitro label-ling studies mitochondria were isolated [3] and inc-ubated as described in table 4. For in vivo labelling the conditions were as described in table 6. Initially, the isolated labelled mitochondria were fractionated by the method developed by Richardson et al. [11] for structural protein isolation, modified by collect-ing fractions precipitating at 12%, 18%, 23% and 100% saturation with ammonium sulphate. A second fractionation was developed by modifying a method of Tsagaloff (personal communication), and the de-tails are shown in fig. 1. Cytochrome estimations were made from difference spectra using the extinc-tion coefficients for mammalian cytochromes [12].

Electron microscopy

The cells were fixed for 30 min at room tempera-ture in ten volumes of 2% potassium permanganate, water washed, suspended for 2 hr in ten volumes of 1% OsO_4, 0.06 M potassium phosphate pH 7.2, water washed then dehydrated through a graded series of aqueous acetone solutions. The dehydrated cells were embedded in araldite, polymerized by incubation at 65°C for 3–4 days, sectioned with glass knives and examined. The thin sections were stained with 1% uranyl acetate in 50% aqueous ethanol for 30 min, then with lead citrate [13] prior to examination in an Hitachi 11A electron microscope.

Results and discussion

Effects of antibiotics on protein synthesis in vivo

The contribution of the mitochondrial protein synthetic system has been quantitated by measure-ment of (^{14}C) leucine incorporation into total mito-chondrial protein in normal yeast in the presence and absence of chloramphenicol or erythromycin, both of which specifically inhibit the mitochondrial system. The effect of glucose repression on the magnitude of the mitochondrial contribution was measured by performing such experiments on both repressed and derepressed cells. These experiments were carried out using strain L410, sensitive to both chloramphenicol and erythromycin at concentrations of 0.1 mg/ml, and strain L411, which is a cytoplasmically deter-mined erythromycin-resistant mutant derived from L410 [4,8].

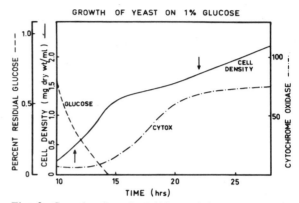

Fig. 2. Growth of strain L410 on 1% glucose medium. Inoculum 0.02 mg dry wt/ml medium of a culture grown on 5% glucose medium. Cytochrome oxidase activity expressed as first order rate constant/g protein of whole cell homo-genate.

Fig. 2 shows the course of growth and derepres-sion of this strain in batch culture on a medium initially containing 1% glucose; the inoculum was grown on 5% glucose medium. There is a relatively rapid growth phase lasting until the glucose in the medium is exhausted, during which period the cells remain subject to glucose repression; there follows an interim period of variable duration when the culture grows very slowly while mitochondrial function in-creases markedly, and then a second, slow growth phase occurs, supported by oxidative metabolism. The level of cytochrome oxidase, an index of mito-chondrial development, is low in the glucose re-pressed phase and increases about tenfold during derepression. The repressed cells at a culture density of 0.5 mg dry weight/ml were pulsed with (^{14}C) leucine for 30 min in the presence or absence of antibiotic (chloramphenicol 0.5 mg/ml or erythro-mycin 1–8 mg/ml). The derepressed cells at a culture density of 1.7 mg dry weight/ml were pulsed with (^{14}C) leucine for 100 min in the presence or absence of antibiotic. The duration of the pulse was chosen in each case to correspond to an increase in cell mass of about 15%.

Table 1 shows the mean specific activities of the isolated cytoplasmic and mitochondrial proteins from antibiotic inhibited cells expressed as a percentage of those isolated from the control cells. It is apparent that there was no significant alteration in specific activity of the cytoplasmic protein in either dere-

Table 1

Effect of inhibition of mitochondrial protein synthesis on incorporation of (^{14}C) leucine into cytoplasmic and mitochondrial protein in vivo.

Cell type	Cytoplasmic protein (C)	Mitochondrial protein (M)	C−M [a]
	(specific activity as % of control)		
Repressed	98.2 ± 1.5 (15)	93.7 ± 1.4 (15)	4.5 ± 2 (15)
Derepressed	102.2 ± 3.2 (9)	89.7 ± 2.6 (9)	12.5 ± 1.3 (9)

The figures given are mean ± standard error of mean, the number of experiments is shown in brackets.

[a] The standard error of the mean of C−M was calculated from the set of individual (C−M) values obtained, considering each experiment as giving a single such value.

pressed or repressed cells, in keeping with the lack of inhibition of cytoplasmic ribosomes in vitro [3]. In the derepressed cells the mean specific activity of mitochondrial protein was some 13% lower than that of the cytoplasmic proteins, which would indicate that in such cells the mitochondrial system synthesises 13% of the organelle protein. In repressed cells the mean decrease in labelling of the mitochondrial protein resulting from the incubation with the antibiotics was of the order of only 4%. This would indicate that in the repressed cells the mitochondrial contribution per se is of the order of only 4% of the total organelle protein, a figure significantly lower than in the derepressed cells and in keeping with the poorly developed mitochondria of repressed cells. In order to confirm that the mitochondrial protein synthetic process was fully derepressed at the cell density of 1.7 mg/ml chosen for the above experiments, cells derepressed by chemostat culture as described later were also examined, and also showed a decrease in the mitochondrial protein specific activity of 13% in the antibiotic treated cells. In the erythromycin resistant strain L411 there was no effect of erythromycin on the labelling of cytoplasmic or mitochondrial protein, even at levels as high as 8 mg/ml of the antibiotic.

The antibiotic cycloheximide has an action complementary to chloramphenicol and erythromycin, as in vitro it strongly inhibits cytoplasmic ribosomes but is completely without effect on the yeast mitochondrial protein synthetic system. Accordingly, it was anticipated that incubation of yeast cells with (^{14}C) leucine and cycloheximide should enable a direct measure to be made of the magnitude of the contribution of the mitochondrial system. The effects of incubation of yeast cells for a period of 30 min with various levels of cycloheximide are shown in table 2. There is a graded inhibition of labelling of cytoplasmic protein by cycloheximide, so that a concentration of 50 μg/ml of the compound is required for complete inhibition. The inhibition of labelling of mitochondrial protein is almost as strong as that of cytoplasmic protein. Thus with repressed cells at a cycloheximide concentration of 10 μg/ml the residual cytoplasmic labelling was 2% of the control while the residual mitochondrial labelling was only 3% of the control. However, the inhibition of labelling of mitochondrial protein by cycloheximide was proportionately less when shorter incubation times were used. Table 3 shows that with a 5 min incubation period the mitochondrial labelling was of the order of 15% of that in the control, while the cytoplasmic labelling was still almost completely inhibited; with longer incubation periods the residual mitochondrial labelling decreased. Similar results have been reported for a liver mitochondrial system [14] in experiments using lowered temperature to reduce the metabolic rate of the tissue so that short pulses can be studied. It seems therefore that there is an interlock between the two protein synthesizing systems, such that when

Table 2

Effect of cycloheximide on incorporation of (^{14}C) leucine into cytoplasmic and mitochondrial protein in vivo.

Cell type	Concentration of cycloheximide μg/ml	Cytoplasmic protein	Mitochondrial protein
		(specific activity as % of control cells)	
Repressed	2	9	14
	10	2	3
Derepressed	2	12	17
	10	2	4
	50	0.2	2

The results shown are those of a typical experiment at each growth stage.

Table 3
Effect of time of incubation on cycloheximide inhibition of incorporation of (^{14}C) leucine into cytoplasmic and mitochondrial protein in vivo.

Time of incubation (min)	Cytoplasmic protein	Mitochondrial protein
	(specific activity as % of control cells)	
5	3	15
15	1.4	6
30	1.3	5

The results shown are those for a typical experiment at a cycloheximide concentration of 20 μg/ml.

the cytoplasmic ribosomal contribution is rendered unavailable the mitochondrial protein synthesizing system becomes inhibited after a short time. The short term experiments, however, yield an estimate of the magnitude of the mitochondrial contribution similar to that obtained from the chloramphenicol and erythromycin experiments. These results are in agreement with those of Mahler and coworkers [15,16] that the major part of the yeast mitochondrial protein is made by the cytoplasmic ribosomal system, which is closely integrated with the mitochondrial system. Similar conclusions have been reached for liver mitochondria [17].

Nature of product of mitochondrial protein synthesis

Isolated mitochondria were labelled by incubation in vitro with (^{14}C) leucine in the presence of an external energy source, and then fractionated by the modified method of Richardson et al. [11] for structural protein preparations. Table 4 shows that the fraction precipitating between 0 and 12% ammonium sulphate saturation, corresponding to the structural protein fraction, contained the (^{14}C) labelled protein at twice the specific activity of the starting material, while it had no cytochrome a, little cytochrome c_1, and a moderate amount of cytochrome b. The fraction precipitating between 18 and 23% saturation of ammonium sulphate contained all of the cytochrome a and moderate amounts of cytochromes b and c_1, but the specific activity with respect to the radioactive label was only about three quarters of that of the whole mitochondria. These results suggest that the product of in vitro mitochondrial protein synthesis is not associated with the membrane bound cytochromes to any major degree, in confirmation of work of Kadenbach [18]. However, the cytochromes b and c_1 are spread throughout the fractions by this method, and a more satisfactory separation of cytochromes was achieved by the deoxycholate-KCl method shown in fig. 1. Table 5 shows that essentially all the cytochrome a and almost half of the cytochromes b and c_1 were separated into a protein

Table 4
Distribution of protein, radioactivity and cytochromes a, b and c_1 after fractionation of mitochondria by method of Richardson et al. [11].

Fraction	Percent recovered protein	Percent recovered cpm	Percent cytochromes in fractions		
			Cyt. a	Cyt. b	Cyt. c_1
Soluble protein	17	4	0	0	0
Insoluble in cholate-deoxycholate	21	8	0	19	0
Ammonium sulphate fractions					
0–12%	25	54	0	41	17
12–18%	10	18	0	19	4
18–23%	18	13	100	28	39
23–100%	18	4	0	4	21

The results shown are those for a typical fractionation of mitochondria labelled in vitro following isolation from derepressed yeast. The incubation conditions were as follows: 40 mM Tris-HCl (pH 7.4); 100 mM KCl; 13 mM MgCl$_2$; 100 mM sorbitol; 5 mM PEP; 1.5 mM ATP; pyruvate kinase, 5 enzyme units/ml; 20 μCi (^{14}C) leucine (48 μCi/μmole); in a final volume of 40 ml containing 40 mg mitochondrial protein. Incubation was for 20 min at 30°.

Table 5

Distribution of protein, radioactivity and cytochromes a, b and c_1 after deoxycholate + KCl fractionation of mitochondria labelled in vitro.

Fraction	Percent recovered protein	Percent recovered cpm	Percent cytochromes in fractions		
			Cyt. a	Cyt. b	Cyt. c_1
1	52	54	0	23	47
2	17	25	0	21	16
3	17	9	94	48	31
4	14	12	6	8	6

For details of experimental conditions see table 4. Fractionation as shown in fig. 1.

fraction whose specific radioactivity was only one half that of the starting material, and whose absolute content of label was only 9% of the total.

Wheeldon and Lehninger [19] showed with liver mitochondria that about half of the total counts incorporated were released into solution by a subsequent period of incubation with puromycin, and this finding was confirmed with the yeast system. Fractionation studies showed that the radioactivity remaining bound after puromycin treatment was distributed in essentially the same way as shown in table 5. Again, we have considered that various cofactors, especially porphyrins, may have been unavailable to isolated mitochondria in vitro, some of the enzymes of the biosynthetic pathway being located in the cytoplasm. In the absence of pophyrin only cytochrome apoprotein could be made, and would not be recognizable by our assays. Accordingly incubations of isolated mitochondria were performed with the addition of a mixture of porphyrin a, protoporphyrin, haem a and haem, together with ferrous and cupric ions and glutathione. There was however no change in either the extent or distribution of the incorporated labelled amino acid in comparison with the unsupplemented mitochondria.

The experiments in vivo with cycloheximide already described have shown that normal functioning of the mitochondrial protein sythesizing system is dependent on the simultaneous operation of the cytoplasmic ribosomal system. Accordingly the interpretation of in vitro labelling experiments with yeast mitochondria must be interpreted with great caution.

For example, even if a small number of complete polypeptide chains were formed, it is highly likely that they would not be integrated into their correct membrane loci in the absence of cytoplasmically synthesized proteins and lipids, and so may well not appear in the expected fractions after the fractionation procedures used. Accordingly, whole cells were incubated in the presence of cycloheximide (20 μg/ ml) and (^{14}C) leucine for a short period and the mitochondria isolated from these cells were fractionated by the deoxycholate-KCl method, with results as shown in table 6. It can be seen that essentially the entire cytochrome complement has now appeared in the fraction 3, containing about 30% of the incorporated radioactivity. Essentially the same pattern of distribution of label among the several fractions was found with incubation periods from as short as 5 min to as long as 2 hr, although the rate of incorporation of amino acid into mitochondrial protein became progressively slower with time, as already discussed. Presumably a small cytoplasmic protein pool rapidly became almost exhausted.

Fraction 3 as isolated in this study must contain at least the equivalent of complexes III and IV; thus it includes cytochromes a, b and c_1 and the other associated non-cytochrome proteins such as core proteins, non-haem iron proteins etc. [20]. The previous estimate that the organelle manufactures about 13%

Table 6

Distribution of protein, radioactivity and cytochromes a, b and c_1 after deoxycholate + KCl fractionation of mitochondria labelled in vivo.

Fraction	Percent recovered protein	Percent recovered cpm	Percent cytochromes in fractions		
			Cyt. a	Cyt. b	Cyt. c_1
1	37	20	0	0	0
2	16	23	2	10	13
3	25	32	96	77	78
4	22	25	2	11	9

Yeast was grown to a cell density of 1.7 mg/ml on 1% glucose medium, harvested and resuspended in 600 ml 1% ethanol medium, at a cell density of 8 mg/ml. Cycloheximide was added to a final concentration of 20 μg/ml and 2 min later 20 μCi (^{14}C) leucine added. After 30 min incubation at 30°, 1 mg/ml carrier leucine was added and the suspension cooled to 0° and mitochondria isolated and fractionated.

of its protein may be combined with the present result that about 30% of the mitochondrially synthesized protein becomes located in the fraction 3 to indicate that the mitochondrially synthesized component or components of fraction 3 make up about 4% of the total mitochondrial protein. In derepressed cells of this type the total insoluble cytochrome content is of the order of 2 nmoles/mg protein. Assuming an average molecular weight of 25,000 for the cytochrome apoproteins, this represents a total of 50 μg cytochrome/mg protein; thus the 4 insoluble cytochromes together make up of the order of 5% of the mitochondrial protein. This is of a similar order of magnitude to the estimate that the mitochondrial contribution to fraction 3 represents about 4% of the mitochondrial protein. Thus this mitochondrial contribution would represent sufficient protein to comprise the cytochrome content of the mitochondria, or some of the core proteins, but evidently not both. The remainder of the mitochondrially synthesized protein presumably becomes localized in other membrane constituents, which are found in the other fractions following our proceudre. These experiments support the conclusions of other workers that the products of mitochondrial protein synthesis become localized in the inner mitochondrial membranes [21,22]. Tuppy et al. [23,24] have suggested that the cytochrome $a\,a_3$ apoproteins are actually made by the cytoplasmic ribosomes, and that the mitochondrial system synthesizes a structural type of protein; this conclusion is based on the apparent occurrence of the apoprotein of cytochrome oxidase in the mitochondria of the cytoplasmic petite mutant. Experiments reported by Chen and Charalampous [31] also indicate that the cytochrome oxidase complex receives contributions from both cytoplasmic and mitochondrial protein synthetic systems, and that a small pool of precursor material may be present in the cell, as has also been suggested by experiments from our laboratory [32].

Effect of derepression and antibiotics on mitochondrial morphology

The estimate of the contribution of the mitochondrial protein synthetic system obtained in the present experiments appears to be small in view of the profound morphological effects resulting from growth of cells in the presence of chloramphenicol

[2]. In such cells the mitochondria show extensive loss of internal membrane. However, these cells are respiratory incompetent, due to the loss of the membrane bound cytochromes, and their growth in batch culture therefore requires the continued presence of a fermentable carbohydrate as energy source; they are thus inevitably subject to a concomitant severe catabolite repression. To avoid this complication we have grown cells in a chemostat with glucose as the limiting substrate, under which conditions its steady state concentration is of the order 2×10^{-4} M. For these experiments we have used the diploid M strain yeast, to allow comparison with the previous work from this laboratory. Derepression has been assessed by study of the cytochrome content; cells grown in batch culture under conditions of maximal derepression, using a non-fermentable substrate such as lactate or ethanol, manifest a very high total cytochrome content with a relatively high proportion of cytochromes $a + a_3$ [25]. Normal cells grown in the chemostat develop over a period of 20–30 hr a cytochrome spectrum approaching this maximal derepression picture. Cells grown in the presence of inhibitory levels of chloramphenicol, or petite cells manifest only cytochrome c in batch culture; the trace of a b type cytochrome in such cells is not demonstrable in whole cell spectra. In such cells the concentration of cytochrome c becomes much higher in the chemostat culture, attaining a plateau after about 30–40 hr at a concentration about three times that of the batch grown cells.

In table 7 are shown the cell yields obtained in the chemostat, calculated on the basis of mg dry weight of cell material per mmole of glucose utilized. As the concentration of chloramphenicol in the medium is increased, there is a progressive decrease in yield, reflecting the gradual decrease in activity of the oxidative pathway and the inability to obtain energy by combustion of the ethanol produced by fermentation. With the strain used the yield reaches a plateau at 28 mg/mmole glucose at a chloramphenicol concentration of 0.5 mg/ml; this concentration is also that at which the membrane bound cytochromes have disappeared and the cyanide sensitive respiration is no longer demonstrable. Further increase in chloramphenicol concentration up to 4 mg/ml medium is without further effect on the yield.

Fig. 3 shows a section through a normal cell

Table 7
Effects of chloramphenicol and petite mutation on cell yield
in chemostat.

Strain	Chloramphenicol concentration mg/ml	Cell yield (mg dry weight/mMole glucose utilized)
$M\rho^+$	0.0	120
	0.1	70
	0.25	38
	0.4	30
	0.5	28
	0.6	28
	0.75	28
	4.0	28
$M\rho^-$	0.0	19
	4.0	19

Cells were cultured for not less than 45 hr as described in
Methods.

derepressed in the chemostat. The mitochondria are
numerous, well differentiated from the cytoplasm,
and they contain considerable amounts of well defined cristae. Figs. 4, 5 and 6 show sections of cells
grown in the chemostat in the presence of concentrations of chloramphenicol of 0.4, 0.75 and 4.0 mg/ml.
These are respectively a concentration at which the
insoluble cytochromes occur in only trace amounts, a
concentration at which they are undetectable, and a
great excess of the antibiotic. With increasing concentration of the antibiotic there is a progressive loss of
definition of cristae without visible effect on the
outer membrane; the overall appearance of the mitochondria in fig. 4 is very similar to that of normal
mitochondria while those in fig. 6 show very marked
changes.

There are at least two specific effects of chloramphenicol distinguishable in yeast in vivo. At relatively
low concentrations, the exact level depending on the
sensitivity of the strain studied, there is progressive
loss of the insoluble cytochromes a, a_3, b and c_1 as
the culture grows, and the cyanide sensitive respiratory activity is ultimately lost [2]. At higher concentrations of the antibiotic we have found an immediate
inhibition of respiration in normal non-growing cells,
reaching 40% with the present strain at a chloramphenicol concentration of 4 mg/ml medium.
Firkin and Linnane [26] have shown a similar situation with HeLa cells growing in tissue culture, where

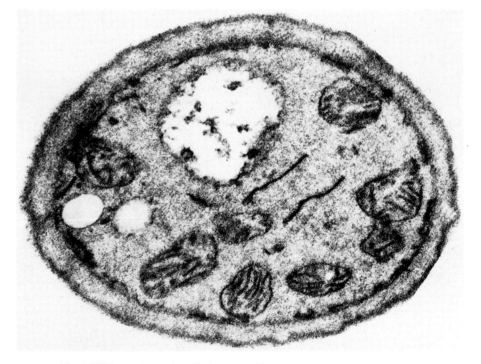

Fig. 3. Wild-type M strain cell, derepressed by growth in chemostat. 31,500 X.

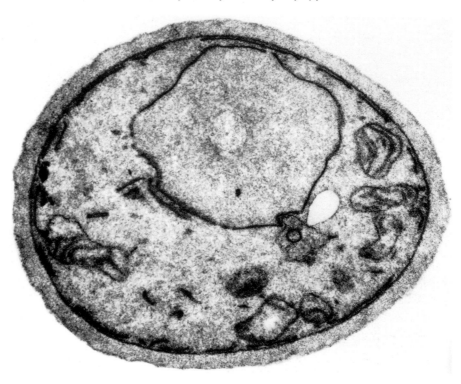

Fig. 4. Wild-type M strain cell, derepressed by growth in chemostat in presence of 0.4 mg chloramphenicol/ml medium. 31,500 ✕.

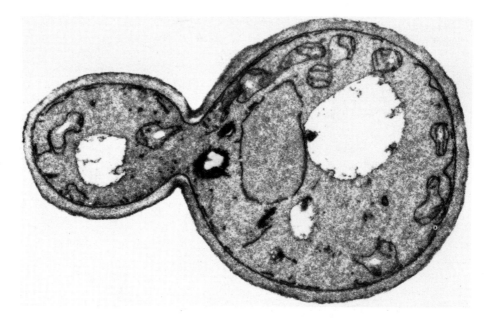

Fig. 5. Wild-type M strain cell, derepressed by growth in chemostat in presence of 0.75 mg chloramphenicol/ml medium. 17,100 ✕.

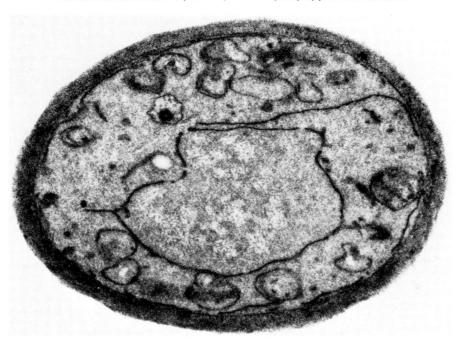

Fig. 6. Wild-type M strain cell, derepressed by growth in chemostat in presence of 4.0 mg chloramphenicol/ml medium. 24,300 X.

at low concentrations of chloramphenicol there is a progressive dilution of cytochromes a and a_3 during growth, while at high concentrations there is an immediate inhibition of growth and inhibition of respiration.

Confirmatory evidence that the additional morphological changes in yeast cells resulting from growth in the presence of very high levels of chloramphenicol are due to a secondary effect of the drug, and that the mitochondrial protein synthetic system is not required for some development of the cristae, may be obtained from study of the cytoplasmic mutant petite yeast. We have already drawn attention to the close resemblance between chloramphenicol grown wild type cells and petite cells [2]. Both lack the insoluble cytochromes a, a_3, b, and c_1, have reduced levels of several other mitochondrial enzymes and have poorly developed internal membranes when grown in batch culture. Wintersberger and Viehhauser [27] have reported that mitochondria isolated from batch grown petite cells have a total RNA content of the order of 10 μg/mg protein and lack the high molecular weight RNA which presumably belongs to the mitochondrial ribosome. Bowers et al. [28] have reported that mitochondria of petite cells grown

under conditions of derepression in a chemostat have normal cristae.

We have searched carefully for evidence of protein synthesis by mitochondria isolated from petite cells of several types grown under various conditions of glucose repression or following chemostat derepression. Table 8 shows that mitochondria isolated from petite cells of either the diploid or the haploid strain did not incorporate (^{14}C) leucine into protein in the presence of an ATP regenerating system, GTP and an amino acid mixture. In all cases, rapid exhaustion of the energy source by ATPase activity was excluded as an explanation for the inability to demonstrate protein synthesis, as PEP was still present at the end of the incubation. The soluble cytochrome c was regularly present in the mitochondria isolated from strain M petite cells, at levels varying from 0.3–1.1 nmoles/mg protein depending on the degree of glucose repression, suggesting that there had been no extreme damage to the organelles during isolation. We found that mitochondria isolated from petite cells derepressed by growth in the chemostat may contain about 3–4 times as much total RNA as those described by Wintersberger and Viehhauser [27] but such mitochondria were also quite inactive in protein synthesis.

Table 8

Protein synthesis by mitochondria from wild type and petite yeast.

Strain	Incorporation (cpm/mg protein)
Mρ^+ (batch)	2500
Mρ^- (batch)	< 10
Mρ^- (chemostat)	< 10
L411ρ^+ (batch)	3800
L411ρ^- (batch)	< 10
L411ρ^- neutral (chemostat)	< 10
L411ρ^- 30% supp. (chemostat) [a]	< 10
L411ρ^- 80% supp. (chemostat)	< 10

Each 1 ml incubation mixture contained 40 mM Tris-HCl (pH 7.4), 5 mM potassium phosphate, 100 mM KCl, 8 mM MgCl$_2$, 100 mM sorbitol, 5 mM PEP, 1.5 mM ATP, 1 mM GTP, amino acid mixture excluding leucine, 30 μg (5 E.U.) pyruvate kinase and 0.2 μCi of (^{14}C) leucine (40 μCi/μMole). Incubations were for 20 min at 30°. Under these conditions a specific activity of 2500 cpm/mg protein corresponds to the incorporation of 45 pMoles leucine/mg protein.

[a] This petite strain retained the erythromycin resistance determinant. The term supp. refers to the degree of suppressiveness of the strain.

It is not commonly appreciated that the cytoplasmic petite mutant is not a unitary mutation and that certain petite strains still retain demonstrable amounts of genetic information, as shown in our laboratory [8,29,30]. Based on these findings additional parameters become available to characterize petite cells, in particular their status with respect to retention of the erythromycin determinant and their degree of suppressiveness. Accordingly, several petite strains derived from the erythromycin resistant cytoplasmic mutant L411 were examined for the presence of mitochondrial protein synthesis; some of the strains retained the resistance determinant while others had lost it, and their degree of suppressiveness ranged from 0–80%. Table 8 shows that mitochondria isolated from chemostat cultures of all of these strains completely lacked the ability to synthesize protein. We conclude therefore that there is a general loss of the mitochondrial protein synthesizing ability associated with the cytoplasmic petite mutation.

Fig. 7 shows a section of a petite cell derepressed by growth in the chemostat; it can be seen that mitochondria are numerous with well differentiated

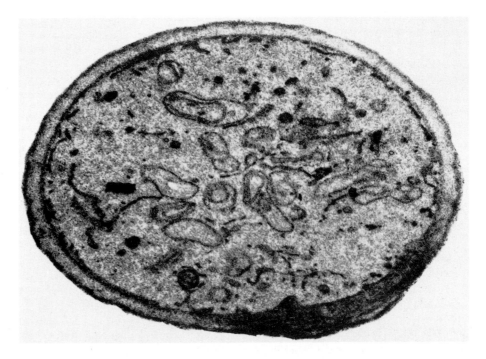

Fig. 7. Petite cell derived from strain M, derepressed by growth in chemostat. 24,300 X.

outer membranes and that many of the profiles contain internal membrane, often concentrically related to the outer membrane as described by Yotsuyanagi [33] and Federman and Avers [34]. While this appearance is certainly qualitatively different from the normal cell (fig. 3), quantitative differences are more uncertain and clearly a very significant amount of inner mitochondrial membrane may be formed under conditions of complete lack of mitochondrial protein synthesis. The results indicate that our initial equation of the petite cell with the chloramphenicol grown cell, although a convenient approximation at the time, is really an oversimplification, and that there are a number differences. Firstly, at the morphological level, the arrangement of cristae shown in fig. 7 is different from that shown in figs. 4 and 5. Secondly, at the metabolic level, the cell grown in the presence of chloramphenicol on a glucose medium has a generation time indistinguishable from the same cell in the absence of chloramphenicol, while the petite cell has a longer generation time than the corresponding wild type cell. Again there is a significant difference in the cell yields obtained in the chemostat; as illustrated in table 7, a plateau of 28 mg/mmole glucose for chloramphenicol inhibited wild type cells was reached, but the cell yield was only 19 mg/mmole glucose for petite cells; in keeping with the lack of mitochondrial protein synthesis, addition of chloramphenicol to petite cell cultures even at 4 mg/ml, was without effect on the cell yield.

It is relevant to consider the finding of Davey et al. [35] in our laboratory, that mitochondria isolated from yeast grown for many generations in the presence of high concentrations of chloramphenicol and lacking demonstrable membrane bound cytochromes, are capable of amino acid incorporation into protein. This finding has been interpreted to indicate that while the RNA of the mitochondrial ribosome is transcribed from the mitochondrial DNA [36], all the proteins of the mitochondrial protein synthesizing system including the ribosome are themselves manufactured in the cytoplasm. Kuntzel [37] has demonstrated in *Neurospora crassa* that the protein of the mitochondrial ribosome appears to be synthesized by the cytoplasmic ribosomes. Thus the chloramphenicol grown cell would lack only the products of the mitochondrial protein synthetic system, but

the system itself is intact; the petite cell lacks both the system and its products. The exact localization and nature of the mitochondrial ribosome is still uncertain; Henson et al. [15] have suggested that it is found in a relatively insoluble fraction of the mitochondrion. As the RNA content of normal yeast mitochondria is of the order of 40 μg/mg protein, this would suggest that the mitochondrial ribosome represents about 80 μg total mass/mg mitochondrial protein; if localized in the cristae this would represent a significant proportion and its absence could well explain the morphological changes seen in the petite cell in fig. 7.

The interpretation of our results is summarized in fig. 8. The outer membrane and matrix enzymes including soluble Krebs cycle enzymes, are manufactured by the cytoplasmic ribosomes. The cytochrome complexes would have contributions from both cytoplasmic and mitochondrial systems; the remaining inner membrane proteins would likewise have a dual origin, as inner membrane of some type can be formed in the petite cell yet its function is deficient as judged by the lack of oligomycin sensitivity of its ATPase [38]. The mitochondrial ribosome derives its RNA from the mitochondrial DNA and its proteins from the cytoplasmic ribosomes, although our genetic evidence suggests that the information for some of these proteins is encoded in the mitochondrial DNA. The difference between the petite and the chloramphenicol grown cell would then result from the additional lack in the petite cell of the mitochondrial ribosomes and some of the information in the mitochondrial DNA. There is as yet no unequivocal explanation of the significant differences in growth rates and total cell yields of petite and chloramphenicol-grown wild type cells. However, several suggestions can be offered. It may be that there is overactivity in vivo of the disorganized ATPase which is known to have lost its oligomycin sensitivity, so that the full yield of fermentation energy is not available to the petite. Again, a deficiency in some structural protein necessary not only for mitochondrial membranes but also for other cell membranes may occur, similar to the findings of Woodward and Munkres [39,40] in *N. crassa*. Finally, there may be lack of a mitochondrial messenger RNA which has some regulatory function in the cytoplasm, similar to the suggestions of Attardi [41] for HeLa cells.

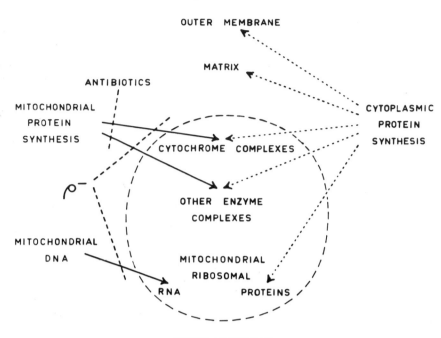

Fig. 8. Summary of data on origin of mitochondrial components. Components of the inner membrane are enclosed by the circular broken line.

Summary

The present paper is concerned with the proportion of the total mitochondrial protein synthesized by yeast mitochondria. Results will be presented which indicate that in catabolite derepressed cells the mitochondria synthesize about 15% of their protein, while in catabolite repressed cells as little as 5% of mitochondrial protein is made by the mitochondria. The products of the mitochondrial system have been examined following labelling both in vitro and in vivo; in agreement with previous work very little label is incorporated into the cytochrome rich fraction in vitro but in vivo about 30% of the label is found in this fraction.

The cytoplasmic mutant petite has lost the capacity for mitochondrial protein synthesis. Petite cells when grown in batch culture show complete loss of the cytochromes a, a_3, b and c_1, and extensive loss of internal mitochondrial membrane. Also, as we have reported earlier chloramphenicol inhibited cells grown in batch culture approach a phenocopy of the petite cell. These extensive changes in mitochondrial architecture appear paradoxical in view of the above estimates of the magnitude of the mitochondrial protein synthesis. We have therefore re-examined the morphology of cells derepressed by growth in a chemostat with glucose limitation, and now find that chloramphenicol inhibited cells, while still completely respiratory deficient, possess mitochondria of almost normal appearance. Although petite cell mitochondria under these conditions have considerable internal membrane they are never morphologically normal. It is apparent that the petite cell, which lacks the mitochondrial protein synthetic apparatus, is more severely altered than the chloramphenicol grown cell in which we have previously shown the apparatus to be still present. Further evidence for this difference is provided by the slower growth rate of petite cells than chloramphenicol inhibited cells, suggesting that the normal mitochondrial genome has some function even when mitochondrial protein synthesis is inhibited.

Acknowledgement

The authors wish to thank Dr. D.B. Morell, of Royal North Shore Hospital, Sydney, Australia, for preparing the porphyrin and haem compounds described in this paper.

References

[1] G.D. Clark-Walker and A.W. Linnane, Biochem. Biophys. Res. Commun. 25 (1966) 8.

[2] G.D. Clark-Walker and A.W. Linnane, J. Cell Biol. 34 (1967) 1.

[3] A.J. Lamb, G.D. Clark-Walker and A.W. Linnane, Biochim. Biophys. Acta 161 (1968) 415.

[4] A.W. Linnane, in: Biochemical Aspects of the Biogenesis of Mitochondria, ed. E.C. Slater, J.M. Tager, S. Papa and E. Quagliariello (Adriatica Editrice, Bari, 1968) p. 333.

[5] G.M. Kellerman and J.E. Hansby, Proc. Australian Biochem. Soc. (1968) 95.

[6] J.E. Hansby and G.M. Kellerman, Proc. Australian Biochem. Soc. (1969) 16.

[7] P.G. Wallace, M. Huang and A.W. Linnane, J. Cell. Biol. 37 (1968) 207.

[8] A.W. Linnane, G.W. Saunders, E.B. Gingold and H.B. Lukins, Proc. Natl. Acad. Sci. U.S. 59 (1968) 903.

[9] I. Malek and Z. Fencl, Theoretical and Methodological Basis of Continuous Culture of Microorganisms (Academic Press, New York, 1966).

[10] Worthington Biochemicals 'Glucostat' Reagent.

[11] S.H. Richardson, H.O. Hultin and S. Fleischer, Arch. Biochem. Biophys. 105 (1964) 254.

[12] R.W. Estabrook and A. Holowinsky, J. Biochem. Biophys. Cytol. 9 (1961) 19.

[13] E.S. Reynolds, J. Cell. Biol. 17 (1963) 209.

[14] D.S. Beattie, J. Biol. Chem. 243 (1968) 4027.

[15] C.P. Henson, C.N. Weber and H.R. Mahler, Biochemistry 7 (1968) 4431.

[16] C.P. Henson, P. Perlman, C.N. Weber and H.R. Mahler, Biochemistry 7 (1968) 4445.

[17] M.A. Ashwell and T.S. Work, Biochem. Biophys. Res. Commun. 32 (1968) 1006.

[18] B. Kadenbach, Biochim. Biophys. Acta 134 (1967) 430.

[19] L.W. Wheeldon and A.L. Lehninger, Biochemistry 5 (1966) 3533.

[20] D.E. Green, N.F. Haard, G. Lenaz and H.I. Silman, Proc. Natl. Acad. Sci. U.S. 60 (1968) 277.

[21] W. Neupert, D. Brdiczka and T. Bücher, Biochem. Biophys. Res. Commun. 27 (1967) 488.

[22] D.S. Beattie, R.E. Basford and S.B. Koritz, Biochemistry 6 (1967) 3099.

[23] H. Tuppy, P. Swetly and I. Wolff, Europ. J. Biochem. 5 (1968) 339.

[24] H. Tuppy and G.D. Birkmayer, Europ. J. Biochem. 8 (1969) 237.

[25] H.B. Lukins, D. Jollow, P.G. Wallace and A.W. Linnane, Australian J. Exptl. Biol. Med. Sci. 46 (1968) 651.

[26] F.C. Firkin and A.W. Linnane, Biochem. Biophys. Res. Commun. 32 (1968) 398.

[27] E. Wintersberger and G. Viehhauser, Nature 220 (1968) 699.

[28] W.D. Bowers, D.O. McClary and M. Ogur, J. Bacteriol. 94 (1967) 482.

[29] E.B. Gingold, G.W. Saunders, H.B. Lukins and A.W. Linnane, Genetics 62 (1969) 735.

[30] G.W. Saunders, E.B. Gingold, K. Trembath, H.B. Lukins and A.W. Linnane, this volume.

[31] W.L. Chen and F.C. Charalampous, J. Biol. Chem. 244 (1969) 2767.

[32] A.W. Linnane, in: Biochemical Aspects of the Biogenesis of Mitochondria, ed. E.C. Slater, J.M. Tager, S. Papa and E. Quagliariello (Adriatica Editrice, Bari, 1968) p. 437.

[33] Y. Yotsuyanagi, J. Ultrastruct. Res. 7 (1962) 141.

[34] M. Federman and C.J. Avers, J. Bacteriol. 94 (1967) 1236.

[35] P.J. Davey, R. Yu and A.W. Linnane, Biochem. Biophys. Res. Commun. 36 (1969) 30.

[36] H. Fukuhara, Proc. Natl. Acad. Sci. U.S. 58 (1967) 1065.

[37] H. Küntzel, Nature 222 (1969) 142.

[38] G. Schatz, J. Biol. Chem. 243 (1968) 2192.

[39] D.O. Woodward and K.D. Munkres, Proc. Natl. Acad. Sci. U.S. 55 (1966) 872.

[40] K.D. Munkres and D.O. Woodward, Biochim. Biophys. Acta 133 (1967) 143.

[41] G. Attardi and B. Attardi, Proc. Natl. Acad. Sci. U.S. 61 (1968) 261.

BIOSYNTHESIS OF MITOCHONDRIAL CYTOCHROMES

Bernhard KADENBACH

Institut für Physiologische Chemie und Physikalische Biochemie
der Universität München, Germany

The steady state of a living cell is characterized by a continuous and constant rate of synthesis and degradation of the components of the membranes. To maintain this highly organized state, a specific and directed mechanism of transport of high molecular weight compounds is required, which transfers the molecules or molecular complexes involved in membrane functions from their site of synthesis to their site of action.

Such a transfer of proteins and phospholipids from microsomes – their site of synthesis – into mitochondria – their site of action – has been shown in vitro by Kadenbach [1,2]. In further studies the synthesis and transfer of cytochrome c, an enzyme exclusively located in the mitochondria, was demonstrated [3,4]. However, in contrast to simple proteins, the biosynthesis of the cytochromes is a complex process involving various steps: (1) the synthesis of the apoprotein; (2) the synthesis of the prosthetic group; (3) the binding of both together to form the active holoenzyme. The pathway of biosynthesis is further complicated by different topological sites of synthesis of the apoprotein and the prosthetic group in the cell, by the requirement of a special mechanism for the transfer of apoprotein from the site of synthesis to the site of action, and, finally, by a combined requirement for nuclear and cytoplasmic genes, both of which seem to be involved in controlling cytochrome synthesis.

Concerning the requirement for both nuclear and cytoplasmic genes, only cytochrome c seems to be an exception. Sherman et al. [5] have shown that the biosynthesis of cytochrome c is controlled only by nuclear genes. This finding is in agreement with the observation that isolated mitochondria cannot incorporate labelled amino acids into soluble proteins or into cytochrome c [1]. Also, kinetic studies of the incorporation of ^{14}C-lysine in vivo into cytochrome c by Cadavid and Campbell have shown that the endoplasmic reticulum represents the site of cytochrome c synthesis [6]. These authors concluded furthermore, that cytochrome c is synthesized at the endoplasmic reticulum in toto, including the attachment of the heme group. The same conclusion was drawn from studies of incorporation of ^{14}C-δ-aminolevulinic acid into mitochondrial and microsomal cytochrome c in vivo by Davidian et al. [8,7] and of ^{59}Fe incorporation into cytochrome c by Kadenbach [4].

From quantitative studies of the kinetics of incorporation of ^{14}C-lysine into mitochondrial and microsomal cytochrome c of rat liver in vivo, we were able to calculate the pool size and turnover time of the microsomal cytochrome c [9] according to the method of Zilversmit [10]. Fig. 1 shows the time dependency of the labelling of cytochrome c and free lysine in the rat liver. It is evident that a precursor-product relationship between free lysine and microsomal cytochrome c can be postulated. According to the theory, at the maximum of the specific activity of microsomal cytochrome c, its specific activity must be equal to that of the precursor, e.g., free lysine. However, the measured specific activity of microsomal cytochrome c was found to be lower than the specific activity of free lysine by a factor of 11.5. This discrepancy can be easily explained by the assumption that the cytochrome c extracted from isolated microsomes consists mainly of mitochondrial cytochrome c, which has a much lower specific activity and which, during homogenization of the liver, is redistributed into all sub-cellular fractions. This redistribution of cytochrome c has been described by several authors [8,11,12].

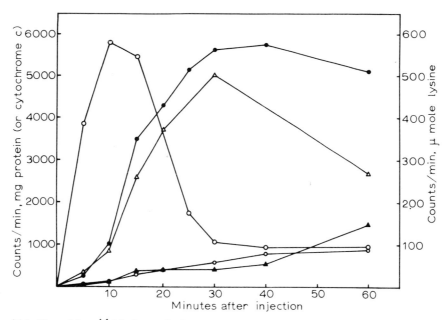

Fig. 1. Kinetics of labelling of free ^{14}C-lysine and of cytochrome *c* and protein in rat liver after injection of ^{14}C-lysine. Taken from ref. [9]. ○—○ Free lysine, ●—● microsomal protein, △—△ microsomal cytochrome *c*, ○—○ mitochondrial protein, ▲—▲ mitochondrial cytochrome *c*.

Our data are summarized in table 1. This table indicates that cytochrome *c*, after synthesis at the ribosomes, which may take less than 1 min, is attached to the endoplasmic reticulum for more than 1.5 hr. We interpret this long time of attachment to mean that the formation of a protein-phospholipid complex precedes transfer into the mitochondrion and that transfer into the organelle is effected by chance contact between the complex and individual mitochondria. The table shows, furthermore, that the

half-life of total rat liver cytochrome *c* is about 13 days. This agrees with data from the literature of about 10 days which, however, was estimated from the decay of specific radioactivity in vivo [13,14].

The third result shown in the table concerns the pool-size of microsomal cytochrome *c*. Many efforts were made to estimate the content of cytochrome *c* in rat liver microsomes in vivo [8,11,12]. In these studies the possibility that a large percentage of the cytochrome *c* extracted from the isolated microsomes

Table 1
Pool-size, turnover-time and half-life of rat liver cytochrome *c*. Taken from ref. [9].

	Pool-size mμmoles/g protein	Pool-size mμmoles/g total rat liver protein	Turnover-time	Half-life
Mitochondrial cytochrome *c*	217	63.0 *	19 days	13.2 days
Microsomal cytochrome *c* (measured)	11.7	2.7 **		
Microsomal cytochrome *c* (calculated)	1.02	0.235 **	102 min	71 min

* Assuming 29% mitochondrial protein from total protein.
** Assuming 23% microsomal protein from total protein.

actually originated from the mitochondria was never completely ruled out. We calculated that only 1/10 of the cytochrome *c* extracted from our purest microsomal fraction represented newly-synthesized cytochrome *c*. This means that the main part of the so-called 'microsomal cytochrome *c*' is not of microsomal but of mitochondrial origin. The conclusion drawn by Cadavid and Campbell [6] that the highly labelled compound extracted from microsomes represents in fact microsomal cytochrome *c* was not proven therefore. It seemed equally possible that a different compound, which during chromatography behaves similarly to cytochrome *c* and therefore occurs in the cytochrome *c* fraction, carries the label. To test this hypothesis we followed the distribution of radioactivity along the chromatogram for the separation of cytochrome *c* on CM-cellulose, as shown in fig. 2. The peak of radioactivity of the mitochondrial cytochrome *c* coincides with the peak of absorption at 280 and 450 mμ. In contrast, the peak of the microsomal cytochrome *c* does not coincide with the peak of radioactivity, which appears just before.

This result indicates that the labelled compound eluted with the isolated microsomal cytochrome *c* is not identical with cytochrome *c*. However, the close proximity to cytochrome *c* in the elution pattern indicates that it may possess a related chemical structure. The concentration of the labelled compound must have been very low, since no corresponding absorption at 280 or 405 mμ was observed. This is to be expected since the content of newly synthesized 'microsomal cytochrome *c*' was calculated to be only 8% of the total microsomal cytochrome *c* [9].

The results of fig. 2 suggest that the labelled compound extracted with the microsomal cytochrome *c* may represent apocytochrome *c*; this compound was prepared by a chromatographic modification of the method described by Sano and Tanaka [16]. In fig. 3 a chromatogram of apocytochrome *c* together with cytochrome *c* on CM-cellulose is presented. At 405 mμ two peaks were observed which correspond to reduced (appearing first) and oxidized cytochrome *c*. At 280 mμ 2 peaks were also found, the first corresponding to reduced cytochrome *c*. The second 280 mμ peak was higher than the correspond-

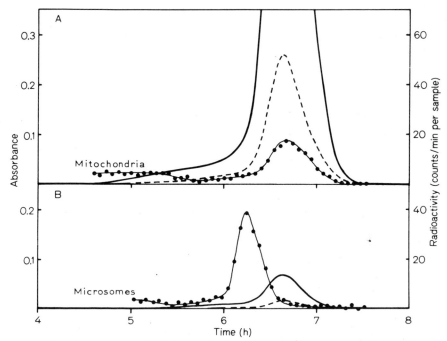

Fig. 2. Chromatography of cytochrome *c* of rat liver mitochondria and microsomes labelled in vivo with ^{14}C-L-lysine on CM-cellulose columns. Rats were killed 20 min after intraperitoneal injection. For details see ref. [15]. —— 405 mμ, – – – 280 mμ, ●——● [14]C radioactivity.

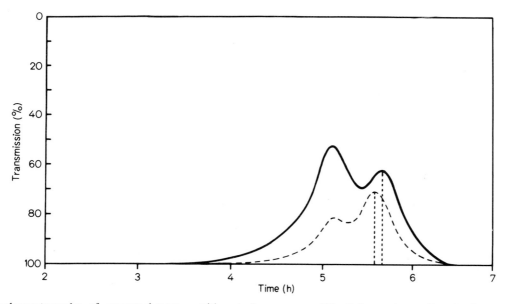

Fig. 3. Cochromatography of apocytochrome c with cytochrome c on CM-cellulose column. For details see ref. [15]. —— 405 mμ, – – – 280 mμ.

ing peak for oxidized cytochrome c, due to apocytochrome c, which only gives an absorption at 280 mμ. The second peak at 280 mμ was shifted, so that it appeared before oxidized cytochrome c. This chromatographic behaviour is similar to that observed with the labelled compound extracted from the microsomes.

If the heme group of cytochrome c is attached to the apoprotein within the mitochondria after the transfer from the endoplasmic reticulum, no labelling of the microsomal compound — which for convenience will be called 'precytochrome c' — should be observed after injection of ^{59}Fe. Therefore the incorporation of ^{59}Fe into mitochondrial and microsomal cytochrome c was studied. 40 min after injection of ^{59}Fe (100 μc) into rats, chromatograms of both mitochondrial and microsomal cytochrome c gave peaks of radioactivity which coincided with the cytochrome c peaks (fig. 4). A quantitative analysis of the specific radioactivities of mitochondrial and microsomal cytochrome c gave identical values for each (table 2). This result indicates that the microsomal cytochrome c is labelled at the same rate as the mitochondrial cytochrome c. Therefore it cannot be a precursor of mitochondrial cytochrome c. Furthermore, no labelled peak was observed at the position

of precytochrome c, which indicates that this compound contains no iron.

Sano and Tanaka have synthesized cytochrome c by the non-enzymatic reaction of apocytochrome c with protoporphyrinogen, followed by autoxidation and insertion of iron [16]. Because the reaction occurred at room temperature and in aqueous solution, the authors suggested that this pathway may represent the biosynthetic sequence of cytochrome c. Therefore the possibility was considered that protoporphyrinogen is attached to the apoprotein of cytochrome c at the microsomes and that only the

Table 2

Specific radioactivities of isolated cytochrome c labelled with ^{59}Fe in vivo (fig. 4).

	Cyto-chrome c extracted μg	Total radioactivity of cyto-chrome c cpm	Specific radioactivity of cyto-chrome c cpm/mg
Mitochondria	108.6	443	407
Microsomes	2.6	11	423

See fig. 4 and ref. [15] for details.

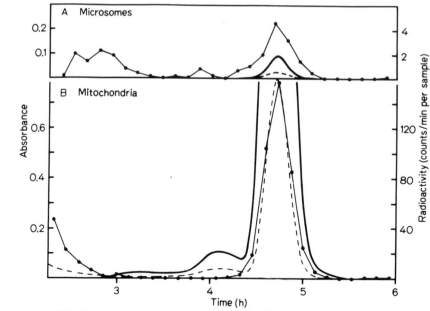

Fig. 4. Chromatography on CM-cellulose columns of cytochrome c of rat liver mitochondria and microsomes labelled in vivo with
[59]Fe. Rats were killed 40 min after injection of 100 μc ^{59}Fe into the vena femoralis. For details see ref. [15]. ——— 405 mμ,
– – – – 280 mμ, ●———● radioactivity.

insertion of iron occurs within the mitochondria after the transfer. We therefore studied the incorporation of ^{14}C-δ-aminolevulinic acid, the biosynthetic precursor of protoporphyrinogen, into cytochrome c in vivo. However, as shown in fig. 5, this possibility must be excluded. In chromatograms of microsomal and mitochondrial cytochrome c, radioactivity was only associated with the cytochrome c peaks. This behaviour is similar to the incorporation of [59]Fe (fig. 4). Also, the specific radioactivities of microsomal and mitochondrial cytochrome c were the same (table 3) and no peak of radioactivity was found at the position of 'precytochrome c'.

· Further studies were undertaken to demonstrate that 'precytochrome c' may in fact be the precursor of cytochrome c by measuring the kinetics of its labelling with ^{14}C-lysine in vitro. Rat liver slices were incubated for 60 and 120 min with purified ^{14}C-lysine. A mitochondrial fraction was isolated between 750 and 17,000 g which contained both mitochondria and some attached microsomes. The CM-cellulose chromatogram again showed 2 labelled peaks (fig. 6). The experimental data are given quantitatively in fig. 7. It is apparent that the kinetics of labelling of

both radioactive peaks resembles that of mitochondrial and 'microsomal cytochrome c' described previously [6,9]. This result is consistent with the idea that the radioactive peak appearing just before cytochrome c on CM-cellulose chromatography represents the precursor of cytochrome c.

Altogether these data seem to prove that the attachment of the heme group to the apoprotein of cytochrome c occurs within the mitochondria.

If we now compare the biosynthesis of cyto-

Table 3
Specific radioactivities of cytochrome c labelled with ^{14}C-δ-aminolevulinic acid in vivo (fig. 5).

	Cytochrome c extracted μg	Total radioactivity of cytochrome c cpm	Specific radioactivity of cytochrome c cpm/mg
Mitochondria	133	3515	264
Microsomes	4.2	107	255

See fig. 5 and ref. [15] for details.

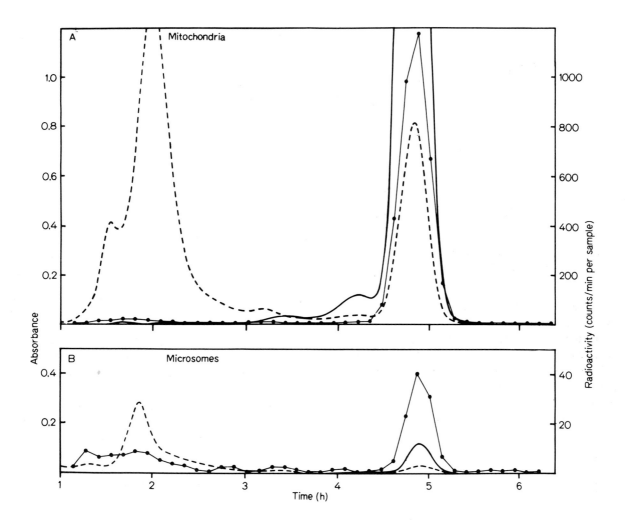

Fig. 5. Chromatography on CM-cellulose columns of cytochrome *c* of rat liver mitochondria and microsomes labelled in vivo with ^{14}C-δ-aminolevulinic acid. Rats were killed 40 min after intraperitoneal injection. For details see ref. [15]. ——— 405 mμ, − − − 280 mμ, ●———● radioactivity.

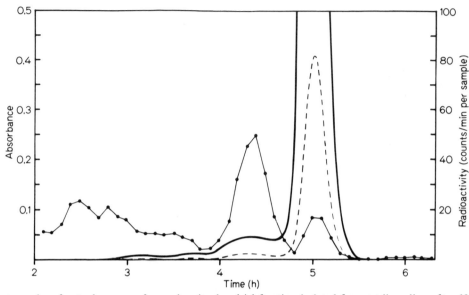

Fig. 6. Chromatography of cytochrome c of a crude mitochondrial fraction isolated from rat liver slices after 60 min incubation with ^{14}C-L-lysine. For details see ref. [15]. ——— 405 mµ, – – – 280 mµ, ●———● radioactivity.

chrome c with that of the cytochromes b, c_1 and $a + a_3$, it can be shown that cytochrome c differs from the others in several respects: (1) it can be easily extracted from mitochondria with salt solutions; (2) the heme group of cytochrome c, and also that of cytochrome c_1, is covalently linked to the apoprotein

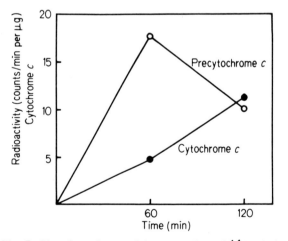

Fig. 7. Time-dependency of incorporation of ^{14}C-L-lysine into cytochrome c and precytochrome c of mitochondria from rat liver slices. The radioactivity of precytochrome c was related to the amount of cytochrome c extracted. The data at 60 min were taken from fig. 6.

by two thioether linkages; (3) its biosynthesis in yeast [17], HeLa cells [18] and rat liver [19] has been shown by Linnane and coworkers to be uninhibited by chloramphenicol, a specific inhibitor of mitochondrial protein synthesis, which inhibits the synthesis of the other cytochromes. The latter finding has been taken by Clark-Walker and Linnane to indicate that the synthesis of the structurally-bound cytochromes b, c_1 and $a + a_3$ occurs within the mitochondria [17].

However, Tuppy and Birkmayer have recently been able to demonstrate the occurrence of apocytochrome oxidase in mitochondria of the petite mutant of yeast [20]. This indicates that the cytoplasmic-inherited respiratory deficiency of the petite mutant, which in many respects resembles that of yeast cells grown in the presence of chloramphenicol, concerns only the formation of the holoenzyme and is not connected with the synthesis of the apoprotein. The question is then raised as to which factors regulate the synthesis of the structurally-bound cytochromes. In this connection it is of interest to consider the localization of the cytochromes at the inner mitochondrial membrane. The last few years have brought a detailed insight into the localization of structurally-bound enzymes at the inner mitochondrial membrane. Klingenberg and coworkers have shown that it is only the inner mitochondrial membrane which represents a barrier to the transport of substrates and ions into the mitochondria [21]. In this context it is of particular interest to know whether an enzyme is located at the outer or inner side of the inner mitochondrial membrane. It is only in the latter case that a specific translocase is required to transport the substrate to the site of its corresponding enzyme. Fig. 8 summarizes the current knowledge on the localization of some of the mitochondrial enzymes bound at the inner mitochondrial membrane [21–29]. This figure shows that both cytochromes c and c_1, which have a covalent linkage between the apoprotein and the prosthetic group, are situated at the outer side of the inner membrane. After synthesis, apocytochrome c has to be transported through only one membrane to be bound at its functional site. On the other hand, the apoproteins of the cytochromes b and $a + a_3$ have to pass through two membranes after synthesis, if we make the assumption that they are synthesized at the endoplasmic reticulum as well.

Our further studies were designed to compare the pathway of synthesis of one of the structurally-bound cytochromes with that of cytochrome c. We chose cytochrome $a + a_3$ since a simple procedure, developed by Jacobs and Andrews [30], was available for its isolation. We developed a method to isolate, in parallel, samples of pure cytochrome c and cytochrome $a + a_3$ from one rat liver. The scheme for this procedure is shown in fig. 9. The last step in the

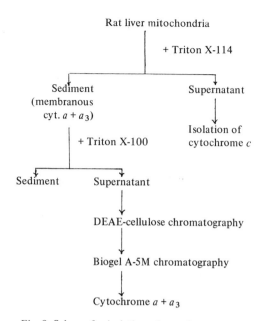

Fig. 9. Scheme for isolation of cytochrome $a + a_3$.

purification of cytochrome $a + a_3$ on Biogel A-5m in the presence of cholate gives a chromatogram shown in fig. 10. Cytochrome $a + a_3$ is separated from the detergent Triton X-100 and appears as a single peak shown by the parallel absorption at 280 nm, which

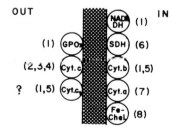

Fig. 8. Localization of structurally-bound enzymes at the inner mitochondrial membrane. 1. Klingenberg and Von Jagow [22]; 2. Lenaz and McLennan [23]; 3. Lee and Carlson [24]; 4. Muscatello and Carafoli [25]; 5. Tyler [26]; 6. Klingenberg [27]; 7. Palmieri and Klingenberg [28]; 8. Jones and Jones [29].

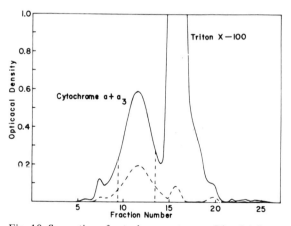

Fig. 10. Separation of cytochrome $a + a_3$ on Biogel A-5m. To the fraction of cytochrome $a + a_3$ obtained after DEAE-cellulose chromatography, a solution of saturated ammonium sulfate was added to give 30% saturation. The precipitated oxidase collected on top of the tube after centrifugation was dissolved in a buffer containing 20 mM phosphate, pH 7.7 and 20 mM Na-cholate, and chromatographed on Biogel A-5m (0.6 × 45 cm). —— 280 nm, – – – 405 nm.

records the protein, and at 405 nm, which records the heme group. Cytochrome oxidase obtained by this procedure is enzymatically active and can be activated 3 to 4 times by the addition of Tween 80. The preparation results in spectroscopically pure cytochrome $a + a_3$, as shown in fig. 11. A small shift of the alpha and gamma band is observed, which may be due to the treatment with cholate, since before chromatography on Biogel A-5m peaks are observed at 605 and 445 nm.

A further separation of the oxidase was performed by disc electrophoresis on polyacrylamine-gel in the presence of 4 M urea, according to the method of Birkmayer [31]. Fig. 12 shows a photograph of such a disc after staining with amido-black. One main band can be clearly seen, but in addition there are 6 further bands of varying intensity. A densitogram of this disc is shown in fig. 13. This figure also gives the percentages of intensity of the various bands. The main band, 3, contains about half of the total protein of the cytochrome oxidase preparation. Tuppy and Birkmayer have shown that cytochrome oxidase from yeast mitochondria contains two bands after electrophoresis on polyacrylamide-gel [20]. From the electrophoretic pattern of rat liver cytochrome oxidase it may be assumed that only one band is due to the oxidase. This would mean that cytochrome oxidase consists of one single protein.

Using the method described, we have investigated the labelling of protein fractions and of the isolated cytochromes of rat liver mitochondria after intraperitoneal injection of ^{14}C-labelled lysine and leucine together. In addition, the effect of cycloheximide and chloramphenicol on the labelling pattern was studied. Table 4 gives the specific activities of the various fractions and enzymes after injection of the label and

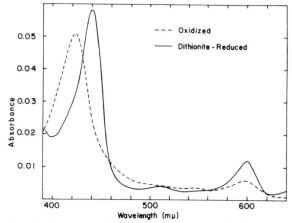

Fig. 11. Spectrum of cytochrome $a + a_3$ purified on Biogel A-5m. The spectrum of the diluted oxidase was measured with a highly sensitive recording split-beam spectrophotometer constructed by Prof. M. Klingenberg.

the inhibitors at various times. The percentage inhibition caused by cycloheximide and chloramphenicol is also shown. A marked difference in the specific radioactivities of cytochrome c and cytochrome $a + a_3$ was found. Whereas the specific activity of cytochrome c, on a mg protein basis, was higher than the specific activity of the corresponding soluble mitochondrial proteins, the specific activity of cytochrome $a + a_3$ was lower than that of the corresponding insoluble mitochondrial proteins by a factor of 10. This result indicates either a lower turnover time of cytochrome $a + a_3$ compared to the other mitochondrial proteins, or a much longer half-life of the apoprotein of cytochrome oxidase compared to the half-life of cytochrome c, which we estimated to be more than 1.5 hr.

Cycloheximide   blocked   the   incorporation   of

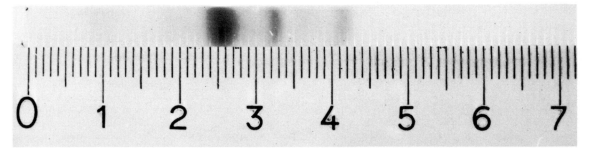

Fig. 12. Disc electrophoresis of cytochrome $a + a_3$ on polyacrylamide gel in 50% acetic acid–5 M urea. After electrophoresis the gel was stained with amido black.

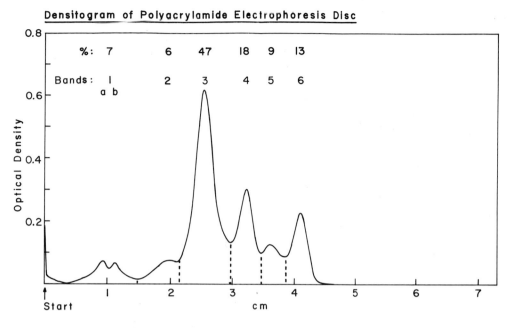

Fig. 13. Densitometer trace of the disc shown in fig. 12 at 578 nm.

Table 4

Effect of cycloheximide and chloramphenicol on the incorporation of ^{14}C-lysine and ^{14}C-leucine into protein fractions and isolated cytochromes of rat liver 60 min after injection.

		Microsomes	Soluble cytoplasmic protein	Mitochondrial soluble protein	Mitochondrial insoluble protein	Cyt c	Cyt $a + a_3$
Control: cpm/mg protein		3410	1681	781	819	1070	87
cpm/nmoles heme						13	23
	Inhibitor mg/kg body wt			%			
Control	–	100	100	100	100	100	100
Cycloheximide	50 [a]	4.5	1.8	1.7	12	2	45
Chloramphenicol	300 [a]	158	145	129	132	134	40
Chloramphenicol	200 [a] + 300 [b]	138	154	111	99	163	51
Chloramphenicol + cycloheximide	300 [a] 50 [a]	6.8	1.0	2.6	5.2	–	15

[a] Injected 5 min prior to the label.
[b] Injected 120 min prior to the label.

labelled amino acids into all protein fractions almost completely. Only the insoluble mitochondrial fraction gave a lower percentage of inhibition, and this is clearly due to mitochondrial protein synthesis which is not influenced by cycloheximide. The percentage inhibition of the labelling of cytochrome c was the same as that of the other soluble proteins, indicating that its apoprotein is synthesized at the endoplasmic reticulum. But the labelling of the cytochrome $a + a_3$ preparation was inhibited by cycloheximide by only 55%.

In contrast, chloramphenicol had the opposite effect to that of cycloheximide and caused an increased labelling of all protein fractions, with one exception. Only the labelling of the isolated cytochrome oxidase was inhibited by chloramphenicol and in this case only by 60%. The inhibition of labelling of cytochrome $a + a_3$ by both cycloheximide and chloramphenicol may be explained in one of two ways: since only 50% of the protein in this preparation is contained in band 3, which may correspond to cytochrome $a + a_3$, it may be that the labelling of this protein band is inhibited by cycloheximide, due to the synthesis of apocytochrome c at the cytoplasmic ribosomes. The other protein bands or some of them may be synthesized within the mitochondria and therefore show an inhibition of labelling by chloramphenicol. The other explanation would be that the labelling of cytochrome $a + a_3$ involves two steps: (1) the synthes of the apoprotein at the cytoplasmic ribosomes, which is inhibited by cycloheximide, and (2) transfer of the apoprotein across the inner mitochondrial membrane, accompanied by binding to a protein which is synthesized within the mitochondria and which in addition is necessary for the binding of the prosthetic group to the apoprotein. The latter step would be inhibited by chloramphenicol.

To decide between these two possibilities we estimated the specific radioactivities of the protein bands in the electrophoresis discs. The bands were cut out, ground in formic acid and counted in butyl-PBD-scintillator [32]. To our surprise, the radioactivity recorded in the single bands was extremely low. Therefore we could only accurately determine the radioactivity in the main band 3. As shown in table 5 the specific radioactivity of band 3 of the control was only 1/5 of that of the total cytochrome

Table 5
Inhibition of the incorporation of ^{14}C-leucine and ^{14}C-lysine into band 3 of cytochrome $a + a_3$ (fig. 13) in vivo by cycloheximide and chloramphenicol.

	cpm/mg protein	per cent of control
Control	16	100
Cycloheximide	5.0	30
Chloramphenicol	3.4	20
Chloramphenicol + cycloheximide	1.8	10

$a + a_3$ fraction (table 4). However, it is clear that the labelling of band 3 is inhibited by both cycloheximide and chloramphenicol. This result supports the view that the apoprotein of cytochrome $a + a_3$ is synthesized at the cytoplasmic ribosomes, while the transport and/or the attachment of the prosthetic group within the mitochondrion involves binding to a protein which is synthesized by the mitochondria. This view is different from that given by Vary et al. [33], who postulated that cytochrome $a + a_3$ consists of several protein components of which some are synthesized at the cytoplasmic and some at the mitochondrial ribosomes.

Acknowledgements

I gratefully acknowledge the technical assistance of Miss Angelika Kalthegener. My thanks are due to Dr. G. Birkmayer for performing the disc electrophoresis and to Dr. M. Weidemann for his help and discussions in preparing the manuscript. This work was supported by a grant from the Deutsche Forschungsgemeinschaft.

References

[1] B. Kadenbach, Biochim. Biophys. Acta 134 (1967) 430.
[2] B. Kadenbach, in: Mitochondria – Structure and Function, FEBS-symposium, Prague, 1968, p. 179.
[3] B. Kadenbach, Biochim. Biophys. Acta 138 (1967) 651.
[4] B. Kadenbach, in: Biochemical Aspects of the Biogenesis of Mitochondria, eds. E.C. Slater, J.M. Tager, S. Papa and E. Quagliariello (Adriatica Editrice, Bari, 1968) p. 415.

[5] F. Sherman, J.W. Stewart, E. Margoliash, J. Parker and W. Campbell, Proc. Natl. Acad. Sci. U.S. 55 (1966) 1498.

[6] W.F. Gonzalez-Cadavid and P.N. Campbell, Biochem. J. 105 (1967) 443.

[7] R. Penniall and N. Davidian, FEBS Letters 1 (1968) 38.

[8] N. Davidian, R.Penniall and W.B. Elliott, Arch. Biochem. Biophys. 133 (1969) 345.

[9] B. Kadenbach, Europ. J. Biochem. 10 (1969) 312.

[10] D.B. Zilversmit, C. Entenman and M. Fishler, J. Gen. Physiol. 26 (1943) 325.

[11] H. Beinert, J. Biol. Chem. 190 (1951) 287.

[12] N.F. Gonzalez-Cadavid, M. Bravo and P.N. Campbell, Biochem. J. 107 (1968) 523.

[13] M.J. Fletscher and D.R. Sanadi, Biochim. Biophys. Acta 51 (1961) 356.

[14] B. Kadenbach, Biochim. Biophys. Acta 186 (1969) 399.

[15] B. Kadenbach, Europ. J. Biochem. 12 (1970) 392.

[16] S. Sano and K. Tanaka, J. Biol. Chem. 239 (1964) 3109.

[17] G.D. Clark-Walker and A.W. Linnane, J. Cell Biol. 34 (1968) 1.

[18] F.C. Firkin and A.W. Linnane, Biochem. Biophys. Res. Commun. 32 (1968) 398.

[19] F.C. Firkin and A.W. Linnane, Exptl. Cell Res. 55 (1969) 68.

[20] H. Tuppy and G.D. Birkmayer, Europ. J. Biochem. 8 (1969) 237.

[21] M. Klingenberg and E. Pfaff, in: Regulation of Metabolic Processes in Mitochondria, eds. J.M. Tager, S. Papa, E. Quagliariello and E.C. Slater (BBA Library Vol. 7, Elsevier Publishing Co., Amsterdam, 1966) p. 180.

[22] M. Klingenberg and G. von Jagow, in: Electron Transport and Energy Conservation, eds. J.M. Tager, S. Papa, E. Quagliariello and E.C. Slater (Adriatica Editrice, Bari) in press.

[23] G. Lenaz and D.H. McLennan, J. Biol. Chem. 241 (1966) 5260.

[24] C.P. Lee and K. Carlson, Fed. Proc. 27 (1968) 828.

[25] U. Muscatello and E. Carafoli, J. Cell Biol. 40 (1969) 602.

[26] D.D. Tyler, in: Biochemistry of Mitochondria, Warwick Meeting 1969, in press.

[27] M. Klingenberg, in: Biochemie des Sauerstoffs, eds. B. Hess and Hj. Staudinger (Springer-Verlag, Berlin/ Heidelberg/New York, 1968) p. 131.

[28] F. Palmieri and M. Klingenberg, Europ. J. Biochem. 1 (1967) 439.

[29] M.S. Jones and O.T.G. Jones, Biochem. J. 113 (1969) 507.

[30] E.E. Jacobs and E.C. Andrews, Federation Proc. 26 (1967) 455.

[31] G.D. Birkmayer, personal communication.

[32] M.J. Weidemann and H.A. Krebs, Biochem. J. 111 (1969) 69.

[33] M.J. Vary, C.L. Edwards and P.R. Stewart, Arch. Biochem. Biophys. 130 (1969) 235.

THE NATURE OF THE PROTEINS AND NUCLEIC ACIDS
SYNTHESIZED BY ISOLATED CHLOROPLASTS

D. SPENCER, P.R. WHITFELD, W. BOTTOMLEY
and A.M. WHEELER *

Division of Plant Industry, CSIRO, Canberra, Australia

It is now well established that chloroplasts have the capacity to synthesize DNA, RNA and protein. In a qualitative way they thus seem to have the *potential* for a high degree of autonomy. The central question being asked at this time concerns the *actual* contribution of this synthetic machinery to the assembly of the component macromolecules of the chloroplast. In other words, we are now at the stage of asking about the nature of the information contained in chloroplast DNA. We would like to know the particular classes of RNA which are transcribed from this DNA, and the particular proteins that arise from the subsequent translation of some of this RNA. Bearing on this question is the relationship between the information carried in the chloroplast DNA and that in the nuclear DNA.

This paper presents a summary of the current status of the work in our laboratory which relates to this overall question of chloroplast autonomy. Aspects of DNA, RNA and protein synthesis in isolated chloroplasts from higher plants will be discussed with particular reference to the nature of the products synthesized by chloroplasts in vitro.

DNA synthesis

The DNA polymerase system of isolated chloroplasts has, in general, the same characteristics as similar systems both in procaryotic cells and in the nuclei of eucaryotes [1–3]. It shows a similar sensi-

* Commonwealth Post Graduate Scholar in the Research School of Biological Sciences, Australian National University, Canberra.

tivity to inhibitors, dependence on deoxynucleoside triphosphates and response to relatively high concentrations of inorganic salts such as KCl [3].

The product of the in vitro reaction is fairly heterogeneous in size having an average sedimentation value of 18 S. It is double-stranded, as shown by its behaviour when treated with alkali and, furthermore, the denatured product renatures rapidly when held under annealing conditions. This ability to renature is characteristic of chloroplast DNA and serves to distinguish it from nuclear DNA [1,4]. Degradation of the in vitro product by digestion with snake venom phosphodiesterase releases the incorporated ^{3}H-thymidylic acid progressively and in parallel with the release of 260 nm absorbancy showing that the ^{3}H-TTP has been built into a regular polynucleotide structure [3].

As would be expected the radioactive DNA product synthesized by isolated spinach chloroplasts hybridizes with spinach chloroplast DNA. However it is of particular interest that this product also shows a significant degree of hybridization with spinach nuclear DNA (table 1). This interaction was greater than the non-specific interaction of product with *Escherichia coli* DNA. Tobacco nuclear DNA showed some interaction with the radioactive, in vitro product synthesized by spinach chloroplasts, but much less than did spinach nuclear DNA. The extent to which spinach chloroplast product DNA hybridized to spinach nuclear DNA is considerably greater than could be attributed to the presence of DNA synthesized by nuclei contaminating the chloroplast preparation. It would thus appear from these DNA-DNA hybridizations that there is some degree of relatedness between chloroplast DNA and the homologous

Table 1

Hybridization of radioactive chloroplast product DNA to chloroplast and nuclear DNA

DNA on filter	Radioactive product DNA in solution			
	Experiment 1 [a]		Experiment 2 [a]	
	% counts bound	µg DNA bound	% counts bound	µg DNA bound
Blank	1 [b]	0.076 [b]	0	0
Spinach chloroplast (10 µg)	23	1.77	29	0.58
Spinach chloroplast (10 µg)	20	1.58	27	0.54
Spinach nuclear (10 µg)	4.5	0.35	5.7	0.11
Spinach nuclear (10 µg)	5.0	0.38	5.9	0.12
Tobacco nuclear (20 µg)	–	–	1.3	0.03
E. coli (14 µg)	0.4	0.04	0.4	0.01

[a] In expt. 1 there were 7.6 µg (2300 cpm) product DNA in solution; in expt. 2 there were 2 µg (2300 cpm) product DNA in solution.

[b] All other values in expt. 1 have been corrected for the amount of DNA bound to the blank filter. See ref. [3] for further experimental details.

nuclear DNA, suggesting the duplication of some parts of the total cell genome in both nucleus and chloroplasts.

RNA synthesis

Endogenous RNA synthesis

Isolated chloroplasts from a wide range of plants carry out a classical RNA polymerase reaction which is similar in its overall characteristics to the DNA-dependent RNA-synthesizing systems of both higher and lower organisms. In isolated chloroplasts the associated chloroplast DNA serves as the template for RNA transcription. The RNA-synthesizing system of isolated spinach chloroplasts has been characterized and the newly-synthesized RNA fractionated on sucrose gradients [5]. The RNA synthesized by isolated spinach chloroplasts was found to be very polydisperse, ranging in size from about 5 S to more than 25 S with a peak around 12 S. It was concluded that this was the equivalent of the rapidly labelled RNA fraction of chloroplasts and therefore was presumably indicative of a heterogeneous population of messenger RNA molecules of a wide range of sizes. This conclusion was based on the fact that short-term in vivo incorporation of ^{3}H-uridine into spinach

plants, followed by isolation and fractionation of the newly-formed chloroplast RNA, yielded a similar polydisperse pattern. In contrast, long term in vivo incorporation of ^{3}H-uridine resulted in radioactivity predominantly in chloroplast ribosomal and transfer RNA [5].

Recently we have re-investigated the nature of the RNA synthesized by isolated chloroplasts using polyacrylamide gel electrophoresis to fractionate the newly-formed RNA. This was done in the expectation that the superior resolving power of polyacrylamide gel electrophoresis, compared with that of linear sucrose gradients, would enable us to see whether a significant fraction of the newly-made RNA was associated with the major chloroplast RNA species viz. ribosomal and transfer RNA. The effect of a number of factors on the nature of the RNA synthesized by isolated chloroplasts has been examined. These include leaf age, plant species and state of preservation of the chloroplasts.

Chloroplasts were isolated and freed of all but a trace of contaminating nuclei as described earlier [5]. After incubation with ^{3}H-ATP and non-radioactive CTP, GTP and UTP, the newly synthesized ^{3}H-RNA was isolated by extraction with phenol-SDS in the presence of added carrier RNA which additionally served as a marker during subsequent fractionation on polyacrylamide gels. The gel system used was essentially that of Loening [6] with the addition of agarose as suggested by Peacock and Dingman [7]. This kind of experiment has now been carried out many times with chloroplasts from spinach leaves at various stages of development and from greening pea shoots. Incubation times have varied from 10 min to 1 hr. Although there were small variations from one preparation to another the RNA made by the endogenous system has shown a very consistent distribution pattern on polyacrylamide gels. Fig. 1 shows a typical fractionation of the RNA synthesized by isolated spinach chloroplasts. The chloroplasts used in this experiment were isolated from very young spinach leaves, one inch and less in length, and incorporation of ^{3}H-ATP was allowed to proceed for 20 min. Very similar distribution patterns have been obtained with different incubation periods and with chloroplasts from older leaves. The overall result has essentially confirmed the earlier work [5], namely that the product of RNA synthesis by isolated chloroplasts comprises a heterogeneous population of molecules, ranging in size up to about 2×10^6 molecular weight.

For several reasons we do not consider that the polydisperse pattern is indicative of degradation. Firstly, the pattern is remarkably reproducible from one experiment to another and with repeated fractionations of aliquots of the same product after prolonged storage. This reproducibility also holds between carefully preserved, intact chloroplasts and osmotically shocked chloroplasts, and even between chloroplasts of different species namely spinach, peas and tobacco. Secondly, long-term incubations tend to shift the size of the product slightly upwards rather than downwards [5] and thirdly, as discussed earlier, the in vitro product has a similar size distribution to the rapidly-labelled RNA fraction of chloroplasts in the intact cell.

There do appear to be discrete components within the heterogeneous population of RNA molecules which are synthesized in vitro (fig. 1), and occasionally a peak of radioactivity does coincide with the chloroplast ribosomal RNA regions at 0.55 and 1.1×10^6 molecular weight. The larger molecular weight species of chloroplast ribosomal RNA has been shown by Ingle [8] to be extremely unstable and to readily break down initially to two fragments of 0.7 and 0.4×10^6 molecular weight, and sub-

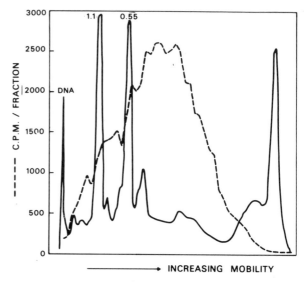

Fig. 1. Polacrylamide gel fractionation of RNA synthesized by isolated spinach chloroplasts. Solid line = densitometer tracing of gel following staining with Azure 1. *E. coli* RNA was added as carrier after incubation of chloroplasts and prior to RNA extraction. Broken line shows distribution of newly synthesized, radioactive RNA. In figs. 1–4 molecular weight ($\times 10^6$) of the two ribosomal RNA components of the carrier RNA is indicated on the densitometer tracing.

sequently to smaller fragments. However, even with these breakdown products of chloroplast ribosomal RNA in mind we have not been able to persuade ourselves that radioactive species of RNA coincident with chloroplast ribosomal RNA make up a major fraction of the RNA made by isolated chloroplasts from either spinach or greening peas. It is, of course, entirely possible that chloroplast ribosomal RNAs are among the species being transcribed by isolated chloroplasts, since they are well within the size range of the fairly continuous spectrum of species found in the in vitro product, but they do not appear to constitute a major fraction of this product. It is hoped that hybridization experiments now underway will resolve this question of the ribosomal RNA component of the in vitro product.

There is evidence that isolated chloroplasts may be capable of synthesizing transfer RNA. Fig. 2 shows the fractionation of ^{3}H-RNA synthesized by isolated chloroplasts from 9-day-old, dark-grown pea shoots which were illuminated with white light for 10 hr prior to harvest. In general the product is similar to

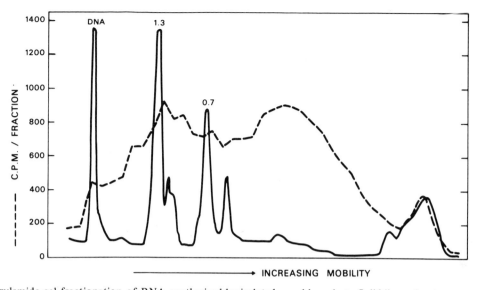

Fig. 2. Polyacrylamide gel fractionation of RNA synthesized by isolated pea chloroplasts. Solid line = densitometer tracing of gel following staining with Azure 1. Pea root RNA was added as carrier after incubation of chloroplasts and prior to RNA extraction. Broken line shows distribution of newly synthesized radioactive RNA.

that synthesized by spinach chloroplasts with sizes ranging up to 2×10^6 molecular weight. However, in addition these plastids very consistently yield a radio-active product which is coincident with transfer RNA (fig. 2). Once again no evidence was obtained for chloroplast ribosomal RNA as a major product.

The failure to detect chloroplast ribosomal RNA synthesis in vitro by simple fractionation of the newly-transcribed RNA is not surprising. Ribosomal RNA cistrons make up only 1 to 2% of the chloroplast genome [9,10]. If a significant fraction of the total chloroplast DNA is being transcribed in vitro then the proportion accounted for by ribosomal RNA will be small. Whilst ribosomal RNA does eventually become the dominant RNA species in vivo, the mechanisms whereby this is brought about, namely the conservation of ribosomal RNA in a stable form in intact ribosomes and the selective degradation of messenger RNA, almost certainly do not operate in the isolated chloroplast system.

Interaction with E. coli RNA polymerase

The aim of the experiments in the previous section was to learn something about the nature of the RNA that is currently being transcribed from chloroplast DNA by the endogenous polymerase at the time of

isolation of the chloroplasts. This endogenous system does not respond to exogenous DNA [5], suggesting that chloroplast RNA polymerase is limiting the overall reaction, and that the in vitro reaction represents largely the completion of RNA chains already initiated at the time of chloroplast isolation. The potential amount of information in a higher plant chloroplast is large (about 5×10^{-15} g DNA), and it seems reasonable to assume that this information is not all used all the time. In other words some sort of control is presumably exercised over gene expression. It is relevant to ask whether transcription of DNA is still controlled in isolated chloroplasts. To approach this question, exogenous RNA polymerase (from *E. coli*) was added to isolated chloroplasts and the interaction of exogenous polymerase with the endogenous system was studied. It was found that RNA synthesis by isolated spinach chloroplasts was enhanced up to ten-fold by the addition of *E. coli* RNA polymerase. The RNA synthesized under these conditions was then isolated and fractionated on polyacrylamide gels to determine whether the extra RNA whose synthesis was elicited by the *E. coli* polymerase resembled the product formed by the endogenous system or whether it was qualitatively different. Fig. 3 shows the result of such an experiment in

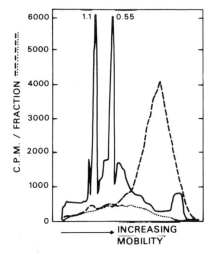

Fig. 3. Polyacrylamide gel fractionation of RNA synthesized by isolated spinach chloroplasts in the presence (dashed line) and absence (dotted line) of added *E. coli* RNA polymerase. The respective RNA fractions were electrophoresed on separate gels but have been plotted on a single graph for direct comparison. In both cases *E. coli* RNA was added as carrier. Solid line = densitometer tracing of gel following staining with Azure 1.

which separate aliquots of the same chloroplast preparation were incubated under standard RNA synthesizing conditions either with or without added *E. coli* RNA polymerase, and the newly-formed radioactive RNA isolated and fractionated separately on acrylamide gels. For conveneince the distribution of the radioactive RNA from the two reaction mixtures is plotted on a common densitometer tracing. The RNA synthesized by the interaction of the *E. coli* RNA polymerase with isolated chloroplasts is qualitatively different from that which was transcribed in the endogenous system. The additional RNA is more homogeneous and is around 6 to 7 S in size, in contrast to the heterogeneous and, in general, much larger product of the endogenous system. This suggests that either the 'controls' have been removed from the chloroplast DNA during chloroplast isolation, or that the *E. coli* enzyme does not recognize the same controls as does the chloroplast RNA polymerase.

For comparison, the nature of the product formed by the transcription of purified native spinach chloroplast DNA by *E. coli* RNA polymerase was also examined. This reaction yields a product which is

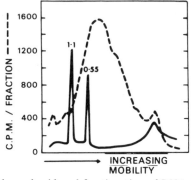

Fig. 4. Polyacrylamide gel fractionation of RNA synthesized by *E. coli* RNA polymerase with purified native spinach chloroplast DNA as the template. Time and temperature of incubation were the same as for fig. 3. *E. coli* RNA was added as carrier. Solid line shows the densitometer tracing of the gel at 254 nm. Broken line shows the distribution of the newly synthesized radioactive RNA.

remarkably similar in its size distribution to that synthesized by the endogenous chloroplast system in the absence of *E. coli* RNA polymerase (fig. 4). It ranges in size from 2×10^6 down to 25,000 with an average molecular weight around 0.4×10^6. In other words, in terms of size only, the endogenous product appears to be fairly representative of the total chloroplast genome. The additional RNA transcribed by *E. coli* RNA polymerase in the chloroplast system is clearly not representative of the total genome (fig. 3), and the intriguing question arises as to what this RNA fraction represents. We are now in a position to learn more about this extra RNA which is elicited by *E. coli* RNA polymerase in the endogenous chloroplast system. It is planned to determine its base composition and, by means of hybridization experiments, to assess the proportion of the chloroplast genome that it represents and whether it is related to any of the major chloroplast RNA species. One possible speculation is that it is a fraction of RNA with a control function.

Rifampicin

In many respects the nucleic acid and protein synthesizing apparatus of the chloroplast appears to have the characteristics of the corresponding bacterial systems rather than those of the cytoplasm of eucaryotes. It was of interest then to test the effect on RNA synthesis by isolated chloroplasts of the new

RNA polymerase inhibitor, rifampicin, which selectively inhibits bacterial (and some viral) RNA polymerases. Rifampicin is known to inhibit chain initiation in sensitive bacteria by reacting with that subunit of RNA polymerase which is involved in the reaction of the DNA enzyme complex with the first nucleotide of the new RNA chain. Once the initiation step is concluded rifampicin is without effect on subsequent chain elongation [11]. It was found that rifampicin has no effect on endogenous plastid RNA polymerase even at very high concentrations (100 μg/ml). Plastids from etiolated peas, greening peas, from corn and spinach were all equally insensitive. Control experiments with E. coli RNA polymerase showed 100% inhibition at 1μg rifampicin/ml. This was taken to indicate that either no chain initiation was occurring in vitro, or that chloroplast RNA polymerase was insensitive to this antibiotic. We are inclined to favour the latter explanation because rifampicin also had no effect on the incorporation of ^{3}H-uridine into chloroplast RNA of young corn leaves in vivo. Excised stems of 12-day-old, light-grown corn seedlings were stood in a solution of rifampicin (400 μg/ml) and when this had been taken up the shoots were supplied with ^{3}H-uridine (100 μCi) together with more rifampicin (50 μg/ml). After 5.5 hr the chloroplasts were isolated from the leaves, and the radioactivity in the cold TCA-insoluble fraction was measured. Considerable incorporation of ^{3}H-uridine was found (46,000 cpm/mg chlorophyll) but this was not affected by rifampicin. A similar lack of effect was found when the incorporation period with ^{3}H-uridine in the presence of rifampicin was extended to 20 hr. This result suggests that the RNA polymerase of higher plant chloroplasts is not sensitive to this potentially useful inhibitor. However in view of the conflict with findings presented elsewhere in this Conference (Bogorad and Woodcock, this symposium), further work is needed to examine the possibility that the mature plastids of light-grown corn seedlings, in contrast to the plastids of etiolated corn, are impermeable to rifampicin.

Protein synthesis

It is now well established that isolated chloroplasts from a range of higher plants have the capacity to incorporate amino acids into polypeptides. The size of chloroplast ribosomes and their component RNA species, their requirement for high Mg^{++} concentrations (10 mM), and the presence of N-formylmethionyl-tRNA [12] are all characteristics of a 'bacterial' type of system, and are in contrast to the conventional eucaryotic type of system found in the cytoplasm of plant cells.

One of the central questions in the field of chloroplast autonomy now concerns the specific identity of proteins synthesized by the chloroplast. Two main lines of approach to this question have been used. One has made use of the selective inhibitors, chloramphenicol and cycloheximide, on intact organisms, and the other has been to see whether particular proteins, known to be localized in the chloroplast, become labelled during amino acid incorporation by isolated chloroplasts. This paper reports our attempts to assess the usefulness of the latter approach. The same approach has been used by several workers, in particular Margulies and Parenti [13] have claimed to have found amino acid incorporation into fraction 1 protein of isolated bean chloroplasts, and Jagendorf and colleagues [14] into the photophosphorylation coupling factor in isolated wheat chloroplasts.

We have chosen to use spinach chloroplasts isolated from very young leaves (less than 1.5 inches) because they yield the highest activities of a number of chloroplast sources tested. Incorporation rates of 50–100 $\mu\mu$moles of amino acids per mg of chlorophyll per 30 min are commonly obtained. A 4 sec blending step now replaces the razor chopping method described earlier [15,16]. This procedure has the virtue of greater speed and gives a high yield of intact chloroplasts with high amino acid incorporating activity. Assay conditions were as described earlier [16] with the exception that the concentration of MgCl$_2$ in Honda medium was increased to 10 mM and a reconstituted mixture of either ^{14}C or ^{3}H-labelled amino acids was used. Under these conditions, incorporation of radioactive amino acids results in hot TCA-insoluble radioactivity in all the major fractions (table 2). A 30 min incubation was followed by a 5 min chase with puromycin after which the reaction mixture was fractionated as described in table 2. A significant proportion of incorporated amino acids was found in the soluble fraction, the ribosome

Table 2

Distribution of incorporated amino acids into chloroplast fractions.

Fraction	% of total radio-activity incorporated
Total reaction mixture (crude chloroplasts)	100
144,000 g supernatant	37
144,000 g pellet [a]	62
Nuclei [b]	9
Ribosomes	22
Lamellar fraction (Triton-soluble)	26

[a] After incubation with ^{14}C amino acids the reaction mixture was diluted with 0.01 M tris pH 7.8 and 0.01 M MgCl$_2$ and centrifuged at 144,000 g for 1 hr, to yield pellet and supernatant fractions.

[b] The 144,000 g pellet was treated with 4% Triton X-100 and centrifuged at 1,000 g for 5 min to yield a nuclear pellet. The resultant supernatant was centrifuged at 144,000 g for 1 hr to yield a pellet (ribosomes) and a supernatant (lamellar fraction).

fraction and the 'lamellar' (Triton soluble) fraction. To date, an attempt has been made to find incorporated amino acids in specific proteins of the ribosome and soluble fractions.

Ribosomes were isolated from the reaction mixture by Triton X-100 treatment followed by 2 cycles of low and high speed centrifugation. They were then dissociated with urea and LiCl after which the solution of ribosomal proteins was fractionated on polyacrylamide gel at pH 4.5 in the presence of 8 M urea by the procedure of Gesteland and Staehelin [17]. The ribosomal proteins were resolved into about 19 distinct species but on no occasion was there any radioactivity associated with any of these proteins. The radioactivity remained at the top of the gel in all cases, so clearly it was not associated with chloroplast ribosomal proteins. It probably was associated with other proteins which came through the ribosome isolation procedure and then were denatured by the low pH of the fractionation which is designed for the rather basic ribosomal proteins. These results suggest that chloroplast ribosomal protein is not synthesized by the chloroplast. This conclusion has perhaps been made more plausible by the recent finding that mitochondrial ribosomal protein is synthesized in the

cytoplasm of yeast and not within the mitochondrion [18]. It is worth noting that Lyttleton [19] has shown that in spinach the spectrum of proteins which make up the chloroplast ribosome is different to that found for cytoplasmic ribosomes. If chloroplast ribosomal proteins are made in the cytoplasm there has to be a highly specific mechanism for seeing that they become associated with the right RNA and end up in the chloroplast ribosomes, and not in the cytoplasmic or mitochondrial ribosomes.

With the aim of seeking amino acid incorporation into specific proteins, we have also attempted to fractionate the incorporated radioactivity in the 144,000 g supernatant (see table 2) derived from chloroplasts after in vitro amino acid incorporation. The proteins of the soluble fraction (144,000 g supernatant) were concentrated by 100% ammonium sulphate precipitation and the resultant pellet dissolved in a small volume of buffer, clarified by centrifugation and fractionated on polyacrylamide gel at pH 8.4. After electrophoresis the gels were stained with Coomassie Blue and the pattern recorded on a densitometer. They were then sliced and the slices were extracted 5 times with hot TCA containing non-radioactive amino acids before being counted for radioactivity. The resultant distribution of protein and radioactivity is shown in fig. 5. Incorporated radioactivity was found to be distributed more or less throughout the entire length of the gel with 2 major peaks obvious. One of these is approximately, but not precisely, coincident with the major protein-staining band, which is presumed to be fraction 1 protein. The other peak is associated precisely with a fast-moving, brown protein which is coloured brown prior to staining. This fractionation was carried out with a large pore gel, namely 3.5% acrylamide. If the 144,000 g supernatant fraction is subdivided into 20–60% and 60–100% ammonium sulphate fractions and then fractionated on gels consisting of a 3.5% acrylamide layer plus a 7.5% acrylamide layer, it is seen (fig. 6) that the two major peaks in incorporated radioactivity seen in fig. 5 are clearly separated. Fig. 6B shows the distribution of radioactivity in the 20–60% ammonium sulphate fraction. It can be seen that the components associated with the major part of the incorporated radioactivity have failed to enter the 7% gel and only a small proportion of the incorporated counts is associated with the major

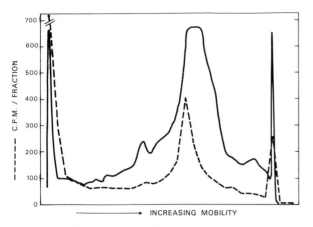

Fig. 5. Fractionation on 3.5% polyacrylamide gel of the proteins from the 144,000 g supernatant derived from spinach chloroplasts after in vitro incorporation of radioactive amino acids. The 144,000 g supernatant was concentrated by 100% ammonium sulphate precipitation prior to the electrophoresis. Solid line shows the densitometer tracing of the gel following staining with Coomassie Blue. Broken line shows distribution of TCA-insoluble radioactivity on the gel.

protein staining band which is presumed to be fraction 1. In the 60–100% ammonium sulphate fraction (fig. 6A) the radioactivity is almost entirely localized in a single fast-moving fraction. This band is a very obvious brown colour and the possibility that it may be due to ferredoxin, which is both small and acidic, immediately suggested itself. When purified ferredoxin was co-electrophoresed with the 60–100% ammonium sulphate fraction under the same conditions as shown in fig. 6A, it was found to be exactly coincident with the coloured, radioactive band.

Ferredoxin is commonly prepared by ion exchange chromatography on DEAE-cellulose [20]. Being rather acidic it binds strongly to DEAE-cellulose and is eluted only at high salt concentrations. To further test the possibility that ferredoxin becomes labelled during the incorporation of radioactive amino acids by isolated chloroplasts, a total 144,000 g supernatant from an incubated reaction mixture was mixed with crude carrier ferredoxin and passaged through

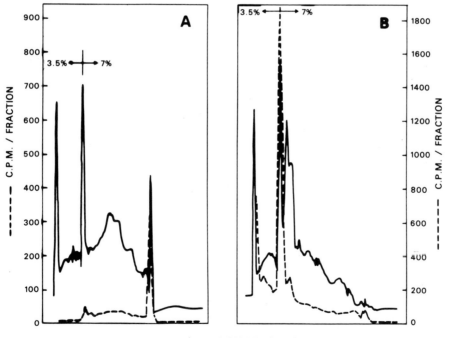

Fig. 6. Fractionation of proteins of the 144,000 g supernatant derived from spinach chloroplasts after in vitro incorporation of radioactive amino acids. Proteins were fractionated on gels consisting of 3.5% and 7% polyacrylamide. Fig. 6A and 6B respectively show the fractionation of the 60–100% and 20–60% ammonium sulphate fractions of the 144,000 g supernatant. Solid lines show the densitometer tracing after staining with Coomassie Blue. Broken lines show the distribution of TCA-insoluble radioactivity.

DEAE-cellulose equilibrated with 0.15 M chloride. A major portion of the incorporated counts passed straight through the column which was then eluted with salt solutions of increasing molarity (0.20 to 0.26 M) and finally the reddish brown ferredoxin band was eluted with 0.8 M salt. Fractions were collected and the content of ferredoxin and of incorporated radioactivity was determined (fig. 7). Clearly there were TCA-precipitable counts associated with the ferredoxin fraction. When this peak fraction was subjected to gel electrophoresis it was found (fig. 8) that a major portion of the incorporated amino acids failed to enter the gel. However virtually all the radioactive product which entered the gel was associated with the ferredoxin band.

Incorporated amino acids have thus been shown to remain associated with ferredoxin during ammonium sulphate and acrylamide gel fractionation, and during DEAE-cellulose chromatography followed by acrylamide gel fractionation. However, the evidence for the synthesis of radioactive ferredoxin by isolated spinach chloroplasts is at best suggestive. Far more rigorous proof is required, but this is difficult to provide because of the limitations of the endogenous

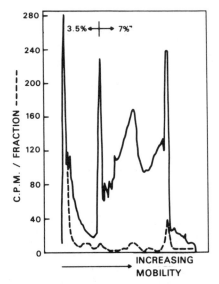

Fig. 8. Electrophoresis on polyacrylamide gel of an aliquot of the peak fraction (fraction 5) shown in fig. 7. The gel consisted of 3.5% and 7% polyacrylamide. Solid line shows the densitometer tracing of the gel after staining with Coomassie Blue. Broken line shows the distribution of TCA-insoluble radioactivity.

protein synthesizing system of the isolated chloroplasts.

There appear to be two main drawbacks inherent in the endogenous chloroplast systems which render the approach of looking for the in vitro synthesis of specific protein products likely to be an unprofitable one. Firstly, the total amount of synthesis of radioactive protein achieved by practicable amounts of chloroplasts and radioisotopes is very low, and secondly, the fact that at any one point in time the chloroplast is synthesizing a wide range of proteins. This is consistent with our observations of the extremely heterogenous nature of the RNA which is transcribed by isolated chloroplasts. If this is messenger RNA, as we suggest, then it implies that a wide spectrum of proteins is being synthesized at this time. The combination of these two factors means that it will be very difficult to obtain a purified labelled product which is sufficiently radioactive to enable tryptic digestion and identification by fingerprinting of radioactivity in the appropriate peptides. It seems that this latter would provide the most rigorous proof, and one that there is little prospect of achieving, at least with the chloroplast systems we have

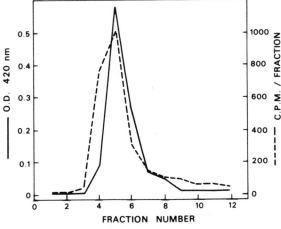

Fig. 7. Elution from DEAE-cellulose of the 144,000 *g* supernatant derived from spinach chloroplasts after in vitro incorporation of radioactive amino acids. The total 144,000 *g* supernatant together with crude carrier ferredoxin was applied to a DEAE-cellulose column and eluted successively with solutions containing 0.20 M, 0.23 M, 0.26 M and 0.8 M chloride. The fig. shows the distribution of ferredoxin (as indicated by absorbancy at 420 nm) and TCA-insoluble radioactivity in fractions collected during elution with 0.8 M chloride.

examined. We believe that we must go beyond the endogenous system and look to a reconstituted system in which some components are borrowed from more active systems such as *E. coli*. In particular we would like to see a protein synthesizing system dependent on chloroplast DNA. Since the basic question we are asking concerns the information content of the chloroplast DNA it is not too important where the other components of the system come from as long as they will translate into protein the information in the chloroplast DNA. On this point it is worth remarking that because a certain protein is made on 70 S ribosomes in the endogenous systems of isolated chloroplasts we cannot conclude that the messenger RNA which specified that protein was transcribed on chloroplast DNA. We must allow for the possibility that messenger RNA made elsewhere in the cell may be imported into the chloroplast.

Relation of macromolecule synthesis to chloroplast structure

It is of interest that all three systems responsible for the synthesis of macromolecules in isolated spinach chloroplasts are tightly attached to the chloroplast membranes, and are not free in the stroma (table 3). Neither the chloroplast DNA nor the endogenous systems which synthesize RNA and protein are removed by repeated hypotonic washes. In the case of the protein-synthesizing system it is, of course, only the ribosomes plus endogenous mes-

Table 3
Membrane-bound nature of DNA, and the RNA- and protein-synthesizing systems of chloroplasts.

Number of washes	DNA [a] (w/w of chlorophyll)	RNA [b] synthesis (cpm/mg chl)	Protein [c] synthesis (cpm/mg chl)
None	2.33×10^{-2}	47,500	2,380
1	2.38×10^{-2}	38,200	2,950
2	2.38×10^{-2}	45,150	2,810
3	–	46,600	2,890

[a] See ref. [3].
[b] See ref. [22].
[c] See ref. [16].

Table 4
Endogenous and poly U-stimulated phenylalanine-^{14}C incorporation by spinach chloroplast fractions.

Fraction	^{14}C-phenylalanine incorporation (cpm/ml) [a]	
	Endogenous	+ poly U
A. Non-washed 1000 *g* pellet [b]	1670	4600
B. Once-washed 1000 *g* pellet	1197	2875
C. Supernatant from wash of 1000 *g* pellet	168	2390

[a] Activities are expressed as cpm/ml of original extract or equivalent volume of the various fractions.
[b] A is the standard non-washed 1,000 *g* pellet; B and C are the 1,000 *g* pellet and supernatant resulting from one wash of A in Tris-MgCl$_2$-mercaptoethanol. The pellet B was resuspended in a separate aliquot of 1,000 *g* supernatant which had been centrifuged at 144,000 *g* for 2 hr.

senger RNA which remain attached to the chloroplast membranes. An aliquot of the original supernatant fraction minus liberated ribosomes was restored to the membrane fraction after each washing. This constancy of amino acid incorporating activity despite repeated washing does not imply that there are no free ribosomes — indeed electron micrographs indicate that the majority of ribosomes are not attached to membranes. However it does imply that those ribosomes which are actively translating messenger RNA are bound to the chloroplast membranes. This is borne out by the fact that the 'free' chloroplast ribosomes, i.e., those readily washed out by hypotonic treatment, have a very low endogenous activity, but are relatively responsive to polyuridylic acid, while the reverse is true for the bound ribosomes (table 4).

The fact that the DNA together with RNA polymerase and endogenous protein-synthesizing activities are all firmly attached to the chloroplast membranes suggests that these components are all physically closely linked to one another in a complex similar to that proposed for *E. coli* [21]. In other words the evidence is quite consistent with a very close physical coupling between the acts of transcription and translation.

References

[1] K.K. Tewari and S.G. Wildman, Proc. Natl. Acad. Sci. U.S. 58 (1967) 689.

[2] D. Spencer and P.R. Whitfeld, Biochem. Biophys. Res. Commun. 28 (1967) 538.

[3] D. Spencer and P.R. Whitfeld, Arch. Biochem. Biophys. 132 (1969) 477.

[4] P.R. Whitfeld and D. Spencer, Biochim. Biophys. Acta 157 (1968) 333.

[5] D. Spencer and P.R. Whitfeld, Arch. Biochem. Biophys. 121 (1967) 336.

[6] U.E. Loening, Biochem. J. 102 (1967) 251.

[7] A.C. Peacock and C.W. Dingman, Biochemistry 7 (1968) 668.

[8] J. Ingle, Plant Physiol. 43 (1968) 1850.

[9] K.K. Tewari and S.G. Wildman, Proc. Natl. Acad. Sci. U.S. 59 (1968) 569.

[10] N.S. Scott and R.M. Smillie, Biochem. Biophys. Res. Commun. 28 (1967) 598.

[11] A. Sippel and G. Hartmann, Biochim. Biophys. Acta 157 (1968) 218.

[12] G. Burkard, B. Eclancher and J.H. Weil, FEBS Letters 4 (1969) 285.

[13] M.M. Margulies and F. Parenti, Plant Physiol. 43 (1968) 504.

[14] M. Ranalletti, A. Gnanam and A.T. Jagendorf, Biochim. Biophys. Acta 186 (1969) 192.

[15] D. Spencer and S.G. Wildman, Biochemistry 3 (1964) 954.

[16] D. Spencer, Arch. Biochem. Biophys. 111 (1965) 381.

[17] R.F. Gesteland and T. Staehelin, J. Mol. Biol. 24 (1967) 149.

[18] H. Kuntzel, Nature 222 (1969) 142.

[19] J.W. Lyttleton, Biochim. Biophys. Acta 154 (1963) 145.

[20] M. Losada and D.I. Arnon, in: Modern Methods of Plant Analysis, eds. H.F. Linskens, B.D. Sanival and M.V. Tracey (Springer-Verlag, Berlin, 1964). Vol. VII, p. 595.

[21] R. Byrne, J.G. Levin, H.A. Bladen and M.W. Nirenberg, Proc. Natl. Acad. Sci. U.S. 52 (1964) 140.

[22] J. Semal, D. Spencer, Y.T. Kim and S.G. Wildman, Biochim. Biophys. Acta 91 (1964) 205.

ORIGIN AND SYNTHESIS OF CHLOROPLAST RIBOSOMAL RNA AND PHOTOREGULATION DURING CHLOROPLAST BIOGENESIS

N. STEELE SCOTT, Rana MUNNS, D. GRAHAM and Robert M. SMILLIE

*Plant Physiology Unit, CSIRO Division of Food Preservation, Ryde
and School of Biological Sciences, University of Sydney, Australia.*

The discoveries that chloroplasts contain their own specific DNA, RNA, ribosomes and protein synthesizing apparatus (for recent reviews see [1,2]) have led the concept of the chloroplast having some degree of autonomy within the photosynthetic cell. Isolated chloroplasts have been shown to synthesize DNA [3–5] and RNA [2,6] and hybridization experiments suggest that chloroplast DNA can code for chloroplast ribosomal RNA [7,8]. In this paper we present further evidence to show the existence of cistrons for chloroplast ribosomal RNA in the chloroplast DNA of *Euglena gracilis* strain Z.

Up to 70% of the total leaf nitrogen may be found in the chloroplasts [6] and studies which use chloramphenicol and cycloheximide as inhibitors of protein synthesis on chloroplast and cytoplasmic ribosomes, respectively [9–12], suggest that many of the proteins of the chloroplasts are synthesized on the chloroplast ribosomes. We have investigated the synthesis of chloroplast ribosomes in *E. gracilis* and the leaves of the pea (*Pisum sativum* var. Green Feast) and wheat (*Triticum aestivum* var. gamut) in an attempt to further define the role of chloroplast ribosomes in chloroplast biogenesis.

Ribosomes in Euglena gracilis

There are several conflicting reports of the size and composition of ribosomes in *E. gracilis*. We have found that the chloroplast ribosomes of this organism are very sensitive to changes in temperature and Mg^{++} concentration. The sedimentation values shown in

Table 1
Sedimentation coefficients of ribosomes and ribosomal RNA in *Euglena gracilis*.

	Cytoplasm		Chloroplast	
Ribosome	88		70	
Sub-units	67	46	50	30
RNA	26,24,22	22	23.5	16.5

The sedimentation coefficients of ribosomes from *E. gracilis* were obtained by analytical centrifugation and centrifugation on sucrose gradients [13], and those of the ribosomal RNA by polyacrylamide gel electrophoresis [14,15].

table 1 were obtained by isolating both the chloroplast and chloroplast ribosomes at 0–2° in 10 mM $MgCl_2$ and 40 mM KCl. The large species of ribosomal RNA were characterized by polyacrylamide gel electrophoresis and fig. 1 shows that chloroplast ribosomal RNA has 23 S and 16 S species very similar to the procaryotic bacterial cell [15]. The cytoplasmic ribosomal RNA shows several species of RNA and, as isolated, presumably contains degraded forms. We have not investigated the 5 S ribosomal RNA species of the chloroplast or cytoplasm.

Although electron micrographs of *E. gracilis* proplastids show some ribosomes [16], we have been unable to detect chloroplast-type ribosomes or ribosomal RNA in dark-grown cells of *E. gracilis* either directly or by hybridization experiments. Thus, when dark-grown cells are exposed to light and develop chloroplasts, one would expect them to make chloroplast ribosomal RNA.

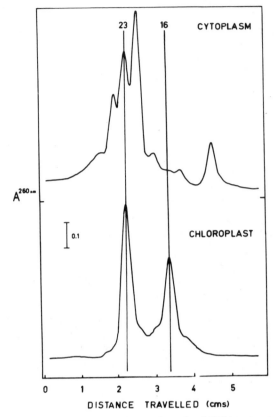

Fig. 1. Polyacrylamide gel electrophoresis of RNA extracted from the ribosomes of *E. gracilis*. Top, cytoplasm; bottom, chloroplasts. When RNA from *E. coli* and the cytoplasm of *E. gracilis* are mixed and run together, the middle peak of the latter RNA runs slightly behind the 23 S peak of *E. coli* RNA and a value of 24 S has been assigned to it (see Table 1).

Synthesis of ribosomal RNA in Euglena gracilis

The standard conditions used for following the formation of chloroplasts in *E. gracilis* (greening) are shown in fig. 2. Dark-grown cells were starved in a carbon-free medium for three days and during this time the number of cells approximately doubled while the DNA content per cell remained constant. The cells were exposed to light and the formation of chloroplasts was followed by the appearance of chlorophyll over the next four days. Under these conditions of low light intensity and high cell density there was no increase in cell number during the light period. During the dark period when the cells divided there was a drop in the RNA per cell. Over the whole

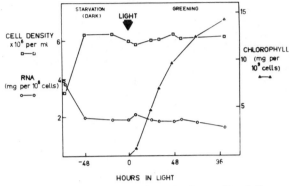

Fig. 2. Chloroplast development in *E. gracilis*. Cells were grown heterotrophically in the dark, transferred to a carbon-free medium [17] for 3 days and then illuminated (white fluorescent light, 120–150 ft-candles) for 4 days. □, cell number; ○, RNA per cell; △, chlorophyll per cell.

period of the experiment there was no net synthesis of RNA in the culture.

Total chloroplast RNA

Chloroplast fractions were isolated from greening cells and the total amount of RNA in them measured chemically [18]. The RNA was characterized by gel electrophoresis and contamination with cytoplasmic RNA was found only in the 12 hr samples. Table 2 shows that the chloroplast RNA in the cells reaches a maximum of 7% of the total after 24 hr of greening, rather less than the 20 to 25% content of chloroplast RNA found in autotrophically growing cells.

Table 2
Production of chloroplast RNA during chloroplast development.

Length of illumination (hours)	Chloroplast RNA (% of total cellular RNA)
12	2 [a]
16	6
24	7
48	6
72	8
Autotrophic cells	20–25

[a] Corrected for cytoplasmic RNA contamination (see text).

The appearance of chloroplasts in *E. gracilis* was followed as described in fig. 2 and the amount of chloroplast RNA determined as described in the text.

Incorporation of ^{14}C-orotic acid into RNA

Fig. 3 shows the distribution of ^{14}C-RNA centrifuged on a sucrose gradient after the RNA was isolated from illuminated cells which had been exposed to ^{14}C-orotic acid for 16 hr. The orotic acid was taken up into both the cytoplasmic and the chloroplast RNA but the specific activity of the chloroplast ribosomal RNA was five times that of the cytoplasmic ribosomal RNA. This experiment was repeated at different times during the greening period and fig. 4 shows the results expressed as the ratio of the specific activity of the chloroplast to the cytoplasmic RNA. At all times the specific activity of the chloroplast RNA was higher than that of the cytoplasmic RNA whether the comparison was between the total or the ribosomal RNA.

Further studies of the rates of uptake of precursor into RNA are needed to define more closely the time of maximum synthesis of chloroplast ribosomal RNA. However, these experiments show that an increase in the amount of chloroplast RNA occurs during the first 24 hr of the greening period and that the rate of chloroplast RNA synthesis relative to cytoplasmic RNA synthesis is at its highest during this same period. Since there was no net synthesis of RNA per culture in these experiments, the synthesis of chloroplast RNA may be at the expense of cytoplasmic RNA. Studies with an inhibitor of ribosomal RNA

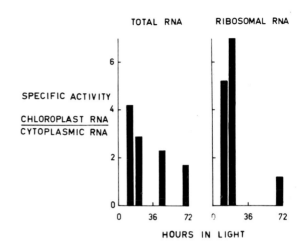

Fig. 4. The ratio of the specific activity of chloroplast RNA to cytoplasmic RNA isolated from E. gracilis incubated with ^{14}C-orotic acid. The experimental conditions were as described for fig. 2. ^{14}C-orotic acid was present for the following periods of illumination and the cells were harvested immediately following the period of incubation with ^{14}C-orotic acid: 0–16 hr, 4–24 hr, 32–48 hr, 56–72 hr. The specific activity of the total RNA was determined after KOH extraction and the ribosomal RNA after phenol extraction and sucrose gradient centrifugation as described in fig. 3.

synthesis, 5-fluorouracil, showed that it only inhibited chloroplast development in E. gracilis if added in the first 14–15 hr of the greening period [19].

The origin of chloroplast ribosomal RNA

DNA prepared from the chloroplasts of E. gracilis contains typically two species of DNA when centrifuged on a CsCl gradient as shown in fig. 5. When the DNA is recycled through CsCl, pure chloroplast DNA ($\rho = 1.686$) can be obtained. Nuclear DNA ($\rho = 1.707$) can similarly be purified from whole cell DNA by CsCl gradient centrifugation. These species of DNA, as well as nuclear DNA from bleached mutants of E. gracilis have been used in hybridization experiments with chloroplast ^{32}P-ribosomal RNA. Fig. 6 shows the distribution of chloroplast ^{32}P-ribosomal RNA centrifuged on a sucrose gradient. The hybridization procedure used was that of Gillespie and Spiegelman [20].

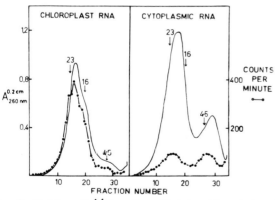

Fig. 3. Uptake of ^{14}C-orotic acid into chloroplast and cytoplasmic RNA in E. gracilis. The experimental conditions were the same as in fig. 2. Cells were incubated with orotic acid (20 μCi, 0.8 mCi/mmole) as described in the text, the RNA extracted with phenol, precipitated with cetyltrimethylammonium bromide and centrifuged on a sucrose density gradient [13].

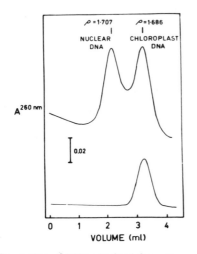

Fig. 5. Separation of DNA obtained from *E. gracilis* chloroplasts on a CsCl gradient. Top, DNA isolated from chloroplasts. Bottom, chloroplast DNA after two passages through a CsCl gradient.

Hybridization of chloroplast DNA with chloroplast ribosomal RNA

Fig. 7 shows that chloroplast DNA does hybridize with chloroplast ribosomal RNA. By progressively increasing the amount of ^{32}P-ribosomal RNA added to hybridization mixtures containing a constant amount of DNA (2.88 μg), it is seen that at saturation 1.6% of the DNA has hybridized with the RNA. About 20% of the DNA bound to the filter was lost during the incubation procedure and so the true saturation figure is close to 2.0%. This value is higher than the 1% figure reported previously [7], when longer incubation times resulted in more DNA being lost from the filter.

To test the specificity of the hybridization between chloroplast DNA and chloroplast ribosomal RNA, various amounts of unlabelled chloroplast ribosomal RNA were included in hybridization mixtures

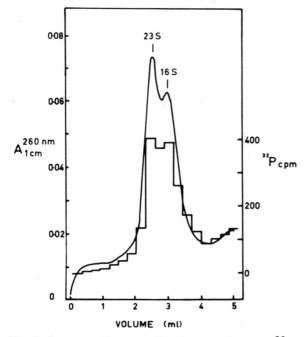

Fig. 6. Sucrose gradient centrifugation of chloroplast ^{32}P-ribosomal RNA. *E. gracilis* ^{32}P-ribosomal RNA was prepared [15] and after centrifugation on a sucrose gradient [13] pumped through a flow cell, the absorbance recorded (continuous line), fractions collected and the radioactivity of the fractions measured (histogram).

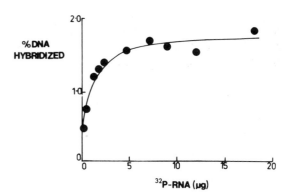

Fig. 7. Hybridization of chloroplast DNA and chloroplast ribosomal RNA (for description see text).

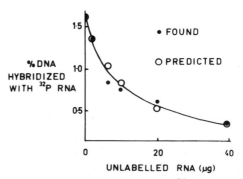

Fig. 8. Competition between chloroplast ^{32}P- and unlabelled-ribosomal RNA (for description see text).

Table 3
Specificity of binding of chloroplast DNA to chloroplast ribosomal RNA.

Source of unlabelled RNA	Counts/min in hybrid	% inhibition
None	82	–
Euglena gracilis		
Chloroplasts	42	49
Dark-adapted heterotrophs	91	0
Bleached mutant	90	0
Escherichia coli	85	0

3.0 μg of chloroplast DNA was incubated with 5.0 μg of chloroplast ^{32}P-ribosomal RNA and 5.0 μg of unlabelled ribosomal RNA from various sources.

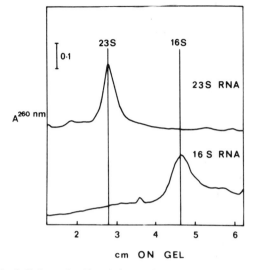

Fig. 9. Polyacrylamide gel electrophoresis of 23 S (top graph) and 16 S (bottom graph) chloroplast ribosomal RNA purified as described in the text.

containing constant amounts of chloroplast DNA (2.78 μg) and chloroplast ^{32}P-ribosomal RNA (10.6 μg). Increasing the amount of unlabelled ribosomal RNA lowered the amount of ^{32}P-ribosomal RNA bound to the DNA and good agreement was obtained between the observed and predicted values (fig. 8). However, the addition of unlabelled ribosomal RNA from sources other than chloroplasts had no effect (table 3). Ribosomal RNA from *Escherichia coli* did not compete with chloroplast ribosomal RNA, neither did cytoplasmic ribosomal RNA from *E. gracilis.* Cytoplasmic ribosomal RNA did not hybridize with chloroplast DNA.

Hybridization of 23 S and 16 S chloroplast RNA with chloroplast DNA

When chloroplast ribosomal RNA is thoroughly extracted from chloroplasts with phenol and sodium dodecyl sulphate, the 23 S and 16 S ribosomal RNA species are sufficiently stable to enable separation and purification on sucrose gradients, although some breakdown may occur upon storage [15]. Fig. 9 shows the polyacrylamide gel electrophoresis of 23 S and 16 S chloroplast ribosomal RNA purified by recycling through sucrose gradients. In an experiment similar to that shown in fig. 8 the addition of increasing amounts of unlabelled 23 S ribosomal RNA from chloroplasts lowered the amount of chloroplast ^{32}P-ribosomal RNA bound to the chloroplast DNA (fig. 10). The amount was lowered still further when chloroplast 16 S ribosomal RNA was added. Again there is good agreement between the experimental

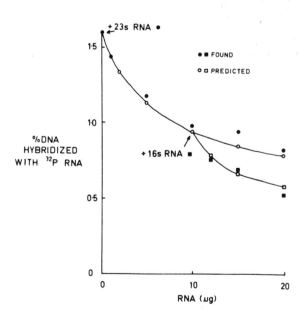

Fig. 10. Competition between chloroplast ^{32}P-ribosomal RNA and unlabelled 23 S and 16 S chloroplast ribosomal RNA. Chloroplast DNA (2.88 μg) was incubated with chloroplast ^{32}P-ribosomal RNA (10.2 μg) and increasing amounts of chloroplast 23 S ribosomal RNA (circles) added. In some incubations after 10 μg of 23 S ribosomal RNA had been added, increasing amounts of 16 S ribosomal RNA (squares) were added.

and the predicted results. This experiment shows that the 23 S and 16 S ribosomal RNA from chloroplasts have distinct cistrons on the chloroplast DNA.

The amount of chloroplast DNA in *E. gracilis* has been reported to be about 2–5% of the total DNA [21]. We have measured the amount several times and have obtained values of from 4–9%. From this figure and from the saturation curve obtained in fig. 7 we can calculate that there are about 230–500 cistrons for chloroplast ribosomal RNA in the chloroplasts of *E. gracilis*. If there are about ten chloroplasts in a mature cell this means about 23–50 cistrons per chloroplast.

Hybridization of chloroplast ribosomal RNA with nuclear DNA

We have also found that chloroplast ^{32}P-ribosomal RNA hybridizes with nuclear DNA (fig. 11) and in this case saturation occurs when 0.2% of the DNA is annealed. Similar results were obtained with DNA isolated from two bleached mutants of *E. gracilis* (ZUV-1 and BEN-A28, obtained by treating normal cells with UV light and benadryl, respectively) which do not form chloroplasts and in which we cannot detect chloroplast DNA. The mutants were obtained from Dr. H. Lyman, State University of New York, Stony Brook, N.Y. Competition experiments with cytoplasmic ribosomal RNA and ribosomal RNA from *E. coli* (table 4) showed that the hybridization was specific for chloroplast ribosomal RNA. The saturation curve shown in fig. 11 predicts 300 cistrons for chloroplast ribosomal RNA in the nucleus of *E. gracilis*.

Table 4
Specificity of binding of nuclear DNA to chloroplast ribosomal RNA.

Source of unlabelled RNA	Counts/min in hybrid	% inhibition
None	21	0
Euglena gracilis		
Chloroplasts	8	60
Dark adapted heterotrophs	22	0
Bleached mutant	19	0
Escherichia coli	19	0

2.72 μg of nuclear DNA was incubated with 5.0 μg of chloroplast ^{32}P-ribosomal RNA and 5.0 μg of unlabelled RNA from various sources.

It should be emphasized there are several problems in interpreting the results of hybridization with nuclear (or eukaryotic) DNA [22–24] and further experiments are required before it can be safely concluded that cistrons for chloroplast ribosomal RNA are present in the nucleus of *E. gracilis*.

The occurrence of cistrons for chloroplast ribosomal RNA in the chloroplast DNA suggests a definite function for the chloroplast DNA in the synthesis of chloroplast ribosomes. If cistrons for the chloroplast ribosomal RNA do in fact exist in both the nucleus and the chloroplast it is not clear why this should be so. Possibly both sets of cistrons are needed for use on separate occasions during the development or senescence of chloroplasts. Alternatively the chloroplasts may contain more than one type of ribosome and the chloroplast and nuclear cistrons may be different. This possibility has been proposed to explain other results obtained in our laboratory [10].

Photoregulation of chloroplast ribosomal RNA synthesis in pea stem apices and wheat leaves

In *E. gracilis* the synthesis of most of the chloroplast ribosomal RNA occurs after exposure of dark-grown cells to the light. In most higher plants, as in *E. gracilis*, light is required for chlorophyll synthesis, but less is known about the light requirement in higher plants for the formation of chloroplast proteins and nucleic acids. To what extent are these compounds

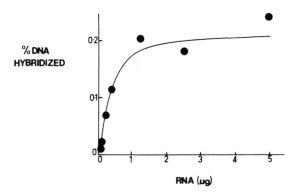

Fig. 11. Hybridization of chloroplast ^{32}P-ribosomal RNA with nuclear DNA. Nuclear DNA (2.72 μg) was incubated with increasing amounts of chloroplast ^{32}P-ribosomal RNA.

synthesized in etiolated plants? Does light stimulate their synthesis and if so, what is the nature of the photoacceptor? Does chloroplast ribosome synthesis precede chloroplast protein synthesis? These questions have been examined by studying the effects of various light treatments on the synthesis of chloroplast ribosomal RNA in dark-grown pea stem apices and wheat leaves.

Ribosomal RNA in pea stem apices

The sedimentation values of ribosomal RNA from chloroplast ribosomes (approximately 23 S and 16 S) are different from those of plant cytoplasmic ribosomes (approximately 25 S and 18 S) [2,14] and the RNA species can be separated by polyacrylamide gel electrophoresis.

In the experiments reported here, RNA was extracted from dark- or light-grown pea stem apices and four species of RNA were found (fig. 12). By analogy, the 23.4 S and 17.3 S RNA are chloroplast or etioplast [6] ribosomal RNA and the 26.1 S and 19.0 S RNA are cytoplasmic ribosomal RNA. Fig. 12 shows that in the dark-grown pea stem apex the etioplast ribosomal RNA is less than 10% of the total

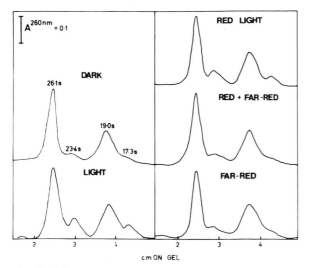

Fig. 12. Ribosomal RNA in pea stem apices. Pea seedlings were germinated and grown in the dark and subsequently exposed to brief illumination with red light, far-red light or red light immediately followed by far-red light as described in the text [25]. The RNA was extracted from the stem apices, fractionated by polyacrylamide gel electrophoresis [14] and the gels scanned in a spectrophotometer [15].

while chloroplast ribosomal RNA is as much as 27% of the total ribosomal RNA in the light-grown peas.

Fig. 12 also shows the effect of various treatments with red light on the ribosomal RNA content of the pea stem apices. Four-day-old seedlings were exposed to red light (661 nm) for 3 min daily on each of 3 consecutive days and then returned to the dark. On the fourth day they were harvested and the ribosomal RNA extracted. The amount of etioplast ribosomal RNA in these non-green seedlings was about the same as the amount of chloroplast ribosomal RNA found in the green plant. If the red light treatment was immediately followed by treatment with far-red light (733 nm) for 10 min, the increase in etioplast ribosomal RNA was less, as was the case if the plants were treated with far-red light alone.

Chloroplast proteins

Table 5 shows the effects of red light on various parameters of growth of pea stem apices. There was a 5- to 7-fold increase in fresh weight and protein, RNA and DNA contents after red light treatment and about a 3-fold increase in the activities of cytoplasmic enzymes. These effects were all partly reversed by far-red light and were not nearly so marked if the plants were irradiated with far-red light only. Increases in photosynthetic enzymes and Fraction I protein in response to red light were much greater, ranging from 11- to 91-fold. The activities of two plastid enzymes, ribulose-1,5-diphosphate carboxylase and alkaline fructose-1,6-diphosphatase in plants treated with red light were 40% and 20% respectively of those found in plants grown in white light [25,34].

These changes in red-irradiated plants took place over a much longer period [25,34] than the changes in etioplast ribosomal RNA which were essentially completed after 3 days of irradiation.

There is a discrepancy between the increase in the amount of Fraction I protein (11-fold) and its associated enzymic activity ribulose-1,5-diphosphate carboxylase (91-fold) after red light treatment (table 5). This may be due to several contributing factors affecting the measurements. (1) Irradiation with red or white light can cause a 2-fold activation of the carboxylase enzyme in dark-grown pea seedlings (D. Graham, A.M. Grieve and R.M. Smillie, unpublished results) and white light is known to activate the

Table 5
Effect of light treatments on various parameters in pea stem apices.

	(1) Red	(2) Red/ far red	(3) Far red
Growth			
Fresh weight	5.1	2.5	1.6
Protein	5.4	2.7	1.7
DNA	7.6	3.3	2.3
RNA	5.0	2.5	1.7
Non-photosynthetic enzymes			
Isocitric dehydrogenase	3.5	1.9	1.6
Malic dehydrogenase	3.6	1.9	1.3
Enolase	3.3	1.9	1.4
Photosynthetic enzymes			
NADP-glyceraldehyde-3-phosphate dehydrogenase	35	18	6
Alkaline fructose-1, 6-diphosphatase	15	3	2
Ribulose-1,5-diphosphate carboxylase	91	36	17
Fraction I protein	11	5	3

Seedlings grown for 7 days in the dark were exposed to (1) red light (661 nm, 72 $\mu W/cm^2$) for 5 min, or (2) red light for 5 min followed by far-red light (733 nm, 20 $\mu W/cm^2$) for 20 min or (3) far-red light for 20 min, on each of 5 consecutive days and returned to the dark. Controls were maintained in the dark. Stem apices were excised 24 hr after the final irradiation treatment and various parameters of growth, enzymic activities and fraction 1 protein were measured as previously described [25]. Values are expressed as multiples of the dark values and are based on data per 100 apices.

enzyme in maize [26]. (2) Low levels of Fraction I protein, such as are found in dark-grown pea tissues, may be over-estimated by the polyacrylamide gel technique [25] because of significant absorption by other proteins in the Fraction I area of the gel. (3) The kinetic properties of ribulose-1,5-diphosphate carboxylase in dark-grown seedlings may be different from those in irradiated plants [27]. This last point is probably not important in our experiments. We use assay conditions similar to those reported by other laboratories [28–32] and have found no difference between properties of the enzyme in extracts of dark-grown and irradiated seedlings thus far. These considerations do not invalidate the general conclusion that red irradiation results in disproportionately large increases in chloroplast enzymes compared with

the increases in non-photosynthetic enzymes and other parameters of cellular growth and development.

The increases in etioplast RNA and protein after treatment of the etiolated seedlings with red light and the inhibition of these increases by far-red light are similar to many responses produced by the photomorphogenetic pigment-protein, phytochrome [33].

Possible ways in which phytochrome may act as a regulator in chloroplast development in higher plants are discussed by Smillie and Scott [2].

Ribosomal RNA in wheat leaves

The leaves of wheat seedlings expand in the dark and brief irradiation with red light had little effect on fresh weight or the activities of respiratory enzymes of the terminal 5 cm of the leaves. However, the activities of several chloroplast enzymes in these leaves increased about 2-fold compared with leaves of unirradiated plants [34] which already had about 30% of the activity found in the green leaves of plants grown in white light. On the other hand, as shown in fig. 13, there was no difference between the relative amount of plastid and cytoplasmic ribosomal RNA in leaves from plants grown in continuous darkness or white light. Treating the dark-grown plants with red,

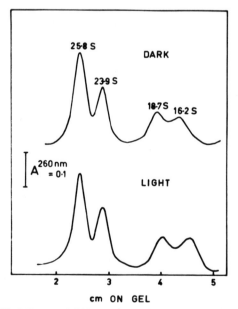

Fig. 13. Ribosomal RNA in wheat leaves. Ribosomal RNA was extracted from the terminal 5 cm of wheat leaves from 8-day-old seedlings as described in fig. 12.

red/far-red, or far-red light did not affect this ratio. Thus the level of ribosomal RNA in the etioplasts is already high in dark-grown wheat leaves and while light could conceivably stimulate the rate of synthesis of plastid ribosomes in young expanding leaves, it does not appear to influence the final levels attained in fully expanded leaves. Expanded dark-grown leaves of the bean plant (*Phaseolus vulgaris*) also contain about the same amount of plastid ribosomes as green bean leaves [35].

Except for photoconversion of part of the existing protochlorophyll to chlorophyll there was no significant synthesis of chlorophyll and consequently no photosynthetic capacity in either the pea or wheat plants as a result of the treatments with red light.

Conclusions

(1) The DNA-RNA hybridization studies reported here and in other publications [7,8,24] show that cistrons for chloroplast ribosomal RNA are found in chloroplast DNA. There is also the possibility that further cistrons for chloroplast ribosomal RNA exist in the nucleus. Competition experiments show that the 23.5 S and 16.5 S chloroplast ribosomal RNA have separate cistrons on the chloroplast DNA.

(2) Chloroplast ribosomal RNA synthesis largely precedes the synthesis of other major chloroplast constituents in *E. gracilis*. Thus the low levels of ribosomal RNA in the plastids of cells grown in darkness increase during the first 24 hr of white illumination and then remain constant while chlorophyll and chloroplast protein continue to accumulate for at least 4 days.

(3) In dark-grown pea stem apices, which remain relatively undeveloped unless illuminated, a phytochrome-mediated system regulates the synthesis of both plastid ribosomal RNA and plastid protein. As in *E. gracilis*, RNA accumulation precedes the synthesis of most of the plastid protein. Dark-grown wheat leaves, which have largely completed their expansion in the dark, apparently contain their full complement of plastid ribosomal RNA and much of their plastid protein. Brief illumination with red light or continuous illumination with white light causes no further change in the ribosomal RNA but leads to an increase in protein content of the plastids. Many

lower plants form chlorophyll in the dark [6], and presumably contain chloroplasts that have developed to a large extent in the absence of light. Thus the stage of plastid development reached in the dark can vary widely in different species of plants. Where chloroplast development is not completed in the dark, light absorbed by protochlorophyll, phytochrome, or other photoacceptors [36] may induce further development. It is apparent that the extent and mode of the photoregulation of chloroplast biogenesis varies greatly throughout the plant kingdom.

Acknowledgement

The authors wish to thank Miss Ruth Burley for excellent technical assistance.

References

[1] J.T.O. Kirk, in: Biochemistry of Chloroplasts, ed. T.W. Goodwin (Academic Press, London and New York, 1966) Vol. 1, p. 319.

[2] R.M. Smillie and N.S. Scott, in: Progress in Molecular and Subcellular Biology, ed. F.E. Hahn (Springer-Verlag, 1969) vol. I, p. 136.

[3] D. Spencer and P.R. Whitfeld, Biochem. Biophys. Res. Commun. 28 (1967) 538.

[4] K.K. Tewari and S.G. Wildman, Proc. Natl. Acad. Sci. U.S. 58 (1967) 689.

[5] N.S. Scott, V.C. Shah and R.M. Smillie, J. Cell Biol. 38 (1968) 151.

[6] J.T.O. Kirk and R.A.E. Tilney-Basset, in: The Plastids: Their Chemistry Structure and Inheritance (W.H. Freeman and Co., London and San Francisco, 1967).

[7] N.S. Scott and R.M. Smillie, Biochem. Biophys. Res. Commun. 28 (1967) 598.

[8] K.K. Tewari and S.G. Wildman, Proc. Natl. Acad. Sci. U.S. 59 (1968) 569.

[9] R.M. Smillie, D. Graham, M.R. Dwyer, A. Grieve and N.F. Tobin, Biochem. Biophys. Res. Commun. 28 (1967) 604.

[10] R.M. Smillie, N.S. Scott and D. Graham, in: Comparative Biochemistry and Biophysics of Photosynthesis, eds. K. Shibata, A. Takamiya, A.T. Jagendorf and R.C. Fuller (University of Tokyo Press, Tokyo, 1968) p. 332.

[11] J.K. Hoober, P. Siekevitz and G.E. Palade, J. Biol. Chem. 244 (1969) 2621.

[12] R.M. Smillie, D.G. Bishop, G.C. Gibbons, D. Graham, A.M. Grieve, J.K. Raison and B.J. Reger, this volume.

[13] H. Noll, Nature 215 (1967) 360.

[14] U.E. Loening and J. Ingle, Nature 215 (1967) 363.

[15] N.S. Scott and R.M. Smillie, Currents Mod. Biol. 2 (1969) 339.

[16] J.A. Schiff, this volume.

[17] S.H. Hutner, M.K. Bach and G.I.M. Ross, J. Protozool. 3 (1956) 101.

[18] R.M. Smillie and G. Krotkov, Can. J. Bot. 38 (1960) 31.

[19] R.M. Smillie, W.R. Evans and H. Lyman, Brookhaven Symp. Biol. 16 (1963) 89.

[20] D. Gillespie and S. Spiegelman, J. Mol. Biol. 12 (1965) 829.

[21] G. Brawerman and J.M. Eisenstadt, Biochim. Biophys. Acta 91 (1964) 477.

[22] B.J. McCarthy, Bact. Rev. 31 (1967) 215.

[23] E. Stutz, this volume.

[24] J. Ingle, R. Wells and J.V. Possingham, this volume.

[25] D. Graham, A.M. Grieve and R.M. Smillie, Nature 218 (1968) 89.

[26] S. Chen, D. McMahon and L. Bogorad, Plant Physiol. 42 (1967) 1.

[27] T. Sugiyama, N. Nakayama and T. Akazawa, Arch. Biochem. Biophys. 126 (1968) 737.

[28] A. Weissbach, B.L. Horecker and J. Hurwitz, J. Biol. Chem. 28 (1956) 795.

[29] E. Racker, Arch. Biochem. Biophys. 69 (1957) 300.

[30] N.G. Pon, B.R. Rabin and M. Calvin, Biochem. Z. 338 (1963) 7.

[31] P.W. Trown, Biochemistry 4 (1965) 908.

[32] J.M. Paulsen and M.D. Lane, Biochemistry 5 (1966) 2350.

[33] H.W. Siegelman and W.L. Butler, Ann. Rev. Plant Physiol. 16 (1965) 383.

[34] A.M. Grieve, Synthesis of proteins during chloroplast development. Ph.D. Thesis, The University of Sydney, 1969.

[35] N.K. Boardman, Exptl. Cell Res. 43 (1966) 474.

[36] K.W. Henningsen, in: Biochemistry of Chloroplasts, ed. T.W. Goodwin (Academic Press, London and New York, 1967) vol. I, p. 453.

THE ORIGINS OF CHLOROPLAST RIBOSOMAL-RNA

J. INGLE, R. WELLS, J.V. POSSINGHAM and C.J. LEAVER

Botany Department, University of Edinburgh, U.K.

The presence of a unique DNA within the chloroplast [1] poses the problem of the function of this genetic material in relation to the biosynthesis of chloroplast components and the autonomy of the organelle. Differential sensitivity to inhibitors, such as chloramphenicol versus cycloheximide, indicates that certain proteins are synthesised within the chloroplast [2,3], but it does not necessarily follow that the information for this synthesis is present in the chloroplast DNA. Many of the typical chloroplast components appear to be under nuclear control when analysed genetically [4]. The site of transcription of the chloroplast ribosomal-RNA should be more easily determined than that of protein-specific messenger-RNA, since the ribosomal-RNAs represent initial, stable, gene products. Consequently homologies between these RNAs and various cellular DNAs may be studied by the hybridisation technique. The chloroplast ribosomal-RNAs, with molecular weights of 1.1×10^6 and 0.56×10^6, differ in size from those of the cytoplasm, 1.3×10^6 and 0.7×10^6, and are readily distinguished from the latter by electrophoretic fractionation on polyacrylamide gels [5]. The properties of the chloroplast ribosomal-RNAs, such as their stability and synthesis, may therefore be conviently studied using preparations of total RNA from the leaf. Such preparations must also contain mitochondrial ribosomal-RNAs, the sizes of which are not known with any certainty for higher plants. However, since the total RNA preparation of root shows no significant components other than the two cytoplasmic ribosomal-RNAs, the mitochondrial RNAs must either represent a very small amount of the total RNA, less than 1%, or be of similar size to the cytoplasmic RNAs. In either case they are not likely to complicate the study of the chloroplast RNAs, which constitute about 30% of the total ribosomal-RNA.

A variety of experiments provide indirect evidence that chloroplast and cytoplasmic ribosomal-RNAs are synthesised in different places in the cell, and since the site of cytoplasmic RNA synthesis is in the nucleus, this indicates that chloroplast RNA is made somewhere other than the nucleus, presumably in the chloroplast. There is, for instance, a differential effect of light on the synthesis of chloroplast and cytoplasmic RNAs [6]. The accumulation of cytoplasmic and chloroplast RNAs in radish cotyledons during growth in light and dark is shown in fig. 1. Whereas light has little effect on the accumulation of the cytoplasmic RNA, resulting in only 10% more RNA in the light than in the dark, the accumulation of chloroplast RNA is very light dependent. These re-

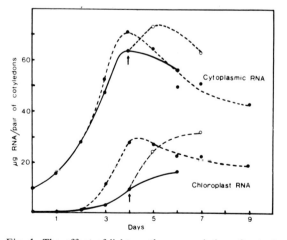

Fig. 1. The effect of light on the accumulation of cytoplasmic and chloroplast RNA. Gel fractionations of total RNA were quantitated after making corrections for breakdown of the 1.1×10^6 M RNA species [6]. The amounts of both chloroplast ribosomal RNA species and both cytoribosomal RNA species were calculated during growth in the dark (•———•), light (•----•) and after transfer from dark to light (○----○) as indicated by the arrow.

sults show that there is no absolute light requirement for the synthesis of chloroplast RNA, but that the rate of accumulation is much lower, about one third, in the dark than in the light. Differences in the synthesis of chloroplast and cytoplasmic RNAs are also indicated from inhibitor experiments [6]. In 3-day-old radish cotyledons both cytoplasmic and chloroplast RNAs are synthesised, as measured by the incorporation of ^{32}P-orthophosphate, the specific activity of the chloroplast RNA being about twice that of the cytoplasmic (fig. 2A). Chloramphenicol completely prevents the incorporation of radioactivity into the chloroplast RNA while having no inhibitory effect on incorporation into cytoplasmic RNA (fig. 2B). A similar selective inhibition of the chloroplast RNA occurs in the presence of cycloheximide (fig. 2C). 5-Fluorouracil, on the other hand, inhibits the synthesis of cytoplasmic RNA more than chloroplast RNA. We do not want to try to interpret these inhibitions, but just use them to show the selective inhibition of either the cytoplasmic or chloroplast RNAs. These kinds of studies suggest, therefore, that chloroplast and cytoplasmic RNAs are made in different places in the cell.

A more direct approach to this problem is to study the hybridisation of chloroplast and cytoplasmic RNAs to chloroplast and nuclear DNAs. To do such hybridisation experiments one requires the correct DNAs, the RNAs and suitable conditions for hybridisation, and we would like to consider these requirements briefly before describing the experiments. The problems concerned in identifying chloroplast DNA have been admirably discussed by Dr. Kirk [1], and all we need do is to define what we call chloroplast DNA. We have looked at a series of plants in which the density of the nuclear DNA varies considerably, from 1.691 g/cm^{-3}, which is equivalent to 30% G–C, to 1.702, 42% G–C. Throughout this range of nuclear DNA density, we have found the density of chloroplast DNA to be remarkably constant at 1.697 ± 0.001. Similarly the mitochondrial DNA is constant with a density of 1.706. The chloroplast and mitochondrial DNAs were prepared by differential centri-

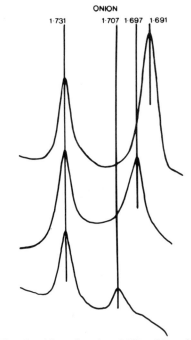

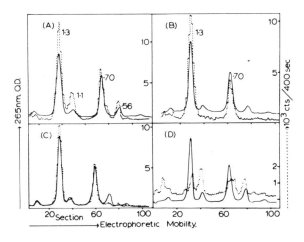

Fig. 2. The inhibition of chloroplast and cytoplasmic RNA synthesis. Cotyledons from 3 day radish seedlings were excised and incubated in the absence (A) and presence of 100 μg/ml chloramphenicol (B), 0.5 μg/ml cycloheximide (C) and 162 μg/ml 5-fluorouracil (D) for 1 hr. ^{32}P-orthophosphate was then added, and the incorporation during the following 6 hr was used as a measure of RNA synthesis. Total RNA was prepared and fractionated by gel electrophoresis in EDTA buffer for 3.5 hr [6].

Fig. 3. The densities of onion DNAs. Microdensitometer tracings of photographs obtained after analytical CsCl-density-gradient centrifugation of the DNA. The DNA samples were adjusted to a density of 1.720 g cm^{-3} with CsCl and centrifuged at 44,770 rpm for 20 hr at 25°C. The marker DNA from *Micrococcus lysodeikticus* has a density of 1.731 g cm^{-3}, nuclear DNA, 1.691 g cm^{-3}; chloroplast DNA, 1.697 g cm^{-3}; mitochondrial DNA, 1.707 g cm^{-3}.

fugation of the organelles followed by deoxyribonuclease treatment to remove all DNA other than that present in intact organelles [7]. This method does not rely on preparing chloroplasts completely free from nuclear contaminants. Onion, which contains nuclear DNA with the low density of 1.691, has chloroplast and mitochondrial DNAs with densities of 1.697 and 1.707 respectively (fig. 3). Swisschard represents the more usual situation in that the density of the nuclear DNA, 1.694, is more similar to that of the chloroplast DNA, 1.697 (fig. 4). In wheat the density of the nuclear DNA, 1.702, is greater than that of the chloroplast DNA, 1.698 (fig. 5). The chloroplast DNA used throughout these experiments has been the component with the density of 1.697–8.

Although it is easy to separate the chloroplast ribosomal-RNA from cytoplasmic ribosomal-RNA by electrophoresis, there is still one difficulty – the instability of the large chloroplast component. The range of stability of this 1.1 M RNA (the RNA components are referred to in terms of their molecular weight in millions, i.e. 1.1 M RNA = RNA with a

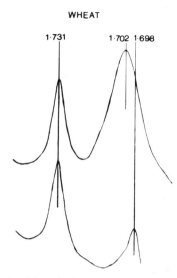

WHEAT

Fig. 5. The densities of wheat DNAs. Nuclear DNA, 1.702 g cm^{-3}; chloroplast DNA, 1.698 g cm^{-3}.

molecular weight of 1.1×10^6), is shown in fig. 6. In *Pinus sylvestris* the chloroplast RNAs are present in a ratio of 2:1, indicating very little degradation. In a tissue such as spinach the ratio is around 1:1, which is the more common situation. An extreme case of instability is that of swisschard, where the ratio is down to 0.2. The instability of the 1.1 M RNA from radish cotyledons has been studied in more detail. Three points may be concluded from the fractionation of RNA prepared from 6-day-old cotyledons (fig. 7). Firstly there is only a small amount of 1.1 M chloroplast RNA, resulting in a low chloroplast ratio, 1.1 M/0.56 M, of 0.69, secondly there is an additional RNA component of molecular weight 0.4 M, and thirdly the ratio of the cytoplasmic components, 1.3 M/0.7 M, is also low, with a value of 1.16 instead of the expected 1.86 These data can be interpreted by assuming that the 1.1 M RNA is breaking at a single point to give the 0.4 M RNA and a 0.7 M component. This latter molecule would be identical in size to the normal cytoplasmic 0.7 M ribosomal-RNA, and would account for the low 1.3 M/0.7 M cytoplasmic ratio. If a correction is made for such breakdown from the size of the 0.4 M peak, then the ratios for both the cytoplasmic and the chloroplast components are close to 2 (table 1). The correction is not fortuitous for this particular preparation, since it

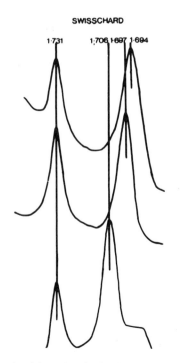

SWISSCHARD

Fig. 4. The densities of swisschard DNAs. Nuclear DNA, 1.694 g cm^{-3}; chloroplast DNA, 1.697 g cm^{-3}; mitochondrial DNA 1.706 g cm^{-3}.

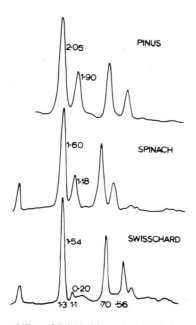

Fig. 6. The stability of 1.1 M chloroplast RNA from different tissues. Total nucleic acid was prepared from the shoots of 6-week-old *Pinus sylvestris* seedlings by the chloroform-detergent procedure. Spinach leaf and swisschard leaf total nucleic acid was prepared by the phenol-detergent method [14]. Preparations were fractionated by electrophoresis in EDTA buffer on 2.4% polyacrylamide gels for 3.5 hr. The RNA components are referred to in terms of their molecular weight in millions, i.e. 1.3 M RNA refers to an RNA of size 1.3×10^6 daltons. The value next to the 1.3 M peak is the ratio of the cytoplasmic components, 1.3 M/0.7 M, and that next to the 1.1 M peak is the chloroplast ratio, 1.1 M/0.56 M.

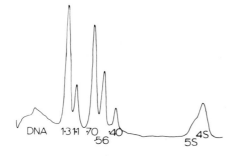

Fig. 7. The stability of radish RNA. Total nucleic acid was prepared from the cotyledons of 6-day-old seedlings and fractionated on 2.4% polyacrylamide gel in EDTA buffer for 2 hr [8]. The numerals specify the RNA species as described in table 6.

has been successfully applied to dozens of different radish RNA fractionations [8]. With radish RNA, then, the 1.1 M component appears to be split into two pieces, 0.7 M and 0.4 M.

It was therefore necessary to try to prevent this breakdown. The preparations of RNA were routinely made in the presence of bentonite, but the omission of bentonite gives slightly more of the 1.1 M component. This is surprising since, if the breakdown has anything to do with nuclease action, bentonite would improve the situation. We then realised that the bentonite preparation contained EDTA, and found that EDTA without bentonite similarly gives a very low yield of 1.1 M RNA. The obvious experiment, with added magnesium instead of EDTA, increases the yield of the 1.1 M RNA. This indication that the magnesium/EDTA level is important, necessitates a reconsideration of the fractionation procedure, since the electrophoresis buffer that was being used contains EDTA. We found that if a buffer containing magnesium was used, a much higher yield of 1.1 M RNA results. When RNA is extracted from radish cotyledons in the presence of EDTA, and fractionated in the EDTA buffer, the ratio of the chloroplast components is 0.5 (fig. 8). If the same preparation is run in the magnesium buffer, then more 1.1 M RNA

Table 1

Fractionation of radish RNA, and correction for the breakdown of the 1.1 M RNA. The fractionation of RNA shown in fig. 7 was quantitated by measurement of the areas of the peaks, which were expressed as a percentage of the total area and related to the µg nucleic acid loaded onto the gel [8]. The RNA components are referred to in terms of their molecular weight in millions.

RNA component	µg RNA		
	Determined from scan	Correction from 0.4 M peak	Corrected values
1.3 M	5.7	–	5.7
1.1 M	1.8	+ 3.0	4.8
0.7 M	4.9	– 1.9	3.0
0.56 M	2.6	–	2.6
0.4 M	1.1	– 1.1	0.0
Ratio			
1.3 M/0.7 M	1.16		1.90
1.1 M/0.56 M	0.69		1.85

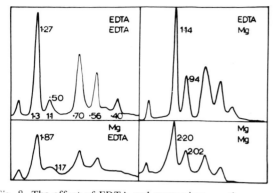

Fig. 8. The effect of EDTA and magnesium on the preparation and fractionation of total RNA. Total nucleic acid was prepared from 6-day-old radish cotyledons by the detergent-phenol procedure [14] in the presence of either 10 mM EDTA or 10 mM $MgCl_2$. The preparations were then fractionated by gel electrophoresis using either the normal EDTA buffer [36 mM–Tris, 30 mM–NaH_2PO_4, 1 mM EDTA (Na_2 salt), pH 7.7–7.8 at 20°C] or the magnesium buffer [30 mM–Tris, 16 mM–HCl, 0.1 mM–EDTA, 2 mM magnesium acetate, pH 8.1 at 20°C]. RNA prepared in the presence of EDTA was fractionated by EDTA electrophoresis (EDTA/EDTA) and by magnesium electrophoresis (EDTA/Mg). RNA prepared in the presence of magnesium was fractionated by EDTA electrophoresis (mg/EDTA) and by magnesium electrophoresis (Mg/Mg).

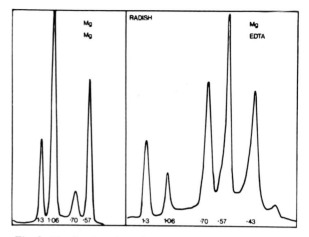

Fig. 9. The fractionation of RNA prepared from chloroplasts isolated from radish. Chloroplasts were isolated from 6-day-old radish cotyledons [15]. The chloroplasts were resuspended and solubilised with Triton X-100 (2% final concentration). After centrifugation at 10,000 g for 10 min at 2–4°C, total nucleic acid was prepared from the supernatant by shaking it with an equal volume of double strength detergent (12% 4-amino salicylate, 2% tri-isopropyl naphthalenesulphonate, 50 mM KCl, 10 mM $MgCl_2$ and 10 mM Tris, pH 7.4), followed by phenol deproteinisation. This preparation was fractionated by magnesium electrophoresis (Mg/Mg) and by EDTA electrophoresis (mg/EDTA).

is present, the ratio being 0.94. When the RNA is extracted in the presence of magnesium, and fractionated in the magnesium buffer, the ratio of the chloroplast RNAs is close to 2. It is therefore possible to get 2:1 ratios for both the cytoplasmic and the chloroplast RNAs from radish if magnesium is present all the time.

The effect of magnesium on the stability of the 1.1 M RNA is seen more clearly if RNA is prepared from isolated chloroplasts, rather than from the total tissue. When RNA is prepared from radish chloroplasts in the presence of magnesium, and fractionated in the magnesium buffer, the two chloroplast RNAs are nicely separated with a ratio of 2:1, and the cytoplasmic components are present as minor contaminants (fig. 9). Fractionation of the same RNA in the EDTA buffer results in the almost complete loss of the 1.1 M chloroplast component, with increases in the amount of 0.7 M and 0.4 M RNAs. This particular pattern of breakdown of the 1.1 M RNA is fairly specific for radish, and rather different patterns are obtained with other tissues. When chloroplasts from

broad bean are extracted in the presence of magnesium, and the RNA is fractionated in the magnesium buffer, the two chloroplast components are present with a 2:1 ratio (fig. 10). When this RNA is run in the EDTA buffer, there is considerable breakdown of the heavy component with the production of 0.66 M and 0.52 M RNAs. The stabilization, or holding together, of the 1.1 M RNA is not specific to magnesium, since fractionation in a calcium buffer gives similar results (fig. 10). Furthermore, EDTA does not have to be present in the electrophoresis buffer to promote the breakdown. The electro potential applied during the fractionation appears to be sufficient to remove magnesium from the RNA, resulting in breakdown (fig. 10).

We do not know whether this initial break in the 1.1 M RNA occurs in vivo in the ribosome, or whether it happens during extraction, and there is no time to consider this now. We would just like to make the point that degradation of the chloroplast RNA certainly occurs, and that a knowledge of the state of the RNA is essential before hybridisation experiments can be attempted.

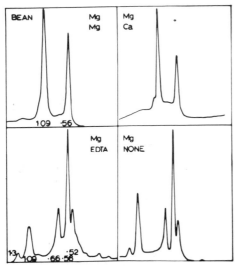

Fig. 10. The effects of magnesium, calcium and EDTA on the stability of RNA prepared from broad bean chloroplasts. Chloroplasts were isolated and total nucleic acid was prepared as described for fig. 9. The preparation was fractionated by magnesium electrophoresis (Mg/Mg), calcium electrophoresis [10 mM Ca^{++} replaced the Mg^{++}] (Mg/Ca), EDTA electrophoresis (Mg/EDTA) and by electrophoresis in a buffer containing neither magnesium nor EDTA (Mg/none).

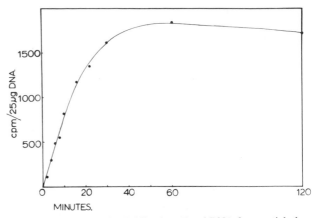

Fig. 11. Kinetics of hybridisation. Total DNA from artichoke tuber was fixed to filters and hybridised with cytoplasmic ribosomal-RNA (1.3 M plus 0.7 M in 2:1 mixture, specific activity of 630,000 cpm/μg) as described in table 2.

The production of ^{32}P labelled-chloroplast RNA with the high specific activity necessary for hybridisation, is further complicated by the limited period of chloroplast RNA synthesis during development of the leaf [8,9]. For these experiments small plants were grown in liquid culture in the presence of ^{32}P-orthophosphate for 7 days, and total RNA was prepared from those leaves which had been produced during this growth period. In this way chloroplast and cytoplasmic RNAs with specific activities of 100,000 cpm/μg of RNA are obtained. The total RNA is fractionated by electrophoresis and the individual components are eluted from the gel.

For the hybridisation experiments the alkali-denatured DNA, fixed to Millipore filters [10], is incubated with the various ribosomal-RNAs at 3 μg/ml, in 6 × SSC at 70°. Under these conditions the rate of hybridisation is very fast, and saturation is reached in 30—60 min (fig. 11). This contrasts with the much longer periods e.g. 16 hr that are often employed for hybridisation [11]. A hybridisation period of 2 hr has been used routinely in these experiments.

The results of a typical hybridisation experiment are shown in table 2. Cytoplasmic RNA hybridises to 0.3% of the nuclear DNA. This corresponds to 6,000 cistrons in a diploid nucleus containing 2.5×10^{-12} g of DNA. The chloroplast RNA hybridises to a much larger percent of the chloroplast DNA, 1.5%. This corresponds to 1 cistron per chloroplast genome of 150×10^6 daltons [7], or around 30 cistrons per chloroplast. Since an average swisschard cell contains around 200 chloroplasts, this in turn corresponds to approximately 6,000 cistrons per cell. Hybridisation of the chloroplast DNA with cytoplasmic RNA re-

Table 2

Hybridisation of cytoplasmic and chloroplast RNAs to nuclear and chloroplast DNAs. 25 μg DNA was alkali-denatured and fixed to a millipore filter [10]. The filters were incubated with RNA at 3 μg/ml [a] in 6 × SSC (standard saline citrate) at 70°C for 2 hr. The filters were then washed once with 6 × SSC, 3 times with 2 × SSC, incubated with 10 μg/ml of ribonuclease at 25°C for 15 min in 2 × SSC, dried and counted. The DNA content of the filters was checked by acid hydrolysis after counting [13].

	% DNA hybridised
Nuclear DNA + cytoplasmic RNA	0.3
Chloroplast DNA + chloroplast RNA	1.5
Chloroplast DNA + cytoplasmic RNA	0.3
Nuclear DNA + chloroplast RNA	0.3

[a] See fig. 13 for details of the RNA specific activities.

sults in only 0.3% hybridisation, relative to the 1.5% in the homologous system, so that the cytoplasmic RNA is only about 20% as effective for the hybridisation as is the chloroplast RNA. Chloroplast RNA, on the other hand, hybridises to 0.3% of the nuclear DNA i.e. the cytoplasmic and chloroplast RNAs hybridise to nuclear DNA to the same extent. These data are very similar to those of Tewari and Wildman [11], and from such data it has been concluded that cistrons for making chloroplast RNA are in the nucleus as well as in the chloroplast. We have tried to take these experiments one step further and see if it is possible to distinguish between the nuclear cistrons and the chloroplast cistrons. From previous work we knew that the cistrons for making cytoplasmic ribosomal-RNA move out from the main DNA band on caesium chloride centrifugation (fig. 12). Hybridisation of cytoplasmic RNA to the DNA fractions obtained from the caesium chloride gradient results in a peak at a density of 1.705–6. This has been confirmed with four plant tissues. The cytoplasmic ribosomal-RNA cistrons therefore move out from the main nuclear DNA band to a characteristic position on the gradient. The position on the gradient of the

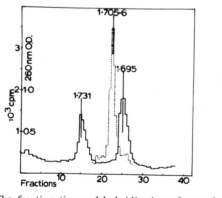

Fig. 12. The fractionation and hybridisation of cytoplasmic ribosomal-RNA cistrons from artichoke DNA. Total DNA from artichoke tuber was fractionated by preparative CsCl-density-gradient centrifugation. The DNA solution was adjusted to 1.720 gcm^{-3} and centrifuged at 44,000 rpm for 72 hr at 25°C in an M.S.E. 50 rotor. *Micrococcus lysodeikticus* DNA, of density 1.731 gcm^{-3}, was included as a marker. Five-drop fractions were collected, and the 260 mμ O.D. determined after the addition of 0.4 ml of 0.1 X SSC. The DNA in each fraction was alkaline denatured, fixed to millipore filters and hybridised with cytoplasmic ribosomal-RNA, as described in table 2 and fig. 11.

chloroplast ribosomal-RNA cistrons was therefore determined. The caesium chloride gradients for the hybridisation combinations described in table 2 are shown in fig. 13. Cytoplasmic RNA hybridises to nuclear DNA with a density of 1.706. Chloroplast RNA hybridises to nuclear DNA to the same extent, and also at an identical position in the gradient, 1.706. The hybridisation of chloroplast RNA to chloroplast DNA occurs essentially within the main band of the chloroplast DNA, peaking on the heavy side with a density of 1.699 relative to the main band peak at 1.697. Hybridisation of the chloroplast DNA with cytoplasmic RNA is much lower, approximately 20% that of the homologous chloroplast system, but the hybridisation again peaks at a density of 1.699. Cistrons are therefore present at two density positions in the gradient, 1.706 in nuclear DNA and 1.699 in chloroplast DNA. The fact that cistrons occupy two different positions on the gradient does not necessarily mean that there are two different types of cistrons. The position of the cistron on the gradient depends on the number of cistrons present, on the integration of the cistrons with the rest of the DNA and on the size of the DNA that is fractionated. We wanted to know whether there were two types of cistrons present in the cell, and if so, whether only one type was present at a density of 1.706 on the gradient and one at 1.699, or were both present in both positions. The latter question i.e. are there two types of cistrons at a density of 1.706, can be readily answered from competition experiments. The competition between cytoplasmic and chloroplast RNA for nuclear DNA is shown in table 3. Hybridisation with cytoplasmic RNA at a concentration of 3 μg/ml results in about 80% saturation of the system. With chloroplast RNA at 2.1 μg/ml the system is again saturated to about 80%, and in both cases the cytoplasmic and chloroplast RNAs bind to identical levels. When nuclear DNA is challenged with a mixture of cytoplasmic plus chloroplast RNA the presence of two different cistrons in the DNA, one of which binds cytoplasmic RNA, the other chloroplast RNA, should result in the sum of the two individual hybridisations i.e. in this experiment a binding of approximately 4,600 counts. If only one type of cistron is present, which will bind either cytoplasmic or chloroplast RNA, then the combination of RNAs should result in a binding of between 2,700 and

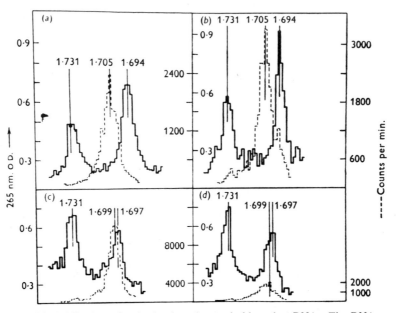

Fig. 13. The fractionation and hybridisation of swisschard nuclear and chloroplast DNAs. The DNA samples were fractionated and the fractions hybridised with RNA as described in fig. 12 and table 2. Nuclear or chloroplast DNA was hybridised with cytoplasmic ribosomal-RNA, a 2:1 mixture of 1.3 M and 0.7 M with a specific activity of 130,000 cpm/μg RNA, and with chloroplast ribosomal-RNA, a 2:1 mixture of 1.1 M and 0.56 M with a specific activity of 110,000 cpm/μg RNA. (a) Nuclear DNA and cytoplasmic RNA; (b) nuclear DNA and chloroplast DNA; (c) chloroplast DNA and chloroplast RNA; (d) chloroplast DNA and cytoplasmic RNA.

3,100 counts. The results therefore indicate that only one type of cistron is present at a density of 1.706 on the gradient, and that this cistron binds both types of RNA with equal efficiency. Once it has been established that there is just one type of cistron at 1.706, then the cistron at 1.699, in the chloroplast DNA,

Table 3
Competition between cytoplasmic and chloroplast RNA for nuclear DNA from swisschard. Hybridisation was carried out as described in table 2.

RNA		cpm/10 μg nuclear DNA
Source	μg/ml	
Cytoplkasmic	3.0	2415
0.7×10^6	6.0	3146
Chloroplast	2.1	2243
0.56×10^6	4.2	2770
Cytoplasmic	3.0	
+ chloroplast	+2.1	2800

must be different, since it binds chloroplast RNA 5 to 10 times more efficiently than cytoplasmic RNA.

Can we conclude therefore, from the observation that chloroplast RNA hybridises with nuclear DNA, that the nucleus contains the cistrons for chloroplast RNA? We think the type of experiment described shows that we cannot really make this conclusion. On the caesium chloride gradients it is possible to get at least an apparent resolution of two types of cistrons, at densities of 1.706 and 1.699, and the cistrons at 1.706 appear to bind cytoplasmic and chloroplast RNA equally well. This rather low specificity of hybridisation is not altogether unexpected, since it has been known for several years that the hybridisation between nuclear DNA and cytoplasmic RNA depends only on the origin of the DNA, being independent of the RNA [12]. The hybridisation of cytoplasmic RNA from 4 plants to total DNA from 5 species is shown in table 4. It can be seen that as far as wheat DNA is concerned, artichoke RNA is just as acceptable as wheat RNA. We can therefore ask the question, in relation to the binding of chloroplast

Table 4

Specificity of cytoplasmic RNA — total DNA hybridisation. Hybridisation was carried out as described in table 2.

[32]P ribosomal-RNA (cytoplasmic)	Hybridisation (% total DNA)				
	Wheat DNA	Swisschard DNA	Onion DNA	Artichoke DNA	Cucumber DNA
Wheat	*0.069*	0.22	0.076	0.035	0.56
Swisschard	0.073	*0.30*	0.078	0.043	0.75
Onion	0.083	0.29	*0.098*	0.048	0.69
Artichoke	0.069	0.21	0.081	*0.042*	0.83

RNA to nuclear DNA; is the difference between swisschard cytoplasmic and chloroplast RNA necessarily any greater than the difference between swisschard cytoplasmic and onion cytoplasmic RNA? From our present knowledge I think we have to say no. Swisschard chloroplast RNA is probably as similar to swisschard cytoplasmic RNA as is onion cytoplasmic RNA. We think therefore that until the specificity of the hybridisations is evaluated, by determining the stability of the hybrids via their melting curves, or until such experiments as we have described have been repeated under more stringent conditions i.e. conditions which will only allow the formation of truly complementary hybrids, one cannot really conclude whether the cistrons for chloroplast RNA are present in the nucleus as well as in the chloroplast.

Acknowledgements

We thank Dr. M. Birnstiel for discussions during the course of the work, and Professor R. Brown for his interest and support of the project. One of us (J.I.) thanks the University of Edinburgh, the Royal Society and the International Union of Biochemistry for a travel grant which enabled him to attend the symposium.

References

[1] J.T.O. Kirk, this symposium.
[2] R.M. Smillie, D.G. Bishop, G.C. Gibbons, D. Graham, A.M. Grieve, J.K. Raison and B.J. Reger, this symposium.
[3] U.W. Goodenough, R.P. Levine, S.J. Surzycki and J.J. Armstrong, this symposium.
[4] J.T.O. Kirk, in: Biochemistry of chloroplasts, Vol. 1, ed. T.W. Goodwin (Academic Press, London, 1966) p. 327.
[5] U.E. Loening and J. Ingle, Nature 215 (1967) 363.
[6] J. Ingle, Plant Physiol. 43 (1968) 1850.
[7] R. Wells and M. Birnstiel, Biochem. J. 112 (1969) 777.
[8] J. Ingle, Plant Physiol. 43 (1968) 1448.
[9] J. Ingle, J.V. Possingham, R. Wells, C.J. Leaver and U.E. Loening, in: The Development and Inter-relationship of Cell Oganelles, Symp. No.24 of Soc. Exptl. Biol. (1969), ed. P.L. Miller (Cambridge University Press, Cambridge).
[10] D. Gillespie and S. Spiegelman, J. Mol. Biol. 12 (1965) 829.
[11] K.K. Tewari and S.G. Wildman, Proc. U.S. Natl. Acad. Sci. 59 (1968) 569.
[12] K. Matsuda and A. Siegel, Proc. U.S. Natl. Acad. Sci. 58 (1967) 673.
[13] D.D.Brown and C.S.Weber, J. Mol. Biol. 34 (1968) 661.
[14] J. Ingle and R.G. Burns, Biochem. J. 110 (1968) 605.
[15] W. Cockburn, D.A. Walker and C.W. Baldry, Plant Physiol. 43 (1968) 1415.

AN APPROACH TOWARDS ASCERTAINING THE FUNCTION
OF CHLOROPLAST DNA IN TOBACCO PLANTS

S.G. WILDMAN

Department of Botanical Sciences, Molecular Biology Institute,
University of California, Los Angeles, California, USA

Biologists now possess a remarkably comprehensive knowledge of how biological information encoded within DNA is utilized for the growth and reproduction of procaryotic organisms, in particular, *E. coli.* Now, the important question is being asked as to how far this knowledge will extend to account for the behavior of more complex organisms belonging to the animal and plant kingdoms. Instead of a single bacterium displaying minimal differentiation of its living matter and possessing only a single molecule of DNA with no redundancy in its informational content, the highly differentiated protoplasm of cells of higher organisms is composed of a system of particulate organelles, of which nuclei, mitochondria and chloroplasts are all known to contain DNA. The different DNAs are also different in size as well as redundancy. The very fact that DNA is present in mitochondria and chloroplasts begs the question as to how much of the information required for the total synthesis of such organelles is contained in their own DNAs. Furthermore, how do organelles other than nuclei maintain continuity during sexual reproduction of higher organisms? Must a preformed organelle be transmitted via egg or sperm, or is it sufficient that only the DNA of the organelle be the agent which is physically transferred from generation to generation? The purpose of this essay is to point out that an especially favorable opportunity exists for discovering the precise role of extranuclear DNA in the growth and reproduction of organelles because of the unique organization of higher plant chloroplasts. In the interests of briefness, I will only summarize the salient information necessary for my point of view and register an unseemly parochialism by restricting the material mainly to work from my laboratory on

chloroplasts from tobacco leaves. Also, I will leave the experimental details to be found in the original literature. Several reviews can provide access to this material [1−4].

Organization of higher plant chloroplasts

The appearance of chloroplasts, as seen in living mesophyll cells of spinach leaves by phase microscopy, is shown by the photomicrographs in fig. 1. Chlorophyll is located in the nearly spherical regions distinguished by the presence of numerous grana. The spherical part of the chloroplast does not change position in the cell during long periods of continuous observation and is therefore called the *stationary component*. The stationary components are surrounded by translucent structures called the *mobile phase,* the reason for this designation becoming apparent when the same chloroplasts in the two photomicrographs, taken only about two minutes apart, are compared. In the case of the chloroplast marked 'a', almost the entire shape of the mobile phase has changed, and this is also noticeable with the chloroplast marked 'b', particularly because a protuberance emerged from the mobile phase in this short interval of time. Often, the mobile phase will be seen to embrace two stationary components as in 'c'. However, continued observation might reveal the separation of the mobile phase to yield two chloroplasts as in 'a' and 'b'. Or, conversely, the single mobile phases surrounding two stationary components might be seen to fuse via ameboid movements to yield the condition shown in 'c'. In addition to its constant dynamic changes in shape, the mobile phase also

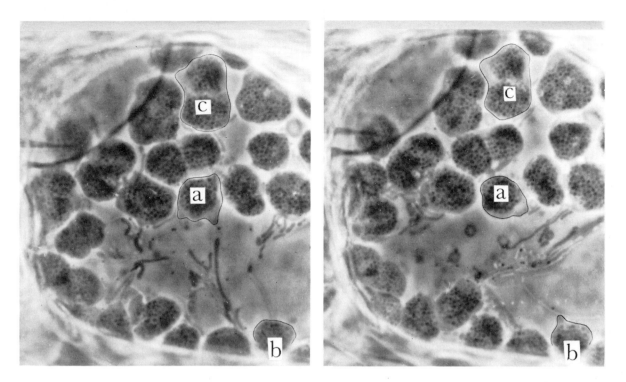

Fig. 1. The appearance and behavior of the mobile phase and stationary component of chloroplasts in a living cell in mesophyll tissue of a spinach leaf. Observed by phase microscopy using 100 X objective; flash photography. Two photomicrographs taken about 2 min apart. (a) change in shape of mobile phase; (b) emergence of protuberance from mobile phase; (c) two stationary components enclosed by a single mobile phase. Stationary component of 'a' is 10 μ in diameter.

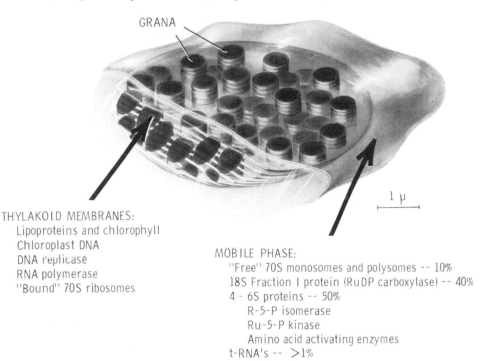

GRANA

THYLAKOID MEMBRANES:
   Lipoproteins and chlorophyll
   Chloroplast DNA
   DNA replicase
   RNA polymerase
   "Bound" 70S ribosomes

MOBILE PHASE:
   "Free" 70S monosomes and polysomes -- 10%
   18S Fraction I protein (RuDP carboxylase) -- 40%
   4 - 6S proteins -- 50%
      R-5-P isomerase
      Ru-5-P kinase
      Amino acid activating enzymes
   t-RNA's -- >1%

1 μ

Fig. 2. Model of higher plant chloroplast and location of various macromolecular components within the chloroplast. A section through the model has exposed a typical low magnification view of the thylakoid membrane system as seen by electron microscopy, whereas the upper face is that seen as the stationary component as in fig. 1.

displays fascinating interactions with the mitochondria and endoplasmic reticulum which have been described [5–7] and will not be further dwelt on here. What has been described for spinach chloroplasts also applies to the appearance and behavior of chloroplasts in living leaf cells of several species of tobacco as well as a variety of other kinds of herbaceous plants commonly used for biochemical research [8].

The drawing in fig. 2 is a model depicting a current view of the general organization of a higher plant chloroplast utilizing information derived from both light and electron microscopy [9] and our present concept of the location of macromolecular components in tobacco chloroplasts. The essential feature of the model is the probable organization of the intergrana thylakoid membranes in the form of a stack of flattened vesicles interspersed with stacks of grana membranes, the entire system of thylakoids being surrounded and interpenetrated by the mobile phase. The outer surface of the latter displays osmotic properties [10] and therefore constitutes a semipermeable barrier delimiting the biphasic chloroplast from the cytoplasm of the cell.

Macromolecular composition of mobile phase

Methods have been developed for obtaining isolated chloroplasts with intact mobile phase [11] and then to permit separation of the mobile phase constituents from the thylakoid system of membranes [12]. Analysis of the macromolecular composition of the mobile phase of tobacco chloroplasts has revealed the following components:

(1) *70 S ribosomes,* constituting about 10% of the macromolecules. The ribosomes are present both as monomers and polymers, the latter ranging in size from dimers to octomers [13]. The monomers closely resemble 70 S monomers found in bacteria and blue-green algae in regard to physical and morphological properties [14]. They furthermore contain RNAs of the 16 S and 23 S varieties.

(2) *Fraction I protein,* making up about 40% of the macromolecules. This protein, when purified to a high degree, has an $S_{20,w}^0$ of ca. 18.5 S, a molecular weight of about 525,000, and displays RuDP carboxylase activity [3]. This is the enzyme which catalyzes

the first step in CO_2 fixation during photosynthesis whereby ribulose diphosphate is converted into two molecules of 3-phosphoglyceric acid.

(3) *A mixture of proteins,* comprising about 50% of the total macromolecules, ranging in size up to about 6 S. The mixture can be resolved into six distinct components by polyacrylamide gel electrophoresis [15]. Enzyme activities necessary for the operation of the Calvin cycle are found in this mixture, in particular, R-5-P isomerase and Ru-5-P kinase. In addition, enzymes required for activation of amino acids, acylation of tRNAs and peptide synthesis are found among this mixture of proteins [12].

(4) *tRNAs,* very small in amount, but a sufficient mixture to provide for protein synthesis within the chloroplast [2].

Components associated with thylakoid membranes

Upon removal of the mobile phase constituents from isolated tobacco chloroplasts, the following components still remain firmly associated with the thylakoid membranes:

(1) *DNA,* amounting to about 10^{-14} g per chloroplast, possessing a density in CsCl of 1.700 ± 0.001, a length of 30–150 μ, renatures completely after melting, and has a genome size of $11.4 \pm 1.1 \times 10^7$ daltons as ascertained from renaturation kinetics [2]. Chloroplast DNA, in contrast to nuclear DNA, is notable for the absence of 5-methylcytosine [16]. Features which serve to distinguish chloroplast DNA from other tobacco leaf DNAs are summarized in table 1. Hybridization experiments reveal there to be at least one cistron for each of the 23 S and 16 S species of chloroplast ribosomal RNA. No regions of the chloroplast DNA are complementary with the 26 S and 18 S species of cytoplasmic ribosomal RNA [17]. Chloroplast DNA hybridizes with tRNA to a level sufficient to code for 20–30 tRNAs having molecular weights of about 25,000. Only about 1% of the chloroplast DNA genome appears to be required to code for chloroplast ribosomal and tRNAs [2].

(2) *DNA replicase activity.* Only the gross properties of this enzymatic activity are known but they seem to conform to a Kornberg type enzyme [18,19]. Efforts to release the enzyme from the

Table 1
Properties which distinguish between nuclear, chloroplast, and mitochondrial DNAs of tobacco leaves.

Property	Chloroplast	Nuclear	Mitochondria [32]
% total leaf DNA	9	90	>1
DNA per organelle	9×10^{-15} g	100×10^{-15} g	?
Density in CsCl	1.700 ± 0.001	1.697 ± 0.001	1.724
Genome size			
Electron microscope	$30-150 \mu$	?	$2-12 \mu$
Renaturation kinetics	$11.4 \pm 1.1 \times 10^7$ d.	?	?
Renaturation	complete	incomplete	complete
5-methyl cytosine	absent	2–5%	?
Histones	absent	present	?
Transcription rate:			
$\mu\mu$moles RNA/hr/μg DNA	40	1.2	?

thylakoid membranes and chloroplast DNA have been unsuccessful. Hybridization experiments indicate that the in vitro product synthesized by the crude enzyme resembles chloroplast DNA [15]. Thus, chloroplasts appear to have the enzymatic machinery necessary for replication of their own DNA.

(3) *RNA polymerase activity.* This activity can be concentrated to a large degree but not to the extent where separation of the enzyme from thylakoid membranes and chloroplast DNA has been achieved. No condition has been obtained where added template DNA has significantly stimulated the enzymatic activity. However, the crude preparations are highly active, most likely as a consequence of the absence of histones in chloroplasts, and cause synthesis of heterogeneous RNAs ranging in size from about 4 to 30 S [20]. That the in vitro synthesized RNA products have the characteristics of mRNAs has been shown by hybridization experiments which indicate that as much as 21% of the chloroplast DNA genome can be transcribed in vitro by the RNA polymerase [2].

(4) *Bound 70 S ribosomes* [15]. Identification of 70 S ribosomes tightly bound to thylakoid membranes has come about from studies of the radioactive products formed during amino acid incorporation into peptides by isolated tobacco chloroplasts. About 50% of the radioactive peptides are found largely in the form of nascent peptides attached to the polysomes and monosomes of the mobile phase and are thus readily removed from the thylakoid membranes. The other 50% is not released from the thylakoids

even by the most strenuous washing procedures which also result in extensive swelling and partial disruption of the thylakoids. The radioactivity can however be released in a form where it no longer sediments with chlorophyll containing materials by treating the thylakoids with deoxycholate (DOC). Ribosomes are simultaneously released from the thylakoids by DOC and the radioactive product is now found primarily in the form of nascent peptides still attached to the ribosomes. As estimated by RNA analysis, about 50% of the total chloroplast ribosomes are firmly-bound to the thylakoid membranes, with the other 50% 'free' in the mobile phase. The 'bound' ribosomes by themselves are capable of inducing amino acid incorporation into peptides but require the addition of part of the mobile phase in the form of a soluble extract containing tRNA and amino acid activating enzymes for this activity.

Function of chloroplast DNA

The view which emerges from the analyses described above is that all of the molecular biological ingredients necessary for a complete system to replicate, transcribe and translate the information in chloroplast DNA are found within the isolated chloroplast. The gross molecular biological picture of isolated chloroplasts seems to be entirely consistent with the much earlier view of chloroplast autonomy derived from genetical investigations which had demonstrated the non-Mendelian mode of chloroplast inheritance. Moreover, the DNA is of sufficient size to provide coding information for more than 1000 different proteins containing as many as 200 amino acids in their primary structures, as well as information for chloroplast ribosomal and tRNAs. Allowing for ca. 60 proteins as making up the structure of the chloroplast ribosomes, perhaps 30 proteins for the Calvin cycle, something of these orders of magnitude for enzymes for fatty acid and starch synthesis, it does not seem likely that the total number of individual proteins to be eventually found in chloroplasts will exceed the potential capacity of chloroplast DNA to code for all of them. If chloroplast DNA does code for the entire chloroplast macromolecular structure, it would seem that the DNA has suffered few mutations which have survived during the course of its evolutionary history in higher plants. As judged by observation of chloroplasts in living cells of a dozen different herbaceous plants, chloroplasts display a remarkable constancy in their phenotypic appearance. Perhaps chloroplasts in bundle sheaths of some plants represent a non-lethal mutation of chloroplast DNA since these chloroplasts do not have well-defined grana as part of their thylakoid system. However, the question now remains as to how it can be demonstrated that chloroplast DNA does, in fact, serve as the source of information during synthesis and reproduction of chloroplasts. According to the precepts of molecular biology, a rigorous demonstration would have to show, as a minimum, an arrangement of base sequences in chloroplast DNA that would be predicted from a knowledge of the arrangement of amino acids in a primary peptide chain of chloroplast protein. Needless to say, we are still a long way from meeting this objective in the first place because of our ignorance of the arrangement of amino acids in chloroplast proteins. Lacking this fundamental knowledge, reliance must be placed on correlative information in our quest for understanding the function of chloroplast DNA.

That chloroplast DNA does function in the metabolism of chloroplasts finds support from experiments on tobacco leaves infected with tobacco mosaic virus (TMV). Recent studies indicate that TMV in the process of rapid multiplication interferes with the macromolecular metabolism of tobacco chloroplasts [21]. The studies showed that the presence of TMV strongly inhibited synthesis of the 16 S and 23 S ribosomal RNAs of the 70 S chloroplast ribosomes without affecting synthesis of 18 S and 26 S RNAs of the 80 S cytoplasmic ribosomes. Chloroplast protein synthesis, most noticeably fraction 1 protein synthesis, was also strongly inhibited when TMV infected leaves were compared to healthy leaves. Further experiments indicated that mRNA synthesis in chloroplasts was also greatly reduced while TMV was undergoing maximum rate of reproduction. Of more significance was the finding that the reduced amount of chloroplast mRNA synthesis also resulted in mRNA having different AMP/GMP and UMP/GMP ratios than mRNA synthesized by chloroplasts from uninfected plants. The results suggest that both rRNA and mRNA production in chloroplasts was affected by some undefined regulator produced during the grand period of TMV reproduction, the reduction and change in mRNA composition being reflected in a greatly reduced amount of protein synthesis within the chloroplasts. Regulation apparently occurred at the chloroplast DNA transcription level and these results can be taken to signify that chloroplast DNA serves as a source of information for macromolecular synthesis by chloroplasts.

Nuclear and chloroplast cistrons for chloroplast ribosomal RNA

Another approach towards ascertaining the function of chloroplast DNA has been to investigate whether cistrons complementary to chloroplast ribosomal RNAs are exclusively located on chloroplast DNA. Cross-hybridization experiments have shown that nuclear DNA hybridizes to a significant extent with chloroplast rRNAs. Whereas chloroplast DNA hybridizes with chloroplast rRNA to 0.5% when heat denatured DNA is used [17] and to 1.5% when the DNA is denatured by alkali [2], nuclear DNA hybridizes with chloroplast rRNA to the extent of slightly more than 0.1%. Since nuclear DNA constitutes about 90% of the total DNA in tobacco leaves compared to about 9% for chloroplast DNA, the hybridization values lead to an estimate that the DNA in a single nucleus contains about 600 regions with base sequences complementary to chloroplast rRNAs compared to only about 16 regions for chloroplast rRNA on the DNA contained in a single chloroplast [17]. Thus, the question needs to be asked as to whether the chloroplast rRNA cistrons in nuclear DNA also have a function in supplying information for the synthesis of higher plant chloroplasts. In this connection, there is suggestive evidence that chloroplast DNA itself also has the capacity to hybridize with nuclear DNA [18].

One notion that has been advanced to explain the abundance of chloroplast RNA hybridization sites on nuclear DNA is that they are simply non-functional relics from the past evolution of chloroplasts. Suppose that evolution of the higher plant chloroplast took the direction depicted in fig. 3. In this conception, single thylakoids bearing chlorophyll first appeared in unicellular blue-green algae whose present day progenitors do not contain organelles and whose DNA is present as a single molecule as in bacteria. Primeval chloroplast DNA is taken to be a specific region of nucleotides in the DNA molecule which provides information for synthesis of the thylakoid proteins and RNA components. As evolution advanced to the level of complexity displayed by a green algal cell such as modern day *Chlorella,* differentiation of the protoplasm into discrete organelles

occurred. The process was accompanied by encapsulation of the naked DNA within a nucleus, as well as encapsulation of thylakoids into the organelle we call a chloroplast. Perhaps, encapsulation of thylakoids also required replication and detachment of that portion of the DNA molecule which coded for thylakoids while at the same time, that piece of the DNA became an integral part of the chloroplast while the entire parent DNA molecule was sequestered into the nucleus. We imagine that chloroplast DNA now leads an independent existence and serves as the entire source of information for the manufacture of chloroplasts. However, its parent is still present as a small portion of the total nuclear DNA but is now merely a non-functional redundancy. An argument against this notion is that so many mutants have been detected which affect chlorophyll biosynthesis as well as chloroplast metabolic functions, all of which are inherited in a Mendelian manner and therefore require nuclear DNA as the source of this kind of information [22]. Furthermore, work on a cytoplasmic mutant of tobacco persuades me that a more attractive hypothesis may be one that invokes nuclear DNA as containing some of the information required for the total synthesis of the chloroplast.

Macromolecular composition of defective chloroplasts in a cytoplasmic mutant of tobacco [23]

A spontaneous mutant tobacco plant was recognized because of the presence of variegated leaves displaying random and bizarre patterns of green and white areas. The variegation character is only transmitted through the maternal line. The cause of the variegation is the presence of defective chloroplasts intermingled with normal chloroplasts within the same cells of the leaf. When the ratio of defective to normal chloroplasts is high, the tissue is white in appearance; when low, green. The defective chloroplasts are devoid of chlorophyll and grana and hence appear to have suffered a great reduction in thylakoids. The defective chloroplasts have never been seen in wild-type tobacco plants. While other work on the tobacco mutant has suggested that the defective chloroplasts arise by degeneration of normal chloroplasts [24], our observations of large numbers of defective chloroplasts in living cells examined during

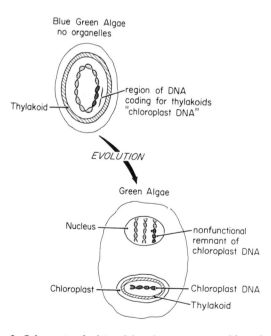

Fig. 3. Scheme to depict origin of autonomous chloroplast DNA during evolution resulting in organelles within cells.

several stages of leaf growth do not support this view. We have been unable to detect what we would consider as intermediate stages between normal and defective chloroplasts. Rather, we are left with the impression of an all-or-none situation: the organelle is either entirely defective, or entirely normal in appearance. In living cells, defective chloroplasts appear to be not much more than equivalent to the mobile phase of wild-type chloroplasts, the defective chloroplasts displaying the same type of continual ameboid changes in shape as well as interactions with the mitochondria. In addition, the frequent presence of starch grains in defective chloroplasts implies that the enzymes of the Calvin cycle resulting in carbohydrate synthesis could be present in the defective chloroplasts.

Analysis of isolated defective chloroplasts showed the presence of a DNA in amounts approximately equivalent to those found in isolated wild-type chloroplasts. DNA polymerase activity could also be detected. Defective chloroplasts were also found to contain 70 S chloroplast ribosomes in amounts about equal to the 'free' 70 S ribosomes contained in the mobile phase of wild-type chloroplasts. Fraction 1 protein was also present, but was reduced to about 25% of the amount found in mature, wild-type chloroplasts although the protein had the same specific RuDP carboxylase activity as fraction 1 protein from wild-type chloroplasts. The presence of starch together with RuDP carboxylase is an indication that most, if not all of the other enzymes required for operation of the Calvin cycle may be found in defective chloroplasts. In contrast to normalcy in amounts of DNA, 70 S ribosomes, and catalytic activity of fraction 1 protein, RNA polymerase activity was almost undetectable in defective chloroplasts although a nuclear RNA polymerase was as active in isolated nuclei obtained from cells containing a great preponderance of defective chloroplasts as in nuclei obtained from wild-type tobacco leaves. Evidently, the factor responsible for the low activity is specific to defective chloroplasts. Experimental evidence was also obtained to suggest that something had occurred which impaired the catalytic properties of the RNA polymerase.

A plausible interpretation of the findings from the cytoplasmic mutant is that a mutation occurred in chloroplast DNA and this resulted in formation of defective chloroplasts. Lacking photosynthetic ability, the defective chloroplasts survive what would otherwise appear to be a lethal mutation when they are in the presence of a sufficient number of wild-type chloroplasts. In fact, some seedlings of the mutant cannot grow beyond the earliest stages of germination; they contain too few normal chloroplasts to satisfy the photosynthetic needs of the plant to grow. The presence of active DNA polymerase in defective chloroplasts is a sign that mutated chloroplast DNA and defective chloroplasts could be perpetuated by replication of the DNA. Possibly, the mutation could have amounted to little more than production of a defective RNA polymerase. Repression of transcription of chloroplast DNA as a mechanism to account for defective chloroplasts seems very unlikely because defective chloroplasts are found together with normal chloroplasts in the same cell and it is hard to imagine how a repressor substance could select some chloroplasts as targets for action and not the others. Rather, it would seem that defective chloroplasts arise because of mutation of the chloroplast DNA, the phenotypic manifestation of the mutation being an incomplete and defective system of thylakoid membranes resulting in an organelle having the appearance of being composed mainly of mobile phase. Recalling that about one-half of the ribosomes in wild-type chloroplasts are firmly bound to thylakoid membranes, we may further speculate that it is the 'bound' ribosomes which perform the function of producing thylakoid proteins after having been messaged from chloroplast DNA via the chloroplast RNA polymerase. In this view, the formation of a mobile phase could still occur in the near absence of thylakoid membranes because information for synthesis of mobile phase macromolecules was provided by nuclear DNA.

Further research on wild-type tobacco chloroplasts has brought forth evidence which is consistent with the conjecture that total synthesis of tobacco chloroplasts requires information from nuclear as well as chloroplast DNAs. Radioactive peptides synthesized by 'bound' ribosomes appear to have a property which is different from peptides synthesized by 'free' ribosomes.

Nature of peptides synthesized by isolated tobacco chloroplasts

After incorporation of radioactive amino acids into hot acid insoluble products by isolated tobacco chloroplasts, about one-half of the incorporated radioactivity is readily removed with mobile phase constituents away from the thylakoid membranes. Most of the mobile phase radioactivity remains associated with the 'free' 70 S monosomes and polysomes as nascent peptides; only a small fraction of the peptides appears as soluble protein. The radioactivity which remains with the thylakoids is not removed by strenuous washing but is solubilized from chlorophyll bearing materials by DOC, which also releases 'bound' ribosomes. Nearly all of the DOC released radioactivity still sediments with the ribosomes apparently as nascent peptides. The nascent radioactive peptides associated with the 'free' mobile phase ribosomes are completely discharged into soluble form (i.e. no longer sedimenting with ribosomes) by puromycin. In contrast, puromycin did not result in such solubilization of the nascent radioactive peptides associated with the ribosomes bound to thylakoid membranes. However, when puromycin was followed by DOC, the detergent not only released the 'bound' ribosomes from the membranes but also solubilized the radioactive peptides so that they no longer sedimented with the released ribosomes. We take these results to mean that puromycin had discharged the radioactive peptides from the 'bound' ribosomes but the discharged product was so hydrophobic in character that it re-bound to membranes and could only then be 'solubilized' by partial disruption of the membranes with DOC [15].

Thus, the suspicion arises that 'free' ribosomes of the mobile phase are concerned with manufacturing peptides which are different from those manufactured by 'bound' ribosomes of the thylakoid membranes and that messages from different sources of DNA could be required for this division of labor. Two messages may also be required for synthesis of fraction 1 protein, this deduction arising out of inspection of the amino acid composition of fraction 1 proteins obtained from three genera of plants belonging to two families.

Peptide structure of fraction 1 proteins

As first shown for fraction 1 protein from spinach leaves by Rutner and Lane [25] and now confirmed and extended to the protein obtained from tobacco [26] and *Beta vulgaris* [27] leaves, fraction 1 proteins can be dissociated into two distinct and separable subunits. From a knowledge of the number of tryptic peptides together with the total amino acid composition, the minimum chemical molecular weight of the larger subunit of tobacco and spinach fraction 1 protein was calculated to be in the range of 20,000 and 10,000 for the smaller subunit, suggesting that native fraction 1 protein is composed of 16 larger and 16 smaller subunits to produce a particle of around 500,000 molecular weight [28]. Amino acid analysis of the two subunits as they are obtained from fraction 1 proteins from three different genera of plants are shown in table 2 [3]. Tobacco (*Solanaceae*), spinach (*Chenopodiacea*) and *Beta vulgaris (Chenopodiacea)* are widely separated plants in phylogenetic terms. Within the limits of experimental error, no difference in quantity of 16 individual amino acids is found when the compositions of the larger subunits are compared. In striking contrast, nearly all 16 amino acids are different when the compositions of the smaller subunit are compared. Fingerprints of tryptic peptides of the larger and smaller subunits of tobacco and spinach fraction 1 proteins are compared in fig. 4 [28]. With the larger subunit, 19 and 20 distinct tryptic peptides were obtained and only two or three were different in amino acid composition. In the case of the smaller subunit, 7 out of 9 tryptic peptides were different in amino acid composition.

The relatively slight change in primary structure of the larger subunit and the very extensive changes in the smaller subunit encountered in the comparison between fraction 1 proteins isolated from two plants widely separated phylogenetically seems to me to argue for the possibility that two DNA cistrons code for fraction 1 protein. My view is that fraction 1 protein is of very ancient origin and probably first appeared on the evolutionary scene at the time that the photosynthetic process had evolved from bacterial origins into the O_2 evolving system utilizing chlorophyll *a* which is now found in such an enormous variety of photosynthetic organisms ranging from

Table 2
Amino acid composition of subunits obtained from different species of Fraction I protein.

Plant source	Larger fragment				Smaller fragment			
	Spinach	Spinach	Tobacco	Spinach beet	Spinach	Spinach	Spinach	Spinach beet
Dissociation and separation	SDS G-100	SDS G-100		Urea G-200	SDS G-100	SDS G-100		Urea G-200
Phenylalanine	1.00 [a]	1.00	1.00	1.00	1.00	1.00	1.00	1.00
Lysine	1.18	1.04	1.07	1.15	1.21	*1.19*	*1.77*	1.44
Histidine	0.67	0.66	0.62	0.62	0.45	*0.53*	0.42	*0.21*
Arginine	1.44	1.42	1.44	1.26	*0.99*	0.96	0.96	*0.58*
Aspartate	2.18	2.35	2.27	2.00	2.14	2.31	*2.88*	*1.44*
Threonine	1.75	1.63	1.45	1.41	1.19	1.29	*1.60*	*0.77*
Serine	0.81	0.87	0.86	0.93	*0.77*	0.78	*1.51*	0.91
Glutamate	2.20	2.42	2.52	2.30	2.26	2.77	*4.14*	*2.17*
Proline	1.13	1.17	1.13	1.06	1.56	1.56	1.64	1.43
Glycine	2.10	2.43	2.44	2.22	*1.15*	1.28	*2.33*	1.58
Alanine	2.16	2.21	2.23	2.13	*0.86*	0.98	*1.63*	1.09
Valine	–	1.52	1.49	1.57	–	1.15	1.36	1.40
Methionine	0.42	0.35	0.36	0.33	0.46	0.43	0.41	0.30
Isoleucine	0.88	0.69	0.76	0.88	0.56	*0.47*	*0.84*	0.70
Leucine	2.09	2.19	2.16	1.91	1.63	1.81	*2.10*	*1.54*
Tyrosine	0.92	0.91	0.85	0.76	1.57	1.43	*1.92*	*0.99*
Reference	Rutner & Lane [25]	Kawashima [26]		Moon & Thompson [27]	Rutner & Lane [25]	Kawashima [26]		Moon & Thompson [27]

[a] These numbers are calculated as relative molar ratios to phenylalanine. Italicized numbers indicate differences greater than normal error of method.

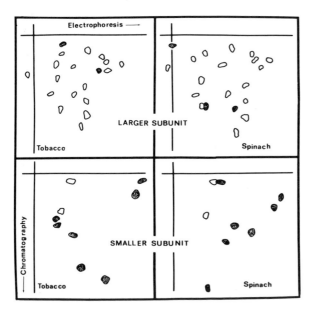

Fig. 4. Comparison of the fingerprints of tryptic peptides obtained from the larger and smaller subunits of Fraction I protein isolated from tobacco and spinach leaves. Peptides which are different indicated in black.

blue-green algae to higher plants. It is clear that RuDP carboxylase of green sulfur and purple non-sulfur photosynthetic bacteria is a much smaller molecule than Fraction I protein [29]. Perhaps much of the history of the origin of Fraction I protein is still preserved in the primary structure of the larger subunit, the coding DNA having suffered a minimum of mutations as more and more photosynthetic organisms evolved and acquired the protein. However, acquisition of the foreign protein also required accommodation by a partial change in its structure to make it more acceptable in an immunological sense. Accommodation occurred by altering the primary structure of the smaller subunit which also came from repeated mutation of its coding DNA. A recent experiment by Kawashima [30] on in vivo synthesis of fraction 1 protein can also be interpreted as indicating that two DNA cistrons are involved in synthesis of the protein.

In vivo synthesis of fraction 1 protein

After tobacco leaves had incorporated $^{14}CO_2$ by photosynthesis, fraction 1 protein was isolated, the two subunits were resolved, and the amino acids composing the two subunits were separated. ^{14}C was found in 8 of the 16 amino acids in each subunit. As shown in table 3 [30], the striking result was that the specific radioactivity of each of the 8 amino acids in the larger subunit was significantly higher than that of the same amino acid obtained from the smaller subunit. The difference in specific activities could be explained by assuming that a pool of non-radioactive smaller subunits was already present when the $^{14}CO_2$ was provided. The pool diluted the radioactivity of this portion of the fraction 1 protein molecule as it was formed from union with the newly-synthesized larger subunits whose radioactivity was not so diluted. Alternatively, one could envision that the two subunits were synthesized on ribosomes which were spatially separated from each other, and had been coded by two different DNA cistrons. I believe that we have now reached a level of knowledge concerning the macromolecular composition of tobacco chloroplasts where it will be possible to test these speculations experimentally and locate the source of DNA coding for various chloroplast proteins.

Locating the source of coding information for chloroplast proteins

The genus *Nicotiana* may prove uncommonly useful for testing whether or not chloroplast DNA contains all of the information for the entire synthesis of the proteins of the chloroplast. A great variety of viable inter-specific crosses can be produced within this genus. The genus consists of 60 well-defined species, most of which can be induced to cross and produce viable seeds for F_1 generation plants [31]. Furthermore, markers are available to differentiate between cytoplasmic and Mendelian inheritance. Therefore, if fraction 1 proteins from different species of *Nicotiana* can be shown to be different by as little as a single amino acid replacement as ascertained by fingerprint analysis, analysis of the primary structure of fraction 1 protein from the F_1 hybrid plants should surely reveal whether the difference is transmitted via Mendelian (nuclear DNA) inheritance, or only as a cytoplasmic inheritable factor as would be the case for chloroplast DNA inheritance. The same genetical approach could also reveal the location of DNA coding for the 'free' and 'bound' ribosomes of the chloroplast, the marker in this instance being any

Table 3

Comparison of the specific radioactivity of aminoacids contained in the larger and smaller subunits of in vivo synthesized tobacco fraction 1 protein. From Kawashima [30].

	Larger fragment			Smaller fragment		
	Amount (μmole)	(cpm)	Sp. act. (cpm/μmole)	Amount (μmole)	(cpm)	Sp. act. (cpm/μmole)
Asp.	1.78	710	399	2.16	600	230
Ser.	0.69	1590	2304	1.43	2040	1420
Gly.	1.98	2992	1511	2.40	2580	1075
Ala.	1.80	2040	1333	1.91	1585	830
Val.	1.03	430	417	1.46	440	301
Met.	0.28	1078	3850	0.54	1163	2154
Tyr.	0.68	1025	1505	2.36	2600	1016
Phe.	1.03	1265	1622	1.06	1355	1075

change in the electrophoretic properties of the 40—50 ribosome subunits which can be resolved by poly-acrylamide gel electrophoresis. I would even be so optimistic as to suppose that it will prove possible to reproducibly fingerprint the proteins of thylakoid membranes and thereby complete the picture of the sources of coding information for most of the macro-molecular structure of the biphasic higher plant chloroplast.

Acknowledgement

The tobacco plant has been the main center of attraction in my laboratory since 1950. What has been learned and recorded in that time is the result of the efforts of an assortment of extremely talented graduate students and post-doctoral associates. My function has been to insist that they use the tobacco plant in their experiments. While disclaiming any responsibility on their part for any unrealistic conjectures presented in this paper, I would nevertheless have had nothing to say without access to the more recent research performed by Drs. Nobumaro Kawashima, K.K. Tewari, Jane Chen, Catherine Liao, William Burton, Atsushi Hirai, Tasani Hongladarom-Honda, Shigeru Honda, R.I.B. Francki, N.K. Board-man, and Donald Spencer. I am also grateful for the financial support received from the U.S. Atomic Energy Commission, U.S. Public Health Service, and National Science Foundation.

References

[1] S.G. Wildman, Proc. XXXth Biology Colloquium, Oregon State Univ. Press (1969), in press.
[2] K.K. Tewari and S.G. Wildman, Symp. Soc. Exptl. Biol. 24 (1970) 147.
[3] N. Kawashima, and S.G. Wildman, Ann. Rev. Plant Physiol. 21 (1970) 325.
[4] S.G. Wildman, in: Biochemistry of Chloroplasts, ed. T.W. Goodwin (Academic Press, London, 1966) vol. 11, p. 295.
[5] S.G. Wildman, T. Hongladarom and S.I. Honda, Science 138 (1962) 434.
[6] T. Hongladarom, S. Honda and S.G. Wildman, Appearance of Organelles in Living Plant Cells. 26 min., natural color, 16 mm cine-photomicrographic film with commentary on sound track. Extension Media Center, University of California, Berkeley, California (1965).
[7] S.I. Honda, T. Hongladarom and S.G. Wildman, in: Primitive Motile Systems in Cell Biology, eds. R.A. Allen and N. Kamiya (Academic Press, New York, 1964) p. 485.
[8] T. Hongladarom, A Study of Living Plant Cells and Their Organelles in Vivo and In Vitro. Thesis, University of California, Los Angeles (1964).
[9] D. Spencer and S.G. Wildman, Australian J. Biol. Sci. 15 (1962) 599.
[10] T. Hongladarom and S.I. Honda, Plant Physiol. 41 (1966) 1686.
[11] S.I. Honda, T. Hongladarom and G. Laties, J. Exp. Bot. 17 (1966) 460.
[12] R.I.B. Francki, N.K. Boardman and S.G. Wildman, Biochemistry 4 (1965) 865.
[13] Jane L. Chen and S.G. Wildman, Science 155 (1967) 1271.
[14] N.K. Boardman, R.I.B. Francki and S.G. Wildman, J. Mol. Biol. 17 (1966) 470.
[15] Jane L. Chen and S.G. Wildman, Biochim. Biophys. Acta 209 (1970) 207.
[16] K.K. Tewari and S.G. Wildman, Science 153 (1966) 1260.
[17] K.K. Tewari and S.G. Wildman, Proc. Natl. Acad. Sci. U.S. 59 (1968) 569.
[18] K.K. Tewari and S.G. Wildman, Proc. Natl. Acad. Sci. U.S. 58 (1967) 689.
[19] D. Spencer and P.R. Whitfield, Biochem. Biophys. Res. Comm. 28 (1967) 538.
[20] K.K. Tewari and S.G. Wildman, Biochim. Biophys. Acta 186 (1969) 358.
[21] A. Hirai and S.G. Wildman, Virology 38 (1969) 73.
[22] J.T.O. Kirk and R.A.E. Tilney-Bassett, in: The Plastids (W.H. Freeman & Co. Ltd., 1967).
[23] Catherine L. Liao, Macromolecular Composition of Defective Chloroplasts From a Cytoplasmic Mutant of Tobacco. Thesis, University of California, Los Angeles (1968).
[24] D. von Wettstein and G. Eriksson, in: Genetics Today, ed. S.J. Geerts (Pergamon Press, Oxford, 1965) vol. III, p. 591.
[25] A.C. Rutner and M.D. Lane, Biochem. Biophys. Res. Commun. 28 (1967) 531.
[26] N. Kawashima, Plant and Cell Physiol. 10 (1969) 31.
[27] K.E. Moon and E.O.P. Thompson, Australian J. Biol. Sci. 22 (1969) 463.
[28] N. Kawashima and S.G. Wildman, Abst. XIth Intl. Bot. Congr. (1969).
[29] L. Anderson, G.B. Price and R.C. Fuller, Science 161 (1968) 482.
[30] N. Kawashima, Biochem. Biophys. Res. Commun. 38 (1970) 119.
[31] T.H. Goodspeed, in: The Genus Nicotiana (Chronica Botanica Press, Waltham, Mass., 1954).
[32] W.G. Burton, Localization and Physical Characterization of Tobacco Mitochondrial DNA. Thesis, University of California, Los Angeles (1969).

SYNTHESIS OF CHLOROPLAST tRNA SPECIES
DURING PLANT SEED EMBRYOGENESIS AND GERMINATION

L.S. DURE III and W.C. MERRICK

Department of Biochemistry, University of Georgia, Athens, Georgia, USA

We have been investigating the levels of isoaccepting tRNA species for several amino acids in the cotyledons of cotton embryos during embryogenesis and germination. A necessary aspect of these studies involved ascertaining if any of the isoaccepting tRNA species for a specific amino acid might be localized in chloroplasts, and, hence, be of chloroplast origin.

We would like to present here our data concerned with the identification of tRNA species for valine and isoleucine that appear to represent chloroplast-affiliated species; and our data concerning the levels of these species in cotyledons during the embryogenesis, germination and greening processes of differentiation.

In brief, our data suggest that chloroplast-specific tRNA species do exist. They exist in low levels in the cotyledons of very young embryos, of dry seeds, and also in the roots of germinating seeds. These species increase markedly in the cotyledons during the first five days of germination, and this increase is observed in etiolated cotyledons to the same extent as in cotyledons which have been allowed to green.

Methods

The general protocol for these studies is as follows: partially purified but unfractionated preparations of mixed aminoacyl-tRNA synthetases and preparations of purified tRNA were made from the cotyledons of (a) cotton embryos that are one-fifth their final weight, (b) dry mature seeds, (c) 5-day dark-germinated seedlings, (d) 5-day germinated seedlings that have been allowed to green by exposure to light for the last 2 days. Similar preparations were obtained from isolated chloroplasts that had been prepared by the non-aqueous method of Stocking [1], and from the roots of 3-day germinated seedlings.

The tRNA from these sources was then acylated in a cell-free reaction mixture with a radioactive amino acid, the acylated tRNA isolated from the reaction mixture, and the number and relative levels of the isoaccepting species of tRNA for the amino acid existing in the preparation determined by a combination of Freon and DEAE chromatography.

Controls

There are a number of ways that spurious information can be collected in comparative studies of this sort of which the investigator may be unaware. We would like to illustrate the criteria that we have used in characterizing our tRNA preparations in an attempt to insure that the chromatographic profiles of isoaccepting tRNA species are indicative of the true number and relative levels that exist within the tissues. Fig. 1 outlines our procedures for characterizing our tRNA preparations.

A major problem is that of being certain that all of the tRNA of the tissue is extracted rather than some non-random sample of tRNA that could very well vary with the age of the tissue. Our extraction procedure employs 1% sodium deoxycholate in the homogenization medium (which we have found to be an absolute requirement for total extraction) followed by phenol extraction and this followed by isolation of the tRNA, 5 S ribosomal RNA and DNA from a DEAE column. These elute from the column with 1 M NaCl after the column is extensively washed with 0.3 M NaCl to elute the majority of the proteins that are not denatured by phenol.

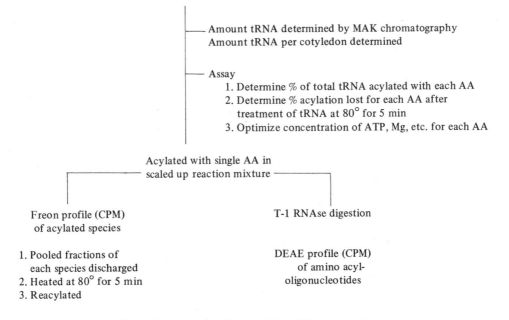

Fig. 1. Procedures for characterizing tRNA preparations.

The amount of tRNA in this fraction is determined by methylated albumin Kieselguhr (MAK) chromatography of a small aliquot which separates the tRNA from the 5 S component and the DNA. This value for tRNA is put on a per cotyledon basis and compared with a maximum value for the tRNA content per cotyledon that we have previously determined from preparations of total nucleic acids obtained by exhaustive extraction procedures employing a number of detergents. Only preparations of tRNA that represent this total extraction were utilized.

Another necessity, of course, is the proof that the tRNA preparation is essentially undamaged i.e. that the bulk of the molecules can be acylated with their specific amino acid. It is possible for tRNA molecules to have undergone RNAse hydrolysis of phosphodiester bonds which are critical for acylation but that do not result in fragmentation of the molecules because of their secondary structure, and hence does not alter their MAK elution profile. The loss of the CpCpA terminus of tRNA molecules is a particular hazard. In order to determine the acylation capability of each preparation, we routinely enzymatically aminoacylate aliquots of the preparation individually with 16 amino acids and total the percentage of the tRNA molecules that have been acylated. We utilize preparations for subsequent experiments in which 70–80% of the tRNA molecules can be acylated with the 16 amino acids.

It is possible to acylate tRNA molecules that have sustained RNAse hydrolysis of phosphodiester bonds which are not critical for enzyme recognition and acylation. These acylated although damaged tRNA molecules frequently behave on Freon columns as individual species of tRNA, and consequently are a common source of artifactual information regarding the number and levels of isoaccepting species. Damaged molecules which still can be aminoacylated can be detected by melting the tRNA (80°C for 5 min) and subsequently running the acylation reaction for the amino acid with the melted and cooled aliquot. Transfer RNA preparations that show a loss of acceptance for a given amino acid after melting and cooling are not used in these studies.

In addition, the concentrations of the components of the acylation reaction are optimized with aliquots of the tRNA and synthetase preparation, and these concentrations utilized in the scaled up reaction mixture used to produce enough acylated tRNA for the subsequent chromatography. The aminoacylated tRNA (along with the unacylated) is purified from the reaction mixture by DEAE batchwise chromatography (0.3–1.0 M NaCl fraction taken) and one half of this material fractionated by each of the two chromatographic procedures.

Freon chromatography

The Freon column [2] separates individual whole tRNA molecules on the basis of their base composition and tertiary structure in response to a salt gradient. It is very sensitive to small differences in base composition or structure, and consequently specific peaks of radioactivity in the Freon column elution profile are thought to represent distinct species of tRNA.

By assaying for tRNA species that do not record a loss of amino acid acceptance after heating and cooling, we reduce the possibility of 'false' species appearing in the elution profiles. However, there is the possibility of RNAse damage to tRNA molecules occurring during the acylation reaction itself which will generate 'false' species just prior to the Freon chromatography.

To determine the intactness of the radioactive tRNA in each elution peak from the Freon column, the individual fractions comprising each peak are pooled and concentrated, the tRNA 'discharged', melted, cooled and reacylated. Peaks that represent RNAse-produced 'false' species do not reacylate after this procedure. The Freon profiles shown today contain no 'false' species as revealed by this procedure.

T-1 RNAse digestion and DEAE chromatography

Although the Freon column is extremely sensitive to small overall differences in tRNA species, not all isoaccepting species for a given amino acid are resolved. For this reason an additional procedure was utilized to determine the number and relative levels of isoaccepting tRNA species in these tissues.

This procedure involves the enzymatic digestion of acylated tRNA at pH 5.5 by Takadiastase T-1 ribonuclease. This enzyme specifically cleaves the phosphodiester linkages of RNA at guanosine residues producing oligonucleotides having guanosine 3'-phosphate at its 3' terminus. Treating tRNA acylated with a radioactive amino acid will produce radioactive amino-acyl-oligonucleotides whose nucleotide number and composition is determined by the location in the chain of the guanosine nearest to the CpCpA-amino acid terminus.

The position of this G may differ in some of the isoaccepting tRNA species and if so, different sized radioactive oligonucleotides will be produced from the different isoaccepting species. The relative position of this G may not differ in isoaccepting species but the nucleotide composition between this G and the CpCpA-amino acid terminus may differ. Both of these types of differences in radioactive oligonucleotides can be resolved by DEAE chromatography of these oligonucleotides at pH 4.5 in the *absence* of urea. This pH is used so as to produce about ½ of a full positive charge on cytosine residues, and a slight positive charge on adenine residues. Urea, which is often utilized in the chromatography of oligonucleotides, is omitted so as to allow maximum non-electrostatic interaction between the nucleoside residues and the column, thus promoting separation of oligonucleotides that differ in nucleotide composition but not in nucleotide number.

Let me emphasize that this T-1 RNAse digestion followed by DEAE chromatography will separate isoaccepting tRNA species that differ in the position of the guanosine residue nearest the CpCpA-amino acid end or in the nucleotide composition between this guanosine and the amino acid terminus but it will *not* distinguish isoaccepting species that differ in nucleotide composition elsewhere in the chain, for the T-1 RNAse digestion will generate the same radioactive fragment from species that differ only in other areas of the polynucleotide chain. Thus there may be more isoaccepting tRNA species for a given amino acid than are revealed by this procedure, but the different species that are demonstrable reflect a concrete difference in nucleotide composition and hence may be more trustworthy than species revealed by Freon chromatography alone.

Results

Fig. 2 represents the elution from a DEAE column of radioactive amino-acyl-oligonucleotides produced by T-1 RNAse digestion of tRNA from the tissues indicated acylated with valine. As can be seen from the first column of profiles, there is very little difference in the number and levels of tRNA^{Val} species obtained from very young embryo cotyledons, dry seed cotyledons or germinating roots. Three major species and several minor species are indicated by this procedure.

However, tRNA obtained from green cotyledons shows a marked increase in the level of one of the minor components. The profile obtained with tRNA from etiolated cotyledons shows the same increase in the level of this species. In tRNA obtained from non-aqueously prepared chloroplasts the species that increases during germination is seen to be the dominant species contained in this tRNA preparation.

If it is assumed that non-aqueously prepared chloroplasts are somewhat contaminated with adhering cytoplasmic components, these profiles suggest that chloroplasts contain a tRNA^{Val} that exists in small amounts in young embryos, in dry seeds and in germinating roots, and that this species increases in cotyledons during the first 5 days of germination relative to the other species. Furthermore, this increase is not mediated by light. (The levels of the presumed chloroplast tRNA^{Val} fragments is accentuated in each profile with stippling.)

Fig. 3 represents the DEAE radioactivity profiles obtained with tRNA^{Ile} prepared from the various tissues. As with tRNA^{Val} the number and levels of

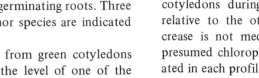

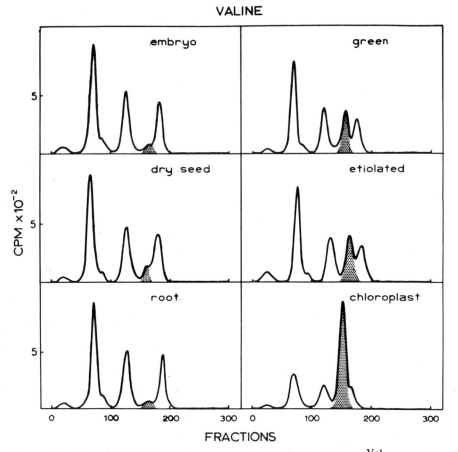

Fig. 2. Radioactivity profiles from DEAE-column chromatography obtained with tRNA^{Val} prepared from chloroplasts and several tissues.

ISOLEUCINE

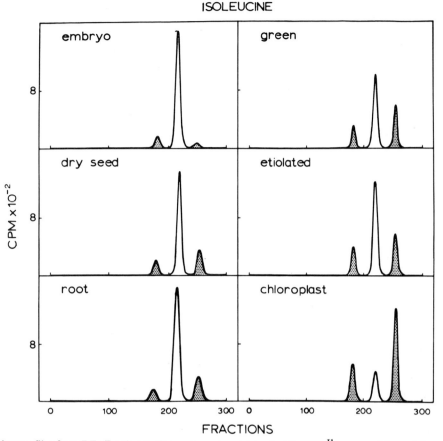

Fig. 3. Radioactivity profiles from DEAE-column chromatography obtained with tRNA[Ile] prepared from chloroplasts and several tissues.

isoaccepting species appear very similar in tRNA prepared from young embryos, dry seeds and germinating roots. However, upon germination two species in the cotyledons increase in amount relative to the third species, and this increase is observed in etiolated cotyledons to the same extent as it is in green cotyledons. The profile of tRNA[Ile] from purified chloroplasts shows these two species to be apparent chloroplast species in this preparation, and shows that the third tRNA[Ile] species is reduced to the level of a cytoplasmic contaminant. These two apparent chloroplastic tRNA[Ile] species are always present in roughly the same relative amounts, one to the other, in all tRNA preparations except that from young embryos.

These DEAE column profiles would seem to indicate that the chloroplastic tRNA species exist in young embryos, dry seeds and roots, and increase in the cotyledons during germination. However, since these profiles are of the aminoacyl-oligonucleotide fragments of tRNA species, there is an alternate interpretation; namely, that new and hitherto non-existant species arise during germination in proplastids that happen to produce the same aminoacyl-oligonucleotide fragments as are produced by pre-existing species found in young embryos, dry seeds and roots. As we shall see, the Freon column elution profiles of intact aminoacyl-tRNA molecules rule out this alternate interpretation.

Fig. 4 represents some of our Freon column elution profiles for tRNA[Val] and tRNA[Ile]. The top two profiles are of tRNA[Val] species from dry seed cotyledons and from green cotyledons. As can be seen, the Freon column does not resolve as many species of tRNA[Val] as are predicted by the DEAE

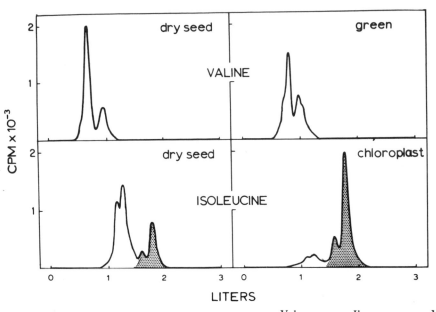

Fig. 4. Radioactivity profiles from Freon columns obtained with tRNA^{Val} and tRNA^{Ile}. Top: tRNA^{Val} from dry seed cotyledons and green cotyledons. Bottom: tRNA^{Ile} from dry seed cotyledons and from purified chloroplasts.

chromatography. The profile obtained with dry seed cotyledon tRNA^{Val} shows only two major components, and this profile is also obtained with young embryos and root tRNA^{Val}. The profile of tRNA^{Val} from green cotyledons, which has the increased amount of the apparent chloroplastic tRNA^{Val} species, does show different levels of the two major components, and the shoulders on these elution peaks indicate a heterogeneous population of species in each peak. Nevertheless very little can be ascertained about tRNA^{Val} species from the Freon column run under our conditions.

The bottom two profiles of tRNA^{Ile} are more informative. The profile of tRNA^{Ile} obtained with dry seed cotyledons is easily equated with the DEAE elution profile of this preparation. The large double peak eluting first from the Freon column represents the same relative amount of the radioactivity that is found in the large single, aminoacyl-oligonucleotide observed in the DEAE profile of this preparation. This would indicate that two cytoplasmic species of tRNA^{Ile} produce the same aminoacyl-oligonucleotide fragment upon T-1 RNAse digestion. Thus they elute as a single species from the DEAE column but are revealed as two species on the Freon column. The remaining two species seen on the Freon profile of

this dry seed cotyledon tRNA preparation are in the proper proportions to represent the two smaller species observed in the DEAE profile, and are considered to be chloroplastic species.

The Freon column profile of tRNA^{Ile} obtained with purified chloroplast tRNA demonstrates that the two apparent chloroplastic species of tRNA^{Ile} seen on this profile and on the DEAE elution profile of this preparation are, indeed, the same two species that are found in the dry seed cotyledons, young embryos and roots. Thus, it would seem that the chloroplast species that increase during germination do pre-exist in young embryos and dry seeds, and occur in germinating roots.

Table 1 gives quantitative data on the relative levels of the tRNA^{Val} and tRNA^{Ile} species, both cytoplasmic and chloroplastic, derived from these elution profiles and others not shown. The top two lines indicate the percentage of the total tRNA^{Val} and tRNA^{Ile} extracted from each of the tissues that may be considered to be chloroplastic tRNA^{Val} and tRNA^{Ile}. The levels of chloroplastic tRNA^{Val} existing in the embryonic and dry seed cotyledons and in roots is difficult to assess from these profiles, but it appears to be between 5 and 10% of the total tRNA^{Val} in these tissues. During germination it

Table 1
Chloroplast tRNA: % of tRNA^{-aa}.

	Embryo cotyledons	Dry seed cotyledons	Root	Green cotyledons	Etiolated cotyledons	Chloroplasts
Valine	5−8	5−10	5−8	20	20	60
Isoleucine	14	28	20	52	48	82
Amino acid acceptance: % of total tRNA						
Valine	8.0 (0.56) [a]	8.0 (0.56) [a]	7.2 (0.50) [a]	7.0 (1.4)	6.3 (1.3)	7.6 (4.56)
Isoleucine	3.5 (0.50)	2.4 (0.67)	2.2 (0.62)	3.2 (1.6)	4.4 (2.2)	4.8 (4.0)
Aspartic acid				5.3		5.2

Numbers in parentheses are values for chloroplast species.
[a] Calculated at 7% chloroplast contribution (see text).

increases until it comprises about 20% of the tRNA^{Val}. It represents only 60% of the tRNA^{Val} in the chloroplast preparation indicating a large contamination with the cytoplasmic species.

Chloroplastic tRNA^{Ile} on the other hand constitutes a larger proportion of the total tRNA^{Ile} in all the tissues and is only 18% contaminated with cytoplasmic tRNA^{Ile} species in the chloroplast preparation.

These figures would seem to indicate a large difference in the amounts of chloroplast tRNA^{Val} and tRNA^{Ile} per plastid in all of the tissues examined. However, when the difference in the relative levels of cytoplasmic tRNA^{Val} and tRNA^{Ile} are taken into consideration, a different picture emerges. The bottom two lines of table 1 give the percentage of the total tRNA of each preparation that accepts valine and isoleucine. These figures show that there is appreciably more tRNA^{Val} than tRNA^{Ile} in all of these tissues. Consequently, when the amount of chloroplast tRNA^{Val} and tRNA^{Ile} is calculated (numbers in parentheses) in each preparation, the relative amounts of chloroplast tRNA for each amino acid appear to be roughly the same. In both cases there appears to be about a three-fold increase in the amount that chloroplast tRNA contributes to the total tRNA during germination.

Valine and isoleucine chloroplastic tRNA species produce aminoacyl-oligonucleotide fragments that are different from those produced by cytoplasmic species for these amino acids. It would seem likely that in the case of some amino acids, both cytoplasmic and chloroplastic tRNA species would produce the same aminoacyl-oligonucleotide fragment. Such apparently is the case for aspartic acid as is shown in fig. 5. This figure gives the DEAE radioactivity elution profile of aspartyl-oligonucleotides produced from aspartyl-tRNA^{Asp} prepared from green cotyledons and from purified chloroplasts. The chloroplast profile has been normalized by making the radioactivity of the small peak correspond to that of the small peak from the green cotyledon profile. A comparison of these two profiles indicates that the tRNA^{Asp} species whose level is presumably increased in the chloroplast preparation produces the same aminoacyl-oligonucleotide fragment as does a cytoplasmic species, and thus cannot be visualized but only inferred from the DEAE column data.

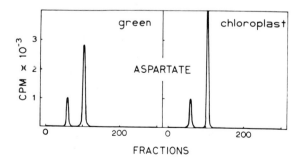

Fig. 5. Radioactivity profiles from DEAE-column chromatography obtained with tRNA^{Asp} prepared from green cotyledons and from purified chloroplasts.

Discussion

Chloroplast tRNA and chloroplast protein synthesis

That there appear to be tRNA species that are localized in chloroplasts should not be surprising. There is an extensive literature concerning the existence of all the cellular machinery for protein synthesis in chloroplasts. The machinery employs ribosomes that are quite distinct from those of the non-organelle cytoplasm both in their chemical and biological properties.

However, this chloroplast protein synthesizing apparatus could conceivably utilize tRNA species that are common to the entire cytoplasm in the translation process. Yet in view of the uniqueness of the chloroplast ribosomes and their apparent origin from the chloroplast DNA, it would not seem unlikely that chloroplast DNA codes for chloroplast specific tRNA as well.

These studies in no way indicate a chloroplast DNA origin for these apparent chloroplast tRNA species but merely demonstrates (1) an increase in the levels of tRNA species that is associated with proplastid development in cotyledons, (2) that these tRNA species appear to be concentrated in chloroplasts, (3) that these tRNA species exist in embryonic and non-green tissue. The demonstration of the specific hybridizability of these tRNA species with chloroplast DNA should constitute proof of their origin from chloroplast DNA.

If we assume that these apparent chloroplast tRNA species do arise from the chloroplast DNA, a number of further questions become of interest: (1) Are chloroplast tRNA species recognized solely by chloroplast specific aminoacyl-tRNA synthetases? (2) Do chloroplast tRNA species participate solely in the synthesis of chloroplast proteins? Do cytoplasmic tRNA species participate also? (3) How much of the chloroplast DNA is taken up with encoding the components of the chloroplast protein synthesizing apparatus? (4) Does the chloroplast protein synthesizing system utilize the entire genetic code?

If the 'wobble' possibilities in the recognition of codon by anticodon as described by Crick [3] are utilized to their theoretical maximum, there should be a minimum of 31 chloroplast tRNA species if all code words are utilized (not including initiation and termination specific tRNA species). There would be a necessity for at least two tRNA^{Val} species if all four valine code words are used. The fact that we find only one apparent chloroplast tRNA^{Val} species in the DEAE profile does not preclude the possibility that the DEAE valyl-oligonucleotide that appears to be chloroplastic is generated from two chloroplast species that produce identical valyl-oligonucleotides upon T-1 RNAse digestion.

Isoleucine on the other hand, having three code words, could be incorporated into protein by means of only one tRNA species, should this species have UAI as its anticodon, i.e., make use of the maximum 'wobble' possibility. Yet on both DEAE and Freon chromatography we find evidence for two chloroplast tRNA^{Ile} species. This may indicate that not all of the 'wobble' possibilities are utilized by these chloroplasts. On the other hand, both species could read all three isoleucine code words and have differences in base sequences in trivial areas of the polynucleotide chain. These differences could have arisen as neutral mutations. The ratio of the levels of the two chloroplast tRNA^{Ile} species appears to be 2:1 in most all of the tissues studied. Since it is hard to imagine identical point mutations in several cistrons, it may be that further study of the isoleucine chloroplast species will reveal that there are three cistrons for tRNA^{Ile} on the chloroplast DNA, one of which has experienced a mutation.

Chloroplast tRNA and chloroplast biogenesis

The existence of chloroplast tRNA species in very young embryos is demonstrated by these data. The levels of these species relative to cytoplasmic species do not appear to increase during the subsequent embryonic growth of the cotyledons, and hence the dry seed cotyledons show roughly the same relative levels of chloroplast tRNA to cytoplasmic tRNA. It would be of interest to know if chloroplastic tRNA exists in even younger embryos. It would seem a priori that the chloroplast protein synthesizing system must always be totally in existence in order to synthesize more of itself. Otherwise, its protein components would initially have to be synthesized by the cytoplasmic system at some point in the life cycle.

The chloroplastic tRNA species increase in amount relative to cytoplasmic species in cotyledons during the first 5 days of germination, and this increase is not in response to a light activated stimulus. This increase is coincident with the development of pro-plastids which have been shown in another species to contain an independent protein synthesizing system [4] . Since there is no further cell division in cotton cotyledons after late embryogenesis, this increase in chloroplastic tRNA represents an increase per cell. It would be interesting to know if this represents an increase in the number of chloroplasts per cell or an increase in tRNA molecules per chloroplast. Chloro-plast number per cell has been shown to increase when etiolated tobacco leaves are exposed to light [5] , but proplastid development during early germi-nation may represent different phenomenon.

If the increase in chloroplast tRNA during early germination solely reflects an increase in the proplas-tid number per cell, perhaps it will be shown that plastids have a fixed amount of tRNA at any stage of development much like bacteria.

Of related interest would be some knowledge of the life time of chloroplastic tRNA species and other components of the chloroplastic protein synthesizing system. Is one set of components provided for each chloroplast which must be utilized throughout its life time or are these components 'turning over'?

The finding of chloroplastic tRNA in root tissue should not be surprising to those of us familiar with the recent book *The Plastids* [6] . The evidence supporting a common origin of amyloplasts and chloroplasts has been elegantly collated in the last chapter. It is not too unreasonable to visualize amylo-plasts in roots as chloroplasts whose cistrons for producing the light-harnessing apparatus have been repressed or in some cases lost. The starch bio-synthetic enzyme complement may prove to be the same proteins in each type of plastid. Our data suggests that a protein synthesizing system similar to that found in chloroplasts should be found someday in the amyloplasts of roots.

References

[1] C.R. Stocking, Plant Physiol. 34 (1959) 56.
[2] J.F. Weiss and A.D. Kelmers, Biochemistry 6 (1967) 2507.
[3] F.H.C. Crick, J. Mol. Biol. 19 (1966) 548.
[4] M.M. Margulies and C. Brubaker, Plant Physiol., in press.
[5] R. Boasson and W.M. Laetsch, Science 166 (1969) 749.
[6] J.T. Kirk and A.E. Tilney-Bassett, in: The Plastids (W.H. Freeman, London, 1967) pp. 572–583.

DETERMINATION OF THE SITES OF SYNTHESIS OF PROTEINS AND LIPIDS OF THE CHLOROPLAST USING CHLORAMPHENICOL AND CYCLOHEXIMIDE

Robert M. SMILLIE, D.G. BISHOP, G.C. GIBBONS, D. GRAHAM,
A.M. GRIEVE, J.K. RAISON and B.J. REGER *

*Plant Physiology Unit, CSIRO, Division of Food Preservation, Ryde
and School of Biological Sciences, University of Sydney, Australia*

In 1963, Smillie et al. [1] first reported that the antibiotic chloramphenicol inhibited chloroplast development. Since, at least in *Euglena gracilis,* the development of chloroplasts can be inhibited by chloramphenicol without any apparent inhibition of cytoplasmic protein synthesis, chloramphenicol has proved to be a useful antibiotic for investigating whether individual proteins of the chloroplast are synthesized on 70 S ribosomes of the chloroplast, or alternatively, on 80 S ribosomes of the cytoplasm. In this paper evidence will be presented that chloramphenicol can inhibit specifically protein synthesis in chloroplasts. In *E. gracilis* this inhibition can be obtained without inhibiting the synthesis of non-chloroplast proteins. Using this experimental system, a study has been made of the production of various lipids and proteins of the chloroplast in cells treated with chloramphenicol.

The fact that the synthesis of a particular chloroplast protein is inhibited by chloramphenicol is not sufficient evidence in itself to establish that the protein is made on the chloroplast protein-synthesizing system, since inhibition of this system could possibly result, through feed-back or other mechanisms, in repression of the synthesis of chloroplast proteins in the cytoplasm, even though synthesis of the normal cytoplasmic constituents is not affected. For this reason the action of chloramphenicol on chloroplast development has been compared with that of cycloheximide, an inhibitor of cytoplasmic protein synthesis in higher organisms.

* Graduate School of Biomedical Sciences, Oak Ridge National Laboratory, Oak Ridge, Tennessee.

Specificity of chloramphenicol inhibition of protein synthesis in photosynthetic eucaryotes

The use of chloramphenicol for determining the cellular sites of synthesis of chloroplast proteins is based on the premise that experimental conditions can be found in which chloramphenicol specifically inhibits protein synthesis within the chloroplast, but not elsewhere in the cell, including the mitochondria. We shall summarize evidence showing that (1) chloramphenicol inhibits chloroplast development by interfering with the protein-synthesizing system of chloroplasts and (2) under the conditions that we use for studying chloroplast development in *Euglena gracilis,* strain Z, chloramphenicol does not inhibit protein synthesis outside of the chloroplast nor the cytoplasmic energy-producing processes of the cell.

Inhibition of photoautotrophic growth

In general, the growth of eucaryotic organisms is not very sensitive to chloramphenicol. Algae grown photoautotrophically constitute a notable exception. Chloramphenicol inhibits growth or chlorophyll synthesis in several strains of *Chlorella* [2–6], *Polyhedriella helvetica* [6], *Chlamydomonas reinhardi* [6,7], *Scenedesmus quadricauda* [8], *Ankistrodesmus braunii* [4], *Coelastrum proboscideum* [4] and *Euglena gracilis* var. *bacillaris* [9] and the Z strain [10–12]. Concentrations of 10 to 100 times that necessary to inhibit the growth of bacteria are usually required. Thus there is ample evidence of inhibition by chloramphenicol of the growth of eucaryotic organisms where the growth is dependent upon photosynthesis.

Comparative effects of chloramphenicol on photo-autotrophic and heterotrophic growth of E. gracilis

A variety of carbon sources, including glucose, ethanol and acetate, support the growth of *E. gracilis*. Chloramphenicol at 0.5 mg/ml severely inhibits the growth of cells cultivated photoautotrophically, but heterotrophic growth, either in light or in darkness, is not significantly inhibited [1]. The most obvious explanation of this result is that chloramphenicol inhibits the synthesis of chloroplast protein and consequently, the cells are unable to produce functional chloroplasts. The synthesis of other cellular proteins is not inhibited and if a carbon source whose utilization is not dependent on chloroplast metabolism is provided, then growth is not prevented by chloramphenicol. However, there are other possible

explanations. Chloramphenicol might directly inhibit photosynthesis, but, as is shown below, this does not appear to be the case. We have, however, observed that at high light intensities (e.g., 800–2000 ft–candles), chloramphenicol has a toxic effect on green cells of *E. gracilis* and also on detached green leaves of bean which results in rapid loss of chlorophyll and death of the cells. The reason for this toxicity is not known, but it is possible that under conditions favouring high rates of photosynthesis the mitochondrial system becomes sensitive to chloramphenicol. All experiments on chloroplast development reported below have been carried out at low light intensities (100–150 ft–candles) where this toxic effect of chloramphenicol has not been observed.

While chloramphenicol does not inhibit hetero-

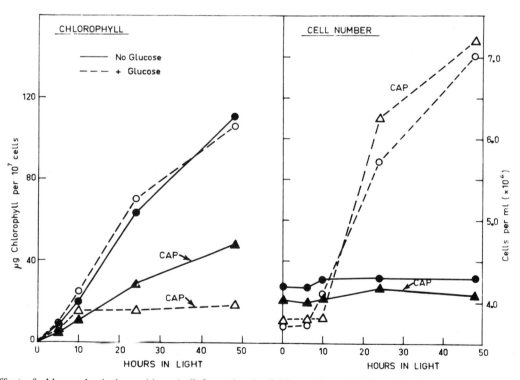

Fig. 1. Effect of chloramphenicol on chlorophyll formation in dividing and non-dividing cells of *E. gracilis*. The experimental conditions for studying chloroplast development in *E. gracilis*, strain Z, were as follows. Cells were cultivated in the dark in Hutner's organic growth medium [13] until the cell density had reached $2-4 \times 10^6$ cells/ml. The cells were harvested by centrifugation and washed twice in Hutner's medium from which all carbon compounds, other than thiamine and vitamin B_{12}, and EDTA, had been omitted. Cells were suspended in this medium to $3-4 \times 10^6$ cells/ml and shaken in the dark for 1–2 days. Transfers were carried out using a green safe-light [14]. Chloroplast development was initiated by illuminating the culture with continuous light (white fluorescent tubes, 120–150 ft-candles). At the cell densities and light intensity used there was no cell division during a 3-day greening period. To study chloroplast development in dividing cells, glucose (22.5 g/l) was added just prior to illuminating the cells. Chloramphenicol was also added immediately before commencing the illumination. ○,●, control cells; △,▲, chloramphenicol (CAP) 1.0 mg/ml; solid lines, no glucose; broken lines, plus glucose.

trophic growth of *E. gracilis* in the light, it does inhibit chlorophyll synthesis. The formation of chlorophyll in *E. gracilis* can be conveniently studied by illuminating a culture of dark-adapted cells which contain no measurable amounts of chlorophylls. If the cells are transferred to an inorganic medium and kept in the dark for 24 hr or longer before being illuminated, chlorophyll formation can be studied in non-dividing cells. If glucose is added at the beginning of illumination, both chlorophyll formation and cell division will commence after a lag of 6 to 12 hr. Fig. 1 shows that chloramphenicol at 1 mg/ml inhibits chlorophyll formation in a non-dividing culture and even more severely in a dividing culture. Cell division is not inhibited.

The inhibition of chlorophyll synthesis by chloramphenicol varies with the light intensity. Maximum inhibition occurs at 80–120 ft-candles and at higher intensities inhibition progressively decreases.

It is well known that the chloroplasts of *E. gracilis* are not essential for the growth of the cells and it is relatively easy to obtain mutants which have lost both chloroplast DNA and the capacity to produce chloroplasts. In many other algae, and especially those which do not normally become etiolated when grown in the dark, the role of the chloroplast may not be confined to photosynthesis. If in these organisms some of the plastid enzymes were to function in heterotrophic metabolism, even in the dark, then it would not be unexpected to find inhibition of growth by chloramphenicol whether the cells are grown as heterotrophs or photoautotrophs.

Comparison of the action of chloramphenicol and cycloheximide on chlorophyll formation and growth in procaryotic and eucaryotic cells

In order to contrast the effects of chloramphenicol on chlorophyll synthesis and growth in *E. gracilis* and its action in bacterial cells, we have measured growth and bacteriochlorophyll formation by the photosynthetic bacterium, *Rhodospirillum rubrum* in the presence of various concentrations of chloramphenicol. The results are summarized in table 1. Two differences are apparent in the comparison between *R. rubrum* and *E. gracilis:* (1) chlorophyll synthesis in *E. gracilis* is much less sensitive to chloramphenicol than is the synthesis of bacteriochlorophyll in *R. rubrum* and (2) chloramphenicol inhibits both pigment syn-

Table 1

Effect of chloramphenicol and cycloheximide on pigment formation and growth of *Rhodospirillum rubrum* and *Euglena gracilis*.

R. rubrum			
		Bacterio-chlorophyll	Growth
Chloramphenicol	0.25–1.0 µg/ml	+	+
Cycloheximide	500 µg/ml	–	–

E. gracilis			
		Chlorophyll	Growth (on glucose)
Chloramphenicol	20–1000 µg/ml	+	–
Cycloheximide	2.0 µg/ml	–	+

+ inhibits, – no inhibition.

thesis and growth in the case of *R. rubrum*, but only the former in the case of *E. gracilis*. From experiments with cell-free systems (see below), the different sensitivities of the chloroplast and bacterial systems to chloramphenicol probably reflect differences in the properties of their respective ribosomal systems rather than a differential uptake of the antibiotic.

Cycloheximide has a very different effect than chloramphenicol. Concentrations as high as 500 µg/ml affect neither pigment formation nor the growth of *R. rubrum*. In contrast, cell division of *E. gracilis* is prevented by cycloheximide at 1–2 µg/ml. This concentration does not inhibit the rate of chlorophyll synthesis except for an initial lag of a few hours. Higher concentrations (10–20 µg/ml), however, are inhibitory.

Lack of inhibition of respiration and photosynthesis by chloramphenicol

High concentrations of chloramphenicol have been reported to inhibit oxidative phosphorylation by isolated plant mitochondria [15] and respiratory processes in mammalian cells [16–18]. Concentrations of chloramphenicol which are inhibitory to chloroplast development in *E. gracilis* grown heterotrophically apparently do not affect respiration adversely since the growth rates are not depressed (fig. 1). On the other hand, anaerobiosis or respirato-

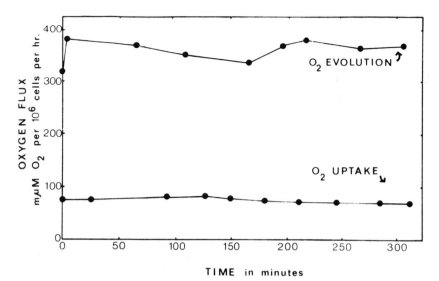

Fig. 2. Photosynthesis and respiration of E. gracilis in the presence of chloramphenicol. Autotrophic cells were placed in an oxygen electrode chamber and oxygen evolution or uptake were measured in three light/dark cycles. The mean values of these measurements represent the zero time values on the graph. Chloramphenicol at 1 mg/ml was added at zero time and further measurements of oxygen evolution or uptake were made in light/dark cycles. Oxygen evolution was measured during illumination with white light (250 ft-candles) and O_2 uptake was measured in the dark.

ry poisons severely inhibit chloroplast development [1]. Fig. 2 shows that in autotrophic cells neither photosynthesis nor respiration is inhibited by chloramphenicol at 1 mg/ml. In other experiments, chloramphenicol failed to affect the rate of photo-oxidation of cytochrome-552 in autotrophic cells or the photo-reduction of NADP by chloroplasts isolated from E. gracilis.

Inhibition by chloramphenicol of amino acid incorporation by isolated chloroplasts and lack of inhibition by cycloheximide

The incorporation of amino acid into protein by chloroplasts isolated from a wide range of organisms is sensitive to chloramphenicol (table 2). Cytoplasmic ribosomes from E. gracilis [27] or tobacco leaves [22] are comparatively insensitive. Fig. 3 shows the inhibition by chloramphenicol of amino acid incorporation into protein by plastids isolated from E. gracilis and from leaves of wheat and pea plants. The activities of chloroplasts and etioplasts are affected similarly, 50% inhibition occurring at about 5 μg/ml. Thus while protein synthesis by isolated chloroplasts is inhibited by chloramphenicol it is less sensitive to this antibiotic than is protein synthesis by bacterial cell-free systems.

Cycloheximide, on the other hand, has little effect on amino acid incorporation by isolated chloroplasts (fig. 3). When cycloheximide and chloramphenicol were added together, inhibition was the same as with chloramphenicol alone.

Table 2
Inhibition by chloramphenicol of amino acid incorporation by isolated chloroplasts.

	Chloramphenicol (μg/ml)	% inhibition	Reference
Pea leaves	235	89	[19]
Spinach leaves	20	58	[20]
	200	78	
Tobacco leaves	200	70	[21]
	300	77	[22]
Tomato leaves and cotyledons	1540	38	[23]
Wheat leaves	40	77	[24]
Acetabularia mediterranea	5	50	[25]
E. gracilis	100	89	[26]
	210	45	[27]

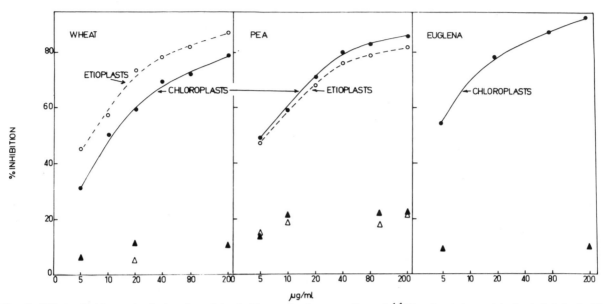

Fig. 3. Effect of chloramphenicol and cycloheximide on the incorporation of ^{14}C-leucine into protein by isolated plastids. Plastids were isolated from young etiolated or green leaves of pea and wheat seedlings and from autotrophic cells of *E. gracilis*. The isolation medium consisted of 10 mM HEPES buffer pH 8.0, 0.33 mM sucrose, 1 mM dithiothreitol, 15 mM $MgCl_2$, 100 mM KCl and 0.2% (w/v) bovine serum albumin. Amino acid incorporation was determined using the filter paper disc method of Mans and Novelli [28,29]. Triangles represent treatment with cycloheximide, circles represent treatment with chloramphenicol. Open symbols, etioplasts; closed symbols, chloroplasts.

Binding of chloramphenicol by ribosomes from chloroplasts

Vasquez [30] has shown that bacterial ribosomes can bind chloramphenicol, whereas ribosomes from the cytoplasm of higher organisms do not. Anderson

Table 3
Binding of ^{14}C-chloramphenicol by ribosomes.

Source	Chloramphenicol bound (mµg/mg RNA)
Escherichia coli	88
Anabaena cylindrica	55
E. gracilis	
Chloroplasts	13
Cytoplasm	6
Pea leaf	
Chloroplasts	22
Cytoplasm	7
Wheat leaf	
Chloroplasts	28
Cytoplasm	11

Data from Anderson and Smillie [31].

and Smillie [31] showed that chloroplast ribosomes also bind chloramphenicol, although not as strongly as ribosomes from bacteria or blue-green algae (table 3).

Specific inhibition of chloroplast development by the D-threo isomer of chloramphenicol

Besides the naturally occurring D-*threo* isomer, three other stereoisomers of chloramphenicol, the L-*threo*, D-*erythro* and L-*erythro* isomers, are known.

Inhibition of amino acid incorporation by bacterial ribosomes is specific for the D-*threo* isomer, the L-*threo* and L-*erythro* isomers showing about 10% of the activity of D-*threo* chloramphenicol [32]. Similarly, chloroplast development is inhibited specifically by the D-*threo* isomer, whereas other processes in plants which are inhibited by chloramphenicol such as salt uptake, root elongation and oxidative phosphorylation are inhibited by more than one isomer (table 4). Fig. 4 shows that chloroplast development in *E. gracilis* is inhibited much more strongly by the D-*threo* isomer than by the corresponding L-isomer.

Table 4
Inhibition of various processes in bacteria and plants by stereoisomers of chloramphenicol.

Process	Organism	Inhibition by isomer				Reference
		D-*threo*	L-*threo*	D-*erythro*	L-*erythro*	
Protein synthesis	Bacteria	+	–		–	[32]
Development of photosynthetic capacity	Bean leaves	+	–			[33]
Increase in chlorophyll and protein	Bean leaves	+	–			[33]
Amino acid incorporation by isolated chloroplasts	Tobacco leaves	+	–	–		[22]
Salt uptake	Pea, beetroot, carrot	+	+			[34]
Root elongation	Lupin	+	+			[35]
Oxidative phosphorylation by isolated mitochondria	Corn shoots	+	+			[36]

+ inhibition; – no effect.

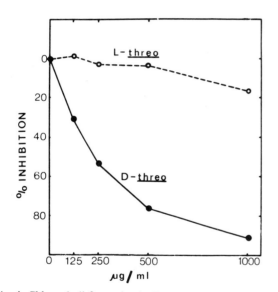

Fig. 4. Chlorophyll formation in *E. gracilis* in the presence of D-*threo* chloramphenicol and L-*threo* chloramphenicol. Experimental conditions (non-dividing cells) are described in fig. 1. Chlorophyll values were determined after the cells had been illuminated for 69 hr.

Inhibition of chlorophyll synthesis by cycloheximide

Although cycloheximide is an inhibitor of cytoplasmic protein synthesis, it also inhibits the synthesis of chlorophyll in *E. gracilis* [37,38]. The kinetics of chlorophyll synthesis in the presence of various con-

centrations of cycloheximide are shown in fig. 5. The addition of cycloheximide at a concentration of 2.5 μg/ml results in an initial lag in chlorophyll synthesis. This lag is followed by a period in which the rate of synthesis far exceeds that of untreated cells. The lag period is extended by doubling the concentration of cycloheximide and when the concentration of inhibitor was further increased to 10 μg/ml, there was almost complete inhibition of chlorophyll synthesis during 72 hr of illumination. Cell division is inhibited by all of the concentrations used. Inhibition of both chlorophyll synthesis and cell division is reversible.

The effect of chloramphenicol and cycloheximide on the synthesis of chloroplast lipids

Fig. 6 compares the effects of chloramphenicol and cycloheximide on the production of chlorophyll and galactolipids during chloroplast development in *E. gracilis*. Small amounts of galactolipid are found in the dark-grown cell. Their intracellular location has not been determined accurately, but it may be the proplastid. Exposure of the dark-grown cells to light for 48 hr results in a 4-fold increase in lipid-bound sugar. As the accumulation of chlorophyll is progressively inhibited by increasing amounts of chloramphenicol, an inhibition of galactolipid accumulation also takes place, although the degree of inhibition of galactolipid accumulation is not as great as that of

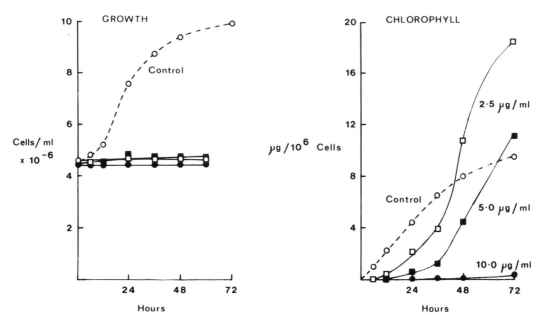

Fig. 5. Effect of cycloheximide on cell division and chlorophyll synthesis. The experimental conditions (dividing cells) are described in fig. 1. Cycloheximide concentrations: ○, none; □, 2.5 μg/ml; ■, 5.0 μg/ml; ●, 10.0 μg/ml.

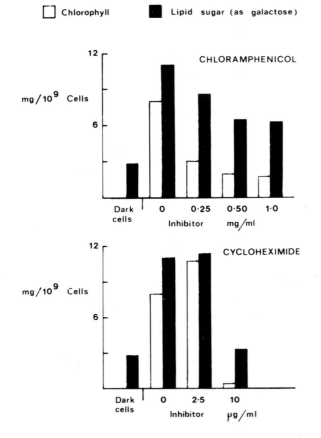

chlorophyll. At the lower concentration of cycloheximide shown in fig. 6 (2.5 μg/ml), the amount of chlorophyll and galactolipid per cell are approximately equal to that of the control. At this stage of growth (48 hr illumination) the initial lag period of chlorophyll synthesis in the presence of 2.5 μg/ml of cycloheximide has been overcome (fig. 5). In contrast, galactolipid accumulation is markedly inhibited by the higher concentration of cycloheximide which inhibits chlorophyll production (10 μg/ml) [39].

The production of galactolipids by *E. gracilis* in the presence of cycloheximide exhibits a lag phase similar to that shown by chlorophyll (fig. 7). Cycloheximide also causes a change in the molar ratio of

Fig. 6. The effect of chloramphenicol and cycloheximide on the production of chlorophyll and galactolipids during chloroplast development in *E. gracilis*. Lipid was extracted with chloroform:methanol (2:1) from cells exposed to light for 48 hr under the standard greening conditions (fig. 1). Lipid-bound sugar was estimated on a sample of the supernatant fluid obtained after hydrolysis of a lipid sample by 2N H_2SO_4 in the presence of silica gel HR (Merck) [40]. □, Chlorophyll; ■, lipid sugar (as galactose).

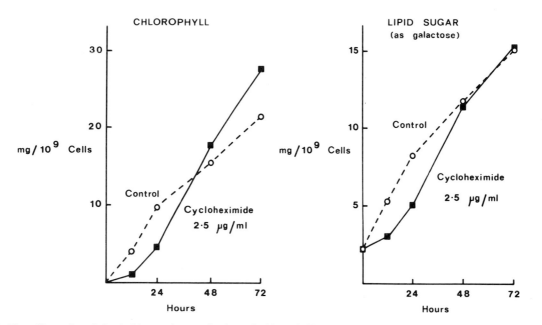

Fig. 7. The effect of cycloheximide on the production of chlorophyll and galactolipids during chloroplast development in *E. gracilis.* For experimental procedures see fig. 6.

monogalactosyl diacylglycerol to digalactosyl diacylglycerol in the cell. This ratio in dark-grown cells is about 4.3 and on illumination of the cells, the ratio decreases within 12 hr to about 2.5, remaining constant thereafter until at least 72 hr after the beginning of illumination. In cycloheximide-treated cells (2.5 μg/ml), the ratio changes within the first 12 hr of illumination to 3.5 and thereafter remains constant. In cells treated with chloramphenicol the ratio falls to the same level as that in the control [39]. These changes in the molar ratio of monogalactosyl diacylglycerol to digalactosyl diacylglycerol result in the apparent conservation of galactose in cycloheximide-treated cells, even though the total number of moles of galactolipid synthesized in cycloheximide-treated cells is greater than that in control cells. The availability of galactose or a closely related compound may be restricted in cycloheximide-treated cells, perhaps due to inhibition of its synthesis in the cytoplasm, and the chloroplast responds by synthesizing less digalactosyl diacylglycerol.

The fatty acid composition of the galactolipids produced by cells grown in the presence of cycloheximide differs markedly from that of control cells. Illumination of control cells results in a marked reduction of the chain length of the constituent fatty acids of monogalactosyl diacylglycerol. The monogalactosyl diacylglycerol of cells grown in the dark contains fatty acids, over 40% of which contain 20 or 22 carbon atoms. After illumination for 72 hr, this amount is reduced to 6%. A similar reduction in the chain length of the constituent fatty acids appears in the monogalactosyl diacylglycerol of cycloheximide-treated cells. These reductions in the proportions of fatty acids containing 20 or 22 carbon atoms is accompanied by an increase in the fatty acids containing 16 or 18 carbon atoms. However, the fatty acid composition of the monogalactosyl diacylglycerol of cycloheximide-treated cells also shows a significant increase in the content of polyunsaturated fatty acids (fig. 8). The amounts of dienoic and trienoic fatty acids containing 16 or 18 carbon atoms after 72 hr of illumination are substantially greater than in the monogalactosyl diacylglycerol of control cells, in which monoenoic acids are the major components. Similar changes are found in the fatty acid composition of the digalactosyl diacylglycerol [39]. The significance of these changes in galactolipid fatty acids is not yet apparent.

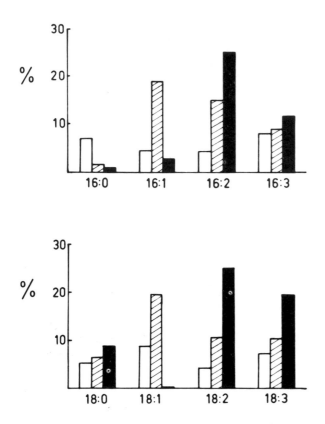

Fig. 8. The effect of cycloheximide on the fatty acid composition of monogalactosyl diacylglycerol during chloroplast development in *E. gracilis*. Lipid was extracted from whole cells with chloroform:methanol (2:1). The monogalactosyl diglyceride fraction was isolated from the total lipid by column chromatography on acid-washed Florisil [41]. Fatty acid methyl esters were prepared by transesterification with boron trifluoride-methanol and analyzed by gas-liquid chromatography. □ Dark-grown cells. □ Control cells after exposure to light for 72 hr. ▪ Cells exposed its light for 72 hr in the presence of cycloheximide (2.5 µg/ml).

The effect of chloramphenicol and cycloheximide on the synthesis of chloroplast proteins

Fraction I protein

During chloroplast development in *E. gracilis* there is an increase in the activity of ribulose-1,5-diphosphate carboxylase and the amount of Fraction I protein [38]. Our previous studies have demonstrated that these increases are prevented by chlorampheni-

col, in both non-dividing and dividing cells [38,42, 43]. While this result is consistent with the site of synthesis of Fraction I protein being the 70 S ribosome of the chloroplast, it does not provide conclusive proof since inhibition of the synthesis of some chloroplast proteins within the chloroplast could conceivably lead to repression of synthesis of other chloroplast proteins in the cytoplasm. If it can also be shown that the synthesis of Fraction I protein continues when cytoplasmic protein synthesis is inhibited, then this, taken in conjunction with the results of the experiments with chloramphenicol, would provide evidence for the in vivo synthesis of Fraction I protein on chloroplast ribosomes.

In the presence of concentrations of cycloheximide as high as 15 µg/ml, ribulose-1,5-diphosphate carboxylase and Fraction I protein continues to increase in illuminated cells of *E. gracilis* [38,42]. This concentration of cycloheximide is 15–30 times higher than is necessary for inhibition of cell division, and cytoplasmic protein synthesis is almost certainly inhibited. Similar or lower concentrations of cycloheximide inhibit the incorporation of amino acids into the cells [44], the uptake of glucose from the medium [42] and the ethanol-induced synthesis of isocitrate lyase (unpublished experiments). The synthesis of chlorophyll (fig. 5) and chloroplast cytochromes is also inhibited [38].

As was shown in fig. 5, lower concentrations of cycloheximide (1–5 µg/ml) produce only an initial lag in chlorophyll synthesis and this is followed by a period in which the rate of synthesis is higher than in untreated cells. The synthesis of Fraction I protein is also stimulated. Fig. 9 contrasts the effects of cycloheximide at 3 µg/ml with that of chloramphenicol (500 µg/ml) on the light-induced production of Fraction I protein. While inhibition with chloramphenicol was almost total, the rate of synthesis with cycloheximide was 2- to 3-fold higher than the rate in control cells.

At both the higher and lower ranges of cycloheximide concentrations used in these experiments protein synthesis in the cytoplasm is severely curtailed. Evans et al. [44] have shown that cycloheximide at 2 µg/ml inhibits the incorporation of valine into the protein of *E. gracilis* by 92% during the first 12 hr of illumination, and we have obtained similar results. While it seems unlikely that the accelerated

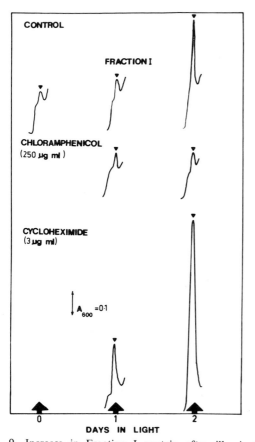

Fig. 9. Increase in Fraction I protein after illuminating dark-grown *E. gracilis*. Experimental conditions are described in fig. 1, except that the light intensity was 60–70 ft-candles. Chloramphenicol was added at 250 μg/ml and cycloheximide at 3 μg/ml. At the times indicated, samples of the cells were broken in a French Pressure Cell and centrifuged at 34,000 g for 30 min and proteins in the supernatant fluid were separated by electrophoresis at pH 8.3 on polyacrylamide gel. After staining with amido black, the gel was scanned using a densitometer. The traces show the region of gel occupied by Fraction I protein. The results are comparable on a per cell basis.

Table 5
The effect of chloramphenicol and cycloheximide on the incorporation of ^{14}C-amino acids into proteins of *E. gracilis*.

Protein	Specific activity		
	Control	Cyclo-heximide	Chlor-amphenicol
Fraction I protein	2.7	9.1	0.2
Cytochrome *f* (cytochrome-552)	1.35	0.13	0.24
Cytochrome *C* (cytochrome-556)	28	0.2	

Autrotrophic cells were grown to a cell density of 2.5×10^6 cells/ml, and chloramphenicol (1.25 mg/ml) or cycloheximide (5 μg/ml) were added. Two hours later, ^{14}C-protein hydrolyzate was added (7.5 μc/l; 0.2 mc/mg) and the culture was shaken in the light for a further 4 hr. Proteins were purified by column chromatography and electrophoresis on polyacrylamide gel and their specific activities were determined. The amount of Fraction I protein was estimated following gel electrophoresis as described in fig. 9. The dye adsorbed to Fraction I protein was extracted and its absorbancy at 600 nm was measured. Specific activity is given as cpm per absorbancy of 100 at 600 nm. Specific activities for the other proteins are expressed as cpm per mg protein. Control cells were not treated with inhibitor.

synthesis of Fraction I protein in the presence of cycloheximide could be due to some residual activity of the cytoplasmic protein-synthesizing system, we have tried to eliminate this possible explanation by comparing the rates of incorporation of amino acids into individual chloroplast and cytoplasmic proteins in the presence and absence of cycloheximide. Cells were fed ^{14}C-amino acids for 4 hr, proteins were extracted and purified and their specific activities

determined. In order to have sufficient quantities of chloroplast proteins for purification, green autotrophic cells were used. Table 5 shows that cycloheximide at 5 μg/ml inhibits the incorporation of label into mitochondrial cytochrome *C*. Presumably, this protein is synthesized on cytoplasmic ribosomes as is the case in several other organisms [45–47]. Synthesis of the chloroplast cytochrome-552 is also inhibited. This is not unexpected since in greening cells, there is an initial inhibition of chlorophyll and chloroplast cytochrome synthesis by cycloheximide. Most significantly, the incorporation of label into Fraction I protein is accelerated while incorporation into cytochrome *C* is inhibited. Chloramphenicol inhibits incorporation into both of the chloroplast proteins. From these studies we conclude that in *E. gracilis*, Fraction I protein and its associated enzyme, ribulose-1,5-diphosphate carboxylase, are synthesized on the 70 S ribosomes of the chloroplast.

Proteins of the electron transfer pathway of chloroplasts

Chloramphenicol inhibits the synthesis of ferredoxin-NADP-reductase and chloroplast c-type and b-type cytochromes [38]. The incorporation of Mn^{++}, an essential component of the chloroplast electron transfer system, is also inhibited by chloramphenicol [1].

The action of cycloheximide on the synthesis of these electron transfer proteins is similar to its effect on chlorophyll synthesis, that is, their synthesis is actually stimulated by low concentrations of cycloheximide after an initial lag (cf. fig. 5). However, in order to ascertain whether the continued synthesis of these proteins is independent of protein synthesis in the cytoplasm it will be important to establish if, under these conditions of accelerated chlorophyll synthesis, cytoplasmic protein synthesis remains inhibited.

Conclusions

Chloramphenicol, which inhibits protein synthesis in bacteria, also inhibits growth of photoautotrophic cells. In those green cells in which growth is not dependent upon plastid metabolism (such as in the case of *E. gracilis* which can be grown as bleached mutants lacking chloroplasts and chloroplast DNA), it is possible to inhibit chloroplast development with chloramphenicol without inhibiting growth. Neither photosynthesis nor respiration is directly inhibited. In contrast, in the photosynthetic bacterium, *R. rubrum,* both pigment synthesis and growth is inhibited by chloramphenicol.

The site of action of chloramphenicol in the photosynthetic eucaryotic cell is the 70 S ribosome of the chloroplast. This was indicated by the inhibition by chloramphenicol of amino acid uptake by isolated chloroplasts, by the binding of chloramphenicol to chloroplast ribosomes and by the specificity of the D-*threo* isomer of chloramphenicol in inhibiting chlorophyll synthesis.

The comparison is frequently drawn between the similarities of the properties of bacterial and chloroplast ribosomes. Both are similar in size, in the size of their RNA components (16 S and 23 S), in their requirement of Mg^{++} ions for stability and in the inhibition of their activity for protein synthesis by certain antibiotics. However, there appears to be a definite difference in the sensitivities of the two ribosomal systems towards chloramphenicol as evidenced by the higher concentration of chloramphenicol required for inhibition of both chloroplast development in intact cells and amino acid incorporation by isolated chloroplasts compared with bacterial protein synthesis, and the weaker binding of chloramphenicol to chloroplast ribosomes compared with bacterial ribosomes.

Cycloheximide, in contrast to chloramphenicol, does not inhibit the synthesis of bacteriochlorophyll or the growth of *R. rubrum.* Nor is amino acid incorporation by isolated chloroplasts inhibited by cycloheximide. Concentrations of 1 μg/ml or more inhibit growth of *E. gracilis.*

The results obtained for Fraction I protein provide a clear-cut illustration of how these two inhibitors can be used to determine the site of synthesis of a chloroplast protein. The net increase in Fraction I protein and the increase in activity of ribulose-1,5-diphosphate carboxylase in greening cells of *E. gracilis* are inhibited by concentrations of chloramphenicol which do not inhibit heterotrophic growth, and are stimulated by concentrations of cycloheximide which inhibit growth. In shorter-term labelling experiments using autotrophic cells, the synthesis of Fraction I protein is similarly inhibited by chloramphenicol and stimulated by cycloheximide, although the cycloheximide inhibits synthesis of both chloroplast cytochrome-552 and mitochondrial cytochrome *C* (cytochrome-556). We conclude that Fraction I protein is synthesized in vivo largely within the chloroplasts.

The production of chlorophyll, galactolipids and photosynthetic electron transfer proteins is inhibited by chloramphenicol. In the case of chlorophyll and galactolipids, synthesis is stimulated by cycloheximide, but only after a lag period whose length varies with the concentration of cycloheximide present. The lag in synthesis of these components of the chloroplast lamellae may be due to a requirement for the synthesis in the cytoplasm of some components of the lamellar membranes as has already been suggested [42]. Alternatively, interference with the regulatory mechanisms for the synthesis of chloroplast lamellar components could have resulted indirectly from the drastic inhibition of cytoplasmic protein synthesis, if it is assumed that metabolites common to both systems are involved in regulating protein synthesis in the chloroplasts and cytoplasm.

While low concentrations of cycloheximide (2–5 μg/ml) can accelerate chloroplast development in *E. gracilis* after the initial lag, the composition of the chloroplast is not the same as that of untreated cells. The molar ratio of monogalactosyl diacylglycerol to digalactosyl diacylglycerol is higher in cycloheximide-treated cells than in untreated cells and the proportion of polyunsaturated fatty acids in the galactolipids of cycloheximide-treated cells is much higher than in untreated cells. What these changes mean in terms of structure and function of the chloroplast lamellae is being investigated.

References

[1] R.M. Smillie, W.R. Evans and H. Lyman, Brookhaven Symp. Biol. 16 (1963) 89.

[2] R.A. Galloway and R.W. Krauss, Amer. J. Bot. 46 (1959) 40.

[3] H. Tamiya, Y. Morimura and M. Yokota, Arch. Mikrobiol. 42 (1962) 4.

[4] F.C. Czygan, Arch. Mikrobiol. 47 (1964) 251.

[5] S. Aokim J.K. Matsubara and E. Hase, Plant Cell Physiol. 6 (1965) 475.

[6] P. Echlin and I. Morris, Biol. Rev. 40 (1965) 143.

[7] G.A. Hudock, G.C. McLeod, J. Moravkova-Kiely and R.P. Levine, Plant Physiol. 39 (1964) 898.

[8] F.J. Taylor, Nature 207 (1965) 783.

[9] Y. Ben-Shaul and Y. Markus, J. Cell Sci. 4 (1969) 627.

[10] A.W. Linnane and P.R. Stewart, Biochem. Biophys. Res. Commun. 27 (1967) 511.

[11] B.G.T. Pogo and A.O. Pogo, J. Protozool. 12 (1965) 96.

[12] S. Aaronson, B.B. Ellenbogen, L.K. Yellen and S.H. Hutner, Biochem. Biophys. Res. Commun. 27 (1967) 535.

[13] S.H. Hutner, M.K. Bach and G.I.M. Ross, J. Protozool. 3 (1956) 101.

[14] R.B. Withrow and L. Price, Plant Physiol. 32 (1957) 244.

[15] J.B. Hanson and T.K. Hodges, Nature 200 (1963) 1009.

[16] W. Godchaux and E. Herbert, J. Mol. Biol. 21 (1966) 537.

[17] K.B. Freeman and D. Haldar, Biochem. Biophys. Res. Commun. 28 (1967) 8.

[18] F.C. Firkin and A.W. Linnane, Biochem. Biophys. Res. Commun. 32 (1968) 398.

[19] N.M. Sissakian, I.I. Filippovitch, E.N. Svetailo and K.A. Aliyev, Biochim. Biophys. Acta 95 (1965) 474.

[20] D. Spencer, Arch. Biochem. Biophys. 111 (1965) 381.

[21] D. Spencer and S.G. Wildman, Biochemistry 3 (1964) 954.

[22] R.J. Ellis, Science 158 (1969) 477.

[23] T.C. Hall and E.C. Cocking, Biochim. Biophys. Acta 123 (1966) 163.

[24] M.S. Bamji and A.T. Jagendorf, Plant Physiol. 41 (1966) 764.

[25] A. Goffeau and J. Brachet, Biochim. Biophys. Acta 95 (1965) 302.

[26] A. Gnanam and J.S. Kahn, Biochim. Biophys. Acta 142 (1967) 493.

[27] J.M. Eisenstadt and G. Brawermen, J. Mol. Biol. 10 (1964) 392.

[28] R.J. Mans and G.D. Novelli, Arch. Biochem. Biophys. 94 (1961) 48.

[29] R.M. Smillie, in: Methods in Enzymology: Photosynthesis and Nitrogen Fixation, ed. A. San Pietro (Academic Press, New York, 1970), in press.

[30] D. Vazquez, Nature 203 (1964) 257.

[31] L.A. Anderson and R.M. Smillie, Biochem. Biophys. Res. Commun. 23 (1966) 535.

[32] R. Rendi and S. Ochoa, J. Biol. Chem. 237 (1962) 3711.

[33] M.K. Nikolaeva, O.P. Osipova and Yu.V. Krylov, Doklady Akad. Nauk SSSR 175 (1967) 487.

[34] R.J. Ellis, Nature 200 (1963) 596.

[35] F. Rønnike, Physiol. Plant. 11 (1958) 421.

[36] J.B. Hanson and W.A. Krueger, Nature 211 (1966) 1322.

[37] J.T.O. Kirk and R.L. Allen, Biochem. Biophys. Res. Commun. 21 (1965) 523.

[38] R.M. Smillie, D. Graham, M.R. Dwyer, A. Grieve and N.F. Tobin, Biochem. Biophys. Res. Commun. 28 (1967) 604.

[39] D.G. Bishop and R.M. Smillie, Arch. Biochem. Biophys. 139 (1970) 179.

[40] P.G. Roughan and R.D. Batt, Anal. Biochem. 22 (1968) 74.

[41] K.K. Carroll, J. Amer. Oil Chem. Soc. 40 (1963) 413.

[42] R.M. Smillie, N.S. Scott and D. Graham, in: Comparative Biochemistry and Biophysics of Photosynthesis, eds. K. Shibata, A. Takamiya, A.T. Jagendorf and R.C. Fuller (University of Tokyo Press, Tokyo, 1968) p. 332.

[43] R.M. Smillie, N.S. Scott, D. Graham and B.D. Patterson, in: The Productivity of Photosynthetic Systems, part II: Theoretical Foundations of Optimization of the Photosynthetic Productivity (IBP Symposium, Moscow, Sept. 1969), ed. A.A. Nichiporovich (1970), in press.

[44] W.R. Evans, R. Walenga and C. Johnson, Ann. Res. Report C.F. Kettering Res. Lab. Yellow Springs, Ohio, 1967, p. 99.

[45] D.S. Beattie, R.E. Basford and S.B. Koritz, Biochemistry 5 (1966) 926.

[46] G.D. Clark-Walker and A.W. Linnane, J. Cell Biol. 34 (1967) 1.

[47] K.B. Freeman, D. Haldar and T.S. Work, Biochem. J. 105 (1967) 947.

STUDIES ON THE METABOLISM OF NUCLEIC ACID AND PROTEIN ASSOCIATED WITH THE PROCESSES OF DE- AND RE-GENERATION OF CHLOROPLASTS IN *CHLORELLA PROTOTHECOIDES*

EIJI HASE

Division of Biosynthesis, The Institute of Applied Microbiology,
University of Tokyo, Tokyo, Japan

One of the features of biogenesis of chloroplasts is that it is markedly affected by the environmental conditions, especially light and nutritional conditions, under which plants or algae are placed. Following the early work by Kruger [1] with *Chlorella protothecoides,* striking changes of pigmentation have been reported to be induced in several algae when grown on media containing organic sources of carbon [2–4]. Recently, investigations have been made by a research group in our laboratory on profound degeneration and regeneration of chloroplasts in *Chlorella protothecoides* induced by the control of nutritional as well as light conditions. This article summarizes the results so far obtained on the metabolism of nucleic acid and protein associated with the processes of de- and re-generation of chloroplasts in these algal cells. When this alga is grown in media rich in glucose or other metabolizable organic carbon compounds and poor in a nitrogen source (ammonium salt, urea or amino acid), apparently chlorophyll-less cells with markedly degenerated plastids containing neither 'discs' nor lamellae (called 'glucose-bleached' cells) are produced either in the light or in darkness [5–8]. When, on the other hand, it is cultured in media rich in a nitrogen source and poor in an organic carbon source, normal green cells with fully organized chloroplasts are obtained in the light, while in darkness pale green cells with only partially organized plastids containing a few discs (called 'etiolated' cells) are produced. These three different types of algal cells are transformed into each other as schematically shown in fig. 1. When the green cells are incubated in a medium containing a high concentration of glucose

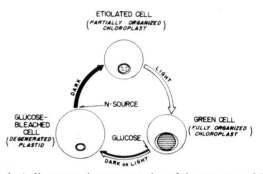

Fig. 1. A diagrammatic representation of the processes of de- and re-generation of chloroplasts in *Chlorella protothecoides.*

or other organic carbon source, they are relatively rapidly bleached [5,9] becoming eventually the glucose-bleached cells, either in the light * or in darkness ('bleaching process'). During this process chloroplast lamellae are first loosened and then broken into smaller pieces which are subsequently lost completely [5,8]. When glucose-bleached algal cells are incubated in a medium containing a nitrogen source but no organic carbon source, they are turned into normal green cells under illumination ('greening process') and into the etiolated cells in darkness [7,10]. These processes of bleaching and greening have been shown to proceed independent of cell multiplication [9,11].

---

* Light delayed the start of bleaching [5]. All the bleaching experiments described in this article were, therefore, performed in darkness, unless otherwise mentioned.

Metabolism of RNA, protein and chlorophyll during the process of chloroplast degeneration

RNA metabolism

Experiments showed that the RNA content of algal cells remained nearly constant on a per-culture basis during the process of bleaching [9]. However, when changes in ribosomal RNA (rRNA) were studied, a marked and relatively rapid degradation of chloroplast rRNA (16 S and 23 S) was found to occur to give rise to an increase in polynucleotides of lower sedimentation values [12,13]. This was most probably due to a sharp rise in ribonuclease activity observed after the incubation of green cells with glucose [13]. In contrast, the cytoplasmic rRNA (16 S and > 23 S) seemed to remain intact without being degraded at least during the degradation of chloroplast rRNA.

Protein metabolism

The bulk protein content of algal cells was also found to remain almost constant on a per-culture basis during the bleaching process [9], but there were indications that certain proteins are degraded while others are being synthesized. Fig. 2 shows that ribulose-1,5-diphosphate (RuDP) carboxylase, whose activity resides on fraction 1 protein, was significantly decreased, resulting in a parallel decrease of the activity of photosynthetic CO_2-fixation of the algal

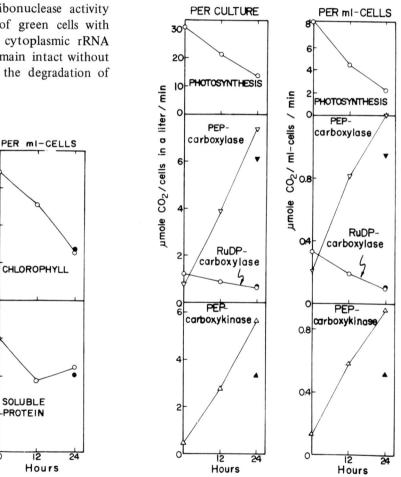

Fig. 2. Decrease of the activity of RuDP carboxylase and increases of the activities of PEP-carboxylase and -carboxykinase during the bleaching process. Green cells were incubated, in darkness, in a medium containing glucose (initial conc.: 1%). Methods of determination of chlorophyll content [30] and photosynthetic CO_2-fixation activity [7] of cells have been described previously. Sonicated algal materials were centrifuged at 27,000 g for 20 min. The protein content of the supernatant was estimated ('soluble protein'). An aliquot of the supernatant was passed through a Sephadex G-25 column and used as the enzyme preparation. Further details and the methods of assay for enzyme activities will be described elsewhere [14]. Solid circles and triangles represent the results obtained in the presence of chloramphenicol (10^{-2} M).

cells [14]. As shown later, little or no RuDP carboxylase activity was detected in glucose-bleached cells. This is in agreement with the observation made by Fuller and Gibbs [3] that extracts from bleached cells of *Chlorella variegata* obtained by growing the alga on organic source of carbon in the light showed no activity of RuDP carboxylase. On the other hand, the activities of phosphoenolpyruvate (PEP)-carboxylase and -carboxylkinase were conspicuously increased when green cells were incubated with glucose. By examining the distribution of carboxylating enzymes among non-aqueously separated subcellular fractions, evidence was obtained suggesting that PEP carboxylase is mainly present outside of chloroplast in *C. protothecoides,* while RuDP carboxylase is localized in chloroplast [14]. A smaller part of PEP carboxylase activity seemed to reside in chloroplast. The increase of the chloroplast PEP carboxylase seems to be inhibited by chloramphenicol (fig. 2).

The bleaching process was found to be severely inhibited by a low concentration of cycloheximide [15], but not by chloramphenicol [9]. The former antibiotic has been shown to inhibit protein synthesis in cell-free systems containing ribosomes of the 80 S-type [16,17], but has no effect on protein synthesis in systems containing the 70 S-type ribosomes. On the other hand, the latter antibiotic has been shown to block protein synthesis on ribosomes of the 70 S-type [18,19]. There is evidence suggesting that *C. protothecoides* cells contain the 70 S-type ribosomes in their chloroplasts and the 80 S-type ones in the cytoplasms [20], like higher plant cells [21–24]. Cycloheximide was shown to strongly inhibit the assimilation of added glucose or acetate into lipids (fatty acids) that was induced in green cells on incubation with glucose or acetate and that proceeded very actively during the process of bleaching [25–27]. These results suggested that a certain enzyme(s) participating in fatty acid synthesis is inductively formed in the cytoplasm on the addition of an organic source of carbon, and that the fatty acid synthesis may be causally related to bleaching.

Chlorophyll degradation

Chlorophyll *a* was found to be degraded more rapidly than chlorophyll *b* during an early phase of the bleaching process [28]. In parallel with the disappearance of chlorophyll, red pigments were found to be excreted into the medium [29,30]. The red pigments are not porphyrins but seem to be products of degradation of chlorophyll with oxidative rupture of tetrapyrrole ring. Chiba et al. [31] found that in parallel with the degradation of chlorophyll, hydrolytic activity of chlorophyllase decreased.

Differential disintegration of photosynthetic electron transport chain

Experiments showed that there occurs differential decreases of the activities for the phenazine methosulphate (PMS)- and flavin mononucleotide (FMN)-catalyzed photophosphorylations and the Hill reaction during the bleaching process [28]. The decrease of the activity for the PMS-photophosphorylation proceeded approximately in parallel with that of chlorophyll content, and was slower than the decreases in other photosynthetic activities. The rate of decay of the Hill reaction activity was significantly slower with 2,6-dichlorophenol-indophenol than with quinone. It was suggested that systems I and II may be relatively rapidly cut off at a certain site(s) of the photosynthetic electron transport chain during the bleaching process, the disconnected systems, however, retaining activities for partial reactions of photosynthesis for some time.

Formation of RNA, protein and chlorophyll during the process of chloroplast regeneration

Syntheses of bulk RNA and protein

The first event detected after incubation of glucose-bleached algal cells in a medium containing a nitrogen source but no organic carbon source was the start of an active RNA synthesis [10]. The RNA synthesis was found to be significantly enhanced by light. Following the RNA synthesis, there started a marked formation of protein which in the light was accompanied by an equally active synthesis of chlorophyll. The syntheses of protein and chlorophyll proceeded in parallel and were greatly enhanced by light. When the greening algal cells were transferred from light to darkness, the rate of protein synthesis was rapidly reduced to the dark control rate [7]. When etiolated algal cells containing RNA and protein at appreciable levels but very small amounts of chlorophyll were illuminated, a large amount of chlorophyll was formed with further increase in the protein content and little increase in the RNA level [10].

Syntheses of chloroplast and cytoplasmic ribosomal RNA

Glucose-bleached algal cells contained only traces of rRNA, although they contained soluble RNA at appreciable levels [13,32]. When they were incubated in a nitrogen-enriched medium, both cytoplasmic and chloroplastic rRNAs were actively synthesized [32]. The synthesis of chloroplast rRNA was found to be accelerated more by light than that of cytoplasmic rRNA. The start of active formation of soluble RNA lagged behind that of rRNA.

Effects of actinomycin and 5-fluorouracil

The greening of glucose-bleached algal cells was found to be inhibited by actinomycin (C complex) [33]. On applying the antibiotic at different times during the chloroplast regeneration, the inhibitory effect was found to be markedly reduced with the progress of the developmental process. This indicates that DNA-directed RNA synthesis is important for the chloroplast regeneration in glucose-bleached cells, at least during its early phase *. 5-Fluorouracil (5-FU) was also found to inhibit the greening process strongly when applied at early stages, especially before the provision of a nitrogen source, but the suppressive effect decreased with the delay of its application [34]. No inhibitive effect was observed when 5-FU was applied later than the time around which chlorophyll formation started. 5-FU seems to inhibit the greening of glucose-bleached cells by interfering with the ribosome production occurring during an early phase of the chloroplast regeneration. Smillie has observed strikingly similar inhibitory effects of 5-FU on the light-induced greening of dark-grown *Euglena* cells [35]. As discussed later, cytoplasmic ribosomes as well as chloroplast ribosomes seem to be important for the syntheses of proteins associated with the chloroplast development in *Chlorella* cells.

---

* It is highly probable that the permeability of algal cell membranes change during the developmental process and that the antibiotic was unable to enter the cells at later stages of development.

Effects of cycloheximide and chloramphenicol on the greening process

An inhibitory effect of chloramphenicol has been observed in a previous study [34]. Recently comparative studies were performed on the effects of cycloheximide and chloramphenicol [36]. As shown in fig. 3, it was disclosed that cycloheximide strongly inhibited an early, essentially dark process as well as the subsequent light-dependent process of the greening of glucose-bleached cells (fig. 3b). Chloramphenicol inhibited a light-dependent process immediately following the early dark process and lasting for a comparatively short period, but it had no effect upon the subsequent light-dependent process of greening (figs. 3(a) and 3(c)). These effects of inhibitors of protein synthesis will be discussed later in connection with the effect of light on the greening process.

Synthesis of RuDP carboxylase

The development of RuDP carboxylase, a major soluble protein in chloroplast, was followed during the greening process [14]. Figs. 4 and 5 reproduce the results obtained. Glucose-bleached algal cells contained barely detectable amounts of fraction I protein when examined by polyacrylamide gel-electrophoresis and by Sephadex G-200 column chromatography. It is clearly seen from the figures that the synthesis of RuDP carboxylase was dependent on light, while the formation of PEP carboxylase existing outside of chloroplast was independent of light. The development of photosynthetic CO_2-fixation activity of algal cells paralleled the formation of RuDP carboxylase. When the algal cells that were developing RuDP carboxylase in the light were transferred to darkness, the formation of the enzyme was rapidly stopped, a result which is in agreement with the previous observation that the photosynthetic CO_2-fixation activity stopped increasing under similar conditions [7]. The light-dependent formation of RuDP carboxylase was strongly inhibited by both chloramphenicol and cycloheximide.

Chlorophyll synthesis

When glucose-bleached algal cells were incubated in a nitrogen-enriched medium under illumination, the formation of chlorophyll started after a lag period, as mentioned earlier. Chlorophyll synthesis also began in dark-incubated algal cells, but it soon

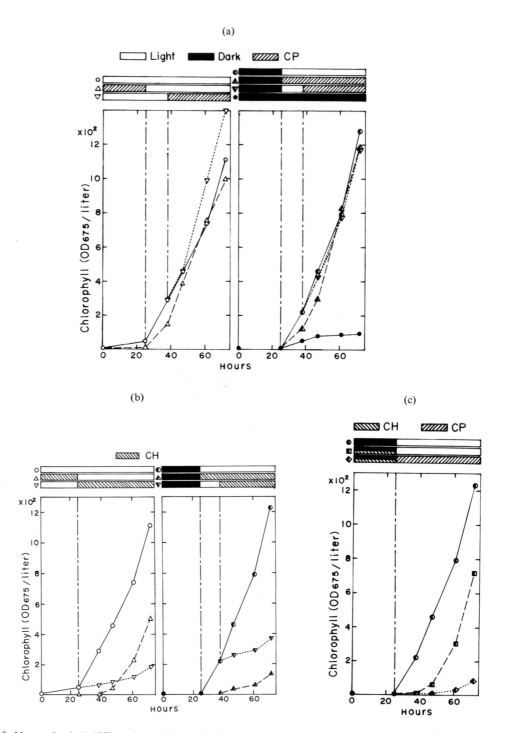

Fig. 3. Effects of chloramphenicol (CP) and cycloheximide (CH) on the greening process [36]. CP: 10^{-2} M, CH: 15 μg/ml. Glucose-bleached algal cells were incubated in a medium containing glycine (5×10^{-2} M) as nitrogen source. Cultures were illuminated by white fluorescent lamps at a light intensity of 2,000 lux. Chlorophyll content was determined by measuring light absorption at 675 mμ of cell suspension with a Shimadzu-MPS-50 with corrections for absorbance at 750 mμ.

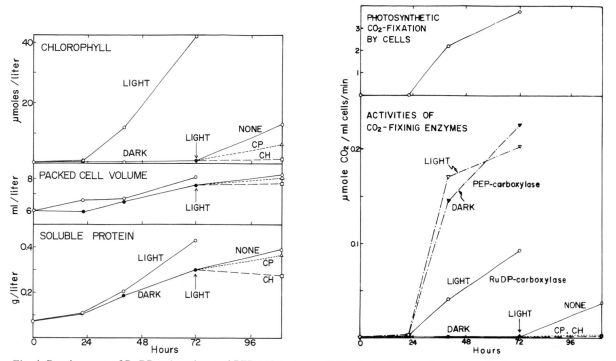

Fig. 4. Development of RuDP carboxylase and PEP carboxylase activities during the greening process. Culture conditions were the same as in fig. 3. CP: chloramphenicol, 10^{-2} M; CH: cycloheximide, 15 μg/ml.

ceased or continued at a very slow rate. The synthesis of chlorophyll *a* commenced before that of chlorophyll *b* [7]. No detectable amount of protochlorophyll(ide) was found in algal cells throughout the greening process either in darkness or in light. As can be seen from fig. 6, when the algal cells that were greening in light were transferred to darkness, chlorophyll synthesis continued at the same rate for a short time [37]. This was followed by a slow but continuing formation of chlorophyll. The rate although slow, was distinctly higher than that in algal cells incubated in continuous darkness. It was suggested that there are two different light effects on the chlorophyll formation: a short-term effect disappearing in a few hours in darkness after the illumination of etiolated cells and a long-term effect lasting for many hours after the illumination. The addition of δ-aminolevulinic acid (ALA) caused an increase of the synthesis of chlorophyll * in etiolated algal cells in darkness

* The pigment produced under these conditions seemed to be chlorophyll(ide), not protochlorophyll(ide) [cf. 38,39], as judged from its absorption spectrum observed using cell suspensions.

as well as in light. The increased rate of the dark synthesis of chlorophyll, however, was much lower than the rate in light without added ALA [37]. Succinic acid and glycine could not be substituted for ALA in the enhancement of dark formation of chlorophyll, but in the light ALA could be replaced by these two substances. Fig. 7 illustrates a tentative scheme for the effects of light on the chlorophyll formation in etiolated algal cells. It was suggested that the light effect I in the scheme may be the long-term effect and the second one the short-term effect. Evidence was obtained suggesting that when etiolated algal cells are illuminated, the production of ALA at first limits the over-all rate of chlorophyll synthesis for a relatively short period, and that light reaction II subsequently becomes rate-limiting. As was shown earlier, chloramphenicol suppressed only an early phase of the light-induced greening of etiolated algal cells, and cycloheximide inhibited all phases of the greening process. It appears likely that the light reaction I is linked with the formation of a certain enzyme(s) participating in the production of ALA

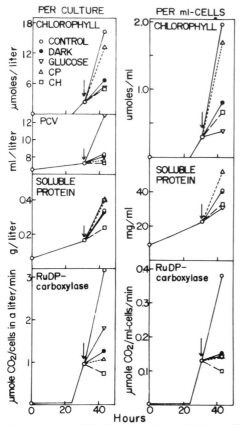

Fig. 5. Development of RuDP carboxylase activity as affected by dark-incubation of greening algal cells and by the addition of glucose or inhibitors of protein synthesis. Glucose: initial conc. 0.1%; CP: chloramphenicol 10^{-2} M; CH: cycloheximide 15 μg/ml. The time of addition is indicated by the arrows.

Fig. 7. A tentative scheme for enhancing effects of light and suppressive effects of glucose on the chlorophyll formation in etiolated algal cells.

from succinic acid and glycine, and that the enzyme formation is inhibited by chloramphenicol. Recently Steer and Gibbs [40] demonstrated that the illumination of etiolated bean leaves caused an increase in the activity of succinyl coenzyme A synthetase. The light reaction II seems to be related with the production of certain protein(s) associated with an intermediate step(s) of chlorophyll synthesis; the protein synthesis in this case is inhibited by cycloheximide but not by chloramphenicol.

Effects of light and CO_2 on the formation of RNA, protein and chlorophyll

The process of chloroplast development in algal cells in light has been demonstrated to be virtually unaffected by 3-(*p*-chlorophenyl)-1,1-dimethylurea (CMU) [41,42], which completely inhibited photosynthetic CO_2-fixation. This indicates that some light-dependent process other than photosynthesis is

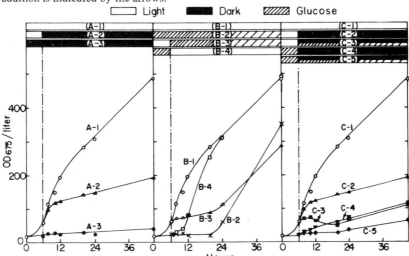

Fig. 6. Effects of light and glucose on the greening process. A: chlorophyll formation in darkness after a 6 hr illumination of etiolated algal cells. B: effects of added glucose (initial conc. 0.1%) on the chlorophyll formation proceeding under continuous illumination. C: effects of added glucose on the dark synthesis of chlorophyll after the 6 hr illumination. The starting etiolated cells were obtained by incubating glucose-bleached cells in a medium containing urea (0.5%) in darkness.

involved in the chloroplast development. To obtain information about the nature of this light-dependent process, effects of coloured light on the greening process have been investigated [42–44]. In earlier experiments [42,43], in which the cells were kept under CO_2-free air to remove effect of photosynthetic CO_2-fixation, blue light was found to be most effective for the chlorophyll formation in glucose-bleached cells, the effect of yellow, green and red light following in the decreasing order. Light did not effect the changes in total RNA or protein content. Recently, CO_2 was found to be important for the development of chloroplasts in algal cells [44]. Fig. 8 clearly shows a markedly enhancing effect of CO_2 on the greening process in the presence of CMU. Fig. 9 indicates that not only the formation of chlorophyll, but also synthesis of RNA and protein were considerably enhanced by CO_2. It is also seen in fig. 9 that the lag period before the start of synthesis of chlorophyll was remarkably shortened in the presence of CO_2. In fig. 10 are illustrated the effects of coloured

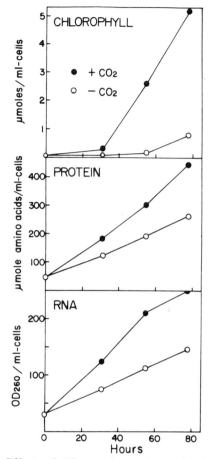

Fig. 9. Effects of CO_2 on the syntheses of chlorophyll, protein and RNA during the greening process. Glucose-bleached algal cells were incubated, under illumination with white fluorescent lamps (light intensity: 2,000 lux at the level of culture), in a medium containing glycine (5×10^{-2} M) as nitrogen source and CMU (5×10^{-5} M). Aerated with CO_2(1%)–air. Protein and RNA were determined by the Schmidt-Thannhauser method [cf. 10].

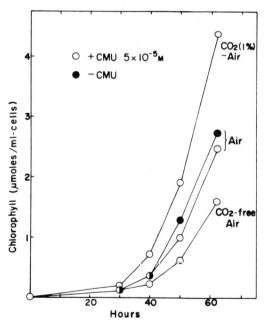

Fig. 8. Effect of CO_2 on the greening process. Glucose-bleached algal cells were incubated, under illumination, in a medium containing glycine (5×10^{-2} M) as nitrogen source, and aerated with air containing the indicated concentrations of CO_2. CMU has been shown to completely inhibit photosynthetic CO_2-fixation by algal cells at 5×10^{-5}M [42].

lights of low intensity on chlorophyll formation with and without provision of CO_2. In the absence of CO_2, blue light was much more effective than green or red light, a result which is in agreement with earlier observations [42]. Where CO_2 was provided, however, green light as well as blue light enhanced chlorophyll formation, except during its early phase (the lag phase), in which period green light and red light were less effective than blue light. It was found that the enhancement of chlorophyll formation by green and blue light under the provision of CO_2 occurs in

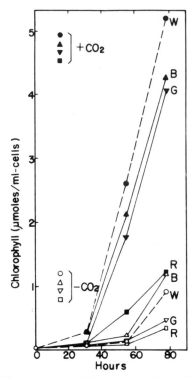

Fig. 10. Effects of coloured light on the chlorophyll formation in glucose-bleached algal cells under aeration with
CO_2-enriched (1%) and CO_2-free air. This experiment was
performed simultaneously with that in fig. 9. W: white light,
2,000 lux (this has been shown to be a saturating light
intensity for the greening of glucose-bleached cells [42]. B:
blue, G: green, R: red light (100 ergs/cm^2/sec). The spectral
energy distribution of coloured light has been shown in a
previous paper [42]. CMU was added to each culture to give
a final concentration of 5×10^{-5}.

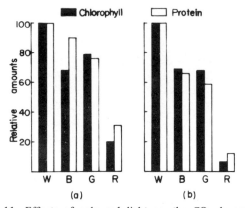

Fig. 11. Effects of coloured light on the CO_2-dependent
synthesis of chlorophyll and protein. (a) The results obtained
at the 78th hr in the experiment shown in fig. 10. (b) Those
obtained at the 78th hr in the experiment shown in fig. 12.
The ordinate represents the differences between the levels of
chlorophyll or protein in the algal cells with and without
provision of CO_2, in terms of percent of the values obtained
in white light.

parallel with the acceleration of protein synthesis, as
may be seen from fig. 11a. This suggests that green
and blue light may primarily enhance the synthesis of
protein which is closely associated with chlorophyll
formation. It was further found that blue light en-
hances RNA synthesis where CO_2 is provided, and
green and red light are similarly less effective. As
mentioned earlier, the synthesis of RNA (mostly
rRNA) is a main event occurring during an initial
phase of chloroplast regeneration in glucose-bleached
algal cells, and this is followed by a period of protein
synthesis concomitant with chlorophyll formation.
To explain the above-described results, it may be
assumed that under CO_2-deficiency (aeration with

CO_2-free air), the CO_2-dependent and blue light-
enhanced synthesis of a certain species of RNA (most
probably rRNA) is limiting the overall rate of chlorophyll synthesis throughout the process of chloroplast
regeneration; the synthesis of this RNA species proceeds in competition with light-independent synthesis
of other species of RNA, resulting in no light enhancement of the increase in total RNA content of
the algal cells [43]. Under the provision of ample
CO_2, the green and blue light-dependent synthesis of
a certain protein(s) may be assumed to be rate-
limiting chlorophyll formation during the later light-
dependent phase of chloroplast development. Then,
one would expect that if the etiolated algal cells
which have been incubated in darkness under provision of CO_2-enriched air for a sufficiently long period
of time are illuminated, the enhancement of chlorophyll formation by blue light through its acceleration
of RNA synthesis will be reduced or disappear, since
sufficient RNA synthesis would have taken place
during the dark incubation. This was borne out by
the experiments reproduced in figs. 12 and 11(b). In
the experiments shown in fig. 12 [44], the starting
cells were the etiolated ones obtained under CO_2-
enriched air. With these cells, both green and blue
light were much more effective for promoting chlorophyll synthesis than red light, either with or without

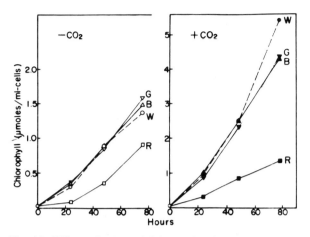

Fig. 12. Effects of coloured light on the chlorophyll formation in etiolated algal cells with and without provision of CO_2. The starting etiolated cells were obtained by incubating in darkness, glucose-bleached cells in a medium containing glycine (5×10^{-2} M) as nitrogen source for 24 hr under CO_2-enriched air. Other conditions were the same as in fig. 10.

provision of CO_2. Fig. 11(b) illustrates that the CO_2-dependent synthesis of chlorophyll proceeded closely in parallel with the CO_2-dependent protein synthesis under these conditions, as was the case in the experiments shown in fig. 10. However, no significant enhancement by light was recognized with the increase in the RNA content of cells under these conditions. As shown before, chloramphenicol suppresses only an early phase of the light-induced greening of etiolated algal cells, while cycloheximide inhibits all phases of the greening process. This suggests that the site of synthesis of the protein associated with chlorophyll synthesis may be on cytoplasmic ribosomes.

Suppressive effects of glucose on the chloroplast development

Concentrations of glucose or other metabolizable organic carbon compounds that are sufficiently low not to cause bleaching of green cells can nevertheless inhibit the greening process [37,45]. Fig. 6 shows the results of experiments, in which glucose was applied under different conditions. Added glucose was found to stop immediately and completely chlorophyll synthesis in the light as well as in darkness, either with or without added ALA. On the basis of these and other

results, a tentative scheme is presented in fig. 7 for the suppressive effects of glucose on the chlorophyll synthesis in algal cells. As seen from fig. 5, glucose was found to suppress also the development of RuDP carboxylase, but not of PEP carboxylase (fig. 2). An electron microscopic study revealed that starch-like granules accumulated in the plastids of algal cells provided with glucose, like in the bleaching algal cells, and that the development of chloroplast lamellae was suppressed [46]. These results suggest that glucose suppresses multifarious aspects of the chloroplast development in algal cells.

Development of photosynthetic electron transport chain

Etiolated algal cells containing a small but appreciable quantity of chlorophyll were found to show detectable activities for the PMS- and FMN-catalyzed photophosphorylations and the Hill reaction [7]. During the process of chloroplast development in the light, these activities increased. The activities for the PMS- and FMN-photophosphorylations were measurable at an early stage of greening of glucose-bleached algal cells, in which the 'discs' were developed but no complete lamellae were observed. Further increases in these photic activities were studied in relation to chlorophyll formation under continuous illumination and under light-dark conditions [7]. The PMS-photophosphorylation activity always developed in parallel with chlorophyll formation. In contrast, the development of the FMN—photophosphorylation activity measured at the different light intensities stopped immediately after the dark incubation. The development of the Hill reaction activity measured at the low light intensity (under the light-limiting conditions) continued for some time during incubation in the dark, while that of the activity measured at the high light intensity ceased rapidly. These and other observations lead us to the conclusion that the process of development of photosynthetic electron transport chain consists of two successive phases: an early one which is limited by the formation of chlorophyll (or the pigment system involved in the primary 'light-process' of electron transport), and the subsequent one in which light-dependent syntheses of other components are necessary and rate-determining for the development of the complete electron transport chain.

Concluding remarks

An attempt is made in fig. 13 to correlate the above-described experimental results and inferences. As may be seen, the main event during an initial phase of the chloroplast regeneration in glucose-bleached algal cells is the synthesis of ribosomes, and the synthesis of rRNA is enhanced by blue light as well as by the provision of CO_2. During the subsequent process of active greening of algal cells, the synthesis of protein which is dependent on both green and blue light as well as on CO_2, is the rate-limiting step. The light-dependent development of RuDP carboxylase also actively proceeds during this phase of chloroplast regeneration. Judging from the effects of coloured light and other results, the photoreceptors of the light reactions involved in the chloroplast regeneration in glucose-bleached algal cells seem to be different from chlorophyll, protochlorophyll(ide) [39] or phytochrome [47–51], which has been shown to be a photoreceptor acting in the light-induced development of chloroplast in higher plants. There have been indications that some other light reaction(s) may be involved in the chloroplast development in various plants [cf. 38]. It was reported by Bjorn [52] and Bjorn and Odheluis [53] that in excised roots of wheat, cucumber and pea,

chlorophyll was accumulated in blue light, but was practically absent in red light. Kasemir and Mohr [54] reported that blue light stimulates the chlorophyll formation in fern sporeling with concurrent enhancement of protein synthesis. Miller and Zalik [55] showed that seedlings of a mutant of Gateway barley accumulated most chlorophyll under green light. On the other hand, the role of CO_2 is morphogenesis seems to be universal and of fundamental importance, although little is known about its precise action [cf. 56]. Recently Ogasawara and Miyachi in our laboratory [57] found that blue light of a low intensity stimulated the incorporation of $^{14}CO_2$ by *Chlorella ellipsoidea* cells into aspartate, glutamate and malate in the presence of CMU at concentrations completely suppressing the ordinary photosynthetic CO_2-fixation. Whether or not this blue light effect is related with the blue light effect on the greening process of glucose-bleached algal cells described above is an interesting problem to be worked out.

Glucose (and other organic carbon sources) exerts dual suppressive actions on chloroplasts in *C. protothecoides*: one is the disintegration of existing chloroplasts induced by relatively high concentrations and the other the suppression of development of plastids into fully organized chloroplasts at lower concentrations. Recently organic sources of carbon have been shown to inhibit chlorophyll synthesis in *Euglena* [58–61] and in higher plant [62] as well as formation of the Calvin cycle enzymes [58,63].

The marked responses of the algal cells to light and carbon sources as described above might be related to mechanisms for their adaptation and survival in changing environmental conditions.

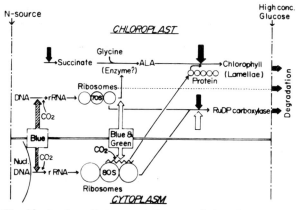

Fig. 13. A schematic representation correlating observations and inferences made of the formation and degradation of chloroplasts in cells of *C. protothecoides*. White arrows represent the light-dependent processes, cross-hatched ones the light-enhanced but essentially dark processes, and black ones the processes inhibited by lower concentrations of glucose, or the degradation processes induced by high concentrations of glucose.

Summary

Profound degeneration and regeneration of chloroplasts have been demonstrated to be induced in cells of *Chlorella protothecoides* by the control of nutritional as well as light conditions. Suppressive effects of glucose and enhancing effects of light and CO_2 on the chloroplast development in the algal cells were summarized and discussed with special reference to the metabolism of nucleic acid, protein and chlorophyll. Glucose at high concentrations was found to induce degradation of chloroplast ribosomal RNA and RuDP carboxylase as well as chlorophyll. At lower concentrations, it markedly inhibited not only the formation of RuDP carboxylase and chlorophyll but also the development of lamellar structure. Light and CO_2 significantly enhanced chlorophyll formation and the syntheses of RNA and protein in the presence of CMU that completely inhibited photosynthetic CO_2-fixation. The RNA synthesis was enhanced most markedly by blue light and, to appreciably lesser extents, by green and red light. The protein synthesis, on the other hand, was strikingly accelerated by green as well as blue light, the effect of red light being much less. It was inferred that the blue light-enhanced and CO_2-dependent production of RNA (probably ribosomal RNA) was limiting the overall rate of the greening of algal cells during an early phase, while during the subsequent phase in which the most active greening of algal cells occurred, synthesis of protein induced by green or blue light and dependent on CO_2 was the rate-limiting process. The protein synthesis in question seemed to be inhibited by cycloheximide, but not by chloramphenicol. The development of RoDP carboxylase was also found to be greatly dependent on light, but was inhibited by both cycloheximide and chloramphenicol.

Modes of disassembly and re-formation of the photosynthetic electron transport chain during the de- and re-generation of chloroplasts were discussed on the basis of differential degradation and development of various photosynthetic activities.

References

[1] W. Kruger, Beitrage z. Morph. u. Phys. Nied. Org. Leipzig 4 (1894) 69.

[2] H. Nakano, J. Coll. Sci. Tokyo Imperial Univ. 40 (1917) Art. 2.

[3] R.C. Fuller and M. Gibbs, Plant Physiol. 34 (1959) 324.

[4] I. Shihira and R.W. Krauss, in: Chlorella — Physiology and Taxonomy of Forty-One Isolates, (Port City Press, Baltimore, 1963).

[5] I. Shihira-Ishikawa and E. Hase, Plant Cell Physiol. 5 (1964) 227.

[6] K. Takashima, I. Shihira-Ishikawa and E. Hase, ibid. 5 (1964) 321.

[7] T. Oh-hama, I. Shihira-Ishikawa and E. Hase, ibid. 6 (1965) 743.

[8] S. Yumoto, unpublished.

[9] S. Aoki, M. Matsuka and E. Hase, Plant Cell Physiol. 6 (1965) 487.

[10] S. Aoki and E. Hase, ibid. 5 (1964) 473.

[11] S. Aoki and E. Hase, ibid. 6 (1965) 347.

[12] Y. Oshio and E. Hase, ibid. 9 (1968) 69.

[13] S. Mihara, K. Kimura and E. Hase, ibid. 9 (1968) 87.

[14] Y. Oshio and E. Hase, in preparation.

[15] M. Matsuka and E. Hase, Plant Cell Physiol 10 (1969) 277.

[16] M.R. Siegel and H.D. Sisler, Biochim. Biophys. Acta 87 (1964) 83.

[17] H.L. Ennis and M. Lubin, Science 146 (1964) 1474.

[18] A.D. Wolfe and F.E. Hahn, Biochim. Biophys. Acta 95 (1965) 146.

[19] R.M. Smillie, N.S. Scott and D. Graham, in: Comparative Biochemistry and Biophysics of Photosynthesis, eds. K. Shibata, A. Takamiya, A.T. Jagendorf and R.C. Fuller (University of Tokyo Press, Tokyo, 1968) p. 332.

[20] S. Mihara and E. Hase, Plant Cell Physiol. 10 (1969) 465.

[21] J.W. Lyttleton, Exptl. Cell Res. 26 (1962) 312.

[22] M.F. Clark, R.E.F. Mathews and R.K. Ralph, Biochim. Biophys. Acta 91 (1964) 289.

[23] N.K. Boardman, R.I.B. Francki and S.G. Wildman, J. Mol. Biol. 17 (1966) 470.

[24] E. Stutz and H. Noll, Proc. Natl. Acad. Sci. U.S. 57 (1967) 774.

[25] M. Matsuka, H. Otsuka and E. Hase, Plant Cell Physiol. 7 (1966) 651.

[26] M. Matsuka, S. Miyachi and E. Hase, ibid. 10 (1969) 513.

[27] M. Matsuka, S. Miyachi and E. Hase, ibid. 10 (1969) 527.

[28] T. Oh-hama, M. Matsuka and E. Hase, in: Comparative Biochemistry and Biophysics of Photosynthesis, eds. K. Shibata, A. Takamiya, A.T. Jagendorf and R.C. Fuller (University of Tokyo Press, Tokyo, 1968) p. 279.

[29] Y. Oshio and E. Hase, Plant & Cell Physiol. 10 (1969) 41.

[30] Y. Oshio and E. Hase, ibid. 10 (1969) 51.

[31] Y. Chiba, I. Aiga, M. Idemori, Y. Satoh, K. Matsushita and T. Sasa, ibid. 8 (1967) 623.

[32] S. Aoki and E. Hase, ibid. 8 (1967) 181.

[33] S. Aoki and E. Hase, ibid. 5 (1964) 485.

[34] S. Aoki, J. Khan Matsubara and E. Hase, ibid. 6 (1965) 475.

[35] R.M. Smillie, Canad. J. Bot. 41 (1963) 123.

[36] S. Ochiai, M. Matsuka and E. Hase, in preparation.

[37] S. Ochiai and E. Hase, Plant Cell Physiol. 11 (1970) 663.

[38] L. Bogorad, in: Chemistry & Biochemistry of Plant Pigments, ed. T.W. Goodwin (Academic Press, London and New York, 1965) p. 29.

[39] N.K. Boardman, in: The Chlorophylls, eds. L.P. Vernon and G.R. Seely (Academic Press, New York, 1966) p. 437.

[40] T. Steer and M. Gibbs, Plant Physiol. 44 (1969) 775.

[41] M. Matsuka and E. Hase, Plant Cell Physiol. 7 (1966) 149.

[42] Y. Sokawa and E. Hase, ibid. 8 (1967) 495.

[43] Y. Sokawa and E. Hase, ibid. 8 (1967) 509.

[44] T. Oh-hama, in preparation.

[45] I. Shihira-Ishikawa and E. Hase, Plant Cell Physiol. 6 (1965) 101.

[46] S. Ochiai, T. Osafune and E. Hase, in preparation.

[47] L. Price and W.H. Klein, Plant Physiol. 36 (1961) 733.

[48] A. Marcus, ibid. 35 (1960) 126.

[49] J.L. Mego and A.T. Jagendorf, Biochim. Biophys. Acta 53 (1961) 237.

[50] D. Graham, A.M. Grieve and R.M. Smillie, Nature 218 (1968) 89.

[51] B. Filner and A.O. Klein, Plant Physiol. 43 (1968) 1587.

[52] L.O. Bjorn, Physiol. Plant 18 (1965) 1130.

[53] L.O. Bjorn and I. Odheluis, ibid. 19 (1966) 60.

[54] H. Kasemir and H. Mohr, Planta 67 (1965) 33.

[55] R.A. Miller and S. Zalik, Plant Physiol. 40 (1965) 569.

[56] W.J. Nickerson and S. Bartniki-Garcia, Ann. Rev. Plant Physiol. 15 (1964) 327.

[57] N. Ogasawara and S. Miyachi, Plant Cell Physiol. 11 (1970) 1.

[58] A.A. App and A.T. Jagendorf, J. Protozool. 10 (1963) 340.

[59] D.E. Buetow, Nature 213 (1967) 1127.

[60] J.T.O. Kirk and M.J. Keylock, Biochem. Biophys. Res. Commun. 28 (1967) 927.

[61] R.C. Harris and J.T.O. Kirk, Biochem. J. 106 (1968) 34p.

[62] R. Achantz, H. Duranton and M. Peyriere, Comp. Rend. Acad. Sci. Paris, Ser. D, 265 (1967) 205.

[63] D. McMahon and L. Bogorad, Plant Physiol. 43 (1968) 188.

DEVELOPMENT AND ENVIRONMENT STUDIES ON CHLOROPLASTS
OF AMARANTHUS LIVIDUS

J.W. LYTTLETON, J.E.M. BALLANTINE * and B.J. FORDE

Applied Biochemistry Division and Plant Physiology Division,
D.S.I.R., Palmerston North, New Zealand

An alternative pathway of carbon dioxide fixation to that demonstrated by Calvin and co-workers and believed to operate in most higher plants [1] has been found, first in sugar cane [2] and now in many allied grasses as well as a few dicotyledons of tropical origin [3]. Photosynthesis in these species produces initially 4-carbon dicarboxylic acids as a result of carboxylation of phosphopyruvate, whereas in most higher plants the initial product is phosphoglycerate which is produced by carboxylation of ribulose diphosphate. The two types of carbon dioxide fixation are now commonly described as the C4 and C3 pathways respectively.

The plants which photosynthesise by the C4 pathway share a particular anatomical feature in the existence of a well-defined layer of cells, known as the bundle sheath cells, around the vascular bundle of the leaf. These cells contain a high concentration of chloroplasts which are generally morphologically distinct from the chloroplasts of the surrounding mesophyll tissue. In the case of maize and sugar cane, for instance, the bundle sheath chloroplasts contain no recognisable grana, but are filled with extensive lamellae, in contrast to the normal grana-containing structure found in the mesophyll chloroplasts [4,5]. This absence of grana is not found in the bundle sheath of all C4 plants, as some such as certain *Amaranthus* species can contain grana in both types of chloroplasts [6].

In C4 plants the bundle sheath chloroplasts are often more elongated, and accumulate starch more readily than the corresponding mesophyll chloro-

plasts. They are also normally concentrated in a group at one end of the cell, in contrast to those of the mesophyll, which are distributed more uniformly around the cell.

A further feature of chloroplasts of C4 plants is the presence of a peripheral reticulum, particularly in the mesophyll [6]. This is in contrast to chloroplasts of C3 plants, where such a structure occurs only occasionally in a rudimentary form.

The roles of the two types of chloroplast in the photosynthetic process carried out by C4 plants is naturally of great interest, and studies of the enzyme complement of chloroplasts isolated in a non-aqueous system have been made [7]. Use was made of the fact that bundle-sheath chloroplasts accumulate starch preferentially, giving them a greater effective density when banded by density gradient centrifugation in organic solvents. The separation thus achieved indicated that the enzymes concerned with the primary carboxylation were located in the mesophyll chloroplasts, while those concerned with later steps in the carbon cycle were located in the bundle sheath.

Although non-aqueous techniques are valuable in determining the location of enzymes in cell organelles, the external membrane of such organelles is usually destroyed in the isolation, and they can no longer be expected to function in vitro as integrated units. It would be desirable to have methods for separating the two classes of chloroplast of C4 plants in an aqueous system, so that more might be learned of their individual roles in the fixation of carbon dioxide.

We have been studying the chloroplasts of *Amaranthus lividus* from two points of view: (1) the influence of different environments on the morpholo-

* Present address: National Vegetable Research Station, Wellsbourne, Warwickshine, England.

gy of the two classes of chloroplast, (2) devising a method of separating the two classes of chloroplast in an aqueous system.

Effect of environment on chloroplasts of Amaranthus lividus

Plants were grown in pots in a glasshouse in which the light intensity was somewhat below full daylight, at temperatures 18–23°C. When the plants were 3–4 in. high, some were left in the glasshouse, and were considered as in medium light intensity. Others were shaded for ten days by two layers of gauze in the glasshouse, and thus received low intensity light. To expose plants to higher light intensity, some plants were transferred to growth cabinets. At the highest intensity, (293 W/m² with 56 W/m² in ex-

cess of 700 nm) growth was very poor, and starch accumulation so severe that details of chloroplast structure were obscured. At approximately half this intensity growth was vigorous, and plants were grown for ten days in such a cabinet for chloroplast studies at high light.

The main conclusion that can be drawn from electron microscope studies of sections of leaf tissue taken from expanding leaves of plants grown under the three light conditions outlined above is that chloroplasts of both mesophyll and bundle sheath cells exhibit very different degrees of granal content under different levels of illumination.

Under high light, the granal structure is almost absent, and both classes of chloroplast are filled with lamellae which appear similar to those described for bundle sheath chloroplasts in maize and other tropical grasses [4,5]. There are, however, occasional

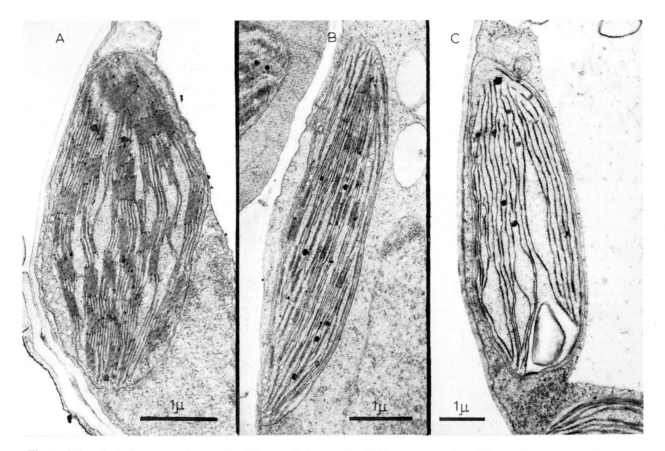

Fig. 1. Chloroplasts from mesophyll cells of leaves of *Amaranthus lividus,* grown under different degrees of illumination: (A) low light, (B) medium light, (C) high light.

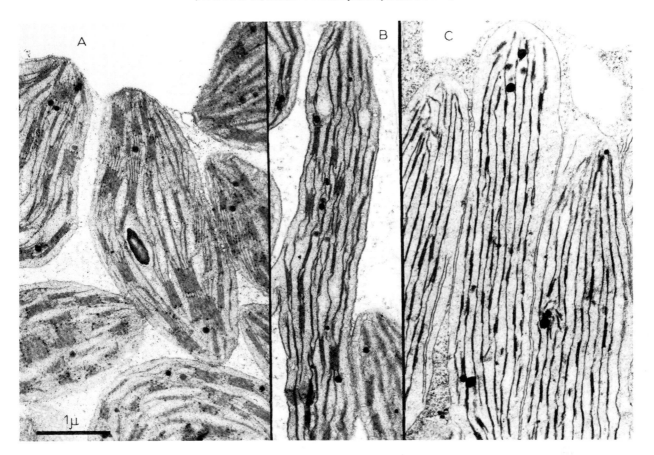

Fig. 2. Chloroplasts from bundle sheath cells of leaves of *Amaranthus lividus*, grown under different degrees of illumination: (A) low light, (B) medium light, (C) high light.

regions of overlap of the extended thylakoids which traverse the chloroplast, so some few rudimentary grana might be said to be present even at high light.

With decreasing level of illumination, more obvious granal stacking is found, and at the lowest level investigated, in shaded conditions in the glasshouse, in both classes of chloroplast the grana were well developed.

This phenomenon of granal modification by light is not peculiar to C4 plants, for it has also been observed in leaves of the soya bean, where under high intensity illumination the grana of the mesophyll chloroplasts almost disappeared [8]. In this case the loss of grana was shown to be reversible, in that return of the plants to low light produced increased granal formation in the same leaves that had previously contained chloroplasts from which grana were absent.

Another feature of chloroplast structure in *A. lividus* which appeared to be influenced by light intensity was the degree of development of the peripheral reticulum. The results are not so clear as with the granal structure, but it appears that in mesophyll chloroplasts the reticulum is more extensive under low light than high light, when it is sometimes replaced by peripheral lamellae. Fig. 3 shows well defined cases of this, but examination of large number of electron micrographs shows that under either light condition a considerable range of reticulum development can be observed.

Within the bundle sheath chloroplasts the peripheral reticulum is less well developed, and its extent seems not to be affected by light intensity in the same way as in the mesophyll.

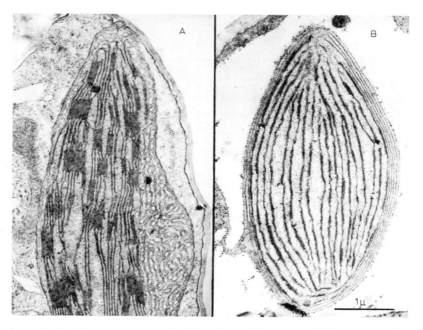

Fig. 3. Appearance of peripheral reticulum in mesophyll chloroplasts of *Amaranthus lividus* grown under different conditions of illumination: (A) low light — reticulum very marked, (B) high light — reticulum replaced by closely spaced lamellae.

Fractionation of chloroplasts in an aqueous system

In order to fractionate chloroplasts from *A. lividus* into those derived from bundle sheath and from mesophyll cells, the density difference arising from the higher starch content of the bundle sheath chloroplasts was exploited in an aqueous system suitable for isopycnic density gradient centrifugation. For this purpose a colloidal silica suspension of high density and low viscosity, commercially available as 'Ludox HS' (E.I. du Pont de Nemours & Co., Wilmington, Del.) was used. When added to 0.4 M sucrose, which provides a satisfactory osmotic environment for chloroplasts, Ludox will readily give densities up to 1.25. At a density of 1.22, equivalent to that of a 50% sucrose solution, its viscosity is 4 cp at 25°, compared with 12.5 cp for the sucrose solution. In addition, the osmotic effect of the colloidal silica particles, which have a diameter of 14—15 nm, is negligible over the concentration range used, so that chloroplasts do not move into a hypertonic medium as they sediment through the density gradient.

We have used Ludox to produce either a continuous or a discontinuous density gradient, both in swing-out centrifuge tubes and in a zonal rotor (Type A, Measuring & Scientific Equipment Ltd, London), usually with a density range of 1.08 to 1.22, which appears to span possible chloroplast densities in aqueous suspension.

As an illustration of the relative performance of Ludox and sucrose step density gradients, a preparation of chloroplasts from spinach, isolated in 0.4 M sucrose, was layered over a 4-step gradient of Ludox in buffered 0.4 M sucrose, and over a gradient of similar density steps containing only buffered sucrose at higher concentrations. Fig. 4 shows the result of centrifuging these gradients at low (300 *g*) and high (2500 *g*) speeds for 15 min. In the case of the Ludox gradient, the low speed spin has separated the chloroplasts into three classes on the basis of density, while in the sucrose gradient the chloroplasts have moved very little past the first density step. The high speed spin brought about some redistribution and broadening of the bands in the Ludox system, and in the sucrose produced only two bands, as well as some material which sedimented to the bottom of the tube. The more rapid resolving power of the Ludox system

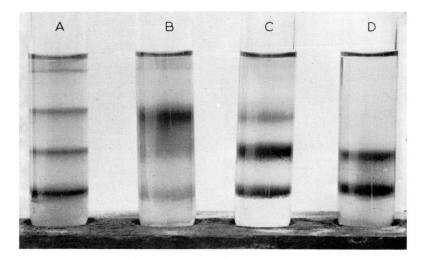

Fig. 4. Fractionation of spinach chloroplasts in discontinuous density gradients of Ludox and of sucrose. (A) and (C) Ludox gradients in buffered 0.4 M sucrose. 2 ml volumes of densities 1.22, 1.17, 1.13 and 1.08, over which was layered 0.5 ml of chloroplast suspension. (B) and (D) Buffered sucrose gradients, 2 ml volumes of same densities as in (A) and (C), with 0.5 ml chloroplast suspension. (A) and (B) centrifuged at 300 *g* for 15 min. (C) and (D) centrifuged at 2500 *g* for 15 min.

is probably due to its lower viscosity, while the shift of chloroplasts to the denser regions in sucrose appears to be due to shrinkage in a hypertonic medium. Microscopic observation of the chloroplasts recovered from the sucrose gradient supports this hypothesis.

In preliminary experiments with the zonal rotor, filled to its capacity of 1300 ml with a continuous gradient of Ludox in buffered 0.4 M sucrose, spinach chloroplasts gave an unsatisfactory diffuse distribution on reaching isopycnic equilibrium, and recovery of the chloroplasts after pumping the gradient out of the rotor was made difficult because of the large volumes involved and the long time (up to 60 min) to recover the rotor contents. A more effective separation resulted from loading the rotor with 4 × 150 ml volumes of densities 1.08, 1.13, 1.17 and 1.22, and filling the remainder of the rotor volume with undiluted Ludox. When spun at 4,000 rpm, the chloroplasts reached isopycnic equilibrium in about 5 min and the resulting sharp bands at the density steps could be collected much more rapidly in relatively small volumes.

Phase contrast microscopy of the recovered spinach chloroplasts showed the band between densities 1.08 and 1.13 to contain mainly chloroplasts which

were osmotically intact, while that between 1.17 and 1.22 contained only chloroplasts which had lost their outer membrane.

Application of this system to chloroplasts isolated from *A. lividus* gave satisfactory resolution between those with high starch content, which collected between densities 1.17 and 1.22, and those almost free of starch, which were found between 1.08 and 1.13. A mixed population of chloroplasts was found between densities 1.13 and 1.17.

To date, only poor recoveries of enzyme activity have been obtained from chloroplasts of *A. lividus* fractionated in this manner. The fault seems to lie in the isolation procedure rather than in any undesirable effect of the gradient system. A number of different extractant media have been tested, but it appears both from enzyme studies and from electron microscopy of the isolated chloroplasts that it is considerably more difficult to preserve the chloroplast membrane when dealing with C4 plants than when dealing with leaves such as spinach, broad bean or pea, which are commonly used as sources of chloroplasts. It is tempting to speculate that the presence of a peripheral reticulum in the C4 chloroplasts may offer special problems in their isolation in an intact form.

Conclusion

The ease with which the internal structure of the chloroplasts of *A. lividus* can be modified by varying the intensity of illumination, and the intrinsic interest of having two classes of chloroplast with apparently different functions in the same leaf, make it highly desirable to obtain these chloroplasts in an intact, fully active form for study of their reactions in vitro. It is possible that the photosynthetic light reactions may be able to be studied with the chloroplasts fractionated as described above, but full exploitation of the system must await isolation of truly intact chloroplasts.

References

[1] M. Calvin and J.A. Bassham, in: The Photosynthesis of Carbon Compounds (W.A. Benjamin Inc., New York, 1962).

[2] H.P. Kortschak, C.E. Hartt and G.O. Burr, Plant Physiol 40 (1965) 209.

[3] H.S. Johnson and M.D. Hatch, Phytochemistry 7 (1968) 375.

[4] A.J. Hodge, J.D. McLean and F.V. Mercer, J. Biophys. Biochem. Cytol. 1 (1955) 605.

[5] W.M. Laetsch, D.A. Stetler and A.J. Vlitos, Z. Pflanzen-physiol. 54 (1965) 472.

[6] W.M. Laetsch, Am. J. Botany 55 (1968) 785.

[7] C.R. Slack, Phytochemistry 8 (1969) 1387.

[8] J.E.M. Ballantine and B.J. Forde, Am. J. Botany, in press.

PROTEIN AND RNA SYNTHESIS DURING
AGEING OF CHLOROPLASTS IN WHEAT LEAVES

C.J. BRADY, B.D. PATTERSON, HENG FONG TUNG
and Robert M. SMILLIE

*Plant Physiology Unit, CSIRO Division of Food Preservation, Ryde,
and School of Biological Sciences, University of Sydney, Australia*

It has been a general experience that chloroplasts isolated from young, expanding leaves can synthesize ribonucleic acid (RNA) and protein in vitro, but that chloroplasts recovered from mature leaves have little or no activity [1–4]. The relatively poor synthetic ability of chloroplasts from older leaves does not appear to result from the isolation procedure [4], so that chloroplasts in older leaves may synthesize little RNA and protein.

Protein synthesis on the chloroplast-located 70 S ribosomes [5] is much more active in vitro than is synthesis on the 80 S ribosomes of the cytoplasm [6]. There is no certainty, however, that the full activity of the cytoplasmic ribosomes is expressed after isolation [7]. In consequence, the relative contributions of the chloroplast and cytoplasmic systems to protein synthesis in the leaf remain undefined [2].

If most of the chloroplast proteins are made within the chloroplast [8], and chloroplast ribosomes are capable of the high synthetic rates found in bacterial systems [9], the contribution of the chloroplast to protein synthesis in the expanding leaf will be large. However, since the ability of isolated chloroplasts to incorporate amino acids into protein decreases greatly when the growth rate declines, the contribution of the chloroplast to protein turnover in the mature leaf may be small. As leaves age, the cytoplasmic system may be responsible for an increasing proportion of the total protein synthesis.

There are difficulties in obtaining reliable estimates of protein turnover in non-growing leaf tissues, and in long term experiments with detached leaves a contribution by contaminating micro-organisms is difficult to avoid. However, a number of different methods have yielded turnover rates for protein in mature leaves in the range 0.2 to 1.5% per hr [10–13]. This rate declines as leaves age, but remains significant as long as the leaves are green. Protein turnover also occurs in non-green plant storage organs [14,15], and active systems of protein synthesis, presumably based on cytoplasmic ribosomes have been derived from these [16,17].

Chloroplasts from young wheat leaves incorporate amino acids into protein in vitro; those from older leaves do not [3,4]. We have used wheat leaves of different ages to measure in vivo precursor incorporation into a number of chloroplast proteins, and into chloroplast and cytoplasmic rRNA.

Ribosomal RNA in developing leaves

When wheat (*Triticum aestivum* L var Timgalen) is grown under standard conditions (12 hr day at 22°, 12 hr night at 16°) the first leaf grows to its maximum size in seven days from sowing (fig. 1). Total RNA increases in proportion to growth, but as soon as growth stops the RNA content declines (fig. 2). There is no decline in protein or chlorophyll in the first few days after growth stops.

Leaf rRNA can be separated by gel electrophoresis into the subunits from the cytoplasmic ribosomes (25 S and 18 S RNA) and the subunits from the chloroplast ribosomes (23 S and 16 S RNA) [19,20]. The method can be quantitated spectrophotometrically, and we have used it to measure the amounts of RNA derived from the chloroplast and cytoplasmic ribosomes. Because the smaller subunits (16 S chloro-

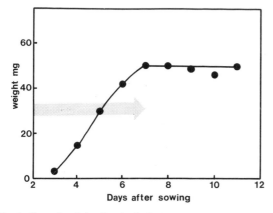

Fig. 1. Growth of the first leaf of wheat. Wheat was grown in vermiculite with a 12-hr day at 22°C and a 12-hr night at 16°C. Light intensity was about 9,000 lux. The leaf blade only was taken for measurements. The shaded arrow shows the period of growth.

plast and 18 S cytoplasmic RNA) are more stable during the isolation from the leaves [21], these have been used as a measure of rRNA from chloroplasts and cytoplasm. Fig. 3 shows that the amounts of both chloroplast and cytoplasmic rRNA are greatest seven days after sowing, when the leaves reach their maximum size. About 60% of the chloroplast and cytoplasmic RNA is lost from each system in the three days after day seven.

Can the decline in the amount of rRNA in the fully grown leaf be correlated with a reduction of RNA synthesis? To determine whether the synthesis of rRNA continues in the non-growing leaf, we have

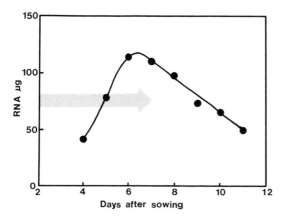

Fig. 2. Changes in total RNA in the first leaf of wheat. RNA was estimated by the method of Smillie and Krotkov [18]. The shaded arrow shows the period of growth.

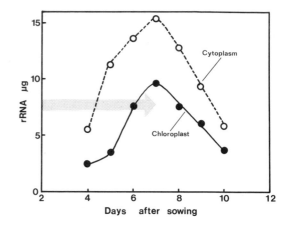

Fig. 3. Changes in rRNA in the first leaf of wheat. RNA was extracted [20] and separated by gel electrophoresis [19]. The amount of RNA in the 18 S (cytoplasmic) and 16 S (chloroplast) subunits was determined by scanning the gels at 260 nm. The shaded arrow shows the period of growth.

measured the incorporation of radioactive precursors into rRNA.

Leaves were detached from plants eight days old and fed (^{14}C)-uracil in the light for 6 hr. The rRNA subunits were separated by electrophoresis [19] through gels of rectangular cross section [22] and revealed by staining [23]. To define the position of the 23 S and 16 S components, rRNA from *Escherichia coli* was run in parallel. In these 8-day leaves which had finished growing (fig. 1), (^{14}C)-uracil was incorporated into the cytoplasmic subunits of rRNA (25 S and 18 S), but scarcely at all into the chloroplast subunits of rRNA (23 S and 16 S) (fig. 4).

When the incorporation of (^{3}H)-uracil into rRNA was followed in this manner into leaves of different ages, both cytoplasmic and chloroplast fractions were substantially labelled as long as the leaves were growing (fig. 5). When growth stopped and there was a loss of both cytoplasmic and chloroplast rRNA, incorporation into chloroplast rRNA dropped to a low level. In contrast, incorporation into the cytoplasmic rRNA continued at a significant rate (fig. 5), indicating that some synthesis of cytoplasmic rRNA continues at this time even though there is a net loss of this component. From a study of RNA in radish cotyledons, Ingle [21] concluded that synthesis of cytoplasmic rRNA occurred at a time when there was no further synthesis of chloroplast rRNA.

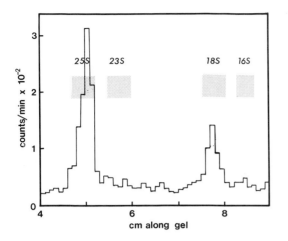

Fig. 4. Incorporation of (^{14}C)-uracil into rRNA of the first leaf eight days after sowing. Wheat leaves were washed with 0.1% cetyltrimethylammonium bromide and sterile distilled water and incubated in the light for 6 hr with (^{14}C)-uracil. The labelled RNA was separated on flat gels [22] which were stained [23] and dried before cutting into 1 mm fractions for counting. RNA from *E. coli* was run in parallel as a marker for the 23 S and 16 S bands. The position of the 25 S, 23 S, 18 S and 16 S stained bands is shown by the shaded bars.

Our results with wheat leaves can be interpreted in a similar way, but an alternative explanation for both sets of experiments is possible. In the bacterium, *Escherichia coli,* exogenous radioactive precursors are

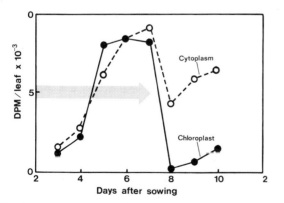

Fig. 5. Incorporation of ^{3}H-uracil into leaves of different ages. Leaves were fed (^{3}H)-uracil for 6 hr and the RNA then extracted and separated as described in fig. 4. ^{3}H activity in the 16 S (chloroplast) band and the 18 S (cytoplasmic) bands were measured, and incorporation calculated on a per leaf basis. The shaded arrow shows the period of growth of the leaves.

incorporated into RNA only when there is a net synthesis of RNA [24]. In *E. coli,* the entry of labelled uracil into the immediate precursor pool for RNA synthesis is not greater than the net requirement of uracil and cytosine for RNA synthesis [25]. When there is no net synthesis of RNA, breakdown products from RNA turnover are preferentially used for new synthesis, and exogenous radioactive precursors are not incorporated [24,26]. In addition, feedback controls [27] may limit the entry of precursors such as purines, pyrimidines and phosphate into the immediate RNA precursor pool.

Similar controls may well function in chloroplasts in which a net loss of RNA is occurring. It should therefore be stressed that both our results and those of Ingle [21] indicate either an exclusion of uracil and phosphate from the chloroplast rRNA precursor pool, or an absence of turnover of chloroplast rRNA once growth ceases. Neither explanation is particularly favoured by our results, although some experiments [28] with isolated chloroplasts have suggested that RNA-nucleotidyltransferase activity decreases in chloroplasts as leaves age.

Protein synthesis on chloroplast ribosomes

Synthesis of fraction 1 protein in growing and non-growing leaves

Fraction 1 [29] represents a large portion of the chloroplast protein. Present evidence indicates that at least part of this protein is made on the chloroplast ribosomes, as its synthesis is inhibited by inhibitors which bind to 70 S ribosomes [30,31], and isolated chloroplasts are reported to incorporate amino acids into fraction 1 [32]. Chloroplast ribosomes which are capable of making fraction 1 can be accepted as functional. The opposite does not, of course, follow, for a failure to observe synthesis of fraction 1 does not necessarily indicate that other chloroplast proteins are no longer made.

The amount of fraction 1 in developing wheat leaves increases in proportion to the fresh weight (fig. 6). For the few days immediately following full expansion, the amount of fraction 1 per leaf neither increases nor decreases. Growing leaves readily incorporate (^{3}H)-leucine into fraction 1 (fig. 6), but the amount of incorporation falls rapidly when growth

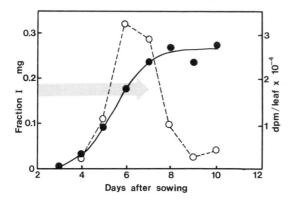

Fig. 6. Changes in fraction 1 protein of the first leaf of wheat, and incorporation of (^{3}H)-leucine. Fraction 1 protein was extracted from leaves by homogenising in 0.01 M Trischloride buffer (pH 8.0) containing 5 mM dithiothreitol. The extract was centrifuged at 18,000 g for 20 min and portions of the supernatant separated by electrophoresis through 5% (w/v) acrylamide gels at pH 8. The gels were scanned in a spectrophotometer at 280 nm, and the amounts of fraction 1 calculated assuming an extinction coefficient of E^{280} nm = 14.1 [33]. To incorporate (^{3}H)-leucine into fraction 1, leaves were fed (^{3}H)-leucine (100 μCi/ml, 50 mCi/mM) as described for the incorporation of labelled uracil in fig. 3. Fraction 1, after separation on polyacrylamide gels, was stained with Amido Schwarz 10B, and fractions (0.5 mm) cut from the frozen gels. The profile of radioactive incorporation was correlated with the dyed band of fraction 1, and the ^{3}H activities incorporated per leaf calculated from the results. The shaded arrow shows the period of growth.

stops. Two days after full expansion, there is little incorporation into fraction 1. While dilution by endogenous leucine may increase with age, the decline in incorporation into fraction 1 coincides with the decrease in chloroplast rRNA and therefore may reflect an overall decrease in the capacity for protein synthesis within the chloroplast.

Synthesis of fraction 1 in detached leaves

To study further the synthesis of fraction 1 and other chloroplast proteins, experiments were performed with leaves in which protein catabolism was accelerated by detaching leaves from the plant [34, 35]. In this system responses are uncomplicated by the relation of the leaf to the rest of the plant; precursors can be fed without further disturbance of the leaf, and the demand for a continuing supply of carefully matched samples is lessened.

Experiments with detached leaves were made with

the second leaf of the primary stem, using blades detached at one of two stages of development. Leaves described as 'prior to full expansion' were removed from the plant two days after the leaf blade was fully emerged (tables 1, 2 and 4, figs. 7–9). At this stage while most of the cells were fully expanded, some expansion growth persisted in the basal portions of the blade. 'Fully expanded leaves' (tables 5–11, figs. 10, 11) were detached seven to nine days later when the blades of the second leaf had assumed a horizontal position, and the fifth leaf was emerging.

When leaves are detached a day before full expansion and kept in a flowing stream of humid air in the light [36] their contents of chlorophyll and protein remain constant for about two days and then decline rapidly (fig. 7). Similar changes and a corresponding decline in the RNA content have been shown in detached leaves of other monocotyledons [39–41]. Wood et al. [35] concluded from studies of detached leaves of a number of species that cytoplasmic and chloroplast proteins were degraded at comparable

Table 1
Synthesis of different soluble proteins in detached leaves.

	Specific activity (dpm × 10^{-3}/mg)	
	Day 0	Day 3
Total protein [a]	268	138
Sephadex peak 1 [b]	564	598
Sephadex peak 2 [b]	211	28
Sephadex peak 3 [b]	125	187
Fraction 1 protein [c]	99	4

[a] Protein extract was prepared from leaves exposed to $^{14}CO_2$. The extraction of 'total protein' and separation on a column of Sephadex G-200 is described in fig. 9. Total protein refers to the sample loaded onto the Sephadex column. Specific activity in all fractions was determined after washing TCA (5%, w/v) precipitates exhaustively with ethanol-ether (1/1, v/v) and 5% TCA at 90°C. Protein samples were hydrolyzed in 6N HCl (24 hr, 105°C), and separated into basic and non-basic components by ion-exchange chromatography. The specific activity relates ^{14}C in the basic components to total protein.

[b] Bulked fractions from peak 1 (fractions 4 and 5), peak 2 (fractions 9 and 10), and peak 3 (fractions 18 to 22) from fig. 9.

[c] Purified by iso-electric focusing on a 110 ml column. Fraction 1 was precipitated in the pH 5.4 zone.

Table 2
Synthesis of fraction 1 protein in detached leaves.

	A Soluble protein [a]	B Fraction 1 protein [b]	A/B [c]
Day 0	36.7	11.2	0.33
Day 1	25.9	6.5	0.25
Day 2	20.7	2.0	0.10
Day 3	40.7	1.5	0.04
Day 4	32.2	0.0	0

Leaves detached before full expansion were kept turgid and in the light. DL-(^{3}H)-valine (G) was fed for 2 hr immediately after detaching the leaves (day 0), and then to further batches of leaves at daily intervals. The amount of (^{3}H)-valine given to the leaves was progressively increased from 5 μCi on day 0.

[a] Soluble protein was prepared as in fig. 7.
[b] The amount of fraction 1 protein was measured after gel electrophoresis by the method detailed in fig. 5. (^{3}H)-valine in fraction 1 was measured in formamide extracts of appropriate zones of the polyacrylamide gels.
[c] Specific activity of fraction 1 protein as a fraction of the specific activity of the soluble protein.

rates. However, the methods available to them for distinguishing cytoplasmic and chloroplast proteins were inadequate.

In our detached wheat leaves, the content of fraction 1 declines within one day and before any loss

Table 3
Synthesis of fraction 1 and fraction 2 proteins in attached leaves.

Leaf number	Specific activity fraction 2/fraction 1
1	12.5
2	14.1
3	11.0
4	11.9
5	9.0
6	4.3
7	0.7

Wheat plants without tillers were fed $^{14}CO_2$ for 1 hr. Proteins from leaves in positions 1 to 7 were separated by gel electrophoresis and the specific activity of fraction 1 and one of the fraction 2 bands determined. To measure specific activity, radioactivity in basic compounds released by acid hydrolysis (6N HCl, 105°C, 24 hr) was related to adsorbed dye.

Table 4
Synthesis of fraction 1 and fraction 2 proteins in detached leaves.

	^{3}H/^{14}C [a]
Fraction 1 protein	4.0
Fraction 2 proteins	
Band A	0.9
B	1.2
C	1.2
D	1.1
E	1.3
F	1.0

Leaves detached prior to full expansion were kept 2 days in kinetin solution (1 × 10^{-4} M) or in water. DL-(^{3}H)-valine (G) was fed to leaves in kinetin and L-(^{14}C)-valine (U) to control leaves. After 2 hr, the two lots of leaves were mixed, extracted in Tris-borate-EDTA buffer containing mercapto-ethanol (1 mM), pH 8.5, and the soluble proteins separated by gel electrophoresis. Proteins were visualized by staining with Amido-Schwarz 10B. Protein zones were hydrolyzed in acid (6N HCl, 105°C, 48 hr) and the amounts of ^{3}H and ^{14}C measured.

[a] The ^{3}H/^{14}C ratio for each protein band is related to the ^{3}H/^{14}C ratio of band F.

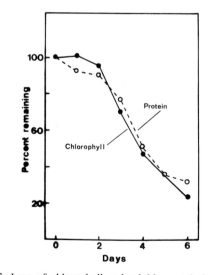

Fig. 7. Loss of chlorophyll and soluble protein in detached leaves. Chlorophyll was determined by the method of Arnon [37]. Portein was extracted by grinding leaves (400 mg fresh weight) in Tris (0.005 M)–glycine buffer, pH 8.0 (3.6 ml) containing dithiothreitol (1 mM). The homogenate was centrifuged at 27,000 g for 60 min and TCA added to the supernatant to a concentration of 5% (w/v). Protein was measured in the precipitate [38].

of chlorophyll and protein can be measured (fig. 8). The fact that the loss of the quantitatively important fraction 1 protein is not recorded as a loss of soluble protein suggests that fraction 1 initially dissociates into its subunits [43,44], or is degraded to polypeptide fragments which precipitate with trichloracetic acid. The ribulose diphosphate carboxylase activity which is associated with fraction 1 [33] also declines within the first day. After 4 days the enzymic activity has declined to a greater extent than has protein as fraction 1.

Since little synthesis of fraction 1 could be demonstrated in expanded leaves in which rRNA degradation had commenced (fig. 6), it was to be anticipated that synthesis of fraction 1 would decline rapidly once leaves were detached. To measure this, the incorporation of ^{14}C into fraction 1 in leaves detached for different times was followed. To avoid the possibility that access of introduced amino acids to the chloroplast was limited in detached leaves which have enlarged pools of free amino acids, leaves were

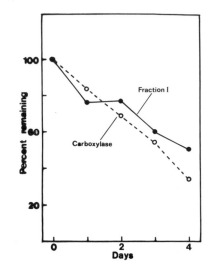

Fig. 8. Loss of fraction 1 protein and ribulose diphosphate carboxylase activity in detached leaves. Fraction 1 protein in extracts made with Tris-glycine buffer was separated by gel electrophoresis and measured as absorption at 280 nm (fig. 6). Ribulose diphosphate carboxylase was assayed [42] in the leaf homogenates.

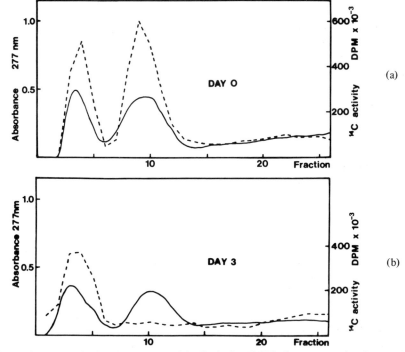

Fig. 9. Incorporation into soluble protein fractions separated by Sephadex G-200 chromatography. Leaves close to full expansion were detached and either kept turgid in the light for 3 days before exposing to $^{14}CO_2$ (2 mCi per 4.8 g fresh weight for 1 hr) or exposed to $^{14}CO_2$ (1 mCi per 7.1 g fresh weight, for 1 hr) immediately. Proteins were extracted in Tris–borate-EDTA buffer, pH 8.5 [46], with mercaptoethanol (1 mM) and glycerol (10%, v/v), and passed through a column of Sephadex G-200 (30 × 2.5 cm) under nitrogen. Equilibration and elution of the Sephadex was with the extraction buffer. Absorption at 277 nm (solid line) and radioactivity in protein (dashed line) was measured in 3.5 ml fractions by the methods described in table 1.

Table 5
Turnover of valine in protein of detached leaves.

Days after detaching	% turnover/hr	
	Experiment 1	Experiment 2
0	1.92	1.87
1	0.53	0.37
2	0.22	0.27
3	0.20	0.29

Leaves were detached after full expansion and kept turgid in water in the light. At daily intervals, L-(^{14}C)valine was fed to a batch of leaves via the transpiration stream. The specific activity of valine in protein and of free valine was measured 20 min and 30 min after feeding commenced. Turnover was calculated by relating valine entering protein in this 10-min interval to the mean specific activity of free valine at these times.

labelled by exposure to radioactive carbon dioxide in the light [45]. The distribution of ^{14}C in protein was measured after fractionation on Sephadex G-200 (fig. 9) and iso-electric focusing [47] (table 1) columns. Fraction 1, which was concentrated in the second peak from the Sephadex column, was the major protein in extracts from leaves either just detached, or detached and kept in water for three days. Contaminants of fraction 1 in the Sephadex peak were removed by iso-electric focusing, when fraction 1 formed a precipitate about pH 5.4. It was dissolved in phenol—acetic acid—water, and its purity

Table 6
Specific activity of proteins leached from chloroplasts.

Protein [a]	Specific activity [b]
Fraction 1	2.5 ± 1.6
Fraction 2	
Band A [c]	61.1 ± 3.2
Band B	38.8 ± 5.7
Band C	202.1 ± 2.0
Band F [d]	275.3 ± 9.0

[a] Chloroplasts were prepared and proteins separated as in fig. 11b. Protein and dye was extracted from the minced gel with formamide.
[b] Dpm per unit of absorption at 620 nm.
[c] Band on the leading edge of fraction 1.
[d] Band slightly behind the buffer front. The spectral characteristics of this band are those of ferredoxin.

Table 7
Synthesis of proteins extracted or not extracted by buffer.

Source of label	Time from harvest (hr)	Specific activity [a]	
		Soluble [b]	Insoluble
L-valine	0	313	463
	24	224	327
Acetate [d]	0	169	366
	48	185	448
	96	162	468
CO_2 [d]	0	64	128
	48	32	59

[a] Dpm × 10^{-3} per mg protein. The amount of precursors fed was increased with time from harvest, so comparisons of specific activities from one time to another are without meaning.
[b] Extracted by homogenizing in Tris-borate-EDTA buffer containing sucrose (0.3 M) and mercaptoethanol (1 mM), and remaining in the supernatant after centrifuging at 30,000 g for 60 min.
[c] Not extracted as in (b) but extracted by phenol—acetic acid—water (2:1:1, w:v:v).
[d] Protein fractions recovered after feeding 2-(^{14}C)-acetate or $^{14}CO_2$ were acid hydrolyzed, and the hydrolyzate divided into basic and non-basic components by ion exchange chromatography. ^{14}C in the basic components was used to measure specific activity.

Table 8
Location of synthesized proteins in cell fractions.

Fraction	Protein mg	Specific activity [a] dpm/mg
500 g, 10 min	4.53	840
1,500 g, 10 min	4.07	853
20,000 g, 20 min	0.37	1,407
105,000 g, 60 min	0.55	1,013
Supernatant	6.00	610

[a] Fully expanded leaves were detached and fed L-(^{14}C)-valine (U) for 1 hr. The tissue was homogenized [58] and fractionated. Each fraction was washed once in the homogenizing medium and then washed exhaustively with TCA (5%, w/v), ethanol/ether (l/l, v/v) and TCA (5%) at 90°C. Protein was dissolved in alkali (1N) and re-precipitated with TCA. Protein and ^{14}C activity were measured in each fraction.

Table 9

Synthesis of lamellae proteins in detached leaves.

	Protein mg	Specific activity [a] dpm $\times 10^{-3}$/mg
Buffer extracted protein [b]	7.44	28.0
Lamellae protein [c]		
A. Acetic acid extractable		
(1) Soluble at pH 6.5	1.28	2.3
(2) Insoluble at pH 6.5	3.23	26.4
B. Acetic acid insoluble	1.33	14.5

[a] Fully expanded leaves were detached and fed $^{14}CO_2$ for 60 min. Protein fractions were prepared and specific activities measured by the method detailed in table 8.

[b] The supernatant fraction in the initial homogenate. In chloroplast washes the specific activity increased progressively from 12.2×10^3 to 21.9×10^3 dpm/mg.

[c] Lamellae were recovered from osmotically disrupted chloroplasts and washed with water. Pigments were extracted with acetone at $-15°C$, and the residue extracted 3 $\times$ with 1.4N acetic acid at $0°C$. The extract was adjusted to pH 6.5 and after 1 hr at $0°C$, centrifuged (20,000 g, 20 min). The deposit was washed with water and dissolved in 0.5N acetic acid.

Table 10

Change in the synthesis of lamellae proteins with time from leaf detachment.

Fraction	Specific activity [a] dpm $\times 10^{-3}$/mg		
	Days from detachment		
	0	1	2
Total protein	42.5	20.6	20.8
Buffer extracted protein	24.0	12.4	16.3
Lamellae protein [b]			
(1) Soluble at pH 6.5	12.3	7.1	3.0
(2) Insoluble at pH 6.5	46.0	13.3	8.1

[a] Fully expanded leaves were detached and fed $^{14}CO_2$ for 60 min, immediately after detachment, and one and two days later. In the final sample (day 2), the amount of isotope used per leaf was doubled. Protein fractions were prepared and specific activities measured by the methods detailed in table 8.

[b] Acetic acid-extractable lamellae proteins were prepared as in table 9. The yield of each fraction was about the same portion of total leaf protein on each day.

Table 11

Relative synthesis of different proteins in detached leaves.

	Time from detachment, days			
Protein fraction	0		2	
	Protein mg	Specific activity [a]	Protein mg	Specific activity
Total	124.5	70.7	85.6	36.4
Supernatant	34.5	58.7	23.2	28.2
Fraction 1 [b]		14		3
Fraction 2				
Band C		240		204
Band F		220		410
Triton-treated lamellae [c]				
A. 30,000 g supernatant	18.9	71.0	14.4	8.3
B. 30,000 g deposit	2.8	59.3	1.3	81.4

[a] Specific activity is expressed as dpm $\times 10^{-3}$ per mg for the supernatant, lamellae and total protein fractions. For fraction 1 and fraction 2 proteins, ^{14}C in protein (dpm) is related to absorption at 620 nm as in table 6.

[b] Samples of the washes of chloroplasts in Tris-glycerol solution (fig. 11) were concentrated by pressure dialysis and fractionated by gel electrophoresis.

[c] Lamellae were prepared as in table 9. The lamella preparation contained 36.6% of the chlorophyll of the leaves in the day 0 sample, and 28.9% of the chlorophyll in the day 2 sample. The samples were treated with 1% (v/v) Triton X-100 in a medium containing Tris (0.005 M), Mg acetate (5 mM) and mercaptoethanol (0.1 mM). After standing in Triton solution at $0°C$ for 60 min the samples were centrifuged (30,000 g) for 60 min. The supernatant fluid was freeze-dried and extracted with acetone. Protein in both supernatant fluid and deposit was extracted with 1N NaOH, and reprecipitated with TCA. Specific activity was measured as in table 6.

checked by electrophoresis through polyacrylamide equilibrated with the same solvent.

Although 40% of the radioactivity in soluble proteins was associated with fraction 1 in freshly detached leaves, virtually no incorporation into this plastid protein was detected in leaves which had been detached for three days (table 1). However, ^{14}C was still incorporated into other soluble proteins. One explanation for the preferential curtailment in the synthesis of some soluble proteins is that protein synthesis ceases on chloroplast ribosomes while synthesis on cytoplasmic ribosomes persists. A decline in the synthesis of fraction 1 relative to other soluble

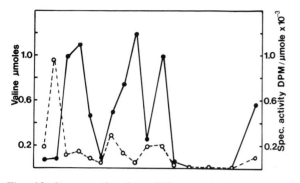

Fig. 10. Incorporation into different soluble proteins in detached leaves. Leaves were detached after full expansion and fed (^{14}C)-valine (U) for 2 hr. The soluble proteins were separated by gel electrophoresis and the stained bands hydrolyzed with acid to determine the amount (—) and specific activity (– – –) of protein-bound valine.

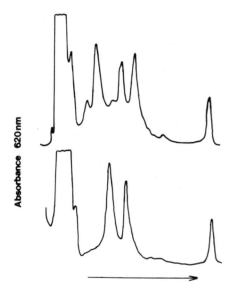

Fig. 11. Soluble proteins leached from chloroplasts. Chloroplasts isolated from fully expanded leaves were washed three times in the isolating medium [58], twice in a tris (0.01 M)–Mg (3 mM)–mercaptoethanol (1 mM) buffer (pH 7.8), then repeatedly in Tris (0.001 M)–glycerol (0.05 M)–mercaptoethanol (0.1 mM). Proteins in the original tissue homogenate (top) and those concentrated by pressure dialysis from the third Tris-glycerol-wash (bottom) were separated by gel electrophoresis. Gels were stained with Amido Schwarz 10B and scanned at 620 nm. Movement of proteins was from left to right.

proteins has also been shown for leaves of *Perilla frutescens* [48,49].

To measure how quickly the synthesis of fraction 1 declines in detached leaves, incorporation of (^{3}H)-valine into fraction 1 purified by gel electrophoresis is related to incorporation into soluble protein (table 2). The incorporation into fraction 1 relative to total protein decreases daily from harvest until day 4, when incorporation into fraction 1 cannot be measured. The fall in the synthesis of fraction 1 will be higher than indicated by the declining ratios in table 2 since the absolute rate of synthesis of the soluble proteins also declines (table 5). The decrease in incorporation into fraction 1 is more rapid than is the loss of fraction 1 protein (table 2, fig. 8).

To eliminate the possibility that low incorporation into fraction 1 in older leaves is caused by a particular sensitivity of this system to leaf detachment, synthesis in attached leaves of different ages was measured. ^{14}C was introduced as carbon dioxide to growing plants, and incorporation into fraction 1 in each leaf was related to incorporation into a lower molecular size protein whose synthesis was well maintained in older detached leaves (table 3). Incorporation into fraction 1 is relatively low in all except young expanding leaves.

Maintenance of fraction 1 synthesis in detached leaves treated with kinetin

The loss of RNA, and in particular rRNA, which occurs rapidly in attached leaves as they age (figs. 2, 3) also occurs, though more slowly, in detached leaves [50]. If the detached leaves are treated with kinetin, the loss of RNA is delayed [51]. There is no evidence as to whether both cytoplasmic and chloroplast ribosomes are maintained by kinetin treatment.

If fraction 1 is made on chloroplast ribosomes, the maintenance of the synthesis of fraction 1 as a result of kinetin treatment would be evidence that the hormone maintains protein synthesis within the chloroplast. There is already ample evidence that kinetin, and other hormones retarding leaf senescence, limit the loss of chloroplast components [52]. Since the synthesis of fraction 1 has been shown to decline particularly rapidly in detached leaves, the hypothesis that kinetin maintains its synthesis can be tested by comparing amino acid incorporation in fraction 1 and the smaller fraction 2 [29] proteins in detached leaves treated or not treated with kinetin. Kinetin treatment has a far larger influence on incorporation into fraction 1 than on incorporation into any of a number of fraction 2 proteins separated by gel electrophoresis (table 4). Thus, the synthesis of fraction 1 protein, and by inference the synthesizing capacity of the chloroplast ribosomal system, is maintained by kinetin treatment. Other experiments have shown that incorporation into fraction 1 protein relative to incorporation into total soluble protein is similar in freshly detached leaves and in leaves detached and treated with kinetin solution for two days. Consequently, kinetin appears to maintain protein synthesis in the state existing in the leaves when treatment commences. Synthesis of fraction 1 is not stimulated beyond that level, but the decline in its synthesis is prevented. Kinetin treatment also prevents the fall in the amount of fraction 1 and in its ribulose diphosphate carboxylase activity.

Among the lower molecular weight proteins, only minor differences are found in the relative incorporation rate with and without kinetin (table 4). The rates of synthesis of these proteins appear to change to similar extents during ageing.

Synthesis of chloroplast soluble proteins in detached leaves

A comparison of the distribution of radioactive precursors in different proteins in leaves aged after detachment (fig. 9, tables 1, 4) shows that synthesis of some proteins is relatively well maintained as leaves age. That the synthesis of some of these proteins is high relative to that of fraction 1 protein when leaves are aged on the plant is shown in fig. 10. In this experiment, leaves were left on the plant for seven days after full expansion, and incorporation into soluble proteins was made by a method which allowed a valid direct comparison between protein fractions. Incorporation into fraction 1 is very low, but some other soluble proteins attain a high specific activity. In these fully expanded leaves, the absolute rate of protein synthesis is quite high, though it declines when the leaves are detached (table 5).

In considering the possibility that the decline in synthesis of fraction 1 protein in older leaves reflects the reduced capability of the chloroplast ribosome system rather than the specific repression of the synthesis of fraction 1, it is of particular interest to know where the rapidly synthesized fraction 2 proteins are located in the leaf cells. Ridley et al. [53] have described the release of some fraction 2 proteins during the washing of chloroplasts from spinach beet in hypotonic media. Our examination of the proteins in washes of chloroplasts prepared from fully expanded wheat leaves produced clear evidence that a number of the fraction 2 components readily separated by gel electrophoresis were located in the chloroplast (fig. 11). The spectral characteristics of the fastest moving of these bands indicates that it contains ferredoxin.

Incorporation into some of these bands relative to incorporation into fraction 1 protein is high (table 6). No measurements of specific activity of these bands by a method which allows a comparison between different proteins have been made. However, in terms of dye absorption they are all of very high specific activity relative to fraction 1. While low affinity for the dye may contribute to this, no smaller protein is present in sufficient concentration to permit detection by their absorption at 280 nm, though fraction 1 is readily detected by this means. Their concentration is then low compared to fraction 1 and consequently, since the amount of incorporation is relatively great, their specific activities must be high.

Incorporation of precursors into these proteins can be demonstrated two days after the fully expanded leaves have been detached. There is no direct evidence concerning the site of synthesis of these proteins. However, many soluble chloroplast proteins appear to be synthesized on the chloroplast system [31]. If this is true of these chloroplast-located soluble proteins, then this is evidence that not all protein synthesis in the chloroplast is curtailed at the time when the synthesis of fraction 1 protein declines.

Synthesis of insoluble chloroplast proteins

In the experiments outlined above, except when the decline in the absolute rate of protein synthesis in detached leaves was reported (table 5), incorporation into proteins extracted from leaves by aqueous buffers was measured. However, about half the leaf protein is not extracted by buffers, and the amount of this 'insoluble' protein decreases in detached leaves. A study of changes in the pattern of protein synthesis with leaf age obviously must include this protein fraction. In incorporation experiments, it has repeatedly been observed that the specific activity of 'insoluble' proteins exceeded that of the proteins extracted by buffer (table 7). A contribution of micro-organisms to the activity in insoluble proteins is difficult to eliminate, but the proportion of label in insoluble proteins is not less when chloramphenicol (0.3 mM) is introduced with the precursor, or when short term (5 min) exposure to $^{14}CO_2$ is used to label protein and this argues against a large contribution by micro-organisms. The portion of total incorporation into insoluble proteins does not increase with time of leaf detachment.

Studies of ageing leaves by electron microscopy reveal that organization is lost in the endoplasmic reticulum and in the chloroplasts before any changes in mitochondria are detected [54–56]. This suggests that incorporation into insoluble proteins in fully expanded leaves no longer making fraction 1 protein may be largely in mitochondria. Synthesis of mitochondrial proteins has been reported to be rapid in mature tobacco leaves [57].

When the proteins of fully expanded wheat leaves are labelled by exposing leaves to $^{14}CO_2$, fractions enriched in mitochondria have the highest specific activity (table 8). However, other fractions, and particularly those in which the chloroplasts are concentrated, contain much more protein. The largest amount of newly synthesized protein is in the chloroplast enriched fractions. From the results in tables 7 and 8, it is clear that most of the newly synthesized 'insoluble' protein must be in the fractions containing the chloroplasts (500 g and 1,500 g fractions).

Synthesis of chloroplast 'insoluble' protein has been studied by measuring the specific activity of proteins in chloroplast preparations from which soluble proteins have been removed in hypotonic buffer. In fully expanded leaves, in which little synthesis of fraction 1 occurs (table 6), the insoluble proteins of the chloroplast preparations are significantly labelled (table 9).

A crude fractionation of these proteins may be achieved by extraction with acetic acid solution [59] after removal of the pigments with acetone. By this means three protein fractions are derived (table 9); all three are heterogeneous by gel electrophoresis [60,61]. In the acetic acid extract, the average specific activity of proteins precipitating at neutral pH is much greater than that of the proteins remaining in solution. This has been so in all preparations made from leaves at full expansion.

This result indicates that some of the chloroplast lamellae proteins are replaced within chloroplasts in which the synthesis of fraction 1 protein has virtually ceased. This is not due to contaminating bacteria, for these would not decrease in density during the hypotonic washes. Chloroplast soluble protein obtained from the initial hypotonic wash, when denatured by freezing and thawing and acetone treatment cannot be dissolved in acetic acid. Proteins denatured during recovery of the lamellae are therefore likely to contribute to the residue insoluble in acetic acid.

Incorporation into the chloroplast proteins which are insoluble in acetic acid decreases progressively after the detachment of expanded leaves (table 10). Incorporation into this fraction decreases more rapidly than does incorporation into soluble protein, or into the total leaf protein. The extent to which the synthesis of chloroplast lamellae proteins contributes to incorporation in the insoluble proteins of the leaf (table 7) appears to fall rapidly once the leaf is detached.

Pigments and proteins of the chloroplast lamellae are readily dispersed by treatment with detergents [62–64]. Treatment with Triton X-100 is also useful in distinguishing amino acid incorporation into chloroplasts from that into nuclei and bacteria [2,28]. In fully expanded leaves, most of the protein of the lamellae preparations, and most of the incorporated radioactivity is dispersed by treatment with Triton X-100 (table 11). If the leaves are detached, the amount of protein in Triton X-100 supernatants of lamellae preparations falls a little during two days, but incorporation relative to the overall incorporation into protein falls by about 75% (table 11). In detached leaves a marked and rapid fall in the synthesis of lamellae proteins is indicated. The results in table 11 also show how the pattern of protein synthesis changes as leaves age. In these expanded leaves, there is little incorporation in fraction 1 protein when the leaves are detached, but incorporation into the Triton-dispersed lamellae proteins and into some fraction 2 proteins continues. Two days after detachment, incorporation into fraction 2 proteins continues, but there is now little incorporation into the major lamellae fraction. Incorporation appears to be reduced selectively, first into fraction 1, then into the major lamellae fraction, and then into the fraction 2 components.

Conclusion

When leaf growth stops, the content of both cytoplasmic and chloroplast ribosomes declines greatly. From this it can be presumed that the role of each is reduced when growth ceases. In the cytoplasm, rRNA undergoes some turnover in the non-growing leaf. Whether chloroplast rRNA is also turned over is not clear from our evidence. But bacteria are known to use the products of turnover of RNA in preference to exogenous precursors, as discussed above. In view of the many similarities between chloroplasts and free-living procaryotes, there is room to doubt whether lack of incorporation of labelled uracil necessarily implies lack of synthesis of rRNA in chloroplasts.

In newly-expanded leaves protein turnover, which involves some chloroplast proteins, occurs at an appreciable rate. Of the chloroplast proteins replaced some are associated with membranes, and some are soluble. Among both membrane bound and soluble proteins there are large differences in turnover rate. The polypeptides of fraction 1 show only a very low turnover rate.

After full expansion the rate of protein synthesis gradually declines. The synthesis of some proteins declines more rapidly than the synthesis of others. The synthesis of some chloroplast components appears to continue at a time when the loss of chloroplast structure has commenced.

The gradual reduction in the complexity of the protein population being made within ageing leaves indicates that synthesis is reduced in a step-wise fashion, and that a number of distinct controls are activated or de-activated. Loss of function by the degradation of chloroplast ribosomes could be one of these steps.

References

[1] D. Spencer and S.G. Wildman, Biochemistry 3 (1964) 954.
[2] D. Spencer, Arch. Biochem. Biophys. 111 (1965) 381.
[3] M.S. Bamji and A.T. Jagendorf, Plant Physiol. 41 (1966) 764.
[4] M. Ranalletti, A. Gnanam and A.T. Jagendorf, Biochim. Biophys. Acta 186 (1969) 192.
[5] J. Lyttleton, Exptl. Cell Res. 26 (1962) 312.
[6] N.K. Boardman, R.I.B. Francki and S.G. Wildman, J. Mol. Biol. 17 (1966) 470.
[7] U. Heber, Nature 195 (1962) 91.
[8] J.T.O. Kirk, in: The Biochemistry of Chloroplasts, ed. T.W. Goodwin (Academic Press, London, 1966) vol. 1, p. 319.
[9] T.C. Hall and E.C. Cocking, Biochim. Biophys. Acta 123 (1966) 163.
[10] D. Racusen and M. Foote, Arch. Biochem. Biophys. 90 (1960) 90.
[11] J.A. Hellebust and R.G.S. Bidwell, Can. J. Botany 41 (1963) 969.
[12] J.A. Hellebust and R.G.S. Bidwell, Can. J. Botany (1964) 1.
[13] J.A. Hellebust and R.G.S. Bidwell, Can. J. Botany 42 (1964) 357.
[14] R.E. Click and D.P. Hackett, Proc. Natl. Acad. Sci. U.S. 50 (1963) 243.
[15] C.J. Brady, P.B.H. O'Connell, J.K. Palmer and R.M. Smillie, Phytochemistry 9 (1970) 1037.
[16] R.J. Ellis and I.R. MacDonald, Plant Physiol. 42 (1967) 1297.
[17] R.J. Ellis and I.R. MacDonald, Planta 83 (1968) 248.
[18] R.M. Smillie and G. Krotkov, Can. J. Botany 38 (1960) 31.
[19] U.E. Loening, Biochem. J. 102 (1967) 251.
[20] U.E. Loening and J. Ingle, Nature 215 (1967) 363.
[21] J. Ingle, Plant Physiol. 43 (1968) 1448.
[22] P. Akroyd, Anal. Biochem. 19 (1967) 399.
[23] A.C. Peacock and C.W. Dingman, Biochemistry 6 (1967) 1818.
[24] D.P. Nierlich, Proc. Natl. Acad. Sci. U.S. 60 (1968) 1345.
[25] B.J. McCarthy and R.J. Britten, Biophys. J. 2 (1962) 35.
[26] D.P. Nierlich, Science 158 (1967) 1186.
[27] R.D. Berlin and E.R. Stadtman, J. Biol. Chem. 241 (1966) 2679.
[28] K.K. Tewari and S.G. Wildman, Biochim. Biophys. Acta 186 (1969) 358.
[29] S.G. Wildman and J. Bonner, Arch. Biochem. Biophys. 14 (1947) 381.
[30] M.M. Margulies, Plant Physiol. 39 (1964) 579.
[31] R.M. Smillie, D. Graham, M.R. Dwyer, A. Grieve and N.F. Tobin, Biochem. Biophys. Res. Commun. 28 (1967) 604.

[32] M.M. Margulies and F. Parenti, Plant Physiol. 43 (1968) 504.

[33] P.W. Trown, Biochemistry 4 (1965) 908.

[34] A.C. Chibnall, in: Protein Metabolism in the Plant (Yale University Press, New Haven, Conn., 1939).

[35] J.G. Wood, D.H. Cruikshank and R.H. Kuchel, Australian J. Exptl. Biol. Med. Sci. 21 (1943) 37.

[36] Y. Fuchs and M. Lieberman, Plant Physiol. 43 (1968) 2029.

[37] D.I. Arnon, Plant Physiol. 24 (1949) 1.

[38] O.H. Lowry, N.J. Rosebrough, A.L. Farr and R.J. Randall, J. Biol. Chem. 193 (1951) 265.

[39] M. Shaw, P.K. Bhattacharya and W.A. Quick, Can. J. Bot. 43 (1965) 739.

[40] B.E.S. Gunning and W.K. Barkley, Nature 199 (1963) 262.

[41] B.I.S. Srivastava, Intern. Rev. Cytol. 22 (1967) 349.

[42] C.R. Slack and M.D. Hatch, Biochem. J. 103 (1967) 660.

[43] A.C. Rutner and M.D. Lane, Biochem. Biophys. Res. Commun. 28 (1967) 531.

[44] T. Sugiyama and T. Akazawa, J. Biochem. (Tokyo) 62 (1967) 474.

[45] J.A. Hellebust and R.G.S. Bidwell, Can. J. Botany 41 (1963) 985.

[46] S. Raymond, Clin. Chem. 8 (1962) 455.

[47] O. Vesterberg and H. Svensson, Acta Chem. Scand. 20 (1966) 820.

[48] H.W. Woodhouse, Symp. Soc. Exptl. Biol. XXI (1967) 179.

[49] C.G. Kannagra and H.W. Woolhouse, New Phytologist 67 (1968) 533.

[50] B.I.S. Srivastava and C. Arglebe, Plant Physiol. 42 (1967) 1497.

[51] B.I.S. Srivastava, Arch. Biochem. Biophys. 110 (1965) 97.

[52] A.E. Richmond and A. Lang, Science 125 (1957) 650.

[53] S.M. Ridley, J.P. Thornber and J.L. Bailey, Biochim. Biophys. Acta 140 (1967) 62.

[54] R.D. Butler, J. Exptl. Botany 18 (1967) 535.

[55] R. Barton, Planta 71 (1966) 314.

[56] S.J. Roux and J.T. McHale, Phyton (Buenos Aires) 25 (1968) 113.

[57] B. Parthier, Biochim. Biophys. Acta 72 (1963) 503.

[58] W. Cockburn, D.A. Walker and C.W. Baldry, Biochem. J. 107 (1968) 89.

[59] S. Fleischer, W.L. Zahler and H. Ozawa, Biochem. Biophys. Res. Commun. 32 (1968) 1031.

[60] E.M. Jordan and S. Raymond, Anal. Biochem. 27 (1969) 205.

[61] K. Takayama, D.H. MacLennan, A. Tzagoloff and C.D. Stoner, Arch. Biochem. Biophys. 114 (1966) 223.

[62] J.S. Kahn, Biochim. Biophys. Acta 79 (1964) 234.

[63] J.P. Thornber, R.P.F. Gregory, C.A. Smith and J.L. Bailey, Biochemistry 6 (1967) 391.

[64] J.S.C. Wessels, Biochim. Biophys. Acta 153 (1968) 497.

THE CONTROL OF MITOCHONDRIAL PROLIFERATION IN THE FACULTATIVE ANAEROBE, SACCHAROMYCES CEREVISIAE

A.J.S. BALL and E. Reno TUSTANOFF

Department of Pathological Chemistry, University of Western Ontario,
London, Ontario, Canada

The current literature has a plentitude of information on the varying aspects of the biosynthesis of various cellular components. Investigations of one of these components, mitochondria, have recently been centred on the facultative anaerobe, *Saccharomyces cerevisiae.* The reason this organism has been singled out for this particular study has been well documented [1,2]. This organism can be grown readily and reproducibly under relatively simple conditions. By manipulating the growth environment, the organism can be induced to alter the number if functioning oxidative organelles and thus present a discrete biological system which lends itself to the study of the assembly and organization of these sub-cellular particles.

The control mechanisms which are responsible for altering the enzymic machinery in a viable organism or cell after an alteration in its environment is under the control of the cell's own genetic apparatus. Recent reports have established fairly incontrovertibly that yeast mitochondria possess all the characteristics of self-replicating organisms, i.e., DNA [3], RNA [4], tRNA [5] and the various nucleic acid polymerases [6]. It has been also established that these constituents are functional in terms of RNA [7] and protein synthesis [8], with the resulting proteins being incorporated into mitochondrial structures [9]. In addition, the characteristics of the mitochondrial system are quite different from the nucleo-cytoplasmic system in yeast and are similar to those possessed by bacteria [3–8]. This difference not only adds credibility to the proposed independent functioning of the mitochondrial replicating system [10], but also allows one to employ these differences to preferentially alter the functional aspects of the mitochondrial systems without directly affecting those of the nucleo-cytoplasmic system [11].

The primary investigations on the development of mitochondria in yeast established the fact that the number of mitochondrial profiles per cell are altered in response to changes in their gaseous environment, i.e., aerobic versus anaerobic [12,13] and in response to different fermentable carbon sources [14]. Adaptation experiments, in which anaerobically grown yeast cells were exposed to oxygen have been carried out by a number of investigators [14,15] as well as studies on the repression phenomenon, wherein the carbon energy source was varied [16,17]. In order to rationalize our approach to the control aspect of this broad problem we have focused our attention specifically on the effects of the fermentative carbon source and the energy-carbon source transitions. Initially, experiments were carried out with stationary phase cells which were then exposed to repressing concentration of fermentative sugars (usually glucose) in a fresh and often different medium. They were then allowed to utilize this carbon source until a non-repressive environment [16,17] was reached and during this phase, the synthesis of mitochondrial components were studied [18]. During this type of experimental approach, yeast would undergo an initial adaptive lag phase, growth by fermentation and then a further adaptive lag phase before growth would resume on the products of fermentation. Inherent in this type of experiment are a number of other physiological phenomena; e.g. adaption to a different medium, which would superimpose themselves on the primary effects of the carbon source and/or its subsequent transitional effect. In the experiments to be presented here, we have chosen our conditions with these conclusions in mind and suggest that our results are due solely to the carbon source transition while acknowledging the basic limitation of batch culture technique.

It is the purpose of this paper to describe the aerobic growth characteristics of yeast and to further show how these characteristics are modified by the fermentative carbon source and to attempt to elucidate a regulating mechanism for the genesis of yeast mitochondria.

In this present report we are not primarily concerned with the nature of the assembly of mitochondria, but rather in the manner in which this assembly is controlled by the milieu of the cell.

Methods

Microorganism and culture procedures

Saccharomyces cerevisiae, strain 77 of the National Collection of Yeast Cultures (Brewing Industries Research Foundation, Nutfield, Surrey) was maintained as previously described [19]. Cells were grown aerobically in shake flasks at 30° containing the complete anaerobic medium supplemented with ergosterol and Tween 80 [14]. This medium is sufficient to support full growth on 2.7% hexose. Growth was initiated by the addition of sufficient inoculum to achieve a density of 0.40 mg of cells (dry wt) per g of added fermentable carbon source. In transfer experiments, cells grown aerobically on 2.7% galactose medium were harvested in their mid-fermentative phase, where the repressive carbon source was still in excess, and then transferred without washing into fresh galactose medium [11]. Preliminary experiments had shown that cells harvested at 0° and washed with cold water reverted physiologically into initial lag phase cells. This was probably due to cold shock and consequently these harvesting steps were omitted in order to minimize the physiological shock caused by the transfer.

Harvesting of cells

Cells were harvested sequentially during their growth periods according to standard procedures. Dry weight determinations were carried out on the twice washed cell pellets after heating at 115° for 24 hr.

For enzyme determination yeast pellets were subjected to fractionation by passage through a chilled French pressure cell (23,000 p.s.i.). Cell-free extracts were obtained by removing the cell debris by centrifugation (1,500 g for 5 min).

Analytical procedures

Respiratory ability of the cultured cells was measured on samples removed from the growth chamber and then aerated for 1.0 min prior to monitoring their oxygen uptake polarographically using a Clark-type electrode. Protein [20], glucose [21], galactose [22], ethanol [23] and DNA [24] concentrations were determined by standard procedures reported in the literature. Apparent absolute low temperature spectrums of whole yeast cells were determined with a Cary Model 11 Spectrophotometer. Thick suspensions of cells (5–10 mg dry wt/ml) in 50% glycerol were reduced with a few grains of sodium dithionite, frozen in liquid nitrogen and compared to a reference cuvette containing Kemwipe tissue [25].

Enzyme assays

Pyruvate decarboxylase [26], hexokinase [27], malic dehydrogenase [28], succinic dehydrogenase [29], isocitric dehydrogenase [30], alcohol dehydrogenase [31], NADH-cytochrome c reductase [32] and cytochrome c oxidase [33] activities were determined spectrophotometrically using standardized enzymic procedures. Malic dehydrogenase isoenzymes were resolved on a DEAE-A50 Sephadex column [34] and these activities were further visualized by polyacrylamide gel electrophoresis.

Expression of results

The kinetics of cellular proliferation and increase in enzymic activity are expressed in three different modes. Specific activity as a function of time is the most popular method, however it fails to elucidate the prime cause of an effect since it is a relative rate. One may only observe alterations in the relative rate of enzymic synthesis when the total cellular synthesis has changed. If both rates are changing at the same moment it is impossible to decide which is the prime effect (cf. figs. 5A, 9A and 9B).

The isometric plot first described by Monod et al. [35] where the increase in the specific activity is plotted against the increase in cell mass of the culture has also been used. A change in the slope of this curve signifies a rate change in the synthetic activity. A basic assumption of this type of plot is that the total rate of cell growth remains constant during the time in which the rate of enzymic synthesis changes. If both the rate of cellular growth and the slope (total activity per cell mass) change at the same time, it is difficult to interpret the prime cause of this change. This is illustrated by comparing figs. 5A and 9B.

The last type of representation and the one in which the majority of our data is expressed is a semi-logarithmic plot of total activity against time. This type of kinetic representation permits a comparison of the absolute rate of synthesis of various cellular parameters with a change in total cell mass. Those activities whose synthesis is coupled to the total growth rate give rise to curves which are parallel to those given by the total cell synthesis. Where activity curves deviate from this parallelism, then one must conclude that the effect is due either to a change in the expression of existing enzyme(s) or to a change in the mode of gene expression. The semi-logarithmic mode was used because these experiments give rise to biphasic growth curves and both phases are logarithmic in nature.

Results and observations

Effect of fermentable carbon source on aerobic growth characteristics

Fig. 1 illustrates the rate of increase in dry wt with time of yeast grown aerobically on three different carbon sources. Cells grown on fructose and sucrose manifested growth curves which were superimposable on the glucose-growth curve. These observations were also valid for ethanol production and respiratory potential (cf. figs. 2 and 3). These observations were valid when cells were either cultured on glucose slopes and grown in sucrose medium or cultured on sucrose and grown in sucrose medium; the yeast strain apparently being constitutive for α amylase. All growth curves demonstrated a biphasic or diauxie growth cycle, the first phase of which is dependent on energy from a fermentative sugar while the second phase utilizes the products of this fermentation for further growth, the length of the lag between phases being conditioned by the primary carbon source.

Fig. 2 shows the production and utilization of ethanol during these experiments. The nature of the sugar determines both the rate of accumulation as well as the maximum production of ethanol. The

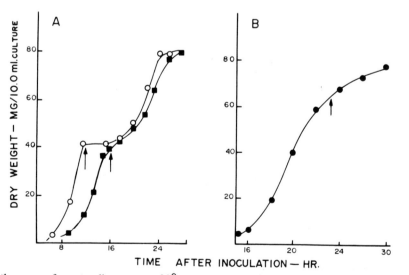

Fig. 1. Aerobic growth curves of yeast cells grown at 30°. A: yeast grown on 2.7% glucose medium —○———○—; yeast grown on 2.7% maltose medium —■———■—. B: yeast grown on 2.7% galactose medium —●———●—. Arrows indicate disappearance of primary carbon source from media.

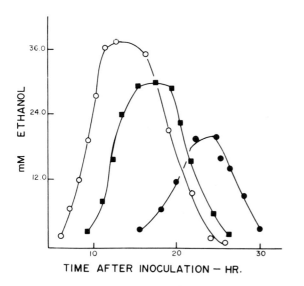

Fig. 2. Production and utilization of ethanol by aerobically grown yeast. Analyses were carried out on the supernatants obtained from samples used for dry wt determinations in fig. 1. Ethanol values for cells grown on 2.7% glucose medium −○——○−; on 2.7% maltose medium −■——■−; on 2.7% galactose medium −●——●−.

formation of ethanol terminates approximately half an hour after the primary carbon source is consumed and utilization begins just prior to the commencement of the secondary growth phase. This is clearly illustrated in the glucose-grown cells, where a long intermediary lag phase (fig. 1A) is associated with a broad plateau in ethanol formation (fig. 2). It would appear from this data, that the carbon energy source, during the secondary growth phase, is predominantly ethanol (cf. Polakis et al. [18]), and that the ethanol produced is completely utilized by the end of the experiment. Under these conditions further growth may occur at the expense of acetate [18].

The Q_{O_2} data from these cultures are illustrated in fig. 3A. It will be noted that the degree of repression of respiratory ability per cell is inversely proportional to the fermentative capacity of the yeast cells: glucose > maltose > galactose. The yeast endowed with the greatest respiratory capacity, accumulates the least amount of ethanol (cf. figs. 2 and 3A). These Q_{O_2} data also illustrate the phenomenon of derepression [16]; once the primary carbon source becomes exhausted, the Q_{O_2} increases rapidly, the

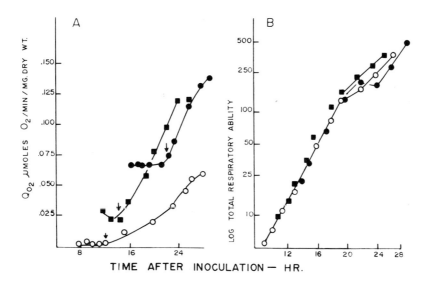

Fig. 3. Respiratory ability of yeast cells grown on various sugars. A: Q_{O_2} data of yeast cells samples at various times from their corresponding growth curves (cf. fig. 1). B: Log of the total respiratory ability (μmoles O_2 consumed per min per 5.0 ml of culture) of these same cells. Cells grown on 2.7% glucose medium −○——○−; cells grown on 2.7% maltose medium −■——■−; cells grown on 2.7% galactose medium −●——●−.

effect being similar for each carbon source. The final Q_{O_2} attained by these cells is conditioned by the nature of the primary carbon source, and not by the physiology of growth which at this stage are similar for each culture.

These results are in accord with our earlier report [36]. However, when this respiratory data is represented on a semi-log plot (fig. 3B) as a function of the total respiration against time, the rate of increase of the total respiratory ability is the same in each culture and the three sets of data fall progressively on the same straight line until late exponential phase conditions are reached. This rate change occurs at the same time in all cultures, which might suggest that this is the point at which these cells finally become dependent on mitochondrial energy production and therefore the point at which fermentative utilization of the primary carbon source finally ceases. Similar results were obtained with a variety of different strains of *S. cerevisiae* in our laboratory [37]. These data demonstrate that the rate of increase in the respiratory ability is the same in each culture and it is unaffected by the nature of the fermentable carbon source. The Q_{O_2} can be defined

as μmoles O_2/hr/mg dry wt; consequently if the amount of respiration (μmoles O_2/hr) is not conditioned by the carbon source (cf. fig. 3B), but the Q_{O_2} is (fig. 3A), then the effect of the carbon source must be exerted as a result of the increase in dry wt; i.e., the absolute cellular growth rate (figs. 1A and B).

A number of theoretical curves, Q_{O_2} versus time are shown in fig. 4A. These curves were calculated by assuming a constant logarithmic rate of increase in total respiration and calculating the ratios obtained when a change in the logarithmic rate of cell mass occurs. Fig. 4(a) shows a transition form similar rates to a relatively very fast growth rate (i.e., glucose, early exponential), (b) shows the same transition for a relatively fast growth rate (maltose, early exponential), (c) shows the transition from a fast growth rate to a very slow one (glucose, plateau phase), and (d) shows the transition from a very slow growth rate to a much faster one (glucose, end of plateau phase). All of these curves are asymptotic, as one would expect, and the various transitions, coded according to the relative rate of dry wt increase, when compared to the constant rate of total respiration increase, produce curves which bear a remarkable

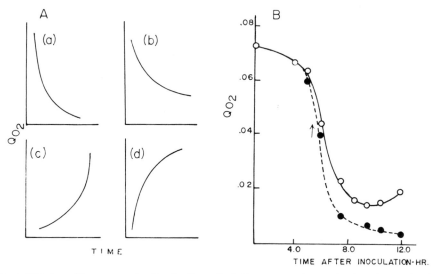

Fig. 4. Respiratory quotient profiles on yeast cells. A: theoretical Q_{O_2} curves demonstrating changes in the rate of Q_{O_2}/time when various changes occur in the growth rate during the maintenance of a constant rate of increase in total respiratory ability. (a) Similar to much faster growth rate, (b) similar to faster, (c) fast to slow and (d) slow to fast. B: lag phase respiration of aerobic glucose cells. Growth conditions were similar to those in fig. 1. These observations illustrate the change in Q_{O_2} between the onset of glucose utilization and the mid-fermentative phase −o——o−. The dilution curve was calculated on the premise that there would be no further increase in the total respiratory ability after 4.0 hr −●−−−●. Arrow indicates the commencement of glucose utilization.

resemblance to those observed experimentally (fig.3A). A transition of this kind is expected to occur during the lag phase of yeast cells grown aerobically on glucose. The results shown in fig. 4B show that the Q_{O_2} of the inoculum is maintained at a high value until glucose utilization is initiated and rapid growth results. Thus the changes observed in this experiment were what one would expect from the theoretical curves shown in fig. 4A[(a)]. The dilution curve (fig. 4B) was calculated by assuming that there would be no further synthesis of respiratory ability once glucose fermentation was initiated. Obviously, despite the large loss of respiratory ability per cell, there has been a continued synthesis of respiratory ability as predicted by fig. 3B.

One may conclude from these experiments that the fermentative carbon source has a profound effect on the ratio, total respiration/cell mass, while at the same time acknowledging that some respiratory limiting substance(s) is being synthesized at a constant rate independent of the carbon source in all three cultures. One is tempted to conclude that these substances are integral parts of the respiratory chain, and the following experiments were carried out in an attempt to characterize the substance(s).

Effect of D(–) chloramphenicol on various kinetic parameters of de-repressed cells

To exploit the possibility that some substance within the mitochondrion was being synthesized at a rate which is unaffected by the fermentative carbon source or its metabolite(s) the following series of experiments were carried out. Yeasts were grown aerobically on 2.7% galactose medium and transferred to fresh galactose medium or fresh galactose medium containing 1.0 mg of D(–) chloramphenicol per ml as described in the methods section. Since yeasts demonstrate a relatively high Q_{O_2} while fermenting galactose (partial de-repression) and because the galactose-ethanol transition is accomplished gradually and with a minimal intermediary lag phase (cf. figs. 1B and 3B) galactose was chosen as the carbon source. D(–) chloramphenicol has been shown to be a specific inhibitor of yeast mitochondrial protein synthesis [13], thus it was incorporated in this series of experiments to test its effects on various kinetic parameters which are altered during carbon source de-repression.

The increase in dry wt and total DNA after yeast cells have been transferred to new media are shown in figs. 5A and B respectively. The control growth curve

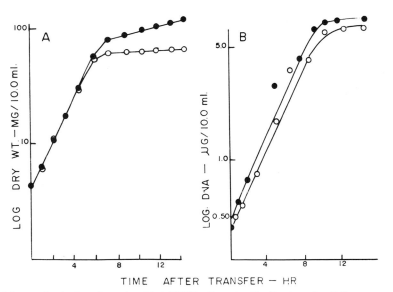

Fig. 5. The effect of chloramphenicol on dry wt and total DNA of yeast cells after transfer. Cells were grown aerobically at 30.0° in 2.7% anaerobic galactose medium, harvested by centrifugation and re-inoculated (4.0 g of wet wt cells per litre of medium) into fresh 2.7% galactose media, with and without 1.0 mg/ml of D(–) chloramphenicol. A: dry wt. B: total DNA. Control –●———●–; medium supplemented with chloramphenicol –○———○–.

is logarithmic and biphasic as is the DNA synthetic curve. Chloramphenicol has very little effect on the fermentative phase of growth, but it does abolish the secondary phase of cell growth. DNA synthesis is virtually abolished during this secondary phase, and this correlates with the viable cell count.

As expected, chloramphenicol inhibits the synthesis of respiratory ability (fig. 6A) and causes the Q_{O_2} to gradually decrease. This decrease in Q_{O_2} is a result of a decreased rate of synthesis of respiratory ability (cf. fig. 6B) while the growth rate remains unchanged (fig. 5A). The slope of the control curve indicates the doubling time for the synthesis of the respiratory apparatus was 2.0 hr and this value was identical with that observed when these cells were grown on the other sugars (cf. fig. 3B).

Fig. 7A shows the rate of synthesis of hexokinase, a cytoplasmic enzyme, during these transfer experiments. The rate of synthesis of this enzyme initially is logarithmic and then gradually it slows down to the point where fermentation ceases (galactose exhausted) when it completely stops. Chloramphenicol has no effect on this enzyme. Fig. 7B shows similar data for isocitric dehydrogenase. Again, chloramphenicol has no effect on the synthesis of this enzyme during the fermentative phase. This enzyme exhibits the first observed induction phenomenon in our system, a change in rate occurring 2 hr after transfer, during

which time the rate of cell growth remains constant. Synthesis then continues at this rate until the exhaustion of galactose causes a decrease to a second steady rate. The kinetics for succinic dehydrogenase synthesis is illustrated in fig. 8A. The synthesis of this enzyme follows a pattern similar to that of isocitric dehydrogenase, showing biphasic kinetics, the transition point coinciding with that of the growth curve and the changeover from galactose fermentation to ethanol metabolism. The effect of chloramphenicol on this enzyme system is not immediate, but results in a gradual inhibition, culminating in a 20% inhibition. The data for pyruvate decarboxylase activity (fig. 8B) is similar to that shown for succinic dehydrogenase. However, the two control curves differ in that the synthesis of pyruvate decarboxylase stops after the transition point. The effect of the antibiotic on these two enzymes is similar.

Fig. 9A and B illustrate the data for two enzymes of the respiratory chain, cytochrome c oxidase and NADH cytochrome c reductase. The kinetics of synthesis of these two enzymes differs radically from the other enzyme activities monitored during these transfer experiments. These activity curves are monophasic straight lines, and the rate of synthesis is unaffected by the growth rate transition. Chloramphenicol causes an immediate inhibition of the synthesis of these enzymes, inhibiting cytochrome c

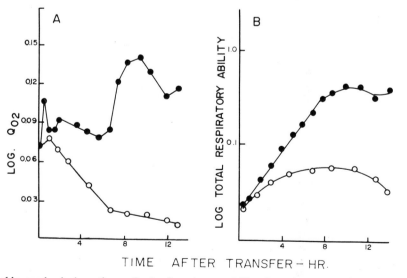

Fig. 6. The effect of chloramphenicol on the synthesis of respiratory ability. Experimental conditions are identical to fig. 5. Q_{O_2} data. B: log of the total respiratory ability. Control —●———●—; medium supplemented with chloramphenicol —○———○—.

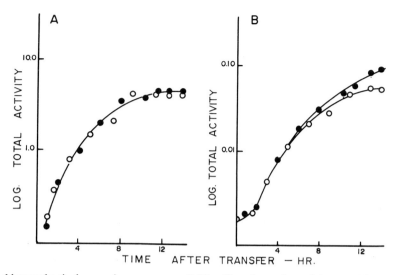

Fig. 7. The effect of chloramphenicol on various enzyme activities. Experimental conditions are identical to fig. 5. A: log total activity of hexokinase after transfer. B: the log of total activity of isocitric dehydrogenase. Control —●———●—; medium supplemented with chloramphenicol —○———○—.

oxidase 50% and the reductase 25%. The kinetics, however, remained logarithmic in the presence of the antibiotic until the transition point is attained when energy starvation prevents the synthesis of these and other enzymes. The doubling time for the oxidase was 2 1/4 hr and the reductase 2 hr, which are in accord with doubling time for the respiratory ability

of transferred cells and cells grown on only one carbon source. The doubling time for the inhibited oxidase was 6.0 hr and the inhibited reductase 3.0 hr. In figs. 10A and B the data for NADH cytochrome c reductase is represented by the other two methods described in the methodology section. The first (fig. 10A) illustrates specific activity of the enzyme

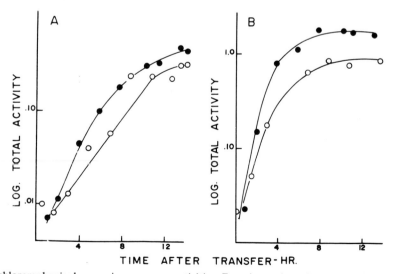

Fig. 8. The effect of chloramphenicol on various enzyme activities. Experimental conditions are identical to fig. 5. A: log total activity of succinic dehydrogenase after transfer. B: the log of total activity of pyruvate decarboxylase. Control —●———●—; medium supplemented with chloramphenicol —○———○—.

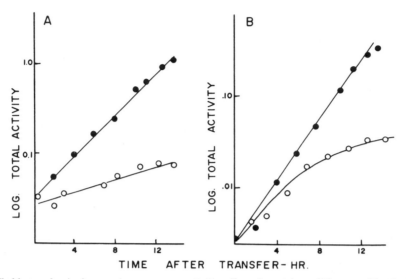

Fig. 9. The effect of chloramphenicol on various enzyme activities. Experimental conditions are identical to fig. 5. A: log total activity of cytochrome c oxidase after transfer. B: the log of total activity of NADH cytochrome c reductase. Control —●———●—; medium supplemented with chloramphenicol —○———○.

against time. This method shows that the amount of enzyme per cell remains constant up to the point of transition when this ratio rapidly changes. These rate changes are also reflected when this data is replotted using the isometric plot (fig. 10B). The constant specific activity is shown to be due to a constant relative rate of synthesis (i.e., 0 to 50 mg/ml) followed

by an accelerating relative rate which generates a continually increasing specific activity of this catalytic protein. Chloramphenicol, which inhibits the synthesis of this enzyme, brings about a decrease in specific activity and consequently a reduced slope of this plot.

Apparent absolute low temperature spectrums of

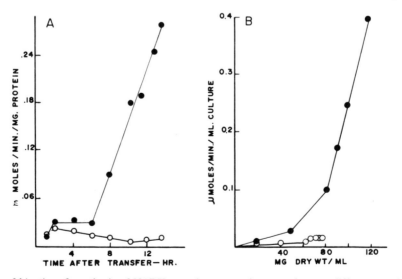

Fig. 10. Comparison of kinetics of synthesis of NADH cytochrome c reductase using two different graphic representations. Data from fig. 9B has been replotted A, as specific activity and B, as an isometric plot. Control —●———●—; medium supplemented with chloramphenicol —○———○—.

these galactose-transferred cells are shown in fig. 11. The spectrum of the control cells A and B (7 and 14 hr after transfer) show the classical α-absorption bands at 602, 558, 553 and 547.4 mμ which correspond to cytochromes a-a$_3$, b, c$_1$ and c respectively. The chloramphenicol grown cells E and D (7 and 14 hr after transfer) reveal one major peak at 550 mμ and a suggestion of a minor one at 556 mμ and a β peak region at about 522 mμ. For comparison absorption spectrum of anaerobically grown yeast cells are shown, C. These latter cells are devoid of cytochromes a-a$_3$, b and c$_1$. These results support the observation by Clark-Walker and Linnane [19a] that chloramphenicol completely inhibits the formation of cytochrome a-a$_3$, b and c$_1$.

Effect of D(−) chloramphenicol on malic dehydrogenase isoenzyme activity

The total activity curve for malic dehydrogenase (fig. 12A) shows that the response of this enzyme differs from that of all the others measured. The initial rate for the control transfer was logarithmic, ceasing when ethanol became the prime energy source, and then resuming again after a lag phase of 2.0 hr. This secondary synthetic phase was suppressed

by chloramphenicol. Since Holzer et al. [28] initially characterized two isoenzyme activities for this enzyme, experiments were undertaken to isolate these activities by DEAE-Sephadex chromatography (cf. fig. 13A). When the total activity of this enzyme was resolved in its constituent isoenzyme pattern (mitochondrial and cytoplasmic), a differential synthesis of these isoenzymes occurred during the transitional lag phase. This data is illustrated in table 1. The ratio of these isoenzymes in the control remains approximately constant for 10 hr and then abruptly changes during the transitional phase in favour of the mitochondrial isoenzyme. The secondary synthesis of the mitochondrial enzyme is abolished by the antibiotic. These isoenzymes were further resolved by disc electrophoresis (fig. 13B). The electrophoretic pattern of 36 hr grown culture exhibited a greater portion of cytoplasmic isoenzyme activity than the 24 hr grown cells. This activity is retained in the spacer gel in our system. Four bands appear in the running gel and have been shown to be mitochondrial in origin since isolated rat liver mitochondria showed a similar pattern. This latter activity from the mammalian mitochondria appeared only when they were traumatically treated to release their bound malic dehydrogenase

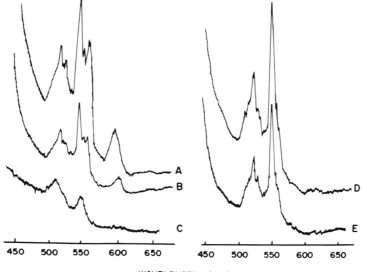

Fig. 11. Apparent absolute low temperature spectrums of galactose transferred cells. A: control cells 7 hr after transfer. B: control cells 14 hr after transfer. C: anaerobic cells. D: chloramphenicol treated cells 7 hr after transfer and E: chloramphenicol treated cells 14 hr after transfer.

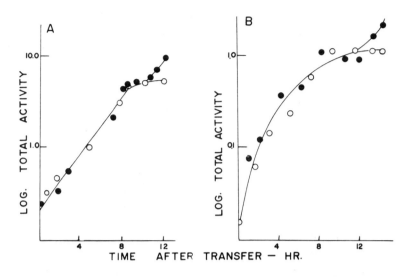

Fig. 12. The effect of chloramphenicol on various enzyme activities. Experimental conditions are identical to fig. 5. A: log total activity of malic dehydrogenase after transfer. B: the log of total activity of alcohol dehydrogenase. Control —●———●—; medium supplemented with chloramphenicol —○———○—.

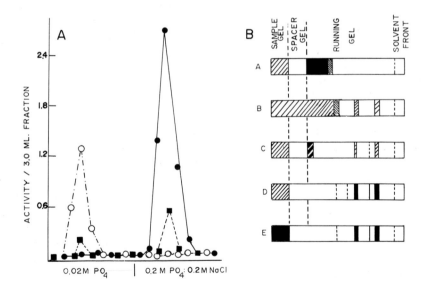

Fig. 13. Column and electrophoretic separation of malic dehydrogenase isoenzymes. A: resolution of malic dehydrogenase isoenzymes from rat liver mitochondrial —○———○; from a crude yeast extract, —■———■—; from yeast mitochondria —●———●—, by DEAE Sephadex A ion-exchange chromatography. B: disc electrophoresis of malic dehydrogenase carried out at room temperature for 1.5 hr at 4 mA/tube. (a) Rat liver mitochondria, (b) Rat liver mitochondria treated with Triton X-100, (c) Rat liver mitochondria diluted 1:10 and frozen and thawed three times, (d) 24 hr aerobic yeast extract, and (e) 36 hr aerobic yeast extract.

Table 1

Yeast cells were transferred from slightly repressed conditions (2.7% galactose medium) to fresh 2.7% galactose medium. Isoenzymes were resolved on a DEAE-A50 Sephadex column according to Schmidt et al. [34], and assayed spectrophotometrically [28]. Values expressed are the ratio of cytoplasmic:mitochondrial enzyme activities.

	Time in fresh medium (hr)			
	0.0	4.0	10.0	14.0
Control	1:4.7	–	1:4.7	1:10.7
D(–) chloramphenicol	–	1:5.5	–	1: 5.5

(gels B and C). These results would appear to confirm Holzer's findings [38] that there are several mitochondrial isoenzymes for malic dehydrogenase. Similar kinetics were observed for the synthesis of total alcohol dehydrogenase activity (fig. 12B). This activity has also been shown to be divisable into two isoenzymes, one cytoplasmic and the other mitochondrial [39,40].

Discussion

The growth experiments on different fermentable sugars document the effects of the fermentative carbon source on some of the parameters of aerobic growth. The carbon source determines the length of the initial lag phase and the growth rate of the fermentative phase (figs. 1A and B), but apparently does not affect the rate of the synthesis of a substance(s) which endows the yeast cells with their respiratory potential (fig. 3B). The alteration in Q_{O_2} (fig. 3A), can be explained in terms of these two parameters. The final ratio of the total respiration/cell mass depends not on the growth phase, but on the relative rate of increase in these two values. Thus the final respiratory quotients are determined by the point in time at which the measurement is made, and the determining factor is the varying growth rate which in turn is conditioned by the fermentative carbon source. In contrast to the conclusion drawn by Tustanoff and Bartley [41], it would appear that the fermentative carbon source does not control the rate of synthesis of the respiratory ability, but controls the rate of growth of the nucleo-cytoplasmic system.

The data in fig. 4B shows that the decrease in Q_{O_2} during the early fermentative phase is dependent on cellular growth, and in combination with the information shown in fig. 3B demonstrates that glucose is not repressing the synthesis of the respiratory-limiting substance(s). The effect of the lag phase in this case is minimal and is a function of the cell's fermentative ability. The logarithmic synthetic rate is modified independently in each culture after 20 hr of growth and this change is independent of the carbon source (fig. 3B). This observed effect bears no apparent relationship to ethanol utilization (cf. figs. 2 and 3B) but, considering the batch culture method used in these experiments, these results are probably due to limiting aeration rate.

It must be stressed that the linearity of the logarithmic plot (total respiratory ability against time) is not affected by the transition from fermentation to ethanol utilization. This effect is common to both experimental approaches; i.e., total growth on one carbon source and a transfer from one to another. These experiments suggest that the production of some rate-limiting substance, possibly a component of the respiratory chain, is being synthesized at the same rate in all the different cultures, and that this rate is unaffected by the changes in the gross synthetic rates of the whole cell. The most likely source of such a substance(s) would be an independent DNA-RNA protein synthesizing system, which is protected from changes in the external environment of the yeast. Such a compartmentalized system is found inside the mitochondrion.

The transfer experiments relate the respiration data directly to both mitochondrial and cytoplasmic enzyme activity. The only enzyme activity curves which parallel those of the total respiratory ability are cytochrome c oxidase and NADH cytochrome c reductase. These two activities are affected by the "petite" mutation and are also inhibited by chloramphenicol. There are good grounds for suggesting that the activities of these two enzymes are dependent upon some product of mitochondrial DNA [42,43]. The nature of these two enzymes, cytochrome c oxidase (complex IV), and NADH cytochrome c reductase (complex I and III), militates against a system of independent control mechanisms. This would not be inconceivable, if all the components of these systems originated within the confines of the mitochondrion. Unfortunately, many of the components of these haem-lipid-protein complexes originate extramitochondrially [44] and would therefore be subjected to the kinds of inductions and repressions characteristic of nucleo-cytoplasmically derived substances (cf. figs. 6–10). A more attractive hypothesis is one in which the organizing of these complexes within the mitochondrion is dependent upon some "polarizing" or "organizer" substance which itself could be derived from mitochondrial DNA. This would be analogous to mitochondrial structural protein [45]. Many of the so-called structural proteins have on further analysis proved to be complex mixtures [46]. However, Lusena et al. [46] have shown that only one component of a structural protein extract is affected by "petite" mutations, and is therefore dependent upon a wild-type mitochondrial DNA for its production. Thus, a single protein or organizer molecule, synthesized independently of any direct nuclear control by mitochondrial DNA and not affected subsequently by the external environment, would explain our result. The organizer would be responsible for generating the activities of both the oxidase and the reductase complexes and thus be responsible for the independent, rate-limited synthesis of total respiratory ability. The other enzymic data substantiates this hypothesis. Enzymes such as succinic dehydrogenase, which are possibly less dependent upon the structural integrity of the mitochondrion for their activity, are less affected by the reduced concentration of the organizer than are those enzymes such as complex IV (cf. figs. 8A and 9). Enzymes which are even less dependent upon the integrity of the mitochondria for activity (i.e., malic and isocitric dehydrogenase) are even less affected by the proposed inhibition of the organizer substance by chloramphenol. These and similarly derived enzymes (e.g. hexokinase) are subjected to induction and repression in response to the environment of the yeast as amply illustrated by the data in figs. 6–8. That the two respiratory chain enzymes differ from this group is made explicit by the results shown in figs. 9A and B and emphasized by the malic dehydrogenase data (cf. fig. 12A and table 1). The latter data were included in order to show that although malic dehydrogenase is bound to the mitochondrial mem-

branes, and can be differentially released (fig. 13B) from the mitochondrion, the synthesis of its isoenzymes are directly influenced by changes in cellular metabolism. Whether this effect is due to a differential gene induction (cf. Munkres [47]), or whether it is due to a greater availability of mitochondrial membrane for the soluble component to bind to, is not clear in this case. However, the pre-treatment of the mitochondrial preparations also gives rise to a variety of electrophoretic patterns (cf. fig. 13B) for this enzyme and these artifacts must be considered.

Very little is known regarding the assembly of sub-cellular organelles or how this assembly is controlled. The experiments described above suggest that there is no direct link between the two different DNA systems in yeast cells. Ball and Tustanoff [11] have previously suggested that the synthesis of tricarboxylic acid cycle enzymes which occurs at the end of the fermentative phase is induced by a change in the internal nucleotide concentration resulting from the cessation of fermentation. This conclusion is consistent with the data presented above and adds credence to the hypothesis that the nucleo-cytoplasmic and mitochondrial synthetic systems are independent.

In terms of biogenesis of mitochondria, this independence would lead to an uncoordinated form of auto-assembly. The result would be a population of mitochondria which might approach homogeneity under constant culture conditions, but which under changing conditions should be heterogeneous.

Adaption (anaerobic to aerobic) may be an example of such a condition and evidence for mitochondrial heterogeneity already exists [45,48]. Lloyd et al. [49], studying the distribution pattern of organelles from anaerobic whole yeast extracts, isolated two particles by zonal centrifugation which sedimented at a density of 1.15 and 1.21. Since these particles were present under anaerobic conditions, it was possible to follow the enzymic changes associated with these organelles during oxidative adaptation. During adaptation, cytochrome c oxidase activity initially appeared mainly in the 1.2 particle and after 30 min in the 1.15 particle and after 1 hr in particles < 1.15, indicating that for at least one enzyme of the respiratory chain, synthesis and binding into a mitochondrial particle are not synchronizid. Similar transitions were also observed for the oligomycin-sensitive and insensitive ATPases.

The role of oxygen in this system is not clear. Both Criddle and Schatz [45], as well as Watson et al. [50] have observed mitochondrion-like particles in anaerobic yeast, and have shown that the presence of oxygen is necessary for the elaboration of a true respiratory chain and its resulting structures.

Anaerobically-grown yeast do not become "petite", so one must conclude that mitochondrial DNA is passed from cell to cell during anaerobic growth. It seems reasonable to suggest that it is embedded in some form of particle which would serve to protect it from its cellular surroundings during this passage. Indeed Criddle and Schatz [45] found mitochondrial DNA associated with one of their mitochondrion-like particles. From an analogy with bacterial systems [51], one might expect the replication of the mitochondrial DNA to be in some way associated with the concentration of its gene products. Thus the organizer substance might well be produced under all growth conditions, partly to ensure that the DNA was protected from the yeast cytoplasm, and partly to ensure its replication, by triggering the appropriate feed-back mechanisms. The role of oxygen therefore would be to provide nacent mitochondria with a terminal electron acceptor, such that the functioning of the as yet embryonic mitochondria would then influence the cytoplasmic milieu in such a manner as to favour the production of larger amounts of mitochondrial precursors. Cohen-Bazire et al. [52] have suggested that the oxidation-reduction potential of the haem groups plays a role in the control of haem synthesis in *Rhodoseudomonas* and a similar role for oxygen could be imagined in the mitochondrial system.

Since this discussion is highly speculative, one may conclude that the lack of coordination between the nucleo-cytoplasmic and the mitochondrial DNA-RNA protein synthesizing systems demonstrated in this paper does at least set a limit to the possible ways mitochondriogenesis might be controlled in yeast.

Acknowledgements

This investigation was supported by a grant (MA-1460) from the Medical Research Council of Canada given to E.R.T. We wish to thank Dr. W.B. Elliot, Department of Biochemistry, State University of New York at Buffalo, for the use of this instrument which was purchased on a grant awarded him by the U.S. Public Health agency (Grant GM06241).

References

[1] P.P. Slonimski and F. Sherman, Biochim. Biophys. Acta 90 (1964) 1.
[2] K.K. Tewarti, W. Votsch and H.R. Mahler, Biochem. Biophys. Res. Commun. 21 (1965) 141.
[3] E. Winterberger and G. Viehause, Nature 220 (1968) 679.
[4] E. Winterberger, in: Regulation of Metabolic Processes in Mitochondria, eds. J.M. Tager, S. Papa, E. Quagliariello and E.C. Slater (Elsevier, Amsterdam, 1966), p. 439.
[5] T.S. Work, J.L. Coote and M. Ashwell, Federation Proc. 27 (1968) 1174.
[6] S.A. Leon and H.R. Mahler, Arch. Biochem. Biophys. 126 (1968) 305.
[7] M. Ashwell and T.S. Work, Biochem. J. 116 (1970) 25P.
[8] A.W. Linnane, A.J. Lamb, C. Christodoulou and H.B. Lukins, Proc. Natl. Acad. Sci. U.S. 59 (1968) 1288.
[9] G.M. Kellerman, D.E. Griffiths, J.W. Hansby, A.J. Lamb and A.W. Linnane, these proceedings.
[10] A.J.S. Ball and E.R. Tustanoff, FEBS Letters 1 (1968) 255.
[11] A.J.S. Ball and E.R. Tustanoff, Biochim. Biophys. Acta 199 (1970) 476.
[12] P.P. Slonimski, Soc. Chem. Ind., Monograph 3, 1958.
[13] A.W. Linnane, D. Biggs, M. Huang and G. Clark-Walker, in: Some Aspects of Yeast Metabolism, ed. R.K. Mills (Blackwell Scientific Press, Oxford, 1968) p. 217.
[14] E.R. Tustanoff and W. Bartley, Canad. J. Biochem. 42 (1964) 651.
[15] P.P. Slonimski, La Formation des Enzymes Respiratories chez la Levure, (Mason et Cie, Paris, 1956).
[16] M.F. Utter, E.A. Duell and C. Bernofsky, in: Some Aspects of Yeast Metabolism, ed. R.K. Mills (Blackwell Scientific Press, Oxford, 1968) p. 197.
[17] A. Castelli, G. Parenti-Castelli, E. Bertoli and G. Lenaz, Ital. J. Biochem. 38 (1969) 35.
[18] E.S. Polakis, W. Bartley and G.A. Meek, Biochem. J. 97 (1965) 303.
[19] E.S. Polakis, W. Bartley and G.A. Meek, Biochem. J. 90 (1964) 369.
[19a] G.D. Clark-Walker and A.W. Linnane, J. Cell Biol. 34 (1967) 1.
[20] O.H. Lowry, N.J. Rosebrough, A.L. Farr and R.J. Randal, J. Biol. Chem. 193 (1951) 265.
[21] J.D. Teller, Abstracts of papers, 130th Meeting A.C.S. (1965) 69C.
[22] Worthington Biochemical Corp. Technical Bulletin, Freehold, N.J. U.S.A.
[23] R. Bonnichsen and H. Theorell, Scand. J. Clin. Lab. Invest. 3 (1951) 58.

from 6 to 4 hr. After synchronization the cells were stored at a density of 5 mg dry wt per ml (about 2×10^8 cells per ml) of starvation medium. For as long as 7 days such cells retained the capacity to give rise to cultures with a high degree of synchrony, once inoculated into the appropriate growth medium.

Culture system

Cultures were grown in a semisynthetic medium [12] containing 3% glucose or galactose as the principal carbon source. Sufficient inoculum was added to yield an initial cell density of 10×10^6 cells per ml, after which 3 to 6 liters of culture were grown in a MicroFerm fermentor (New Brunswick Sci. Co., New Brunswick, New Jersey) at either 20.0 ± 0.1 or $25.0 \pm 0.1°C$. A circulation system was devised which permitted the culture to be pumped out of the fermentor vessel continuously, at a rate of about 400 ml per min. The effluent was passed through a flow cell (5 mm lightpath) in a Bausch & Lomb Spectronic-20 colorimeter, which was coupled by an adapter to a linear recorder, and the turbidity of the culture was recorded continuously at 515 nm. Samples could be removed when necessary as the culture passed through a two-way valve during the return flow to the fermentor vessel.

Control cultures consisted of asynchronous cell populations which, in every other respect, were handled identically with the synchronous experimental materials.

Procedures to monitor cell synchrony

Initially, synchrony was monitored by determining the percentage of budding cells and by direct microscope cell counts, according to the criteria established by Williamson and Scopes [17]. In addition, whole cell DNA [18], whole cell protein [19], and culture turbidity routinely were assayed throughout the growth period. A comparison of the automatic trace of culture turbidity with the microscope observations of cells and budding revealed that the shape of the turbidity trace provided a precise measure of the synchrony achieved in a culture, and always was in accord with the microscope determinations. For later experiments, therefore, culture turbidity alone was used as a major means to monitor synchrony in a growing culture, along with measurements of whole cell DNA and protein content as independent checks of culture synchrony.

Oxygen uptake of whole cells

Both glucose and galactose grown cultures, at either 20 or 25°C, were assayed using three different polarographic methods. In every case the rates of oxygen uptake were measured on a unit-of-culture-volume basis. One polarographic method involved withdrawing portions of the culture at 10-min intervals, and measuring oxygen uptake of these samples by means of a Clark-type electrode and a YSI Biological Oxygen Monitor (Yellow Springs Instr. Co., Yellow Springs, Ohio). The rate of oxygen consumption was determined from the slope of the current-versus-time curve as the saturated oxygen concentration declined from 100% to zero. A second method permitted continuous recording of oxygen saturation at one electrode inside the fermentor vessel and at another electrode placed near the end of the circulation system, the latter having adjustments to remove all air bubbles from the culture being withdrawn from the fermentor as well as being constructed of materials which ensured negligible oxygen diffusion. By measuring the percent oxygen saturation of the culture, the percent saturation after flow through the closed circulation system, and the time of flow between electrodes, the rate of oxygen uptake could be determined. The third method was similar to one described by Scopes and Williamson [12] in that oxygen uptake was measured while aeration to the culture was interrupted at regular but brief intervals. In this system, Q_{O_2} determinations were based upon the rate of loss of oxygen from the culture and the rate of diffusion of oxygen into the culture.

Enzyme assays

Eighty ml samples were removed at 15-min intervals throughout growth, and were stored in an ice bath until the end of the growth period. The cells then were sedimented and resuspended in 50 mM Tris buffer (pH 7.4), after which the suspension was passed once through a French pressure cell at 18,000 psig. The intact cells and debris were sedimented at $4,000 g$ for 10 min, and the resulting cell-free supernatant was stored overnight at $-20°C$. Tests indicated that none of the enzymes which were assayed had lost activity during storage. All enzyme activities were measured spectrophotometrically: cytochrome c oxidase by the method of Avers et al.

[16], L-malate dehydrogenase according to the ox-aloacetate method described by Ferguson et al. [20], and alkaline phosphatase using Torriani's procedure [21]. Each assay was performed in duplicate on all samples collected during the growth period.

Since enzyme activities were expressed per ml of culture, it was necessary to take account of the variability of cell breakage efficiency and, therefore, to obtain a reasonable estimate of the breakage achieved for each sample. The percentage of cell breakage for each sample was estimated by measuring the amount of protein released into the cell free brei and then comparing this value to the whole cell protein determination. In this fashion, activities then could be expressed as an extrapolation to the level that would be obtained per milliliter of culture had there been 100% broken cells.

Ultrastructure analysis

Twenty ml portions of the culture were removed at 10 min intervals during growth, and immediately mixed with 20 ml of chilled 5% cacodylate-buffered glutaraldehyde (pH 7.2). Then, the cells were sedimented and resuspended in Karnovsky's fixative [22] for 2 hr, in an ice bath. The cells then were spun down, washed four times with distilled water, and resuspended at room temperature for 70 to 90 min in 5% sodium permanganate buffered with 0.1 M cacodylate buffer (pH 7.2). Cells embedded in warm 2% agar then were immersed for 1 hr in a 1:1 mixture of saturated uranyl acetate and 100% acetone. After this, the cells were dehydrated in 90% acetone and in three changes of 100% acetone, for 0.5 to 1 hr in each. The samples were embedded in Luft's Epon 812 [23] mixture, and the resin was permitted to polymerize at 60°C for 48–72 hr. Sections were cut with diamond knives using the LKB or Porter-Blum MT-2 ultramicrotomes, and were floated onto parlodion-coated, carbon-stabilized grids for scanning and photography with the RCA-3G electron microscope. Most sections were left unstained. Photographs usually were taken at initial magnifications of 3,500 to 6,000 ×. Tabulations were made of mitochondrial numbers per cell section using no fewer than 30 sections of different cells for intervals during synchronous growth.

Results

Duration of the cell cycle

Cultures which were grown at 25°C on either glucose or galactose containing media completed a synchronous cell cycle in about 100 min, but the range was 85 to 120 min in different experiments. When identical cultures were grown at 20°C, the duration of one cell cycle was extended to 165 min in each experiment performed at this temperature.

Whole cell protein and culture turbidity

A typical pattern of the kinetics of increase in turbidity and in protein content of a single synchronous cell culture when grown at 25°C is shown in fig. 1. The first round of bud formation was observed by light microscopy to occur 55 min after the initiation of growth, and a second burst of budding occurred after 135 min of growth. According to the turbidity readings, the first increase appeared after 55 min and, after 135 min, the rate of increase

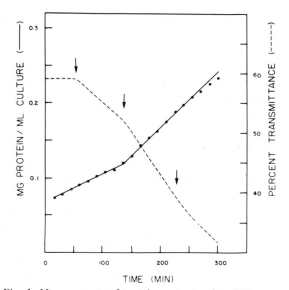

Fig. 1. Measurements of protein content and turbidity were made during synchronous cell growth of a single culture grown at 25°C on 3% galactose medium. Doubling in rate of turbidity increase (shown as decrease in transmittance) occurred at times indicated by arrows, which also coincided with the onset of budding cycles as seen with the microscope. The linear rate of increase in total protein doubled at the same time as the onset of the second budding cycle and the doubling in rate of turbidity increase.

essentially doubled. The linear rate of protein synthesis also doubled at 135 min, as determined from whole cell protein content. However, according to these measurements there was a constant rate of protein synthesis prior to the appearance of the first set of buds at 55 min. Both sets of values (fig. 1) show that synchrony begins to decay after the completion of the second cell cycle, which we found to be due primarily to depletion of the carbon source. In contrast with these data, both culture turbidity and protein content of the asynchronous control cultures showed essentially exponential increases during the growth period.

DNA synthesis

Synchronous cultures which were grown at 25°C on media containing 3% glucose (3% GLU) showed a distinct stepwise increase in DNA content at about the beginning of bud formation, as well as the suggestion of a minor increment in each cycle which occurred 20–30 min before budding (fig. 2b). To examine this situation in greater detail, synchronous cultures were grown on the same medium but at 20°C rather than 25°C. To verify that such cultures were entirely equivalent to others grown at the higher temperature, except for the expected reduction in metabolic rate, careful tests were made of the kinetics and rates of increase of the major parameters in cultures grown at each temperature. We found that turbidity, whole cell protein, whole cell DNA, oxygen uptake, and enzyme activity increases were entirely equivalent except that the cultures at 20°C exhibited a general reduction in rates of increase of those constituents which were measured. This observation permitted us to exploit the system at the lower temperature so that we could monitor events on an expanded time-scale, so to speak, and examine events more thoroughly than might be possible for all parameters and conditions at 25°C. The data from the 20°C cultures clearly indicated the occurrence of a distinct stepwise increase in DNA before budding and prior to the doubling of the bulk DNA of the cell (fig. 2c). To enhance the minor-DNA stepwise increase, on the assumption that it represented mitochondrial DNA, we also monitored synchronous cultures in media containing 3% galactose (3% GAL) and grown at 25°C. Since both 3% GLU and 3% GAL cultures were grown at 25°C, the time relationships

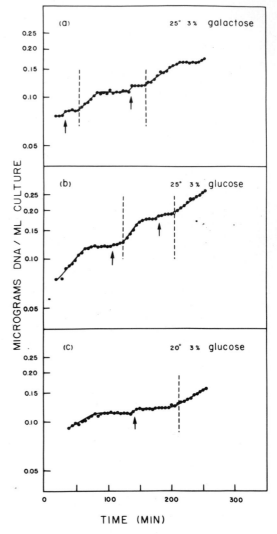

Fig. 2. DNA synthesis for three single cultures grown on media containing 3% glucose at 20 and 25°C and on 3% galactose media at 25°C. Note the enhancement of a small stepwise increase in DNA content in the galactose culture (a) as compared with the glucose culture (b) at 25°C, and the extended cycle duration at 20°C (c) compared with 25°C (b) for the two cultures grown on glucose media.

between the two stepwise increases in DNA would remain essentially the same. But we would expect an enhancement of mitDNA synthesis on galactose media, just as was found by Smith [24] using *Saccharomyces lactis*. As we had predicted, the stepwise increases were sharper in the 3% GAL cultures and the minor DNA increment consisted of about 20% of the total DNA in a cell cycle (fig. 2a).

Oxygen uptake

There were no stepwise increases in oxygen uptake (fig. 3) regardless of the polarographic method used. Stepwise increases were absent in synchronous cultures grown at 20 or 25°C, and on media containing either glucose or galactose. In each case there was either continuous or very frequent sampling, so that it is unlikely that these increments would have been overlooked if they had occurred at all. The exponential increases in rates of oxygen uptake were identical in the synchronous and the asynchronous cell cultures under all conditions which were tested.

Enzyme activities

All three enzymes showed stepwise increases in activity rate (fig. 4), but whereas alkaline phosphatase showed two steps of increase per cell cycle, both cytochrome *c* oxidase and malate dehydrogenase exhibited only single stepwise doublings during each

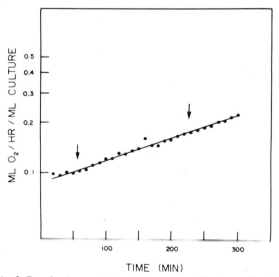

Fig. 3. Results from a single synchronous cell culture showing the exponential increase in rate of oxygen uptake during the entire growth period. Arrows indicate the time of onset of bud formation as was determined by microscopy, in cells grown on galactose media at 20°C. Readings of oxygen consumption were determined polarographically at 10 min intervals using the YSI Biological Oxygen Monitor and a Clark-type electrode.

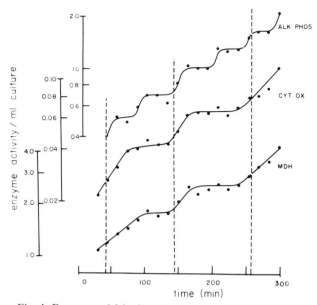

Fig. 4. Enzyme activities in a single synchronous cell culture grown at 25°C on galactose media. Increase in activity of alkaline phosphatase was measured as micromoles nitrophenol produced per min per ml culture; cytochrome oxidase activity as the first-order reaction rate constant, K (sec^{-1}), per ml of culture, and malate dehydrogenase activity as nanomoles NADH oxidized per min per ml of culture. The onset of budding in each cell cycle is indicated by a vertical dashed line.

cell cycle. The latter two enzymes showed increased activities at approximately the same time during a cell cycle, but the rate of increase was lower for malate dehydrogenase than for cytochrome *c* oxidase. The relationships of onset and termination of the several

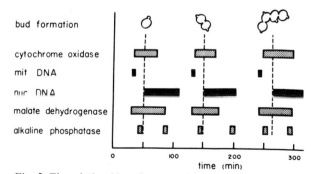

Fig. 5. The relationships of onset and termination of increase in DNA content and of enzyme activities are shown relative to the time of budding in each cell cycle (shown by vertical dashed lines). These values all were derived from a single synchronous cell culture.

biochemical activities measured during a single experiment are shown in fig. 5. Enzyme activity assays also were performed on asynchronous control cultures and in all cases essentially exponential increases in activity were observed throughout growth.

Ultrastructure analysis

Thus far we have examined in some detail samples from only one cell cycle of the three which were collected. All cells contained elongate and bizarrely shaped mitochondria (fig. 6). There was evidence from light microscopy and from electron micrographs (fig. 7) which indicated mitochondrial movement within the common cytoplasm of parent cell and bud. The mean number of mitochondria per cell section varied from 5.1 to 8.8, depending upon the time of sampling during the cell cycle. As can be seen from fig. 8, there was essentially a linear rate of increase in mitochondrial numbers per cell section in the interval corresponding to one cell cycle. There was a direct correlation between this increase in mitochondrial number and the increase in total cytoplasmic volume of the budding cell, with time (figs. 8 and 9). The parent cell and its bud were considered to be a single nucleocytoplasmic unit until the wall completely divided the complex into two separate cells.

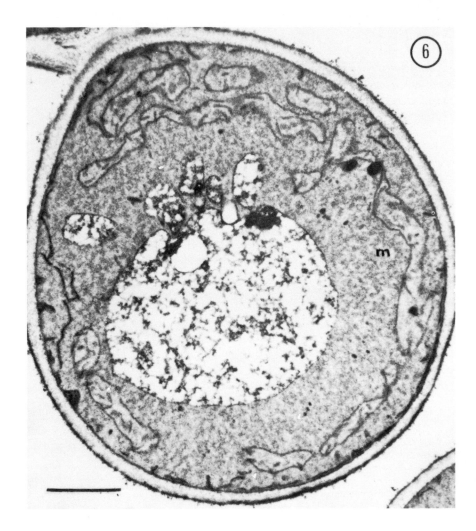

Fig. 6. Cell section showing the mitochondrial (m) morphology which is characteristic of exponentially-growing cultures, in contrast with organelles usually encountered in stationary-phase cells [16]. The bar is equal to 1μ.

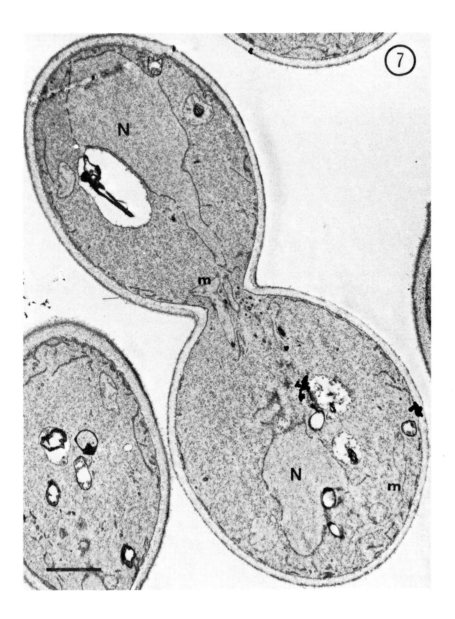

Fig. 7. Budding cell showing movement of nucleus (N) and mitochondria (m) within the common cytoplasm of parent and bud. The bar is equal to 1μ.

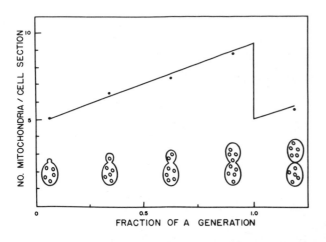

Fig. 8. Relationship between counts of mitochondrial profiles per cell section and the time during the cell cycle. A minimum of 30 whole cell sections were included to obtain the mean values plotted for mitochondrial numbers at each point shown in the cycle.

One constant feature of these exponentially growing cells was the abundance of cytoplasmic membranes of various lengths (fig. 9). These membranes usually were found to be closely associated with mitochondria, rather than occurring randomly in the cell.

Discussion

In these studies as in others which have been reported [15], many simple events appear to increase in rate in discrete steps during the cell cycle of synchronous cultures whereas complex developmental events more often demonstrate continuous change patterns. Also, stepwise increases may be obscured in randomly changing populations as we observed in every assay using asynchronous cultures as controls. Indeed, such asynchronous systems demonstrated the same exponential patterns of increase as were found to characterize phenomena based upon several or many separate components, all of which together contribute to one total measurable activity. For example, we found that identical patterns of increase occurred both in synchronous and in asynchronous cell cultures for changes in oxygen uptake rate. This

phenomenon almost certainly is a function of a number of individual events which have not yet been separated successfully [12].

The three enzymes which were assayed all showed step increases in synchronous cultures, but continual exponential rates of increase in the asynchronous controls. Cytochrome c oxidase is a uniquely mitochondrial enzyme which is structurally incorporated into the inner membrane of the organelle [25]. The abrupt onset of increase in enzyme activity coincided with the time of mitochondrial DNA (mitDNA) replication, and continued to rise until activity was doubled after about one-third of a cell cycle interval. Similar patterns of stepwise doublings characterized the other two enzymes, too, but there were differences in duration and number of steps for alkaline phosphatase, and for malate dehydrogenase there was a difference with respect to time of replication of the genome exercising control. Both alkaline phosphatase (AlkPh) and malate dehydrogenase (MDH) are considered to be under nuclear control exclusively [26,27]. We developed an assay based upon substrate inhibition phenomena [28] which showed that approximately 85–90% of the MDH activity was due to the cytoplasmic enzyme and only 10–15% was a function of mitochondrial MDH activity. Thus, if we consider the onset and duration of the stepwise increases relative to nuclear DNA replication, we find that MDH activity begins about 20–30 min after the cessation of nuclear DNA synthesis, but continues to increase for a significant portion of the succeeding cell cycle until the activity reaches a plateau value. AlkPh, on the other hand, showed two distinct and very abrupt doubling steps in each cell cycle, as has been reported by others [29]. Since we are observing activity and not synthesis of the enzyme proteins, we cannot determine whether any lag is due to a transcriptional or a translational control mechanism [30], or perhaps to some other form of regulation.

The failure to detect stepwise increases in the rates of oxygen uptake is in accord with observations made by Muller and Dawson [31] using synchronous phased cultures of *Candida utilis,* but contradicts the data reported by Scopes and Williamson [12] using *Saccharomyces cerevisiae* and that of Osumi and Sando [13] for synchronous cell cultures of the fission yeast *Schizosaccharomyces pombe.* Although there might be some question concerning the data for

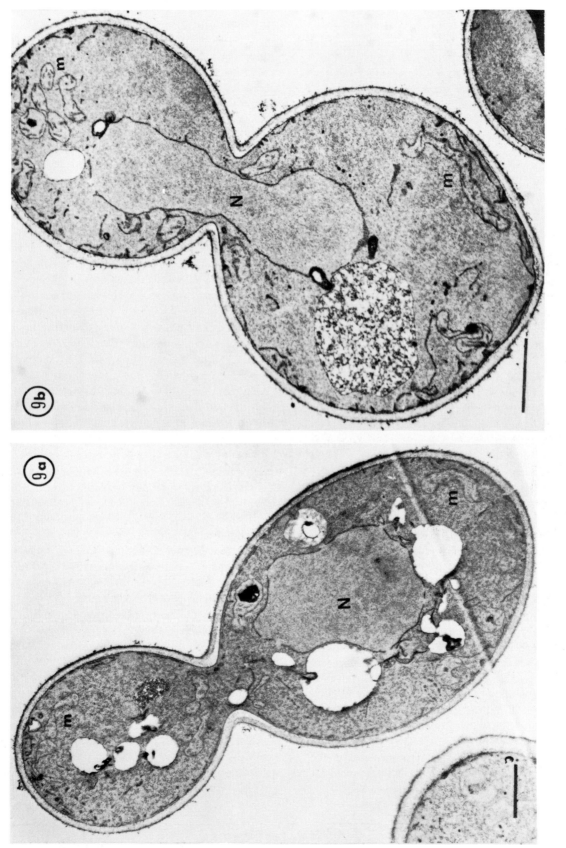

Fig. 9. Views through budding cells from different times during a cell cycle. Note the relationship between mitochondrial number per budding cell section and the relative size of the bud. The size of the parent cell remains relatively constant throughout the cell cycle.

S. pombe since readings were made at 30 min intervals during cell cycles lasting only 90 min, we cannot explain the differences between our results and those of Scopes and Williamson, since essentially identical conditions and methods were used for the same species of yeast. Since oxygen consumption is a concomitant of various metabolic processes in yeast, it does seem unlikely that all activities would occur sufficiently close together in time to yield a stepwise increase in the rate of oxygen uptake. One explanation for the conflicting results might be that the different synchronous cultures contained mitochondrial populations of different degrees of synchrony.

Throughout this report we have referred to mitDNA replication during the cell cycle. The identification of the "minor" DNA component was based only on preliminary lines of evidence and upon comparisons with other studies [7,24]. Firstly, the percentage of the "minor" DNA relative to whole cell DNA varied according to the physiological state of the culture. Thus, the "minor" component comprised about 12% of the total DNA when 3% GLU media were used, and increased to about 20% for cells grown on 3% GAL. Such an increase in a nuclear DNA moiety would be unlikely in response to a repressing versus a nonrepressing carbon source, but would be expected for mitDNA which is known to vary with growth conditions of a culture [32]. Similar increases in the proportion of mitDNA have been found by Smith [24] in studies of *Saccharomyces lactis* grown on different carbon sources at different concentrations in the medium. A second line of evidence to support the assumption that the small, rapid burst of synthesis is due to mitDNA replication is provided by the observations in yeast [7] in which mitDNA clearly was shown to replicate at a specific time during the cell cycle which was distinct from that for nuclear DNA. In a single preliminary experiment in which we purified whole cell DNA and isolated nuclear (ρ = 1.700) and mitochondrial (ρ = 1.684) DNA after analytical ultracentrifugation in CsCl density gradients [33], we calculated the percentage of mitDNA relative to total DNA of the sample, from the areas under the peaks of the densitometer tracings. We found that mitDNA comprised about 6% of the total DNA in a sample collected shortly before the step increase of the "minor" component, but that mitDNA accounted for

10–12% in each of two different samples collec... ...at separate times in the interval after the "minor" DNA increase but prior to nuclear DNA synthesis of the succeeding cell cycle. These preliminary results are in accord with our interpretations and, also, coincide very closely with results from similar studies of *Saccharomyces lactis* [6,7]. Clearly, we plan to repeat the experiment to include more than one cell cycle. We also have initiated pulse labeling experiments to obtain independent data concerning the identity of the DNA components which double in amount at two separate and repeatable periods of the cell cycle.

Although the ultrastructure analysis has just begun, we can discern the occurrence of a linear increase in mitochondrial numbers per cell section throughout a synchronous cell cycle. There were about five mitochondria per cell section at the onset of the cycle, and a steady increase to about ten organelles per section near the end of the budding cycle. Upon completion of the wall between bud and parent, each of the resulting cells was observed to contain approximately five mitochondria in thin-section, thus indicating periodicity. Accurate tabulations of mitochondrial populations were rendered difficult because of the contorted morphology of the organelles. The data to be examined from the entire sequence of three cycles will be essential to the final determination of the organelle population pattern. But, we may interpret the available information as showing that all the mitochondria of a cell developed as a population group, in synchrony with other intracellular mitochondrial populations in the culture. The occurrence of synchronous cycles of increase in mitDNA and cytochrome oxidase activity, along with the regular increase in mitochondrial numbers per cell, all provide support for this interpretation.

We may examine the pattern of mitochondrial increase in relation to the possible modes of biogenesis [34], on the basis of the interpretation that there was synchronous development of intracellular mitochondrial populations during the cell cycle. The simultaneous growth and division of organelles would evoke a pattern in which there was an abrupt increase and subsequent decrease to a base level, on a per cell basis. Even if division of mitochondria occurred somewhat randomly, but only after a period of synchronous organelle growth, then we would expect a pattern of unchanging numbers early in the cycle, with some sort of increase pattern confined to the end of the growth cycle. Other possibilities exist, but they are more remote than those proposed. Considering the pattern of steady increase, which paralleled the increase in cytoplasmic volume of the budding cell during the cycle, it is likely that some mechanism other than growth and division was responsible. A simple suggestion is that a new mitochondrion was added to the population unit of each cell in the synchronous culture, at regular intervals during the cell cycle. The precise mode of formation is unknown but since de novo formation has been eliminated in most recent considerations of the subject [35,36], we are left with the third major alternative, namely, formation from non-mitochondrial membranous precursors. Whether or not the abundant cytoplasm membranes observed in all the cells are participants in such a biogenetic process remains to be determined.

We believe that stepwise increases both in mitochondrial and in extramitochondrial systems provide the basis for testing a number of possible regulation mechanisms in the cell, and of comparing mechanisms which control activities in different parts of the same cell. Because of the amplification which is effected using synchronous cell cultures, such systems should continue to provide us with new insights into the ordered events which are characteristic of the growth and development of the cell and its component parts.

Acknowledgements

This study was supported in part by a grant from the U.S. Public Health Service (AI-07262) and contract AT(30–1)-3997 from the U.S. Atomic Energy Commission. We are most grateful to Mr. H.P. Hoffmann for his excellent contributions to the electron microscopy phase of the investigation.

References

[1] J. Engelberg and H.R. Hirsch, in: Cell Synchrony, eds. I.L. Cameron and G.M. Padilla (Academic Press, New York, 1966) p. 14.
[2] D.H. Williamson and A.W. Scopes, Nature 193 (1962) 256.
[3] J.M. Mitchison, Science 165 (1969) 657.
[4] J.E. Cummins and H.P. Rusch, J. Cell Biol. 31 (1966) 577.
[5] K.-S. Chiang, Proc. Natl. Acad. Sci. U.S. 60 (1968) 194.
[6] P. Tauro, E. Schweizer, R. Epstein and H.O. Halvorson, in: The Cell Cycle, eds. G.M. Padilla, G.L. Whitson and T.L. Cameron (Academic Press, New York, 1969) p. 101.
[7] D. Smith, P. Tauro, E. Schweizer and H.O. Halvorson, Proc. Natl. Acad. Sci. U.S. 60 (1968) 936.
[8] J. Koch and E.L.R. Stokstad, Europ. J. Biochem. 3 (1967) 1.
[9] R. Braun and T.E. Evans, Biochim. Biophys. Acta 182 (1969) 511.
[10] J.R. Cook, J. Cell Biol. 29 (1966) 369.
[11] K.-S. Chiang and N. Sueoka, Proc. Natl. Acad. Sci. U.S. 57 (1967) 1506.
[12] A.W. Scopes and D.H. Williamson, Exptl. Cell Res. 35 (1964) 361.
[13] M. Osumi and N. Sando, J. Electron Micr. 18 (1969) 47.
[14] P. Tauro, H.O. Halvorson and R.L. Epstein, Proc. Natl. Acad. Sci. U.S. 59 (1968) 277.
[15] G.M. Padilla, G.L. Whitson and I.L. Cameron (ed.), The Cell Cycle (Academic Press, New York, 1969).
[16] C.J. Avers, M.W. Rancourt and F.H. Lin, Proc. Natl. Acad. Sci. U.S. 54 (1965) 527.
[17] D.H. Williamson and A.W. Scopes, Exptl. Cell Res. 20 (1960) 338.
[18] K. Burton, Biochem. J. 62 (1956) 315.
[19] E.R. Tustanoff and W. Bartley, Canad. J. Biochem. 42 (1964) 651.
[20] J.J. Ferguson, M. Boll and H. Holzer, Europ. J. Biochem. 1 (1967) 21.
[21] A. Torriani, Biochim. Biophys. Acta 38 (1960) 460.
[22] M.J. Karnovsky, J. Cell Biol. 27 (1965) 137A.
[23] J.H. Luft, J. Biophys. Biochem. Cytol. 9 (1961) 409.
[24] J.D. Smith, Ph.D. Thesis (Univ. Wisconsin, 1967).
[25] A.L. Lehninger, The Mitochondrion (W.A. Benjamin, Philadelphia, 1964).
[26] J. Gorman, P. Tauro, M. LaBerge and H.O. Halvorson, Biochem. Biophys. Res. Commun. 15 (1964) 43.
[27] G.P. Longo and J.G. Scandalios, Proc. Natl. Acad. Sci. U.S. 62 (1969) 104.
[28] I. Witt, R. Kronau and H. Holzer, Biochim. Biophys. Acta 128 (1966) 63.
[29] P. Tauro and H.O. Halvorson, J. Bacteriol. 92 (1966) 652.
[30] H.O. Halvorson, R.M. Bock, P. Tauro, R. Epstein and M. LaBerge, in: Cell Synchrony, eds. I.L. Cameron and G.M. Padilla (Academic Press, New York, 1966) p. 102.
[31] J. Muller and P.S.S. Dawson, Canad. J. Microbiol. 14 (1968) 1127.
[32] E. Moustacchi and D.H. Williamson, Biochem. Biophys. Res. Commun. 23 (1966) 56.
[33] F.E. Billheimer and C.J. Avers, Proc. Natl. Acad. Sci. U.S. 64 (1969) 739.
[34] D.B. Roodyn and D. Wilkie, Biogenesis of Mitochondria (Methuen Ltd., London, 1968).
[35] H. Plattner and G. Schatz, Biochemistry 8 (1969) 339.
[36] D.J.L. Luck, J. Cell Biol. 24 (1965) 461.

MITOCHONDRIAL SPECIFICATION OF THE RESPIRATORY CHAIN

H.R. MAHLER, Philip S. PERLMAN and Bam D. MEHROTRA

Chemical Laboratories, Indiana University, Bloomington, Indiana, USA

. On the last day of a symposium devoted to the autonomy of mitochondria and chloroplasts — and in a year that has seen the appearance of several reviews devoted to this topic at various levels of penetration and critical rigor [1—4] — it is probably redundant to emphasize once again that these organelles possess a unique genetic system responsible in some manner for their replication and continuity.

There are two aspects to mitochondrial self-specification that are logically distinct, even though they have at times been confounded, perhaps due to the identity or close juxtaposition of the responsible systems in a space confined by the inner membrane of the particle. The scope of the problem becomes evident from fig. 1.

(1) What is the nature of the message, of the entities, specified by the mitochondrial DNA, i.e. inscribed in its DNA and transcribed into various species of RNA? We assume transcription to have taken place in situ, since we shall not admit, for the sake of simplicity, the possibility of the export of mit-DNA (mitochondrial-DNA) into the NC (nucleo-cytoplasmic) system, but at the same time we must admit the possibility of some of the mit-transcripts finding their way to this site.

(2) What is the nature of the entities translated by the mitochondrial system responsible for polypeptide synthesis? And here, as a corollary, we shall have to admit the possibility that some of the transcripts — not just tRNA, but also perhaps rRNA, and more to the point within the present context, some of the mRNA found in the mitochondrion — had its origin in the nucleus.

In this contribution I will try to bring together and discuss — I hope critically — the efforts of just one laboratory concerned with these problems, dealing in the main with just one organism, the facultative anaerobe *Saccharomyces cerevisiae*. Most of the experiments have taken advantage of one or more of three well-known paradigms:

(1) When this organism is grown on glucose, even aerobically, the synthesis of its respiratory apparatus is repressed; upon exhaustion of the fermentable substrate (fig. 2) there then ensues, coincident with the shift from aerobic glycolysis to active respiration, a rapid elaboration of functional mitochondria. This sequence of events may be monitored by following the increases in a number of parameters, such as oxygen uptake, various enzymes in the different mitochondrial compartments, and even of the mitochondrial transcribing enzyme [5—11].

(2) Studies on the specificity of two inhibitors of protein synthesis, namely chloramphenicol (CAP) and cycloheximide (CYCLO) in vitro suggested an absolute dichotomy of their site of action: the former blocks exclusively the mitochondrial, the latter the NC system (i.e. the appropriate ribosomal components) [12—19]. The implications of these findings to cell physiology were first explored by one of our hosts, who found that only derepression but not

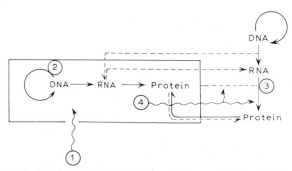

Fig. 1. Possible interactions between the nucleo-cytoplasmic and mitochondrial genetic systems. The area within the full line block represents the mitochondrion.

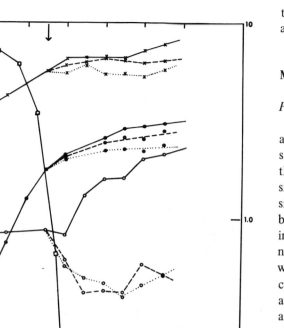

Fig. 2. Course of derepression of respiratory activity in *S. cerevisiae* and effects thereon of CAP (4 mg/ml) and CYCLO (0.5 μg/ml) as a function of time and glucose concentration. Time = 0 at moment of inoculation. Parameters and symbols shown are: mg protein per ml of growth medium X10 min [X]; turbidity at 600 nm [●]; oxygen uptake (nmoles min^{-1} mg protein^{-1}) X 10^{-2}, [○]; glucose concentration in the growth medium (mg/ml), [□].

tion of the mitochondrial apparatus to the two aspects of self-specification mentioned initially.

Mitochondrial protein synthesis in vivo

Pools, kinetics and inhibitors

Since most of the results of these studies have already been published [30] a brief summary will suffice. Initially we were interested in demonstrating the existence, and to arrive at some estimate of the significance, of the intramitochondrial system responsible for polypeptide synthesis. This was to be done by means of experiments on whole cells and employing appropriate pulse labeling and pulse-chase techniques. To this end we had to establish first of all whether the acid-soluble pools of precursors in the cytosol and the mitochondria became filled rapidly and were freely accessible to exchange with unlabeled aminoacids added to the medium, and whether the use of CAP and CYCLO as selective inhibitors was justified. Using phenylalanine as a tracer we were able to show that (a) derepressing cells and their mitochondria incorporate this precursor into protein with a lag of less than one minute in a linear fashion for periods ranging up to sixty minutes; (b) this incorporation can be diluted in the manner predicted by theory by the addition of unlabeled aminoacid; and (c) the specific activity of the acid-soluble material becomes constant within ≤ 90 sec.

The incorporation shows the following characteristics with regard to the two inhibitors (table 1); CYCLO at microgram levels inhibits within 1 min at all physiological stages; maximum inhibition is 98% at 2 μg/ml. The inhibition is reversible. CAP does not inhibit incorporation by repressed cells; this is in accord with the findings in Linnane's [17–20] and our own laboratory [11] that the mitochondrial system is dispensable for glycolysing cells, at least as judged by these criteria. CAP *does* inhibit incorporation by derepressing (and derepressed) cells, thus providing the chemical basis for the physiological and enzyme data previously reported and those to be described in a subsequent section of this report. This inhibition reaches 22% at a level of 4 mg/ml of medium, a saturated solution; this represent a concentration capable of blocking derepression of respiratory capacity completely (see fig. 2). For whatever it

growth dependent on aerobic glycolysis was interfered with by CAP, and that the former process appeared relatively insensitive to CYCLO [17–20].

(3) The existence of cytoplasmically determined mutants (ρ⁻ petites) which appear completely devoid of respiration and the membrane-bound cytochromes [21–24], coincident with major derangements in their mit-DNA [25–29].

The appropriate superposition of these three phenomena then can be made use of for the design of experiments to assess, at least in theory, the contribu-

Table 1
Sensitivity of incorporation of labeled phenylalanine into yeast cell proteins.

State of cells	CAP (4 mg/ml)	% Inhibition of linear rate by			
		CYCLO (μg/ml) 0.5	1	5	CYCLO (5 μg/ml) + CAP (4 mg/ml)
Repressed	< 2	76	80	>98	
Depressing	22	30		91	96

For experimental details see [30]. Inhibition by CYCLO reaches its maximal level within one minute and remains constant thereafter except for concentration ≤ 1 μg/ml when it may increase with time beyond 10 min.

is worth the magnitude of this inhibitory effect particularly when analyzed on the level of sub-cellular and sub-mitochondrial fractions, taken in conjunction with its non-additivity when superposed on the inhibition by CYCLO was interpreted in terms of some form of coupling between the two protein-synthesizing systems.

Properties of the mitochondrial system

We were able to demonstrate that to analyze the phenomenon it was necessary to sub-fractionate mitochondria. Provided this was done one could by means of pulse and pulse-chase experiments identify a fraction, associated with the most insoluble proteins of the inner membrane, that resembled the membrane-bound, ribosomal system of the cytoplasm in its kinetics in this sense: it appeared to function as a source while all other sub-cellular and sub-mitochondrial fractions exhibited the properties expected of sinks. Unlike the ribosomal system, however, which showed the opposite specificity, it was resistant to CYCLO and sensitive to CAP, but only partly so. Chase experiments also indicated, when performed in the presence of the inhibitors, export from the mitochondrial site into other fractions derived from the inner membrane, particularly the one containing cytochromes (a and a_3); export from the cytoplasmic ribosomes coupled to import into all mitochondrial fractions, including the one also containing the mitochondrial synthetic site was also observed.

The mitochondrial contribution

The least soluble fraction, by our scheme, accounts for 30% of the total (inner) membrane protein or about 15% of the total protein of the organelle. Its contribution to its own synthesis is sensitive at the most to a 50% inhibition by CAP. Conversely, it is about half as sensitive as is the incorporation into total homogenate or soluble cytosol proteins to inhibition by 5 μg CYCLO/ml; furthermore half or more of its protein appears to be derived from the outside, when its own synthetic capacity is blocked by CAP. Thus we estimate that the pool contributed by this fraction does not account for more than about 7.5% of the total mitochondrial protein and may even be less than that. These considerations would tend to rule out the majority membrane proteins (the "structural protein fraction" of others, see below) as having been synthesized in situ under our conditions. It is, of course, possible that these proteins were already present at the beginning of the experiment (i.e. in repressed cells) and thus their synthesis would have escaped detection. The inability of CAP to interfere with mitochondrial development (or that of their analogues) in repressed cells makes this an unlikely possibility.

The inability to push experiments of this sort to a satisfactory resolution concerning the nature and origin of *individual* proteins led us to a concomitant examination of the effect of inhibitors on the elaboration of enzyme activities during derepression.

Inhibition of enzyme formation by CAP and CYCLO

Following up the initial observations of Linnane et al. [17–20], we have attempted to ascertain the effects of these site-specific inhibitors of protein synthesis on the development of enzymes, localized in different intra-mitochondrial compartments, in the course of derepression [11,31]. We have used the following markers: NADH:cytochrome c reductase (NADH:c), succinate dehydrogenase SDH, succinate: cytochrome c reductase (SC), and cytochrome oxidase (cyt ox) for the inner-membrane-cristae and its enzyme complexes; L-malate dehydrogenase (MDH) as an example of a strongly repressible enzyme, which in its various iso-enzymic forms [32,33] is localized in both, the extractable compartment of the mitochondrion as well as the cytosol, but is almost

certainly entirely responsive to the NC apparatus [34,35]; and NAD:L-glutamate dehydrogenase (GDH) as a reference, since, in this strain at least, it is localized exclusively in the cytosol [32] and as such originates in the NC system. *

* Occasionally we have used NADP:isocitrate dehydrogenase and citrate synthase as markers for the easily extractable compartment (matrix) of the mitochondrion.

Table 2a
Long term effects of chloramphenicol added at various stages in depression.

| Drug added when | Increase in enzyme units* | | | | | | | | | |
| | OD = 0.665 | | OD = 1.2 | | OD = 1.5 | | OD = 1.8 | | OD = 2.5 | |
	Control	CAP (% of control)	Control	% of control	Control	% of control	Control	% of control	Control	% of control
Growth (mg/ml)	0.375	79	0.368	70.5	0.285	53.5	0.187	85.6	0.123	87
Cyt ox	35.2	<0	39.1	<0	89.6	0.7	38.7	7.5	44.7	4.7
NADH: c	13.5	16.3	23.8	15	35.7	17.3	25.2	73.8	20.5	102
SDH	18.5	49.6	23	83.5	†	†	20.7	133.0	17.1	125
Succ: c	16.2	26.8	18.9	56	23.1	40.1	10.05	109.0	15	211
GDH	47	46.8	58.5	31.8	133	36.9	82.6	65	52.6	98.5
MDH	1157	33.2	1196	50	†	†	967	122.0	727	171

* Increase in nmoles/min/ml medium during 6 hr of incubation starting at OD_{600} shown (fig. 2). Chloramphenicol (CAP) was added at a concentration of 4 mg/ml. Cytochrome oxidase (cyt ox), NADH cytochrome c reductase (NADH:c), succinate cytochrome c reductase (succ:c), L-glutamate dehydrogenase (GDH), malate dehydrogenase (MDH).
† Not done.

Table 2b
Long term effects of cycloheximide added at various stages of depression.

| Cyclo added when (t_0) | OD = 0.56 | | OD = 0.87 | | OD = 1.5 | | OD = 2.4 | |
	Control	Cyclo	Control	Cyclo	Control	Cycle	Control	Cyclo
Growth	3.74	1.11	2.66	1.27	1.95	1.27	1.54	1.18
Cyt ox	19.4	1.99	24.8	2.09	28.4	5.62	12.7	2.7
NADH:c	40.0	1.48	20.8	2.34	10.7	3.04	2.32	1.35
Succ:c	196.0	1.11	109.0	1.61	55.6	4.92	13.6	3.4
GDH	52.5	1.37	64.3	2.07	25.0	1.65	10.5	1.5
MDH	56.0	0.99	38.4	1.2	19.3	1.7	9.3	1.47

Enzyme units, nmoles/min/ml of medium. Cycloheximide (cyclo) was added at a concentration of 0.5 μg/ml. Abbreviations as given in table 2a.

Long-term effects

Our earlier studies covered periods that either occupied the total time required for derepression or a considerable portion thereof. Under our conditions 10 hr are required for the total process, starting in late exponential phase at a protein concentration of about 0.2 mg/ml or an A_{600} of about 0.5, when most dehydrogenases begin to increase in activity. Derepression of cytochrome oxidase is the last to occur and begins about 2 hr later [10,11]. Maximal activity for all enzymes is reached within about 8 hr of the start of ¹erepression. I shall now present a recent exampl of this type of experiment (table 2) in order to emphasize and enlarge on some of the points raised previously [11,31]. The synthesis of *all* markers, not just of the soluble ones, is susceptible to severe inhibition by CYCLO, regardless when the drug is added during derepression, but the exact extent of inhibition varies with the activity measured. Cyt ox, and to a certain extent also NADH:*c* and SC appear to be significantly more resistant than the other activities.

Conversely, inhibition by CAP is by no means restricted to the elaboration of the mitochondrial respiratory chain. Furthermore here the time of addition of the drug becomes an additional crucial parameter. Conditions can be so chosen, it turns out, as to restrict CAP sensitivity to the formation of cytochrome oxidase, and in fact, this activity appears to behave in a manner qualitatively distinct from that of all others regardless of the time of addition. In any event the net result of the superposition of these various effects appears to the observer as a highly non-coordinated and idiosyncratic pattern.

Short-term effects

Further elaboration on the kind of study just presented did not seem to be particularly attractive for the following reasons: it appeared to us just as likely a priori, that the effects and patterns observed were due to the "sequential" interference with certain proteins required either catalytically or stoichiometrically for the synthesis or integration of several mitochondrial (and extra-mitochondrial) activities, rather than to more direct, "parallel" effects on enzyme synthesis proper. Furthermore it appeared difficult, if not impossible, to rule out even more indirect effects such as interference with the requisite

energy supply or with essential components, such as RNA. In order to minimize these difficulties we have turned more and more to an investigation of the *kinetics* of the changes observed after relatively short intervals subsequent to the addition of the drugs. Figs. 3 and 4 show some of the results obtained with CAP and CYCLO respectively. Our conclusions are:

(1) Integrated over a period of some three hours, the only activity severely affected by CAP is cytochrome oxidase, regardless of the time of drug addition, thus confirming the results of the long-term studies.

(2) All other enzymes (with the possible exception of NADH:*c*) when the drug is added early are made in amounts and over a time course not differing significantly from that of the control.

(3) The kinetics of the formation of cytochrome oxidase under conditions when CAP is added during the exponential phase of cell growth ($A_{600} = 0.665$, see also table 2 and fig. 2) is anomalous; it appears to be produced in amounts exceeding those in the control for about two hours and destroyed thereafter, as if it were synthesized by a CAP insensitive system, but required synthesis of an entity susceptible to this agent for continued function. This result has been confirmed with additions of the inhibitor at A_{600} equal to 0.665, 1.20 and 1.50 with the effect diminishing in that order.

(4) The sudden cessation of further synthesis of all enzymes four hours after addition of the drug in the "early" experiments ($A_{600} \leqslant 1.5$) appears to be due to an exhaustion of the energy supply. This is still fermentative in nature in this set since addition of antimycin A at a corresponding stage permits normal enzyme formation for precisely this length of time, in spite of the fact that this inhibitor rapidly and completely abolishes all respiration.

(5) In the presence of CYCLO (at a level of either 0.5 or 2 μg) the following pattern obtains. If the inhibitor is added at any time prior to stationary state ($A_{600} \leqslant 1.5$) the activity of membrane-bound enzymes increases, after an initial lag, for several hours at rates about 50% those of the control. The drug-resistance effect appears to be maximal at a cell density corresponding to $A_{600} = 1.2$ and declines rapidly soon thereafter. It is observed for NADH:*c*, SC, SDH and cyt ox and totally absent for MDH, IDH and GDH. Its explanation may lie in a relatively slow

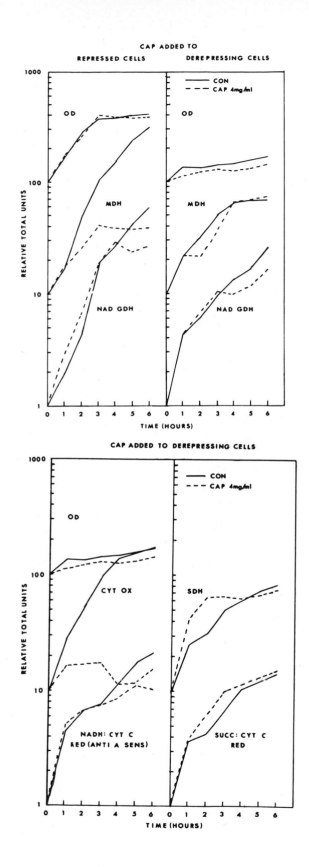

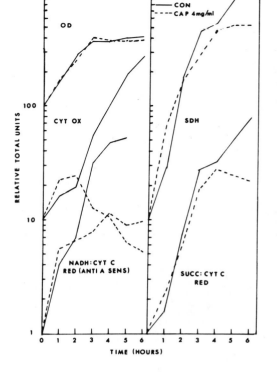

Fig. 3. Kinetics of the effects of CAP on the derepression of several marker enzymes. Data are expressed as relative total units – (units at t)/(units at t_0): units in nmoles/min/ml; t_0 is time of addition of drug; for absolute time scale see fig. 2. Determinations at times shown: A: Comparison between fully repressed (A_{600} = 0.665) and derepressing (A_{600} = 1.84) cells – soluble enzymes. For the first set t_0 values for MDH and GDH are 192 and 4.03 units; for the second, 920 and 17.7 units. B: Continuation of A for enzymes of the inner membrane; repressed cells (A_{600} = 0.665) t_0 values for cyt ox NADH:c, SDH and SC are 6.75, 0.985, 0.870 and 1.04 respectively. C: Continuation of B; derepressing cells (A_{600} = 1.84); t_0 values for cyt ox, NADH:c, SDH and SC are 13.2, 6.81, 15.7 and 7.85 respectively.

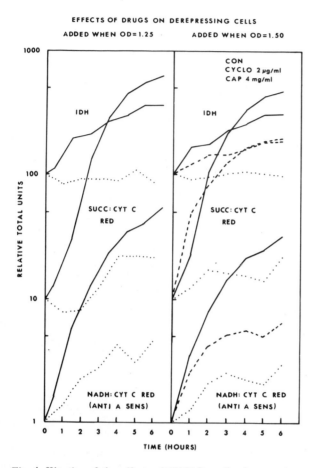

Fig. 4. Kinetics of the effects of CYCLO on the derepression of several marker enzymes. Initial values for A_{600}, IDH, SC and NADH:c are 1.25, 16.8, 0.375 and 0.709 respectively for the left-hand panel and 1.50, 19.63, 0.497 and 1.133 respectively for the right-hand panel. Cell growth typical for this experiment is presented in fig. 2; in the presence of CAP there was an increase in cellular protein slightly less than that of the control cells; there was no net protein synthesis in cyclo treated cells. Drugs were added at t = 0. ——— control, ... cycloheximide, – – – – CAP.

integration into and/or activation by sites on the membranes of a pool of precursors previously synthesized by the NC system. The stringency of the time sequence involved becomes understandable in terms of a shift of the rate-limiting step from the CYCLO-sensitive filling of the pool to its emptying by the CYCLO-insensitive process. This may, for instance, be brought about if we suppose that the latter is regulated by the availability of highly specific attachment sites and that their number increases in the course of derepression.

Phenotype of cytoplasmic petites

Although a full exposition of this topic is beyond the scope of this report, and will be presented elsewhere [36], some of its aspects are highly pertinent within the present context and need some comment.

(1) For a certain sub-class of this class of mutants — which quite in general is characterized by major alterations in its mit-DNA [25–29] — this derangement appears to have taken the rather extreme form of a base composition corresponding to 96% adenine plus thymine, probably arranged in large part in alternating sequence [27,28].

(2) The physiology of the members of this sub-class is very similar to that of all others in the class of ρ^- mutants, including the characteristic lesion leading to an absence of cytochromes b, c_1, a and a_3, with its correlate of a lack of cytochrome c linked activities [22,24].

(3) On the other hand mutants of this sub-class contain mitochondria which, when isolated, share many of the properties of those found in the wild type. These include certain morphological features, including the ability to assume two limiting conformations (fig. 5); in sedimentation coefficient and iso-pycnic density in sucrose gradients (table 3); and in their content of characteristic enzymes such as (fig. 6) MDH, NADP:isocitrate dehydrogenase (IDH) and, as first shown by Kovac and Weissova, and by Schatz [39,40] of the mitochondrial F_1-ATPase (sensitive to Dio-9 and the F_1 inhibitor isolated from beef heart; but insensitive to oligomycin). This enzyme can be considered a marker for inner-membrane, and probably cristae-linked activities.

(4) Finally the deficiency mentioned under (2) is vitiated, at least in part and in some strains, by the presence of certain proteins exhibiting what may be termed partial activities: NADH: and succinate: dye reductases [23,24,41–43], cross reacting material (CRM) related to the cytochrome c binding site of NADH:c [44] and entities exhibiting CRM [45] and apoenzyme activities [46] related to proteins of cytochrome oxidase preparations.

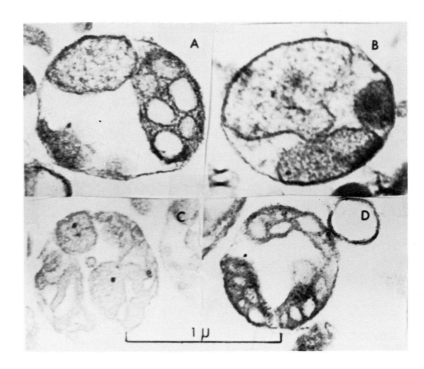

Fig. 5. Morphology of isolated mitochondria from ρ^- and ρ^+ cells. A and B: Mitochondria isolated from spheroplast lysates of the cytoplasmic petite 4D-21. Note the two conformations of the inner membrane: attempts at demonstrating interconvertability as a function of metabolic activity have not been successful. C: Mitochondria from spheroplast lysates of isogenic wild type 4D, fully derepressed after growth on 1% glucose; fixed in "resting state" D: Similar to C, except that mitochondria were actively respiring (20 min at 23° in 1.0 mM KH_2PO_4, 5 mM $MgCl_2$, 10 mM Na succinate, 0.05% BSA; made 0.4 M in sucrose and adjusted to pH 7.0). All samples were fixed with 3% glutaraldehyde in 0.1 M KH_2PO_4, pH 7.4 and 0.4 M in sucrose, post-fixed with OsO_4, embedded in Epon, sectioned, and photographed with a Siemens ELMISKOP at 40 KV.

Table 3
Characteristics of mitochondria of wild type and ∂^- cells.

Strain	$63.000\,g \times t$ Marker	Peak banding position (% Sucrose, 4°) 10 min SDH	a	60 min SDH	MDH IDH	ATPase	Cit synth.	S_{obs} (S)
4D (ρ^+)								
repressed			37	39	39	39		6000 (?)
derepressed	33	33		38	38	38	38	5000
4D-21 (ρ^-)			33b	38	38	38		5×10^3

a MDH, isocitrate dehydrogenase (IDH), citrate synthetase (cit synth), ATPase (F_1).
b Re-calculated from a run at 23° when banding occurs at 38%, isopycnic at 42.5% at that temperature.

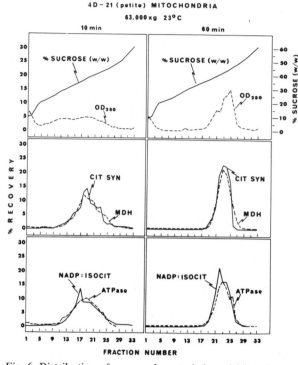

4D – 21 (petite) MITOCHONDRIA
63,000×g   23°C

Fig. 6. Distribution of various characteristic activities when a crude mitochondrial fraction isolated from the ρ^- mutant 4D-21 is placed on a linear (1.059 to 1.258 g/ml) sucrose gradient and centrifuged at 23° at 63.000 g for either 10 or 60 min. Totals loaded were for protein; IDH, ATPase, MDH and citrate synthase (cit. synth):40 (A_{600}); 0.77, 6.0, 20.1 and 11.16 units resp. Recoveries were, after 10 min 85, 117, 79, 87 and 75% respectively, after 60 min 97, 114, 87, 109, and 72% respectively. Equilibrium density in a double labeling experiment in which mixed ρ^+ cells (labeled with ^{3}H) and ρ^- cells (with ^{14}C) were used was at 38% sucrose (1.172 g/ml) at 4°.

Mutagenesis by ethidium bromide

It would be exceedingly useful to have available a tool for interfering, at least in the first instance, specifically with the DNA, rather than the translational machinery of the mitochondrion. Such a tool appears now at hand by appropriate use of the phenanthridine ethidium bromide (EB). This dye has become well known as an intercalating agent for DNA [47–49] and was shown by Slonimski and colleagues [50] to be a highly effective mutagen capable of converting a cell population of *S. cerevisiae* into ρ^-

mutants even in the absence of growth. In theory then, provided that the mutagenesis could be shown to (a) be rapid and quantitative, leaving behind an experimentally insignificant fraction of unconverted wild type cells, and (b) leave unaffected the initial enzyme complement of the mutagenized cells, one can ask the following two questions:

(1) How quickly do the cells thus converted to the ρ^- genotype exhibit the ρ^- phenotype i.e. become unable to respire and utilize a non-fermentable carbon source, such as lactate?

(2) What is the kinetics of the resultant alteration in enzyme complement which in the aggregate accounts for (1); and as a corollary can we

(3) distinguish primary from secondary effects — all triggered of course by some initial event having to do with the exposure of mitochondria and their DNA to EB?

Kinetics of mutagenesis

When cells in the exponential phase of growth, whether under strongly repressing conditions, such as with 4% glucose as the carbon source, or under completely derepressing ones, such as in 3% lactate, are harvested, resuspended in phosphate buffer at a cell density of 2×10^7, starved for 3 hr and then exposed to 2×10^{-5} M EB, conversion to the ρ^- genotype is rapid and quantitative in the sense that (a) the viable cell count (ρ^+ plus ρ^-) remains constant — there is no cell killing; and (b) after an initial lag this conversion follows exponential kinetics down to a level of wild type survivors as low as desired, e.g. less than 0.5% after 100 min. Closer examination of the kinetics by curve fitting using the Indiana CDC 3600 shows that a good fit can be obtained to the function $S = 1 - (1 - e^{-kt})^n$ where S equals the number of wild type cells remaining after a length of time of exposure to EB equal to t, and k and n are constants. This function can be interpreted in terms of a mutagenic event that requires a single interaction (hit) with any one of several equivalent sites (targets) [52]. For this model n then equals the number of targets and kt measures the number of hits applied. (Half-lives, or the length of time required per hit can be determined for the exponential phase of the decay, observed in the limit when t becomes large.) We have already mentioned that the physiological state of the cell appeared without effect and indeed

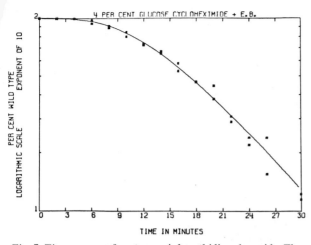

Fig. 7. Time course of mutagenesis by ethidium bromide. The plot is computer generated to fit the function % wild type (= 100−% mutants) = $1-(1-e^{-kt})^n$ for t in minutes and the values of k and n shown in table 4, where other experimental details are described. Glucose-grown cells; CYCLO present.

Table 4

Determination of n and k for ethidium bromide (EB) mutagenesis.

| EB (mM) | Conditions | | k (min^{-1}) | n |
	Growth C-Source	Inhibitor		
2.18×10^{-2}	3% Lactate	Cyclo	0.117	7.93
	Lactate	Eythro	0.164	23.5
	4% Glucose	Cyclo	0.121	5.1
	Lactate	–	0.0922	16.8
1.09×10^{-2}	3% Lactate	Cyclo	0.0485	8.8
2.18×10^{-2}	Lactate	Cyclo	0.109	9.8
4.36×10^{-2}	Lactate	Cyclo	0.140	6.4

Standard conditions: Mid log lactate-grown cells, pre-starved 3 hr in 0.1 M phosphate, pH 6.5, 3 hr. Cell concentration during mutagenesis = 10^7/ml. Fraction of ρ^- cells in population determined as in [50] and checked with tetrazolium agar [51] concentrations of cyclo and erythro were 2 and 100 μg/ml, respectively. Multiple target, single hit; CDC 3600; chi square minimization. Data treatment by subroutine "STEP IT", copyright J.P. Chandler, 1965, distributed by Quantum Chemistry Program Exchange, Indiana University.

(table 4) n and k under the two extreme sets of conditions are indistinguishable, provided that cell and mitochondrial proliferation is inhibited by the addition of CYCLO. If not, then n, but not k increases; on the other hand k, but not n is affected by lowering the EB concentration below the saturation level determined for this many cells; or by increasing the cell population to $> 4 \times 10^7$/ml. All these results are in accord with a single-hit, multiple target model.

Exposure of cells to EB has no effect on enzyme levels

As shown in table 5, exposure of cells to the regime found effective for mutagenesis leaves unaffected the levels of all enzymes tested. We are thus in a position to ask the questions posed earlier.

How long is the phenotypic lag?
On lactate

In several experiments we have grown cells on lactate, mutagenized them with EB and then transferred them back to a lactate medium which permitted normal growth of control cells treated similarly, except for the actual presence of EB. Results of one such experiment are summarized in table 6. It can be seen that the mutagenized cells were

Table 5

Effects of various treatments during EB mutagenesis on enzyme levels.

| Culture Conditions | Enzyme | | | | |
	Cyt ox	SDH	SC	MDH	NADH:c
17 hr in 1.8% glucose	97.3	61	34	2280	47
+ starvation[a] 5 hr	226	49	40	2818	41
+ starvation[a] 10 hr	222	38	34	2130	62
+ EB (2×10^{-5} M)[b]					
0 min	230	53	44	3340	76
60 min	210	52	44	2780	103
5 hr	302	60	55	3340	123
10 hr	213	43	38	2230	58

[a] 0.1 M phosphate, pH 7.4, 1,8% glucose, 30°C;
[b] Produces $\geqslant$ 99% petites in $\leqslant$ 60 min. All activities as nmoles/min/mg homogenate protein. Enzyme abbreviations as given in table 2a.

Table 6
Comparison of enzyme levels in ethidium bromide (EB) treated and control cells growing on 3% lactate.

Treatment	A_{600}	Cell No	% ρ^-	Cyt ox	NADH:c	SC	SDH	MDH	GDH
EB	2.0	1.5	99.0	0.57	1.1	0.76	0.68	1.12	1.03
	4.0	2.5	93.0	1.3	2.2	2.9	5.3	3.5	2.5
Control	2.0		<0.5	1.1	0.86	0.60	0.76	0.44	0.92
	4.0		<0.5	2.8	2.9	2.40	3.2	3.0	6.0

Cells were grown on lactate, mutagenized with EB and then transferred back to lactate, control cells were similarly treated except for the EB exposure. All values relative to those at moment of transfer to lactate: $A_{600} = 0.100$; cell number $= 1.45 \times 10^7$, % $\rho^- = 99.5$. Enzyme abbreviations as given in table 2a.

capable of slow growth for a limited number of generations: the latter was equal to two or less in all experiments. They tell us that exposure of cells to EB leads to a sudden and discontinous alteration of phenotype, characterized by an inability to synthesize active cytochrome oxidase, and of further cell division on a lactate medium; they rule out the possibility that mutagenesis is due to continued action of residual mutagen on buds formed subsequent to the removal of the bulk of the agent and resumption of growth on a nutrient medium.

On galactose

The experiments just presented are, by their very nature, of only limited usefulness in our search for the events that take place in the course of the conversion of an incipient, recently mutagenized, petite into its final, established phenotype. The latter is incapable of growth on anything but fermentable carbon sources, but among them one has a wide choice and one is by no means restricted to sugars producing strong catabolite repression, such as glucose or fructose, — a complication to be avoided. Mellibiose and galactose produce repression in only an attenuated form, and we have used the latter sugar in our experiments.

When mutagenized cells are placed in a liquid medium containing galactose as the carbon source they are found to be capable of continous and exponential growth with a generation time only slightly greater than that of appropriate controls (fig. 8). Over the period of the experiment shown, 8 hr, equivalent to 5.25 generations, the level of mutants never dropped below 95% of the total.

What is the path to the petite phenotype? Evidently, a series of distinct and irreversible events supervenes. With two exceptions all respiratory enzymes tested, including SC and SDH, continue to be synthesized for three generations at rates reminiscent of the control. Beyond this there takes place what appears to be a dismantling of the respiratory apparatus leading, we suppose, to the enzyme distribution characteristic of the established petite. In sharp contrast to the continued synthesis of certain mitochondrial enzymes during the initial metastable state, the cessation of cytochrome oxidase is virtually immediate, while synthesis of NADH:c does appear to take place, but with a rate considerably below that of the other enzymes. The metastable state, which is present for several generations, is thus characterized by an abnormally low ratio of cytochrome oxidase to all other enzymes. In a qualitative sense at least, these results are quite equivalent to those obtained on the lactate medium (table 6).

We interpret these results as follows: Interaction of mit-DNA with EB expresses itself immediately and directly by impeding transcription * of a component required for cytochrome oxidase activity, and perhaps of one having a similar function for NADH:c;

* The postulate of a block in the formation of mitochondrial transcripts — which, incidentally EB is known to be capable of exerting [53,54] — is required, since the alternative which postulates a primary event involving DNA *duplication exclusively* appears to be ruled out: it predicts the presence of unmodified parental strands, available for transcription, both prior and subsequent to the initial round of DNA duplication, post EB treatment, and yet enzyme production is known to cease without delay.

conversely, there is no effect on any of the steps required for the synthesis of all other enzymes, including the complete respiratory chain between succinate and cytochrome *c*. As a corollary to the first statement the mRNA for the component involved in cyt ox activity must have suffered substantial decay during the mutagenic treatment. For the time being we need not be concerned with formulating a precise model for EB mutagenesis proper, in particular with respect to the question whether mutagenesis occurs prior to, coincident with or subsequent to the postulated transcriptional block imposed by EB.

Targets and models

Provided we make the further, not unreasonable, assumption that the development of respiratory and genetic incompetence are closely coupled, then a possible reason can be suggested why incipient petites are incapable of growth on lactate beyond two generations or so. It is, that by this time, on the average each cell (subsequent to redistribution between parent and daughters) contains less than one *fully* competent organelle (or groups of organelles, if a cooperative process is involved). The respiratory insufficiency brought about by the cessation of cytochrome oxidase synthesis may not have been solely responsible, as will become apparent from a consideration of the results obtained when synthesis of this enzyme is blocked by the addition of CAP to cells growing on lactate (see next section).

If this model is accepted it provides us with an independent estimate of *n*, the number of genetically competent targets, which from the data (two generations on lactate, as well as three generations on galactose before there is a break in continuity) would suggest that *n* lies between 4 and 8. This compares favorably with other estimates based either on the kinetics (table 4) or the concentration dependence of EB mutagenesis. Evidently, as already pointed out by Slonimski et al. [50] this number is smaller, perhaps by as much as an order of magnitude, than the number of organelles present under these conditions. However, we have no evidence whatever for the identity of all organelles with regard to their capability of maintaining either genetic continuity or respiratory capacity. The data suggest that they are not, or that several of them must act in concert to

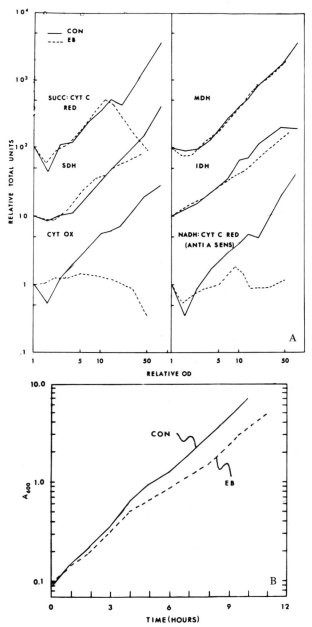

Fig. 8. A: Isometric (relative activity vs cell density on a logarithmic scale) plot of various marker activities in cells originally grown on lactate to mid-exponential phase, exposed to EB for two hours (for conditions see table 4) and then placed on a medium containing galactose as a carbon source. Samples of experimental and control (not exposed to EB) cells were harvested and enzymatic activities determined in each S_{600} [32] at one hour intervals. Initial specific activity (nmoles min^{-1} mg^{-1}) for succ: cyt *c* red, SDH, cyt ox, MDH, IDH and NADH:*c* was (for control cells) 83.6, 91.9, 286, 2780, 74.9 and 134 and (for EBIC) 88.2, 92.8, 281, 2620, 78.6 and 140 respectively. B: growth vs time.

provide a functional unit. They also suggest that certain models for mutagenesis can be ruled out. Since production of mutants can be made more than 99.5% effective in 60 minutes (at a cell concentration of 2×10^6/ml and an EB concentration of 2×10^{-5} M) any model that postulates dependence on and an explanation of the initial lag in terms of, an obligatory synthesis of DNA during actual exposure [2] appears to be ruled out. Furthermore, nalidixic acid, a known inhibitor of DNA synthesis [31–55] has no effect on the kinetics. Alternative models, that postulate a slow accumulation of errors [2], perhaps through the intervention of an altered polymerase [55] appear equally unlikely in view of the rather sudden and discontinuous cessation of growth on lactate. A model that we are currently testing postulates (unidirectional) DNA synthesis affected by a selective and critical interaction of EB interfering with the attachment of DNA to the mitochondrial membrane at the replicating site. As a consequence all DNA regions duplicated before this event has occurred are normal, but all those duplicated subsequent thereto will be formed without close, or perhaps without any, reference to the template.

Translational blocks on derepressed cells

A necessary control for the experiments with EB, i.e. the postulated transcriptional block, is to inquire into the consequence of the substitution of a translational block on similar, fully derepressed cells. Alternatively these experiments constitute an extension of the analogous ones with derepressing cells to their steady state, when that process has been completed, and cells contain the maximal level or respiratory enzymes they are capable of attaining. The paradigm is to look for relative, and differential, rates of dilution of various enzymatic activities as their synthesis is stopped or slowed subsequent to the addition of CAP. Fig. 9 summarizes the results of one

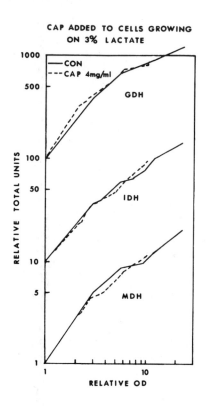

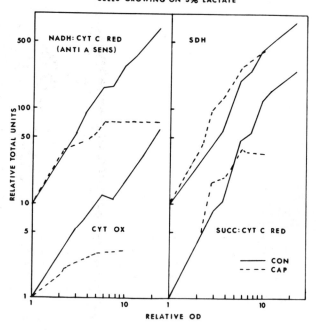

Fig. 9. Isometric plot similar to fig. 8 describing the effect of adding CAP (4 mg/ml) to cells growing exponentially on lactate. A: soluble enzymes; initial values of GDH, IDH and MDH were 341, 107.1 and 2860 nmoles min^{-1} mg^{-1}, respectively. Initial A_{600} was 0.09. B: membrane bound enzymes; control values of NADH:cyt c red, cyt ox, SDH and succ:cyt c red were 35.7, 202, 32.2 and 8.05 nmoles min^{-1} mg^{-1}, respectively.

such experiment. The cells are capable of growth, at a reduced rate for about six generations after the addition of CAP. The enzyme data themselves are of a pleasing simplicity; GDH, IDH and MDH, enzymes that are either present exclusively in the cytosol, or in that compartment as well as the easily extractable ones of mitochondria, are synthesized in normal amounts, and so is the complete, membrane-bound electron transport chain between succinate and cytochrome *c*. NADH:*c* seems to be formed normally for the first generation, at a reduced rate for the second and part of the third, and then it ceases. In contrast, the synthesis of cytochrome oxidase appears sensitive to the inhibitor ab initio, and in this it stands apart from all other enzymes. Qualitatively

these results then suggest that, indeed, the effects of introducing either a translational or trancriptional block into the mitochondrion are similar. The similarity may be very close indeed (fig. 10). In this experiment we added CAP and EB to separate aliquots of cells growing on lactate. If we assume that action of EB is restricted to the nucleic acids of the mitochondrion then we may draw a strong inference with respect to mit specification (in both senses) for some component of the enzymes affected: neither the translation of NC mRNA by mitochondrial ribosomes nor that of mit mRNA by cytoplasmic ribosomes should be affected the same way by both CAP and EB. Under these conditions biosynthesis of *all* enzymes appears equally sensitive to CYCLO: the mit and NC systems are tightly coupled.

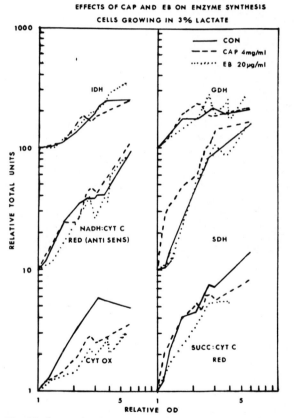

Fig. 10. Isometric plot similar to fig. 9 indicating the similarity between the effects of CAP (4 mg/ml) and EB (2×10^{-5}) when these inhibitors are added to separate aliquots of cells growing exponentially on lactate at an initial A_{600} of 0.486. Initial activity (nmoles $min^{-1} mg^{-1}$) of IDH, NADH:cyt *c* red, cyt ox, GDH, SDH and succ:cyt *c* red was 105, 33.9, 170, 364, 21 and 27.4, respectively.

Discussion

The different paradigms
Concerning ρ^- mutants

In theory, at least, the study of such mutants, particularly those that exhibit an extreme derangement of base composition in their mit-DNA, should permit one to draw strong inferences concerning the extent of mitochondrial specification on the genetic level. The assumption would be that the informational capacity of such a DNA is either severely restricted or absent entirely. Hence *any* activity found in the organelles or any other part of such cells *cannot* have been specified by their mit-DNA. How valid is this assumption? Clearly even 4 mol % (G plus C), the level found by us [27] and by Bernardi et al. [28] in independently isolated strains, can still provide for a considerable level of sense, as long as these bases are retained in wild-type sequences in a small number of cistrons, and in particular if the genome is relatively large. On the basis of the biophysical properties of the DNA we have considered such an arrangement of the remaining G plus C as unlikely. However, much more direct proof seems desirable. We need to know a great deal more about (a) the molecular weight of such mutant DNAs (for wild type and other strains see [58–62]); (b) whether it can be sheared or otherwise degraded to fragments that vary and can provide meaningful base sequences; and (c) whether any of these fragments can be accounted for in terms of primary transcripts, i.e. mit-RNA [63,64]. Such investigations are under way in a number of laboratories, including our own. Pending their outcome, and for the purposes of this discussion we shall maintain the hypothesis just stated concerning complete inability of specification by mit-DNA and even extend it to include translation by the mitochondrial system. This last part rests on the observation that such cells are completely insensitive to the presence of CAP [20]; the original hypothesis is strenghtened so far as transcription is concerned, by the observation that they are also resistant to inhibition by acridines or EB, agents known to be capable of interfering with this event. Yet some caution appears necessary even at this stage, in view of the demonstration that mitochondrial antibiotic-resistance markers can be retained in a ρ^- genome [65]. However, these markers may be clustered within a rather special part of the genome, and it is rather unlikely that their retention extends to the rather extreme form of mutants here under consideration.

Concerning site-specific inhibitors of translation

The usual way for employing agents of this sort is to determine what activities do not increase in their presence. However, a moment's reflection will suffice to convince one that any inference drawn from this kind of experiment can at best be weak. This is, of course, due to the fact, already mentioned, that an effect found to be exerted on a number of activities is at least as likely to have been due to action on a small number of essential steps of a sequence, rather than a simultaneous and parallel one on all the activities affected. The difficulty becomes compounded when experiments extend over a relatively long time span when cumulative effects may be expected to become exacerbated. Furthermore, the best that can be expected is to know that the synthesis of a particular, spectroscopically or catalytically distinct entity appears to be affected. In the absence of additional experiments, this cannot provide the identification of the polypeptide the synthesis of which is blocked. Yet inhibitor experiments can be extremely useful, provided they are designed and interpreted in terms of the only strong inference that appears permissible: the *absence* of inhibition for some activities. This requires in a carefully controlled experiment that (a) inhibitory action can be observed for some

components but not others, hence the inhibition is presumed to be functioning, and (b) its rate and extent can be determined in short-term experiments.

Site of specification of mitochondrial components

With these provisos we are now ready to come to the point of the exercise. The criteria which we shall employ are to examine the presence, or the insensitivity of their synthesis, of a variety of mitochondrial components and activities, under conditions that might be expected to interfere with their successful specification either at the DNA (gene or transcript) or the translational level. In addition to the paradigms already discussed we shall use the additional one of severe glucose repression, where, as already noted, the mitochondrial translational apparatus appears dispensable.

Outer membrane and easily extractable proteins (matrix)

Although we have no good marker for the outer membrane of yeast mitochondria *, the presence of this entity as a distinct, and recognizable morphological feature under conditions when development of normal mitochondrial function is severely impeded or entirely absent, leave little doubt concerning its specification and synthesis by the NC system. The same appears true with regard to all the relatively easily extractable enzymes of the particle, whether repressible or not, which in their aggregate presumably account for the 70% or so of the total mitochondrial protein that appears to behave in a

---

* The particles contain a NADP: and NADPH:cytochrome c reductase which is insensitive to Antimycin A, and whose biosynthesis is blocked by CYCLO and not by CAP. Its precise localization is not known. As a corollary we are certain that the pleiotropic, phenotypic expression of the ρ^- mutation cannot be due to the absence of the bulk structural proteins of mitochondria and other membranous organelles as might have been supposed by analogy to the model proposed for the *mi*-mutations in Neurospora [83]. On isolating enzymes from the inner membrane the polypeptides postulated here may well remain attached by some methods, or else become detached by other, more strongly dissociating or denaturing treatments. A necessary consequence of the model is the presence of the *identical* polypeptide in more than one respiratory enzyme or complex in their active and its absence in their inactive analogue produced by genetic or biochemical manipulation.

manner indistinguishable from the soluble proteins of the cytosol in labeling experiments, and which are similar in their inhibition patterns to the cytosol enzyme GDH.

Inner membrane and cristae

So far the results have been hardly surprising: the postulate that mitochondrial specification does not extend beyond the inner membrane and its attachments appears to be gaining ever wider acceptance [1,2,11,20,30,57,66—72]. However, we feel that application of the criteria just discussed to the results presented in this report allow us to eliminate from further consideration a considerable number of entities localized in precisely the inner membrane and its attachments, namely all those participating in the respiratory chain between succinate and cytochrome c, as well as — and this was already implicit in the work of earlier investigators [39,40] — those that make up the mitochondrial ATPase proper. Thus it would appear that such characteristic proteins of the inner membranes as succinate dehydrogenase (including its non-heme iron moiety), cytochrome b and cytochrome c_1, as well as the proteins comprising the knob-like attachments [73—76] (the headpieces in Green's terminology [77]) all are of extramitochondrial origin.

Let us now turn to the one activity for which we have clearcut evidence to the contrary: cytochrome oxidase. As described, it is the only mitochondrial activity that appears to behave in a qualitatively distinct fashion from all others with respect to the severity and rapidity of the blocks exerted by either EB or CAP. Its synthesis like that of other mitochondrial enzymes is also subject to a block by CYCLO, but, and this constitutes the stronger argument, we can also demonstrate a temporary release from the blocks by CAP and by CYCLO, but with different time constants.

Similar differential kinetics of the sensitivity to these inhibitors of this activity in the course of its elaboration during exposure of anaerobic cells to oxygen have been reported by Chen and Charalampous [78], and a partial sensitivity to CYCLO of the incorporation of precursors into proteins of an enzyme preparation isolated from rat liver has been demonstrated by Schiefer [79].

NADH: cytochrome c reductase appears to share in some, but not all these attributes of cyt ox, but to a reduced extent.

We now need to dispose of a somewhat touchy point: the specification of "structural" or "non-catalytic" proteins of the inner membrane. Without wishing to take a stand on the shifting quicksand [70,80,88] concerning the nature and significance of such entities we can make a fairly strong statement: those proteins that account for the *bulk* of the inner membrane of the yeast mitochondrion appear to be resistant to interference by genetic or physiological blocks with the autonomous mitochondrial system as judged by: (a) morphology as seen on electron micrographs; (b) sedimentation velocity and buoyant density in sucrose gradients; (c) electrophoretic mobility after complete denaturation — in agreement with the results reported by Tuppy et al. [86]. They are specified by the nuclear genes and synthesized by the CAP-insensitive, CYCLO-sensitive system of the NC apparatus. We do *not* exclude, but on the contrary wish to hypothesize, as we have in the past [11], the possibility that the statement just made does *not* apply to a particular class of minority proteins of the inner membrane.

A class of regulatory of morphopoetic proteins

The results with cytochrome oxidase — and to a certain extent with NADH:cytochrome c reductase — can be explained in one of two ways. Either, synthesis of the active enzyme itself, and in this we include the polypeptide chain(s), the prosthetic group, and all other components required for the activity of the enzyme as isolated require the joint and coordinated (in time and space) participation of both the mitochondrial and the NC systems. Or else, the successful integration into the inner membrane/cristae and the attendant expression of the enzymatic activity of an entity otherwise completely specified by the NC system, requires the concurrent, or immediately preceding, synthesis of *another* component wholly indigenous to the particle itself. In theory a choice between the two alternatives will be possible once the protein components of the pure enzyme and their mode of biosynthesis can be established without ambiguity. Meanwhile, we are somewhat inclined to favor the second alternative for the following reasons:

(1) Protein entities related either immunochemi-

cally [45] or enzymologically [46] to active cytochrome oxidase are present in ρ^- mitochondria.

(2) A very similar situation appears to obtain in certain mutants of Neurospora, where the non Mendelian inheritance of an altered cytochrome oxidase has, however, been explained by Woodward et al. [72] in terms of a two-protein model.

(3) The manifestation of a CYCLO and a CAP insensitive phase is both transient and non-overlapping as seen either during release from glucose repression or during respiratory adaptation [78].

(4) During derepression the increase in cytochrome oxidase appears to take place subsequent to the appearance of other respiratory enzymes [10,11] and is known to be regulated by other mitochondrial components [89].

(5) All the effects studied appear to be pleiotropic, but always in a manner that is quantitatively distinct for each of the activities tested. For instance, NADH:c shares some of the attributes of cyt ox — and this too has its counterpart in Neurospora [72] — but usually to a lesser extent. Such a differential effect would be difficult to explain in terms of the alternate model, which would predict that both the enzymes should then be deficient in a component of mitochondrial origin (say a polypeptide, though not necessarily the same) forming an integral part of their structure and essential for their function. Furthermore on the substrate site we have definite evidence that the mitochondrial system is dispensible for the formation of the dehydrogenase, cytochromes b and c_1. One then wonders as to the nature of the component affected since *Saccharomyces* is reported to possess a chain that in any event lacks the non-heme iron protein and/or components required at the coupling site of complex I [75]. Finally, in the case of ATPase, there is definite evidence for the participation and mitochondrial specification of just such a component or components concerned with the integration and regulation of the fully functional enzyme [40].

We therefore wish to propose that the autonomous mitochondrial system, at least of yeast, produces — aside from parts of the translational machinery — a very small number of proteins with an integrating and regulatory function that control the successful elaboration of the inner membrane and its attachments. In thus ascribing an essentially catalytic role in mem-

brane formation to these mitochondrial gene products we can account for their high degree of qualitative effectiveness in spite of their quantitative paucity in number and amount [71,72,87,88].

We also propose that these regulatory features with regard to function find their counterpart in terms of structure: successful development of the proper pattern, infoldings and attachments require the presence of these same proteins as *morphopoetic principles* [11,90], to use Kellenberger's terminology [91]: to construct surfaces more complex than sheets, cylinders or regular polyhedra you need signals at the corners. Finally we may hypothesize that by some negative feedback device or other the systems in the cytosol concerned with synthesis and transport sense the unavailability of sites provided by the regulatory proteins, and shut off further production by the NC system. The latter thus exert controls over the whole cellular and not just the mitochondrial economy.

This model differs in important features from others that have recently been proposed. These include ones that ascribe mitochondrial origin to the insoluble framework of the inner membrane as a whole, with *subsequent* insertion into the latter of soluble components originating in the NC system [69]; those that postulate a separate and distinct origin of inner membranes and cristae [57]; or ones that propose that the synthesis of outer membrane and matrix — presumably under the aegis of the NC system — precede the synthesis of certain inner membrane components such as ATPase — present at first in a soluble form — followed by elaboration of the inner membrane [92]. We envisage normal mitochondrial development during derepression, to proceed somewhat as follows: the formation, entirely by the NC system, of a particle already containing an outer membrane, intermembrane space and matrix, but, with a relatively undifferentiated inner membrane separating the last two compartments. This inner membrane, we propose, already contains the majority proteins of F_1-ATPase and some of the respiratory enzymes, but at low levels relative to the bulk proteins of the membrane. Further differentiation, characterized morphologically by increased cristae formation, and functionally by a complete, highly active respiratory chain, as well as an oligomycin-sensitive ATPase as an integral part of productive coupling of electron to energy flux requires the intervention of the autonomous mitochondrial system. A lesion in the latter, introduced by genetic or biochemical means, can lead to the dissociation or arrest of an otherwise continuous and predetermined sequence of events. The model serves to reconcile observations on such diverse phenomena as the course of derepression and respiratory adaptation in a facultative anaerobe such as *S. cerevisae*; the effects of translational, transcriptional and incipient genetic blocks on the mitochondrial system of such an organism, as well as on an obligate aerobe such as *Candida* or *Hansenula*. A very similar model has been proposed in the case of chloroplast development in *Chlamydomonas reinhardii* by Hoober et al. [93].

The origin of mitochondria

Although it is becoming ever more fashionable [94,95,96] to ascribe the evolutionary origin of mitochondria to the development of obligatory parasitism or symbiosis consequent to some primordial event involving the engulfment of a procaryotic cell by some primitive, amoeboid organism, we wish to advance a different hypothesis. For though it has been stated "that an old adage of parasitology is that the more specialized the form of parasitism, the more simplified or reduced is the parasite's metabolic apparatus, ... this could apply as well to mitochondria" [96] it is difficult to find an example of parasitism where this process of reduction has continued to the point where the parasite, with a genome the size of the mitochondrial one, is incapable of coding either for its own polymerase or its majority proteins.

Instead we wish to propose that the mitochondrion had its origin in a continued series of emergent changes of the bacterial cell membrane, and in particular of the mesosome [98,99] which led, either progressively or simultaneously to (a) increased specialization, culminating in cristae formation; (b) a subdivision of the intracellular space, by means of the membrane, into separate compartments, vesicularization of which culminated in the separation of an intracellular organelle, and (c) the sequestration of some part of the genome (either a completely or a partially duplicated one, thus capable of specification of and perhaps still attached to the membrane [100–102]; or perhaps of a plasmid) into one or more of these compartments. Continued maintenance of this

DNA, first perhaps as an episome, later as an independent entity, but capable of innumerable exchanges with its chromosomal counterpart produced a continously changing pattern of gene sequestration. This, together with further differentiation and segregation of function of the proteins in the organelle, over the course of evolutionary history resulted in the mitochondria of the primitive ancestors of contemporary unicellular eucaryotes. The survival value of the postulated events was three-fold; it further increased the potentialities with regard to specialization of different membranes; it provided for the duplication of crucial genes and their sequestration at the site of their utilization; and it allowed a much closer regulation of the dispensing of crucial gene products at appropriate points in space and time.

Acknowledgements

The research reported was supported by grant GM 12228 from the Institute of General Medical Sciences, of the National Institutes of Health, U.S. Dept. Health, Education and Welfare. It would have been impossible without the cheerful and competent technical assistance of Mrs. Karen Walker and Mrs. Carol Williams. This constitutes Paper 4 in a series entitled Formation of Yeast Mitochondria. H.R. Mahler is the recipient of Research Career Award GM 05060, Institute of General Medical Sciences, currently on leave at the Centre de Genetique Moleculaire, 91 Gif sur Yvette, France. The hospitality of and the many fruitful discussions with Dr. P.P. Slonimski are gratefully acknowledged.

References

[1] S. Nass, Intern. Rev. Cytol. 25 (1969) 55.
[2] P. Borst and A.M. Kroon, ibid., in press.
[3] M.M.K. Nass, Science 165 (1969) 25.
[4] R.P. Wagner, ibid. 163 (1969) 1026.
[5] B. Ephrussi, P.P. Slonimski, Y. Yotsuyanagi and J. Tavlitzki, Compt. Rend. Lab. Carlsberg 26 (1956) 87.
[6] A.W. Linnane, in: Oxidases and Related Redox Systems, eds. T.E. King, H.S. Mason and M. Morrison (Wiley, New York, 1965) p. 1102.
[7] E.S. Polakis, W. Bartley and G.A. Meek, Biochem. J. 97 (1965) 298;
E.S. Polakis and W. Bartley, ibid. 97 (1965) 284.
[8] M.F. Utter, E.A. Duell and C. Bernofsky, in: Aspects of

Yeast Metabolism, ed. R.K. Mills (Blackwell, Oxford, 1967) p. 197.

[9] J. Jayaraman, C. Cotman, H.R. Mahler and C. Sharp, Arch. Biochem. Biophys. 116 (1966) 224.

[10] D. South and H.R. Mahler, Nature 218 (1968) 1226.

[11] C.P. Henson, P. Perlman, C.N. Weber and H.R. Mahler, Biochemistry 7 (1968) 4445.

[12] A.G. So and E.W. Davie, Biochemistry 2 (1963) 132.

[13] R.K. Bretthauer, L. Marcus, J. Chaloupka, H.O. Halvorson and R.M. Bock, ibid. 2 (1963) 1079.

[14] E. Wintersberger, Biochem. Z. 341 (1965) 409.

[15] S.R. de Kloet, Koninkl. Ned. Akad. Wetenschap Proc. 68B (1965) 266.

[16] M.R. Siegel and H.D. Sisler, Biochim. Biophys. Acta 103 (1965) 558.

[17] A.W. Linnane, D.R. Biggs, M. Huang and G.D. Clark-Walker, in: Aspects of Yeast Metabolism, ed. R.K. Mills (Blackwell, Oxford, 1967) p. 197.

[18] H.R. Mahler, K.K. Tewari and J. Jayaraman, ibid., p. 247.

[19] A.J. Lamb, G.D. Clark-Walker and A.W. Linnane, Biochim. Biophys. Acta 161 (1968) 415.

[20] G.D. Clark-Walker and A.W. Linnane, Biochem. Biophys. Res. Commun. 25 (1966) 8. J. Cell Biol. 34 (1967) 1.

[21] B. Ephrussi, Nucleo-Cytoplasmic Relations in Microorganisms (The Clarendon Press, Oxford, 1953).

[22] F. Sherman and P.P. Slonimski, Biochim. Biophys. Acta 90 (1964) 1.

[23] H.R. Mahler, B. Mackler, S. Grandchamp and P.P. Slonimski, Biochemistry 3 (1964) 668.

[24] B. Mackler, H.C. Douglas, S. Will, D.C. Hawthorne and H.R. Mahler, ibid. 4 (1964) 2016.

[25] J.C. Mounolou, H. Jakob and P.P. Slonimski, Biochem. Biophys. Res. Commun. 24 (1966) 218; in Biochemical Aspects of the Biogenesis of Mitochondria, eds. E.C. Slater, J.M. Tager, S. Papa and E. Quagliarello (Adriatica Editrice, Bari, 1968) p. 473.

[26] F. Carnevali, G. Piperno and G. Tecce, Acad. Naz. Lin. Rend. Scien. His. mat. nat. 41 (VIII) (1966) 194.
F. Carnevali, G. Morpurgo and G. Tecce, Science 163 (1969) 1331.

[27] B.D. Mehrotra and H.R. Mahler, Arch. Biochem. Biophys. 128 (1968) 685.

[28] G. Bernardi, F. Carnevali, A. Nicolaieff, G. Piperno and G. Tecce, J. Mol. Biol. 37 (1968) 493.

[29] G. Bernardi, M. Faures, G. Piperno and P.P. Slonimski, ibid., submitted.

[30] C.P. Henson, C.N. Weber and H.R. Mahler, Biochemistry 7 (1968) 4431.

[31] H.R. Mahler, P. Perlman, C. Henson and C. Weber, Biochem. Biophys. Res. Commun. 31 (1968) 474.

[32] P.S. Perlman and H.R. Mahler, Arch. Biochem. Biophys. 136 (1970) 245.

[33] W. Atzpodieu, J.M. Gancedo, W. Duntze and H. Holzer, Europ. J. Biochem. 7 (1968) 58.

[34] K.D. Munkres, N.H. Giles and D.O. Woodward, Arch.

Biochem. Biophys. 109 (1965) 397.
K.D. Munkres and F.M. Richards, ibid. 109 (1965) 457.

[35] Longo and Scandalios, Proc. Natl. Acad. Sci. U.S. 62 (1969) 104.

[36] P.S. Perlman and H.R. Mahler, J. Bioenergetics, 1 (1970) 113.

[37] D.E. Green, J. Asai, R.A. Harris and J.T. Penniston, Arch. Biochem. Biophys. 125 (1968) 684.

[38] C.R. Hackenbrock, J. Cell Biol. 30 (1966) 269, ibid. 37 (1968) 345.

[39] L. Kovač and K. Weissová, Biochim. Biophys. Acta 153 (1967) 55.

[40] G. Schatz, J. Biol. Chem. 243 (1968) 2192.

[41] A.W. Linnane and J. Still, Australian J. Sci. 18 (1956) 165.

[42] D.B. Roodyn and D. Wilkie, Biochem. J. 103 (1967) 3C.

[43] G. Schatz, H. Tuppy and J. Klima, Z. Naturforsch. 186 (1963) 145.

[44] H.R. Mahler, B. Mackler, P.P. Slonimski and S. Grandchamp, Biochemistry 3 (1964) 677.

[45] J.R. Kraml and H.R. Mahler, Immunochemistry 4 (1967) 213.

[46] H. Tuppy and G.D. Birkmayer, Europ. J. Biochem. 8 (1969) 237.

[47] J. Le Pecq, Thèse Université de Paris (1965). J. Le Pecq and C. Paoletti, J. Mol. Biol. 27 (1967) 87.

[48] M.J. Waring, Biochim. Biophys. Acta 114 (1966) 234; J. Mol. Pharmacol. 1 (1965) 1; L.V. Crawford and M.J. Waring, J. Mol. Biol. 25 (1967) 27.

[49] R. Radloff, W. Bauer and J. Vinograd, Proc. Natl. Acad. Sci. U.S. 57 (1967) 1514.

[50] P.P. Slonimski, G. Perrodin and J.H. Croft, Biochem. Biophys. Res. Commun. 30 (1968) 232.

[51] M. Ogur, R.St. John and S. Nagai, Science 125 (1957) 928.

[52] K.C. Atwood and A. Norman, Proc. Natl. Acad. Sci. U.S. 35 (1949) 696.

[53] S. Leon and H.R. Mahler, Arch. Biochem. Biophys. 126 (1968) 305.

[54] E. Zylber, C. Vesco and S. Penman, J. Mol. Biol. 44 (1969) 195.

[55] J.C. Mounolou and G. Perrodin, C.R. Acad. Sci. Paris 267 (1968) 1268.

[56] R.E. Meyer and M.V. Simpson, Biochem. Biophys. Res. Commun. 34 (1969) 238.

[57] G.M. Kellerman, D.R. Biggs and A.W. Linnane, J. Cell Biol. 42 (1969) 378.

[58] M. Guerineau, C. Grandchamp, Y. Yotsuyanagi and P.P. Slonimski, C.R. Acad. Sc. Paris 266 (1968) 1884.

[59] P. Borst, this volume.

[60] K.K. Tewari, W. Võtsch, H.R. Mahler and B. Mackler, J. Mol. Biol. 20 (1966) 453.

[61] L. Shapiro, L.I. Grossman, J. Marmur and A.K. Kleinschmidt, ibid. 33 (1968) 901.

[62] C.J. Avers, F.E. Billheimer, H.P. Hoffmann and R.M. Pauli, Proc. Natl. Acad. Sci. U.S. 61 (1968) 90.

[63] E. Wintersberger and G. Viehhauser, Nature 220 (1968) 699.

[64] H. Fukuhara, M. Faures and C. Genin, Mol. Gen. Genetics 104 (1969) 264.

[65] A.W. Linnane, G.W. Saunders, E.B. Gingold and H.B. Lukins, Proc. Natl. Acad. Sci. U.S. 59 (1968) 903; E.B. Gingold, G.W. Saunders, H.B. Lukins and A.W. Linnane, Genetics 62 (1969) 735; P.P. Slonimski, this volume.

[66] D.S. Beattie, R.E. Basford and S.B. Koritz, Biochemistry 5 (1966) 926; ibid. 6 (1967) 3099.

[67] B. Kadenbach, Biochim. Biophys. Acta 134 (1967) 430.

[68] W. Neupert, D. Brdiczka and Th. Bücher, Biochem. Biophys. Res. Commun. 27 (1967) 488.

[69] D.S. Beattie, J. Biol. Chem. 243 (1968) 4027.

[70] T.S. Work, L.J. Coote and M. Ashwell, Fed. Proc. 27 (1968) 1174; T.S. Work, British Soc. Cell Biol. Symp., in press.

[71] W. Sebald, Th. Hofstötter, D. Hacker and Th. Bücher, FEBS Letters 2 (1969) 177.

[72] D.O. Woodward, D.L. Edwards and R.B. Flavell, Soc. Exptl. Biol. Symp. 24, in press.

[73] H. Fernández-Morán, T. Oda, P.V. Blair and D.E. Green, J. Cell Biol. 22, (1964) 63.

[74] E. Racker, D.D. Tyler, R.W. Estabrook, T.E. Conover, D.F. Parsons and B. Chance, in: "Oxidases and related redox systems", vol. II, eds. T.E. King, H.S. Mason and M. Morrison (Wiley, New York, 1965) p. 1077. E. Racker, Fed. Proc. 26 (1967) 1335.

[75] For the current status see reviews by: H. Lardy, Ann. Rev. Biochem. 38 (1969) 991 and [76].

[76] D.O. Hall and J.M. Palmer, Nature 221 (1969) 717.

[77] D.E. Green and D.H. MacLennan, Bio Science 19 (1969) 213; K. Kopaczyk, J. Asai and D.E. Green, Arch. Biochem. Biophys. 126 (1968) 358.

[78] W.L. Chen and F.C. Charalampous, J. Biol. Chem. 244 (1969) 2767.

[79] H.G. Schiefer, Z. Physiol. Chem. 350 (1969) 235.

[80] R.S. Criddle, R.M. Bock, D.E. Green and H.D. Tisdale, Biochemistry 1 (1962) 821.

S.H. Richardson, H.O. Hultin and D.E. Green, Proc. Natl. Acad. Sci. U.S. 50 (1963) 821.

[81] R.S. Criddle, D.L. Edwards and T.G. Petersen, Biochemistry 5 (1966) 578.

[82] D.H. Mac Lennan and A. Tzogoloff, Biochemistry 7 (1968) 1603.

[83] D.O. Woodward and K.D. Munkres, in: Organizational Biosynthesis, eds. H. Vogel, J.O. Lampen and V. Bryson (Academic Press, New York, 1967) p. 489. D.O. Woodward, Fed. Proc. 27 (1968) 1167.

[84] D.E. Green, N.F. Haard, G. Lenaz and H.I. Sillman, Proc. Natl. Acad. Sci. U.S. 60 (1968) 277.

[85] E.D. Kiehn and J.J. Holland, ibid. 61 (1968) 1370.

[86] H. Tuppy, P. Swetly and I. Wolff, Europ. J. Biochem. 5 (1968) 339.

[87] S. Yang and R.S. Criddle, Biochem. Biophys. Res. Commun. 35 (1969) 429.

[88] G. Schatz, Biochim. Biophys. Acta, in press.

[89] C. Reilly and F. Sherman, Biochim. Biophys. Acta 95 (1965) 640.

[90] C. Cotman, H.R. Mahler and T. Hugli, Arch. Biochem. Biophys. 126 (1968) 821.

[91] E. Kellenberger, in: "Principles of Biomedical Organization" (Ciba Foundation, London, 1965.)

[92] A. Castelli, G. Parenti-Castelli, E. Bertoli and G. Lenaz, Ital. J. Biochem. (1969) 35.

[93] J.W. Hoober, P. Siekevitz and G.C. Palade, J. Biol. Chem. 244 (1969) 2621.

[94] H. Ris and W. Plaut, J. Cell Biol. 13 (1962) 383; H. Ris, Can. J. Gent. Cytol. 3 (1961) 95.

[95] M.W.K. Nass, S. Nass and B.A. Afzelius, Exptl. Cell Res. 37 (1965) 516.

[96] L. Sagan, J. Theoret. Biol. 14 (1967) 225.

[97] S. Nass, Intern. Rev. Cytol. 25 (1969) 101.

[98] M.R. Edwards and R.W. Stevens, J. Bacteriol. 86 (1963) 414.

[99] W. van Iterson, Bacteriol. Rev. 29 (1965) 299.

[100] A. Ryter and F. Jacob, Ann. Inst. Pasteur 107, 384.

[101] A.T. Ganesan and J. Lederberg, Biochem. Biophys. Res. Commun. 18 (1965) 824.

[102] Y. Yotsuyanagi, Compt. Rend. 262 (1966) 1348.